Proceedings in Engineering Mechanics

Research, Technology and Education

This book series publishes the results of meetings dealing with material properties in engineering and science. It covers a wide range of topics, from the fundamentals of materials mechanics and applications for various industries to aspects of scientific training and career development. The volumes in the series are based typically on primary research materials presented at conferences, workshops, and similar scientific meetings, and represent comprehensive scientific and technical studies.

Lucas F. M. da Silva · Robert D. Adams ·
Klaus Dilger
Editors

2nd International Conference on Industrial Applications of Adhesives 2022

Selected Contributions of IAA 2022

Springer

Editors
Lucas F. M. da Silva
Faculty of Engineering
University of Porto
Porto, Portugal

Robert D. Adams
Department of Mechanical Engineering
University of Bristol
Bristol, Avon, UK

Department of Engineering Science
University of Oxford
Oxford, UK

Klaus Dilger
Institut für Füge- und Schweißtechnik
Technische Universität Braunschweig
Braunschweig, Niedersachsen, Germany

ISSN 2731-0221 ISSN 2731-023X (electronic)
Proceedings in Engineering Mechanics
ISBN 978-3-031-11149-5 ISBN 978-3-031-11150-1 (eBook)
https://doi.org/10.1007/978-3-031-11150-1

This Springer imprint is published by the registered company Springer Nature Switzerland AG
The registered company address is: Gewerbestrasse 11, 6330 Cham, Switzerland

Preface

This volume of *Proceedings in Engineering Mechanics—Research, Technology and Education* contains selected papers presented at the 2nd International Conference on Industrial Applications of Adhesives 2022 (IAA 2022), held in Carvoeiro (Portugal) during March 3–4, 2022. The goal of the conference was to provide a unique opportunity to exchange information, present the latest results as well as to discuss issues relevant to industrial applications of adhesives.

This conference is held every two years. The conference is chaired by Lucas F. M. da Silva and co-chaired by R. D. Adams (University of Oxford, UK), Chiaki Sato (Tokyo Institute of Technology) and Klaus Dilger (Technische Universität Braunschweig, Germany). The focus is on applications of adhesive bonding in the industry such as automotive, aeronautic, railway, marine, energy and electronics. The idea is to bring together the adhesive makers and the adhesive users to exchange experiences and facilitate potential synergies and partnerships. Sixty-one papers were presented by researchers from 14 countries.

In order to disseminate the work presented at IAA 2022, selected papers were prepared which resulted in the present volume. A wide range of topics are covered resulting in nine chapters dealing with adhesion assessment between polymers and metals (first chapter), pressure sensitive adhesives (2 chapters), adhesive bonding process optimization (1 chapter), civil applications (2 chapters), adhesive joints in composite materials (2 chapters) and elastic adhesives (last chapter). The book is a state of the art of industrial applications of adhesives and also serves as a reference volume for researchers and graduate students working with adhesive bonding.

The organizer and editor wish to thank all the authors for their participation and cooperation, which made this volume possible. Finally, I would like to thank the team of Springer-Verlag, especially Dr. Christoph Baumann and Ute Heuser, for the excellent cooperation during the preparation of this volume.

Porto, Portugal Lucas F. M. da Silva
lucas@fe.up.pt

Bristol/Oxford, UK Robert D. Adams
Braunschweig, Germany Klaus Dilger
April 2022

Contents

A Method to Evaluate the Interfacial Adhesion Between Polymer and Metal

Achim Frick, Markus Rettenberger, and Marcel Spadaro

Abstract Up to now there is no standardized measuring method to properly quantify the interfacial adhesion between polymer and metal. This work presents a possible approach to this topic based on the quantification of the interfacial tightness of a polymer-metal hybrid. Two types of specimens, a monolithic specimen and a hybrid specimen, were fabricated by injection molding, partially thermally post-treated, and then subjected to tracer gas measurement, using helium as the tracer gas. The recording of the helium gas volume rate over the test duration enables the determination of a tight interface, which is characterized by full-faced contact between polymer and metal. Micro- or sub microscopic gaps between polymer and metal lead to a constant gas volume rate signal, while a tight interface without gaps exhibits a permeation characteristic. The determination of the gas volume rate on both specimens and a subsequent subtraction allows the separation between measurand falsifying permeation rate and actual interfacial gas volume rate. Therefore, tight interfaces without any gaps from different material combinations can be compared regarding their true adhesion performance. On the other hand, microscopic studies of the interface do not provide information on interfacial adhesion, since no interpretable visual differences in the interfaces of demonstrably tight and leaky hybrids could be detected within this study. The newly developed and presented method can enable research into compatible material combinations, effective adhesion modifications and suitable processing conditions for improving interfacial adhesion between polymer and metal.

Keywords Interfacial adhesion · Polymer-metal hybrids · Interfacial tightness · Tracer gas measurement

A. Frick · M. Rettenberger (✉) · M. Spadaro
Institute of Polymer Science and Processing (iPSP), Aalen University, Beethovenstraße 1, 73430 Aalen, Germany
e-mail: markus.rettenberger@hs-aalen.de

A. Frick
e-mail: achim.frick@hs-aalen.de

M. Spadaro
e-mail: marcel.spadaro@hs-aalen.de

L. F. M. da Silva et al. (eds.), *2nd International Conference on Industrial Applications of Adhesives 2022*, Proceedings in Engineering Mechanics,
https://doi.org/10.1007/978-3-031-11150-1_1

1 Introduction

In the automotive sector, the demands for energy-efficient and thus sustainable overall solutions have increased exponentially in recent years. In addition to the switch to alternative drive systems, the lightweight construction concept continues to represent another partial solution to this ambition [1–7]. To realize this concept, monolithic components are preferably replaced by multi-material hybrids, which offer enormous weight savings with identical mechanical load capacities [1, 3, 4]. In addition to load-bearing structures, such as front ends [8], mechatronic systems, such as connectors for signal or power transmission, are nowadays also designed as polymer-metal hybrids (PMHs) [2, 4, 7, 9]. Although the focus here is less on the mechanical load-bearing capacity of the assembly and more on media tightness, both requirements can only be realized as a result of strong adhesion in the interfaces between polymer and metal [1, 4–7, 9–11]. The term adhesion includes all interfacial forces that contribute to the mechanical cohesion between the two joined materials [1, 3, 12, 13]. Several papers have already reported possible interface failures caused by a lack of interactions at the macroscopically finite interfaces of a PMH. These detachments of the polymer component, often titled as so-called gaps, are usually shrinkage [9] or generally strain-induced [4, 14]. Consequently, these gaps indicate incompletely formed adhesion in the interface [6, 15] and therefore offer enormous potential for characterizing these critical regions.

Since the layer thicknesses of the interfaces are in the single-digit nanometer range [12], optical microscopy images are not a satisfactory characterization alternative due to the limited resolution of 0.5 μm [6, 16]. Since interfacial forces cannot be measured directly [6, 12], pull-out or pull-off tests on hybrid specimens are often used, the result of which is the bond strength correlated with the interactions [1–3, 5, 6, 10, 11, 13, 14, 17, 18]. In addition, both the simple feasibility and the high significance speak for the use of this test [3]. However, future applications of PMHs, such as use in electrical conductors [19], often require not only strong but also tight bonds, the majority of which require full-faced contact and thus full-faced adhesion of both materials at the interface [2, 7]. The smallest detachments of the polymer component from the metal surface are not statistically noticeable in the result of the bond strength and are therefore not measurable with the destructive test methods commonly used today. Especially in media-tight applications, these gaps can cause failure of the mechatronic system as a result of a short circuit caused by previous media ingress [1, 2, 4, 7, 9, 19, 20].

In addition, a methodology is to be developed which can quantify different properties between interfaces where full-faced contact is present on the nanoscale and thus offers a possibility to characterize the interfacial adhesion between different materials.

2 Materials and Methods

2.1 *Differential Tightness and Permeation Measurement (DTPM)*

Within the interfacial tightness characterization of PMHs, there is the possibility of a nondestructive tracer gas measurement [14]. The test specimen is sealed, pressurized with the tracer gas and surrounded by an evacuated detector chamber sealed against the environment. Due to the pressure and tracer gas concentration gradient between the test specimen and the detector chamber, there is an unavoidable tracer gas volume flow in the direction of the evacuated detector chamber. As the name implies, the detector chamber is connected to a corresponding detector system, which detects and records the amount of tracer gas escaping from the test specimen over the testing time, thus enabling tracer gas measurement (TGM) according to DIN EN 1779: 1999 B2.1. The test fixture used for TGM in this work is shown in Fig. 1.

Authors propose, that a tight bond between polymer and metal, must show a permeation characteristic when investigating it by means of TGM, while a local detachment of the polymer at the interface leads undoubtedly to an instantaneous tracer gas transfer without any prior dissolution and diffusion mechanism. This allows leaky and tight interfaces to be reliably detected and distinguished. In the case of a tight hybrid, where full-faced contact of the two materials must therefore be present, the time course of the gas volume curve must therefore resemble that of a known permeation curve, which initially exhibits transient behavior until the steady-state permeation level is reached [21–23]. However, assuming there is no gas volume flow through the crystalline structure of the metal [24], this permeation curve is composed of the following gas flows:

1. Permeation through the free volume of the polymer component
2. Permeation through the free volume of the sealings used
3. Permeation through thc necessary sealing contacts to the polymer surface or the test fixture surface.

In order to obtain an unambiguous permeation signal for the actual interfacial gas volume rate, this volume rate and the permeation signals just listed must be separated from each other. The so called "Differential Tightness and Permeation measurement Method "(DTPM) in which the permeation signal of both a hybrid and a geometrically identical, monolithic test specimen made of the same base polymer is recorded by means of TGM, can be used for this purpose. The actual interfacial gas volume rate then can be calculated from the both measured results according to Eq. 1.

$$\dot{q}_{\text{Interface}} = \dot{q}_{\text{Hybrid}} - \dot{q}_{\text{Permeation}} \quad (1)$$

In the equation $\dot{q}_{\text{Interface}}$ means the tracer gas volume rate through the interface, $\dot{q}_{\text{Hybrid}}$ the measured gas volume rate through the hybrid specimen and $\dot{q}_{\text{Permeation}}$ the

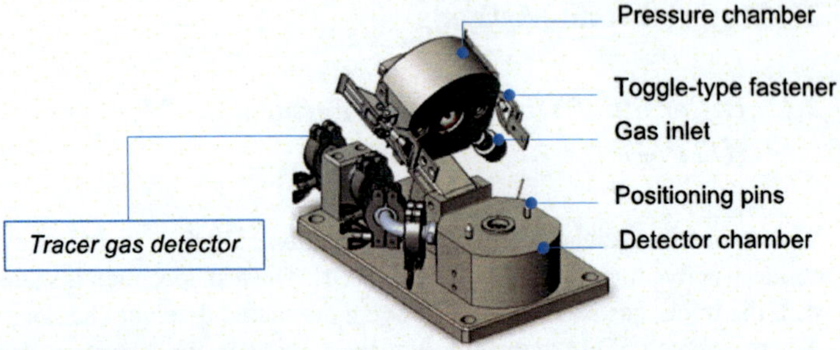

Fig. 1 Possible test fixture necessary for tracer gas measurement (TGM); presented in the opened state without test specimen

measured permeation rate through the monolithic specimen and the sealings of the test fixture. Possible test specimens necessary for the application of the method are presented in the following chapter.

2.2 Test Specimen Geometry

The polymer-metal test specimen investigated was specially developed and consists of an injection-molded polymer blank with a diameter of 30 mm and a wall thickness of 1.2 mm (z-direction), which is penetrated centrally by the metal insert, Fig. 2. The hybrid test specimen is produced in a single-cavity mold by assembly injection molding. In addition, closed, monolithic polymer blanks without metal inserts can be produced by exchangeable inserts in the injection mold, which are then used to measure the permeation rate through the polymer component used.

2.3 Metals

Deoxidized copper (Cu-PHC) with a copper content of 99.95% and pure aluminum (EN AW 1050A) with an aluminum content of 99.50% from the manufacturer Wieland-Werke AG were used as materials for the rectangular, metallic inserts with a sheet thickness of 0.8 mm. The overmolded metal inserts were produced by stamping from semi-finished products and, prior to overmolding, were placed in beakers containing propan-2-ol for 15 min and cleaned in an ultrasonic bath from the manufacturer Elma Schmidbauer GmbH (type: Elmasonic S 10 H) at a water bath temperature of 40 °C. In the following, the aluminum type will be abbreviated as "Al" and the copper type as "Cu".

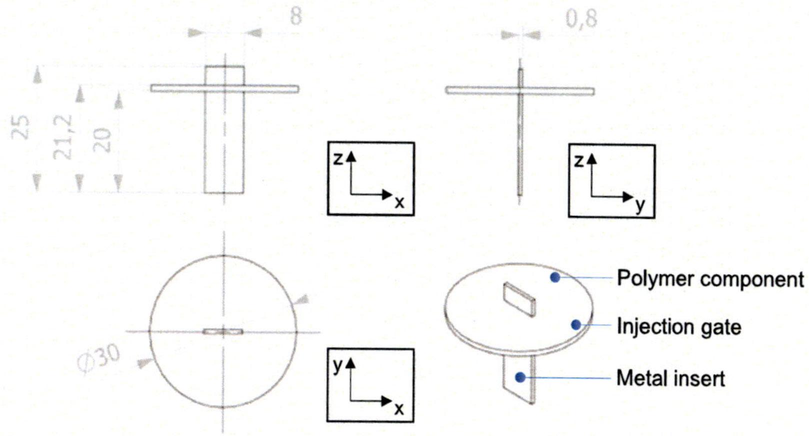

Fig. 2 Hybrid test specimen in different drawing views (dimensions in mm)

2.4 *Polymers*

The polymers used for the production of the test specimens described are, on the one hand, polypropylene (PP) from the manufacturer Ineos Holdings Ltd (type: 100-GA12), whereby two sample batches were adhesion-modified with 20 wt% of a PP grafted with maleic anhydride (PP-g-MAH) from the manufacturer Mitsui Chemicals Europe GmbH (type: Admer 3115E). On the other hand, a commercially available, black colored polyamide 6 (PA6) from the manufacturer LANXESS Deutschland GmbH (type: Durethan B30S) was used, which was dried for 6 h at 80 °C immediately before processing using a dry air dryer from the manufacturer Bierther GmbH (type: DR202 MT), in order to be able to exclude hydrolytic degradation of the material.

2.5 *Specimen Production*

While the monolithic specimens were merely injection molded, the hybrid ones were produced by assembly injection molding. Both specimen variants were produced employing an injection molding machine from the manufacturer Arburg GmbH & Co. KG (type: 220 S Allrounder 150–30) with a screw diameter of 15 mm using a single-cavity mold with spot gating. In order not to produce a different morphology of the two variants within one polymer type with regard to permeation behavior, the injection molding processing conditions were always identical for both specimen variants within one polymer type, Table 1.

The insertion of the metal inserts and the removal of the hybrid specimens were performed without a handling system using nitrile gloves.

Table 1 Injection molding processing conditions for the production of the investigated test specimens

Parameter	Unit	PP	PP + PP-g-MAH	PA6
Nozzle temperature	[°C]	240	240	265
Mold temperature	[°C]	60	60	100
Injection speed	[cm^3/s]	30	30	12
Switchover melt pressure	[bar]	700	700	550
Holding pressure	[bar]	500	500	500
Holding pressure time	[s]	7.5	7.5	9

2.6 Thermal Post-Treatment

In this study, thermal post-treatment means that in an additional process step after assembly injection molding of the PMHs, the areas of the polymer component close to the interface are re-melted by induction heating of the metallic joining partner. On the one hand, this allows the number of intermolecular interactions between the joining partners to be increased [25, 26], and on the other hand, any nanoscale pores present on the metal surface can be better infiltrated by the polymer melt. Both potentially contribute to an increased adhesion effect, regardless of physiochemical [25, 27] or mechanical nature, which should lead to a full-faced contact at the interface of the two materials.

Re-melting of the regions close to the interface was carried out immediately after the injection molding process by means of a mobile induction heating device from the manufacturer EFD Induction GmbH (type: Minac 6/10) with a maximum output power of 10 kW at a frequency range of 10–40 kHz. The temperature of the metal insert during thermal post-treatment was determined by a miniature sheath thermocouple from the manufacturer TC Mess- and Regeltechnik GmbH, and the temperature distribution on the surface of the polymer component was determined using an infrared camera from the manufacturer Optris GmbH. The temperature profile over time was recorded digitally. Possible measurement errors due to a slight inclination of the thermal camera were not investigated further, Fig. 3.

To ensure a high reproducibility of the measurements, the hybrid test specimens were placed and aligned in a holder manufactured for this purpose, consisting of two pneumatically controlled clamping jaws. The material Kelutherm 800 M from the manufacturer KELUX Hochleistungswerkstoffe GmbH, from which the clamping jaws were made, has a low thermal conductivity of only 0.26 W m^{-1} K^{-1}. The thermal conductivity of copper is about 399 W m^{-1} K^{-1}, that of aluminum 220 W m^{-1} K^{-1}, and those of PP and PA6 are 0.22 W m^{-1} $K^{-1}$and 0.31 W m^{-1} K^{-1}, respectively.

During thermal post-treatment, which causes partial melting of the polymer component in the regions near the interface, a joining pressure is generated due to the local volume expansion of the polymer in the region of this zone, which is largely dependent on the expansion coefficient of the respective polymer and its volume increase during the phase transition from the solid to the liquid aggregate state. In

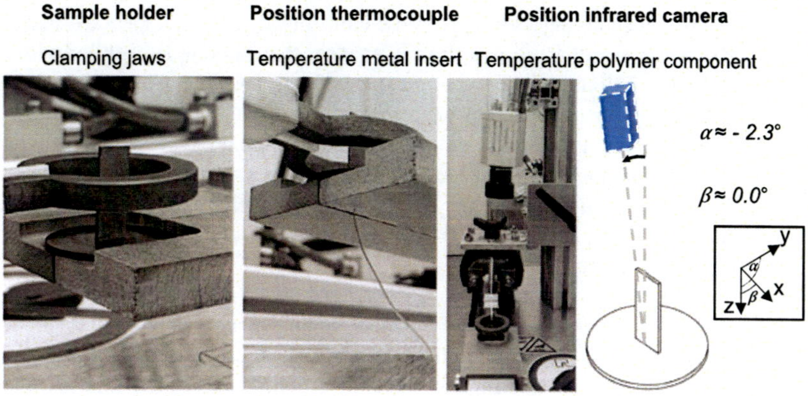

Fig. 3 Overview of the equipment for thermal post-treatment

the present case, however, the joining pressure is limited because the experimental setup ensures free expansion of the polymer volume in the z-direction, Fig. 3.

Both the injection-molded hybrid specimens and the hybrid specimens additionally subjected to thermal post-treatment were stored for at least 48 h under standard atmosphere (23 °C and 50% relative humidity) before further investigations. Thermally post-treated specimens are subsequently referred to as "Thermally post-treated".

2.7 *Tracer Gas Measurement (TGM)*

The employed testing setup for this study is shown schematically in Fig. 4.

In TGM, the test specimen is first placed between two chambers and pressed against the inner sealing, Fig. 4. The toggle-type fasteners and the positioning pins

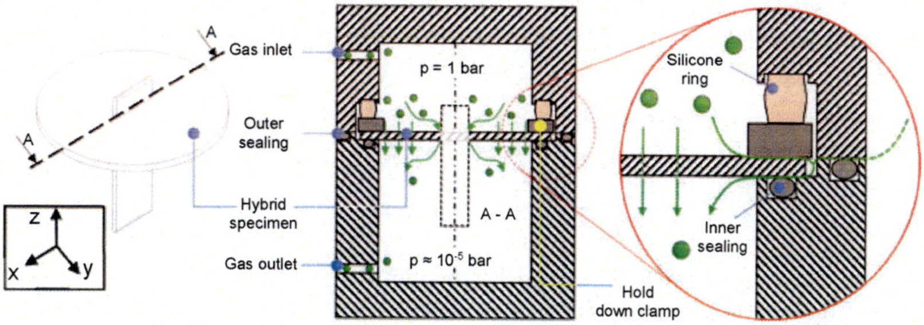

Fig. 4 Full cut along the sectional plane A-A of a hybrid test specimen clamped in the closed test fixture; arrows show possible permeation and leakage paths of the tracer gas

ensure that clamping is always reproducible, Fig. 2. Both chambers are then evacuated in succession so that the pressure chamber can then be completely filled with the tracer gas (here: helium 4.6) to a gas pressure of 1 ± 0.05 bar. The initial concentration gradient and the permanent pressure difference cause a gas volume flow through the test sample towards the detector chamber. Within the present investigations, the detector chamber has been connected to a mass spectrometer of the manufacturer Pfeiffer Vacuum GmbH (type: ASM340), which detects the helium atoms diffusing through the test specimen and the rest of the measurement setup and outputs them as gas volume rate.

Helium is ideally suited as a tracer gas, because it is a light and environmentally friendly gas, which is only present in atmospheric air in low concentrations of five parts per million. In addition, helium can be used to realize fast diffusion and permeation times, ensuring manageable experimental testing times. However, the main reason for using helium is that helium is a noble gas and thus does not react with other substances under normal conditions. This means that a test gas-induced change in the material properties of the polymer component tested and test fixture components can be ruled out.

An O-ring made of nitrile butadiene rubber (NBR) with a durometer hardness of 72 Shore A was used to seal the detector chamber of the tracer gas test fixture, where the inner diameter of the O-ring was 22 mm and the cord thickness was 2.5 mm. In order to generate a uniform permeation characteristic of the measuring system for comparative measurements and thus to be able to record only the permeation through the free volume of the respective polymer component, these sealings were used for all measurements.

In order to generate statistically representative results, five measurements were taken within each material combination. To avoid incorrect measurements caused by adsorption of helium on the chamber sealings during a previous measurement, the O-rings used are changed after each measurement and the complete measurement setup is flushed with compressed air both internally (connection systems) and externally (test fixture). O-rings, which are usually contaminated with helium after only one measurement, are reused under an exhaust hood after 1 h of outgassing at the earliest. It is therefore also not recommended to retest the same test specimen without first depressurizing it and then changing the sealing. All measurements were carried out at room temperature until the steady state level of permeation was reached.

2.8 Light Microscopy (LM) and Scanning Electron Beam Microscopy (SEM)

LM images and SEM images of previously prepared sections in the longitudinal direction (y–z plane) along the sectional plane B-B of the hybrid test specimens

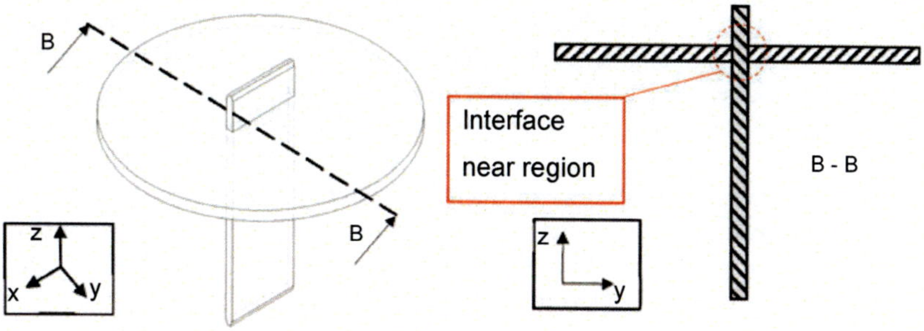

Fig. 5 Schematic representation of the sectional plane for optical measurements

were investigated, Fig. 5. The focus here was on the question of whether statements on interfacial adhesion in PMHs can be made using these established optical measurement methods. The images were compared with TGM results.

The specimens were first coated with a cold-curing resin and ground with increasingly finely grained silicon carbide discs (SiC). Subsequently, the ground sections were polished for 15 min with two diamond suspensions (grain size 1: 3 μm / grain size 2: 1 μm) in succession. Grinding and polishing were performed under constant rotational speed of 150 rpm and low contact forces of maximum 25 N using a rotary grinding machine from Wirtz Buehler GmbH (type: 49–5100-230). While a light microscope from the manufacturer Carl Zeiss AG (type: Axio Imager Vario) with an additional digital camera was used for the light microscopic investigations of the regions near the interface, a SEM from the manufacturer Carl Zeiss AG (type: Leo Gemini 1525), which generates an image of a sample via backscattered secondary electrons, was used for the more detailed investigations. Prior to the SEM investigations, an ultrathin Au/Pt layer was applied to the sample in an argon atmosphere at a current of 20 mA for 40 s in order to increase the conductivity of the polymeric regions of the respective sample.

3 Results and Discussion

This chapter presents the measured results on the interfacial tightness and on the adhesion behavior of hybrid samples manufactured under different processing conditions.

3.1 Tightness Validation of PMHs by TGM Method

When comparing the gas volume rates of injection-molded and adhesion-modified hybrid samples from PP and Cu with and without thermal post-treatment there is a significant difference in tightness behavior observed, Fig. 6. Without thermal post-treatment the samples present a comparatively high gas volume rate, that is almost constant over the whole tested period of thousand seconds. Already with the beginning of the measurement a significant gas volume rate of 1.5 10^{-4} mbar·l s^{-1} is found. Since the gas volume rate remains constant during testing time, some gap formation at the interface of the hybrid samples is expected. In this case the interfacial contact between the modified PP and the copper cannot be intimate, thus the detected gas volume rate should directly correspond with the gas volume which passes the interfacial zone of the hybrid sample. Thus, a measured gas volume rate curve showing a constant gas volume rate (Fig. 6) documents locally missing adhesion in the interfacial joining area of a hybrid sample.

A thermal post-treated injection-molded hybrid sample exhibits a transient gas volume rate curve similar to known and transient permeation curves [21–23], Fig. 6. The post-treatment of the interfacial zone of an injection-molded PMH obviously creates full-faced contact within the interface between polymer and metal, which prevents a sudden gas volume rate at begin of the test for the tracer gas applying TGM method. Then, a certain gas transfer develops during measurement time due to ongoing permeation process of the helium tracer gas in the tested polymer component of PMH and unfortunately in the TGM chamber sealing.

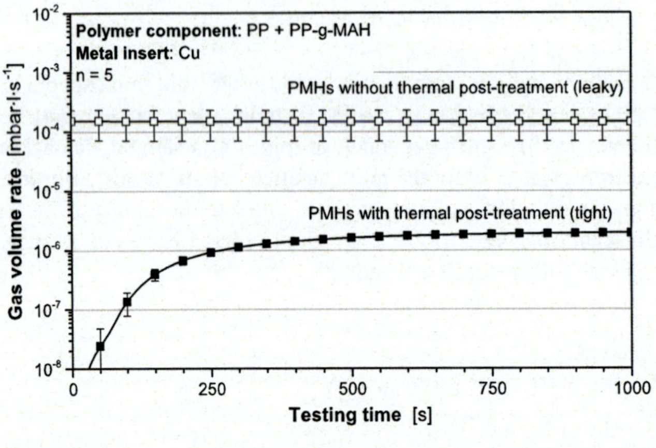

Fig. 6 Gas volume rate curves of hybrid samples from adhesion-modified PP and Cu with and without thermal post-treatment of the interfacial zone; Tightness test (TGM): Tracer gas : Helium 4.6, Pressure gradient: 1 bar, Test chamber seal material: NBR 72 Shore A (Ø22 mm · 2.5 mm), Test temperature: 23 °C; average curves and related standard deviations

3.2 Tightness Validation of PMHs by Optical Methods

Optical inspection of the interfacial zone of PMHs is a widely employed method. The following microscopic study results of injection-molded polymer-metal interfaces shown verify, this method does not provide any information on the tightness of an interface, it is only worth for optical characterization of a sample´s topography.

Figure 7 shows micrographs of the interfacial zone of two different PP/Al hybrid samples sectioned in longitudinal direction (*y*–*z*-plane). The images on the left present a PMH where an adhesion modified PP is used instead of a non-modified one on the right.

Before sectioning the samples and taking micrographs, they were visually inspected in the *x*–*y*-plane for possibly visible gaps between polymer and metal in contact zone and further measured for their tightness by means of TGM. The "tight" sample made of adhesion modified PP component is characterized in TGM by a transient, permeation-like gas volume rate curve, while the "leaky" sample made of a neat PP has presented a leakage curve with a constant gas volume rate of about 10^{-2} mbar·l s^{-1} over the testing time.

Only the use of adhesion-modified PP for PMH manufacture delivers a tight hybrid. Employing unmodified neat PP for PMH with aluminum insert provides a leaky hybrid. The presented micrographs, do not show this evidence, Fig. 7.

3.3 Determination of Interfacial Gas Volume Rate of PMHs Using DTPM

When testing a monolithic polymer material sample (PP) with a wall thickness of 1.2 mm by means of TGM employing helium as tracer gas a permeation curve is measured, Fig. 8. The measured TGM curve of a thermally post-treated PMH with the same wall thickness of its PP blank also exhibits such transient behavior, but at slightly increased level. This implies the gas passes the thermally post-treated PMH similar the monolithic polymer material, namely by permeation. This further means that the measurement of a permeation-like gas volume rate curve applying TGM on a PMH verifies, the interfacial adhesion of the tested sample is pronounced with a full-faced adhesive bonding between polymer and metal.

The interfacial tightness can be quantified using DTPM by calculating the actual gas volume rate at the manufactured interface, e.g. of a post-treated PMH, according to Eq. 1 by subtracting the permeation curve of the monolithic polymer from that of the PMH sample.

A thermal post-treatment of an injection-molded PMH sample enables a significant improvement of the interfacial adhesion and thus tightness in its interfacial zone, proved by a measured, permeation like transient gas volume rate curve, Fig. 6. The resulting very low helium gas volume rate of the hybrid´s interfacial zone, e.g. of a

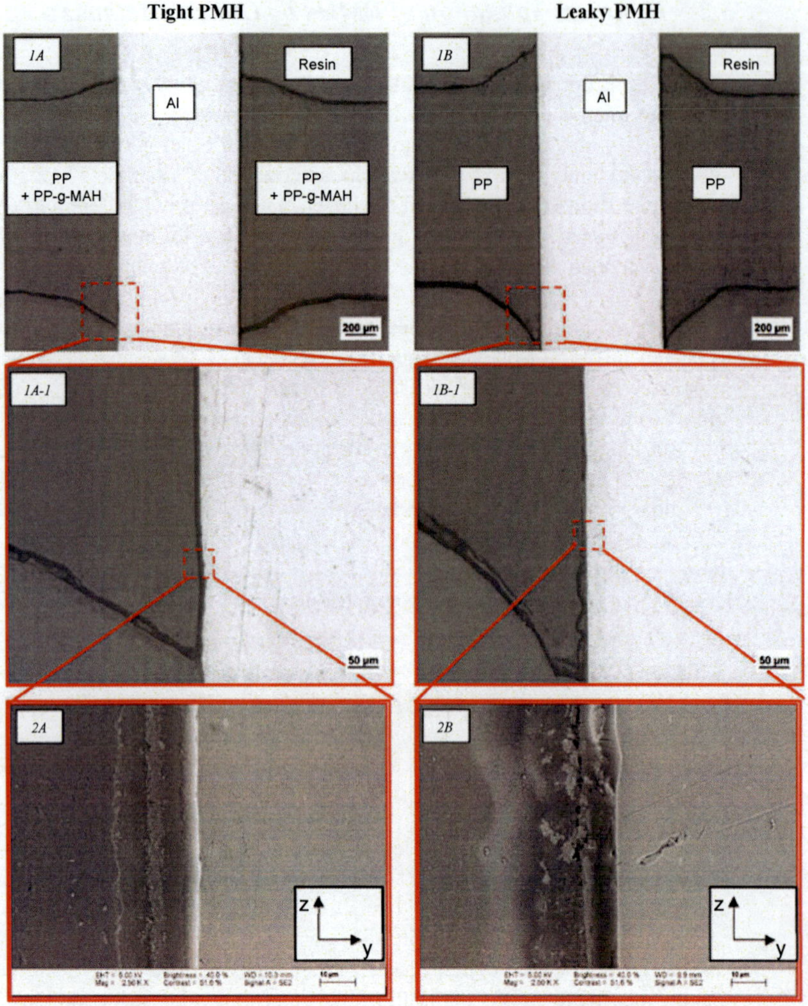

Fig. 7 LM micrographs of longitudinal cross-sections (y–z-plane) of the interface near region of thermally post-treated PMHs prepared from a "tight" and "leaky" sample, i.e. adhesion modified and non-modified PP component both with aluminum insert; SEM images 2A and 2B present enlarged views of the framed areas within the LM micrographs 1A-1 resp. 1B-1

PP/Cu hybrid, is about 10^{-7} mbar l s^{-1} only, when reaching the permeation steady state level.

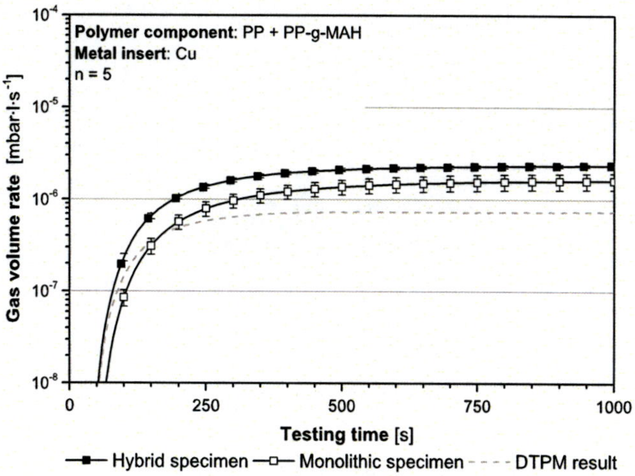

Fig. 8 Differentiation between permeation and interfacial gas volume rate determined by means of DTPM of thermally post-treated, adhesion-modified PP/Cu hybrids; Tightness test (TGM): Tracer gas: Helium 4.6, Pressure gradient: 1 bar, Inner sealing: NBR 72 Shore A (Ø22 · 2.5 mm), Temperature: 23 °C; average curves from five samples

3.4 Quantification of Interfacial Adhesion by DTPM Method for Different PMHs

Figure 9 shows the gas volume rates at steady state level after thousand seconds testing time of different types of thermal post-threated PMHs, all are technically relevant hybrid combinations, established applying DTPM method, as described above. DTPM method allows to characterize quantitatively the possible interfacial adhesion of polymer metal hybrids and to rank them.

As Fig. 9 shows there is practically no difference in the gas volume rate of the interfacial zone, when combining a certain polymer material (either PP or PA6) with aluminum or copper for a PMH. But PP metal hybrids present less tight, i.e. they exhibit higher gas volume rates compared with PA6 metal hybrids.

Further it can be assumed, looking firstly on the combination of a certain polymer with Al or Cu, that in both combinations an almost identical free volume in the interfacial zone might exist between, e.g. PP and the respective bonded metal, enabling comparable diffusion of helium tracer gas atoms.

The measured differences in gas volume rate between PP and PA6 hybrids can be explained by different types or numbers of intermolecular interactions between the polymers and metals surfaces in the interfacial zones.

This result proves PA6/metal hybrids have a tighter adhesive bonded interface caused by a certainly higher number of physicochemical interactions in the polymer-metal interface. The anhydride functional groups based and the -COOH groups based intermolecular interactions are also conceivable in the adhesion-modified PP

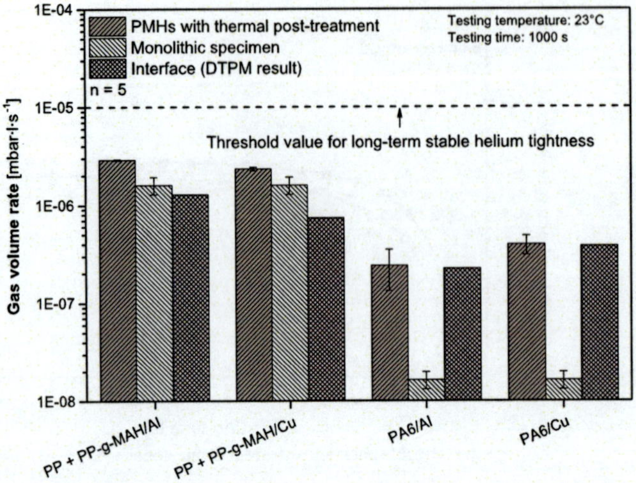

Fig. 9 Gas volume rate of the interfacial zone of various thermally post-treated PMHs using DTPM method; testing time: 1000 s, temperature: 23 °C

[18], thus can be assumed the comparatively pronounced adhesion of the PA6/metal hybrids must origin in a higher physiochemical interaction density.

Despite the measured differences in gas volume rate, all quantified hybrid interfaces are practically permanently helium-tight [28], Fig. 9.

When comparing the measured gas volume rates of the monolithic polymers with those of the related PMHs (Fig. 9) is interesting to find, in case of PP/metal hybrids the difference between the tracer gas permeation rate of the polymer itself and its hybrids is comparatively lower as in case of PA6/metal hybrids, where such difference is distinctly more and about one order of magnitude. Considering that, there is still a further potential to improve the interfacial tightness of injection molded hybrid structures, especially of adhesively bonded polyamide metal hybrids.

Summing up, TGM provides a very suitable method to characterize nano leakage behavior of polymer metal interfaces on the one hand as well as the possible interfacial adhesion of PMHs and allows to rank them quantitatively. Optical inspection of the interfacial zone of hybrids is not a competitive method to the TGM method for characterizing interfacial tightness of PMHs.

DTPM method allows to establish the permeation behavior of an interfacial area of a PMH quantitatively and is a measure for the physicochemical interaction density in the investigated interface.

4 Conclusions

In the present study, variously manufactured polymer-metal hybrid test specimens were investigated for their interfacial tightness properties by means of nondestructive tracer gas measurement (TGM). The suitability of optical methods for characterizing the possible interfacial tightness of these hybrids was also investigated. TGM on monolithic polymer specimens, which have exactly the same geometry and shape as the hybrid specimens, allows the permeation behavior to be quantified under given manufacturing conditions. In the case of full-faced contact, the application of differential tightness and permeation measurement (DTPM), where the measured gas volume rate of the monolithic sample is subtracted from that of the hybrid sample, provides a possibility to quantify the gas diffusion through the free volume of the interface and thus indirectly quantify interfacial adhesion. Further important results of the study are listed below:

(1) If the time-dependent measured curve in TGM resembles a transient permeation curve which strives towards a constant level during testing time, then full-faced contact at the interface exists due to homogeneous dispersed interfacial adhesion between metal surface and surrounding polymer. In this case, the sample represents a tight hybrid.
(2) If there is no full-faced contact between the two dissimilar materials, caused by locally missing adhesion in the interface, the measured gas volume rate presents a constant level from the beginning of the measurement over time. In this case, the sample represents a leaky hybrid.
(3) Neither the light microscopy images nor the scanning electron microscopy images of previously prepared micrographs of the interface near regions of tight and leaky hybrids measured in the TGM revealed interpretable differences in the formation and shape of the interface. Thus, the information content of microscopic images must be seriously questioned in terms of interfacial tightness and adhesion.
(4) In the case of full-faced contact, TGM and a subsequent DTPM enables to detect even small differences in interfacial gas volume rate between thermally post-treated hybrids, such as the discussed PP/metal and PA6/metal hybrids. DTPM is a nanoscopic investigation approach and therefore a worthful and informative analytical method to investigate and characterize polymer-metal interfacial adhesion performance.

It is conceivable to perform TGM at higher temperatures in order to realize economically justifiable measurement times. However, it should be noted that the measurement conditions must not exert any lasting influence on the polymer component, whereby changes in the molecular as well as the super-molecular structure can occur.

The transferability of the method to bonded joints is conceivable with the aid of an adapted test specimen geometry. However, the method is limited when testing so-called "kissing bonds", where the materials are in full-faced contact with each other but there is no mechanical cohesion between metal and polymer [29]. In [19], the authors have already measured the bond strength via pull-out tests of the thermally post-treated hybrids presented in this study and thereby demonstrated that in these hybrids, on the one hand, there is full-faced contact at the interfaces and, on the other hand, there is mechanical cohesion between the polymer and the metal.

Acknowledgements The research results presented in this paper were developed within the framework of a research project "InDiQuant" funded by the German Federal Ministry for Economic Affairs and Energy (BMWi) under the funding code ZF4113815WM8. Furthermore, the authors would like to thank their project partners the companies Gindele GmbH and Dr. Wiesner Steuerungstechnik GmbH for their excellent cooperation collaboration as well as LANXESS Deutschland GmbH, Ineos Holdings Ltd, Wieland-Werke AG and Mitsui Chemicals Europe GmbH for providing of test materials.

References

1. Bonpain, B.: Entwicklung und prozesstechnische Analyse eines festen und mediendichten Kunststoff—Metall—Verbundes. Ph.D. thesis, Technische Universität Dortmund (2017)
2. Kleffel, T., Drummer, D.: Electrochemical treatment of metal inserts for subsequent assembly injection molding of tight electronic systems. J. Polym. Eng. **38**(7), 675–684 (2018). https://doi.org/10.1515/j.polyeng-2017.0432
3. Heckert, A.: Oberflächenstrukturierung von Aluminium mittels gepulster Laserstrahlung für das thermische Fügen an endlosfaserverstärkte Thermoplaste. Ph.D. thesis, Technische Universität München (2019)
4. Heinle, M., Frey, P., Merklein, M., Drummer, D.: Influence of micro formed edges of metal inserts on temperature cycling durability of tight polymer metal hybrids. J. Plast. Technol. **14**(3), 211–232 (2018)
5. Zhao, S., Kimura, F., Yamaguchi, E., Horie, N., Kajihara, Y.: Manufacturing aluminum/polybutylene terephthalate direct joints by using hot water–treated aluminum via injection molding. Int. J. Adv. Manuf. Technol. **107**(11–12), 4637–4644 (2020). https://doi.org/10.1007/s00170-020-05364-0
6. Flock, D.: Heat conduction bonding of plastic-metal hybrid parts. Ph.D. thesis, Rheinisch-Westfälische Technische Hochschule Aachen (2011)
7. Kleffel, T., Frey, P., Merklein, M., Drummer, D.: Analysis of the formation of gap-based leakages in polymer-metal electronic systems with labyrinth seals. J. Polym. Eng. **39**(6), 573–586 (2019). https://doi.org/10.1515/polyeng-2019-0098
8. Ehrenstein, G.W., Amesöder, S., Férnandez, D.L., Niemann, H., Deventer, R.: Werkstoff- und prozessoptimierte Herstellung flächiger Kunststoff-Kunststoff und Kunststoff-Metall-Verbundbauteile. In: Tagungsband zum Berichts- und Industriekolloquium 2003 des SFB 396. Robuste, verkürzte Prozessketten für flächige Leichtbauteile, pp. 149–178, Bamberg, Germany, 15–16 Oct 2003, Meisenbach (2003)
9. Heinle, C., Vetter, M., Brocka-Krzeminska, Z., Ehrenstein, G.W., Drummer, D.: Tight material bonds within mechatronic systems by assembly moulding. J. Plast. Technol. **5**(6), 427–450 (2009)
10. Mitschang, P., Velthuis, R., Emrich, S., Kopnarski, M.: Induction heated joining of aluminum and carbon fiber reinforced nylon 66. J. Thermoplast. Compos. Mater. **22**(6), 767–801 (2009). https://doi.org/10.1177/0892705709105969
11. Ramani, K., Moriarty, B.: Thermoplastic bonding to metals via injection molding for macro-composite manufacture. Polym. Eng. Sci. **38**(5), 870–877 (1998). https://doi.org/10.1002/pen.10253

12. Bischof, C., Bauer, A., Possart, W., Kapelle, R., Schulze, R.D.: Zur Adhäsion in Metall-Polymer-Grenzschichten und ihrer praktischen Nutzung. Acta Polym. **40**(3), 214–221 (1989). https://doi.org/10.1002/actp.1989.010400309
13. Gleich, H.: Zusammenhang zwischen Oberflächenenergie und Adhäsionsvermögen von Polymerwerkstoffen am Beispiel von PP und PBT und deren Beeinflussung durch die Niederdruck-Plasmatechnologie. Ph.D. thesis, Universität Duisburg-Essen (2004)
14. Hallak, A., Mathpati, M., Wilde, J.: Self-healing mechanisms in thermoplastic, media-tight injection-molded housings. In: 2021 44th International Spring Seminar on Electronics Technology (ISSE). 2021 44_{th} International Spring Seminar on Electronics Technology (ISSE), pp 1–7. Bautzen, Germany, 5–9 May 2021, IEEE (2021). https://doi.org/10.1109/ISSE51996.2021.9467648
15. Baldan, A.: Adhesion phenomena in bonded joints. Int. J. Adhes. Adhes. **38**, 95–116 (2012). https://doi.org/10.1016/j.ijadhadh.2012.04.007
16. Frick, A., Stern, C., Muralidharan, V.: Practical testing and evaluation of plastics. Wiley-VCH, Weinheim, Germany (2019)
17. Brostow, W., Chen, I.K., Hnatchuk, N., Hrbacek, J., Hull, A., Pahler, R.H., Srinivasan, A.: Enhanced adhesion of polypropylene to copper substrates. Polym. Test. **63**, 158–162 (2017). https://doi.org/10.1016/j.polymertesting.2017.05.017
18. Chen, M.-A., Li, H.-Z., Zhang, X.-M.: Improvement of shear strength of aluminium-polypropylene lap joints by grafting maleic anhydride onto polypropylene. Int. J. Adhes. Adhes. **27**(3), 175–187 (2007). https://doi.org/10.1016/j.ijadhadh.2006.01.008
19. Frick, A., Rettenberger, M., Spadaro, M.: Highlighting thermal post-treatment for improving long-term media-tightness of polymer-metal hybrids. Academ. J. Polym. Sci. **5**(3) (2021). https://doi.org/10.19080/AJOP.2021.05.555665
20. Byskov-Nielsen, J., Holm, A.H., Højsholt, R., Sá, P., Balling, P.: Testing the permeability and corrosion resistance of micro-mechanically interlocked joints. Appl. Phys. A **104**(3), 975–979 (2011). https://doi.org/10.1007/s00339-011-6473-2
21. Murray, B.R., Leen, S.B., Semprimoschnig, C.O.A., Brádaigh, C.M.Ó.: Helium permeability of polymer materials as liners for composite overwrapped pressure vessels. J. Appl. Polym. Sci. **133** (29) (2016). https://doi.org/10.1002/app.43675
22. Sturm, P., Leuenberger, M., Sirignano, C., Neubert, R.E.M., Meijer, H.A.J., Langenfelds, R., Brand, W.A., Tohjima, Y.: Permeation of atmospheric gases through polymer O-rings used in flasks for air sampling. J. Geophys. Res. **109** (D4), n/a-n/a (2004). https://doi.org/10.1029/2003JD004073
23. Menges, G., Haberstroh, E., Schmachtenberg, E., Michaeli, W.: Menges Werkstoffkunde Kunststoffe, 6th edn. Carl Hanser Verlag, München (2011)
24. Bargel, H.-J., Schulze, G.: Werkstoffkunde. Springer, Berlin Heidelberg (2018)
25. Lamberti, C., Solchenbach, T., Plapper, P., Possart, W.: Laser assisted joining of hybrid polyamide-aluminum structures. Phys. Procedia **56**, 845–853 (2014). https://doi.org/10.1016/j.phpro.2014.08.103
26. Hirchenhahn, P., Al Sayyad, A., Bardon, J., Felten, A., Plapper, P., Houssiau, L.: Highlighting chemical bonding between nylon-6.6 and the native oxide from an aluminum sheet assembled by laser welding. ACS Appl. Polym. Mater. **2**(7), 2517–2527 (2020). https://doi.org/10.1021/acsapm.0c00015
27. Hino, M., Mitooka, Y., Murakami, K., Urakami, K., Nagase, H., Kanadani, T.: Effect of aluminum surface state on laser joining between 1050 aluminum sheet and polypropylene resin sheet using insert materials. Mater. Trans. **52**(5), 1041–1047 (2011). https://doi.org/10.2320/matertrans.L-M2011804
28. Emrich, N., Meyer, S.P., Daub, R.: Ageing behavior of thermally joined hybrids of laser pre-treated metal and thermoplastic polymers. In: Proceedings of the 35th International Conference of the Polymer Processing Society (PPS-35). Proceedings of the 35_{th} International Conference of the Polymer Processing Society (PPS-35), p 20001. Cesme-Izmir, Turkey, AIP Publishing, 26–30 May 2019. https://doi.org/10.1063/1.5142916

29. Désagulier, C., Pérés, P., Larnac, G.: Aerospace Industry. In: Da Silva, L.F.M., Öchsner, A., Adams, R.D. (eds.) Handbook of Adhesion Technology, pp. 1285–1331. Springer International Publishing, Cham (2018)

Development and Characterization of Silicone Pressure-Sensitive Adhesives Modified Sepiolite for Self-Adhesive Tape Applications

Adrian Krzysztof Antosik, Konrad Gziut, Marlena Musik, Piotr Miądlicki, and Mateusz Weisbrodt

Abstract Innovative silicone self-adhesive materials in the form of single-sided tapes with high values of self-adhesive properties were obtained. These materials were obtained by modification of silicone resin with addition of sepiolite. Both pure and hydrochloric acid sepiolite modified was used in this work and characterized by FTIR (Fourier transform infrared), XRD (X-ray diffraction), BET (Brunauer–Emmett–Teller) spectroscopy, among others. The modified adhesive compositions obtained in this way were used to produce silicone adhesive tapes with improved thermal resistance. The adhesive materials obtained with these additives show better cohesion, shrinkage and thermal resistance compared to the composition without modification.

Keywords Silicone pressure-sensitive adhesives · Sepiolite · Acid modification · Cohesion · Thermal resistance

---

A. K. Antosik (✉) · K. Gziut · P. Miądlicki · M. Weisbrodt
Faculty of Chemical Technology and Engineering, Department of Chemical Organic Technology and Polymeric Materials, West Pomeranian University of Technology in Szczecin, Piastów Ave. 42, 71-065 Szczecin, Poland
e-mail: adriankrzysztofantosik@gmail.com

K. Gziut
e-mail: konrad.gziut@zut.edu.pl

P. Miądlicki
e-mail: piotr.miadlicki@zut.edu.pl

M. Weisbrodt
e-mail: mateusz.weisbrodt@zut.edu.pl

M. Musik
Faculty of Chemical Technology and Engineering, Engineering of Catalytic and Sorbent Materials Department, West Pomeranian University of Technology in Szczecin, Piastów Ave. 42, 71-065 Szczecin, Poland
e-mail: marlena.musik@zut.edu.pl

L. F. M. da Silva et al. (eds.), *2nd International Conference on Industrial Applications of Adhesives 2022*, Proceedings in Engineering Mechanics,
https://doi.org/10.1007/978-3-031-11150-1_2

1 Introduction

Pressure-sensitive adhesives (PSA) are a system that actually goes back to the invention of self-adhesive articles in 1935, when R. Stanton Avery manufactured the first coating unit using a wooden cigar box with two holes cut in the bottom. PSA can be defined as a special category of adhesives which, when are in dry form, are permanently tacky at room temperature. Pressure-sensitive adhesives usually exhibit good adhesion to the surface, good resistance to light, oxygen, moisture and plasticizers, adhesion is constant over a very wide temperature range, excellent, low, permanent peel strength and tack, excellent aging performance at room temperature or above. The inherent tackiness of PSAs allows them to wet and quickly adhere under low pressure on a wide variety of substrates. The self-adhesive materials market has been developing very intensively in the last decade. The growing market is the result of expansion in both current and new application areas. Self-adhesive polymers have inherent properties due to their structure which makes them unique. The global market for self-adhesive tapes reached $ 60.4 billion in 2020, expected to grow to $ 79.9 billion by 2025 [1–5].

There are many criteria for the division of PSA, but the most common division is based on their characteristics and purpose, with the basic division based on their chemical basis and method of preparation. Depending on the production process, pressure-sensitive adhesive adhesives can be divided into solvent-based, water-based and solvent-free. The second most popular classification of pressure sensitive adhesives is the breakdown by type of base polymer. In this classification, PSAs are divided into seven groups, respectively: acrylic, rubber, silicone, polyurethane, polyester, polyether and EVA pressure-sensitive adhesive [6–9].

Silicone pressure-sensitive adhesives (Si-PSA) are characterized by high initial adhesion to most surfaces, very high durability, excellent chemical resistance and excellent resistance to environmental conditions (resistance to aging). Silicone PSA are products of specialized use due to specific properties: low glass transition temperature—between −60 and −40 °C, high flexibility of silicone adhesive coatings, low surface tension <20 mN/m, very good adhesion to low-energy substrates, high thermal resistance, hydrophobic, non-toxic. Si-PSA are used for special applications, e.g. for the production of vinyl and aluminum tapes, used for marking traffic routes, fire protection and hazardous places, for joining, sealing and securing cable bundles in cars, and for protecting plastic elements against high temperature and UV radiation [10–14]. Commercially available adhesive tapes based on Si-PSA with high initial adhesion to most surfaces compensate for differences in thermal expansion of the joined materials and are resistant to dynamic stress, thanks to which they ensure a permanent and economical connection of materials. They have been used in the medical industry (e.g. they are included in self-adhesive patches, also patches containing drugs applied trans dermally), in the heavy and automotive industry (as a connector for elements working at elevated temperatures), in the electrical industry (as non-conductive and transdermal tapes). as masking tapes). Currently, tapes characterized by increased thermal resistance are mainly produced on the basis of organic polymers (i.e. polyacrylates, epoxides, rubbers). Despite the higher price, heat-resistant Si-PSA materials are used as insulating materials in district heating and in construction works (when embedding fireplaces); masking in powder coating (baking in furnaces); glass varnishing; thermoprinting or electrical engineering in the insulation of motor windings. Due to the late interest in this group of materials (which was caused by the low chemical reactivity of silicone adhesives themselves), the intensification of research on them can be noted in the last years of the 20th and the beginning of the twenty-first century. Research and development of this group of self-adhesive adhesives are of interest both in the world of science and industry [15–20].

Sepiolite is a low-cost natural clay, magnesium silicate occurring in very small crystals and belongs to the layered silicates [21]. The structure is built of blocks consisting of dimensional silica tetrahedral sheets and one central magnesium octahedral sheet [22]. Sepiolite silica sheets contain some cations like Ca^{2+}, K^{+} and water molecules. Sepiolite has OH functional groups, which strengthens its ability for the adsorption process [23]. Sepiolite has unique microstructure, low density, high specific surface area, fibrous morphology and excellent thermal stability.

Sepiolite modified was used as anti-corrosive pigment for epoxy resin coatings. The modification of sepiolite clay was performed according with different phosphonium-based ionic liquids [24]. Modified sepiolite was further applied for adsorption of oxytetracycline, a common antibiotic in water. The modification involved the use of cetyl trimethylammonium bromide and sodium dodecyl benzene sulfonate to obtain a type of organic sepiolite [25].

In the prepared work, silicone self-adhesive commercial resins were modified with the addition of sepiolite and its acid modifications. For this purpose, sepiolite was modified (acid modification) and characterized by used e.g. FTIR (Fourier transform infrared) spectroscopy, XRD (X-ray diffraction), BET (Brunauer–Emmett–Teller). From the thus obtained modified adhesive compositions silicone pressure-sensitive adhesives tape with improve thermal resistance were obtained. The tapes were tested for typical performance properties such as the adhesion, cohesion, tack, shrinkage and SAFT test.

2 Materials and Methods

2.1 Materials

The materials and reagents used in this study include sepiolite (SEPIFEED), hydrochloric acid (Stanlab, 35–38%, Poland), toluene (Carl Roth, Germany). The Dowsil 2013 was used as a silicone adhesive, while the cross-linking agent was Syl-OFF 4000 (both from The Dow Chemical Company, USA).

Table 1 Symbols of natural and modified sepiolite in this study

Filler	Treatment	Symbols
Sepiolite	Natural, unmodified	SEP
	0.1 M solution of HCl	SEP 0.1
	0.5 M solution of HCl	SEP 0.5
	1.0 M solution of HCl	SEP 1.0

2.2 *Modification of Sepiolite*

Sepiolite was obtained from Sepifeed. Samples of sepiolite were modified with hydrochloric acid (STANLAB, 35–38%) water solutions of different concentrations (0.1–1.0 M). For modification, 100 cm^3 of the appropriate acid solution was used for 10 g of sepiolite sample. Samples were placed in the ultrasonic cleaner for 4 h. The temperature of the water in ultrasonic cleaner was set at 60 °C. The resultant samples were washed with a distilled water until no $Cl-$ ions could be detected in the solution and dried at 100 °C for 24 h. A natural, unmodified sepiolite was labeled as SEP. The names of other obtained samples of modified sepiolite were based on the acid concentration used (e.g., sepiolite modified with 0.1 M solution of HCl was denoted as SEP 0.1).

2.3 *Preparation of One-Side Tapes by Using Silicone Pressure-Sensitive Adhesives*

The pressure-sensitive adhesive (PSA) compositions were based on commercial silicone adhesive (Dowsil 2013) into which a 1.5 wt% of cross-linking agent was introduced. After obtaining a homogeneous mixture, the appropriate filler was added in an amount of 0.5, 1, 3, 5, or 7 parts by weight. The contents of the prepared compositions are shown in Table 1. The obtained adhesive formulations were used to prepare a 30 μm thick coatings on a polyester carrier using an adjustable film applicator. The coated films were then placed in an oven at 150 °C for 10 min to obtain a cross-linked polymer structures. After this process, the obtained silicone pressure-sensitive adhesive films were secured with a polyester films. Finally, strips 2.5 cm wide were cut from the resulting films and used for testing the properties of the adhesive.

2.4 *Methods*

In order to verify the modifications and determine the properties of the samples, various analyses, various analyses were carried out. The Fourier transform infrared spectroscopy (ATR-FTIR, Thermo Scientific Nicolet 380 spectrometer equipped with an ATR diamond plate, Waltham, MA, USA) was implemented within a wavenumber range of 4000–400 cm^{-1} at a resolution of 4 cm^{-1} to identify the chemical bonds. X-ray diffraction (XRD) analyses were performed to determine the structures of the modified montmorillonite samples. The XRD patterns were recorded by an Empyrean PANalytical (Malvern, United Kingdom) X-ray diffractometer with the Cu lamp used as the radiation source in the 2θ range 5–60° with a step size of 0.026. The method of nitrogen sorption at 77 K was performed with the QUADRASORB evo Gas Sorption Surface Area and Pore Size Analyzer in order to determine the textural properties of the montmorillonite samples. The Brunauer–Emmett–Teller (BET) method was applied for the calculation of the specific surface area. The total pore volume (TPV) was estimated utilizing the volume of N2 adsorption at $p/p0 \approx 1$. The density functional theory (DFT) method was utilized to calculate the micropore volume (MPV) and pore size distribution (PSD). The Grain-size distribution was determined in the 0.02–2000 μm range using a laser diffraction particle size analyzer Malvern Mastersizer 2000 (Malvern, UK).

Tack of PSA was examined using universal testing machine (Zwick Roell, Germany) according to international standard Association des Fabricants Europeens de Rubans Auto-Adhesifs AFERA 4015 standard. During the test, the adhesive tape is bent to form a loop and then lowered onto a plate made of steel. This creates an adhesive bond between the tape and the plate with an area of 5 cm^2 (2.5 cm × 2 cm). By tearing the tape off the plate at an angle of 90° (at a speed of 300 mm/min), the machine measures the strength of the joint [13, 26].

Peel adhesion of prepared silicone pressure-sensitive adhesives was also tested with use of Zwick Roell universal testing machine according to (AFERA) 4001 procedures. In order to carry out the test, it was necessary to stick a 2.5 cm wide sample of the tape to a steel substrate with a 2 kg roller. The steel plate with the strip of tape was then placed in the jaws of a testing machine. The tape was peeled off the plate at an angle of 180° at a rate of 300 mm/min. Strength with which the PSA bonded to the substrate, i.e., adhesion, was measured in newtons [N]. Data are presented as the mean value of 10 trials performed on each adhesive [2, 12, 13, 27].

The cohesive strength of the PSA was tested by a shear strength test according to the FTM 8 standard developed by the European self-adhesive labeling industry association (FINAT). A sample of the PSA tape was adhered to an area of 6.25 cm^2 (2.5 cm × 2.5 cm) of a steel plate and loaded with a weight of 1 kg. A machine designed at the Laboratory of Adhesives and Self-Adhesive Materials of the West Pomeranian University of Technology in Szczecin was used to carry out the measurement, which automatically records the time after which the installed weight is no longer held by the adhesive sample that detaches from the steel plate. The test was performed at a temperature of 20 °C and 70 °C [12, 13, 28].

The shrinkage is measured as a percentage or millimeter change in the dimensions of the PVC or PET foil covered with the tested adhesive. After the adhesive has been cross-linked and the foil has been attached to the glass, the sample is conditioned for 8 weeks at 70 °C. Shrinkage greater than 0.5% or 0.5 mm exceeds the allowable value [14, 29].

The shear adhesion failure test (SAFT) was performed to evaluate the heat resistance and shear strength of the PSA films. Failure temperature was measured using a shear tester designed at the Laboratory of Adhesives and Self-Adhesive Materials of the West Pomeranian University of Technology in Szczecin. A 2.5 × 2.5 cm area of PSAT was adhered to stainless-steel plate with use of a 2 kg roller. Then, a weight of 1 kg was hung on the adhesive tape and this system was placed in the machine. The test was carried out in the temperature conditions of 20–250 °C with a temperature increase of 0.5 °C/min. The temperature at which the adhesive failed was recorded as the shear failure temperature [30, 31].

3 Characteristic of Fillers

In sepiolite diffraction spectrum, near 2θ = 7.42°, a strong diffraction peak appears (Fig. 1), which is characteristic of layered two-dimensional lattice structure minerals, proving that sepiolite is a kind of layered fibrous structure mineral similar to concentric circles [32]. The characteristic peaks of sepiolite, according to JCPDS card 13–0595 are shown in the plot. At the concentration of 0.5 M acid, you can notice a significant reduction in the intensity of the peaks corresponding to sepiolite and an increase in the intensity of the peaks corresponding to quartz, as shown by the reflections at 2θ = 26.6°, 36.4°, 42.4°, 20.8° (JCPDS card 85–0930). This is due to the destruction of the crystalline structure of the clay, while quartz is inert to hydrochloric acid.

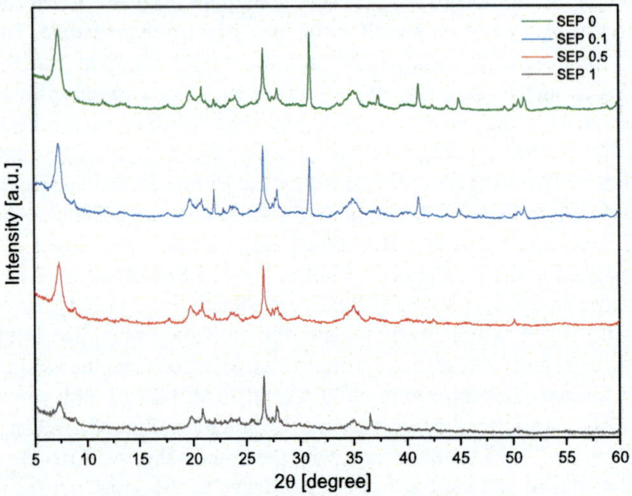

Fig. 1 XRD patterns of sepiolite fillers

Table 2 Surface properties of sepiolite fillers

Sample	BET (m^2/g)	V_{por} (cm^3/g)
SEP 0	101	0.257
SEP 0.1	105	0.26
SEO 0.5	98	0.2461
SEP 1	141	0.29

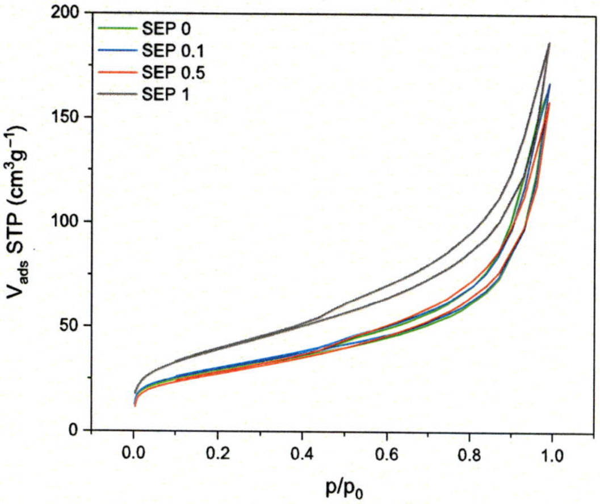

Fig. 2 N_2 adsorption–desorption isotherms of sepiolite fillers

All the isotherms exhibited Type IV isotherm with H3 type hysteresis according to the UPAC classification [33]. The shape of the isotherm indicated unrestricted monolayer-multilayer adsorption and the presence of mesopores and macropores. The shape remains the same after acid treatment. Table 2 presents the surface properties of sepiolite samples. After the acid activation of raw sepiolite, the BET surface area increased from 100 to 140 m^2/g. This could be attributed to the acid leaching of ions from the sepiolite and the formation of silica with central channels [34] (Fig. 2).

The pore size distributions are shown in Fig. 3. All samples have a similar pore distribution. Mainly there are mesopores with a small amount of micropores.

Figure 4 further shows the representative FTIR spectra of raw sepiolite and modified sepiolite. The vibrational bands located in the 4000–400 cm^{-1} range can all be assigned to the characteristic vibrations of clay mineral sepiolite, i.e., the vibrations of the Mg–OH group (3615 cm^{-1}), coordinated water (3566 cm^{-1}), zeolitic water (3384 cm^{-1}, 1652 cm^{-1}); and the bonds of Si–O–Si (1015 and 421 cm^{-1}), Si–O (1200, 1072 cm^{-1} (shoulder), and 977 cm^{-1}) and Mg–O (695 and 647 cm^{-1}) [34]. After the acid treatment the wide band centered at 1015 cm^{-1} (actually consisting of three different bands at 1200, 1072 and 977 cm^{-1}) changed its form and shifted to higher wave numbers, from 977 to 1017 cm^{-1}, indicating the textural changes in the solids. The peaks at 1017, 791, and 457 cm^{-1} agreed with the Si–O–Si bond, confirming that the natural sepiolite transformed into silica after the acid activation.

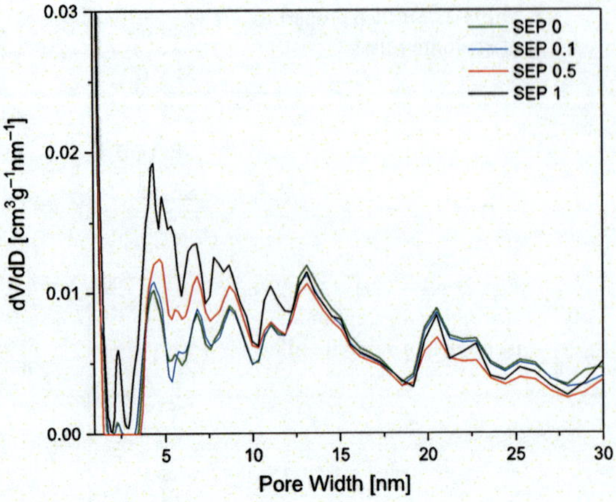

Fig. 3 Pore size distributions of sepiolite fillers

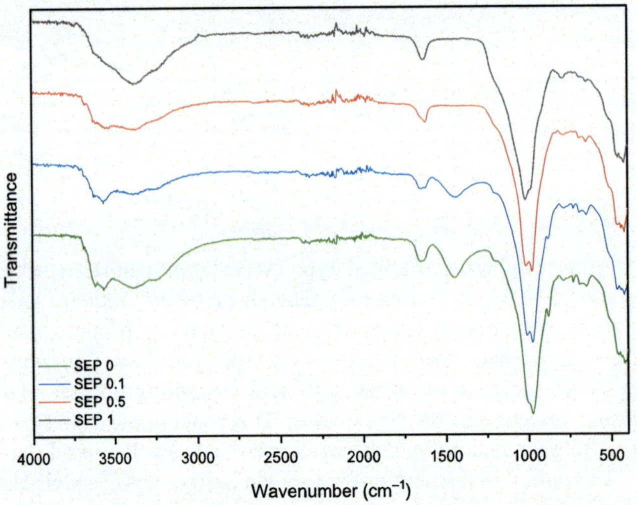

Fig. 4 FTIR spectra of sepiolite fillers

Particle size changes of natural sepiolite before and after acid treatment are shown in Table 3. After acid treatment, the coarsest fraction d(0.5) and d(0.9) of sepiolites were reduced. Changes in the particle size of sepiolite were mainly related to the concentration of the acid solution. After raw sepiolite was treated with 0.1, 0.5, and 1.0 M acid solution, the value of d(0.5) were reduced by 32.3%, 16.9% and 19.1% respectively. On the other hand, after treating the crude sepiolite with the respective acid solutions, the d(0.9) value decreased by 51.0%, 58.7% and 19.1%.

Table 3 Particle size distribution parameters of sepiolite before and after treatment

Sample name	d(0.5)	d(0.9)
SEP 0	237.5	441.7
SEP 0.1	76.8	227.9
SEP 0.5	40.1	259.2
SEP 1	45.4	242.5

4 Results and Discussion

The adhesive properties of the produced tapes are shown in Figs. 5 (adhesion) and 6 (tack).

PSA prepared without filler showed adhesion at the level of 12.7 N/25 mm. Only one filled adhesive showed an increase in adhesion—SEP 0.1 (sepiolite modified with 0.1 M HCl) introduced in the amount of 1 wt%.—13.48 N/25 mm. Moreover, the samples containing SEP 0.1 allowed to obtain adhesives with the highest adhesion for each tested amount of filler. Similar values of adhesion were obtained for the adhesives containing unmodified sepiolite (SEP) and SEP 0.5. However, at higher filler contents (5 and 7 wt%), the adhesion was clearly lower for the samples with SEP 0.5, noting the lowest result among all tested adhesives—9.16 N for a sample containing 5 wt% of the filler). The most unfavorable values of the tested parameter were recorded for SEP_1, for which the adhesion values significantly differed from the adhesive without filler, for each tested amount of additive. With the addition of 0.5 wt%, adhesion of 11.30 N/25 mm was achieved, while with higher contents the values were approx. 10 N/25 mm (the lowest for 3% filler content—9.50 N/25 mm). It should be noted that in the case of unmodified and two modified sepiolites (with

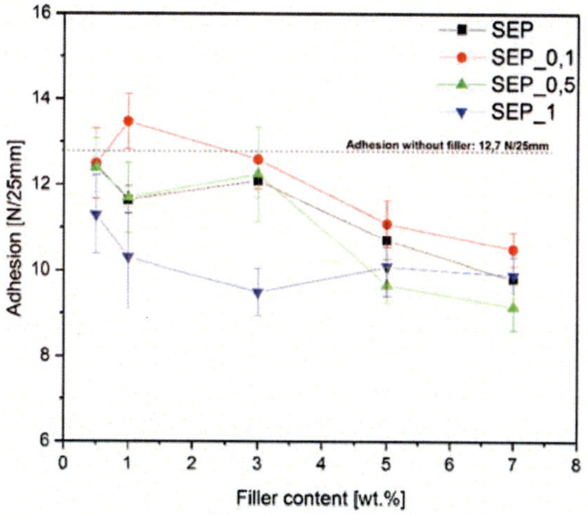

Fig. 5 Adhesion of prepared silicone PSAs with different content of sepiolites

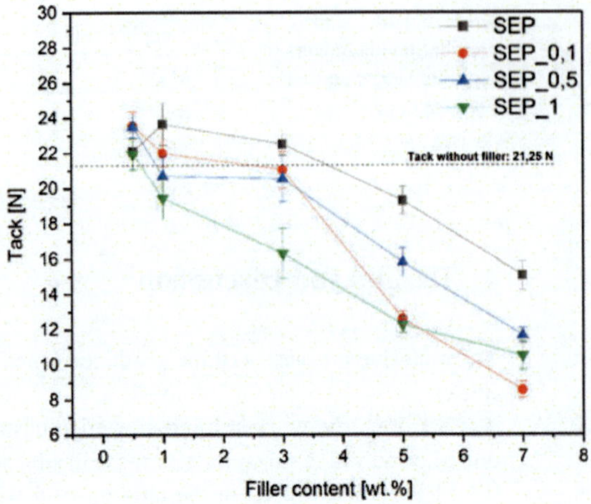

Fig. 6 Tack of prepared silicone PSA with different content of sepiolite

a concentration of 0.1 and 0.5 M HCl), a small addition of the filler (0.5–3 wt%) caused only a slight change in adhesion.

The tack of the adhesive produced without the filler was 21.25 N. Incorporation of each tested filler (modified and unmodified) in the amount of 0.5% by weight cause an increase in tack, especially for samples with SEP 0.1 (23.55 N) and SEP_0.5 (23.5 N). The unmodified sepiolite also caused an increase in tack with the addition of 1 wt% (23.65 N) and 3 wt% (22.5 N). In the case of introducing higher amounts of modified sepiolites (more than 0.5 wt%), only the addition of 1% SEP 0.1 also allowed for higher tack (22 N) than the PSA without filler. With the exception of unmodified sepiolite, whose 1 wt% content caused the greatest tack (23.65 N), the modified fillers caused a decrease in this parameter with the tested additive greater than 0.5 wt%. Moreover, SEP 1.0 exerted the worst influence on tack, with almost every tested additive causing the lowest obtained values—only with the highest tested content (7 wt%) the lowest value was obtained for an adhesive containing SEP 0.1–8.55 N.

Tables 4 and 5 the results of cohesion at room temperature and elevated temperature are collated. It is well known that with increasing adhesives filling, its cohesive properties decrease. This dependence was demonstrated by Si-PSA tapes modified with the addition of sepiolite, which for a filling of 7 wt% both at room temperature and the above temperature showed its relatively low value. The tape peels off during the test after 571 and 10 min, respectively. In the case of modification with the modified filler, the decrease in cohesive properties occurred when tested at elevated temperature.

Results of shear adhesion failure test (SAFT) test determining the maximum operating temperature of the obtained self-adhesive tapes are summarized in Table 6. It was observed that with the increase in the content of sepiolite addition, the resistance of the tested self-adhesive tapes initially increased, depending on the type of filler, between 0.5 and 5 wt% start to fall. The best results were obtained for

Table 4 Cohesion at 20 °C expressed as a time (min) elapsing between the moment burned on the sample until to cohesion/adhesicon failure of one-side tape based on Si-PSA comprising sepiolites

Cohesion at 20 °C (min)				
Content filler in adhesives film (wt%)	Filler			
	SEP	SEP 0.1	SEP 0.5	SEP 1.0
0	>4320	>4320	>4320	>4320
0.5	>4320	>4320	>4320	>4320
1	>4320	>4320	>4320	>4320
3	>4320	>4320	>4320	>4320
5	>4320	>4320	>4320	>4320
7	571	>4320	>4320	>4320

Table 5 Cohesion at 70 °C expressed as a time (min) elapsing between the moment burned on the sample until to cohesion/adhesion failure of one-side tape based on Si-PSA comprising sepiolites

Cohesion in 70 °C (min)				
Content filler in adhesives film (wt%)	Filler			
	SEP	SEP 0.1	SEP 0.5	SEP 1.0
0	>4320	>4320	>4320	>4320
0.5	>4320	>4320	>4320	>4320
1	>4320	>4320	>4320	>4320
3	>4320	>4320	>4320	>4320
5	>4320	>4320	>4320	>4320
7	19	126	124	261

a sample modified with 0.5 wt% SEP 0.5, which means that the modification of sepiolite with 0.5 M hydrochloric acid resulted in the best functional properties.

5 Conclusions

Physical modification of the sepiolite was successfully carried out, obtaining three variants with different properties. The presented fillers have been characterized and their differences between them are presented in graphs and tables. As the acid concentration increased during the modification, the obtained filler was characterized by smaller particle sizes.

The obtained fillers were used to obtain self-adhesive tapes. In all cases, the addition of the filler decreased the shrinkage value, adhesion and stickiness of the adhesive film, and increased the value of cohesion, which is a phenomenon typical for the addition of fillers to pressure-sensitive adhesives [9, 14, 16]. In addition, the value of thermal resistance has increased, which is especially worth it along with

Table 6 Shear adhesion failure test (SAFT) expressed as a temperature (°C) elapsing between the moment burned on the sample until to cohesion/adhesicon failure of one-side tape based on Si-PSA comprising sepiolites

SAFT (°C)				
Content filler in adhesives film (wt%)	Filler			
	SEP	SEP 0.1	SEP 0.5	SEP 1.0
0	150	150	150	150
0.5	214	218	>225	218
1	212	>225	208	>225
3	>225	214	154	220
5	>225	211	140	207
7	174	208	139	172

increasing the amount of acid addition during modification to 0.5 M, with lower filler addition, better properties of the adhesive film are achieved, above this value the properties begin to drop, it may be related to the optimal silicon content in filler particles in correlation to the size of the filler particles and the space they occupy.

In the course of the research, new innovative materials in the form of single-sided tapes with high values of self-adhesive properties were obtained. The new materials may find application in various industries from automotive to aerospace.

Funding This research was funded by the National Center for Research and Development (NCBiR) as part of the LIDER initiative. Project number LIDER/9/0028/L-11/19/NCBR/2020.

References

1. Markets and Marets Raport: Adhesive Tapes Market by Resin Type (Acrylic, Rubber, Silicone), Backing Material (PP, Paper, PVC), Technology (Solvent, Hot-Melt, Water Based), End-Use Industry (Packaging, Healthcare, E&E, Automotive) and Region—Global Forecast to 2025
2. Ebnesajjad, S.: Handbook of Adhesives and Surface Preparation Technology, Applications and Manufacturing 3–13. Elsevier, Oxford (2011)
3. Ciannamea, E.M., Ruseckaite, R.A.: Pressure sensitive adhesives based on epoxidized soybean oil: correlation between curing conditions and rheological properties. J. Am. Oil Chem. Soc. **95**, 525–532 (2018)
4. Czech, Z., Kabatc, J., Bartkowiak, M., Licbarski, A., Mozelewska, K., Kwiatkowska, D.: Novel photoreactive pressure-sensitive adhesives (PSA) based on acrylics containing additionable photoinitiators. Materials **13**, 5151 (2020)
5. Czech, Z., Butwin, A.: New developments in the area of solvent-borne acrylic pressure-sensitive adhesives. J. Adhe. Sci. Tech. **23**, 1689–1707 (2009)
6. Moon, H., Jeong, K., Kwak, M.J., Choi, S.Q., Im, S.G.: Solvent-free deposition of ultrathin copolymer films with tunable viscoelasticity for application to pressure-sensitive adhesives. ACS App. Mater. Inter. **10**, 32668–32677 (2018)

7. Al-Moghazy, M., Mahmoud, M., Nada, A.A.: Fabrication of cellulose-based adhesive composite as an active packaging material to extend the shelf life of cheese. Inter. J. Biol. Macromol. **160**, 264–275 (2020)
8. Athavale, S.P.: Hand Book of Pressure Sensitive Adhesives and Coatings: Pressure Sensitive Adhesives Technology. Notion Press (2018)
9. Benedek, I.: Pressure-Sensitive Adhesives. In Handbook of Adhesive Technology, pp. 471–508. CRC Press (2017)
10. Meng, W., Gao, W., Zhu, L., Liu, M., Qu, X.: Preparation of polyacrylic latex with external crosslinking and its effect on the pressure-sensitive properties. Soft Mater. **18**, 432–442 (2020)
11. Kowalczyk, A., Kowalczyk, K., Weisbrodt, M.: Int. J. Adhes. Adhes. **85**, 286–292 (2018)
12. Antosik, A.K, Ragańska, P, Czech, Z.: Termiczne sieciowanie samoprzylepnych klejów silikonowych nadtelenkami organicznymi. Polimery (2014). https://doi.org/10.14314/polimery.2014.792
13. Antosik, A.K., Bednarczyk, P., Czech, Z.: Jednostronnie klejące taśmy na bazie samoprzylepnych klejów silikonowych—dobór najlepszej kompozycji. Chemik **69**, 95–97 (2015)
14. Antosik, A.K., Bednarczyk, P., Czech, Z.: Aging of silicone pressure-sensitive adhesives. Polym. Bull. (2017). https://doi.org/10.1007/s00289-017-2083-2
15. Han, J., Zhou, Y., Bai, G., Wei, W., Liu, X., Li, X.: Preparation of photo-crosslinkable acrylic copolymer and its debonding property on silicon wafer. J. Adhe. Scie. Tech. 1–13 (2021)
16. Lin, S.B., Durfee, L.D., Ekeland, R.A., McVie, J., Schalau, G.K.: Recent advances in silicone pressure-sensitive adhesives. J. Adhe. Scie. Tech. **21**(7), 605–623 (2007)
17. Lee, K., Tiu, B.D.B., Martchenko, V., Mai, K., Lee, G., Gerst, M., Messersmith, P.B.: A modular strategy for functional pressure sensitive adhesives. ACS App. Mater. Inter. **13**, 3161–3165 (2021)
18. Tong, Y., Yuan, Q., Huang, F.: Preparation and characterization of silicon-containing polyarylacetylene/montmorilloite nanocomposites. J. Macromole. Scie. Part B **58**, 469–488 (2019)
19. Gaom, G., Zhangm, S., Wangm, L., Linm, J., Qim, H., Zhum, J., Chum, M.: Developing highly tough, heat-resistant blend thermosets based on silicon-containing arylacetylene: a material genome approach. ACS App. Mater. Inter. **12**(24), 27587–27597 (2020)
20. Cai, M., Yuan, Q., Huang, F.: Catalytic effect of poly (silicon-containing arylacetylene) with terminal acetylene on the curing reaction and properties of a bisphenol A type cyanate ester. Polym. Int. **67**, 1563–1571 (2018)
21. Demir, I., Savaş, M., Yaprak, H., Doğan, O., Özel, G.: Investigation of effect of sepiolite on the properties of autoclaved aerated concrete, CWB-6 359–365 (2015)
22. Shang, Q., Feng, H., Feng, Z., Chen, N., Tan, L., Qiu, J., Wu, H.: Facile fabrication of sepiolite functionalized composites with tunable dielectric properties and their superior microwave absorption performance. J. Colloid Inter. Sci. **576**, 444–456 (2020)
23. Mortazavi, S., Aghaei, H.: Make proper surfaces for immobilization of enzymes: Immobilization of lipase and α-amylase on modified Na-sepiolite. Int. J Biol. Macromol. **164**, 1–12 (2020)
24. Henriques, R.R., Soares, B.G.: Sepiolite modified with phosphonium ionic liquids as anticorrosive pigment for epoxy coatings. Appl. Clay Sci. **200**, 105890 (2021)
25. Wu, J., Wang, Y., Wu, Z., Gao, Y., Li, X.: Adsorption properties and mechanism of sepiolite modified by anionic and cationic surfactants on oxytetracycline from aqueous solutions. Sci. Total Environ. **708**, 134409 (2020)
26. Duncan, B., Abbott, S., Roberts, R.: Measurement good practice guide no. 26: adhesive tack. National Phys. Lab. (1999)
27. Wilpiszewska, K., Czech, Z.: Citric acid modified potato starch films containing microcrystalline cellulose reinforcement—properties and application. Starch (2014). https://doi.org/10.1002/star.201300093
28. Chruściel, J., Graczyk, T.: Otrzymywanie i właściwości elastomerowych silikonowych warstw antyadhezyjnych na papierze i pergaminie. Polimery **51**, 3 (2006)
29. Stempień, Z., Tokarska, M., Gniotek, K.: UV radiation measurement system for the UV curing of fluids disposed on textiles. Fib. Text. East. Euro. **78**, 59–62 (2010)

30. Tomblin, J., Harter, P., Seneviratne, W., Yang, C.: Characterization of bondline thickness effects in adhesive joints. J. Com. Tech. Res. **24/2**, 332–344 (2002). https://doi.org/10.1520/CTR10562J
31. Yeh, H.-Y., Tandjung, D.: Mixed mode fracture analysis of the lap shear specimen test per ASTM D1002. J. Rein. Plast. Com. (2005). https://doi.org/10.1177/0731684405047771
32. Tang, Q., Wang, F., Liang, M.T.J., Ren, C.: Study on pore distribution and formation rule of sepiolite mineral nanomaterials. J. Nanomaterials, ID 382603 (2012). https://doi.org/10.1155/2012/382603
33. Sing, K.S.W., Everett, D.H., Haul, R.A.W., Moscou, L., Pietrotti, R.A., Rouquérol, J., Siemieniewska, T.: Reporting physisorption data for gas/solid systems with special reference to the determination of surface area and porosity. Pure. Appl. Chem. **57**(4), 603–619 (1985). https://doi.org/10.1351/pac198557040603
34. Yao, Q.-Z., Yu, S.-H., Zhao, T.-L., Qian, F.-J., Li, H., Zhou, G.-T., Fu, S.-Q.: Enhanced potential toxic metal removal using a novel hierarchical SiO_2-$Mg(OH)_2$ nanocomposite derived from sepiolite. Minerals **9**, 298 (2019). https://doi.org/10.3390/min9050298

Influence of Zeolite on the Silicone Pressure-Sensitive Adhesives Properties

Adrian Krzysztof Antosik, Karolina Mozelewska, and Konrad Gziut

Abstract The influence of the filler addition on the self-adhesive properties of silicone adhesives was analyzed. For this purpose, the filler was selected: zeolite and five types of it were tested: zeolite Y, sodium; zeolite Y, hydrogen; zeolite mordenite, sodium; zeolite mordenite, ammonium; zeolite ferrierite, ammonium. The work involved, in the first stage, the modification of a commercial pressure-sensitive silicone adhesive with a crosslinking compound and filler in various concentrations—from 0.1 to 3% by weight. The next stage of the work was to carry out tests involving the production of a self-adhesive material in the form of a one-sided adhesive tape. To this end, the obtained adhesive compositions were coated on a carrier (polyester film), dried to evaporate the solvent, and cut into strips 25 mm wide. Pressure-sensitive adhesive films were tested for self-adhesive properties, i.e. adhesion to steel, tack, shrinkage, cohesion at 20 and 70 °C, and the SAFT test. The main purpose of the presented work is to develop a modification of the silicone pressure-sensitive adhesive in order to increase the functional properties of the obtained tapes.

Keywords Zeolite · Pressure-sensitive adhesives · Silicone based self-adhesives tape · Adhesion

1 Introduction

By the term "adhesives" are meant non-metallic substances capable of holding together at least two surfaces permanently. Adhesives come in many forms, from thin liquids, through light pastes, films, powders, to solids intended to be melted [1].

Industrial adhesives and adhesive sealants have been known for many years, and bonding is considered one of the oldest joining techniques. Until recently, the majority of adhesives and sealants were polymers of natural origin: plant, animal or made of

A. K. Antosik (✉) · K. Mozelewska · K. Gziut
Faculty of Chemical Technology and Engineering, Department of Chemical Organic Technology and Polymeric Materials, West Pomeranian University of Technology, Pulaskiego 10, 70-322 Szczecin, Poland
e-mail: adriankrzysztofantosik@gmail.com

L. F. M. da Silva et al. (eds.), *2nd International Conference on Industrial Applications of Adhesives 2022*, Proceedings in Engineering Mechanics,
https://doi.org/10.1007/978-3-031-11150-1_3

mineral substances. In the early 1900s, synthetic polymer adhesives began to use adhesives made from natural raw materials due to their superior properties, such as high adhesion, greater modifiability and excellent environmental resistance [2, 3].

Due to the raw material base, adhesives can be divided into two groups: adhesives of natural origin and adhesives of synthetic origin. Natural adhesives include so-called animal adhesives (e.g. fish adhesives, animal blood adhesives, casein adhesives), vegetable adhesives (e.g. starch adhesives, soybean adhesives) or adhesives based on minerals or elastomers (e.g. natural rubber adhesives) [4]. On the other hand, adhesives of synthetic origin include, among others, elastomers such as nitrile, butyl, styrene-butadiene rubber or thermoplastic elastomers, polymers (polyesters, polyamides, polyacrylates, vinyl acetate copolymers, polysiloxanes and others) and synthetic resins (e.g. epoxy and polyester resins) [5]. Another criterion for classifying adhesives is their physicochemical properties. Thus, adhesives can be divided into solvent-based adhesives, i.e. adhesives that release volatile solvent compounds (water or organic solvent) during the formation of the joint, and solvent-free, e.g. alloyed or in the form of a film. There is also a classification taking into account the temperature of formation of the adhesive joint. According to this classification, we distinguish between adhesives used cold (below 20 °C) or at room temperature (20–30 °C), adhesives used at elevated temperatures (30–100 °C) and adhesives intended for hot joining at temperatures above (100 °C). Due to the fact that adhesives are available in various forms, for example as liquids, pastes, powders or tapes, they are increasingly used in many industries and in everyday life [6, 7].

Adhesives have been used in areas of life where such effects are desired as the possibility of producing parts and sets with a smooth surface, e.g. production of aircraft structures, joining certain elements of electronic or electrical systems, techniques for joining furniture, kitchen cabinets and other elements related to individual construction [8]. It is also well known to use the adhesive technique in the construction of cars, ships, planes and all those devices that are constantly exposed to strong vibrations and shocks (such as, for example, household appliances: washing machine) [9]. A very important advantage of using adhesives is the ability to connect various types of materials/surfaces, e.g. metals with plastics, wood, ceramics and others. Splicing (gluing) as a joining technique is used in the production of missiles, rockets, ships, while pouring miniature elements of electrical circuits, joining jewelry, strain gauges, etc. in ceilings and walls, in electric motors or in musical instruments, especially stringed ones such as guitar or violin. In addition, adhesives are used to bond tubeless tires, sewers, drinking water pipes, lawn sprinklers, and boat components [10]. In addition, adhesives play a very important role in medicine and dentistry. They are used to protect against the entry of bacteria, in surgery (in bone fractures) and in orthopedics (corrective devices), etc. Adhesives have also been used as insulating or electrically conductive materials, e.g. it is known to use adhesives to protect electrical installation components that pose a risk of electric shock or fire. The possibility of connecting elements of large diameters with thin foils is associated with the use of adhesives in the bookbinding industry, finishing construction works and construction of large-panel and prefabricated elements [11, 12].

The English term "Pressure-sensitive adhesives" from which the PSA abbreviation is derived, often used in professional and German literature, "haftklebstoffe ", perfectly describes the nature and properties of pressure-sensitive adhesives, such as: permanent stickiness and low-pressure adhesion to various surfaces [13]. The main difference between pressure-sensitive adhesives and other types of adhesives is that in the case of PSA, the adhesive surface does not change its properties, both before and after use [14].

Silicones are synthetic organosilicon polymers characterized by high molecular weights and a low TG of approx. −120 °C. Silicone-based pressure-sensitive adhesives are made of two basic components: resin and rubber. The resin is synthesized by reacting a silicone or polysilicon hydrogel with trimethylchlorosiloxane. Rubbers are high molecular weight siloxane polymers [15]. The physical properties of silicone pressure sensitive adhesives depend on the ratio of resin to rubber. Silicone-based pressure-sensitive adhesives are characterized by high thermal (−40 to 300 °C) and chemical resistance (organic solvents, oxidizing and reducing reagents), low surface energy (thanks to which they adhere well to silicone decorative surfaces) and high resistance to external conditions (oxygen, UV radiation, humidity). Silicone PSAs dampen vibrations well and have very good electrical insulating properties [16].

Due to numerous applications, zeolites are one of the dynamically developing materials. These are crystalline aluminosilicates that have significant applications in adsorption, catalysis, in the chemical and petroleum industries. They show good biocompatibility and are non-toxic. Over the last several decades, completely new applications have been found for them: in medicine (as drug carriers, hemostatic agents, antibacterial agents) or as coating materials (for the creation of the so-called zeolite coatings). These types of coatings are used, for example, in orthopedic implants [17]. This large number of applications results in a constant and increasing demand for zeolite materials, which inspires the use of various types of zeolites. In this study, five types of zeolites were investigated to obtain products with the desired properties [18].

The aim of the study was to determine whether and how zeolite materials affect the properties of self-adhesive adhesives. The zeolites differed in the type (zeolite Y, zeolite mordenite, zeolite ferrierite) and the types of ions (sodium; hydrogen; ammonium). Four different concentrations were tested in the study: 0.1, 0.5, 1 and 3 wt%, then one-sided adhesive tapes were produced, tested and conclusions were drawn.

2 Materials

The following materials were used in the tests:

- Zeolite: zeolite Y, sodium; zeolite Y, hydrogen; zeolite mordenite, sodium; zeolite mordenite, ammonium; zeolite ferrierite, ammonium (Alfa Aesar USA),
- Silicone resin of the MQ type—Q-7358 (Dow, USA),

- Toluene (Carl Roth, Germany),
- Crosslinking agent—dichlorobenzoyl peroxide—DClBPO (Gelest, USA).

Q7358 silicone resin was mixed with various types of zeolite and a crosslinker (1.5 wt% 2-4-dichlorobenzoyl peroxide-DCLBPO). The effect of the filler addition was tested by adding the following zeolites to the composition: 0.1, 0.5, 1 and 3 wt% (based on the polymer). The composition was then coated onto a polyester film with a thickness of 50 g/m^2 using a semi-automatic coater (basis weight 45 g/m^2). The adhesive film was then placed in a drying duct for 10 min at 110 °C to cross-link the adhesive film.

3 Methods

The time it takes to increase the viscosity (twice or four times) of the original mixture is potlife. The tests are carried out at room temperature and the test is started immediately after mixing the composition [19].

Another investigated property is tack, which indicates the ability of an adhesive to bond briefly without pressure when in contact with another surface. It can also be defined as the force needed to separate a surface after a short time. The tests were carried out in accordance with the technical standard FINAT FTM9 [20].

Adhesion is one of the most important phenomena in the case of bonding. It is defined as the interaction of surfaces and layers of two bodies or phases. It is closely related to the action of the forces of interfacial attraction on the contact surfaces of both materials. The work test was performed in accordance with the FINAT FTM 1 method [21].

Cohesion is responsible for the strength of the adhesive joint. In addition to adhesion, it is one of the most important properties of self-adhesive tapes and adhesives. Its value is influenced by such factors as: test temperature, concentration, type of cross-linking compounds or the thickness of the adhesive film. It is determined by the FINAT FTM 8 method. The measurement was carried out at room and elevated temperature (70 °C) [22].

The shrinkage of pressure sensitive adhesives is defined as the reduction in the surface area of the tape from its original size. It is an important mechanical and functional property, especially in applications where it can cause surface deformations. In the case of pressure sensitive adhesives, the change in film size is indicated in millimeters or percent. A change of more than 0.5 mm or 0.5% exceeds the allowable shrinkage in the production technology of self-adhesive products [23].

The resistance to elevated temperature was tested using the SAFT test. The test is performed in the same way as in the case of cohesion, but the temperature during the test is not constant, but it is increased by about 2 °C/min to the temperature of 230 °C [24].

This publication does not provide detailed descriptions of research methods due to the fact that they have been described in detail in the previous articles by the authors [25, 26].

4 Results and Discussion

In order to conduct research on the influence of various types of zeolite, tests were carried out on the Q 7358 silicone resin from DOW. The pressure-sensitive adhesives without modification is characterized in the Table 1. The initial pressure-sensitive adhesives shows quite good self-adhesive properties, but in the case of thermal resistance, the temperature measurement result is relatively low (147 °C). Tapes made of this adhesive could not be used in industry. Therefore, it was decided to modify the silicone pressure-sensitive adhesive with zeolite.

The second table presents the characteristics of the fillers used. The highest surface area was characterized by sodium zeolite 900 Y g/m^2, while the lowest was characterized by mordenite sodium zeolite and ferrierite ammonium zeolite. In the presented work, three different types of zeolites were investigated: zeolite Y, zeolite mordenite and zeolite ferrierite. In the case of the first zeolite, two different types were tested: sodium and hydrogen. For zeolite mordenite it was sodium and ammonium, while for zeolite ferrierite only ammonium (Table 2).

The Fig. 1 shows the effect of the type and amount of zeolite on the pot-life of the obtained pressure-sensitive adhesives. In the figures, the viscosity of the adhesive without filler is marked with a black line. For the first type of zeolite, i.e. zeolite Y, the

Table 1 Physical properties of tape prepared with Q2–7358

Tack (N)	Adhesion (N/25 mm)	Cohesion (h)		Maximum work temperature (°C)
		20 °C	70 °C	
6.9	10.2	>72	>72	147

Table 2 Characteristics of the fillers used

Name	Structure	Surface area
Zeolite Y, sodium; (Z-YNa)	SiO_2: Al_2O_3 mole ratio 5.1:1	900 m^2/g
Zeolite Y, hydrogen; (Z-YH)	SiO_2: Al_2O_3 mole ratio 30:1	780 m^2/g
Zeolite mordenite, sodium; Z-MNa)	SiO_2: Al_2O_3 mole ratio 13:1	425 m^2/g
Zeolite mordenite, ammonium; (Z-MNH$_3$)	SiO_2: Al_2O_3 mole ratio 20:1	500 m^2/g
Zeolite ferrierite, ammonium; (Z-FNH$_3$)	SiO_2: Al_2O_3 mole ratio 20:1	400 m^2/g

viscosity values are similar for both variants. In both graphs, the values increase over the course of the test, while for silicone PSA without filler, the values are constant. The highest values are obtained for compositions containing the highest amount of filler. In the figure it can be seen that the initial viscosity of the adhesives after modification with zeolite Y causes a decrease in viscosity in the first days (compared to the original adhesive), while from the third day all concentrations exceed the limit value of 20 Pas, which characterizes the adhesive without modification. Figure 1c and 1d show sodium mordenite zeolite and ammonium mordenite zeolite, respectively. In this case, for the sodium zeolite, the viscosity also exceeds that of the starting adhesive after three days, while for the ammonium zeolite, this viscosity is practically exceeded already on the second day, for all concentrations. As in the previous cases, the viscosity increases with time and is the highest for a concentration of 3%. For sodium zeolite the final viscosity (for all concentrations) is higher than in the other cases (above 30 Pas). Fig. 3e shows a diagram for ammonium ferrierite zeolite. Also for compositions with this filler, practically after the second day, the viscosity exceeds the viscosity of the adhesive without the filler, but reaches a final value above 30 Pas.

The third table shows the effect of zeolite on the shrinkage value of silicone pressure-sensitive tapes made of the Q2-7358 adhesive. Compared to the starting adhesive, i.e. adhesive without filler, all modifications had a positive effect on the shrinkage value, i.e. decreased its value. Comparing the sodium and hydrogen zeolite, better results were obtained for the hydrogen zeolite in practically all the concentrations used. This suggests that the hydrogen ions have a positive effect on the shrinkage value compared to sodium ions. For sodium and ammonium morganite zeolite the values are very similar, while ammonium zeolite has a slightly better effect on shrinkage. The best result was obtained for zeolite ferrierite ammonium for which the concentration of 1 and 3 wt%. are within the acceptable range for adhesive tapes (even after 7 days). It follows that there would be no need to modify them before use (Table 3).

Figure 2 shows the adhesion results of the tapes obtained. For the starting adhesive, it was 10.2 N/25 mm. As can be seen in the diagram, all the fillers used increased the adhesion value, even with the smallest amount of filler added. Sodium modernite zeolite turned out to be the best. In its case, however, a rapid decrease was observed after a concentration of 0.5% to the lowest value, similar to the starting adhesive. For ammonium mordenite zeolite the values are also high, but they do not drop as quickly as for sodium mordenite zeolite. They are more stable over the range of concentrations. For zeolite Y, sodium zeolite had a better effect on adhesion, while zeolite hydrogen turned out to be the worst. The addition of ammonium ferrierite zeolite increases the adhesion value to about 12 N/mm, while it remains at a constant level in the entire range of the tested concentrations.

One of the most important properties of pressure-sensitive adhesives is tack (Fig. 3). For all the tapes obtained, the results were higher than for tapes made of the adhesive without modification. The highest values were obtained for sodium mordenite zeolite (for a concentration of 0.1% by weight), however (as in the case of adhesion) this value decreases intensively and from 1 wt% it shows the lowest values compared to other modifications. It follows that zeolite with sodium ions

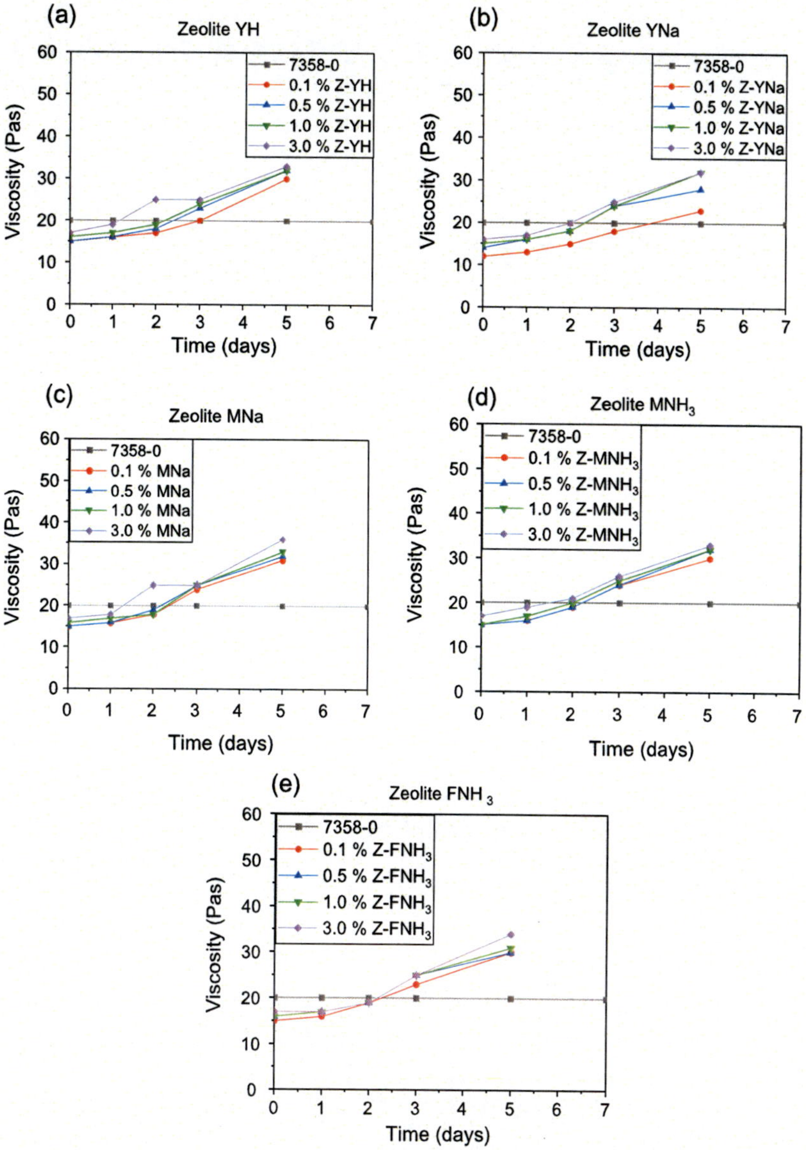

Fig. 1 Influence zeolites on pot-life of Q2-7358 composition: **a** zeolite Y, hydrogen; **b** zeolite Y, sodium; **c** zeolite mordenite, sodium; **d** zeolite mordenite, ammonium; **e** zeolite ferrierite, ammonium

Table 3 Influence zeolites on shrinkage of Q2-7358 adhesives films

Zeolites	Content (% wt)	Viscosity (Pas)											
Time (days)		10 min	30 min	1 h	3 h	8 h	24 h	2 days	3 days	4 days	5 days	6 days	7 days
0.0		0.421	0.412	0.641	0.9	0.96	1.02	1.1	1.14	1.325	1.325	1.325	1.325
Z-YNa	0.1	0.291	0.321	0.378	0.396	0.435	0.501	0.512	0.601	0.687	0.712	0.712	0.712
	0.5	0.284	0.301	0.361	0.387	0.421	0.482	0.503	0.594	0.604	0.696	0.696	0.696
	1.0	0.297	0.301	0.314	0.369	0.408	0.439	0.476	0.503	0.574	0.621	0.659	0.659
	3.0	0.213	0.241	0.264	0.302	0.381	0.402	0.487	0.541	0.601	0.638	0.638	0.638
Z-YH	0.1	0.27	0.31	0.309	0.347	0.425	0.647	0.814	0.961	0.961	0.961	0.961	0.961
	0.5	0.24	0.27	0.298	0.326	0.414	0.502	0.714	0.801	0.847	0.841	0.841	0.841
	1.0	0.2	0.21	0.241	0.301	0.4	0.412	0.621	0.674	0.701	0.738	0.738	0.738
	3.0	0.12	0.147	0.207	0.287	0.314	0.397	0.407	0.412	0.452	0.54	0.55	0.561
Z-MNa	0.1	0.324	0.354	0.389	0.457	0.501	0.541	0.591	0.645	0.697	0.736	0.754	0.754
	0.5	0.301	0.321	0.367	0.423	0.482	0.524	0.589	0.624	0.683	0.701	0.745	0.745
	1.0	0.2	0.21	0.241	0.301	0.421	0.514	0.621	0.674	0.701	0.738	0.738	0.737
	3.0	0.122	0.147	0.207	0.287	0.314	0.397	0.41	0.412	0.452	0.54	0.55	0.621
Z-MNH_3	0.1	0.314	0.324	0.457	0.485	0.496	0.521	0.525	0.525	0.525	0.525	0.525	0.525
	0.5	0.3	0.31	0.441	0.454	0.47	0.499	0.499	0.499	0.499	0.499	0.499	0.499
	1.0	0.214	0.256	0.274	0.321	0.388	0.409	0.452	0.523	0.601	0.638	0.638	0.638
	3.0	0.142	0.177	0.378	0.412	0.433	0.48	0.499	0.517	0.517	0.517	0.517	0.517
Z-FNH_3	0.1	0.2	0.297	0.312	0.368	0.402	0.435	0.535	0.564	0.564	0.564	0.564	0.564
	0.5	0.15	0.273	0.3	0.355	0.392	0.412	0.515	0.531	0.531	0.531	0.531	0.531
	1.0	0.074	0.229	0.245	0.259	0.323	0.346	0.369	0.389	0.401	0.407	0.434	0.434
	3.0	0.124	0.187	0.245	0.301	0.347	0.412	0.414	0.432	0.453	0.453	0.453	0.453

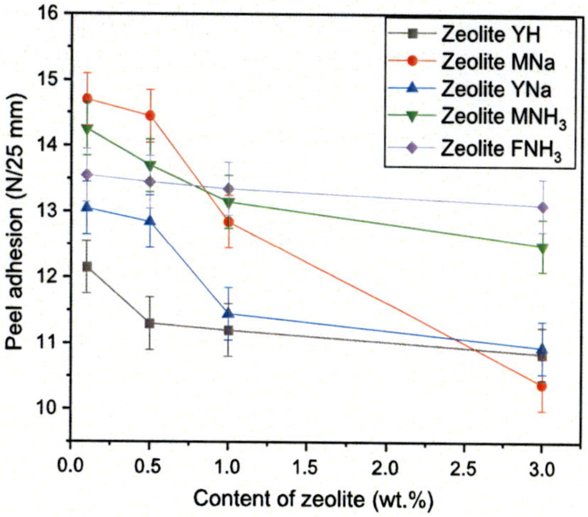

Fig. 2 Influence zeolites on peel adhesion of Q2-7358 composition

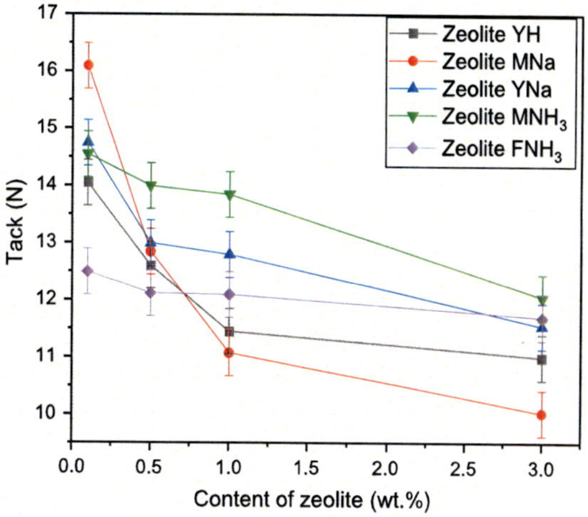

Fig. 3 Influence zeolites on tack of Q2-7358 composition

increases the adhesion and tack to very high values, but only at low concentrations. The ammonium mordenite zeolite shows a lower value at a concentration of 0.1 wt% but from 0.5 wt% these values are the highest. In its case, no intense decline in value was observed either. For the sodium zeolite Y, the values are higher compared to the hydrogen zeolite. Zeolite ferrierite shows intermediate values, as in the case of adhesion, over the entire concentration range. The Y zeolite values are lower compared to the mordenite zeolite. This may be due to the higher surface area of these zeolites.

Tables 4 and 5 show the values of cohesion at room temperature 20 °C and elevated temperature 70 °C. As in the case of the initial adhesive, the values are maximum: over 72 h it means that the modification did not reduce the cohesion value. The cohesion

Table 4 Cohesion at 20 °C expressed as a time (min) elapsing between the moment burned on the sample until to cohesion/adhesion failure of one-side tape based on Si-PSA comprising zeolites

Cohesion at 20 °C (h)					
Content filler in adhesives film (wt%)	Zeolite				
	Z-YNa	Z-YH	Z-MNa	Z-MNH$_3$	Z-FNH$_3$
0	>72	>72	>72	>72	>72
0.1	>72	>72	>72	>72	>72
0.5	>72	>72	>72	>72	>72
1.0	>72	>72	>72	>72	>72
3.0	>72	>72	>72	>72	>72

Table 5 Cohesion at 70 °C expressed as a time (min) elapsing between the moment burned on the sample until to cohesion/adhesion failure of one-side tape based on Si-PSA comprising zeolites

Cohesion at 70 °C (h)					
Content filler in adhesives film (wt%)	Zeolite				
	Z-YNa	Z-YH	Z-MNa	Z-MNH$_3$	Z-FNH$_3$
0	>72	>72	>72	>72	>72
0.1	>72	>72	>72	>72	>72
0.5	>72	>72	>72	>72	>72
1.0	>72	>72	>72	>72	>72
3.0	>72	>72	>72	>72	>72

above 72 h is the limit value for industrial tapes, therefore all tapes produced could be used industrially.

For belts operating at elevated temperatures, the so-called SAFT test. The results for the tapes obtained are shown in Table 6. Practically for all modified tapes, the values were higher compared to the starting adhesive in practically the entire concentration range. The highest values were obtained for the concentration of 0.1 wt% of the filler (above 225 °C), and the lowest for the concentration of 3 wt%. When comparing different types of zeolites, higher results are obtained with mordenite zeolite. For zeolites with sodium ions, the thermal resistance is higher, as in the case of hydrogen and ammonium ions. Zeolite ferrierite shows intermediate values compared to other zeolites.

5 Conclusion

As a result of the research, self-adhesive tapes with good properties compared to the starting adhesive were obtained. Various types of zeolites were successfully applied to the modification: zeolite Y, sodium; zeolite Y, hydrogen; zeolite

Table 6 Shear adhesion failure test (SAFT) expressed as a temperature (°C) elapsing between the moment burned on the sample until to cohesion/adhesion failure of one-side tape based on Si-PSA comprising zeolites

SAFT (°C)					
Content filler in adhesives film (wt%)	Zeolite				
	Z-YNa	Z-YH	Z-MNa	Z-MNH$_3$	Z-FNH$_3$
0	147	147	147	147	147
0.1	>225	>225	>225	>225	>225
0.5	199	189	221	219	191
1.0	191	179	219	197	182
3.0	101	85	219	194	170

mordenite, sodium; zeolite mordenite, ammonium; zeolite ferrierite, ammonium. The best results were obtained for mordenite zeolite in comparison to zeolite mordenite and zeolite ferrierite. It was also checked, how the sodium, hydrogen and ammonium ions contained in zeolites affect the self-adhesive properties when used to modify the pressure-sensitive adhesives. Sodium ions turned out to be the best, as they significantly improve the initial properties of the tape.

The obtained best tapes showing increased thermal resistance could be used in heavy industry for joining elements operating at increased temperature. Regardless of the type of zeolite filler for filling 0.1% wt the highest thermal resistance was obtained each time.

Funding This research was funded by the National Center for Research and Development (NCBiR) as part of the LIDER initiative. Project number LIDER/9/0028/L-11/19/NCBR/2020.

References

1. Mapari, S., Mestry, S., Mhaske, S.T.: Developments in pressure-sensitive adhesives: a review. Polym. Bull. **78**, 4075–4108 (2021). https://doi.org/10.1007/s00289-020-03305-1
2. Keimel, F.A.: Historical development of adhesives and adhesive bonding. In: Pizzi, A., Mittal, K.L. (eds.) Handbook of Adhesive Technology, pp. 10–21. Taylor & Francis Group, LLC, New York (2003)
3. Tenorio-Alfonso, A., Sánchez, M.C., Franco, J.M.: Preparation, characterization and mechanical properties of bio-based polyurethane adhesives from isocyanate-functionalized cellulose acetate and castor oil for bonding wood. Polymers **9**, 132 (2017). https://doi.org/10.3390/polym9040132
4. Dunky, M.: Wood adhesives based on natural resources: a critical review Part I. Protein-based adhesives. Prog. Adhes. Adhes. **6**(3), 199–332, (2020). https://doi.org/10.1002/9781119846703.ch8
5. Cruz, B.A.S., Martinez, A.M.M., et al.: Evaluation of crosslinking reaction in adhesive based styrene-butadiene elastomers using infrared spectroscopy. In: Theophile, T. (ed.) Infrared

Spectroscopy-Anharmonicity of Biomolecules, Crosslinking of Biopolymers, Food Quality and Medical Applications, pp. 117–137
6. Sandberg, D.: Additives in wood products—today and future development. In: Kutnar, A., Muthu, S.S. (eds.) Environmental Impacts of Traditional and Innovative Forest-Based Bioproducts, pp. 105–172, Springer, Singapore
7. Petrie, E.M.: Adhesive bonding of textiles: principles, types of adhesive and methods of use. Joining Text 225–274. https://doi.org/10.1533/9780857093967.2.225
8. Barnes, T.A., Pashby, I.R.: Joining techniques for aluminium spaceframes used in automobiles: part II—adhesive bonding and mechanical fasteners. J. Mater. Process. Technol. **99**(1–3), 72–79 (2000). https://doi.org/10.1016/S0924-0136(99)00361-1
9. Xu, Y.L., Xia, Y.: Monitoring of Temperature Effects in Structural Health Monitoring of Long-Span Suspension Bridges, pp. 177–216. CRC Press
10. Messler, R.W., Jr.: Joining composite materials and structures: some thought-provoking possibilities. J. Thermoplast. Compos. Mater. **17**(1), 51–75 (2004). https://doi.org/10.1177/0892705704033336
11. Yim, M.J., Li, Y., et al.: Review of recent advances in electrically conductive adhesive materials and technologies in electronic packaging. J. Adhes. Sci. Technol. **22**(14), 1593–1630 (2008). https://doi.org/10.1163/156856108X320519
12. Li, Y., Wong, C.P.: Recent advances of conductive adhesives as a lead-free alternative in electronic packaging: materials, processing, reliability and applications. Mater. Sci. Eng. R. Rep. **51**(1–3), 1–35 (2006). https://doi.org/10.1016/j.mser.2006.01.001
13. Pakdel, A.S., Gabriel, V., et al.: A sequential design approach for in situ incorporation of cellulose nanocrystals in emulsion-based pressure sensitive adhesives. Cellulose **27**(18), 10837–10853 (2020). https://doi.org/10.1007/s10570-020-03060-6
14. Jovanović, R., Dubé, M.A.: Emulsion-based pressure-sensitive adhesives: a review. J. Macromol. Sci. Part C: Polym. Rev. **44**(1), 1–51 (2004). https://doi.org/10.1081/MC-120027933
15. Antosik, A.K., Mozelewska, K., Piątek-Hnat, M., et al.: Silicone pressure-sensitive adhesives with increased thermal resistance. J. Therm. Anal. Calorim. (2021). https://doi.org/10.1007/s10973-021-11048-y
16. Antosik, A.K., Mozelewska, K., Czech, Z., et al.: Influence of montmorillonite on the properties of silicone pressure-sensitive adhesives: preparation of a double-sided tape based on the best composition. SILICON **12**, 1887–1893 (2020). https://doi.org/10.1007/s12633-019-00295-2
17. Wang, J., Wang, Z., Guo, S., et al.: Antibacterial and anti-adhesive zeolite coatings on titanium alloy surface. Microporous Mesoporous Mater. **146**, 216–222 (2011). https://doi.org/10.1016/j.micromeso.2011.04.005
18. Zhuopeng, W., Jihong, Y., Ruren, X.: Needs and trends in rational synthesis of zeolitic materials. Soc. Rev. **41**, 1729–1741 (2012). https://doi.org/10.1039/C1CS15150A
19. Si, J., Li, Y., Yu, X.: Curing behavior and mechanical properties of an eco-friendly cold-mixed epoxy asphalt. Mater. Struct. **52**, 81 (2019). https://doi.org/10.1617/s11527-019-1382-5
20. O'Connor, A.E., Willenbacher, N.: The effect of molecular weight and temperature on tack properties of model polyisobutylenes. Int. J. Adhes. Adhes. **24**(4), 335–346 (2004). https://doi.org/10.1016/j.ijadhadh.2003.11.005
21. Baldan, A.: Adhesion phenomena in bonded joints. Int. J. Adhes. Adhes. **38**, 95–116 (2012). https://doi.org/10.1016/j.ijadhadh.2012.04.007
22. Feldstein, M.M., Kireeva, P.E., Kiseleva, T.I., et al.: A new class of pressure-sensitive adhesives based on interpolymer and polymer-oligomer complexes. Polym. Sci. Ser. A **51**(7), 799–814 (2009). https://doi.org/10.1134/S0965545X09070074
23. Czech, Z.: Development in the area of UV-crosslinkable solvent-based pressure-sensitive adhesives with excellent shrinkage resistance. Eur. Polym. J. **40**(9), 2221–2227 (2004). https://doi.org/10.1016/j.eurpolymj.2004.05.012
24. Do, H.S., Park, Y.J., Kim, H.J.: Preparation and adhesion performance of UV-crosslinkable acrylic pressure sensitive adhesives. J. Adhes. Sci. Technol. **20**(13), 1529–1545 (2006). https://doi.org/10.1163/156856106778666462

25. Antosik, A.K., Bednarczyk, P., Czech, Z.: Aging of silicone pressure-sensitive adhesives. Polym. Bull. **75**(3), 1141–1147 (2018). https://doi.org/10.1007/s00289-017-2083-2
26. Antosik, A.K., Weisbrodt, M., Mozelewska, K., et al.: Impact of environmental conditions on silicone pressure-sensitive adhesives. Polym. Bull. **77**, 6625–6639 (2020). https://doi.org/10.1007/s00289-019-03099-x

Optimization of Plasma-Assisted Surface Treatment for Adhesive Bonding via Artificial Intelligence

Jeroen Jordens, Bart Van Doninck, Nasrulloh R. B. Satrio, Alejandro Morales Hernández, Ivo Couckuyt, Inneke Van Nieuwenhuyse, and Maarten Witters

Abstract In this research, Artificial Intelligence (AI) was used to support the optimization of six bonding process parameters for maximal joint strength and minimal production costs. Two industrial bonding processes were investigated, one from electronic potting and another from the manufacturing industry. The focus was on optimizing the plasma treatment of the substrate materials. Two approaches for optimization were compared, namely the traditional approach where the adhesive expert proposes experiments and interpret the results, and an AI approach with Bayesian optimization and Gaussian process models. Similar joint strengths could be achieved via the Bayesian optimization approach with 40% less budget to find the optimum compared to the traditional approach. Additionally, in the electronic potting process, the AI approach resulted in 18% reduction in production cost, while achieving a similar joint strength, compared to the traditional approach. Ageing of the samples

J. Jordens (✉) · B. Van Doninck · M. Witters
Flanders Make, Oude Diestersebaan 133, B-3920 Lommel, Belgium
e-mail: jeroen.jordens@flandersmake.be

B. Van Doninck
e-mail: bart.vandoninck@flandersmake.be

M. Witters
e-mail: Maarten.Witters@flandersmake.be

N. R. B. Satrio · I. Couckuyt
Ghent University-Imec, Technologiepark-Zwijnaarde 126, B-9052 Gent, Belgium
e-mail: nasrulloh.loka@ugent.be

I. Couckuyt
e-mail: Ivo.Couckuyt@UGent.be

A. M. Hernández · I. Van Nieuwenhuyse
Data Science Institute and Research Group Logistics, UHasselt, Agoralaan Gebouw D, B-3590 Hasselt, Belgium
e-mail: alejandro.moraleshernandez@uhasselt.be

I. Van Nieuwenhuyse
e-mail: inneke.vannieuwenhuyse@uhasselt.be

Flanders Make, VCCM, Lommel, Belgium

L. F. M. da Silva et al. (eds.), *2nd International Conference on Industrial Applications of Adhesives 2022*, Proceedings in Engineering Mechanics,
https://doi.org/10.1007/978-3-031-11150-1_4

did not result in a significant drop in joint strength nor changes in failure type or mechanism. This indicates that AI can support adhesive experts to find the optimal bonding process settings and manufacture robust and cost-efficient adhesive bonds.

Keywords Plasma surface treatment · Process optimization · Bayesian optimization · Cataplasma ageing · Artificial Intelligence

1 Introduction

Several industries such as aerospace, automotive, civil construction and manufacturing industry are on the one hand increasingly replacing traditional metallic parts by industrial polymers and on the other hand applying adhesive bonding processes instead of mechanical joining and welding [1–6]. The main reasons are the desired characteristics of polymers and adhesive bonding such as high strength-to-weight ratio, design flexibility, corrosion resistance and fatigue resistance. However, polymers show typically low surface adhesion properties so that achieving good adhesion bonding and efficient joining is a major challenge for several industrial polymers [5, 7–9]. Plasma treatment has been reported as one of the most effective, cost efficient and environmental friendly treatment methods to improve adhesive bonding for polymers [1, 5, 10].

Plasma treatment is believed to induce chemical changes such as the formation of oxygen-containing groups on the surface and changes in surface roughness [5, 7, 8]. These chemical reactions are very complex processes, requiring detailed and extensive experimental campaigns to optimize conditions for plasma treatment for every particular case of selected polymer and type of plasma [11]. The most studied factors impacting the effectiveness of plasma treatment are plasma power, velocity at which plasma treatment is applied, number of passes, distance between nozzle and surface and time between plasma and adhesive application [11–14]. The velocity and number of passes are often converted into one parameter, the plasma exposure time. This plasma exposure time is proportional to the number of passes and inversely proportional to the velocity. All these studies are, however, performed on different substrates and plasma systems and show therefore different optimum settings and sometimes even different trends. Several authors reported for example continuous improvement of the bonding strength till a certain plateau value with increasing plasma exposure times [15–18]. This plateau value was, however, obtained after 1 s, 2.5 s, 6 s, 60 s and 6 min in respectively the papers of Gonzalez et al., Williams and Hicks, Zhang et al., Anagreh et al. and Blackman et al. [14–18]. In addition, other articles report an increase in bonding strength till an optimum, followed by a rapid decline when the plasma exposure time is further enlarged [11, 13]. Also for the other relevant plasma parameters, no generally valid optimum can be defined for all substrates or plasma systems. This indicates clearly the need for a good optimization strategy for optimizing the plasma settings before applying them in industrial bonding processes.

Traditionally, the optimal process parameters are determined through expert knowledge and extensive experimental campaigns [19, 20]. This approach is, however, costly, depends highly on the experience and knowledge of the process designer and generally yields suboptimal results with respect to production costs.

Bayesian Optimization (BO) is one of these optimization techniques, often used for expensive black box problems. BO builds a surrogate model, which is a cheap approximation of the time-consuming objective function. It iteratively refines the surrogate model in search of the global optimum. Some work using Artificial Intelligence (AI), more specifically machine learning for adhesive bonding optimization, was published recently [21–23]. Valdes et al. applied the surrogate-based genetic algorithm approach to define the optimal settings for adhesive type, surface finishing method and overlap length in order to maximize the shear load while minimizing the remaining fiber in the bonding [22]. First a Design of Experiments (DoE) was developed and an accurate global surrogate model build upfront. Afterwards, a genetic algorithm was applied to the surrogate to find the optimum, without updating the surrogate model further. Single-objective and multi-objective BO are proposed by Pruksawan et al. to optimize the bonding strength of an epoxy adhesive in function of the adhesive composition, mixing ratio and curing temperature [21]. It scalarizes the objective functions into one cost function which is optimized, which does not necessarily lead to solutions distributed evenly across the Pareto front [24, 25]. To the authors best knowledge, this single or multi-objective BO technique was not yet applied on optimization of plasma settings during the adhesive bonding process.

In this study, we apply single- and multi-Objective Bayesian Optimization, based on the expected hypervolume improvement [24, 26] for optimization of plasma treatment in two adhesive bonding processes. The optimization is benchmarked against optimal settings defined in a more traditional approach via an adhesive expert and contact angle measurements. The objective for optimization are six plasma settings, namely plasma power, nozzle-surface gap distance, number of passes, plasma velocity, time between plasma and adhesive application and surface cleaning. In addition, similarly to the paper of Pruksawan et al., we employ an ensemble approach by also incorporating an active learning procedure that only accounts for exploration but apply it on plasma optimization instead of adhesive formulation and curing [21]. This active learning procedure will help to search in unexplored regions and make the surrogate model more accurate. This approach adds additional focus on pure exploration which help to get better approximation of the real function landscape in the area within the search domain.

Table 1 Thickness of ABS substrates

Virgin ABS		Recycled ABS	
Thickness (mm)	Tolerance (mm)	Thickness (mm)	Tolerance (mm)
5	0.15	3	0.2
3.5	0.15	4	0.1
		4	0.2
		5	0.15
		5	0.2

2 Materials and Methods

2.1 Materials

Two industrial bonding processes were investigated, one from electronic potting and another from the manufacturing industry. In the former process, two PolyPhenylene Sulfide (PPS) (PPS Ryton R4-200NA, Sirris) where bonded together by a 2-component expoxy adhesive (Araldite 2011, Huntsman). In the latter process, from the manufacturing industry, virgin ABS (ABS sheet PMMA CAP, 001BY0071, Sensoplast) and recycled ABS (recycled ABS virgin capped 001DT0043, Royalite) were bonded via a 2-component PolyUrethane (PU) (ADEKIT A 280/400 BK, Sika). All substrates were machined to dimensions according to the DIN EN 1465 standard. The thickness of the PPS substrates was 2.5 ± 0.04 mm. The thickness of the ABS substrates varied between the different batches as can be seen in Table 1. Within each test condition, substrates of different thickness were mixed for the repetitions to ensure no bias of the substrate thickness on the optimal settings.

2.2 Adhesive Bonding Process

The bonding process consisted of five main steps, namely conditioning of the substrates and adhesives, cleaning of the substrates, plasma treatment, adhesive application and curing. First, the adhesive and substrates were stored overnight at room temperature (23 ± 1 °C) to ensure similar temperatures between the substrates and the adhesive. Next, cleaning was applied by wiping the substrates with a lint-free tissue wetted with 2-propanol (IPA) (>99.5%, Makiba). Plasma treatment was performed by moving an atmospheric plasma torch over the substrates via a robot (HC 10 YR-1-06VXHC10-A00, Yaskawa). Plasma was provided via a Raantec plasma jet G-12P. The conditions for cleaning and plasma treatment were varied among the different tests according to the range and stepsize shown in Table 2. After plasma treatment was completed, the substrates were transferred to Teflon molds to ensure fixation to the right position. The adhesive was applied via a manual glue gun on

Table 2 Process variables

Variable	Range	Stepsize	Unit	Type
Cleaning	[0–1]	1	–	Discrete
Plasma power	[300–500]	10	W	Continuous
Distance plasma nozzle—substrate	[1–20]	0.1	mm	Continuous
Plasma speed	[5–250]	1	mm/s	Continuous
Number of passes	[1–50]	1	–	Discrete
Time between plasma and adhesive application	[1–120]	1	min	Continuous

one of the substrates (PPS or virgin ABS) when the time between plasma and adhesive application was reached. Then, the other substrate was put on top so that an overlap length of 12.5 ± 0.25 mm was obtained and a weight was put on top to fixate the substrates. Spacers are placed in the mold to obtain a bond thickness of 0.2 mm for the PPS substrates and 1 mm for the ABS substrates. Finally, the samples are moved to a curing oven (Heat event 100/150 Weisstechnik) and cured 3 h at 80 °C for the PPS samples and 16 h at 70 °C for the ABS samples. After curing was complete, the samples were cooled down to room temperature before testing started. Five repetitions are conducted for each process setting to ensure reproducible results.

2.3 *Testing*

The contact angle was determined via a goniometer (PGX + model 68–76, TMI) by placing a droplet of distilled water on the substrate surface. Unless otherwise specified, the contact angle was measured 1 min after plasma application, on cleaned substrates with plasma power of 500 W, a speed of 83.3 mm/s and 1 pass. The distance between nozzle and substrate was fixed at 0.85 cm when the number of passes and speed was varied and 0.75 cm when power and time were investigated.

The joint strength was evaluated via standard DIN EN 1465 lapshear tests on a universal test bench (AGS-X-HC Shimadzu) with 20 kN load cell (C05 1/500 AGSX, Shimadzu). Two non shift wedge grips (Shimadzu) were used to clamp the samples in the test bench. Samples were tested at a test speed of 2 mm/min till failure and the maximum break stress and failure mode were determined.

The ageing behavior of the samples were tested according the ISO 9142 standard for cataplasma tests. Samples were wrapped in tissues soaked in deionized water and placed in a sealed plastic bag. Next, the samples are stored in an oven (Heat event 100/150 Weisstechnik) at 70 °C for 1 week. After one week, the samples are taken from the oven and directly placed in a climate chamber at −20 °C (Tensile Event TC/70/40/180 Weisstechnik). The samples are kept in this climate chamber for 16 h at −20 °C. Finally, the samples are removed from the climate chamber and allowed to heat till room temperature (23 ± 1 °C). The aged samples are tested with the same procedure as described in the previous paragraph. As a reference, five samples for each test condition are also stored for the duration of the cataplasma tests at room temperature (23 ± 1 °C) to study the effect of intrinsic ageing. These samples are also tested with the same procedure for lap shear testing.

2.4 Cost Model

To calculate the production cost of the sample, a simplified cost model was generated. This cost models calculates for each step in the production process the cost per piece produced based on the input parameters as can be seen in Table 3. The steps considered in this cost model are planning of the production batch, cleaning, plasma operation, adhesive application and curing. The costs were estimated based on the material costs (e.g. cost of consuming cleaning agent), the equipment cost (e.g. cost to use the plasma device), electricity costs (e.g. cost of electricity consumption by the plasma device) and the salary costs. Material costs were estimated based on the purchase price of the materials and estimations of the amounts consumed per piece. Equipment costs were calculated starting from the purchase price of the equipment, depreciated over 10 years. Next, it was assumed that the equipment could be used 200 days per year and that the equipment was occupied 60% of the time to obtain an equipment cost per hour. Based on the equipment time required per piece, one can hence calculate the equipment cost per sample. Electricity costs were estimated based on average electricity costs for industrial consumers and the estimated power consumption of the equipment per piece [27]. Salary costs were calculated from the average salary of an operator and junior production planner in Belgium and the estimated time needed for each step [28, 29]. It should be noted that this cost model is a rough estimate of the real production cost. It was only created to test the capability of the optimization tool to cope with multi-objective optimization.

2.5 Bayesian Optimization

Bayesian Optimization (BO) consists of two key components: the surrogate model and the acquisition function [30, 31]. The surrogate model approximates the true objective function, while the acquisition function identifies the most informative points to sample next, balancing exploration and exploitation. The main idea is to iteratively evaluate certain areas according to the acquisition function until the optimum is satisfactory or until we run out of the evaluation budget. In other words, BO uses the surrogate model to identify interesting points in the search space to evaluate next. This sequential procedure is a trade-off between evaluating likely optimal points as well as points that could improve the accuracy of the surrogate model.

We employ the Stochastic Gaussian process regression model (SGPR) with the Matern 5/2 kernel [32]. The SGPR is able to handle the heteroscedastic noise which

Table 3 Example of cost model for PPS bonding at 450 W plasma power, with speed of 20 mm/s and 4 passes

Process step	Subcategory	Cost	Unit	Amount required per sample	Unit	Total cost per sample
Production planning	Production planner time	56.2500	€/h	0.01	h	0.750 €
Cleaning	IPA solvent	2.2500	€/L	*0.01*[a]	L	0.023 €
Cleaning	Lint free tissue	0.0275	€/pc	*2*[a]	pc	0.055 €
Cleaning	Cleaning time operator	0.0104	€/s	*30*[a]	s	0.313 €
Plasma	Electricity cost plasma	0.2000	€/kWh	*0.001*[a]	kWh	0.000 €
Plasma	Depreciation plasma device	0.0009	€/s	*5*[a]	s	0.004 €
Plasma	Operator time	0.0104	€/s	*39*[a]	s	0.406 €
Plasma	Yaskawa robot	0.0012	€/s	*5*[a]	s	0.006 €
Gluing	Operator time	0.0104	€/s	30	s	0.313 €
Gluing	Adhesive	0.4416	€/mL	0.42	mL	0.184 €
Gluing	Static mixer	1.7700	€/pc	0.008	pc	0.015 €
Curing	Operator time	0.0104	€/s	2	s	0.021 €
Curing	Electricity cost oven	0.2000	€/kWh	0.4	kWh	0.080 €
Curing	Depreciation oven	0.4155	€/h	0.1	h	0.042 €
Total cost						2.21 €

[a]Values in italic are input variables dependent on the process settings

is inherently present in the adhesive bonding case. This is done by incorporating the replication variance of the experiments into the model.

We propose to use two acquisition functions: the hypervolume-based multi-objective expected improvement (HVEI) and predictive variance [26]. The HVEI is used for optimization purposes while Predictive Variance is used for pure exploration purposes by minimizing the uncertainty of the model.

2.6 Optimization Settings

For the adhesive bonding case, we want to maximize the break stress or minimize the manufacturing cost. Since these two are potentially conflicting objectives, instead of searching for one optimum point, we want to get a set of Pareto front instead [33].

We consider six variables to be optimized for those two objectives. Details of search space are provided in Table 2.

As a benchmark, optimization was initially performed via a "traditional approach" in which optimal settings were defined via two steps. First, contact angle measurements were performed as quick and easy measurements to define the optimal plasma settings. One parameter was changed at a time and the settings resulting in the lowest contact angle of each experimental set were selected in the end. Next, these settings were used in a complete bonding process to generate lapshear samples which were tested for break stress. The budget required to find the optimal solution was also defined based on the required number of experiments, labor time, materials and equipment time.

The BO process is as follows. First, 30 and 18 initial data points are generated for respectively PPS and ABS bonding according to a Latin Hypercube Design [34]. Then, these points are evaluated on the problem with 5 replications. Second, an SGPR model is trained on this initial data set. Third, a new set of query points are identified by maximizing the acquisition function. In particular, the acquisition function is optimized by fine-tuning the best Monte Carlo sample using the gradient-based L-BFGS-B optimizer [35]. New data points are obtained in batches of 6 data points, with 5 repetitions for each condition. Each batch contains one point for single-objective optimization, one for multi-objective optimization and four points for uncertainty reduction. Then, we evaluate these query points on the problem with the same number of replications of the initial data. We then repeat these procedures starting from the second step until we are satisfied with the optimum. The optimization procedure was stopped when the budget for conducting experiments was equal to the budget required to find the optimal solution in the traditional approach. The general workflow of the BO process is illustrated in Fig. 1.

The full Bayesian Optimization approach is implemented in Trieste, a Bayesian Optimization toolbox on python [36]. For fast and easy interaction with the user we also provide an interactive toolbox based on Jupyter Notebook Widgets and Voila-dashboard library.

3 Results

3.1 Benchmark

The contact angle for PPS substrates is shown in Fig. 2 as function of (a) time between plasma and adhesive application, (b) plasma power, (c) number of passes, (d) plasma speed and (e) distance nozzle-substrate. One parameter was changed at a time to limit the number of possible experiments. From each set of experiments, the most optimal setting was selected.

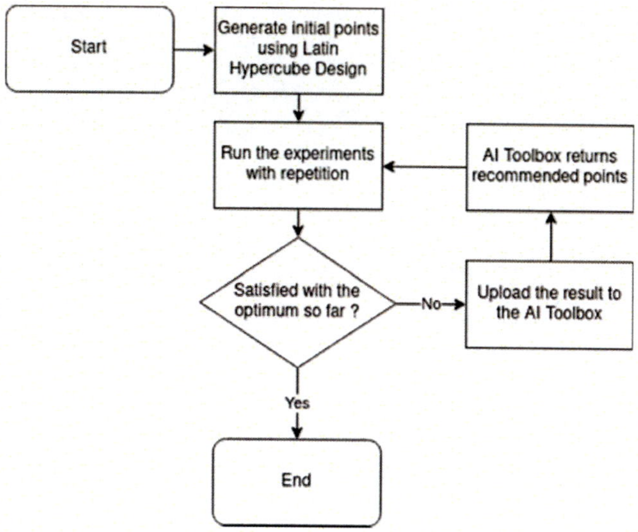

Fig. 1 General flowchart BO process

A contact angle of 77.8° was observed for the untreated PPS substrates which corresponds to values reported in literature for untreated PPS films [15]. A significant drop in contact angle can be observed, in most cases, when plasma is applied. Figure 2a shows for example that the contact angle dropped to 24.5° when the standard settings as defined in Sect. 2.3 are applied and time between plasma application and measurement is 1 min. The most steep incline in contact angle, from 24.5° to 31.4°, is observed in the first 30 min after plasma application. After the first 30 min, the contact angle increases further although at a smaller rate. Similar trends are also observed in literature in which the ageing behavior is related to the restoration of the (polymer) surface to the state before plasma treatment [12, 37, 38]. The results shown in Fig. 2 a indicate that the most optimal setting to apply glue is 1 min after plasma application. As this might not be feasible in practice, 30 min was selected as the most optimal setting for time between plasma and glue application.

In Fig. 2b, the contact angle shows a steady decline with increasing plasma power till 450 W, after which no further decline could be noticed. The effect of plasma treatment increases with higher power levels till a level at which a further increase does not lead to more changes. 450 W was selected as the most optimal power setting as this yielded the lowest contact angle while still consuming less energy compared to 500 W.

A slight decrease in contact angle with the number of passes is shown in Fig. 2c, with a local optimum around 4 passes for the studied range. Also in literature, similar trends are observed which are explained by the fact that longer exposure times to plasma result on the one hand in more polar groups and hence lower contact angles [1, 13]. On the other hand, too long exposure times can degrade the polymer chains

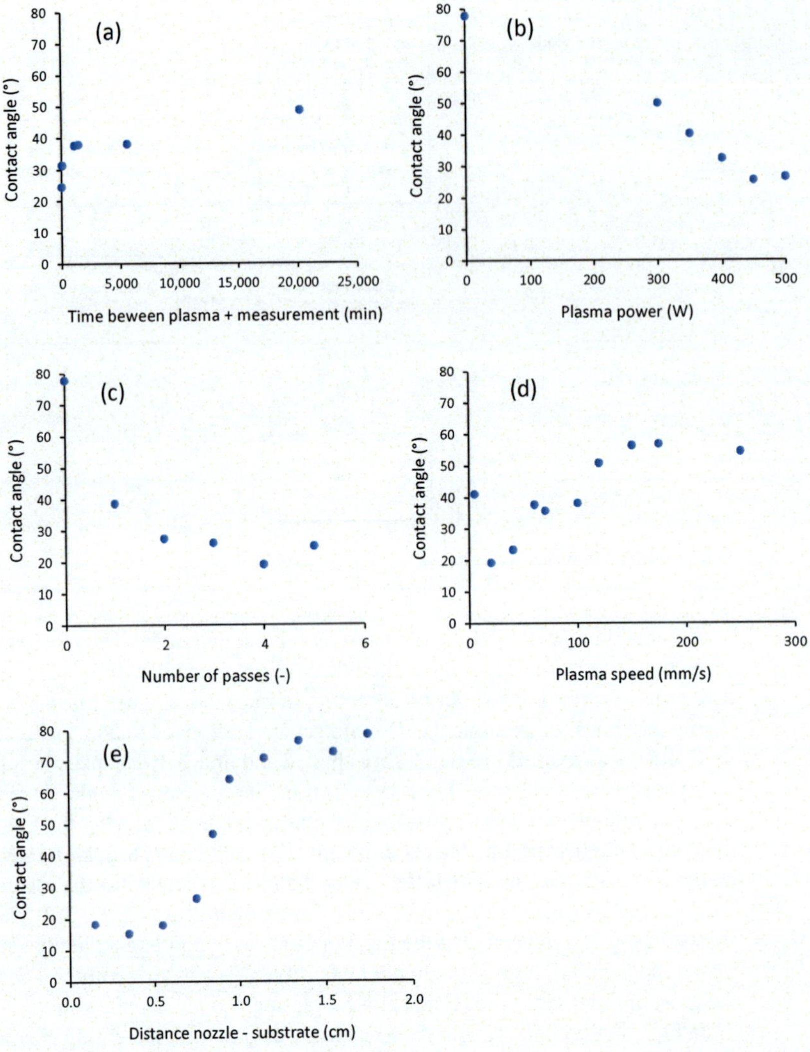

Fig. 2 Contact angle measurements for PPS bonding as function of **a** time between plasma and adhesive application, **b** plasma power, **c** number of passes, **d** plasma speed and **e** distance nozzle-substrate

at the surface and hence increase the contact angle again [37]. These effects also explain why in Fig. 2d an optimum exists around 20 mm/s. At the lowest plasma speed of 5 mm/s, plasma treatment could be too long so that degradation occurs. At speeds above 20 mm/s the exposure time becomes shorter, so that less polar groups are created and hence the contact angle rises again. Therefore, an optimal setting of exposure time and hence number of passes and plasma speed might exist which in our conditions can be defined at 4 passes and 20 mm/s.

Finally, one could notice from Fig. 2e that the distance between plasma nozzle and substrate has a significant effect on the contact angle. The contact angles ranges from 15.4° to 78.4°, which is the largest range among all tested variables. The contact angle stays more or less constant at distances between 0.15 till 0.55 cm. Only from 0.75 cm onwards the contact angle rises with increasing distances till a contact angle around 77°, the value for untreated substrates, is obtained at a distance of 1.35 cm. This again corresponds to literature and is explained by the reduction in plasma density and heat [12]. The lowest contact angle was obtained at 0.35 cm distance.

The most optimal settings obtained from Fig. 2 are shown in Table 4 in column "optimal settings individual". However, when combining these optimal settings, which are obtained by varying one parameter at a time, this can still result in infeasible settings. Black lines were for example observed visually on the substrates when using the settings of Table 4 in column "optimal settings individual". This damage on the substrates is probably caused because of too high temperatures on the substrate surface which could degrade the polymer surface. As substrate damage is unwanted, it was decided to adapt the settings to the ones in the column named "benchmark". The distance between nozzle and substrate was increased to 0.75 cm as the contact angle was still low at this setting in Fig. 2 while the increased distance would decrease the plasma heat reaching the surface. Testing with these benchmark settings resulted indeed in a unburned surface. For the ABS bonding case, optimal settings are retrieved from the equipment supplier as shown in Table 5.

3.2 *Optimization*

Lap shear tests are performed for PPS and ABS bonding according to the benchmark settings defined in Sect. 3.1 and shown in Tables 4 and 5. These results are shown in Fig. 3 by the black squares. A production cost of 2.21 EUR and 2.09 EUR per piece and average lap shear strengths of 9.1 MPa and 5.9 MPa were observed for respectively PPS and ABS bonding. The difference in production cost between PPS and ABS is mainly related to the difference in oven curing time and material costs for adhesive and substrates. No deviations on the productions costs are visible as these are calculated based on a cost model with the same plasma settings as input variables for each repetition. Significant deviations on the measured break stresses could, however, be observed. The standard deviation on the measurements is displayed as error bars in Fig. 3. This standard deviation on the break stress seems to be larger for ABS bonding compared to PPS bonding. This could be due to the fact that recycled ABS is used as one of the substrates in the ABS bonding case. The composition of recycled materials and hence their properties and surface layer deviates typically more compared to virgin materials such as virgin ABS.

Next, the Bayesian optimization routine was applied. After about 60% of the budget for finding an optimal solution was consumed, the single-objective points

Table 4 Optimal settings defined for PPS bonding

Parameter	Optimal settings individual	Benchmark	Single-objective optimization	Multi-objective optimization
Cleaning	Yes	Yes	Yes	No
Plasma power (W)	450	450	330	340
Distance nozzle substrate (cm)	0.35	0.75	0.35	0.35
Plasma speed (mm/s)	20	20	115	60
Time between plasma and adhesive application (min)	30	30	64	10
Number of passes over substrate	4	4	49	10
Average break stress (over 5 repetitions)	–	9.1 MPa	9.6 MPa	9.5 MPa
Standard deviation on break stress	–	0.7	0.4	1.1
Production cost (EUR/pc)	–	2.21	2.28	1.81

Table 5 Optimal settings defined for ABS bonding

Parameter	Benchmark	Single-objective optimization	Multi-objective optimization
Cleaning	No	No	No
Plasma power (W)	500	340	300
Distance nozzle substrate (cm)	0.80	0.30	0.70
Plasma speed (mm/s)	250	65	5
Time between plasma and adhesive application(min)	15	13 min	34 min
Number of passes over substrate	1	1	1
Average break stress (over 5 repetitions) (MPa)	5.9	6.7	6.9
Standard deviation on break stress	1.0	1.1	1.6
Production cost (EUR/pc)	2.09	2.09	2.15

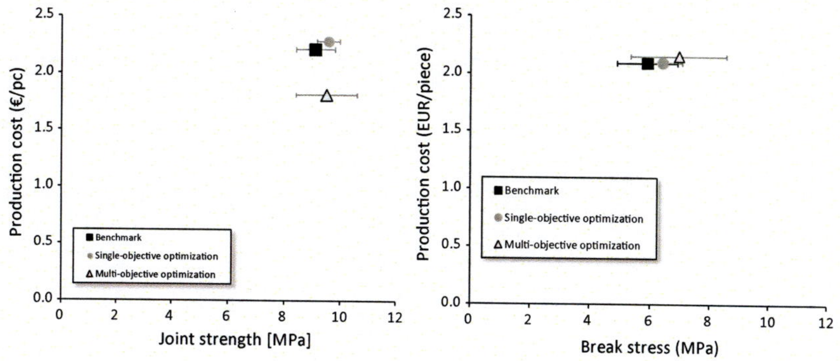

Fig. 3 Production cost as function of the joint strength for the most optimal points obtained via the traditional, single objective and multi-objective approach for PPS bonding (left) and ABS bonding (right). The symbols represent the average value and the error bars the standard deviation

(grey circles) shown in Fig. 3 were obtained. Comparable break stresses and production costs of 2.28 EUR and 2.09 EUR per piece and 9.6 and 6.7 MPa for respectively PPS and ABS bonding were obtained via this Bayesian optimization approach. This indicates that Bayesian optimization is able to find optimal plasma settings at a lower cost to find the optimal solution. In the studied cases, 40% reduction in budget to find the optimal solution could be achieved.

The remainder of the budget was used to complete multi-objective optimization with the goal to maximize the joint strength while minimizing the production cost. The final proposed solutions via multi-objective optimization are presented by grey triangles in Fig. 3. Average joint strengths of 9.5 and 6.9 MPa and production costs of 1.81 EUR and 2.15 EUR per piece were obtained for respectively PPS and ABS bonding. A reduction of 18% in production cost was achieved for PPS bonding via the multi-objective optimization while no significant reduction in production cost was obtained for ABS bonding. More research on the cost composition the processes and exact processes is needed to clarify the exact reasons for this observation. This is, however, outside the scope of this research which wanted to investigate if multi-objective optimization of plasma treatment could be performed via BO. These results indicate that, BO yields comparable break stresses as the traditional approach with lower costs to find the optimal solution and that it can be used to reduce the production cost for some cases.

3.3 *Ageing Tests*

Ageing tests were conducted only for the PPS bonding case to test the reliability of the plasma settings under ageing conditions. Figures 4 and 5 show respectively the failure mode and break stress of these tests. Samples treated with plasma settings obtained

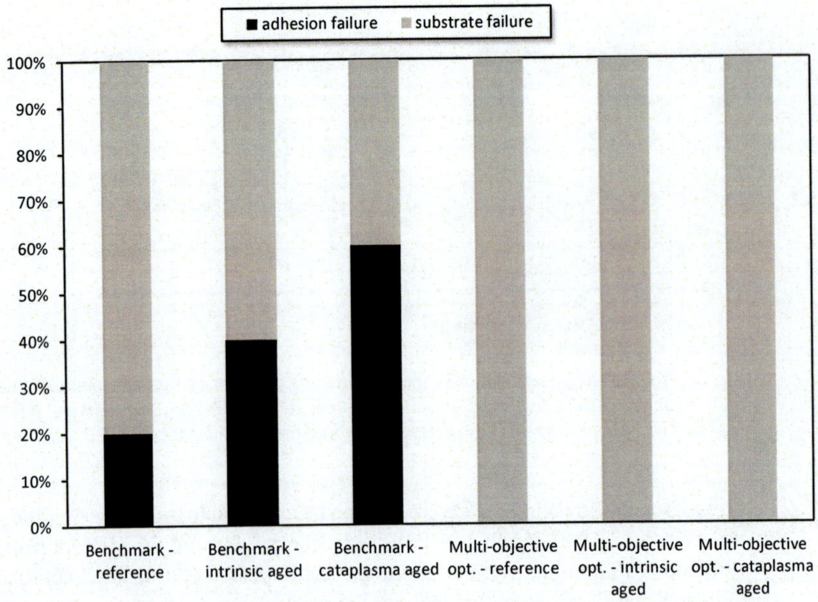

Fig. 4 Percentage of failure mode for samples treated with the plasma settings achieved by the benchmark and multi-objective settings for PPS bonding

via the benchmark and multi-objective settings are displayed respectively on the left and right of each graph. Within each approach, three tests are conducted, namely one reference, another intrinsic aged and cataplasma aged samples. The former are unaged samples tested directly after bonding. The intrinsic aged samples are stored at room temperature for the same duration as the cataplasma tested samples. The latter, cataplasma aged samples, followed an ageing procedure as described in Sect. 2.3.

From Fig. 4, one can observe that the traditional approach resulted in an increase in percentage of adhesively failed samples due to ageing (both intrinsic and cataplasma). The number of tested samples is, however, limited to five so that one should be careful to draw statistical relevant conclusions. The break stress, did increase slightly during cataplasma ageing but the overlapping error bars make it difficult to tell if this change is significant. The multi-objective approach in contrast only showed substrate failure for both aged and unaged samples. In Fig. 5, it can be seen that the break stress did not drop significantly due to cataplasma ageing for both the benchmark and multi-objective settings. These results indicate that reliable plasma treatment can be obtained via the multi-objective BO approach.

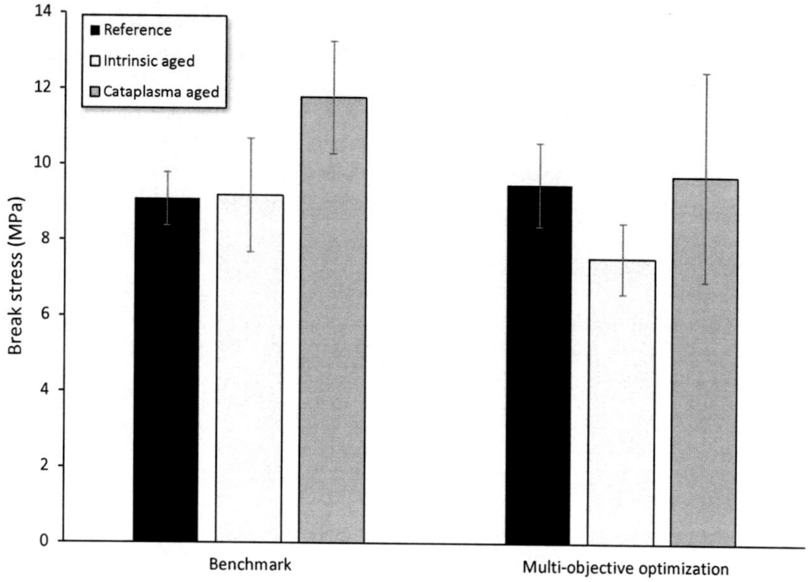

Fig. 5 Break stress of samples treated with plasma settings obtained by benchmark settings (left) and multi-objective settings (right) for PPS bonding. Bars represent the average value over 5 repetitions, error bars the standard deviation on the measurements

4 Conclusion

Optimization of plasma settings was performed for two industrial bonding processes via Bayesian optimization with single-objective optimization for maximal break stress and multi-objective optimization for maximizing the break stress while minimizing the production cost. The obtained results were benchmarked against plasma settings obtained in a more traditional way via contact angle measurements and expert guidance.

Single-objective optimization for the break stress could be achieved with 40% less budget via Bayesian optimization compared to the benchmark. In addition, 18% reduction in production cost could be achieved by multi-objective optimization for electronic potting bonding process. In contrast, no significant reduction in production cost could be achieved for the manufacturing bonding process. This indicates that AI technologies, such as Bayesian optimization, could generate significant reduction in both production and optimization costs. Trade-off analysis to define when it is no longer beneficial to search for an optimum seems, however, be useful to avoid spending optimization budget without significant advantages in the end.

Additionally, cataplasma tests were performed to check the robustness of the proposed settings under ageing conditions. No significant change in type of failure nor break stress could be detected after ageing for the samples treated with the plasma settings obtained via multi-objective optimization. This indicates that reliable process settings can be obtained via Bayesian optimization with lower costs to find the optimal solution or lower production costs.

Future work plans to investigate the application of the Bayesian optimization approach for more production parameters covering multiple bonding process steps. This would increase the complexity of the optimization problem even more and allow to test the optimization approach for optimizing the complete bonding process. In addition, better noise modelling for the surrogate model, which will help to plan when to repeat the measurement is planned.

Acknowledgements This research received funding from the Flemish Government under the "Onderzoeksprogramma Artificiële Intelligentie AI Vlaanderen" program. This research was supported or partially supported by Flanders Make vzw.

References

1. Sim, K.-B., Baek, D., Shin, J.-H., Shim, G.-S., Jang, S.-W., Kim, H.-J., Hwang, J.-W., Roh, J.U.: Enhanced surface properties of carbon fiber reinforced plastic by epoxy modified primer with plasma for automotive applications. Polym. (Basel). **12**, 556 (2020)
2. Martínez-Landeros, V.H., Vargas-Islas, S.Y., Cruz-González, C.E., Barrera, S., Mourtazov, K., Ramírez-Bon, R.: Studies on the influence of surface treatment type, in the effectiveness of structural adhesive bonding, for carbon fiber reinforced composites. J. Manuf. Process. **39**, 160–166 (2019)
3. Min, J., Wan, H., Carlson, B.E., Lin, J., Sun, C.: Application of laser ablation in adhesive bonding of metallic materials: a review. Opt. Laser Technol. **128**, 106188 (2020)
4. Jeevi, G., Nayak, S.K., Abdul Kader, M.: Review on adhesive joints and their application in hybrid composite structures. J. Adhes. Sci. Technol. **33**, 1497–1520 (2019)
5. Sundriyal, P., Pandey, M., Bhattacharya, S.: Plasma-assisted surface alteration of industrial polymers for improved adhesive bonding. Int. J. Adhes. Adhes. **101**, 102626 (2020)
6. Kim, J.Y., Ahn, K., Jeong, S.Y., Jeong, E.D., Jin, J.S., Bae, J.S., Kim, H.G., Cho, C.R.: Enhancement of adhesion between polyphenylene sulfide and copper by surface treatments. Curr. Appl. Phys. **14**, 118–121 (2014)
7. Lin, J., Sun, C., Min, J., Wan, H., Wang, S.: Effect of atmospheric pressure plasma treatment on surface physicochemical properties of carbon fiber reinforced polymer and its interfacial bonding strength with adhesive. Compos. Part B Eng. **199**, 108237 (2020)
8. Pocius, A.V: Adhesion and Adhesives Technology: An Introduction. Carl Hanser Verlag GmbH Co KG (2021)
9. da Silva, L.F.M., Öchsner, A., Adams, R.D.: Handbook of Adhesion Technology. Springer Science & Business Media (2011)
10. Anagreh, N., Dorn, L.: Influence of low-pressure plasma treatment on adhesive bonding between polybutylene terephthalat (PBT) and aluminum. Int. J. Adhes. Adhes. **25**, 165–172 (2005)
11. Krawczyk-kŁys, I., Makowski, P., Wojcik, J., Tyczkowski, J.: Plasma surface modification of commercial SBS rubbers for enhanced adhesive bonding. Mater. Sci. **18**, 132–137 (2012)
12. Kim, M.C., Yang, S.H., Boo, J.-H., Han, J.G.: Surface treatment of metals using an atmospheric pressure plasma jet and their surface characteristics. Surf. Coat. Technol. **174**, 839–844 (2003)
13. Zaldivar, R.J., Kim, H.I., Steckel, G.L., Patel, D., Morgan, B.A., Nokes, J.P.: Surface preparation for adhesive bonding of polycyanurate-based fiber-reinforced composites using atmospheric plasma treatment. J. Appl. Polym. Sci. **120**, 921–931 (2011)

14. II, E.G., Barankin, M.D., Guschl, P.C., Hicks, R.F.: Surface activation of poly (methyl methacrylate) via remote atmospheric pressure plasma. Plasma Process. Polym. **7**, 482–493 (2010)
15. Zhang, S., Huang, G., Wang, X., Huang, Y., Yang, J., Li, G.: Effect of air plasma treatment on the mechanical properties of polyphenylene sulfide/glass fiber cloth composites. J. Reinf. Plast. Compos. **32**, 786–793 (2013)
16. Anagreh, N., Dorn, L., Bilke-Krause, C.: Low-pressure plasma pretreatment of polyphenylene sulfide (PPS) surfaces for adhesive bonding. Int. J. Adhes. Adhes. **28**, 16–22 (2008)
17. Williams, T.S., Yu, H., Hicks, R.F.: Atmospheric pressure plasma activation as a surface pretreatment for the adhesive bonding of aluminum 2024. J. Adhes. Sci. Technol. **28**, 653–674 (2014)
18. Blackman, B.R.K., Kinloch, A.J., Watts, J.F.: The plasma treatment of thermoplastic fibre composites for adhesive bonding. Composites **25**, 332–341 (1994)
19. Budhe, S., Banea, M.D., De Barros, S., Da Silva, L.F.M.: An updated review of adhesively bonded joints in composite materials. Int. J. Adhes. Adhes. **72**, 30–42 (2017)
20. Habenicht, G.: Applied Adhesive Bonding: A Practical Guide for Flawless Results. John Wiley & Sons (2008)
21. Pruksawan, S., Lambard, G., Samitsu, S., Sodeyama, K., Naito, M.: Prediction and optimization of epoxy adhesive strength from a small dataset through active learning. Sci. Technol. Adv. Mater. **20**, 1010–1021 (2019). https://doi.org/10.1080/14686996.2019.1673670
22. Valdés, E., Mosquera-Artamonov, J.D., Cruz-Gonzalez, C., Jaime Taha-Tijerina, J.: Multiobjective optimization on adhesive bonding of aluminum-carbon fiber laminate. Comput. Intell. **37**, 621–634 (2021)
23. Morales-Hernández, A., Van Nieuwenhuyse, I., Gonzalez, S.R., Jordens, J., Witters, M., Van Doninck, B.: Multi-objective simulation optimization of the adhesive bonding process of materials. arXiv Prepr. arXiv2112.06769 (2021)
24. Couckuyt, I., Deschrijver, D., Dhaene, T.: Fast calculation of multiobjective probability of improvement and expected improvement criteria for Pareto optimization. J. Glob. Optim. **60**, 575–594 (2014)
25. Das, I., Dennis, J.E.: A closer look at drawbacks of minimizing weighted sums of objectives for Pareto set generation in multicriteria optimization problems. Struct. Optim. **14**, 63–69 (1997)
26. Yang, K., Emmerich, M., Deutz, A., Bäck, T.: Efficient computation of expected hypervolume improvement using box decomposition algorithms. J. Glob. Optim. **75**, 3–34 (2019)
27. Infogram.com: Evolutie energiefactuur, https://infogram.com/evolutie-energiefactuur-bedrijf-1hdw2jq1dz7p4l0?live
28. Indeed: Operator salary in Belgium, https://be.indeed.com/career/operator/salaries
29. Stepstone GmbH: Junior Planner lonen in België, https://www.stepstone.be/salaris/Junior-Planner.html
30. Močkus, J.: On Bayesian methods for seeking the extremum. In: Optimization Techniques IFIP Technical Conference. pp. 400–404. Springer, Berlin (1975)
31. Frazier, P.I.: A tutorial on Bayesian optimization. arXiv Prepr. arXiv1807.02811 (2018)
32. Ankenman, B., Nelson, B.L., Staum, J.: Stochastic kriging for simulation metamodeling. In: 2008 Winter Simulation Conference. pp. 362–370. IEEE (2008)
33. Miettinen, K.: Nonlinear Multiobjective Optimization. Springer Science & Business Media (2012)
34. Viana, F.A.C.: A tutorial on Latin hypercube design of experiments. Qual. Reliab. Eng. Int. **32**, 1975–1985 (2016)
35. Zhu, C., Byrd, R.H., Lu, P., Nocedal, J.: Algorithm 778: L-BFGS-B: Fortran subroutines for large-scale bound-constrained optimization. ACM Trans. Math. Softw. **23**, 550–560 (1997)

36. Berkeley, J., Moss, H. B., Artemev, A., Pascual-Diaz, S., Granta, U., Stojic, H., Couckuyt, I., Qing, J., Loka, N., Paleyes, A., Ober, S. W., & Picheny, V.: Trieste, https://github.com/secondmind-labs/trieste (2022)
37. Vesel, A., Mozetic, M.: Surface modification and ageing of PMMA polymer by oxygen plasma treatment. Vacuum **86**, 634–637 (2012)
38. De Geyter, N., Morent, R., Leys, C.: Influence of ambient conditions on the ageing behaviour of plasma-treated PET surfaces. Nucl. Instrum. Methods Phys. Res. Sect. B Beam Interact. Mater. Atoms. **266**, 3086–3090 (2008)

Mechanical Performance and Economic Analysis of the Addition of Granite Powder and Fly Ash on the Subsurface Properties of Cementitious Floors Cured Under Different Conditions

Mateusz Moj, Adrian Chajec, and Łukasz Sadowski

Abstract The purpose of this study is to present a comparative analysis of the mechanical and adhesive properties of overlays in cement mortars modified with granite powder or fly ash used in the construction site. The cement replacement ratio analysed with the addition of the supplementary cementitious material was variable from 10 to 30% by weight of cement in the reference sample. The effect of curing conditions on the samples was also analysed. The results of compressive strength and adhesion between subsurface and substrate for mortars modified with the above-mentioned additives were analysed. Then, mechanical performance ratio (MPR) and effective cost ratio (ECR) analyses were performed to show the appropriateness of using granite powder and fly ash to reduce the cost of floor mortars and improve their mechanical properties. Based on the performed research, we concluded that providing appropriate curing conditions for the specimens has a crucial effect on achieving the designed properties of cementitious overlays. We noted that this way of decreasing the amount of cement in cementitious composites allows reducing the carbon footprint of the construction industry and the amount of waste in the environment.

Keywords Mechanical analysis · Economic analysis · Sustainable overlays · Supplementary cementitious materials

---

M. Moj (✉) · A. Chajec · Ł. Sadowski
Department of Materials Engineering and Construction Processes, Wroclaw University of Science and Technology, Wybrzeże Wyspiańskiego 27, 50-370 Wroclaw, Poland
e-mail: mateusz.moj@pwr.edu.pl

A. Chajec
e-mail: adrian.chajec@pwr.edu.pl

Ł. Sadowski
e-mail: lukasz.sadowski@pwr.edu.pl

L. F. M. da Silva et al. (eds.), *2nd International Conference on Industrial Applications of Adhesives 2022*, Proceedings in Engineering Mechanics,
https://doi.org/10.1007/978-3-031-11150-1_5

1 Introduction

Cement is a hydraulic mineral binder, obtained from mineral raw materials (marble or limestone and clay) sintered to clinker in a cement kiln and then grinding the resulting sinter with gypsum, which acts as a setting time regulator. It is used in the preparation of cement pastes, mortars and concrete. It is used in the production of common building materials.

However, cement production has an undesirable impact on the environment. It is responsible for about 7% of total global CO_2 emissions [1]. 900 kg of carbon dioxide is emitted during the production of 1 tonne of this material [2], so partial replacement of cement with another component can contribute to the significant reduction of CO_2 emission to the atmosphere [3]. Therefore, actions are taken to reduce this emission. It has been shown that the reduction of cement production by 10% significantly reduces the annual emission of carbon dioxide [3]. The simplest way may seem to be to reduce the amount of cement in cementitious composites, however, it is necessary to replace the loss of cement with another supplementary cementitious material (SCM). It is estimated that the use of cement mixed with additives derived from industrial waste will reduce CO_2 by 5–20% depending on the type, amount and properties of the used additive [4].

Figure 1 shows the waste hierarchy based on the EU2008/98 EC Waste Framework Directive [5]. From an ecological point of view, it is much better to reuse or recycle material than to landfill it [6]. The way of waste utilization during building materials production contributes to creating new materials often with new or improved properties of standard, unmodified material.

Many researchers are performed research connected with the reusing of industrial wastes in cementitious materials. Awal and Mohammadhosseini [7] noted that carpet waste fibers and palm oil fuel ash can be used to produce cement composites, which

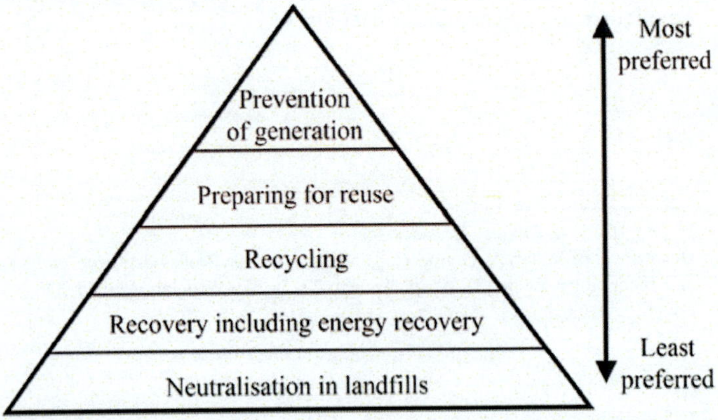

Fig. 1 Hierarchy of desirable waste management practices drawn up by the EU2008/98 EC framework directive

results in increased tensile and flexural strength of the material. Zunino and Lopez 8 observed that rice husk ash influences the strength and durability of concrete and cement mortars. Researchers [9] also stated that it is possible to replace up to 40% of cement with the addition of rice husk ash. This change in the composition of composite does not affect significant changes in the compressive strength of concrete [9]. Many researchers [10, 11] are investigating the impact of the use of glass powder waste as a cement replacement material on the properties of cementitious composites. They [10, 11] stated that replacement of 10–20% of the amount of cement in mixes allows to increase the compressive strength and decrease the water permeability of composite. Moreover, the used additive has a negligible effect on the pull-off strength.

In recent years, there has been a growing interest in granite powder waste used as an additive in cementitious materials [12]. Rudner et al. [13] noted that the addition of fine granite fractions (less than 0.15 mm) allowed for increased compressive strength and adhesive properties of the cementitious composites. Using standards PN-EN 1504 [14], PN-EN 1766:2017-03 [15], PN-EN 206+A2:2021-08 [16], it can be stated that these two parameters are desirable especially for cementitious mortars used for overlays in the construction site. While ensuring compressive strength (min. 20 MPa) for this type of mortar does not generate major problems, ensuring adhesion between overlay and substrate in cementitious materials (min. 1.0 MPa) presents some challenges. The adhesion of the overlay to the substrate is crucial in mortar floors because most of the damage occurs in the subsurface layer. This is due to the exposure of the surface to various destructive conditions and also due to the lack of proper supervision of the floor construction process [17]. Krzywinski et al. [18] and Stawiski [19] have shown that the strengths in the subsurface zone of the substrate can be significantly lower than in other parts of the cement floor. Such changes may be due to the migration of water to the upper part of the overlay, which causes excessive hydration of cement in this zone. Figure 2 shows a cross-section through a layer of such a floor substrate. The blue area indicates the zone of cement sludge. The red line indicates the area of adhesion between the overlay and the substrate.

Fly ash, is commonly used as SCM in such composites [20–22]. Sadowski and Golewski [23] observed that the addition of this filler is beneficial in delaying the destructive processes in cementitious composites. Klabacha and Sypek [24] noted that cementitious industrial overlays with fly ash can achieve better properties than floor substrates using slag cement. Fiedlerová et al. [25] analyzed cementitious substrates using fly ash from three Czech power plants. They emphasized [25] that

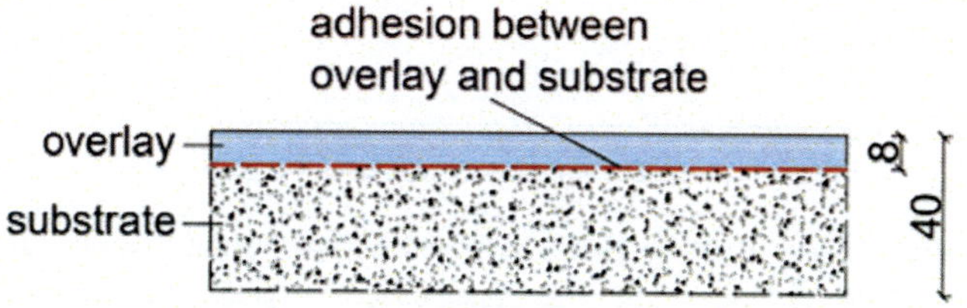

Fig. 2 Cross-section through a layer of the cementitious substrate

the use of fly ash allows for a significant improvement in the strength properties of the overlays.

We have presented our method of using granite powder in the production of such overlays [26] and decided in this paper to compare it with overlays in which fly ash acts as supplementary cementitious material. Chajec [27] analyzed cementitious composites in a similar way. In this study, a novel approach was used because we investigated the mechanical and adhesive properties of modified overlays cured under air-dry and air-humid conditions. This study aims to demonstrate the effect of curing conditions of cement floors on their basic properties. The paper also addresses the issue of economic comparison of given mortars in order to obtain a wider range of information about a given material. This will allow easier selection of a cementitious floor mortar for a particular application.

2 Materials and Methods

2.1 *Ingredients Analysis*

The study focused on cementitious composites for making overlays with cement to the fine aggregate ratio of 1:3 modified according to the filler technology used waste granite powder or fly ash. We decided to investigate the influence of the used curing conditions on its physical and mechanical properties.

CEM I 42.5 (Górażdże, Poland) cement was used to make the mortars, which consisted of 95–100% Portland clinker and 0–5% secondary components. The fine aggregate used in this study has a specific density of approximately 1600 kg/m^3. Granite powder was obtained from a local stone masonry plant in Strzegom, Poland. The silica fly ash used had a specific density of about 2100 kg/m^3, and reactivity after 28 days $\geq$ 75% and after 90 days $\geq$ 85%. Water was taken from the local water supply system. Due to obtaining optimal (dense) consistency of the mixture plasticizers were not used.

Cement, granite powder, fly ash and fine aggregate were sieved according to the procedure given in EN 933-1 [28] to determine the grain size distributions of the components given. The weighed portion of the ingredient was passed through a set of sieves and then the remaining grains were weighed on each sieve. Figure 3 shows the sieve size distribution of used materials.

2.2 *Preparation of Cementitious Overlays*

Seven different cementitious mixtures with different compositions in terms of the amount of granite powder and fly ash used were prepared (Table 1). The first mixture, which is the reference mixture designated REF, contained only Portland cement as a

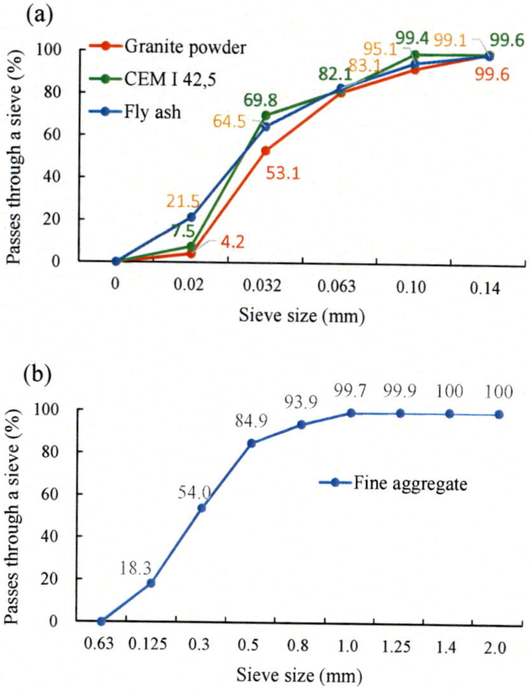

Fig. 3 Parameters of materials used for mortars: **a** grain size of cement, granite powder and fly ash, **b** grain size of dried fine quartz aggregate

binder. In the other mixes, Portland cement was partially replaced with the addition of granite powder or fly ash in a variable of 10%, 20% and 30% (designated GP10, GP20, GP30, FA10, FA20, FA30). The amount of water was assumed constant for all mixes and was determined for a reference composition for a water-cement ratio of 0.5 (−).

The mortars were made into 500 mm × 500 mm × 40 mm sized specimens and cured for the required time of 56 days. All samples were compacted and trowelled manually with a trowel. The samples were cured in air-dry (AIR) and in air-humid conditions (WET), created by applying moist sponges to the surface of the samples (Table 2). The condition of the curing was maintained for 28 days, which in the case of the air-humid conditions consisted in controlling the water saturation of the sponges and re-wetting them if necessary. It was recommended that the sponges were not saturated, so after wetting they were left for 15–30 min to drain excess water.

Table 1 Compositions of the research series

Series (−)	Cement CEM I 42.5R (kg/m^3)	Granite powder (kg/m^3)	Fly Ash (kg/m^3)	Fine aggregate (kg/m^3)	Water (kg/m^3)	Water/Cement ratio (−)	Water/Binder ratio (−)
REF	512.0	0.0	0.0	1536.0	256.0	0.50	0.50
FA10	460.8	0.0	51.2			0.56	
FA20	409.6	0.0	102.4			0.63	
FA30	358.4	0.0	153.6			0.71	
GP10	460.8	51.2	0.0			0.56	
GP20	409.6	102.4	0.0			0.63	
GP30	358.4	153.6	0.0			0.71	

Table 2 Curing conditions of the research series

Curing conditions	Temperature (°C)	Humidity (%)
Air-dry (AIR)	18 ± 3	50 ± 5
Air-humid (WET)	23 ± 3	85 ± 5

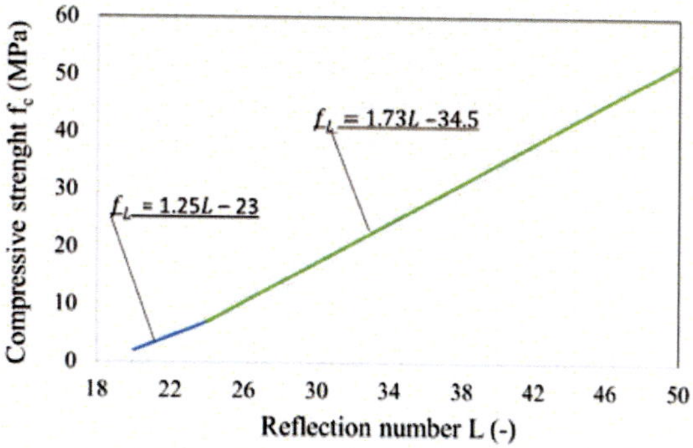

Fig. 4 Base curve according to the standard PN-EN 13791 [30]

2.3 *Schmidt Hammer Compressive Strength Test*

The compressive strength of the specimens was tested using the Schmidt hammer method according to the PN-EN 12504-2 [29] procedure. Thanks to the non-destructive method, it is possible to perform pull-off strength tests at the place where sclerometric tests were performed. This makes it possible to test specimens in one particular place, which eliminates measurement deviations related to changing the measuring location on the specimen. Test measurements were performed on the prepared test subfloors. Three impact tests were performed at each pull-off strength test location. This made it possible to obtain 12 results for each mortar tested. They allow for an approximate determination of the compressive strength by reading the result from the base curve corresponding to a given number of rebounds (Fig. 4). Carrying out the sclerometric test, we obtain the concrete hardness test given as the rebound number indicating the impact weight measurement.

2.4 *Pull-Off Test of the Subsurface Layer*

Four disks were drilled on the prepared plates. The edge of the disc was at least 50 mm from the edge of the specimen. To test the adhesion of the subsurface layer, holes with an inner diameter of 50 mm were drilled to a depth of 8 mm. After all holes were drilled, the surface was vacuumed using a vacuum cleaner. Figure 5 shows how f_h results were obtained during the pull-off test. The test began by adhering steel pucks to the flat surface of the specimen using Poxipol two-component adhesive, first cleaning the specimen and the puck with a solvent that does not compromise the

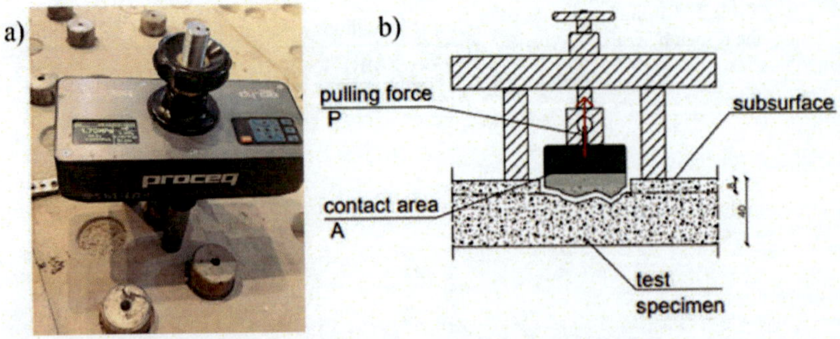

Fig. 5 Pull-off test: **a** photo during the study **b** pull-off test in cross-section

integrity of the coating. The purpose of the pucks is to be able to bond the adhesion force measuring device to the specimen. According to the standard ASTM D4541-17 [31], at least three drill holes had to be drilled to characterize the surface. The distance between each well had to be large enough to allow the device to be positioned. The adhesive was applied to both surfaces and carefully spread over the entire surface. The adhesive used did not affect the surface layer. Once the adhesive was fully cured, the test could proceed. The component connecting the apparatus to the puck had to be carefully screwed into the puck, and the head of the component was attached to the apparatus holder. After the apparatus was leveled, it was programmed so that the test lasted from 10 to 100 s. The loading rate of the apparatus was 0.05 MPa/sec. The result is recorded on the device. A Proceq Dy-216 device was used for the test, which, after entering the appropriate data, automatically calculated the pull-off strength from Eq. (1) and plotted its value as a function of pull-off time.

$$f_h = \frac{P}{A} \quad (1)$$

3 Results

3.1 *Schmidt Hammer Compressive Strength Test*

The specimens were compared to each other because of the amount of granite powder and fly ash added to partially replace the cement, as well as the curing conditions of the specimens. The following graph shows the average compressive strengths obtained when each of the overlays was tested. The approximate strength fc was taken as equal to the f_L value. For the obtained rebound number (L), the compressive strength value was calculated from the base curve (f_L) based on Eqs. (2, 3):

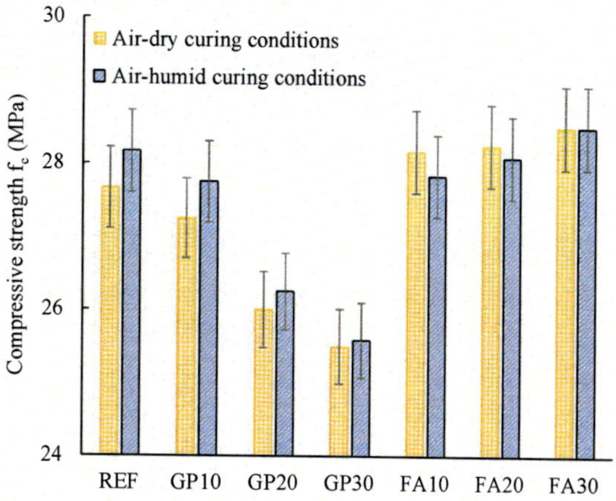

Fig. 6 Average compressive strength f_c of the subsurface layer for particular mortars

$$f_L = 1.25L - 23 \, at \, 20 \le L \le 24 \tag{2}$$

$$f_L = 1.73L - 34.5 \, at \, 24 \le L \le 50 \tag{3}$$

The basic requirement for cementitious floor overlays in terms of compressive strength is to achieve a minimum value of 20 MPa. It can be seen that all of the tested specimens achieved this value. From the graph (Fig. 6), it can be observed that as the amount of granite powder in the mix increases, the compressive strength of the overlay decreases slightly similar to the observation described in studies [26, 27]. The increase in fly ash addition slightly affects the compressive strength (what is widely known). It can be observed that samples containing 10–30% fly ash keep the strength value almost constant. In addition, the differences for specimens cured under air-humid conditions are significantly higher when granite powder addition is used. It should be noted that the results obtained in FA-modified samples are presenting increased compressive strength, although the tests were performed after a longer period (56 days). In the studies [22, 27], Researchers observed that the curing time of the samples especially with the addition of fly ash definitely effected improving the compressive strength.

3.2 *Pull-Off Test of the Subsurface Layer*

The samples were compared to each other due to the amount of granite powder and fly ash added to partially replace the cement, as well as the curing conditions of the samples. Figure 7 presents the average pull-off strengths of each mix.

Especially, a large difference in results can be observed due to the different maturation conditions of the samples. Similarly, the amount of granite powder or fly ash

Fig. 7 Average pull-off strength of the subsurface layer f_h for particular mortars

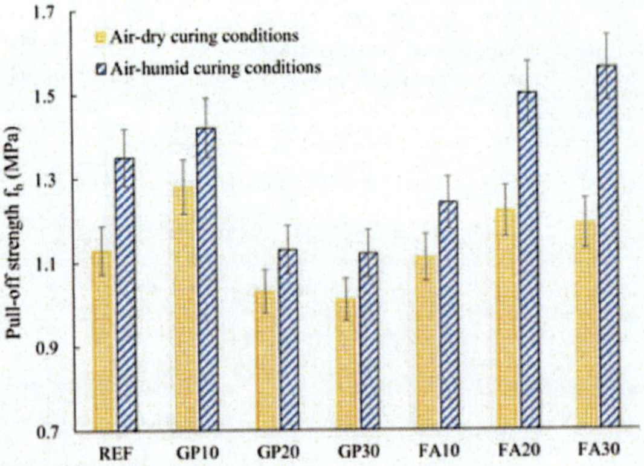

substituted for cement was significant. It can be observed that with the amount of 10% GP filler the results obtained increased which suggests the validity of the use of this waste modern approach to make cement composites. Moreover, comparing the results of air-dry cured samples, it can be concluded that GP10 samples behave best. Similar to Sharma et al. [32] replacing 30% of the cement with granite powder leads to a slight decrease in the pull-off strength by about 10%. Using air-humid curing conditions, larger amounts of cement can be more reliably replaced with one of the additives achieving values as high as 1.56 MPa pull-off strength of the subsurface layer for FA30 specimen. Similarly to [26], it was noted that the curing conditions of the specimens had a greater effect on the pull-off strength obtained than the compressive strength. Moreover, the differences in pull-off strength due to the curing conditions were less for the samples with granite powder additive which may suggest less water requirement in the hydration process for these mortars when comparing them with the reference sample and mortars with fly ash additive.

4 Economic Analysis of the Potential Use of Granite Powder Versus Fly Ash for the Production of Cement Mortars Cured Under Air-Dry and Air-Humid Conditions

The cost of cementitious subfloors currently has a large impact on the choice of material used in their construction. Researchers are striving to optimize the composition of mixes using alternative materials often found locally to minimize the cost associated with sourcing the materials. An important aspect is the consideration of the desired mechanical properties of the mixes. Based on the research conducted, a mechanical and economic analysis is presented to assist in the selection of a specific mix for specific embedded conditions. The analysis was conducted in two parts. The first included an analysis of the cost of the cement mix depending on the type and amount of additive and the calculation of the effective cost ratio (ECR). It should be noted that in the analysis given, it was decided to include the cost associated with providing air-humid curing conditions. In the second part, the mechanical performance ratio (MPR) [33, 34] was calculated based on the mechanical performance analysis performed. Then, a summary was presented comparing the MPR and ECR indices depending on the type and amount of additive.

4.1 *Economic Analysis of the Cementitious Composites*

To determine the cost of the target material it is necessary to know the unit prices of the mixture components. For this purpose of determining the cost of the materials, a survey was made of Polish companies that provide trade services for the specified materials. An important criterion addressed in the paper was the variables of how the samples were cured. Because of the novelty of this issue, it was decided to include the cost associated with providing air-humid conditions for the cementitious mortar floors, taking into account the labor cost associated with controlling and moistening the sponges laid on the floors. After consultation with flooring contractors, reviewing the hourly rate of time needed to provide the above conditions for 1m^2 of completed subfloor and the increased water demand, this cost was estimated at $0.23\$/\text{m}^2$. The cost of making the floor was omitted because it is constant in both cases. The cost of sponges to maintain air-humid conditions was also omitted because they are considered a contractor-owned devices. The assumed price of the materials used was: sand—11.48$/ton, cement—127.46$/ton, granite powder—6.51$/ton, fly ash—27,84$/ton, water—2.11$/ton. To standardize units, the final result in $\$/\text{m}^2$ was adopted, assuming that the thickness of the floor mortar layer is 5 cm. A summary of the cost of making each model composite system along with its strength characteristics is summarized in Table 3.

$$c_{ingredient} = uc_{ingredient} x ingredient\ ratio\ \left(\$/\text{m}^2\right). \tag{4}$$

$$c_{mixes} = c_C + c_{GP} + c_{FA} + c_S + c_W + c_{CE}\ \left(\$/\text{m}^2\right) \tag{5}$$

$$ECR = \frac{c_{mixes}}{c_{ref}} x100\%(\%) \tag{6}$$

To compare the effective cost ratio, the ΔECR parameter was determined as follows:

$$\Delta ECR = \frac{ECR_{mixes}}{ECR_{ref}}(-) \tag{7}$$

Table 3 ECR results for the cementitious composites

Curing conditions	Series	cC	cGP	cFA	cS	cW	cCC	Cmixes	ECR	ΔECR
	(–)	($/m^2)							(%)	(–)
AIR	REF	3.26	0.00	0.00	0.88	0.03	0.00	4.17	100.0	1.00
	FA10	2.94	0.00	0.07				3.92	99	0.94
	FA20	2.61	0.00	0.14				3.66	87.8	0.88
	FA30	2.28	0.00	0.21				3.41	81.7	0.82
	GP10	2.94	0.01	0.00				3.86	92.5	0.93
	GP20	2.61	0.03	0.00				3.55	85.1	0.85
	GP30	2.28	0.04	0.00				3.24	77.6	0.78
WET	REF	3.26	0.00	0.00			0.23	4.40	105.5	1.06
	FA10	2.94	0.00	0.07				4.15	99.4	0.99
	FA20	2.61	0.00	0.14				3.89	93.3	0.93
	FA30	2.28	0.00	0.21				3.64	87.2	0.87
	GP10	2.94	0.01	0.00				4.09	98.0	0.98
	GP20	2.61	0.03	0.00				3.78	90.6	0.91
	GP30	2.28	0.04	0.00				3.47	83.1	0.83

The table above shows that the decrease in the prices of both additives is due to the replacement of cement with additives whose price is about 5 to about 22 times lower. The price differences between mortars with granite powder additives and mortars with fly ash additives differ slightly because the prices of these additives are at similar levels. Nevertheless, it can be said that the use of granite powder allows for a significant reduction in the price of the mortar by up to about 20%.

Figure 8 shows the value of the effective cost ratio (ECR) for a series of air-cured (red) and air-humid cured (blue) cement mortars. There is a noticeable difference here due to the provision of air-humid curing conditions which reinforces the notion that it is appropriate to include this cost in the ECR calculations.

4.2 *Analysis of the Mechanical Performance Ratio of the Cementitious Mortars*

As mentioned in the introduction, providing the pull-off strength is much more complicated than providing the compressive strength of mortars used for the substrates. In order to compare the results of the tests described in Sect. 3, the mechanical performance ratio (MPR) was used, which is as follows:

$$\sum MPR = \frac{2xMPR_{fh} + MPR_{fc}}{3}(\%) \tag{8}$$

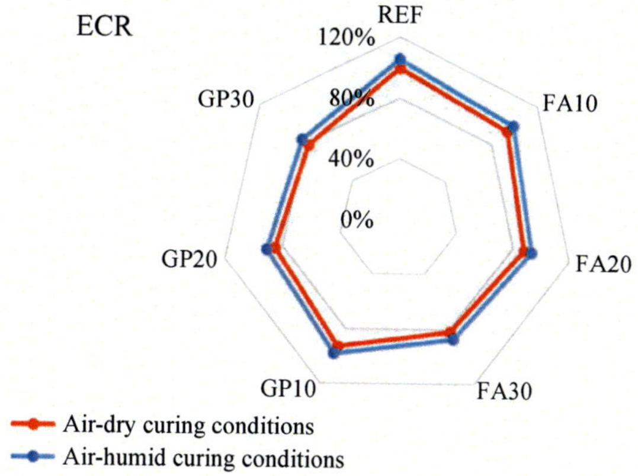

Fig. 8 ECR value with regards to the type of cementitious mortar series

$$MPR_{fh} = \frac{f_{h,mixy}}{f_{h,ref}} x100\%(\%) \tag{9}$$

$$MPR_{fc} = \frac{f_{c,mixy}}{f_{c,ref}} x100\%(\%) \tag{10}$$

To compare the mechanical performance, the ΔMPR parameter was determined as follows:

$$\Delta MPR = \frac{MPR_{seria}}{MPR_{ref}} (-) \tag{11}$$

Figure 9 shows the value of the mechanical performance ratio (MPR) concerning the amount and type of cement replacement additive. The MPR of the air-humid cured samples was lower than that of the reference sample only for GP20 and GP30 mortars. This reinforces the validity of providing appropriate (air-humid) conditions during the curing process of the floor screeds. The differences related to the curing conditions are particularly pronounced for the mortars with fly ash addition. When comparing the samples cured in air-dry conditions, the best result was obtained with sample GP10, which suggests that it is reasonable to use granite powder without the increased costs associated with saturating sponges that hydrate the floor surface.

Table 4 presents a summary of the indicators obtained.

4.3 Combined Economic and Mechanical Performance of the Composites

By combining the economic analysis with the analysis of mechanical properties crucial for cement floors, the results obtained can be correctly interpreted. Table 5

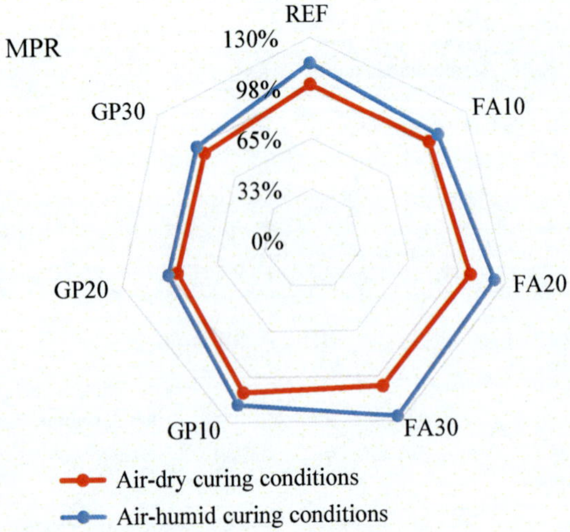

Fig. 9 MPR value with regards to the type of cementitious mortar series

Table 4 MPR results for the cementitious composites

Curing conditions	Series	MPR_{fh}	MPR_{fc}	MPR	ΔMPR
	(–)	(%)	(%)	(%)	(–)
AIR	REF	100.0	100.0	100.0	1.00
	FA10	98.2	101.8	99.4	0.99
	FA20	108.0	102.1	106.0	1.06
	FA30	105.3	103.0	104.5	1.05
	GP10	113.3	98.5	108.3	1.08
	GP20	91.2	94.0	92.1	0.92
	GP30	89.4	92.2	90.3	0.90
WET	REF	119.5	101.8	113.6	1.14
	FA10	109.7	100.6	106.7	1.07
	FA20	132.7	101.5	122.3	1.22
	FA30	138.1	103.0	126.4	1.26
	GP10	125.7	100.3	117.2	1.17
	GP20	100.0	94.9	98.3	0.98
	GP30	99.1	92.4	96.9	0.97

shows the MPR and ECR values for a series of cement mortars with granite powder and fly ash added.

Figures 10 and 11 show the cost dependence of mechanical properties of cement mortars with granite powder and fly ash. Comparing the two additives with each other, it can be concluded that it is more reasonable to use fly ash because the mechanical properties increase as the price of the composite decreases. The addition of granite

Table 5 Mechanical performance ratio and effective cost ratio parameters concerning the series of tested mortars

Curing conditions	Series	MPR	ΔMPR	ECR	ΔECR
	(−)	(%)	(−)	(%)	(−)
AIR	REF	100.00	1.00	100.00	1.00
	FA10	99.42	0.99	93.89	0.94
	FA20	106.01	1.06	87.77	0.88
	FA30	104.54	1.05	81.66	0.82
	GP10	108.34	1.08	92.53	0.93
	GP20	92.09	0.92	85.07	0.85
	GP30	90.31	0.90	77.60	0.78
WET	REF	113.58	1.14	105.51	1.06
	FA10	106.68	1.07	99.40	0.99
	FA20	122.32	1.22	93.29	0.93
	FA30	126.37	1.26	87.17	0.87
	GP10	117.21	1.17	98.05	0.98
	GP20	98.29	0.98	90.58	0.91
	GP30	96.89	0.97	83.12	0.83

powder behaves as expected i.e. the increase in mechanical properties generates an increase in the price of the cement mix. It should be stressed, however, that the graphs are not linear and hasty interpretation of the results may prove to be incorrect in this case. Going deeper into the analysis, it can be seen that sample GP10 maturing in both air-dry and air-humid conditions stands out. Comparing the cost of the REF specimen cured in air-dry conditions and the specimen with 10% granite powder cured under moist sponges, it can be seen that significantly better strength values can be obtained with a similar fabrication cost.

5 Conclusions

This paper describes a comparative analysis of the results of surface layer pull-off strength and compressive strength of two cementitious floor mortars in which 10%, 20%, and 30% of the cement was replaced with granite powder and fly ash.

- An important aspect of the analysis was to compare the mortars concerning their curing conditions. It is important to note that all the samples met the conditions for flooring subfloors, such as a minimum of 1.0 MPa surface layer pull-off strength and a minimum of 20.0 MPa compressive strength.
- The analysis shows that the highest values of pull-off strength of floors cured under air-dry conditions were obtained when 10% cement was replaced with

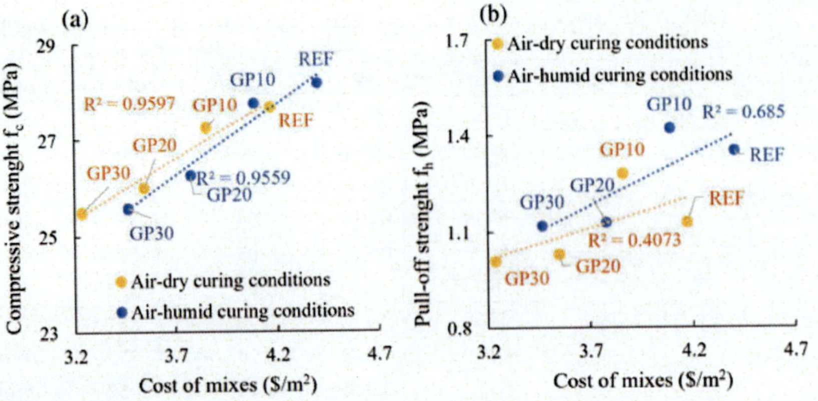

Fig. 10 **a** compressive strength (f_c) and **b** pull-off strength (f_h) compared to the cost of mortars with granite powder added

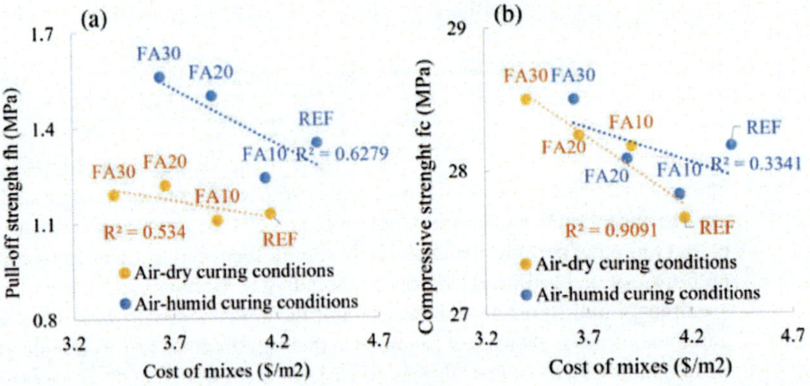

Fig. 11 **a** compressive strength (f_c) and **b** pull-off strength (f_h) compared to the cost of mortars with fly ash added

granite powder in the mortar. The increase was about 13% over the reference sample. The same mixture recorded a slight decrease in compressive strength (about 1.5%).

- The mixes with granite powder obtained noticeably smaller differences in the pull-off strength of the subsurface layer than the mixes with fly ash or the reference mix when comparing the results obtained in air-dry and air-humid conditions. This may be related to the lower water requirement during the hydration process in GP mixes. The above observations indicate the potential of using granite powder as SCM in cementitious floor mortars.
- Providing air-humid conditions increased the pull-off strength of the surface layer by up to 31% (FA30) compared to mortars cured in air. Moreover, the addition

of fly ash slightly improved the compressive strength compared to the reference sample (3%).

- Cost and mechanical analyses performed for the cement mortars used in this paper have shown that the use of fly ash and granite powder as SCMs in floor cement mortars allows, in selected cases, to obtain better values of MPRs while lowering ECRs. Providing air-humid conditions during the curing allows increasing the MPR for floor mortars by up to 12% with an increase in costs by about 5.5% compared to floor mortars cured in the air.

These positive results show that granite powder has the potential to be used in cementitious floor mortars instead of the original materials traditionally used as SCMs. This experimental research has also helped to address the potential use of industrial waste and the reduction of CO_2 emissions associated with cement production. In addition, the issues raised related to the study of the effect of curing conditions indicate the validity of working to ensure adequate moisture content during the construction of cementitious floor screeds. However, further research needs to be conducted to develop the sustainable use of this additive in building materials matured under air-humid conditions and to develop analyses related to the carbon footprint of the materials used.

Funding The authors received funding from the project supported by the National Centre for Research and Development, Poland [grant no. LIDER/35/0130/L-11/19/NCBR/2020 "The use of granite powder waste for the production of selected construction products."

References

1. Ali, M., Saidur, R., Hossain, M.: A review on emission analysis in cement industries. Renew. Sustain. Energy Rev. **15**, 2252–2261 (2011)
2. Duda, J., Tomasiak, J.: Redukcja emisji CO_2 w procesie produkcji cementu (2015)
3. Benhelal, E., Zahedi, G., Shamsaei, E., et al.: Global strategies and potentials to curb CO_2 emissions in cement industry. J. Clean. Prod. **51**, 142–161 (2013)
4. Bosoaga, A., Masek, O., Oakey, J.: CO_2 capture technologies for cement industry. Energ. Procedia **1**(1), 133–140 (2009)
5. Directive of the European Parliament and of the Council 2008/98/WE of 19 November 2008 on waste and repealing certain Directives. Dz.Urz. UE L 312/3 22.11.2008
6. Zając, B., Gołębiowska, I.: Zagospodarowanie odpadów budowlanych. Inż. Ap. Chem. **53**(6), 393–395 (2014)
7. Sivakrishnaad, A., Adesinab, A., Awoyerac, P.O., et al.: Green concrete: A review of recent developments. Materialstoday **27**, 54–58 (2020)
8. Zunino, F., Lopez, M.: Decoupling the physical and chemical effects of supplementary cementitious materials on strength and permeability: a multi-level approach. Cement Concr. Compos. **65**, 19–28 (2016)
9. Al-Khalaf, M.N., Yousif, H.A.: Use of rice husk ash in concrete. Int. J. Cem. Comp. Lightweight Concr. **6**(4), 241–248 (1984)
10. Sadowski, Ł.: Towards the utilization of waste glass powder in sustainable cement based overlays. In: MATEC Web Conferences, vol. 163, Article 03001 (2018)

11. Jain, K.L., Sancheti, G., Gupta, L.K.: Durability performance of waste granite and glass powder added concrete. Constr. Build Mater **252**, Article 119075 (2020)
12. Chajec, A.: Towards the sustainable use of granite powder waste for manufacturing of cementitious composites. In: MATEC Web of Conferences **322**, 01005 (2020)
13. Rudner, M., Chajec, A., Sadowski, Ł.: Mechanical and Adhesive Properties of Cement-Lime Plasters Modified with Waste Granite Powder. In: 6th International Conference on Adhesive Bonding 2021. Springer, Cham (2021)
14. Polish Committee for Standardization, Products and systems for the protection and repair of concrete structures. Test methods. Measurement of adhesion by pull-off, PN EN 1504
15. Polish Committee for Standardization, Products and Systems for the Protection and Repair of Concrete Structures—Test Methods—Reference Concretes for Testing, PN-EN 1766:2017-03
16. Polish Committee for Standardization, Concrete. Properties, Production, Placement and Conformity Criteria, PN-EN 206+A2:2021-08
17. Hoła, J., Sadowski, Ł, Hoła, A.: The effect of failure to comply with technological and technical requirements on the condition of newly built cement mortar floors. Proc. Inst. Mech. Eng. Part L J. Mater. Des. Appl. **233**(3), 268–275 (2019)
18. Krzywiński, K., Sadowski, Ł, Szymanowski, J., et al.: Attempts to improve the subsurface properties of horizontally-formed cementitious composites using tin (ii) fluoride nanoparticles. Coatings **10**(1), 83 (2020)
19. Stawiski, B.: The heterogeneity of mechanical properties of concrete in formed constructions horizontally. Arch. Civ. Mech. Eng. **12**, 90–94 (2012)
20. Diaz-Loya, I., Juenger, M., Seraj, S., et al.: Extending supplementary cementitious material resources: reclaimed and remediated fly ash and natural pozzolans. Cement Concr. Compos. **101**, 44–51 (2019)
21. Lam, L., Wong, Y.L., Poon, C.S.: Effect of fly ash and silica fume on compressive and fracture behaviors of concrete. Cem. Concr. Res. **28**, 271–283 (1998)
22. Golewski, G.L.: Effect of curing time on the fracture toughness of fly ash concrete composites. Compos. Struct. **185**, 105–112 (2018)
23. Sadowski, T., Golewski, G.L.: A failure analysis of concrete composites incorporating fly ash during torsional loading. Compos. Struct. **183**, 527–535 (2018)
24. Klabacha, A., Sypek, M.: Betonowe posadzki przemysłowe wykonane z cementów: popiołowego i żużlowego. Budownictwo Technol. Architektura **1**, 40–41 (2018)
25. Fiedlerová, M., Drochytka, R., Dohnálek, P.: Utilization of fly ash in high performance floor screed based on cement and its influence on physical mechanical properties. Mater. Sci. Forum **865**, 201–205 (2016)
26. Chajec, A., Sadowski, Ł., Moj, M.: The adhesive and functional properties of cementitious overlays modified with granite powder. Int. J. Adhes. Adhes. 103008 (2021)
27. Chajec, A.: Granite powder versus fly ash for the sustainable production of air-cured cementitious mortars. Materials **14**, 1208 (2021)
28. Polish Committee for Standardization, Tests for geometrical properties of aggregates—Part 1: determination of particle size distribution—sieving method, PN-EN, **933**, (1) (2012)
29. Polish Committee for Standardization, Testing of concrete in structures—Part 2: Non-destructive testing—Determination of reflection numer, PN-EN 12504-2
30. Polish Committee for Standardization, Evaluation of compressive strength of concrete in structures and precast concrete products, PN-EN 13791
31. ASTM D4541-17.: Standard Test Method for Pull-off Strength of Coatings Using Portable Adhesion Testers, pp. 1–6. ASTM International: West Conshohocken, PA, USA (06 Feb 2017)
32. Sharma, N.K., Kumar, P., Kumar, S., et al.: Properties of concrete containing polished granite waste as partial substitution of coarse aggregate. Constr. Build Mater. **151**, 158–163 (2017)
33. Kurda, R., De Brito, J., Silvestre, J.D.: Combined economic and mechanical performance optimization of recycled aggregate concrete with high volume of fly ash. Appl. Sci. **8**, 1189 (2018)
34. Kampa, Ł., Sadowski, Ł., Królicka, A.: The use of synthetic and natural fibers in epoxy coatings: a comparative mechanical and economic analysis. Int. J. Adhes. Adhes. 103017 (2021)

Mechanical and Cost Analysis of the Effect of the Addition of Granite Powder Waste on Selected Properties of Cement-Lime Plasters

M. Rudner, A. Chajec, and Ł. Sadowski

Abstract We focus on the effect of the addition of granite powder waste on the mechanical properties of cement-lime plaster. The granite powder was added at 5, 7.5, and 10 percent by weight of the dry plaster mix. A reference test without the additive was also performed. Compressive strength, bending tensile strength tests were conducted for each material configuration. Adhesion properties and mortar consistency were also tested. It was unanimously concluded that a 10% addition of granite powder allows achieving improved strength parameters while keeping the price as low as possible. All tested parameters were improved for this formulation. The 10% addition of granite powder may allow for plasters with greater adhesion, leading to greater durability. This will reduce the problem of falling off and cracking the plasters. The literature currently lacks a cost-economic analysis for such a composite. Thanks to the conducted research, the authors have found new economic possibilities of using the waste material, which is granite powder. This solution is also beneficial from the ecological point of view, it allows to reduce of the production of CO_2 by the cement industry, and in the end, prevents pollution of the environment by dispersion of granite powder to the land near its storage.

Keywords CO_2 emissions · Cement composites · Adhesion of plasters · Granite powder · Cost-economic analysis

M. Rudner (✉) · A. Chajec · Ł. Sadowski
Department of Materials Engineering and Construction Processes, Wroclaw University of Science and Technology, Wybrzeże Wyspiańskiego 27, 50-370 Wroclaw, Poland
e-mail: marlena.rudner@gmail.com

A. Chajec
e-mail: adrian.chajec@pwr.edu.pl

Ł. Sadowski
e-mail: lukasz.sadowski@pwr.edu.pl

L. F. M. da Silva et al. (eds.), *2nd International Conference on Industrial Applications of Adhesives 2022*, Proceedings in Engineering Mechanics,
https://doi.org/10.1007/978-3-031-11150-1_6

1 Introduction

Depending on the purpose, plaster mortar can have a different chemical composition. There are cement, cement-lime, lime or gypsum or clay plasters. They differ in parameters, thanks to which they can be used in various conditions. Nowadays the market is definitely dominated by ready-made plaster mixtures, the preparation of which is easy and fast. The main task of external plaster is to protect the building against the negative effects of weather conditions, or simply to provide an aesthetic appearance of the facade. Internal plasters also provide fire and moisture protection, as well as improve thermal and acoustic insulation of the partition. Incredibly important in plaster mortars is the phenomenon of adhesion. It is the bonding force between two materials that occur at their interface. While there are no major problems with achieving high compressive or tensile bending strengths, good adhesion of mortars is a challenge. Adhesion of plaster depends on the application temperature, among other factors. Francke and Popczyk [1] have noted the influence of a sudden drop in outside temperature during the mortar setting process on the durability of plaster. During plastering works, a possibly constant ambient temperature should be maintained. Also, the work should not be carried out at temperatures below 5 °C and above 25 °C. It leads to uneven drying and discoloration of the plaster. The value of force between the ground and the plaster depends also on the grain size of the mortar itself [2]. Finer mortar grains fill the space more tightly. When the mortar volumetric density is higher, the mortar structure is more compact. Higher adhesion is associated with better mortar packing and a more accommodating contact surface. Omar et al. [3] also noted a relationship between the strength of adhesion and the type and content of the additive added, as well as the substrate and its thickness. This is specifically related to the change in the chemical composition of the mix.

The adhesion of the plaster to the substrate results from its mechanical and chemical bond with the substrate, forming an adhesion surface (Fig. 1b). Mechanically there is the interlocking of the mortar in the irregularities of the substrate, and chemically there is the bonding of the mortar to the substrate. A good combination of these two materials minimizes the risk of problems that occur with plaster. The most common are cracks and mortar falling off the wall (Fig. 1b). A mortar with a good pull-off strength will be able to minimize the risk of damage to the plaster.

The consumption of cement is increasing all over the world, which is quite a problem. Cement is a material with high energy requirements, as its energy demand is about 30 GJ/t [4], and it is also the most important building material. Therefore, scientists from all over the world are striving to find a good eco-friendly substitute for this product [5, 6, 7]. Hospodarova et al. [8] observed a decrease in compressive strength of cement plaster from 26.6 MPa to 14.1–18.0 MPa after the addition of cellulose fibers. Researcher [9] showed an improvement in the adhesive properties of plaster mortar with the addition of granulated slag. Lin et al. [6] evaluated the possibility of using fly ash slag in concrete. They found that this additive can replace up to 20% of the cement content in mortar without compromising quality. An improvement in adhesion properties was also reported in our previous work [10]. As can be

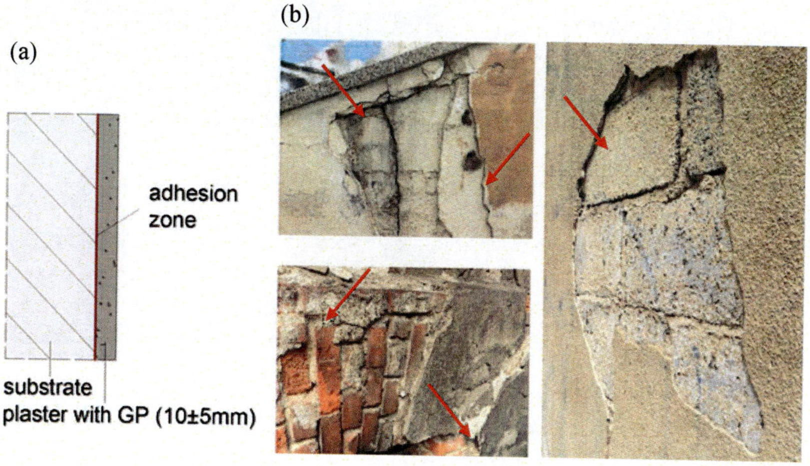

Fig. 1 **a** Plastered wall in cross-section, **b** examples of insufficient adhesion

seen, modification of mortar composition can bring both benefits and deterioration of composite performance. For this reason, it is so important to carry out research on modified mortar.

The main increase in greenhouse gases touts the CO_2 that is produced during the burning of fossil fuels [11, 12]. It is important to note that the cement industry is responsible for more than 25% of global industrial carbon dioxide emissions [13]. The accumulation of CO_2 in the atmosphere causes the capture of infrared radiation emitted from the earth's surface. This is followed by the heating of our planet through the absorption of sunlight. Progressive climate warming has disastrous consequences. It causes violent storms, windstorms, droughts [14], or extremely dangerous floods. To reduce the amount of CO_2 emitted, it is advisable to use local building materials [15]. It allows avoiding additional emissions during transportation. However, the best way is to replace cement in construction products with other additives. This solution solves both the problem of availability of raw material and carbon dioxide emission into the atmosphere.

Taking into account the above ecological and economic aspects, it was decided to use waste granite powder (Fig. 2) as a filler material in cement-lime plaster mortar.

The results of this study describing the effect of granite powder addition to premixed plaster mortar have already been published [10]. Basic mechanical properties such as compressive strength, flexural strength, pull-off strength, and fresh mortar properties were determined. For 10% granite powder addition, there was a significant improvement in all parameters. This paper aims to extend the study, given to show the relationship between the mentioned properties. This addition will contribute new data to the literature. In the further part of the article, the relationship between the cost of plastering and its mechanical properties is analyzed. This is to justify the selection of plaster mortar. An unexpected result is that the price decreases while

Fig. 2 Granite powder

the mechanical properties of the mortar increase. This study is a comprehensive compilation of plaster properties.

2 Materials and Methods

2.1 *Analysis of the Used Ingredients*

The granite powder used was the waste material from the processing and mining of granite blocks (Strzegom, Poland). One type of granite powder was used in this research. The particle size distribution (Fig. 3) of the granite powder was tested by sieve analysis according to PN-EN 933-1:2012 [16]. This testing method separates the material into fractions containing grains of different sizes. This is done by sieving through a set of sieves, which leads to the grains remaining on subsequent sieves with increasingly smaller meshes. The powder was passed through a set of 6 sieves with mesh sizes ranging from 0.02 to 0.14 mm. The volumetric density of used granite powder waste is 1194 kg/m^3.

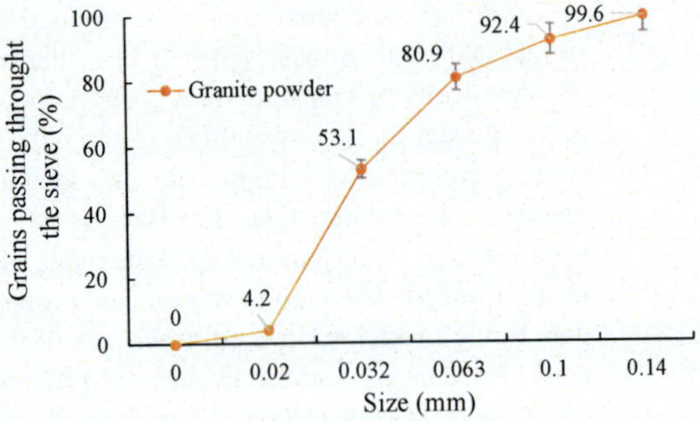

Fig. 3 Particle size distribution of granite powder

Table 1 Mix design proportions of the mortars

Series of mixes	Granite powder	Ready-mix plaster (RMP)	Water
(–)	(kg/m^3)	(kg/m^3)	(kg/m^3)
Ref	0.00	1550	310.00
5%	77.50	1550	325.50
7.5%	116.25	1550	333.25
10%	155.00	1550	341.00

This research involved the determination of the mechanical properties of cement-lime plaster with the addition of waste granite powder. Ready-mix cement-lime plaster (RMP) mortar for manual application, with a grain size of 1 mm, class CS II, was used for the study. It consists of hydrated lime, sand, cement, and other additives. Volumetric density is approximately 1550 kg/m^3. The compressive strength after 28 days was declared to be over 2.5 N/mm^2. Table 1 presents the compositions of the used research series.

2.2 *Composition of Used Research Series and Samples Preparing*

Tests were conducted on two types of specimens (Fig. 4), as described in a previous reference [10]. Compressive strength and bending tensile strength were tested on 40 × 40 × 160 mm rectangular specimens. Pull-off strength was tested on the substrates of the masonry units. Adhesion was investigated on Silka brick class 15, TERMOton P + W ceramic block class 15, and solid brick class 25. The approximate value of water absorption for a ceramic block is 18%, for Silka brick it is about 14%. The highest value of water absorption is for solid brick and it is about 22%.

Fig. 4 Specimens **a** rectangular 40 × 40 × 160 mm, **b** on substrates made of masonry elements

The plaster mortar needed for this study was made using a hand mixer in a plastic vessel. The ingredients were mixed before 3-5 min, according to the instructions in the technical card of the product used. When making the first type of specimens (rectangular samples), it was necessary to grease the molds (Fig. 4) with a release agent beforehand.

After that, the mortar was poured into them and compacted. The substrate of the second type of specimens was first wetted thoroughly and then proceeded to cover the flat surfaces with 10 ± 5 mm thick plaster. All specimens were stored in air-humid curing conditions, covered with foil, at a temperature of about 20 °C, for 28 days.

2.3 Properties of Fresh Cement-Lime Mortars

On the fresh mortar, two tests were performed to determine the mortar consistency and the wet mortar volumetric density. The measurements were repeated for each prescription.

The first test was carried out according to the procedure of PN-EN 1015-3 [17], i.e. the spreading table method. To determine the consistency, a cone-shaped form placed on the spreading table was filled with mortar up to half of its height. This is followed by compaction with 10 strokes of a 4 cm diameter rod. The second layer of mortar is then introduced and compacted again. Excess mortar should be rasped off and the surface of the disc wiped dry. After about 15 s, the form is lifted vertically. The next step is to shake the spreader table 15 times and measure two sample diameters in two perpendicular directions. The arithmetic mean of these measurements is the final result of the test and determines the consistency value of the mortar.

To determine the volumetric density (ρ_m) a vessel made of stainless steel with a volume of $V = 740$ cm^3 was used. Carefully weigh the vessel with an accuracy of 1 g before the measurement to obtain the value (m). The vessel is then filled with the prepared mortar in two layers. They should be properly compacted in each layer. Using a knife, excess mortar is scrapleveledd the upper edge is levelled. The next step is to weigh the mortar vessel with an accurate laboratory balance, reading the weight (m_w). To obtain meaningful results, three measurements are taken and the arithmetic mean is calculated from them. To obtain the volumetric density value, the formula was used:

$$\rho_m = \frac{m_w - m}{V} \tag{1}$$

The study was performed by the procedure in PN-EN 1015-6 [18].

2.4 Properties of Hardened Cement-Lime Mortars

For the determination of the compressive (f_{cm}) and bending tensile (f_{ctm}) strengths, the PN-EN196-1 [19] and PN-EN 1015-11 [20] standards were used, respectively. The research was performed on rectangular specimens. In the bending tensile strength test, the specimens had dimensions of 40 × 40 × 160 mm. After 28 days of manufacturing, they were subjected to testing on a bending machine in the laboratory of Wroclaw University of Science and Technology. During the test, a concentrated force acts on the specimen in the middle of the beam span with a support spacing of 10 cm. The force is applied until the beam breaks and is calculated and displayed on the machine. The bending tensile strength was determined for 3 beams and then the arty-metric mean was drawn from the results, which is the final result. For the compressive strength test, beams (6 halves) that had been previously tested for bending tensile strength were used. They were placed between the compression inserts (Fig. 5) and the press was turned on.

At the time of the destruction of the specimen, the value of the force that acted on it was read. The result obtained is the ratio of the force causing the failure of the beam by the compression area of the beam. All strength results were obtained in the unit of MPa.

The pull-off strength test was performed on three types of masonry units, described in detail in Sect. 2.3. The tests were performed in accordance with PN-EN 1015-12:2016-08E [21]. On each substrate, all series of plaster mortar were laid and space was prepared for at least 3 tests from each recipe. The substrate was thoroughly wetted before applying the mortar, and the samples themselves were stored in air-humid curing conditions for 28 days. Holes were then drilled in the resulting plaster to allow drilling into the substrate material. The centers of the holes, to which the steel discs were glued, were approximately 5 cm from the edge of the substrate. The

Fig. 5 Testing press

Fig. 6 Schematic of the test stand with TERMOton pull-off force measuring device

holes were made with a diamond hole saw. The next stage was a thorough dusting of the surface. Before sticking the disks, it was necessary to wash the plaster and the disk with acetone, in order to obtain a better connection of both materials. To stick the disks, a two-component epoxy glue was used, whose strength is greater than the adhesive strength of the mortar. Excess glue was removed. After the complete set, the actual test was performed. It was necessary to screw a fastener with a spherical end into a steel disc, which was inserted into the machine jaws. After leveling the device, the pull-off rate of 0.01 MPa/s was set. The measurement results in the form of pull-off strength (f_b) were read directly from the pull-off tester. A Proceq by-216 device was used, a photograph of which is shown in (Fig. 6).

3 Results and Analysis

3.1 Properties of Fresh Cement-Lime Mortars

One of the most important properties of plasters is consistency (Fig. 7). It should be dense so that the mortar does not run off the wall, but at the same time, it must be possible to spread it easily.

It is observed that the waste granite powder has a great influence on the value of scattering diameter and also on the bulk density of the mixture. As it increases, the diameter, as tested by the spreading table, decreases.

The smallest spread diameter of 15.2 cm was obtained for 10% granite powder addition. This is 19.79% less than that obtained for the reference sample. The total increase in volumetric density was 10.17%. To conclude, as the amount of granite

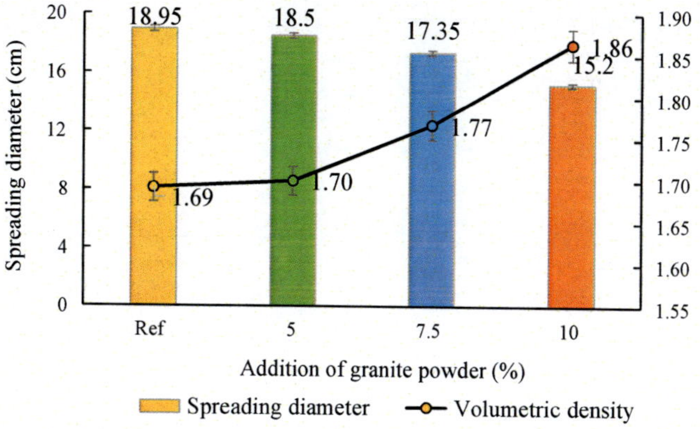

Fig. 7 Value of spreading diameter and volumetric density

powder addition gradually increased, the consistency diameter of the mix decreased and the volumetric density of the mix improved.

The grain size of granite powder is smaller than that of RPM, which causes changes in consistency. This phenomenon leads to the filling of voids, which reduces the porosity of the material. There is an improvement in the packing density of the mortar. The value of volume density of fresh mortar increases, confirming the effect of the additive as a filler. The same conclusion was determined by Li [22] considering granite powder as a filling material.

3.2 *Mechanical Properties of Hardened Cement-Lime Mortars*

Compressive strength and bending tensile strength (Fig. 8), have a common increasing trend. However, it should be noted that among the results obtained, the compressive strength value is identical for the reference test and with the addition of 5% granite powder. However, generalizing, the more granite powder added, the greater the increase in strength. This situation is quite unusual for composite materials. In most cases, adding another material to the mortar results in a decrease in strength [23], and the final choice of the amount of additive is a compromise between quality and price.

Fig. 8 Bending tensile strength and compressive strength after 28 days (MPa)

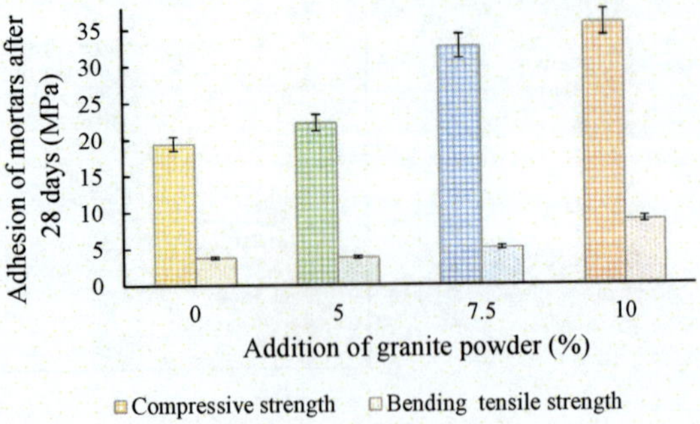

3.3 Adhesion of Hardened Cement-Lime Mortars

The value of adhesion strength (Fig. 9) between the substrate and the plaster mortar was checked on three different substrates. For the solid brick substrate, a continuous decrease in the pull-off strength was noted. It is interesting to note that for 5% and 7.5% granite powder addition, this value was the same on this substrate. A uniform increase was observed for this parameter on the silka brick substrate. It is worth noting that for 5% granite powder, the value of pull-off strength for the substrate of solid brick and silka brick took the same value. The results obtained for the substrate made of TERMOton were surprising. Initially, there was a decrease in the parameter and then a dynamic increase. The final value, for 10% granite powder addition, turned out to be the maximum.

Regardless of the contribution of granite powder, the peeling of the mortar took place at the interface between the mortar and the ground. The increase in the contact area is caused by the tight filling of the space in the plaster mortar. In the literature,

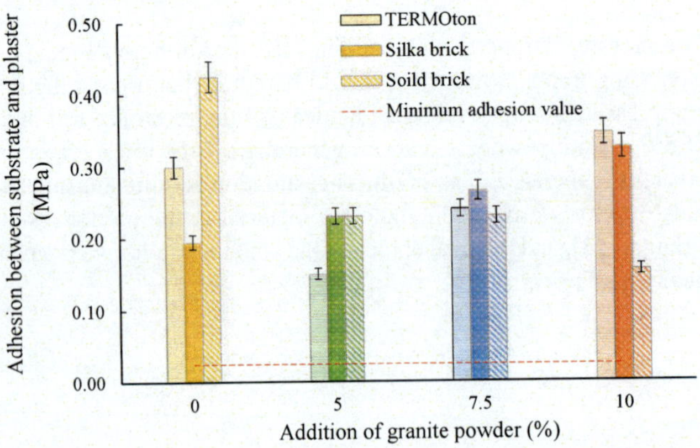

Fig. 9 Pull-off strength value

there is a description of a close relationship between the high strength of plaster and good adhesion of mortar [24]. There is also a better strength on the silicate block floor. The role of lime as a material improving the binding of mortar is not excluded.

3.4 *Influence of Mortar Consistency on Pull-Off Strength Values Concerning the Substrate*

As mentioned earlier, the most important properties of plaster mortars are their consistency and pull-off strength value. These parameters, being at the right level, guarantee good distribution of the compound on the wall, and later prevent cracking or falling off of the finished plaster.

In literature [25], the relationship between the substrate and the consistency of fresh mortars of general purpose is indicated. Based on this, recommended values of mortar consistency depending on the substrate were determined for plaster mortar according to the spreading table method. The consistency for ceramic masonry elements with water absorption of about 22% should be about 16-20 cm. The lower value is predicted for ceramic products with water absorption in the range of 6–21% and silicate elements and it is about 12–16 cm. The approximate value of water absorption for TERMOton ceramic block is 18%, for Silka brick it is about 14%. The highest value of absorption is for solid bricks and it is about 22%.

A correlation was observed between the consistency value and pull-off strength. The adhesion results obtained between the substrate and the plaster as well as the consistency values tested by the spreading table method are shown in the figure below (Fig. 10).

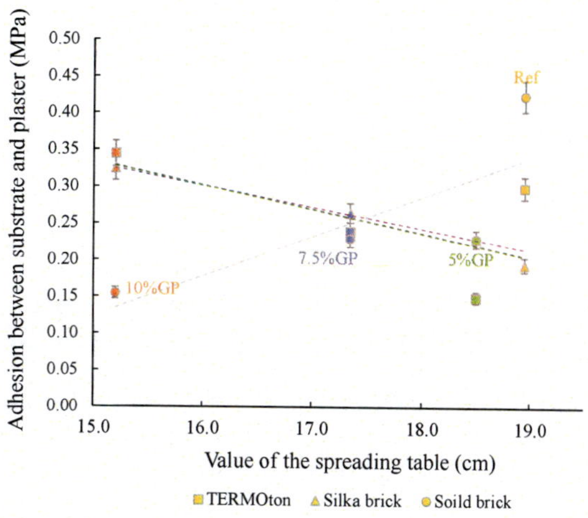

Fig. 10 Change in adhesion of hardened mortars to the substrate concerning the type of substrate and the amount of granite powder added and the consistency value

It was found that when the consistency of the mortar was about 18.95 cm, and therefore within the recommended range for the solid brick substrate, the pull-off strength value was highest precisely on this substrate. This was the case for the reference mortar, without the addition of granite powder. The highest pull-off strength value was obtained on the TERMOton substrate (0.345 MPa) and the Silka brick (0.325 MPa), for 10% granite powder addition. The consistency value was then 15.2 cm, which is also within the recommended values for these sleepers. It is worth noting that the increase in pull-off strength of the cement-lime plaster with an increasing amount of granite powder addition is continuous.

4 Cost-Economic Analysis

This part of the article analyses the dependence of the costs of cement-lime plasterwork on its strength properties. This comparison is important because of the investors' requirements and constantly rising prices. There are many ways to improve or modify the mechanical parameters of plasters, such as pull-off or compressive strength and tensile strength. Almost none of the available publications show the cost of making such plaster. It is natural to strive for the lowest possible price while maintaining the best possible strength parameters. The mortar formulation designer, when selecting a mix, should strive to achieve the best possible price/performance ratio. The following mechanistic and economic analysis related to the addition of granite powder to the mix will help in this assessment. The first step in the evaluation is to analyze the mechanical properties, making them dependent on the percentage of addition (MPR). Then, the cost of the cement-lime mixture with granite powder is determined and the effective cost ratio (ECR) is calculated. The key is to finally compare the MPR and ECR to determine the composition that is the best choice both economically and in terms of mechanical properties.

4.1 Analysis of the Mechanical Properties of a Composite Plaster Mortar

Detailed compressive, bending tensile and pull-off strength values of cement-lime plaster mortar were described in a previous article [10]. To compare the obtained results, the mechanical performance ratio (MPR) was used as follows:

$$\sum MPR = \frac{4x\frac{f_{b,mixy}}{f_{b,ref}} + 2x\frac{f_{cm,mixy}}{f_{cm,ref}} + \frac{f_{ctm,mixy}}{f_{ctm,ref}}}{7} \times 100\%(\%) \quad (2)$$

$$MPR_u = \frac{f_{b,mixy}}{f_{b,ref}} \times 100\%\,(\%) \quad (3)$$

Table 2 Results of analysis of mechanical properties of cement-lime plaster mortar with granite powder added

Series	MPR_b	MPR_{cm}	MPR_{ctm}	MPR	DMPR
	(%)	(%)	(%)	(%)	(-)
Ref	100.00	100.00	100.00	100.00	1.00
5%	117.95	114.43	100.00	114.38	1.14
7.5%	135.90	168.04	133.33	144.72	1.45
10%	166.67	184.33	233.33	181.24	1.81

$$MPR_{cm} = \frac{f_{cm,mixy}}{f_{cm,ref}} \times 100\% \ (\%) \tag{4}$$

$$MPR_{ctm} = \frac{f_{ctm,mixy}}{f_{ctm,ref}} \times 100\% \ (\%) \tag{5}$$

The different mechanical properties were therefore given weights according to the priority given to the parameter. When evaluating the parameter of mortar pu-off value, the results obtained on silica brick substrate were taken into account. This is because it is a material commonly used in large scale construction.

To compare the mechanical performance (MPR), the MPR parameter was determined as follows:

$$\Delta MPR = \frac{MPR_{seria}}{MPR_{ref}} (-) \tag{6}$$

Table 2 shows the results obtained.

A continuous increase in the value of mechanical efficiency is observed. With the addition of 5% granite powder, the value of this ratio increases to 114%. Comparing the mortar with 10% granite powder addition with the reference sample, an increase of nearly 82% is observed. This result is significant. In the literature on cementitious floors, with the same addition, the MPR decreased continuously. Thus, the obtained increase may be due to lime, which is a component of the mix.

5 Economic Analysis of Composite Plaster Mortar

The cost-economic analysis included the direct cost of purchasing granite powder (GP) ready-mix plaster mortar (RMP) and the water (W). We omitted the cost of plaster making as it is the same for all mortars. The price of RMP was about EUR 115.00/1t, while the price of waste granite powder was assumed to be EUR 8.90/1t. For 1 m^3 of water, EUR 1.78 should be paid. The cost breakdown for the different types of plaster is shown in Table 3. To determine the individual material costs, the

Table 3 Economic performance of the properties of the cementitious mortars

Series	c_{GP}	c_{RMP}	c_W	c_{mixes}	ECR	ΔECR
	(€)	(€)	(€)	(€)	(%)	(–)
Ref	0.00	177.73	0.55	178.28	100.0	1.00
5%	0.66	168.85	0.55	170.05	95.4	0.95
7.5%	0.96	164.40	0.55	165.91	93.1	0.93
10%	1.24	159.96	0.55	161.75	90.7	0.91

1. To facilitate the analysis, unit costs were adopted: $u_{GP} = 8.90$ EUR/1000 kg; $u_{RMP} = 114.67$ EUR/1000 kg; $u_W = 1.78$ EUR/1000 kg. 2. Shipping cost was omitted

effective cost ratio (ECR) was used.

$$c_{ingredient} = uc_{ingredient} \times ingredient\,ratio(€) \tag{7}$$

$$c_{mixes} = c_{GP} + c_{RMP} + c_W(€) \tag{8}$$

$$ECR = \frac{c_{mixes}}{c_{ref}} \times 100\%(\%) \tag{9}$$

$$\Delta ECR = \frac{ECR_{mixes}}{ECR_{ref}}(-) \tag{10}$$

ECR value for the mortar with 10% addition of waste material has the lowest value. The index has decreased by almost 9%. From the economic point of view, this is the best mortar. The effect of granite powder addition on the reduction of mix cost was also observed in other literature sources [26, 27].

5.1 *Combined Economic and Mechanical Performance of the Composites*

In order to correctly interpret the results obtained, the economic and strength analysis should be combined. Table 4 shows the MPR and ECR parameter values for a series of plaster mortars.

The relationship between mechanical properties such as compressive strength, bending tensile strength, and pull-off strength and the cost of plaster mortar is shown in Fig. 11. It is observed that a price increase is associated with a decrease in mechanical properties. This is an unexpected result. This trend is opposite to most known in the literature [28]. In mortar with granite powder, the increase in packing-density is well seen, which results in a decrease in the porosity of the material. Which in turn leads to improved mechanical properties of the mortar. Thus, it can be concluded that

Table 4 Mechanical performance ratio (MPR) and effective cost ratio (ECR) parameters with regards to the cementitious mortar series

Series	MPR	DMPR	ECR	DECR
	(%)	(–)	(%)	(–)
Ref	100.00	1.00	100.00	1.00
5%	114.38	1.14	95.38	0.95
7.5%	144.72	1.45	93.06	0.93
10%	181.24	1.81	90.72	0.91

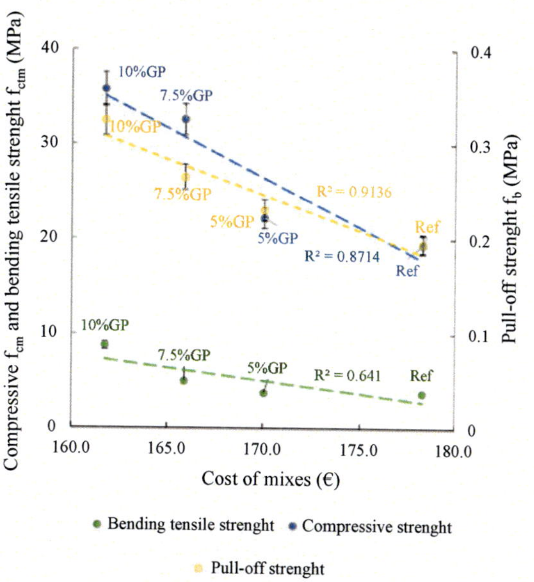

Fig. 11 **a** Compressive strength (f_{cm}), bending tensile strength (f_{ctm}), and pull-off strength (f_b) about the cost of plasters mixes.

cement-lime plaster mortar with 10% waste granite powder is the most economically and mechanically advantageous.

The comparison of the mechanical strength ratios (MPR) and effective cost ratio (ECR) values obtained during the tests confirms the thesis shown earlier.

6 Conclusions

The article describes an analysis of the relationship between the consistency of fresh mortar and its pull-off strength about the substrate. A detailed description of the research and the results obtained are presented in an earlier study by the authors. The substrates analyzed were: solid bricks, silka bricks, and ceramic hollow blocks.

The analyzed ratio of granite powder varied from 5 to 10% about the weight of the plaster mortar. The main conclusions are:

- The analysis shows that the pull-off strength value for the solid brick substrate was the highest for the reference sample.
- The greatest values of pull-off strength were recorded, for the substrate made of silka brick and ceramic hollow brick (materials of close absorption), at 10% addition of granite powder to the plaster mortar.
- All pull-off strengths of the samples were above the minimum value of 0.025 MPa set for plaster mortars.

These conclusions indicate that the absorption value of the primer material has a direct influence on the recommended consistency values, which translates into the pull-off strength results obtained. It is therefore recommended to use cement-lime plaster with granite powder on silka brick and hollow clay brick substrates.

The cost and mechanical analysis carried out for the silka brick substrate shows a clear advantage of the plaster with 10% granite powder addition in each of these aspects. Such a composite allows improved mechanical properties to be obtained while the price decreases. The increase in pull-off strength reaches up to 67% compared to the reference sample.

The results of these analyses indicate a great potential in implementing the solution in the industrial sector. From the results presented, it is clear that the use of waste granite powder improves the mechanical properties of the mortar.

7 Funding Source

The authors received funding from the project supported by the National Centre for Research and Development, Poland grant no. LIDER/35/0130/L-11/19/NCBR/2020 “The use of granite powder waste for the production of selected construction products.”

Acknowledgements The authors would like to thank Mateusz Moj for his help in organizing the study and making the samples and Łukasz Kampa for helping to prepare the samples.

References

1. Francke, B., Popczyk J.: Wpływ nagłego obniżenia temperatury zewnętrznej podczas procesu wiązania na trwałość tynków gipsowych. Materiały Budowlane (2017)
2. Chen, J., Guan, G., Ng, P., et al.: Packing optimization of paste and aggregate phases for sustainability and performance improvement of concrete. Adv. Powder Technol. **32**(4), 987–997 (2021)

3. Omar, H.A., Yusoff, N.I.M., Ceylan, H., et al.: Investigation of the relationship between fluidity and adhesion strength of unmodified and modified bitumens using the pull-off test method. Constr. Build. Mater. **122**, 140–148 (2016)
4. Taylor, M., Tam, C., Gielen, D.: Energy efficiency and CO_2 emissions from the global cement industry. Korea **50**(2.2), 61–7 (2006)
5. Crossin, E.: The greenhouse gas implications of using ground granulated blast furnace slag as a cement substitute. J. Clean. Prod. **95**, 101–108 (2015)
6. Lin, K.L., Wang, K.S., Tzeng, B.Y., et al.: The reuse of municipal solid waste incinerator fly ash slag as a cement substitute. Resour. Conserv. Recycl. **39**(4), 315–324 (2003)
7. Du, H., Dai Pang, S.: High-performance concrete incorporating calcined kaolin clay and limestone as cement substitute. Constr. Build. Mater. **264**, 120152 (2020)
8. Hospodarova, V., Stevulova, N., Junak, J., et al.: Cement materials based on cellulosic fibers for plasters. Sel. Sci. Pap.-J. Civ. Eng. **12**(1), 37–46 (2017)
9. Stevulova, N., Vaclavik, V., Hospodarova, V. et al.: Recycled cellulose fiber reinforced plaster. Materials **14**(11), 2986 (2021)
10. Rudner, M., Chajec, A., Sadowski, Ł.: Mechanical and adhesive properties of cement-lime plasters modified with waste granite powder. In: 6th International Conference on Adhesive Bonding 2021, pp. 33–51. Springer, Cham (2021)
11. Zandalinas, S.I., Fritschi, F.B., Mittler, R.: Global warming, climate change, and environmental pollution: recipe for a multifactorial stress combination disaster. Trends Plant Sci. **26**(6), 588–599 (2021)
12. Bigot, S., Buges, J., Gilly, L., et al.: Pivotal roles of environmental sensing and signaling mechanisms in plant responses to climate change. Glob. Change Biol. **24**(12), 5573–5589 (2018)
13. Chen, C., Xu, R., Tong, D., et al.: A striking growth of CO_2 emissions from global cement industry driven by new facilities in emerging countries. Environ. Res. Lett. (2022)
14. Gray, S.B., Dermody, O., Klein, S.P., et al.: Intensifying drought eliminates the expected benefits of elevated carbon dioxide for soybean. Nat. Plants **2**(9), 1–8 (2016)
15. Bica, S., Roşiu, L., Radoslav, R.: What characteristics define ecological building materials. In: Proceedings of the 7th IASME/WSEAS International Conference on Heat Transfer Thermal Engineering & Environment, pp. 159–164, (2009)
16. PN-EN 933-1:2012: Tests for Geometrical Properties of Aggregates—Part 1: Determination of Particle Size Distribution—Sieving Method
17. PN-EN 1015-3:2000: Methods of Testing Mortars for Masonry—Part 1: Plastering Mortar—Determination of Consistency of fresh Mortar
18. PN-EN 1015-6:2000: Methods of Testing Mortars for Masonry—Part 1: Plastering mortar—Determination of Volume Density of Fresh Mortar
19. PN-EN 196-1:2016: Methods of Testing Cement—Part 1: Determination of Strength
20. PN-EN 1015-11:2001: Methods of Test for Mortar for Masonry—Part 11: Determination of Flexural and Compressive Strength of Hardened Mortar
21. PN-EN 1015-12:2016-08E: Methods of Test for Mortar for Masonry—Part 12: Determination of Adhesive Strength of Hardened Rendering and Plastering Mortars on Substrates
22. Li, L.G., Wang, Y.M., Tan, Y.P., et al.: Filler technology of adding granite dust to reduce cement content and increase the strength of mortar. Powder Technol. **342**, 388–396 (2019)
23. Bicer, A., Kar, F.: Thermal and mechanical properties of gypsum plaster mixed with expanded polystyrene and tragacanth. Therm. Sci. Eng. Prog. **1**, 59–65 (2017)
24. Bochen, J.: Weathering effects on physical-chemical properties of external plaster mortars exposed to different environments. Constr. Build. Mater. **79**, 192–206 (2015)
25. Popek, M.: Wykonywanie zapraw murrskich i tynkarskich oraz mieszanek betonowych WSiP, 42–46 (2014)
26. Chajec, A., Granite powder versus fly ash for the sustainable production of air-cured cementitious mortars. Materials **14**(5), 1208 (2021)

27. Kampa, Ł., Sadowski, Ł.: Engineering technology and preparation of coating systems made with epoxy resin-based bonding agents modified with the addition of granite powder sourced from quarry wastes. In: 6th international conference on adhesive bonding 2021. Springer, Cham, (2021)
28. Kurda, R., De Brito, J., Silvestre, J.D.: Combined economic and mechanical performance optimization of recycled aggregate concrete with high volume of fly ash. Appl. Sci. **8**(7), 1189 (2018)

Evaluating the Influence of Short Fiber Reinforced Thermoplastic Composites Produced by Injection Molding on the Stress Distribution in an Adhesively Bonded Joint Using a Multi-Scale Numerical Modeling Approach

Sofie Verstraete, Frederik Desplentere, and Stijn Debruyne

Abstract As one of the major joining techniques, adhesive bonding has many advantages thanks to its low weight, ability to bond dissimilar materials, preservation of material properties (no heat changes or drilled holes), etc. This last quality is especially important for composite materials. When a bonded joint is loaded in a single lap joint configuration, stress concentrations in both peel and shear at the edge of the overlap may trigger failure. The magnitude of these stresses depends on both the material properties of the adhesive and the material properties of the substrate. Considering the specific case of short fiber composites, local material properties vary according to the local fiber orientation. This work applies a multi-scale modeling approach of the short glass fiber composite substrates with a fiber volume fraction of 10 and 20% to evaluate the effect of homogenized local stiffness variations on stress distribution in the adhesive bond, compared to a macro-scale constitutive transverse isotropic material model. Results show that a significant variability of the fiber orientation within the substrates induce large variation of material properties of the substrate. A reduction in the stress concentrations is observed for an increased fiber volume fraction. This reduction is even more pronounced when the orientation of the short fibers is aligned with direction of the quasi-static load. The comparison when applying the multi-scale modeling approach compared to a global transverse isotropic modeling approach indicates that stress concentrations are underestimated when the

S. Verstraete (✉) · S. Debruyne
Department of Mechanical Engineering, KU Leuven Campus Brugge, Spoorwegstraat 12, 8200 Brugge, Belgium
e-mail: sofie.verstraete@kuleuven.be

S. Debruyne
e-mail: stijn.debruyne@kuleuven.be

F. Desplentere
Department of Materials Engineering, KU Leuven Campus Brugge, Spoorwegstraat 12, 8200 Brugge, Belgium
e-mail: frederik.desplentere@kuleuven.be

L. F. M. da Silva et al. (eds.), *2nd International Conference on Industrial Applications of Adhesives 2022*, Proceedings in Engineering Mechanics,
https://doi.org/10.1007/978-3-031-11150-1_7

local variability in fiber orientation is not considered. In addition, when increasing the overlap length of the adhesive joint, the local material properties located at the edge of the overlap start to be responsible for the influence on these stress concentrations, due to the variation in local bending stiffness. This indicates the necessity of taking these local variations into account when designing an adhesively bonded joint using short fiber composites.

Keywords Short glass fiber polymer (SGFP) · Adhesive bonded joint · Single lap joint (SLJ) · Multi-scale modeling

1 Introduction

Recent trends in industry go towards economical optimization, environmental awareness and increasing mechanical performance of products. Material selection for product design have to result in better performance coupled to a lower production cost. Short fiber reinforced thermoplastic composites in the automotive sector have one of the main advantages that it is a lightweight material with improved mechanical properties leading to a reduction in the fuel consumption [1]. Several studies have shown an increase in mechanical properties for short glass fiber reinforced composites within a polylactic acid (PLA) [2, 3]. These types of composites can be manufactured using a fast production process such as injection molding resulting in a cost reduction. In larger structures, adhesive bonding is widely used in a variety of engineering applications. It offers the advantage of an even weight distributed over a larger bonded surface and also load transfer over that larger bonded surface. Another advantage compared to mechanical joining techniques, especially for composite materials, is that the internal composite structure does not require to be damaged which ensures the performance of the material [4].

Considering single lap bonded adhesive joints, eccentricity in the load causes substrate bending and result in stress concentrations at the edge of the overlap, which may be the initiation of failure of the bond. Several researches studied the effect of the substrate stiffness on the stress distribution and bonding performance in an adhesively bonded joint [5, 6]. They have shown the effect of an increasing substrate stiffness on the shear and peel stress distribution within an adhesively bonded joint. For bonding short fiber composites, it brings also complexities such as a highly anisotropic material behavior and effects of local fiber orientations. Micromechanical models are used to describe the material stiffness depending on the fiber type, volume fraction, shape and aspect ratio [7, 8]. Anisotropy within the material causes a variability in the material stiffness, depending on the orientation of the short fibers with respect to the loading direction of the composite [9, 10]. Using injection molding as a production technique, local variations in fiber orientation depending on the production- and design parameters of the mold results in variations in the local material behavior [11–13].

The main objective of this research is to describe the effect of local fiber orientation on the mechanical material behavior of short fiber reinforced composites. This is done by virtual material modeling of the mechanical behavior of the structure using FEA simulations and show its effect on the stress distribution in the adhesive within a single lap bonded joint.

2 Materials and Methodology

2.1 *Substrate Material: Short Glass Fiber Reinforced Polymer (SGFRP)*

Chopped glass fibers are reinforced in a polylactic-acid (PLA) matrix, supplied by NatureWorks®, to improve the material stiffness of this bio-based polymer. A first drying step of the PLA polymer for 6 h at 80 °C is applied before compounding with the glass fibers using a twin screw Leistritz ZSE18MAXX lab extruder. Another drying cycle of 6 h at 80 °C is applied before the injection molding process, which produces plates of 85 × 85 × 3 mm using an Arburg 320S injection molding machine. Production parameters of the injection molding process are summarized in Table 1.

Three configurations are tested, virgin PLA, PLA with 10% and with 20% fiber volume fraction. The mold is designed in such manner that the fiber orientation in the plate is (mostly) along the direction of the flow. Specimen can be cut out in the longitudinal and transverse direction of the fiber orientation. The nomenclature of the samples used in this paper depends on the volume fraction of the glass fiber reinforced composites: PLA, PLA10 and PLA20.

Mechanical characterization of the test specimen is done on tensile bars according to the ISO 527-1BA standard using a 5 kN tensile bench (Instron 3345) in combination with an extensometer (Instron 2630–102) for several orientations in the plate. The longitudinal and transverse stiffness, the Poisson ratio and the shear modulus of the material for different fiber volume fractions are than used for calibration of the numerical model which is described in a following section.

Table 1 Parameters injection molding process

Process parameter	
Melt temperature	190 °C
Injection volume	30.7 cm^3
Injection pressure	1065 bar
Holding pressure	750 bar
Holding time	15 s
Cooling water temperature	25 °C
Mold temperature	25 °C

Table 2 Material properties 2 k Epoxy adhesive after curing for 48 h at 23 °C

Adhesive	*E*—modulus (Mpa)	$v(-)$	*E*—modulus (Mpa)
Epoxy EA9466	1708	0.4	610

2.2 Adhesive

The adhesive considered in this study is a (2:1) two component epoxy (EA9466), supplied by Loctite®. This adhesive is tested after curing for 48 h at 23 °C for tensile properties using tensile bars according to the ISO 527 standard, in combination with digital image correlation to obtain the Poisson ratio. Shear properties were obtained by iosipescu shear test in combination with digital image correlation, confirming the isotropic material behavior. Material properties are shown in Table 2.

2.3 Setup Numerical Model

A meso-scale approach identifying the local material behavior of the short natural fiber reinforced composite was set up, based on fiber orientation tensor (FOT) data from moldflow simulations of the composite material. This is used to model the material behavior of the substrate in the analysis of the stress distribution in a single lap adhesively bonded joint. Material stiffness according to the fiber volume fraction and aspect ratio is obtained using a local mean field homogenization using the Mori–Tanaka approach. The chopped glass fibers are assumed to as 2D ellipsoidal inclusion, for which the strain concentration tensor is obtained using the Eshelby model. The influence of the fiber orientation on the homogenized stiffness is implemented using the hybrid closure approach. The setup of this model is discussed in the next part of this section. Figure 1 gives a schematic view of the setup of the meso-scale model for the short fiber composite. A local material stiffness matrix is setup for each element in the quadratic C3D20R structural mesh of the composite with a mesh size of 0.5 mm according to the local fiber orientation according to the modeling approach described above. The epoxy adhesive has a thickness of 0.5 mm and the overlap length is varied between 6 and 24 mm. The material behavior is described as a linear elastic material behavior because of its brittle behavior. A C3D20R structural mesh with mesh size 0.125 mm is used, where 4 elements are used trough the thickness.

2.4 Step 1: Mori–Tanaka Generalized Homogenization

The general composite stiffness *C*, according to the Mori–Tanaka generalized homogenization approached, based on the Voigt assumption, can be described according to (Eq. 1) [8, 9].

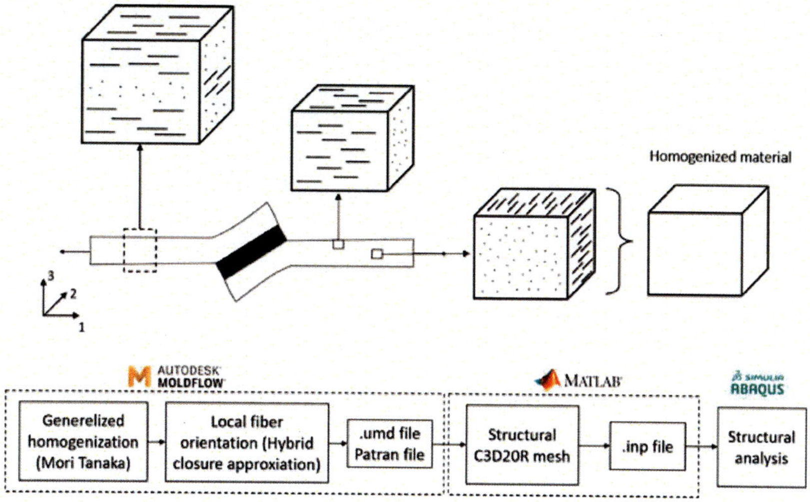

Fig. 1 Schematic view setup meso-scale model

$$C = C_m + V_f(C_f - C_m)A^{Mori-takana} \quad (1)$$

Subscripts m and f reperesent the matrix and fiber respectively. V indicates the volume fraction and $A^{Mori-takana}$ is the strain concentration tensor according to Mori–Tanaka as described in (Eq. 3).

$$A = \left[I + S(C_m^{-1})(C_f - C_m)\right]^{-1} \quad (2)$$

$$A^{Mori-takana} = A\left[(1 - V_f)I + V_f A\right]^{-1} \quad (3)$$

2.5 Step 2: Eshelby Strain Tensor

In Eq. (2), S represents the Eshelby strain tensor, based on an elliptical inclusion with aspect ratio $\rho = {}^{a_1}/_{a_2} = {}^{a_1}/_{a_3}$, where a_1 is the length of the fiber and $a_2 = a_3$ is the diameter of the fiber [14, 15]. The tensor is described according to (Eq. 4), where the aspect ratio has been experimentally determined by measuring the fiber lengths distribution after extraction from the matrix.

$$S = \begin{bmatrix} S_{1111} & S_{1122} & S_{1133} & & & \\ S_{2211} & S_{2222} & S_{2233} & & 0 & \\ S_{3311} & S_{3322} & S_{3333} & & & \\ & & & 2S_{2323} & 0 & 0 \\ & 0 & & 0 & 2S_{1313} & 0 \\ & & & 0 & 0 & 2S_{1212} \end{bmatrix} \tag{4}$$

2.6 Step 3: Fiber Orientation Tensor

Taking into account the effect of the fiber orientation, a 4th order stiffness tensor depending on the micro mechanics model and the local fiber orientation is obtained to predict the material elasticity is described in (Eq. 5) [8].

$$\begin{aligned} T_{ijkl}(p) = {} & B_1\left(a_{ijkl}^{HYB}\right) + B_2\left(a_{ij}\delta_{kl} + a_{kl}\delta_{ij}\right) \\ & + B_3\left(a_{ik}\delta_{jl} + a_{il}\delta_{jk} + a_{jl}\delta_{ik} + a_{jk}\delta_{il}\right) \\ & + B_4\left(\delta_{ij}\delta_{kl}\right) + B_5\left(\delta_{ik}\delta_{jl} + \delta_{il}\delta_{jk}\right) \end{aligned} \tag{5}$$

where a represents the fiber orientation tensor (FOT) obtained from the moldflow simulation, δ is the Dirac delta function and the B-parameters are obtained from the generalized homogenized stiffness matrix obtained via the micromechanics modeling approach as $B_1 = C_{11} + C_{22} - 2C_{12} - 4C_{66}$, $B_2 = C_{12} - C_{23}$, $B_3 = C_{66} + \frac{1}{2}(C_{23} - C_{22})$, $B_4 = C_{23}$, $B_5 = \frac{1}{2}(C_{22} - C_{23})$.

The 4th order orientation tensor a_{ijkl}^{HYB} is obtained from the 2nd order fiber orientation tensor using the hybrid closure approximation as given in (Eq. 6) [8].

$$a_{ijkl}^{HYB} = (1 - f)\left(a_{ijkl}^{LIN} + a_{ijkl}^{QUA}\right) \tag{6}$$

where $f = 1 - 27\det\left(a_{ij}\right)$ and the linear and quadratic 4th order orientation tensor are given in (Eqs. 7, 8) respectively [8].

$$\begin{aligned} a_{ijkl}^{LIN} = {} & \frac{1}{7}\left(a_{ij}\delta_{kl} + a_{il}\delta_{kl} + a_{il}\delta_{jl} + a_{kl}\delta_{ij} + a_{jl}\delta_{ik} + a_{jk}\delta_{il}\right) \\ & - \frac{1}{35}\left(a_{ij}\delta_{kl} + a_{ik}\delta_{jl} + a_{il}\delta_{jk}\right) \end{aligned} \tag{7}$$

$$a_{ijkl}^{QUA} = a_{ij}a_{kl} \tag{8}$$

3 Results

3.1 Material Properties SGFRP

Figure 2 shows the stress–strain curves of the with chopped glass fiber reinforced PLA composites. Due to the brittle nature of the PLA, there is no yielding behavior before material failure. Material properties obtained from these measurements are summarized in Table 3. The composite stiffness increases with an increasing fiber volume fraction, where an increase in stiffness up to 70% for a volume fraction of 20% is measured. On the other hand, elongation at break decreases. Tensile strength increases up to 26% for a fiber volume fraction of 20%, indicating a good interfacial adhesion resulting in an effective load transfer to the reinforced fibers. These material characteristics are used as input parameters to model the composite substrate behavior in the next section of this work.

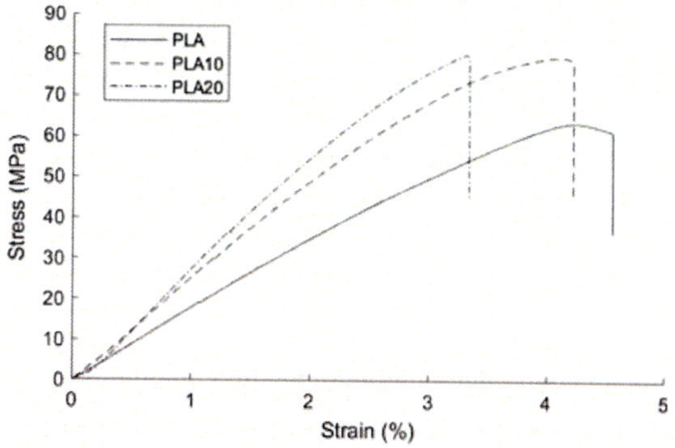

Fig. 2 Stress–strain behavior chopped glass fiber composites for different fiber volume fractions

Table 3 Overview material properties for the chopped glass fiber reinforced composites

Material	PLA	PLA 10% GF	PLA 20% GF
E_{11} (Gpa)	3.8	5.2	6.5
E_{22} (Gpa)	–	3.8	4.1
$\nu(-)$	0.35	0.33	0.32
Tensile strength: $\sigma_{11,max}$ (Mpa)	63.8	77.9	80.5
Tensile strength: $\sigma_{22,max}$ (Mpa)	–	43.8	41.1

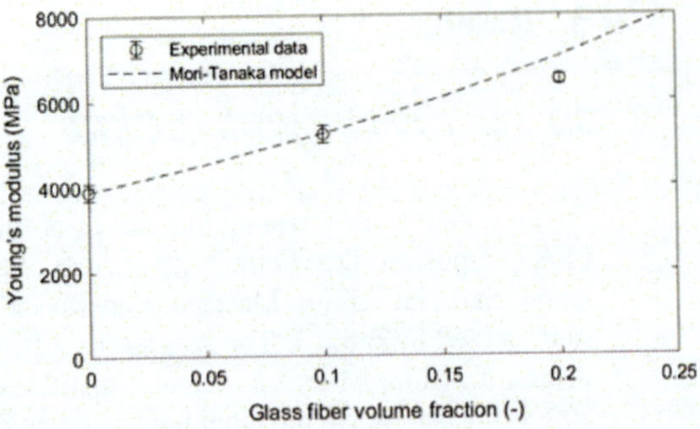

Fig. 3 Comparison between experimental data and the analytical Mori–Tanaka modeling approach

3.2 Analytical Stiffness Prediction SGFRP

In previous section, the analytical Mori–Tanaka model to predict the material stiffness in function of the fiber volume fraction is setup. Results show that there is a high correlation between experimental data and the predicted model (Fig. 3). A deviation of +8.15% on the stiffness is noticeable for the composite with 20% glass fibers. This is due to the fact that for this composite according to the fiber orientation tensor data in the moldflow simulation, not all fibers through the thickness at the location where the tensile bar is produced are aligned with the load, resulting in a lower experimentally obtained material stiffness. This is also discussed in more detail in the next sections.

3.3 FEA Results

High shear and peel stress concentrations at the end the single lap joint (SLJ) overlap are responsible for failure of an adhesively bonded joint. This section studies the main characteristics of injection molded short fiber composites that affect the stress distribution in the adhesive are divided into 3 categories and their effects. For each case, a static load of 1 kN is considered. Figure 4 shows an overview of the setup of the structural mesh with local material properties which is used for structural analysis within Abaqus.

3.4 The Effect of the Fiber Volume Fraction

The most important parameter that is concidered in this work is the fiber volume fraction. As described by several analytical modeling approaches and which is experimentally demonstrated in previous part does an increase in the fiber volume fraction results in an increase of the material stiffness. Within these models, material stiffness

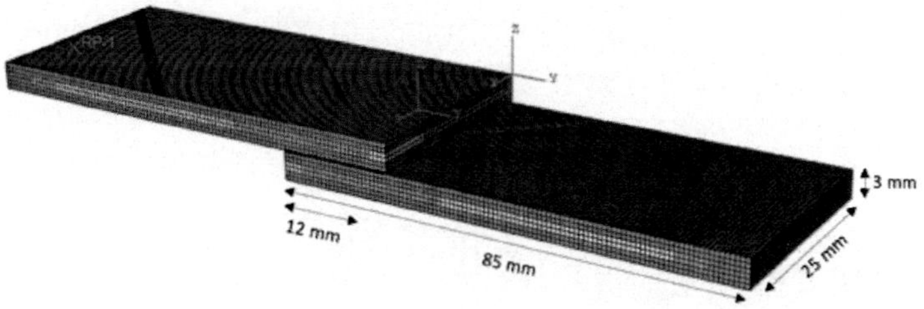

Fig. 4 Structural mesh with local fiber orientation

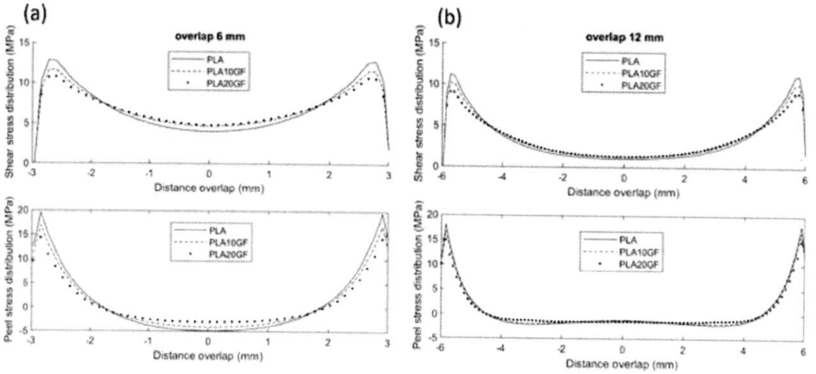

Fig. 5 Shear- and peel stress distribution for an overlap of **a** 6 mm and **b** 12 mm for different fiber volume fractions of the short glass fiber reinforced composite

depends next to the fiber volume fraction also on the fiber type, shape and aspect ratio, which in this work are taken to be constant.

Figure 5 shows results of a FEM simulation where stress distribution in function of the overlap length is plotted for different volume fractions. An increase of the substrate stiffness from 3.8 MPa to 5.2 MPa for a fiber volume fraction of 10% results in a reduction of the peak stress concentrations at the edge of the overlap of 10% in peel and 13% in shear. Concidering composites with 20% fiber volume fraction, the stiffness increases up to a value of 6.5 MPa which leads to a further reduction of the peak stresses of 16% and 27% in shear and peel respectively.

3.5 *The Effect of Fiber Orientation*

The second effect considered in this study is the anisotropic material behavior that occurs due to the fiber orientation. Experimental results indicate a stiffness reduction in the transverse direction on the flow orientation of 27% for a 10% fiber volume fraction and up to 37% for 20% glass fiber volume fraction. Taking this into account

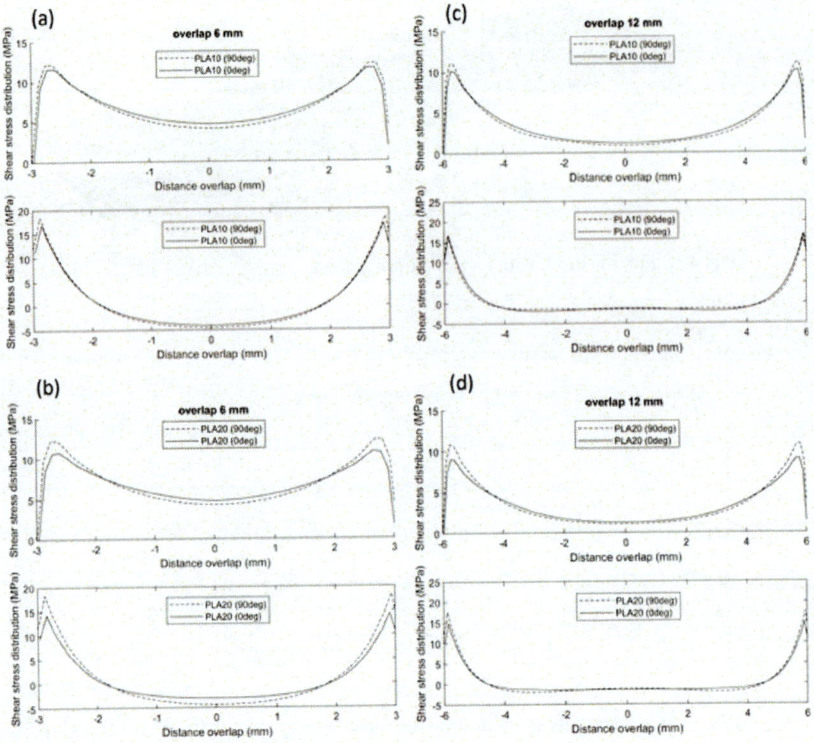

Fig. 6 Shear and peel stress distribution for an overlap of **a–b** 6 mm and **c–d** 12 mm for variation in global fiber orientation along the flow direction

in the FEM simulation for the stress distribution in the adhesive of a single lap joint as seen in Fig. 6, a reduction of the peak stresses up to 17% and 16% for shear and peel stresses respectively could be noticed for 20% fiber volume fraction.

3.6 *Global Transverse Isotropy Versus Multi-Scale Modeling*

The third effect that is considered in this work when concerning injection molded short fiber composites, is the effect of the local variability of the fiber orientations within the composite structure as is schematically shown in Fig. 1. The fiber orientation for each element of the composite is obtained through the fiber orientation tensor (FOT), acquired as output of rheological flow simulations in moldflow. A simple square plate geometry (85 × 85 × 3 mm) is used, which is injected via a fan sprue, designed to optimally orient the fibers in the plate along the direction of the flow. Variations in fiber orientation trough the thickness of the plate are found for

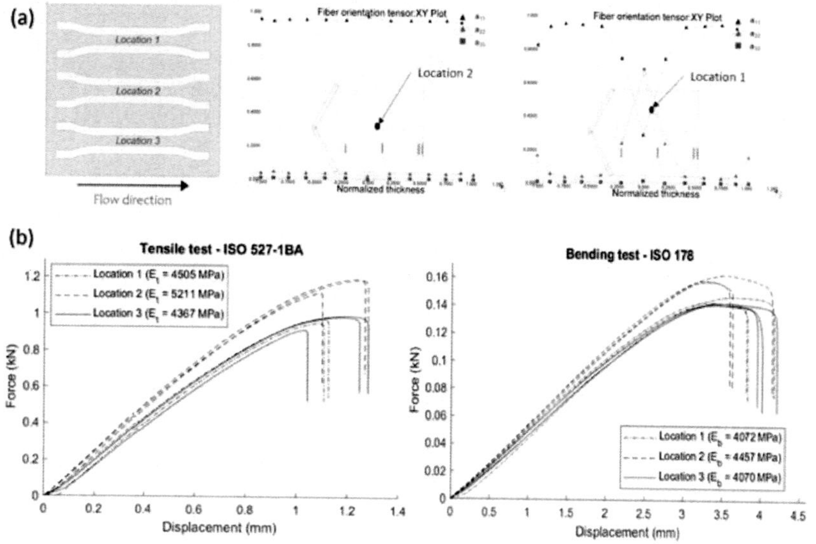

Fig. 7 **a** Fiber orientation tensor (FOT) through the thickness of the plate and results for **b** tensile data and **c** bending data of dog bone samples at different locations of the plate according to the ISO 527-1BA and ISO 178 standard

different locations in the plate as shown in Fig. 7. While characterizing the material at the side of the plate using dog bone samples according to ISO 527-1BA, a reduction in material stiffness of 14.9% and 8.7% for the tensile and bending modulus respectively are measured compared to the center with more aligned fiber orientation for a fiber volume fraction of 10%.

Considering a global transverse isotropic material modeling approach, the fiber orientation is assumed to be aligned with the flow direction through the entire thickness of the plate. Since this fiber orientation varies throughout the thickness of the plate, a lower substrate stiffness occurs, which is overestimated using this approach. A comparative FEM study is performed on the stress distribution in a single lap joint with an overlap of 12 mm, where this global transverse modeling approach is compared to the multi-scale modeling approach presented in this study. Results are shown in Fig. 8, where an underestimate of the stress concentrations at the edge of the overlap up to 5.3% and 3.5% for shear and peel stresses respectively is predicted using the global transverse modeling approach compared to the multi-scale modeling approach.

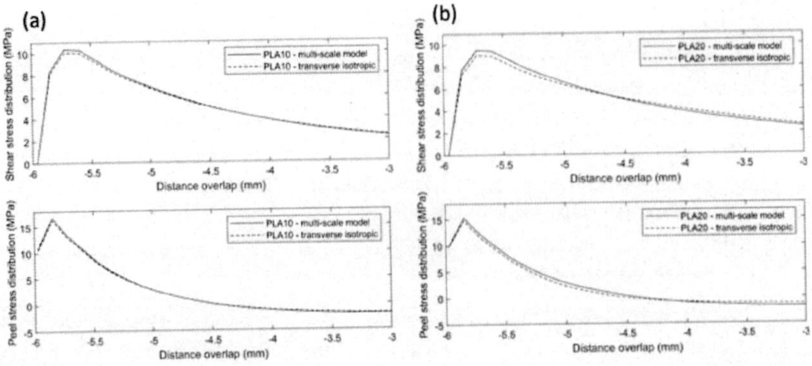

Fig. 8 Shear and peel stress distribution for an overlap of 12 mm using a global transverse isotropic modeling approach compared to the proposed multi-scale approach considering local fiber orientations for a fiber volume fraction of **a** 10% and **b** 20%

3.7 *Effect Joint Design*

Optimization of the bonding performance of short fiber composites can be done by adjusting the composite design to reduce stress concentrations at the edge of the overlap by controlling your local fiber orientation as discussed in the previous section. This section investigates the effect of the overlap length as design parameter on the stress distribution.

Figure 9 shows the effect of the overlap length for substrates with different fiber volume fractions. At a certain overlap length, the load transfer at the center of the overlap decrease to zero. In that case will an increasing overlap length no longer result in a significant increase in the bonding performance. The bonding strength then depends on the stress concentrations at the edge of the overlap, which converge to a constant value, depending on the substrate stiffness. Within the injection molded composites, the local fiber orientation varies throughout the structure resulting in variations of local substrate stiffness. For these composites, the stress concentrations at the edge of the overlap will not converge to a constant, but will depend on these local stiffness at that location.

4 Conclusions

Several studies on adhesive bonding in composites have shown that the location of reinforcements has a large effect on the stress concentration of an adhesive joint. This work investigates the effect of the fiber volume fraction and fiber orientation of short glass fiber composites for composites which are produced by an injection molding process. These parameters have an effect on stress distribution and stress concentrations in an adhesively bonded single lap joint.

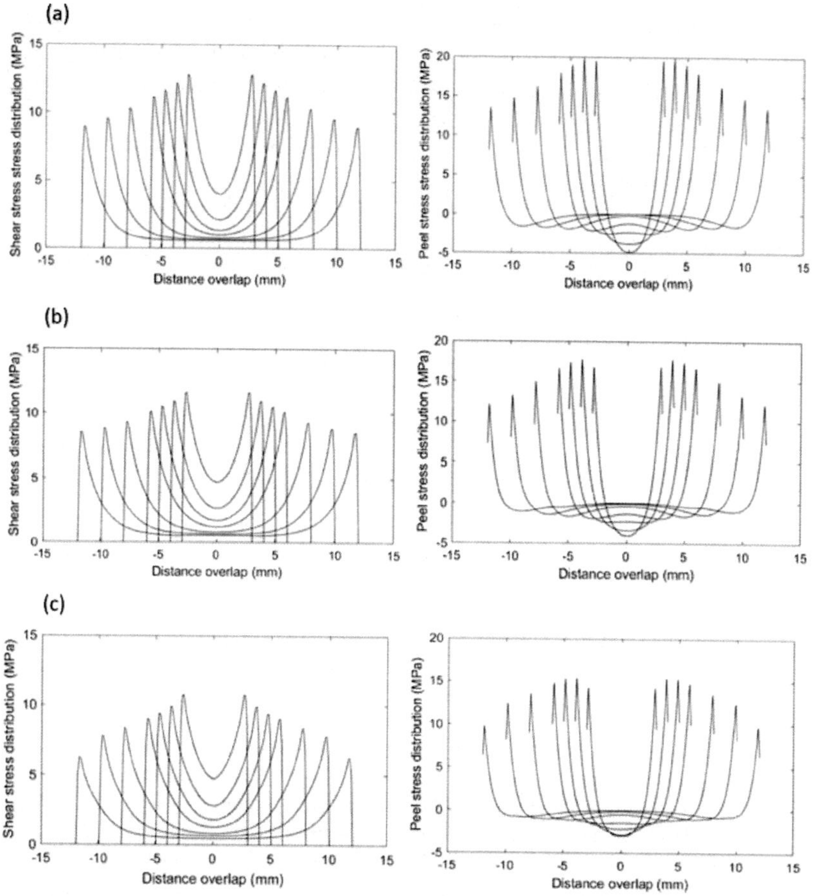

Fig. 9 Shear- and peel stress distribution for **a** PLA, **b** PLA10GF and **c** PLA20GFfor different overlap lengths

An increase in the fiber volume fraction of the short glass fiber composites causes in an increase in substrate stiffness. This results in a reduction of the stress concentrations at the end of the overlap up to 16% in shear and 27% in peel stresses. Further reduces the orientation of the short fibers perpendicular to the direction of the load the material stiffness up to 37% for a fiber volume of 20%. This causes an increase of the stress concentrations at the edge of the overlap up to 17% and 16% for shear and peel stresses respectively compared to substrates where the fibers are oriented in the direction of the load.

This study presents a multi-scale modeling approach for the short glass fiber composites where local fiber orientations in the substrate material are taken into account. The variability in the fiber orientation due to the production process cause an increase in stress concentrations at the edge of the overlap compared to when a global transverse isotropic material modeling approach is applied, where an overestimate of the composite stiffness is assumed. This variability in fiber orientation throughout the structure results in local material stiffness, which effects the bonding performance by effecting the stress concentrations at the edge of the overlap when increasing the overlap length.

Further research on optimization of the bond design can be accomplished by considering effects of the production process parameters, mold design, injection location, etc. to obtain a better, more aligned fiber orientation in the direction of the load at the location where adhesive bonding of the composite will occur.

Acknowledgements This project was funded by the European Community in the frame of the by Interreg V program (France-Wallonie-Vlaanderen) via the ATHENS project.

References

1. Park, H.S., Dang, X.P.: Development of a fiber-reinforced plastic armrest frame for weight-reduced automobiles. Int. J. Automot. Technol. **12**, 83–92 (2011)
2. Rahimizadeh, A., Kalman, J., Henri, R., Fayazbakhsh, K.: Recycled glas fiber composites from wind turbine waste for 3D printing feedstock: effect of fiber content and interface on mechanical performance. Materials (2019)
3. Wang, G., Zhang, D., Wan, G., Li, B., Zhao, G.: Glass fiber reinforced PLA composiete with enhanced mechanical properties, thermal behavior and foaming ability. Polymer (2019)
4. Vinson, J.R.: Adhesive bonding of polymer composites. Polym. Eng. Sci. **29**, 1325–1331 (1989)
5. Her, S., Chang, C.: Interfacial stress analysis of adhesively bonded lap joints. Materials (2019)
6. dos Reis, M.Q., Carbas, R.J.C., Larques, E.A.S., da Silva, L.F.M.: Numerical modelling of functionally graded adherends. Spec. Issue Appl. Adhes. Bond. (2020)
7. Avanti, S., Tucker, C.: The use of tensors to describe and predict fiber orientation in short fiber composites. J. Rheol. (1987)
8. Tucker, C., Liang, E.: Stiffness predictions for unidirectional short-fiber composites: review and evaluation. Compos. Sci. Technol. (1999)
9. Quintana, M.C., Frontini, P.M., Arriaga, A., Plank, B., Major, Z.: Fiber orientation distribution predictions for an injection molded venturi-shaped part validated against experimental micro-computed tomography characterization. Polym. Compos. Mater. (2020)
10. Amir-Rad, A., Wismans, M., Pastukhov, L.V., Govaert, L.E., van Dommelen, J.A.W.: Constitutive modeling of injection-molded short-fiber composites: characterization and model application. J. Appl. Polym. Sci. (2020)
11. Micota, D., Isaincu, A., Marsavina, L.: Micromechanical modeling of glass fiber reinforced plastic material. Mater. Today: Proc. 45. part 5 (4330–4336) (2021)
12. Kugler, S.K., Kech, A., Cruz, C., Osswald, T.: Fiber orientation predictions—A review of existing models. J. Compos. Sci. (2020)
13. Mcnalty, D.: Short fiber orientation and its effect on the properties of thermoplastic composite materials. Polym.-Plast. Technol. Eng. (2008)
14. Jing, X., Wang, Q., Keer, L.M.: A closed-form solution for the Eshelby tensor and the elastic field outside an elliptic cylindrical inclusion. J. Appl. Mech. (2011)
15. Parnell, W.J.: The Eshelby hill: moment and concentration tensors for ellipsoidal inhomogeneities in the Newtonian potential problem and linear elastostatics. J. Elast. (2016)

Investigation of Mechanical Properties of Woven Hybrid Metallic Fabric

Kaan Emre Engin, Ali Ihsan Kaya, and Serif Citil

Abstract In this study, main purpose was to produce hybrid metallic fabrics by using traditional weaving methods. The weaving process was completed by using aluminum wire (Al 1050 H14) and polyethylene fishing line with 0.2 mm thicknesses. Three types of hybrid laminated composite specimens with different orientations were prepared for comparison. Specimens were produced which included hybrid metallic fabrics woven parallel to the weft direction (HMFX), parallel to the warp direction (HMFY), and fishing line fabric (FLF) only, due to the reason that FLF had isotropic structure. Each layer was bonded by using epoxy film adhesive, FM 73 M between Al 5005 H34 sheets. Acquired laminated hybrid composites were cured in a specially designed die by applying pressure and heat using a heat controlled hydraulic press. After the curing process was completed, specimens were cut out according to ASTM D3039 standard for tensile testing. highest values were found in a hybrid metallic fabric with aluminum wires woven parallel to the weft axis (HMFX). Also, all the composites produced were around 63% lighter than the AL 5005 H34 specimen with the same thickness as the composite specimens.

Keywords Hybrid composites · Hybrid metallic fabrics · Woven aluminum and polyethylene

K. E. Engin (✉) · A. I. Kaya · S. Citil
Department of Mechanical Engineering, Engineering Faculty, Adiyaman University, Ataturk Boulvard, Adiyaman, Turkey
e-mail: kengin@adiyaman.edu.tr

A. I. Kaya
e-mail: alikaya@adiyaman.edu.tr

S. Citil
e-mail: scitil@adiyaman.edu.tr

L. F. M. da Silva et al. (eds.), *2nd International Conference on Industrial Applications of Adhesives 2022*, Proceedings in Engineering Mechanics,
https://doi.org/10.1007/978-3-031-11150-1_8

1 Introduction

In today's world, technological advancements bring new type of issues to be dealt with such as environmental degradation, vast increase in the energy consumption levels, the need for new types of materials to meet the mechanical and physical property demands of ever-growing industrial innovations, advanced machinery, and new vehicles [1, 2]. New types of materials hold crucial importance in various application areas where lightness and high strength are required at the same time [3].

Composite materials have growing usage in industrial applications and are preferred over conventional materials because of their high strength, light weight, good strength to weight ratio, dimensional and design flexibility. Apart from the traditional polymer-based composites, textile composites are a relatively new family of materials and are employed in a wide range of lightweight applications, including construction, aerospace and automotive. Their advantageous mechanical qualities, along with low density and weight, also ease of handling and processing, make them very useful [4].

A variety of factors impact the quality of produced textile composites; fibers used in the composite which can be synthetic or natural (metal, carbon, glass, flax and so on), their structure (e.g., roving, yarn), the twist and the fineness of the fibers [5]. The fiber material has dominant effects on the overall strength of the composite. The reinforcing fibers for textile composite should have high tensile strength while remaining as light as possible.

Metallic fibers are an appealing kind of fiber for composite applications due to their superior mechanical characteristics. Metallic fibers are mostly polycrystalline. Due to fewer crystal flaws and greater crystallite alignment in the fiber length direction, a metallic fiber's strength and stiffness qualities are much higher than those of the bulk material. Metal fibers have numerous benefits as reinforcement. They may be easily produced utilizing a variety of fabrication procedures and are more ductile, as well as being less susceptible to surface damage and having excellent strength and temperature resistance. A variety of metals are accessible in fibrous forms such as aluminum, stainless steel, titanium, copper, pure tantalum, nickel, tungsten, etc. However, a major drawback exists for using metallic fibers as reinforcement materials in textile composites which is the addition of weight where lightness is an important asset [5, 6].

But, amongst the other metallic fibers, aluminum fibers present good mechanical properties while being lightweight. Aluminum alloys are used to make the vast majority of aeronautical components. They have low density, high electrical and thermal conductivities, and excellent corrosion resistance. Aluminum fiber composites outperformed glass fiber reinforced composites in terms of mechanical, thermal, and electrical properties Also, aluminum fibers are much available, abundant and relatively affordable compared to other metallic fibers [7–9].

To produce hybrid metallic fabrics, a bonding mechanism should be implemented between the layers. Usage of adhesives are the appropriate method for achieving the bonding. Adhesive joints are of fundamental importance to the many industrial applications including but not limited to textile and composite industry because of their capacity to decrease labor, time, and cost, as well as their excellent corrosion and fatigue resistance, fracture retardation, and superior damping qualities. [10, 11].

There are numerous studies investigating the different aspects of adhesives and aluminum alloys. Single and double acting connection types are the most often utilized adhesive connection types in these kinds of applications. Kweon et al. [12] evaluated the failure loads and mechanisms of adhesive bonding, bolt fastening, and adhesive-bolt hybrid joining for three types of joints. For aluminum and composite bonding, they employed a film-type adhesive FM73 and a paste-type EA9394S, and they compared the strength of the different joint types. The influence of surface treatment on the performance and strength of single lap Aluminum–copper alloys joints for automotive applications was investigated by Boutar et al. [13]. They discovered that single lap joint shear strength is inversely related to surface roughness, and that rougher surfaces have less wettability. Naat et al. [14] examined the effect of surface roughness on fatigue behavior and quasi-static strength in adhesively bonded joints. When bi-functional water-soluble polymers were situated between the adhesive (an epoxy-dicyandiamide) and the aluminum substrate, Layec et al. [15] investigated the influence of bi-functional water-soluble polymers on adhesive bonding of aluminum in transportation. They investigated process parameters like prolonged or brief immersion in polymer-containing treatment baths and formulation with three different concentrations. Then, they performed several mechanical tests like micro-scratch, three-point bending, and single lap shear to evaluate the performances of the BLs. Then, findings were compared to those of previously released products. The influence of oil and water pollution on the binding strength of aluminum-aluminum and aluminum-glass fiber reinforced polymers was examined by Reneckis et al. [16]. They used mechanical and chemical surface preparation methods and found that water contamination was less harmful than mineral oil contamination. Surface pretreatment of metals, unlike composites, was a critical aspect of creating a trustworthy adhesive bond, according to Davis and Brown [17]. Cleaning the metal surfaces is required to provide a good chemical connection between the adhesive and the adherend substance, according to Pethrick [18]. Cavezza et al. [19] and Kuczmaszewski [20] also emphasized the relevance of metal adherent preparation surfaces, which had a significant influence on the chemistry of the substrate's top layer in terms of system durability. They also concentrated on the performance of adhesively bonded joint structures under external tension, corrosive ions, and water effects. Goglio and Rezaei [21] used several surface treatments to investigate the strength deterioration of open-faced single lap joints as a function of age. Among the pretreatments tried, phosphoric acid anodization produced the best results and was sufficient to allow the adhesive layer to fail mainly cohesively. Viana et al. [22] investigated the mechanical behavior of moisture deteriorated bulk tensile adhesive specimens at low and high temperatures, as well as the development of T_g with absorbed moisture. It was discovered that when the adhesive absorbs water, its tensile behavior deteriorates. Using the Taguchi approach, da Silva et al. [23] evaluated the surface treatment, overlap, durability, adhesive thicknesses and adherend, adherend and adhesive effect on the lap shear strength. Kairouz and Matthews [24] studied cross-ply adherends in single-lap bonded joints and found that the surface layer orientation had a direct impact on the failure mechanism and strength of the joints. Citil et al. [25] examined the mechanical properties under tension of double strap joints made of aluminum alloy AA 2024-T3 with a gap and a filled intermediate section, concluding that the intermediate element had no influence on double-strap bonding joints. Citil [26] investigated the novel joint types of AA 2024-T3 aluminum alloy with varied surface geometries and discovered that lap joints with enhanced surface geometry could withstand more load.

As can be seen, there are many studies assessing different cases of adhesive bonding and aluminum material. However, investigations regarding to metallic fabrics are scarce and heavily focused on electrical science. As an example; Yu et al. [27] propound the usage of stainless-steel textiles as a current collector for high-performance graphene-based supercapacitors. They stated that when compared to the commonly used flexible current collectors, stainless steel textiles offer superior characteristics due to the stainless-steel textiles' outstanding conductivity, mechanical flexibility, and great electrochemical stability. They achieved good capacitance results by using stainless-steel textiles. Other studies gravitate towards the manufacturing and researching different aspects of textiles that can be used in capacitors, wearable electronics, and energy storage applications [28–33]. None of the studies investigated hybrid metallic fabrics composed of aluminum and their mechanical properties which create a gap in the literature.

The main purpose of this study is to develop novel hybrid metallic fabric composite materials that might be employed in unmanned aerial vehicles with high strength to weight ratios. The second objective is to overcome the disadvantages that have direct effects on metallic bonding surfaces and the strength of the bonded materials caused by defects and contaminants during the production process of metallic sheets which have drawn attention as an issue and became the topic of many studies. Employing aluminum wires with tiny diameters instead of bulk materials may increase adhesion surface and improve bonding strength.

To achieve this purpose, Hybrid Metallic Fabrics (HMF) were produced by weaving aluminum wires having 0.2 mm diameter and polyethylene fishing lines of the same diameter using traditional rug weaving techniques. Fabrics, only composed of fishing lines (Fishing Line Fabric aka FLF) without aluminum wires were also produced. Then separately, HMF and FLF's were adhesively bonded with aluminum sheets under different orientations by using FM 73 M epoxy film adhesive to constitute hybrid laminated composites and tensile tests were performed to determine the mechanical properties of these laminated hybrid composites.

2 Materials and Method

2.1 *Preparation of Woven Hybrid Metallic Fabrics*

Aluminum wire and polyethylene fishing line with size of 0.2 mm diameters were obtained from local vendors. Both materials were chosen to promote flexibility and proper strength at the same time.

The material of the wire was Aluminum 1050 H14 aluminum alloy which is a wrought alloy with good electrical conductivity, corrosion resistance, and workability. It is frequently used in the electrical and chemical sectors. When compared to more heavily alloyed metals, it has a poor mechanical strength. This alloy was chosen specifically to ease the handmade weaving process [34]. The mechanical properties of Al 1050 H14 were given in Table 1.

The used fishing line "Effe Sumo X8" is made from polyethylene material and the line is leased with 8 thinner lines similar to sailor rope which is referred as X8. The line with 0.2 mm diameter can withstand 18.4 kg of load, according to the producer's data.

The woven hybrid metallic fabric was composed of weft and warps. Weft was made with aluminum wires, while warp was made from polyethylene fishing line.

After the fishing line threads were tied as warp to the conventional weaving equipment, a layer of traditional cotton thread was woven to serve as a foundation for the aluminum wires to be linked as weft, and the aluminum wires were incorporated into the system. The reason behind the choice of fishing line as a warp was to avoid ruptures generated by friction during the compression stage of the wefts, which would have occurred if an aluminum warp had been employed. For hybrid metallic fabrics, a knot density of 55 per 1 cm^2 was used to calculate the warp density of the fishing line threads. The early weaving experiments were the rationale for adopting this density. The woven fabrics can be seen in Fig. 1.

Same weaving procedure was repeated to produce fabrics only composed of polyethylene fishing line carrying the purpose of comparing the change in mechanical properties between HMF and FLF.

2.2 *Preparation of Hybrid Laminated Composites*

Hybrid laminated composites were produced to observe the effects of different weaving directions of hybrid metallic fabrics compared with fishing line fabrics on the strength of materials under tensile loading.

ASTM D3039 tensile test standard was used to perform the tests. The specimens should be prepared according to the dimensional requirements of the standard. The standard states that specimens should have rectangular shapes with 25 ×

Table 1 Mechanical properties of aluminum wire Al 1050 H14 [35]

Alloy and temper	Ultimate tensile strength (MPa) Min–Max	Tensile yield strength (MPa) Min–Max	Elongation (50%) Min–Max	Hardness (Brinell) Min–Max
1050 H14	110–115	90–105	9	35–36

Fig. 1 The woven hybrid metallic fabrics

250 mm dimensions. To ensure dimensional stability throughout the whole lamination process, all specimens first cut with bigger tolerances (30 × 300 mm). Examples of cut fabrics were given in Fig. 2.

Structural adhesives are commonly utilized to join composites because of their low shear stress and increased notch sensitivity [36]. This bonding was achieved by epoxy film adhesive, FM® 73 M by CYTEC. FM 73 M film adhesive is an aerospace epoxy that may be used for a variety of applications. It can operate in temperatures ranging from −67°F to 180°F (−55 °C to 82 °C). FM 73 M film adhesive is designed to join metals and is also ideal for bonding a variety of structural composite systems [37].

This epoxy film adhesive was donated by Turkish Airlines for this study. Properties of FM 73 M epoxy film adhesive was given in Table 2.

Hybrid laminated composites were composed of 5 layers where aluminum sheets resided at the bottom and top layers while the remaining layers were placed between these aluminum sheets. The illustration for the configuration of layers were given in Fig. 3. The material of the sheets was Aluminum 5005 H34 which is a medium-strength alloy with excellent atmospheric corrosion resistance and weldability.

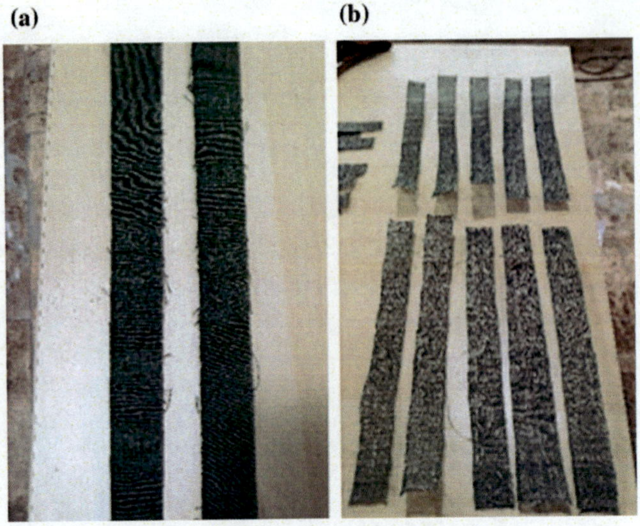

Fig. 2 Cut specimens; **a** polyethylene fishing line fabric, **b** hybrid metallic fabric

Table 2 Properties of FM 73 M epoxy film adhesive [37]

Nominal weight, lb/ft^2 (g/m^2)	Nominal thickness, inches (mm)	Color	Tensile shear at 75 °F (24 °C) for metal to metal psi (MPa)	Floating roller peel at 75 °F (24 °C) for metal to metal lb/in (KN/m)	Blister detection at 75°F (24 °C) for metal to metal psi (MPa)	Climbing drum metal-to-metal peel at 75 °F (24 °C) in-lb/in (Nm/m)
0.045 (210)	0.007 (0.18)	Yellow	6073 (41.9)[a]	67 (11.74)[a]	4870 (33.56)[a]	73 (323.34)[a]

[a] Calculated by interpolation as per data sheet

Fig. 3 Configuration of layers in hybrid laminated composites

Al 5005 is commonly used in building and structural applications, pipes, tubes, chemical and food processing equipment, etc. [38]. The mechanical properties of Al 5005 were given in Table 3.

It is vital to be free of things such as paint, oil, or dust on the surface that the adhesive comes into touch within order to have adequate adhesion between the adhesive

Table 3 Mechanical properties of aluminum wire Al 5005 H34 [39]

Alloy and temper	Ultimate tensile strength (MPa) Min–Max	Tensile yield strength (MPa) Min–Max	Elongation (50%) Min–Max	Hardness (Brinell) Min–Max
5005 H34	145–165	110–145	9	35–36

and the contacting surfaces. To accomplish this, the surfaces of the bonded samples were cleaned and degreased prior to bonding by using a surface cleaning spray consist of Acetal, Naphtha and Prophane.

The arrangement of layers was kept same to obtain an equal condition carrying the purpose of observing the differences in mechanical properties. The aluminum layer had a thickness of 0.5 mm whereas epoxy film adhesive had 0.018 mm. Only the characteristics of the fabrics were changed. Specimens were prepared with hybrid metallic fabrics (HMF) and with fishing line fabrics (FLF). Because the weaving direction is significant and HMF was anisotropic, three distinct specimens were produced: HMF woven in the x direction (woven direction parallel to the weft axis, referred as HMFX), HMF woven in the y direction (woven direction parallel to the warp axis, referred as HMFY), and fishing line fabric (FLF).

FLF's weft and warps were composed of the same material hence admissible as isotropic therefore, there was no need for different specimen preparation including x and y directions. 3 hybrid laminated composite specimens were produced for each fabric characteristic. The layering of HMF and FM 73 M epoxy film adhesive was given in Fig. 4.

Fig. 4 The layering of fabrics and epoxy film adhesives

Fig. 5 The fixation of hybrid laminated composites on to the die

After positioning of all layers, sealing tapes were used to fix the laminates at their terminal points. In order to cure the adhesive, an amount of pressure should be applied on to the specimens under the environmental temperature of 120 °C and should be kept at this position for 2 h.

A manufactured die set consist of upper and lower plates with polished surfaces was employed to make the exertion of necessary pressure possible and prevent sliding during the curing stage. It should be mentioned that before being used, die sets were also degreased and dust purified. Wooden supports with the height of specimens were used to prevent the upper plate to squish the specimens. Pressure was applied by a hydraulic press with 20 kN loading capability.

The hydraulic press also has heating units that can reach to 900 °C. The placement of layered specimens on to the die before the curing stage was given in Fig. 5. The hydraulic press was shown in Fig. 6.

After the curing process was accomplished for 2 h at 120 °C, specimens were held at room temperature for 24 h. Then, specimens having 30 × 300 mm dimensions were cut to 25 × 250 ± 0.2 mm by using wet cut circular saw. During the cutting procedure, each specimen received the same amount of trimming from all sides, resulting in specimens that were evenly aligned with one another. Then, tensile samples weights were scaled.

2.3 *Tensile Testing of the Hybrid Laminated Composites*

Tensile tests were performed using the ASTM D3039 tensile test standard, as previously stated. Tensile tests of hybrid laminated composites were performed in an ambient laboratory environment using a U-Test universal tensile testing equipment (Ankara, Turkey) with 10 kN capacity under 2 mm/min crosshead speed. The tensile

Fig. 6 The hydraulic press used for the curing of specimens

tests were carried out on three bonded samples from each group and the results obtained were evaluated. The experimental setup was shown in Fig. 7.

3 Results and Discussion

The weights of tensile test samples were determined using precision scales with an accuracy of 10^{-4} g before they were subjected to tensile testing. Each specimen was weighed three times on the same precision scales, and the average results were considered. The result of the averaged weights of specimens was given in Table 4.

It was observed that the weight of HMF was slightly lighter than FLF for both x and y directions. Despite the addition of a metal such as aluminum to the fabric, the reduction in weight compared to a composite with a polyethylene structure has the potential to produce lighter materials if employed as textiles.

The weight of the laminated Al 5005 H34 with the identical dimensions and laminate numbers as the composite ones was 88 g, demonstrating that composite materials are well suited for situations where weight is a factor, as mentioned in prior studies [40–43].

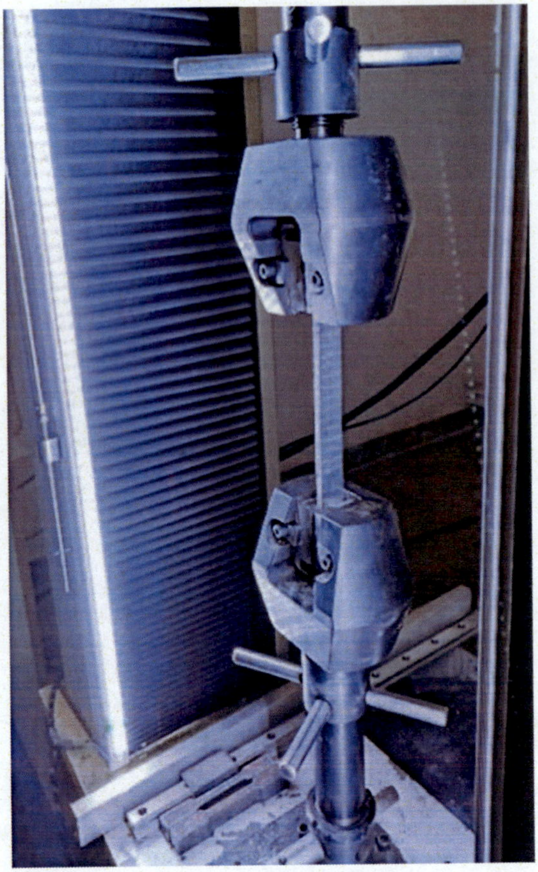

Fig. 7 Universal tensile testing machine used for tensile tests

Table 4 Average weights of specimens for each group

	Hybrid metallic fabrics X-direction (HMFX)	Hybrid metallic fabrics Y-direction (HMFY)	Fishing line fabric (FLF)
Average weights of samples	33.834 gr	31.069 gr	34.973 gr

Tensile test results of three specimens for each of the composites composed of HMFX, HMFY and FLF were shown in Figs. 8, 9 and 10, respectively. The value obtained from test showed that there was nominal difference between the test samples of all composites hence can be considered as the results in correlation with each other. The average of the results for each composite characteristic were calculated and the results were given as a comparison in Fig. 11.

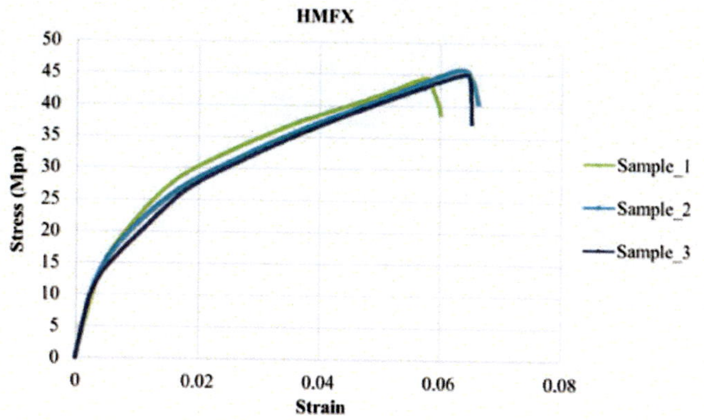

Fig. 8 Tensile test results for composites composed of HMFX

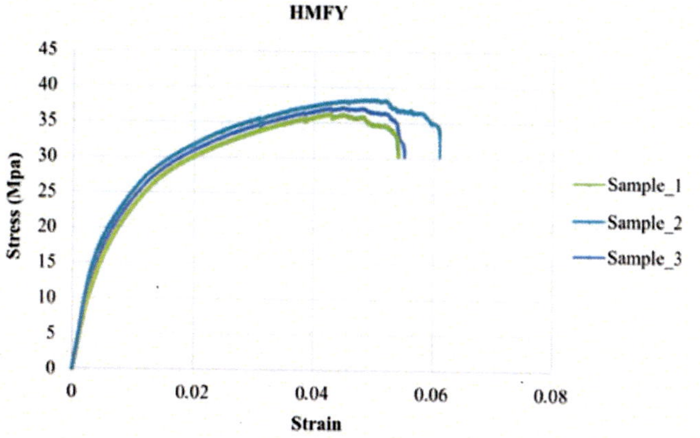

Fig. 9 Tensile test results for composites composed of HMFY

It was observed that the ultimate tensile strength for composites composed of HMFX was found to range between 44 and 47 MPa, with strain values ranging between 0.058 and 0.065 resulting an average modulus of elasticity, $E_x = 4.3$ Gpa.

These results were highest amongst the other specimens, HMFY and FLF. These values were anticipated as result of anisotropic structure of the HMF.

The tensile load was in the direction parallel to axis of the aluminum wires therefore the load is applied predominantly onto the aluminum wires. The tensile strength of the wire was higher than the fishing line and as a result, the composite withstand the highest tensile load.

The influence of the anisotropic structure of HMF manifested itself again in the tensile test results of HMFY. In the instance of HMFY, the weaving direction was

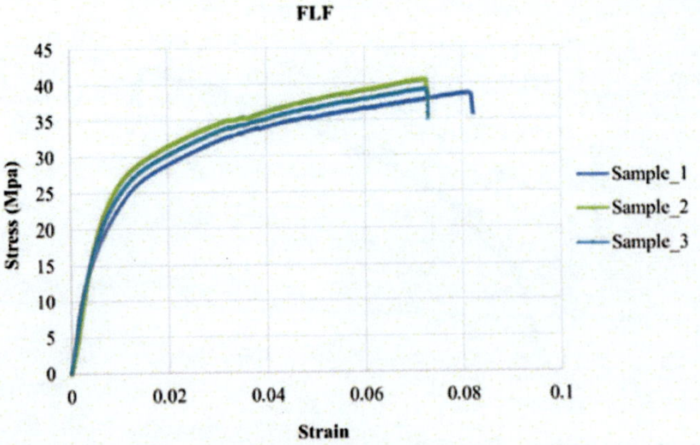

Fig. 10 Tensile test results for composites composed of FLF

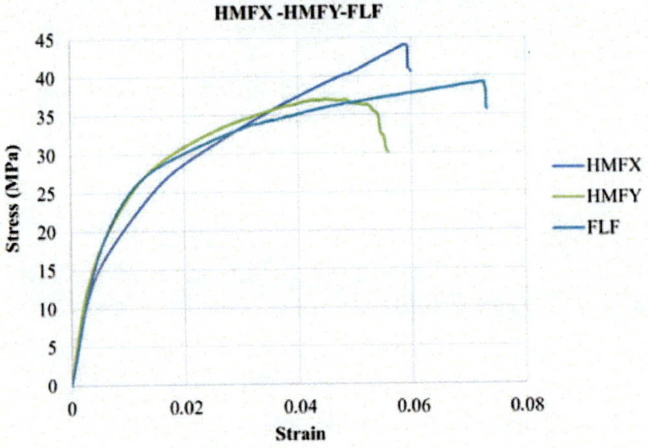

Fig. 11 Comparison of the averaged tensile test values of three type of specimens

parallel to the warp axis, which was the direction in which fishing line were weaved. Fishing line had lower mechanical properties than aluminum wires.

It was observed that the ultimate tensile strength for composites composed of HMFY was found to range between 35 and 38 MPa, with strain values ranging between 0.054 and 0.062 that having an average modulus of elasticity, $E_x = 2.45$ Gpa. These results were lower than the results of X-direction and again proved that fiber orientation has significant effects on the mechanical properties of the composite subjected to eternal forces as stated in other studies [44–46].

The ultimate tensile strength of FLF composites was determined to be between 38 and 41 MPa, with strain values ranging from 0.072 to 0.084 resulting an average modulus of elasticity, $E_x = 3.41$ Gpa. Composites consisted of FLF showed better results compared to the results gathered from composites composed of HMFY.

This situation was the result of weaving density of the wires and lines. The fabrics were made by hand utilizing traditional carpet looms and procedures, as previously described. Aluminum and fishing line have various degrees of elasticity. Weaving fishing line is similar to weaving cotton thread, except weaving aluminum is more challenging. Aluminum was employed as weft in the weaving process, which meant that every row had to be folded on itself and forced downwards to make the row by using special thread combs. During that folding and compression stage, aluminum tended to break. To avoid the break, the compression could only be done to the point below that breakage point. This circumstance resulted in a lack of aluminum wire numbers in some parts, compared to fishing line numbers. FLFs, on the other hand, were created solely from fishing line. In comparison to hybrid fabrics, the number of lines spread throughout the fabric was larger in FLF. This situation promoted to higher strains due to the flexibility of fishing lines. But, not enough to surpass the tensile strength values gathered from HMFX.

After tensile test were performed, layers were examined and found out that failure mechanisms developed in the bonded regions. This mechanism was discovered to be a combination of adhesive and cohesive failure that occurred inside the coating layer, in the adhesive itself, and at the adhesive–substrate contact. As a result, failure mechanisms indicated that the adhesion area's cohesive forces were insufficient, and that the boundary layer was weak [47]. This weakness was more severe with HMF composites than FLF composites. This situation also led to a decrease in the overall tensile strength of the composites if compared to aluminum sheet with same thickness. However, with some minor precautions like increasing the curing pressure and temperature, this bonding weakness issue can easily be eliminated.

4 Conclusion

In this study, Hybrid Metallic Fabrics (HMF) were produced by weaving aluminum wires (Al 1050 H14) with a diameter of 0.2 mm and polyethylene fishing lines with the same diameter using standard rug weaving techniques. Fabrics made entirely of fishing lines (Fishing Line Fabric or FLF) without the use of aluminum wires were also made. Then, using FM 73 M epoxy film adhesive, HMF and FLFs were adhesively bonded with aluminum sheets (Al 5005 H34) in various configurations to construct hybrid laminated composites, and tensile tests were performed to assess the mechanical properties of these laminated hybrid composites. The findings were summarized below:

- It was presented that Aluminum wires and Polyethylene lines can be progressively weaved in typical carpet looms by h and,

- This investigation verified a well-known property of composite materials: weight reduction. All of the composites produced were around 63% lighter than the AL 5005 H34 specimen with the same thickness as the composite specimens.
- In terms of tensile strength, hybrid metallic fabric with aluminum wires woven parallel to the weft axis (HMFX) had the higher values. Because, during the tests, the tensile stress was delivered in a direction parallel to the axis of the aluminum wires which could resist tensile force than fishing line.
- Composites consist of fishing line fabric without aluminum wire (FLF) gave better results than hybrid metallic composed of fishing lines woven parallel to the warp axis (HMFY). The main reason is the weaving of aluminum which was more challenging than fishing line. This led to the lack of aluminum wire numbers in some parts of HMF, compared to fishing line numbers in FLF. High number of materials resulted in higher tensile strength and strain.
- When bonded regions were examined after tensile tests, a mixture of adhesive and cohesive failure was discovered that occurred inside the coating layer, in the adhesive itself, and at the adhesive–substrate interface resulting in the overall decrease in tensile strength of composites. But these weaknesses are easily preventable with some minor changes like increasing pressure and temperature in the curing process.

Acknowledgements This study was supported by coordinatorship of Scientific Research Projects Office of Adiyaman University with the project titled as "Metal Kumas Uretimi ve Bu Kumaslarin Yapistirma Baglantilarinin Mekanik Ozelliklerinin Incelenmesi" with project number of MÜFMAP/2021-0001. In addition, the authors would like to thank to the Turkish Airlines officials for donating the epoxy film adhesive.

References

1. Mann, G.S., Singh, L.P., Kumar, P., Singh, S.: Green composites: a review of processing technologies and recent applications. J. Thermoplast. Compos. Mater. **33**, 1145–1171 (2020). https://doi.org/10.1177/0892705718816354
2. Wang, M., Lawal, A.: Special issue on process and mechanical engineering for carbon capture and transport. P. I. Mech. Eng. E-J Pro. **227**, 87–88 (2013). https://doi.org/10.1177/0954408913485515
3. Liu, C.H., Hsieh, W.H., Chang, Z.Y., Tzou, G.Y., Hanson, S.G., Hwang, Y.L.: Recent advances in precision machinery and manufacturing technology. Adv. Mech. Eng. **6**, 508592 (2014). https://doi.org/10.1155/2014/508592
4. Ravishankar, B., Nayak, S.K., Kader, M.A.: Hybrid composites for automotive applications–a review. J. Reinf. Plast. Compos. **38**, 835–845 (2019). https://doi.org/10.1177/0731684419849708
5. Shabaridharan, K., Bhattacharyya, A.: Metallic fibers for composite applications. In: Rana, S., Fanguerio, R. (eds.) Fibrous and Textile Materials for Composite Applications, pp. 205–230. Springer, Singapore (2016)
6. Winkelmann, J., Shamsuyeva, M., Endres, H.J.: Hybrid fabrics for use in bio-based composites for technical applications. Mater. Today: Proc. **31**, S263–S268 (2020). https://doi.org/10.1016/j.matpr.2019.12.097

7. Suthar, J., Patel, K.M.: Processing issues, machining, and applications of aluminum metal matrix composites. Mater. Manuf. Process. **33**, 499–527 (2018). https://doi.org/10.1080/10426914.2017.1401713
8. Samal, P., Vundavilli, P.R., Meher, A., Mahapatra, M.M.: Recent progress in aluminum metal matrix composites: a review on processing, mechanical and wear properties. J. Manuf. Process. **59**, 131–152 (2020). https://doi.org/10.1016/j.jmapro.2020.09.010
9. Mysiukiewicz, O., Kosmela, P., Barczewski, M., Hejna, A.: Mechanical, thermal and rheological properties of polyethylene-based composites filled with micrometric aluminum powder. Materials. **13**, 1242 (2020). https://doi.org/10.3390/ma13051242
10. Citil, S.: Egrisel yuzeyli yapistirma baglantilarinda malzemenin yapistirici uzerine etkisinin incelenmesi. Dicle Uni. J. Eng. **9**(1), 225–234 (2018)
11. Marques, E.A.S., da Silva, L.F.: Joint strength optimization of adhesively bonded patches. J. Adhes. **84**, 915–934 (2008). https://doi.org/10.1080/00218460802505275
12. Kweon, J.H., Jung, J.W., Kim, T.H., Choi, J.H., Kim, D.H.: Failure of carbon composite-to-aluminum joints with combined mechanical fastening and adhesive bonding. Compos. Struct. **75**(1–4), 192–198 (2006)
13. Boutar, Y., Naïmi, S., Mezlini, S., Ali, M.B.S.: Effect of surface treatment on the shear strength of aluminium adhesive single-lap joints for automotive applications. Int. J. Adhes. Adhes. **67**, 38–43 (2016). https://doi.org/10.1016/j.ijadhadh.2015.12.023
14. Naat, N., Boutar, Y., Naïmi, S., Mezlini, S., Da Silva, L. F. M.: Effect of surface texture on the mechanical performance of bonded joints: a review. J. Adhes. 1–93 (2021). https://doi.org/10.1080/00218464.2021.2008370
15. Layec, J., Ansart, F., Duluard, S., Turq, V., Aufray, M., Labeau, M.P.: Development of new surface treatments for the adhesive bonding of aluminum surfaces. Int. J. Adhes. Adhes. 103006 (2021). https://doi.org/10.1016/j.ijadhadh.2021.103006
16. Reneckis, V., et al.: Investigation of technological factors influencing the strength of bonded Al–alloy. In: Broughton, J., Shaw, S. (eds.) IOP Conference Series Materials. Science and Engineering Vol. 1140, pp. 012042. IOP Publishing (2021)
17. Davis, M., Bond, D.: Principles and practices of adhesive bonded structural joints and repairs. Int. J. Adhes. Adhes. **19**, 91–105 (1999). https://doi.org/10.1016/S0143-7496(98)00026-8
18. Pethrick, R.A.: Design and ageing of adhesives for structural adhesive bonding–a review. Proc. Inst. Mech. Eng. L. **229**, 349–379 (2015). https://doi.org/10.1177/1464420714522981
19. Cavezza, F., Boehm, M., Terryn, H., Hauffman, T.: A review on adhesively bonded aluminium joints in the automotive industry. Metals **10**, 730 (2020). https://doi.org/10.3390/met10060730
20. Kuczmaszewski, J.: Fundamentals of Metal-Metal Adhesive Joint Design. Lublin University of Technology, Poland (2006)
21. Goglio, L., Rezaei, M., Rossetto, M.: Moisture degradation of open-faced single lap joints. J. Adhes. Sci. Technol. **28**, 1382–1393 (2014). https://doi.org/10.1080/01694243.2012.697391
22. Viana, G., Costa, M., Banea, M.D., da Silva, L.F.M.: Behaviour of environmentally degraded epoxy adhesives as a function of temperature. J. Adhes. **93**, 95–112 (2017). https://doi.org/10.1080/00218464.2016.1179118
23. Da silva, L.F.M., Carbas, R.J.C., Critchlowb, G.W., Figueiredo, M.A.V., Brown, K.: Effect of material, geometry, surface treatment and environment on the shear strength of single lap joints. Int. J. Adhes. Adhes. 29, 621–632 (2009). https://doi.org/10.1016/j.ijadhadh.2009.02.012
24. Kairouz, K.C., Matthews, F.L.: Strength and failure modes of bonded single lap joints between cross-ply adherends. Compos. **24**, 475–484 (1993). https://doi.org/10.1016/0010-4361(93)90017-3
25. Citil, S., Ayaz, Y., Temiz, S.: Stress analysis of adhesively bonded double strap joints with or without intermediate part subjected to tensile loading. J. Adhes. **93**, 343–356 (2017). https://doi.org/10.1080/00218464.2015.1075885
26. Citil, S.: Comparison of stepped, curved, and S-Type lap joints under tensile loading. In: da Silva, L.F.M. (ed.) Materials Design and Applications, pp. 377–388. Springer, Cham (2017)
27. Yu, J., Wu, J., Wang, H., Zhou, A., Huang, C., Bai, H., Li, L.: Metallic fabrics as the current collector for high-performance graphene-based flexible solid-state supercapacitor. ACS Appl. Mater. Interfaces. **8**, 4724–4729 (2016). https://doi.org/10.1021/acsami.5b12180

28. Dubal, D.P., Chodankar, N.R., Kim, D.H., Gomez-Romero, P.: Towards flexible solid-state supercapacitors for smart and wearable electronics. Chem. Soc. Rev. **47**(6), 2065–2129 (2018)
29. Guo, X., Zheng, S., Zhang, G., Xiao, X., Li, X., Xu, Y., Xue, H., Pang, H.: Nanostructured graphene-based materials for flexible energy storage. Energy Storage Mater. **9**, 150–169 (2017). https://doi.org/10.1016/j.ensm.2017.07.006
30. Liu, Y.H., Xu, J.L., Gao, X., Sun, Y.L., Lv, J.J., Shen, S., Chen, L.S., Wang, S.D.: Freestanding transparent metallic network based ultrathin, foldable and designable supercapacitors. Energy Environ. Sci. **10**(12), 2534–2543 (2017)
31. Gao, L., Surjadi, J.U., Cao, K., Zhang, H., Li, P., Xu, S., Jiang, C., Song, J., Sun, D., Lu, Y.: Flexible fiber-shaped supercapacitor based on nickel–cobalt double hydroxide and pen ink electrodes on metallized carbon fiber. ACS Appl. Mater. Interfaces **9**, 5409–5418 (2017). https://doi.org/10.1021/acsami.6b16101
32. Liu, Z., Mo, F., Li, H., Zhu, M., Wang, Z., Liang, G., Zhi, C.: Advances in flexible and wearable energy-storage textiles. Small Methods **2**, 1800124 (2018). https://doi.org/10.1002/smtd.201800124
33. Liu, Y.H., Xu, J.L., Shen, S., Cai, X.L., Chen, L.S., Wang, S.D.: High-performance, ultra-flexible and transparent embedded metallic mesh electrodes by selective electrodeposition for all-solid-state supercapacitor applications. J. Mater. Chem. A. **5**(19), 9032–9041 (2017)
34. Gornyakov, V., Ding, J., Sun, Y., Williams, S.: Understanding and designing post-build rolling for mitigation of residual stress and distortion in wire arc additively manufactured components. Mater. Des. **110335** (2021). https://doi.org/10.1016/j.matdes.2021.110335
35. Metalreyonu: Aluminyum 1050 ozellikleri. https://www.metalreyonu.com.tr/icerik/1050// (2022). Accessed 12 Feb 2022
36. Morgado, M.A., Carbas, R.J.C., Marques, E.A.S., da Silva, L.F.M.: Reinforcement of CFRP single lap joints using metal laminates. Compos. Struct. **230**, 111492 (2019). https://doi.org/10.1016/j.compstruct.2019.111492
37. E-aircraftsupply: Cytec FM 73 epoxy film adhesive technical data sheet https://www.e-aircraftsupply.com/MSDS/133433cytec%20FM%2073%20tds.pdf (2022). Accessed 14 Feb 2022
38. Cho, J., Song, S.H.: Modeling hot deformation of 5005 aluminum alloy through locally constrained regression models with logarithmic transformations. Appl. Sci. **12**, 152 (2022). https://doi.org/10.3390/app12010152
39. Metalreyonu: Aluminyum 5005 ozellikleri. https://www.metalreyonu.com.tr/icerik/5005// (2022). Accessed 12 Feb 2022
40. Ramdani, N., Razali, M. S.: Processing, properties, and uses of lightweight glass fiber/aluminum hybrid structures. In: Kumar, K., Babu, B.S., Davim, J.P. (eds.) Handbook of Research on Advancements in the Processing, Characterization, and Application of Lightweight Materials, vol.1, pp. 101–120. IGI Global (2022)
41. Hynes, N.R.J., Vignesh, N.J., Jappes, J.W., Velu, P.S., Barile, C., Ali, M.A., Farooq, M.U., Pruncu, C.I.: Effect of stacking sequence of fibre metal laminates with carbon fibre reinforced composites on mechanical attributes: numerical simulations and experimental validation. Compos. Sci. Technol. **221**, 109303 (2022). https://doi.org/10.1016/j.compscitech.2022.109303
42. Han, Z., Li, H., Xu, X., Wang, H., Li, H., Tao, J.: Crushing characteristics of aluminum/CFRP/aluminum hybrid tubes prepared by spinning forming. Compos. Struct. **249**, 112551 (2020). https://doi.org/10.1016/j.compstruct.2020.112551
43. Ng, L.F., Subramaniam, K., Ishak, N.M.: An overview of the natural/synthetic fibre-reinforced metal-composite sandwich structures for potential applications in aerospace sectors. In: Mazlan, N., Sapuan, S.M., Ilyas, R.A. (eds.) Advanced Composites in Aerospace Engineering Applications, vol. 1, pp. 177–194. Springer, Berlin (2022)
44. Thomas, A.J., Barocio, E., Pipes, R.B.: A machine learning approach to determine the elastic properties of printed fiber-reinforced polymers. Compos. Sci. Technol. **220**, 109293 (2022). https://doi.org/10.1016/j.compscitech.2022.109293
45. Stephen, C., Behara, S.R., Shivamurthy, B., Selvam, R., Kannan, S., Abbadi, M.: Finite element study on the influence of fiber orientation on the high velocity impact behavior of fiber reinforced polymer composites. Int. J. Interact. **1**, 1–10 (2022)

46. Abdallah, M.H., Braimah, A.: Numerical design optimization of the fiber orientation of glass/phenolic composite tubes based on tensile and radial compression tests. Compos. Struct. **280**, 114898 (2022). https://doi.org/10.1016/j.compstruct.2021.114898
47. da Silva, L.F.M., Ochsner, A., Adams, R.D. (eds.): Handbook of Adhesion Technology. Springer, Heidelberg (2011)

Hyperelastic Behaviour of Adhesives and Its Effect on Bonded Joints: An Exploratory Study

I. J. Sánchez-Arce, L. D. C. Ramalho, D. C. Gonçalves, R. D. S. G. Campilho, and J. Belinha

Abstract Ductile adhesives often present a non-linear behaviour, often considered as elastic-plastic, but it can also be hyperelastic. Although adhesive joints are important for modern manufacture, little information about adhesive joints with hyperelastic adhesives exists. The single-lap joints (SLJ) with composite substrates are amongst the most studied. Furthermore, modelling hyperelastic behaviour as elastic-plastic leads to under-predict joint strength, for example, numerical analyses of joints bonded with the SikaForce® 7752 show this effect. This adhesive and the Araldite® 2015 are found in numerous analyses in the literature, and both adhesives show a hyperelastic behaviour corresponding to an Ogden type ($N = 3$). This work aims to evaluate the difference in the joint's mechanical behaviour due to the chosen behaviour for the adhesive layer. SLJs with four different overlaps were evaluated where the two adhesives already mentioned were considered as elastic-plastic and later as hyperelastic. Subsequently, the stress distributions were compared between the two cases, whilst the modelling was done using FEM and compared with experimental data. The hyperelastic models showed larger joint compliance and less element distortion than their elastic-plastic counterparts, whilst their force-displacement curves resembled the experimental ones. Stress distributions obtained with hyperelastic models are similar to those from linear-elastic cases.

Keywords Hyperelastic · Adhesive joints · Ogden model · Carbon-fibre reinforced polymers (CFRP) · Metallic substrates

---

I. J. Sánchez-Arce (✉) · L. D. C. Ramalho · D. C. Gonçalves · R. D. S. G. Campilho · J. Belinha
INEGI, Instituto de Ciência e Inovação em Engenharia Mecânica e Engenharia Industrial, Dr. Roberto Frias 400, 4200-465 Porto, Portugal
e-mail: isidrodjsa@gmail.com; iarce@inegi.up.pt

L. D. C. Ramalho
e-mail: lramalho@inegi.up.pt

J. Belinha
e-mail: job@isep.ipp.pt

R. D. S. G. Campilho · J. Belinha
ISEP, Instituto Superior de Engenharia do Porto, Dr. António Bernardino de Almeida 431, 4249-015 Porto, Portugal

L. F. M. da Silva et al. (eds.), *2nd International Conference on Industrial Applications of Adhesives 2022*, Proceedings in Engineering Mechanics,
https://doi.org/10.1007/978-3-031-11150-1_9

1 Introduction

Adhesive joints present several advantages over other joining methods, amongst these advantages are the reduced stress concentration, lightweight, and ability to join dissimilar materials, making them attractive for applications where lightness is required like aircraft construction [1].

The mechanical behaviour of adhesive joints depends on the individual characteristics of their constituents and on the joint's geometry. Nevertheless, several studies have demonstrated that the mechanical properties of the adhesive have an effect on the final joint strength [2]. Although adhesive joints bonded with flexible adhesives were already studied assuming an elastic-plastic behaviour, more recently, it has been proposed that these adhesives possess a hyperelastic behaviour [3]. Unlike the elastic-plastic behaviour where the material properties are often determined from a uniaxial tensile test, the hyperelastic properties are more difficult to determine; furthermore, several models exist to describe such behaviour, being the Mooney-Rivlin [4], Neo-Hookean [5], Yeoh [6], and Ogden models [7] often used for adhesives and adhesively-bonded joints [3], being the Ogden model suitable for strains up to 700% [8].

Three experimental tests are used for the characterisation of flexible adhesives: the bulk test, planar test, and biaxial test [3, 9]. Then, the hyperelastic parameters are determined by curve-fitting the experimental data to the corresponding hyperelastic strain energy function (W). For numerical simulations, this process is often performed using the built-in capabilities of commercial finite element software. Alternatively, the curve-fitting process can be done using the target W and a least-squares minimisation through custom-written numerical routines, as suggested by Baaser [10] and Rackl [11], but this approach is not suitable for Ogden models, as proven by Rackl [11]. A detailed description of the formulations is provided by Ogden [12].

Adhesive joints bonded with hyperelastic adhesives are often modelled using the Finite Element Method (FEM), mainly with commercial software. Crocker et al. [3] described the experimental methods necessary for the characterisation of these adhesives; moreover, proposed the strain-energy as a failure criterion together with numerical and experimental comparisons for different hyperelastic models and the standard Single-lap Joint (SLJ) geometry; there were in the form of force-displacement ($P - \delta$) curves. Later, Lubowiecka et al. [13] analysed SLJs with metallic and composite substrates; the adhesive behaviour was considered as Neo-Hookean, Mooney-Rivlin, and Arruda-Boyce, a good agreement between numerical and experimental values was attained [13]. Then, Alami et al. [14] modelled SLJs with an acrylic adhesive, which was considered as Mooney-Rivlin. Chiminelli et al. [15] modelled adhesive joints with composite substrates. The adhesive was considered elastic, elastic-plastic, and hyperelastic, a comparison was then made. The hyperelastic constants for the adhesive were obtained by curve-fitting experimental data (tension and compression) using commercial FEM software. The elastic-plastic material model for the adhesive seems to under-predict joint strength, whilst the elastic one over-predicts. Although the hyperelastic behaviour seems to approximate more to the experimental data, differences still exist. In addition, the studied adhesive had a very low strength, whilst the adhesive layers were thick ($t_A = 2$ mm and $t_A = 5$ mm), which is known to negatively affect the overall joint behaviour, reducing its strength [16]. Although the study provides an insight into this type of modelling, the $P - \delta$ curves were shown for a small displacement, and so the non-linearity is not visible. Recently, Srinivasan et al. [17] analysed dual-adhesive joints with Glassfibre Reinforced Polymer (GFRP) substrates. Two stiff adhesives and one hyperelastic were considered. The Exponent Drucker-Prager yield criterion was used for the stiff adhesives, whilst the hyperelastic adhesive was modelled as Marlow [17]. The hyperelastic constants were obtained as described above and then the numerical stress-strain ($\sigma - \varepsilon$) curve was compared against the experimental, with a good agreement; subsequently, the joints were modelled. Nevertheless, for all the literature involving adhesive joints with hyperelastic behaviour, Srinivasan et al. [17] reported some stress distributions (shear and peel) and compared the dual-adhesive joints with mono-adhesive ones, indicating that the former case leads to a better shear distribution and lower peak peel stresses [17]. From these works, it can be observed that most of them compare only $P - \delta$ curves, leaving the stress analysis aside. Furthermore, the joint geometries consisted mostly of the standard SLJ, and the adhesive properties varied amongst the sources.

In consequence, this work aims to explore further how the assumption of a hyperelastic adhesive changes the overall joint behaviour and stress distributions in the bond-line with respect to an elastic-plastic analysis. Moreover, it aims to explore how the increase in overlap length affects those joints with a hyperelastic adhesive. An experimental dataset of SLJs is available from previous work [18], whilst a dataset from the SikaForce® 7752 is also available [19]. The numerical procedure will be based on those data and in the descriptions provided in the literature [3, 15].

1.1 Hyperelastic Behaviour of the Adhesives

The hyperelastic models require material-specific constants that are obtained from curve-fitting experimental data. Starting with the strain-energy function ($W\ (\lambda_1, \lambda_2, \lambda_3)$) for a particular hyperelastic model, its derivatives are calculated to obtain the stresses for uniaxial tension, biaxial tension or pure shear, depending on the type of experimental data available [11]. Subsequently, the curve-fitting is performed to obtain the hyperelastic model's constants [10, 11]. This process leads to similar constants to those obtained using commercial FEM software, except for the Ogden model [11], which requires a more optimised algorithm.

The adhesives employed here are the Araldite® 2015 and SikaForce® 7752, hereafter referred to as A2015 and SF7752, respectively. Their force-displacement ($P - \delta$) curves present a relatively continuous non-linear behaviour, which is observed in hyperelastic materials. Therefore, it was investigated if these two adhesives are in fact hyperelastic. Their material constants were obtained using uniaxial tension experimental data and curve-fitted within the ABAQUS® software (Abaqus

Table 1 Ogden hyperelastic material constants for both adhesives, corresponding to a $N = 3$; constants obtained through the ABAQUS® material module

Constant	Araldite® 2015	SikaForce® 7752
μ_1	−240,842.43	−2956.04
μ_2	146,827.79	1472.70
μ_3	94,678.81	1581.90
α_1	1.99	−3.46
α_2	3.99	−3.40
α_3	−2.00	−5.87

Table 2 Elastic-plastic properties of the adhesives studied

Property	A2015 [2]	SF7752 [19]
E (GPa)	1.85 ± 0.21	0.49 ± 0.09
ν	0.35[a]	0.31[b]
σ_y (MPa)	12.63 ± 0.61	3.24 ± 0.48
σ_f (MPa)	21.63 ± 1.61	11.49 ± 0.25
ϵ_f (%)	4.77 ± 0.15	19.18 ± 1.4
G (GPa)	0.56 ± 0.21	0.19 ± 0.02
τ_y (MPa)	14.6 ± 1.3	5.16 ± 1.14
τ_f (MPa)	17.90 ± 1.8	10.17 ± 0.64
γ_f (%)	43.90 ± 3.4	54.82 ± 6.39

[a] Manufacturer's data
[b] Calculated from the shear modulus (G)

6.17, Dassault Systèmes. RI, USA); it was found that both adhesives correspond to an Ogden ($N = 3$) hyperelastic material. Also, the SF7752 could be considered as a Yeoh ($N = 3$) material; nevertheless, the Ogden model has a better fit.

The stress-strain data corresponding to the A2015 was taken from previous work [2], shown in Fig. 1a. For the SF7752, experimental data corresponding to five bulk tests (from Faneco et al. [19]) were fitted to the Ogden model, as shown in Fig. 1b. The fitting process was performed using the Abaqus® software, which contains the fitting module; this process has some precedence in the literature [3, 15]. Regarding the SF7752, the specimens used correspond to those selected by Faneco et al. [19]; subsequently, the average constants were calculated, which are shown in Table 1.

1.2 Elastic-Plastic Properties of the Adhesives and Substrates

Conversely, the elastic-plastic properties were taken from the literature and are reported in Table 2, which were obtained experimentally, noting that the these are for comparison purposes.

Fig. 1 Experimental stress-strain curves for the Araldite® 2015 and SikaForce® 7752 with their respective Ogden hyperelastic fit

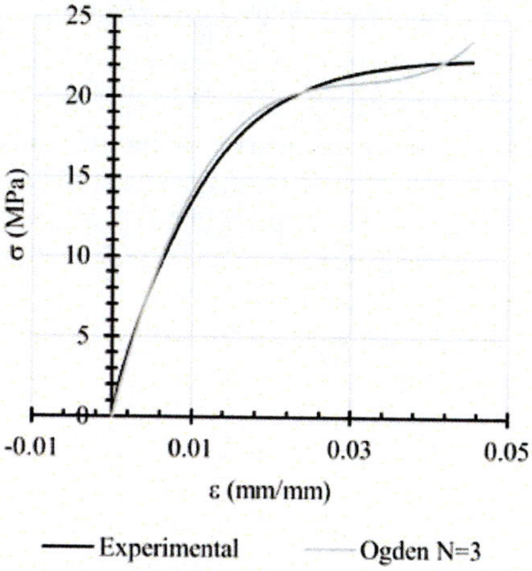

(a) Araldite® 2015

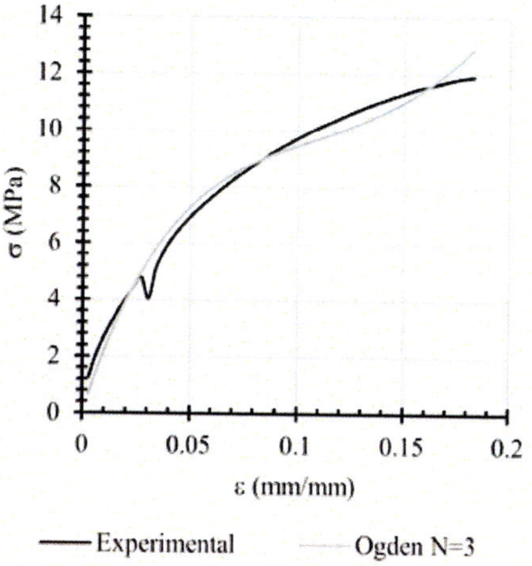

(b) SikaForce® 7752

Metallic substrates were made of an aluminium alloy, A6082-T601, which was used in the experimental work performed by Barbosa et al. [18]. This material was experimentally characterised in previous work [2]. The mechanical properties of this alloy are: Young's modulus (E) of 70.0 GPa or 70,000 MPa, Poisson's ratio of $\nu = 0.30$, yield strength of $\sigma_y = 261.67$ MPa, tensile strength of $\sigma_u = 324.0$ MPa, and failure strain of $\epsilon_f = 21.7\%$ [2].

2 Joint Configurations and Methods

The SLJ geometry corresponds to those experimentally tested by Barbosa et al. [18], whose dataset is used for the numerical comparison. The joints have a total length (L_T) of 170 mm, substrate thickness (t_P) of 3 mm, adhesive layer thickness (t_A) of 0.2 mm, and width (B) of 25 mm. In addition, four overlap lengths (L_O) were evaluated: 12.5, 25.0, 37.5, and 50.0 mm; the schematic of the joint geometry is shown in Fig. 2.

Aiming for consistency, the geometry was created in a parametric form. A similar approach was followed for the meshing, in which all the adhesive layers had a stack of 20 elements. The L_O region (substrates and adhesive) had assigned a double-bias with element size from 0.2 to 1 mm, whilst the remaining substrate had single-bias towards the centre with the same element size. Furthermore, a single-bias was used in the t_P direction, the element size varied from 0.05 to 1 mm. It can be observed that the smallest elements are concentrated towards the adhesive layer free ends, which are high-stress concentration areas (Fig. 3). The models were considered as plane strain two-dimensional cases to reduce the computational cost, this assumption does not compromise the results [20]. Finally, all the models and analyses were run in the Abaqus software using 4-node quadrilateral elements (CPE4RH).

The substrates were modelled as isotropic, elastic-plastic materials with properties corresponding to the aluminium alloy described above. The adhesives, however, were modelled first as elastic-plastic, for comparison, and then as hyperelastic using the

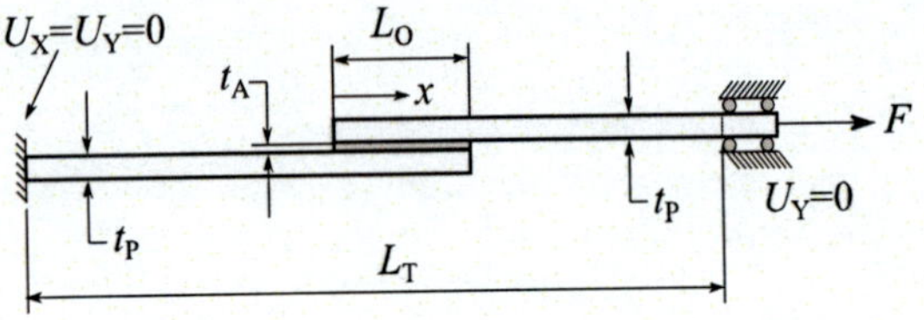

Fig. 2 Schematic representation of the SLJ geometry and involved variables

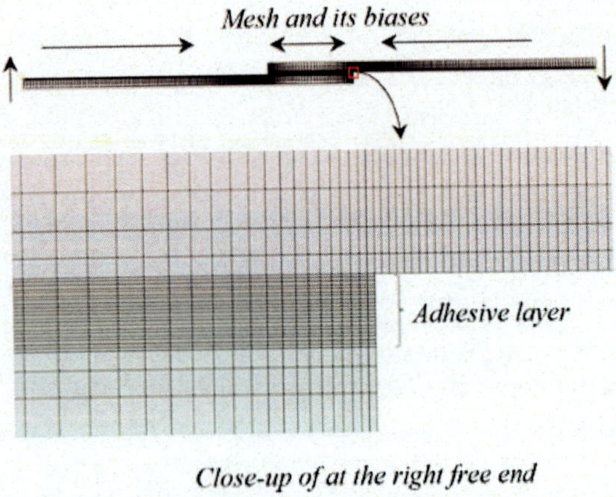

Fig. 3 Mesh on the SLJ and a close-up at its right free end

Table 3 Forces imposed to the force-driven models with aluminium substrates

	Force (N)	
L_O (mm)	A2015	SF7752
12.5	5287.8	3625.6
25.0	9480.7	7175.5
37.5	12,235.2	10,774.1
50.0	15,182.2	14,383.7

Values taken from Barbosa et al. [18]

properties listed in Tables 1 and 2; the hyperelastic adhesives were considered as nearly incompressible [21]. In the elastic-plastic cases, the adhesive was considered to yield under the von Mises criterion.

Regarding the boundary conditions, one end was fixed ($U_X = U_Y = 0$) whilst the other was constrained at the vertical direction ($U_Y = 0$), as shown in Fig. 2. The models were force-driven using the mean experimental load, P_{max} (Table 3), following the approach suggested by Özer and Öz [22].

3 Results and Discussion

Two sets of joint models were run, first with the adhesive layer as elastic-plastic (Table 2); second, as hyperelastic (Table 1), two adhesive types per set (A2015 and SF7752). Subsequently, peel (σ_{yy}) and shear (τ_{xy}) stress distributions were obtained at the adhesive layer mid-thickness and then normalised with respect to the mean shear stress (τ_{avg}) [23], these were retrieved at P_{max}.

The σ_{yy}/τ_{avg} distribution of the joints with A2015 from the elastic-plastic solution are shown in Fig. 4a, whilst the corresponding hyperelastic stress distributions are shown in Fig. 4b.

The elastic-plastic σ_{yy}/τ_{avg} distributions possess the relative standard shape for this type of joint, which has been reported before [24]. Although this adhesive is more suitable for the Exponent Drucker-Prager yield criterion [22, 25], its application to adhesive joints is still scarce; therefore, the von Mises yield criterion was used only for comparison purposes. Furthermore, here the stress distributions are not used for strength prediction but for analysing the effect of using hyperelastic models.

Turning to the hyperelastic σ_{yy}/τ_{avg} distributions, they present a similar shape to the elastic-plastic ones, but with higher peak stresses. Furthermore, the central region of the overlap ($0.3 \leq x/L_O \leq 0.7$) present no stress for all the joints, except for that with $L_O = 12.5$ mm. The stress distribution for the shortest overlap has the shape reported by Lubowiecka et al. [13] for an SLJ with similar dimensions, although the magnitudes cannot be compared because the adhesives are different, being these two orders of magnitude less stiff than the SF7752 reported here. Regardless of this, the peaks on the hyperelastic σ_{yy}/τ_{avg} increase with L_O, as it was also observed on

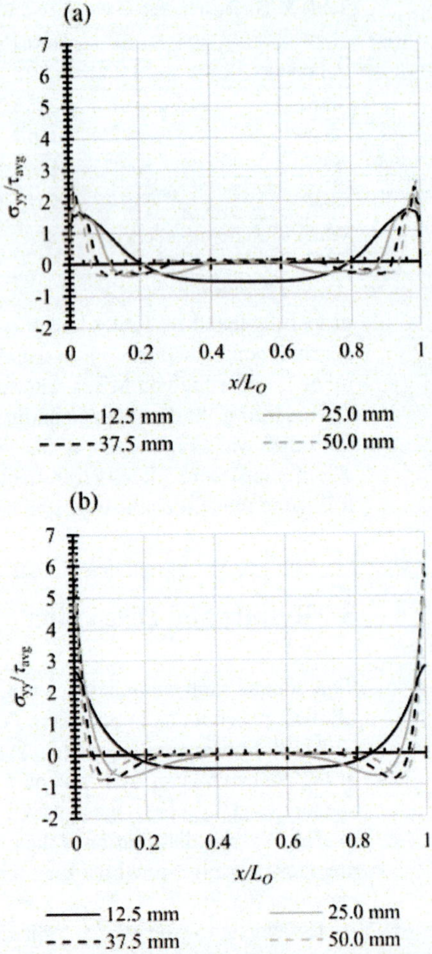

Fig. 4 Normalised peel stress distributions of the SLJ models for the A2015, two adhesive behaviour were compared: **a** elastic-plastic, **b** hyperelastic. Stresses taken at the experimental P_{max}

linear-elastic models of a similar joint [26]. Moreover, the normalised peaks have similar magnitudes, indicating that the approach is reasonable.

Similarly, the normalised shear stress, τ_{xy}/τ_{avg}, distributions for the elastic-plastic models are reported in Fig. 5a, whilst the hyperelastic are in Fig. 5b.

The elastic-plastic τ_{xy}/τ_{avg} present a similar behaviour to those reported by Resende et al. [24], where the two shorter overlaps present a fully developed τ_{xy} pattern, indicating that the load is fully transferred along the bond-line. Conversely, the hyperelastic τ_{xy}/τ_{avg} show patterns and magnitudes similar to those observed on linear elastic models, like those reported by Nunes et al. [26]. This confirms that the hyperelastic behaviour could be related to linear elastic ones.

Fig. 5 Normalised shear stress distributions of the SLJ models for the A2015, two adhesive behaviours were compared: **a** elastic-plastic, **b** hyperelastic. Stresses taken at the experimental P_{max}

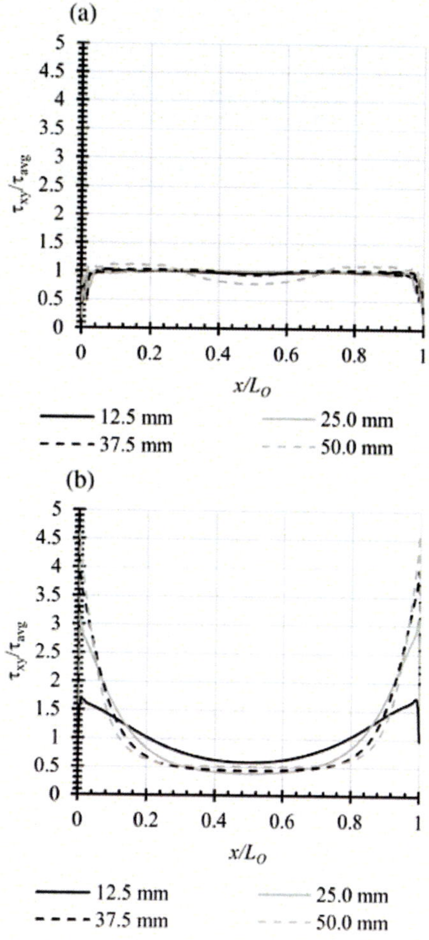

Regarding the joints bonded with SF7752, the σ_{yy}/τ_{avg} distributions of the elastic-plastic models are shown in Fig. 6a, whilst the hyperelastic are shown in Fig. 6b. Similarly, to the joints bonded with the A2015, the peak stresses increase with L_O. However, no previous elastic-plastic studies were found in the literature for this adhesive and joint configuration.

In comparison to the stresses reported in Fig. 4, those corresponding to the SF7752 show a similar overall distribution but with smaller peak stresses. Previous research has found that the peak stresses, for a similar joint configuration, decrease with the adhesive ductility [24]. In consequence, the σ_{yy}/τ_{avg} appears reasonable.

On the other hand, the τ_{xy}/τ_{avg} distributions for the joints bonded with the SF7752 are shown in Fig. 7. The elastic-plastic stress distributions present a similar behaviour to those from the A2015. Furthermore, the joints with the two shorter L_O show a

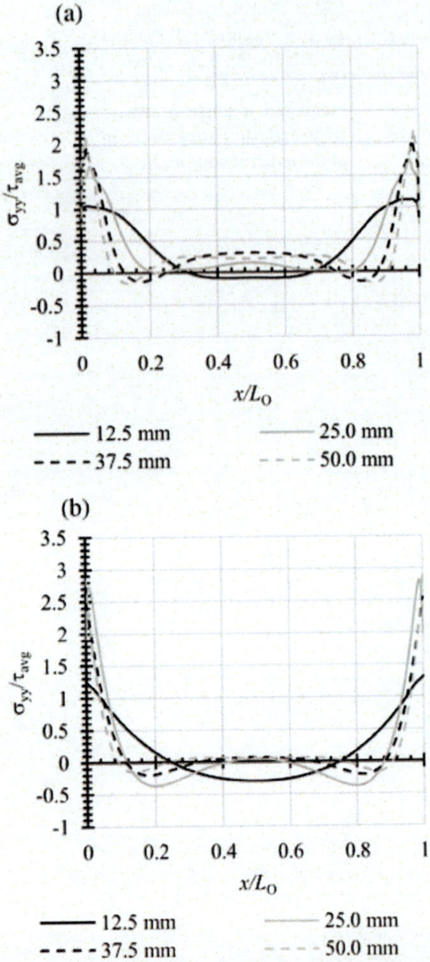

Fig. 6 Normalised peel stress distributions of the SLJ models for the SF7752, two adhesive behaviour were compared: **a** elastic-plastic, **b** hyperelastic. Stresses taken at the experimental P_{max}

fully developed shear distribution, whilst the remaining two were slightly behind. On the contrary, the τ_{xy}/τ_{avg} distributions from the hyperelastic joints exhibit behaviour like that observed on linear-elastic joints. However, the distribution corresponding to the shortest overlap shows a behaviour more developed than its equivalent of A2015.

Although there are no other elastic-plastic nor hyperelastic studies involving this joint configuration with this adhesive, the similarity with those from the A2015, which is more studied, indicates that the results are reasonable. In addition, in a recent numerical study involving a more ductile adhesive, the AF32 from 3M® [17] reported some peel and shear stress curves, which are qualitatively similar to those in Figs. 6b and 7b. Although there are a few studies involving this adhesive with different joint configurations [18, 27, 28], these were linear-elastic and combined

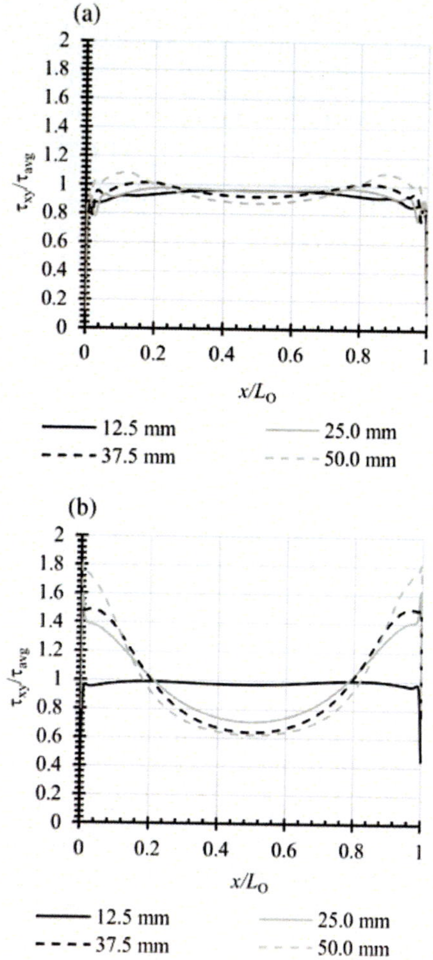

Fig. 7 Normalised shear stress distributions of the SLJ models for the SF7752, two adhesive behaviours were compared: **a** elastic-plastic, **b** hyperelastic. Stresses taken at the experimental P_{max}

with CZM to determine joint strength. More recently, an elastic-plastic analysis of double-L joints included this adhesive [29], in which the yield criterion was the EDP, but the results are not comparable because of the difference between the joint type and the analysis approach.

Finally, the force-displacement ($P - \delta$) curves of the joints bonded with A2015, are shown in Fig. 8, as an example, these curves correspond to the hyperelastic models. The $P - \delta$ curves (Fig. 8) exhibit a clear non-linear behaviour with a convex curvature. Such effect is visible in experimental tests of SLJ with both adhesive systems [18, 26]. Furthermore, in the joints with the two longer overlaps, the curve changes from convex to slightly concave before reaching P_{max}; this effect was also described by Srinivasan et al. [17]. On the contrary, the $P - \delta$ curves, when mod-

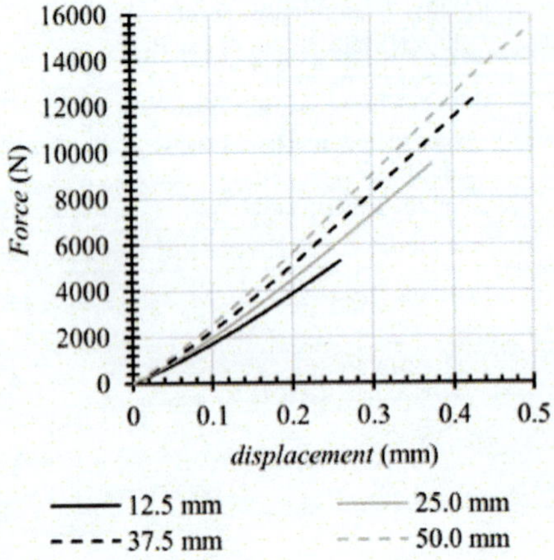

Fig. 8 Numerical force-displacement curves of the joints bonded with Araldite® 2015, where the adhesive was considered as hyperelastic

elled as elastic-plastic show a concave curvature, similar to those shown in Fig. 1, as those reported by Resende et al. [24]. On the other hand, the maximum displacement attained by the joints is significantly smaller than its experimental counterpart, which is a recurrent observation and has been reported before [30]; consequently, this parameter was not further analysed.

4 Conclusions

This study further explores the behaviour of adhesive joints bonded with hyperelastic adhesives. Although hyperelasticity in adhesive joints is scarce in the literature, its relation to flexible adhesives and rubber-like substrates makes it relevant.

For ductile adhesives with moderate strength, the consideration of hyperelasticity on the adhesive layer led to a better representation of the behaviour observed experimentally, in which the relationship between force-displacement is slightly convex. Furthermore, the joints modelled with a hyperelastic adhesive were more compliant than those with an elastic-plastic adhesive, indicating a higher energy absorption.

The stress analyses in the bond-line showed that the joints bonded with hyperelastic adhesives present stress distributions similar to those found in linear-elastic analyses. Furthermore, the joints with the smallest overlap length, bonded with the flexible adhesive, transferred the load along the bond-line more uniformly than their longer counterparts. Although this joint was not the strongest, it exhibit lower peak stresses, which would be relevant for other substrate materials. Moreover, it was observed an increase in the peak stress magnitudes as the overlap was increased. This trend is maintained regardless of the adhesive type, but flexible adhesives lead to lower peak magnitudes with higher or similar strengths.

Although this is an exploratory study, to the authors' best knowledge, this is the first study comparing stress distributions in the bond-line and analysing the effects of overlap length and adhesive type. In addition, this is one of the few studies involving metallic substrates with this adhesive behaviour. Therefore, further research with other flexible adhesives and composite substrates is necessary, but this work shows that, in some cases, hyperelastic material behaviours represent the actual behaviour of adhesive joints.

Acknowledgements The authors thank the following institutions for the funding provided. The Ministério da Ciência, Tecnologia e Ensino Superior through the Fundação para a Ciência e a Tecnologia (from Portugal), under project fundings 'POCI-01-0145-FEDER-028351' and 'SFRH/BD/147628/2019'. Additionally, the authors acknowledge the funding provided by the Associated Laboratory for Energy, Transports and Aeronautics (LAETA), under project UIDB/50022/2020.

References

1. Adams, R.D., Wake, W.D.: Structural Adhesive Joints in Engineering. Elsevier Applied Science Publishers LTD, Essex (1984). https://doi.org/10.1007/978-94-009-5616-2
2. Campilho, R.D.S.G., Pinto, A.M.G., Banea, M.D., Silva, R.F., da Silva, L.F.M.: Strength improvement of adhesively-bonded joints using a reverse-bent geometry. J. Adhes. Sci. Technol. **25**(18), 2351–2368 (2011). https://doi.org/10.1163/016942411X580081
3. Crocker, L.E., Duncan, B.C., Hughes, R.G., Urquhart, J.M.: Hyperelastic Modelling of Flexible Adhesives. Technical Report May. National Physical Laboratory, Teddington, Middlesex, UK (1999)
4. Mooney, M.: A theory of large elastic deformation. J. Appl. Phys. **11**(9), 582–592 (1940). https://doi.org/10.1063/1.1712836
5. Treloar, L.R.G.: The elasticity of a network of long-chain molecules II. Trans. Faraday Soc. **39**, 241–246 (1943). https://doi.org/10.1039/TF9433900241
6. Yeoh, O.H.: Some forms of the strain energy function for rubber. Rubber Chem. Technol. **66**(5), 754–771 (1993). https://doi.org/10.5254/1.3538343
7. Ogden, R.W.: Large deformation isotropic elasticity—on the correlation of theory and experiment for incompressible rubberlike solids. Proc. R. Soc. Lond. A Math. Phys. Sci. **326**(1567), 565–584 (1972). https://doi.org/10.1098/rspa.1972.0026
8. de Souza Neto, E.A., Perić, D., Owen, D.R.J.: Computational Methods for Plasticity, Theory and Applications. Wiley, Sussex, UK (2008)
9. Duncan, B.C., Crocker, L.E.: Characterisation of Flexible Adhesives for Design. Technical Report. National Physical Laboratory, Teddington, Middlesex, UK (2001). www.npl.co.uk
10. Baaser, H.: Development and Application of the Finite Element Method Based on Matlab. Springer, Berlin, Germany (2010). https://doi.org/10.1007/978-3-642-13153-0
11. Rackl, M.: Curve Fitting for Ogden, Yeoh and Polynomial Models. Technical Report May, Ostbayerische Technische Hochschule Regensburg (2015)
12. Ogden, R.W.: Elastic deformations of rubberlike solids, Chap. 14. In: Hopkins, H.G., Sewell, M.J. (eds.) Mechanics of Solids: The Rodney Hill 60th Anniversary Volume, pp. 499–537. Pergamon Press, Oxford, UK (1982). https://doi.org/10.1016/B978-0-08-025443-2.50021-5
13. Lubowiecka, I., Rodríguez, M., Rodríguez, E., Martínez, D.: Experimentation, material modelling and simulation of bonded joints with a flexible adhesive. Int. J. Adhes. Adhes. **37**, 56–64 (2012). https://doi.org/10.1016/j.ijadhadh.2012.01.010
14. Alami, A.H., Bilal, H.: Modelling and verification of an acrylic adhesive as a hyperelastic material. Adv. Mater. Process. Technol. **1**(1–2), 1–12 (2015). https://doi.org/10.1080/2374068X.2015.1095028

15. Chiminelli, A., Valero, C., Lizaranzu, M., López, C.I., Canales, M.: Modelling of bonded joints with flexible adhesives. J. Adhes. **95**(5–7), 369–384 (2019). https://doi.org/10.1080/00218464.2018.1562347
16. Banea, M.D., da Silva, L.F.M., Campilho, R.D.S.G.: The effect of adhesive thickness on the mechanical behavior of a structural polyurethane adhesive. J. Adhes. **91**(5), 331–346 (2015). https://doi.org/10.1080/00218464.2014.903802
17. Srinivasan, D.V., Ravichandran, V., Idapalapati, S.: Failure analysis of GFRP single lap joints tailored with a combination of tough epoxy and hyperelastic adhesives. Compos. Part B Eng. **200**, 108255 (2020). https://doi.org/10.1016/j.compositesb.2020.108255
18. Barbosa, N.G.C., Campilho, R.D.S.G., Silva, F.J.G., Moreira, R.D.F.: Comparison of different adhesively-bonded joint configurations for mechanical structures. Appl. Adhes. Sci. **15**(6), 721–728 (2018). https://doi.org/10.1016/j.promfg.2018.10.122
19. Faneco, T.M.S., Campilho, R.D.S.G., Silva, F.J.G., Lopes, R.M.: Strength and fracture characterization of a novel polyurethane adhesive for the automotive industry. J. Test. Eval. **45**(2), 398–407 (2017). https://doi.org/10.1520/JTE20150335
20. Adams, R.D., Peppiatt, N.A.: Effect of Poisson's ratio strains in adherends on stresses of an idealized lap joint. J. Strain Anal. **8**(2), 134–139 (1973). https://doi.org/10.1243/03093247V082134
21. Simulia User Assistence: Dassault Systèmes Simulia Corp., Providence, RI, USA (2017)
22. Özer, H., Öz, Ö.: The use of the exponential Drucker-Prager material model for defining the failure loads of the mono and bi-adhesive joints. Int. J. Adhes. Adhes. **76**, 17–29 (2017). https://doi.org/10.1016/j.ijadhadh.2017.02.005
23. da Silva, L.F.M., das Neves, P.J.C., Adams, R.D., Spelt, J.K.: Analytical models of adhesively bonded joints—part I: literature survey. Int. J. Adhes. Adhes. **29**(3), 319–330 (2009). https://doi.org/10.1016/j.ijadhadh.2008.06.005
24. Resende, R.F.P., Resende, B.F.P., Sánchez-Arce, I.J., Ramalho, L.D.C., Campilho, R.D.S.G., Belinha, J.: Elasto-plastic adhesive joint design approach by a radial point interpolation meshless method. J. Adhes. 1–27 (2021). https://doi.org/10.1080/00218464.2021.1977633
25. Dean, G.D., Crocker, L.E., Read, B., Wright, L.: Prediction of deformation and failure of rubber-toughened adhesive joints. Int. J. Adhes. Adhes. **24**(4), 295–306 (2004). https://doi.org/10.1016/j.ijadhadh.2003.08.002
26. Nunes, S.L.S., Campilho, R.D.S.G., da Silva, F.J.G., de Sousa, C.C.R.G., Fernandes, T.A.B., Banea, M.D., da Silva, L.F.M.: Comparative failure assessment of single and double lap joints with varying adhesive systems. J. Adhes. **92**(7–9), 610–634 (2016). https://doi.org/10.1080/00218464.2015.1103227
27. Machado, R.M.D., Campilho, R.D.S.G., Rocha, R.J.B., Silva, F.J.G.: Modelling stepped-lap adhesive joints by the extended finite element method. Adv. Transdiscipl. Eng. **7**, 684–692 (2018). https://doi.org/10.3233/978-1-61499-898-3-684
28. Brito, R.F.N., Campilho, R.D.S.G., Moreira, R.D.F., Sánchez-Arce, I.J.: Material and adhesive effect in adhesively-bonded composite stepped-lap joints. Proc. Inst. Mech. Eng. Part G J. Aerosp. Eng. **234**(13), 1967–1979 (2020). https://doi.org/10.1177/0954410020932014
29. Sánchez-Arce, I.J., Ramalho, L.D.C., Campilho, R.D.S.G., Belinha, J.: Meshless analysis of substrate stiffness and its effect on metallic double-L joint strength and stress distributions. Eng. Anal. Bound. Elem. **125**, 190–200 (2021). https://doi.org/10.1016/j.enganabound.2021.01.019
30. Chowdhury, N.M., Chiu, W.K., Wang, J., Chang, P.: Experimental and finite element studies of bolted, bonded and hybrid step lap joints of thick carbon fibre/epoxy panels used in aircraft structures. Compos. Part B Eng. **100**, 68–77 (2016). https://doi.org/10.1016/j.compositesb.2016.06.061

Lecture Notes in Mechanical Engineering (LNME) publishes the latest developments in Mechanical Engineering—quickly, informally and with high quality. Original research reported in proceedings and post-proceedings represents the core of LNME. Volumes published in LNME embrace all aspects, subfields and new challenges of mechanical engineering.

To submit a proposal or request further information, please contact the Springer Editor of your location:

Europe, USA, Africa: Leontina Di Cecco at Leontina.dicecco@springer.com
China: Ella Zhang at ella.zhang@springer.com
India: Priya Vyas at priya.vyas@springer.com
Rest of Asia, Australia, New Zealand: Swati Meherishi at swati.meherishi@springer.com

Topics in the series include:

- Engineering Design
- Machinery and Machine Elements
- Automotive Engineering
- Engine Technology
- Aerospace Technology and Astronautics
- Nanotechnology and Microengineering
- MEMS
- Theoretical and Applied Mechanics
- Dynamical Systems, Control
- Fluid Mechanics
- Engineering Thermodynamics, Heat and Mass Transfer
- Manufacturing
- Precision Engineering, Instrumentation, Measurement
- Tribology and Surface Technology

Indexed by SCOPUS and EI Compendex.

All books published in the series are submitted for consideration in Web of Science.

To submit a proposal for a monograph, please check our Springer Tracts in Mechanical Engineering at https://link.springer.com/bookseries/11693

Lecture Notes in Mechanical Engineering

Anil Kumar · Mohammad Zunaid ·
K. A. Subramanian · Heechang Lim
Editors

Recent Advances in Manufacturing and Thermal Engineering

Select Proceedings of RAMMTE 2022

Springer

Editors
Anil Kumar
Department of Mechanical, Production and Industrial and Automobile Engineering
Delhi Technological University
Delhi, India

Mohammad Zunaid
Department of Mechanical, Production and Industrial and Automobile Engineering
Delhi Technological University
Delhi, India

K. A. Subramanian
Department of Energy Science and Engineering
Indian Institute of Technology Delhi
New Delhi, India

Heechang Lim
School of Mechanical Engineering
Pusan National University
Busan, South Korea

ISSN 2195-4356 ISSN 2195-4364 (electronic)
Lecture Notes in Mechanical Engineering
ISBN 978-981-19-8516-4 ISBN 978-981-19-8517-1 (eBook)
https://doi.org/10.1007/978-981-19-8517-1

This Springer imprint is published by the registered company Springer Nature Singapore Pte Ltd.
The registered company address is: 152 Beach Road, #21-01/04 Gateway East, Singapore 189721, Singapore

Foreword by Prof. J. P. Saini

I am glad to know that an International Conference on *Recent Advances in Materials, Manufacturing and Thermal Engineering* (RAMMTE-2022) has been conducted from 8 to 9 July 2022 at the Department of Mechanical, Production and Industrial and Automobile Engineering and Centre for Energy and Environment, Delhi Technological University (Formerly Delhi College of Engineering), Delhi-110042 (India) in association with Department of Mechanical Engineering, Harcourt Butler Technical University Kanpur-208002 (India). I am grateful to all the delegates and participants across the globe for participating vehemently in the conference and bringing together the national and international talents in one place.

The conference provided an excellent opportunity for the researchers, scientists and industrialists to share and converse on the latest developments in the areas of mechanical engineering and justify the significant role of mechanical engineers in the field of materials design and development, product design and testing, productivity, product quality and safe working environment for the society at large. The publication of the conference proceedings in *Lecturer Notes in Mechanical Engineering*,

Springer, would give a new benchmark and develop deeper insight in R&D initiatives. I also hope that the congregation of eminent experts highlighted the importance of research and innovation to address the rising demand globally.

I congratulate the organizing team of RAMMTE-2022 for the grand success of the conference, and it will help in branding Delhi Technological University as a leading research university among the world's reputed academic institutions.

Prof. J. P. Saini
Vice-Chancellor
Delhi Technological University
Delhi, India

Foreword by Prof. Samsher

It is matter of immense pleasure that International Conference on *Recent Advances in Materials, Manufacturing and Thermal Engineering* (RAMMTE-2022) was organized by the Department of Mechanical, Production and Industrial and Automobile Engineering and Centre for Energy and Environment, Delhi Technological University, Delhi, in collaboration with Department of Mechanical Engineering, Harcourt Butler Technical University (former Harcourt Butler Technological Institute) Kanpur-208002 (India) during 8–9 July 2022 at Delhi Technological University Delhi-110042 (India).

This conference attracted large number of students, academician and researchers all over the world. The conference was very successful in both online and offline modes giving new directions to engineering sciences. Researcher focused their research articles in the areas of computational modelling, simulation, materials, additive manufacturing, automation, biomedical, thermal systems, green manufacturing, etc.

I am thankful to everyone who participated as speakers, authors, and delegates across the world and made this conference very successful. I am aware of that both organizing universities Delhi Technological University Delhi (India) and Harcourt Butler Technical University Kanpur, Uttar Pradesh (India) had worked consistently for one year. I congratulate all team members of organizing teams. I also wish that Lecture Notes on Mechanical Engineering published by Springer will be give new direction to young researcher.

Prof. Samsher
Vice-Chancellor
Harcourt Butler Technical University
Kanpur, Uttar Pradesh, India

Preface

Mechanical Engineering is the application of physical principles of science and engineering in the creation of useful systems, devices, objects and machines. This basic perception of Mechanical Engineering still holds good, while it has evolved into various new fields. Mechanical Engineering is thus rooted deeply in a vast engineering canvas and is in the service of mankind at all levels. Therefore, there is also a need to understand design principles for environment, life-cycle design and sustainable development. It is our pleasure to present academicians and scholars, the proceedings of research papers scheduled for presentation at the 3rd International Conference on "Recent Advances in Materials, Manufacturing and Thermal Engineering (RAMMTE 2022)" from 8–9 July 2022 in the Department of Mechanical, Production and Industrial and Automobile Engineering and Centre for Energy and Environment, at Delhi Technological University (Formerly Delhi College of Engineering), Delhi-110042 (India) in association with Department of Mechanical Engineering, Harcourt Butler Technical University, Kanpur-208002 (India). This conference is in series with the earlier international conference RAME-2020 organized by the Department of Mechanical Engineering, Delhi Technological University (Formerly Delhi College of Engineering), Delhi-110 042 (India). The conference has been very well received by the industry and academia.

The core organizing committee for organizing RAMMTE 2022 is as follows:

Chief Patrons

Prof. Jai Prakash Saini, Hon'ble Vice-Chancellor DTU, Delhi, India
Prof. Samsher, Hon'ble Vice-Chancellor HBTU, Kanpur, India

Patron

Prof. S. K. Garg

Chairman

Prof. Amit Pal

Co-chairman

Prof. Vijay Gautam

Convener

Dr. Anil Kumar

Co-conveners

Dr. Girish Kumar
Dr. Jitendra Bhaskar

Organizing Secretary

Dr. Mohammad Zunaid

Joint Organizing Secretaries

Dr. K. Manjunath
Dr. N. A. Ansari
Dr. N. Yuvraj

The proceeding comprises papers from leading academicians, research scholars and industry experts. The conference papers cover important research areas and the latest trends in the industry. About 188 papers from 350+ authors were presented, and about 450 participants attended the conference. In all, high-quality papers have been selected for presentation during the conference. The main topics of the conference have been classified into the following categories:

- Thermal Engineering
- Manufacturing Processes
- Renewable Energy Technologies
- Industrial and Production Engineering
- Materials Science and Engineering

The technical advisory committee is pleased to mention that the papers have been received on all the topics. Such a voluminous proceeding is only possible with the generous support received from various quarters. The committee would like to record its deep sense of gratitude and appreciation for the persistent efforts of all the reviewers.

We are grateful to all the papers' authors for contributing to enrich this international conference of RAMMTE 2022.

It is our immense pleasure to express our heartfelt gratitude to Prof. Yogi D. Goswami, University of South Florida, USA; Prof. H. C. Lim, Pusan National University, South Korea; Prof. Afzal Husain, Sultan Qaboos University, Oman; Dr. Nitin Upadhaya, University of Modern Science, Dubai, UAE; Dr. Ashish Shukla, Coventry University, UK; Dr. Shyam S. Pandey, Kyushu Institute of Technology, Japan; Prof. Branka Gvozdenac Urošević, University of Novi Sad; Dr. Jan Banout, Czech University of Life Science, Czech Republic, and other colleagues in India and abroad.

Special acknowledgement is made to the support of expert papers and the generous support received from our distinguished alumni of DTU and leading industrialists. We also acknowledge the excellent cooperation received from our colleagues and experts on the review panel for their painstaking efforts in reviewing the papers. We thank the sponsors and extend full credit to the publisher (Springer) for accepting the proposal to publish conference proceedings with the book title *Recent Advances in Manufacturing and Thermal Engineering* under the series title "Lecture Notes in Mechanical Engineering", Springer. Their timely support was not only strength but also an inspiration for the organizers.

We sincerely hope that the engineering fraternity will find this publication a valuable source of knowledge for the application in mechanical engineering.

Delhi, India — Anil Kumar
Delhi, India — Mohammad Zunaid
New Delhi, India — K. A. Subramanian
Busan, South Korea — Heechang Lim

Contents

About the Editors

Dr. Anil Kumar is currently working as Associate Professor at the Department of Mechanical Engineering, Delhi Technological University, India. He has completed his B.Tech. (Mechanical Engineering) from Jamia Millia Islamia, Delhi (India), M.Tech. from Tezpur University, Tezpur (India), and Ph.D. in Solar Energy from the Centre for Energy Studies, Indian Institute of Technology (IIT) Delhi. His nature of experience in teaching and research (science, technology, society, and sustainable development). His areas of specialization are energy technology, renewable energy, solar energy applications, energy economics, heat transfer, natural rubber sheet drying, and environmental issues. He has over 17 years of experience in the field of energy technology. He has published 190 papers in international peer-reviewed journals and 85 papers in international/national conference proceedings. He has developed a thin-layer drying model in 2014. He is the author of 11 books (4 national and 7 international editions). He is Fellow and Chartered Engineer of The Institution of Engineers (India). He appears as the most cited number of Information Systems (IS) researchers featured in the World Ranking of "Top 2% scientists created by Stanford University (2019, 2020 and 2021)".

Dr. Mohammad Zunaid is currently working as Assistant Professor at the Department of Mechanical Engineering, Delhi Technological University, India. He has obtained his B.E. (Mechanical Engineering) and M.Tech. (Thermal Science) from Aligarh Muslim University, Aligarh (India), and his Ph.D. from Delhi Technological University, Delhi (India). He has over 15 years of experience in the field of teaching and research. His areas of specialization are Computational Fluid Dynamics (CFD), computational heat transfer and fluid flow, fluid mechanics, fluid dynamics, refrigeration and air conditioning, power plant engineering, and thermal engineering. He is a life member of The Institution of Engineers, India. He has published more than sixty research papers in national/international journals and conferences. He has guided one Ph.D., fifteen M.Tech. and forty B.Tech. level students for their dissertations.

Dr. K. A. Subramanian is working as Professor and founder Head of the Department of Energy Science and Engineering, Indian Institute of Technology (IIT) Delhi, and former Head of the Centre for Energy Studies. After he completed his Ph.D. from the Department of Mechanical Engineering, IIT Madras, he worked as Scientist at the Indian Institute of Petroleum, Dehradun. He has published more than 75 research articles in international peer-reviewed journals. His main research area includes hydrogen energy, alternative-fueled internal combustion engines/vehicles, hydrogen backfire and combustion, hydrogen fuel cell, oxy-combustion, biofuels and engines, zero-emission vehicles, hybrid energy systems, and integrated energy systems. He is one among the top 2% scientist by Stanford University, 2020, 2021 and 2022, in the world. He recently developed the technology for the utilization of hydrogen in spark-ignition engines for electrical power generation, and dimethyl ether-fueled engines/vehicles with the collaboration of oil and original engine manufacturers. He has also developed alcohol-fueled engines, biogas vehicles, DME-fueled engines/vehicles, dual-fuel engines (hydrogen-diesel, CNG-diesel, biogas-diesel), and energy efficiency improvement using oxy-combustion. He developed many numbers of courses including hydrogen energy, zero-emission vehicles, bioenergy: resources, technologies and assessment, carbon capture and storage, and organic waste to energy conversion technology. He has guided several doctoral scholars and published several research papers in journals of national and international repute.

Dr. Heechang Lim is currently working as Professor at Pusan National University (PNU), South Korea. He obtained his B.S. in Mechanical Engineering from Pusan National University (PNU), South Korea, M.S. and Ph.D. in Mechanical Engineering and Environment Engineering, respectively, from Pohang University of Science and Technology (POSTECH), Pohang, South Korea. He has published more than 100 research papers in international peer-reviewed journals and conferences.

Soy Flour Ash for Adsorption of Cationic and Anionic Dyes from Aqueous Media

Raveena Choudhary, Aayush Gupta, O. P. Pandey, and Loveleen K. Brar

1 Introduction

Increasing world population has resulted in an increase in the requirement for potable water. The expansion of various industries, including textiles and dyeing, paper and pulp, printing, leather treatment, food items, etc., has resulted in an ever-increasing strain on water resources. Synthetic dyes, which are non-biodegradable, mutagenic, and carcinogenic, are mixed in industrial effluents and get dumped into natural water resources at a rate of approximately 800,000 tons each year. These dyes cause serious harm to human and marine life by lowering dissolved oxygen levels in the water, altering the spectral absorption profile of water bodies, and due to the inherent toxic nature of these chemicals [1]. Thus, wastewater treatment is an essential call. There are various techniques available to extract or degrade such organic effluents by physical, chemical, or biological means, e.g., precipitation [2], reverse osmosis [3], ozonation [4], coagulation-flocculation [5], electrochemical techniques [6, 7], and adsorption [2]. Adsorption is the most extensively utilized technique due to its ease of usage and low cost. Clay, zeolite, silica gel, carbon compounds, ashes, and other adsorbents have all been used extensively to remove contaminants. The ash formed by biomass combustion is found to be the most cost-effective and environmentally beneficial of all these adsorbents [8, 9]. In recent years, various works have been reported on the preparation of biomass ash for the adsorption of various pollutants from wastewater. Xu et al. [10] used the biomass ash collected from the power plant for Cu^{2+} ions adsorption from its aqueous solution. Range et al. [11] carried out the adsorption of thiosulphate and polythionates using biomass ash collected from the paper and pulp

R. Choudhary · O. P. Pandey · L. K. Brar (✉)
School of Physics and Materials Science, Thapar Institute of Engineering and Technology, Patiala, Punjab 147004, India
e-mail: brarloveleen@thapar.edu

A. Gupta
Department of Mechanical Engineering, GLA University, Mathura 281406, India

A. Kumar et al. (eds.), *Recent Advances in Manufacturing and Thermal Engineering*,
Lecture Notes in Mechanical Engineering, https://doi.org/10.1007/978-981-19-8517-1_1

industry. Patel et al. [12] synthesized ash using black turmeric rhizome for crystal violet dye adsorption. Rice husk ash was synthesized by Costa et al. [13] to adsorb remazol red dye.

In the present report, soy flour ash was synthesized by heating soy flour at 450 and 600 °C in a muffle furnace without further treatment. TG/DTA/DTG analysis was used to determine the heat treatment temperature. Minerals including calcite, aragonite, arcanite, and fairchildite, as well as carbon, were found in the samples, making them effective adsorbents. For both cationic and anionic dyes [methylene blue (MB) and eriochrome black-T (EBT)], the soy flour heated to 450 °C was shown to be a superior adsorbent. The adsorption data was also fitted with kinetic models to determine the nature of the adsorption process. At equilibrium, 24.1 mg/g of MB and 17.4 mg/g of EBT were adsorbed onto a sample heated to 450 °C.

2 Methodology

2.1 *Synthesis and Characterization*

Soy flour ash was prepared by heating 2 g of soy flour at 450 °C (SM-450) and 600 °C (SM-600) for 4 h in a closed muffle furnace (@ 10 °C/min). Table 1 shows the synthesis parameters and properties of the samples. The as-synthesized samples were used without any further treatment for characterizations and adsorption experiments. The thermal behavior of soy flour was evaluated using TG/DTA/DTG (SII 6300 EXSTAR) analysis with air as carrier gas from room temperature to 700 °C at a heating rate of 5 °C/min. X-ray diffraction (XRD, PANalytical Xpert-Pro diffractometer, $\lambda = 1.5406$ Å) in the range of $2\theta = 5$–$90°$ (step size, $2\theta = 0.013°$) was used for determining the crystalline phase present in as-prepared materials. To understand the morphology of soy flour and heat treated samples, field-effect scanning electron microscope (FE-SEM, Carl Zeiss Sigma 500) was used. Using Fourier transform infrared spectroscopy, the existence of organic and inorganic compounds in the samples was studied (FTIR, Agilent Cary 660, from 400 to 4000 cm^{-1}). X-ray photoelectron spectroscopy (XPS, PHI 5000 Versa Prob II, 1486.7 eV) was used to assess the surface elemental composition.

Table 1 The synthesis parameters and properties of the synthesized samples

Property	SM-450	SM-600
Precursor	Soy flour	Soy flour
Temperature (°C)	450	600
Major minerals	Aragonite, arcanite	Calcite, fairchildite
Carbon content (%)	5.3	3.9
q_e (MB) (mg/g)	24.10	19.15
q_e (EBT) (mg/g)	17.40	14.25

2.2 Dye Adsorption Studies

Batch adsorption tests were carried out to understand the adsorption efficacy of the synthesized materials for cationic and anionic dyes, MB, and EBT (Sigma Aldrich, USA). For adsorption, 1 mg of adsorbent was added to 50 ml (1 mg/L) of each of the dyes solution. The solutions were kept on continuous stirring in dark. After a specified duration, 3 mL solution was extracted and centrifuged to separate the adsorbents. For measuring the amount of dye in the separated solution, UV–Visible spectrophotometer was used.

3 Results and Discussions

3.1 TG/DTA/DTG

To better understand the thermal behavior of soy flour, the TG/DTA analysis was carried out using air (as carrier gas) with 5 °C/min heating rate as shown in Fig. 1a. From ambient temperature to 150 °C, there was a steady weight loss process with an 8% mass loss attributable to the evaporation of volatile matter and the adsorbed moisture content. At 220 °C, soy flour began to lose mass quickly again, lasting until 350 °C, resulting in a mass loss of around 37%. The double DTG peaks between this range of temperatures indicated that this mass loss was caused by the breakdown of cellulose and hemicellulose present in soy flour [14]. The DTA shows a broad exothermic peak for this region indicating that these processes take place simultaneously. A gradual mass loss of 13% occurred from 350 to 450 °C. From 450 to 515 °C, there was again a sudden mass loss of around 27%, and a sharp DTG peak was observed indicating the decomposition of polymeric chains in soy flour. The release of volatiles is accountable for the DTG/DTA peaks at 500 °C. Further, the mass loss of ~ 1.5% occurred from 515 to 700 °C and ~ 9% of mass remained in the sample at 700 °C. On the basis of results obtained from TG/DTA/DTG, the temperature for the combustion of soy flour in the muffle furnace was selected as 450 and 600 °C, immediately before and after the combustion of the carbon content in the soy flour.

3.2 XRD

XRD line profiles are shown in Fig. 1b for SM-450 and SM-600 samples. The major peaks observed for SM-450 and SM-600 were compared with the ICDD database. Analysis confirmed that both the samples possibly consist of calcite ($CaCO_3$, ICDD-01-086-2343), buetschliite ($C_2CaK_2O_6$, ICDD-00-083-1590) aragonite ($CCaO_3$, ICDD-00-041-1475), arcanite (K_2O_4S, ICDD-01-086-2334), and

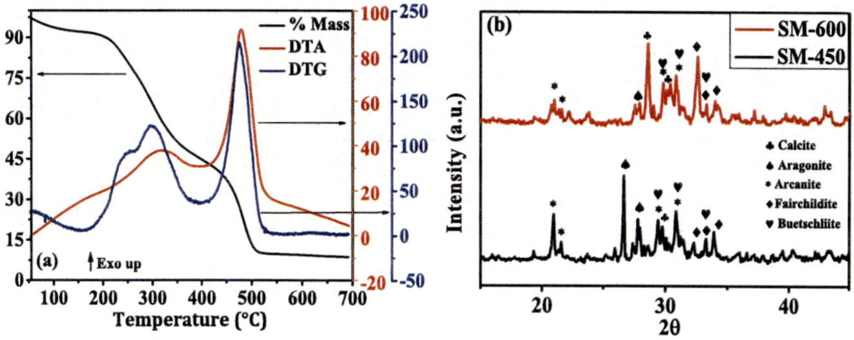

Fig. 1 **a** TG/DTA/DTG of soy flour in air @5 °C/min, **b** XRD of soy flour, SM-450 and SM-600 samples

fairchildite ($C_2CaK_2O_6$, ICDD-00-021-1287). SM-450 sample had the majority of aragonite, arcanite, and buetschliite, while in the SM-600 sample, the content of calcite and fairchildite was higher. All these minerals help in the process of adsorption of dyes via electrostatic interaction [15, 16]. While heating the soy flour in an air atmosphere, we expect to form oxides of the elements present in the majority such as K, Ca. In the present work, we heated the soy flour in a muffle furnace in a confined environment with a limited supply of oxygen. Therefore, the carbon present in the soy flour did not burn out completely and available carbon combined with present elements to form the aforementioned ashes.

3.3 FE-SEM

The influence of heat treatment temperature on the microstructure of samples was examined using FE-SEM imaging of the samples (Fig. 2). Heat-treated samples yielded significantly different morphologies than pure soy flour samples. The growth of the particles was observed with the increasing synthesis temperature of ash. With increasing temperature, the decomposition of organic content present in the soy flour leaves behind the mineral ashes as confirmed by XRD results. With increase in heating temperature, the crystals of the mineral ashes become more prominent. Also, the porosity of the samples improves as the heat treatment temperature rises, as seen in Fig. 2. The images of soy flour, SM-450, and SM-600 samples are shown both at high and low magnifications to visualize the porosity of the sample. The larger pores of uneven size were observed with increasing synthesis temperature. The sample's surface morphology allows for a considerable amount of exposed surface area for the adsorption of dyes.

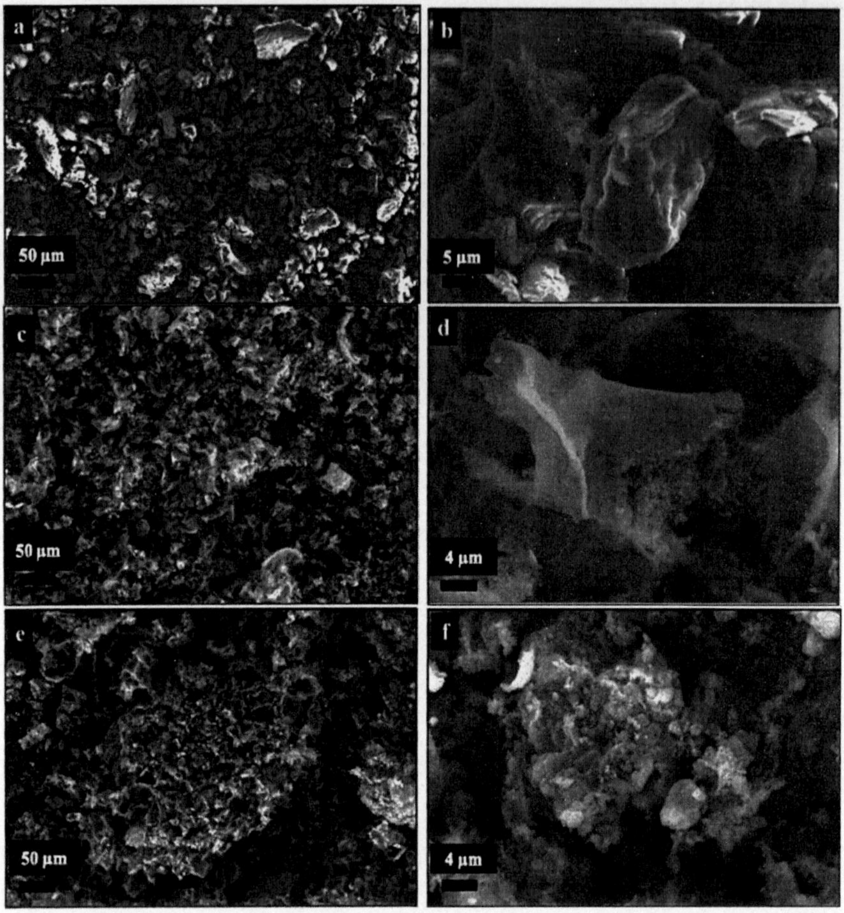

Fig. 2 FE-SEM image of **a**, **b** soy flour, **c**, **d** SM-450 and **e**, **f** SM-600 samples

3.4 FTIR

FTIR spectra for SM-450 and SM-600 samples are shown in Fig. 3a. For both the samples main peaks were observed at 618, 1058, 1620, 2040, and 3400 cm^{-1}, showing the C=C bending group, stretching band of carbonate, C=O stretch, C–C, and O–H stretch groups, respectively [17, 18]. The availability of these functional groups will make the synthesized samples adsorb dye by electrostatic interaction.

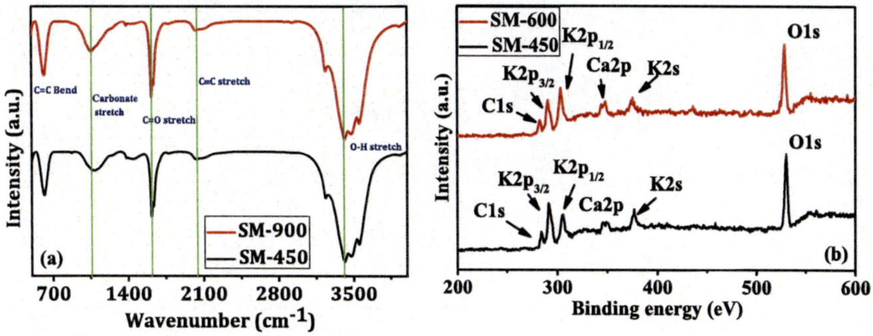

Fig. 3 **a** FTIR, and **b** XPS of SM-450 and SM-600 samples

Table 2 %at of the element species by XPS survey spectra for SM-450 and SM-600 samples

Element species	Total %at	
	SM-450	SM-600
C1s	5.3	3.9
K2p	37.6	39.4
Ca2s	8.3	9.1
K2s	12.9	14.4
O1s	35.9	33.2

3.5 XPS

XPS survey spectra for SM-450 and SM-600 samples are shown in Fig. 3b to confirm the presence of the functionalities observed from XRD and FTIR spectra. XPS survey spectra for both samples consist of peaks between 200 and 600 eV. For both samples, peaks were observed at 285, 305, 348, 375, and 529 eV of C1s, K2p, Ca2p, K2s, and O1s, respectively. The %at of the elements is given in Table 2. With the increase in the heating temperature, the amount of carbon lowered leading to relative increase in content of potassium and calcium in the synthesized samples. The XPS of SM-450 and SM-600 clearly confirm the results observed from the XRD and FTIR spectra.

3.6 Adsorption Study

Adsorption of cationic and anionic dyes MB and EBT (1 mg/L) onto soy flour, SM-450, and SM-600 (20 mg/L) was investigated. The results obtained for adsorption of both the dyes onto soy flour were not significant, i.e., no adsorption took place in 3 h onto soy flour. The MB adsorption onto SM-450 and SM-600 samples is shown in Fig. 4a. Figure 4b shows the % adsorption of EBT onto SM-450 and SM-600 samples with time. In both cases (MB and EBT), the adsorption was better in case

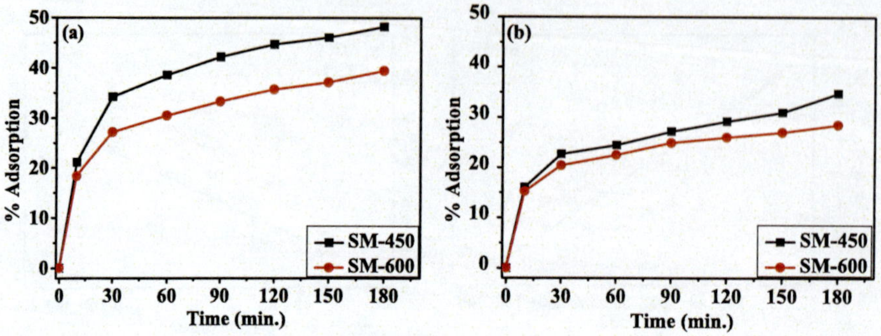

Fig. 4 % Adsorption of **a** MB dye and **b** EBT dye onto SM-450 and SM-600 samples

of SM-450 sample. This is because the content of amorphous carbon has lowered in the SM-600 sample as confirmed by XPS survey spectra. This led to a proportional reduction in the % adsorption of dyes as the availability of amorphous carbon leads to better electrostatic interaction between the dye and available surface functional groups on the sample [19]. Both aragonite and calcite are good dye adsorbents [15, 16]. Therefore, it is the availability of more carbon content in SM-450 samples makes it a better adsorbent for MB and EBT adsorption as compared with SM-600 sample.

3.7 Adsorption Kinetics

Adsorption kinetics were studied using pseudo-first-order (PFO), pseudo-second-order (PSO), and Elovich models for estimating adsorption rates of MB and EBT dyes onto SM-450 sample. Simultaneously, intraparticle diffusion (IPD) model was used to investigate the process of diffusion during adsorption (Fig. 5). The experimentally calculated values of q_e (mg/g) for MB and EBT adsorbed onto the sample at equilibrium, can be given as:

$$q_e = \frac{(C_0 - C_e)V}{m} \quad (1)$$

C_0, C_e, V, m, respectively, represent the initial concentration, equilibrium concentration of dye used in mg/L, solution volume in liters, and adsorbent weight in grams. For MB and EBT dye, the q_e values obtained using SM-450 sample were 24.10 and 17.40 mg/g.

PFO model is considered to be the best-suited model for the low concentrations solutions and it indicates that the adsorption is taking place via the physisorption process with a rate constant k_1, and its equation is:

$$\ln \frac{q_e}{q_e - q_t} = k_1 t \quad (2)$$

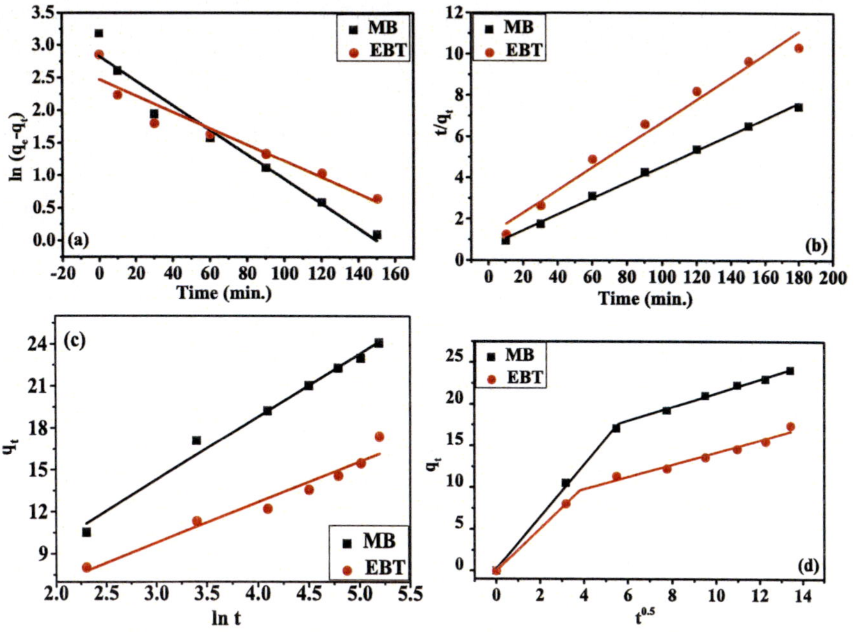

Fig. 5 Kinetic models fitted for MB (1 mg/L) and EBT (1 mg/L) adsorption onto *SM-450* (20 mg/L) using **a** PFO kinetics, **b** PSO kinetics, **c** Elovich kinetics and **d** IPD model

The PSO model implies that the rate at which a pollutant is adsorbed at the adsorbent's surface is proportional to the number of available active adsorption sites and that chemisorption occurs. It can be expressed as:

$$\frac{t}{q_t} = \frac{1}{k_2 q_e^2} + \frac{t}{q_e} \tag{3}$$

The rate constant k_2 was calculated by the intercept and slope of the graph between t/q_t and t [20].

The Elovich model implies that the process of adsorption involves chemisorption via multilayer adsorption and the equation involved is:

$$q_t = \left(\frac{1}{\beta}\right)\ln(\alpha\beta) + \left(\frac{1}{\beta}\right)\ln(t) \tag{4}$$

α, β, and q_t represent the absorption rate, desorption constant, and the dye adsorbed on the adsorbent's surface in mg/g at a specific time t, respectively. The slope and intercept of the curves fitted using experimental data between q_t and ln t were used to calculate these values. Table 3 shows the outcomes of the kinetic model fittings [21].

Table 3 Kinetic models fitting parameters for MB and EBT adsorbed at SM-450 sample

Dye adsorbed	Kinetic model	Parameters	Value	R^2
MB	PFO	k_1 (1/min)	0.0188	0.958
		q_e (mg/g)	7.68	
	PSO	k_2 (g/mg min.)	0.0022	**0.997**
		q_e (mg/g)	25.90	
	Elovich model	α (mg/g min.)	2.177	0.986
		β	0.222	
	IPD	k_{id1}(mg/g min$^{1/2}$)	3.134	0.998
	Model	k_{id2} (mg/g min$^{1/2}$)	0.829	
EBT	PFO	k_1 (1/min)	0.0127	0.905
		q_e (mg/g)	6.72	
	PSO	k_2 (g/mg min.)	0.0024	**0.975**
		q_e (mg/g)	18.21	
	Elovich model	α (mg/g min.)	3.048	0.940
			0.345	
	IPD	k_{id1}(mg/g min$^{1/2}$)	2.545	0.991
	Model	k_{id2} (mg/g min$^{1/2}$)	0.732	

Best followed model is shown in bold for regression coefficient values
(q_e (MB) = 24.10 mg/g and q_e (EBT) = 17.40 mg/g)

The excellent fit with higher R^2 values for the PSO and Elovich models indicated that chemisorption is the mechanism of adsorption. The regression coefficient R^2 for PSO is higher using the kinetic modeling parameters derived for MB and EBT dyes. Also, PSO yields q_e values that are equal to experimentally calculated values. This indicated that MB and EBT adsorption on SM-450 sample followed the PSO adsorption kinetic model, suggesting the presence of physicochemical interaction between the two phases (adsorbent and adsorbate). Thus, it can be concluded both cationic and anionic dyes adsorb via chemisorption on the sample's surface.

To understand the diffusion mechanism between adsorbate and adsorbent, IPD model was employed and its equation can be given as:

$$q_t = k_{id}\sqrt{t} + C \tag{5}$$

The slope of the curve between q_t and $t^{0.5}$ (Fig. 5d) is used to calculate the parameter k_{id} (IPD rate constant), which is given in Table 3. The R^2 values for MB and EBT adsorption on the SM-450 sample obtained > 0.99. Therefore, the adsorption process is completely following the IPD model. The fitting of data points involved a linear curve that went through the origin, indicating that the process as a whole is IPD-driven. Both dyes adsorb onto the SM-450 sample in two steps, with the first phase occurring via boundary layer diffusion and the second phase via

intraparticle diffusion [22]. Boundary layer diffusion occurs for MB and EBT up to 30 and 14.5 min, respectively, after which intraparticle diffusion begins.

4 Conclusions

This study investigated the synthesis of soy flour ash with a simple and cheap method by heating the soy flour in the muffle furnace at 450 (SM-450) and 600 °C (SM-600) without any further physical/chemical modifications. The synthesis temperature was determined by TG/DTA/DTG analysis of soy flour. Morphology of the synthesized samples showed that heat treatment makes the surface porous thereby increasing the surface area. XRD analysis showed the presence of minerals such as calcite, aragonite, arcanite, and fairchildite in SM-450 and SM-600 samples. XRD results were confirmed further by FTIR and XPS. The as prepared samples were used for adsorption of cationic and anionic dyes (MB and EBT) and exhibited good adsorption efficiency. The better adsorption efficiency was observed for SM-450 sample due to the availability of more carbon and aragonite in it, resulting in electrostatic interaction between the dye and accessible surface sites of the adsorbent. The adsorption was found to follow PSO kinetic model indicating chemisorption was taking place with two-step IPD. At equilibrium, 24.1 mg/g of MB and 17.4 mg/g of EBT were adsorbed onto SM-450 sample.

Acknowledgements The authors would like to acknowledge Materials Characterization Facility, TIET for FE-SEM, CIR, IIT Ropar for XRD, SAIF (Panjab University) for FTIR, IIC (IIT Roorkee) for XPS and TG/DTA. The authors are thankful to TIET for funds.

References

1. Berradi M, Hsissou R, Khudhair M, Assouag M, Cherkaoui O, Bachiri A, Harfi A (2019) Textile finishing dyes and their impact on aquatic environs. Heliyon 5:e02711
2. Zhu M, Lee L, Wang H, Wang Z (2007) Removal of an anionic dye by adsorption/precipitation processes using alkaline white mud. J Hazard Mater 149:735–741
3. Bastaki N (2004) Removal of methyl orange dye and Na_2SO_4 salt from synthetic waste water using reverse osmosis. Chem Eng Process 45:1561–1567
4. Sevimli MF, Sarikaya HZ (2002) Ozone treatment of textile effluents and dyes: effect of applied ozone dose, pH and dye concentration. J Chem Technol Biotechnol 77:842–850
5. Moghaddam SS, Moghaddam MRA, Arami M (2010) Coagulation/flocculation process for dye removal using sludge from water treatment plant: optimization through response surface methodology. J Hazard Mater 175:651–657
6. Vlyssides AG, Loizidou M, Karlis PK, Zorpas AA, Papaioannou D (1999) Electrochemical oxidation of a textile dye wastewater using a Pt/Ti electrode. J Hazard Mater B70:41–52
7. Zhao H, Sun Y, Xu L, Ni J (2010) Removal of Acid Orange 7 in simulated wastewater using a three-dimensional electrode reactor: Removal mechanisms and dye degradation pathway. Chemosphere 78:46–51

8. Rattan VK, Purai A, Singh H, Manoochehri M (2008) Adsorption of dyes from aqueous solution by cow dung ash. Carbon Lett 9:1
9. Purai A, Rattan VK (2010) Acid blue 92 (Leather dye) removal from wastewater by adsorption using biomass ash and activated carbon. Carbon Lett 11:1–8
10. Xu L, Cui H, Zheng X, Liang J, Xing X, Yao L, Chen Z, Zhou J (2017) Adsorption of Cu2þ to biomass ash and its modified product. Water Sci Technol 1:115–125
11. Range BMK, Hawboldt KA (2018) Adsorption of thiosulphate, trithionate, tetrathionate using biomass ash/char. J Environ Chem Eng 6:5401–5408
12. Patel A, Soni S, Mittal J, Mittal A, Arora C (2021) Sequestration of crystal violet from aqueous solution using ash of black turmeric rhizome. Desalination Water Treat 220:342–352
13. Costa JAS, Paranhos CM (2019) Evaluation of rice husk ash in adsorption of Remazol Red dye from aqueous media. SN Appl Sci 397:1
14. Liu C, Wang H, Karim AM, Sun J, Wang Y (2013) Catalytic fast pyrolysis of lignocellulosic biomass. Chem Soc Rev 43:7594–7623
15. Suteu D, Bilba D, Aflori M, Doroftei F, Lisa G, Badeanu M, Malutan T (2012) The seashell wastes as biosorbent for reactive dye removal from textile effluents. Clean Soil Air Water 40:198–205
16. Tonge AS, Harbottle D, Casarin S, Zervaki M, Careme C, Hunter TN (2021) Coagulated mineral adsorbents for dye removal, and their process intensification using an agitated tubular reactor (ATR). Chemengineering 5:35
17. Chakrabarty D, Mahapatra S (1999) Aragonite crystals with unconventional morphologies. J Mater Chem 9:2953–2957
18. Liu X, Liao J, Song H, Yang Y, Guan C, Zhang Z (2019) A biochar-based route for environmentally friendly controlled release of nitrogen: urea-loaded biochar and bentonite composite. Sci Rep 9:9548
19. Chen S, Wang Z, Xia Y, Zhang B, Chen H, Chen G, Tang S (2019) Porous carbon material derived from fungal hyphae and its application for the removal of dye. RSC Adv 9:25480
20. Gupta A, Mittal M, Singh MK, Suib SL, Pandey OM (2018) Low-temperature synthesis of NbC/C nano-composites as visible light photoactive catalyst. Sci Rep 8:13597
21. Choudhary R, Pandey OP, Brar LK (2021) High yield glucose assisted carbonization of soy flour for dye removal applications. Mater Chem Phys 260:124174
22. Choudhary R, Pandey OP, Brar LK (2022) Novel ultrasonic pretreatment for HTC carbon nanosphere size control without yield compromise. J Nanopart Res 24:75

Efficacy of ANN and ANFIS as an AI Technique for the Prediction of COF at Finger Pad Interface in Manipulative Tasks

Ashish Kumar Srivastava, Jitendra Singh Rathore, and Sharad Shrivastava

1 Introduction

Skin being the largest and the outermost organ of human body gets in contact with multiple surfaces in day-to-day tasks. Therefore, the tribological (frictional) interaction between finger pad and contact surface is an important factor for the product designs involving human interface. For the tribological behaviour involving human skin (finger pad) multiple theoretical models such as Hertz, Johnson-Kendall-Robert (JKR), and Greenwood Williamson [1] are available.

Optimum friction is crucially important for manipulative tasks [2]. Previous studies have reported the friction between hand and popular handle materials such as aluminium [3–6], rubber [7–9], and steel [5, 7, 10, 11]. Human skin friction irrespective of anatomical region, is affected by multiple intrinsic and extrinsic parameters. Intrinsic parameters include attributes such as age [6, 12], gender [13, 14], and skin hydration [15, 16] while, extrinsic factors constitute normal load [15, 17–19], sliding velocity [19, 20], relative humidity [6, 21], temperature [21, 22] etc. The outcome of the study would provide the model for the frictional interaction at finger pad interface for the better design of the assistive devices, viz. wheelchairs, scooters, sticks etc.

Limited work for the statistical modelling [23] of coefficient of friction at the human skin interface is attempted in the literature. Here, a comparative study of two machine learning methods artificial neural network (ANN) and adaptive neuro fuzzy inference system (ANFIS) is carried out for the modelling of COF at finger pad interface.

In the ambit of artificial intelligence ANN and ANFIS are effectively been used for classification of tasks, model approximations and constrained process controls [24]. With the vide subject wise variation as in human skin tribology domain ANN

A. K. Srivastava (✉) · J. S. Rathore · S. Shrivastava
Department of Mechanical Engineering, Birla Institute of Technology and Sciences, Pilani, Jhunjhunu, Rajasthan 333031, India
e-mail: mechaksrivastava@gmail.com

A. Kumar et al. (eds.), *Recent Advances in Manufacturing and Thermal Engineering*, Lecture Notes in Mechanical Engineering, https://doi.org/10.1007/978-981-19-8517-1_2

and ANFIS can very effectively be applied. ANN works on the principle of human nervous system. On the other hand, ANFIS adapts the features of both ANN and the fuzzy logic. ANN and ANFIS both learn through a set of training data through a mathematical model. ANFIS displays its solution on fuzzy inference system (FIS). ANFIS over ANN has an advantage in determination of the hidden layers and improvement in its prediction ability due to FIS [25]. Ineffective application of ANN parameters remains a challenge for the precision of model in the engineering domain. However, accuracy and efficacy of these techniques depends on the domain of application [26] and input variables. For the selection of the most suitable technique, it is important to evaluate their performance that forms the basis of this study.

2 Methodology

2.1 Experimentation

Test subject A 29-year-old healthy male is taken as a test subject for the study. A written consent was obtained, and subject was well briefed about the experimental protocols. All the experiments were performed using device as depicted in Fig. 1 as per the helsinki declaration.

Experimental setup and test protocol Friction measurements were conducted on the in-house developed rotary-type human skin tribometer. A polished stainless steel material probe is taken as test material for friction measurement of the human skin. All experiments were performed at an ambient temperature of 25 °C and relative humidity of 50 $\pm$ 10%. Subject was instructed to make a continuous contact with the rotating probe at varying normal loads and sliding velocities. Normal load was varied from 2, 4, 6 & 8 N while, and sliding velocity as equivalent to 4, 6, 8 & 10 cm/s of linear speed.

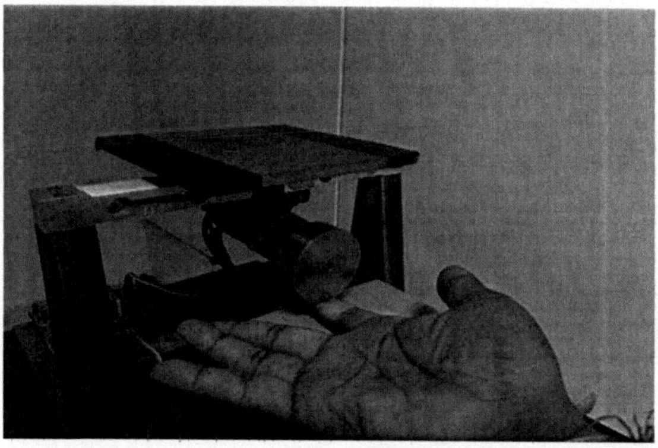

Fig. 1 Device for the finger pad friction measurement

Table 1 Experimental results of the friction measurement

Experimental run	Normal load (N)	Sliding velocity (cm/s)	COF
1	8	6	5.43
2	8	4	5.21
3	6	4	4.42
4	4	8	4.58
5	4	10	2.67
6	6	8	3.19
7	2	8	5.55
8	2	10	3.25
9	6	6	4.65
10	4	4	5.25
11	4	6	6.57
12	6	10	2.23
13	2	6	7.15
14	8	8	2.87
15	8	10	2.29
16	2	4	8.12

Design of experiments Experiments were designed based on Taguchi L16 (4^2) design. In total, 16 experimental runs were performed as shown in Table 1. Input parameters were normal load and sliding velocity. Output parameter was the coefficient of friction between finger pad and steel probe.

2.2 *Artificial Neural Network (ANN)*

ANN draws its inspiration from human nervous system, as learning is acquired by training and prediction/categorization of data. ANN modelling was done using neural fitting toolbox of MATLAB (2021b). Predictions in the ANN are fine tuned by adjusting its parameters such as weights and biases, iteratively unless a minimum error is reached for the inputs. An influence of a suitable network design, algorithms and hyperparameters affects greatly the prediction capability of the network [27]. Neural network performance is measured by its correlation coefficient (R^2) [28]. Trained ANN architecture is shown in Fig. 2.

In this study, data division is done randomly, 70% of experimental data is used for training, 15% each for validation and testing of the developed model. Levenberg–Marquardt (LM), Bayesian Regularization (BR), and Scaled Conjugate Gradient (SCG) are compared for training, validation, and testing.

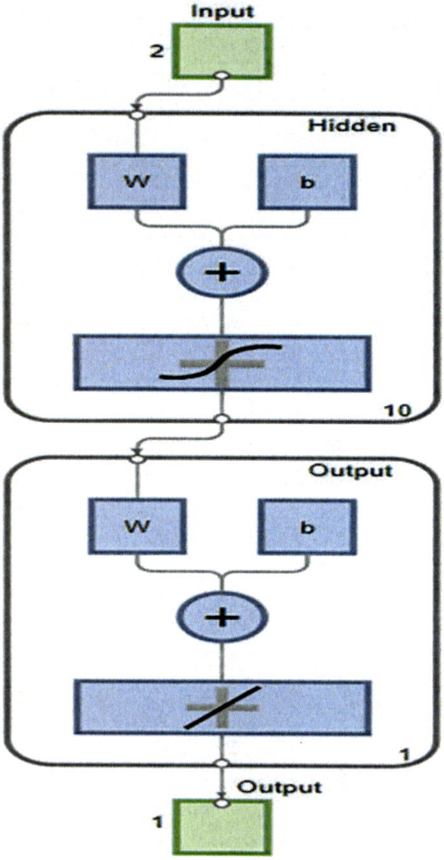

Fig. 2 Network architecture of trained ANN

2.3 Adaptive-Neuro Fuzzy Inference System (ANFIS)

ANFIS modelling is based on the Takagi–Sugeno inference system governed by *if–then* rule. ANFIS model is designed using the neuro-fuzzy designer toolbox of MATLAB (2021b). For the creation of an inference system five different layers viz. fuzzification layer, product layer, normalisation layer, defuzzification layer, and the output layer are required. To briefly, understand the overall functioning of each layer of ANFIS refer Table 2.

Table 2 Layers of inference system with their functioning

Name of the layer	Attribute
Fuzzification layer	Conversion of input data into fuzzy inputs
Product layer	Fuzzy rules are formed by output of the first layer (Algebraic multiplication)
Normalization layer	Normalization of output of second layer
Defuzzification layer	Effect of each part of system's output is determined using fuzzy rules
Output layer	Generates a system output as a numerical variable equal to non-fuzzy part in the fuzzy systems

3 Results and Discussion

3.1 Development of an ANN Model

ANN was trained with the experimental run data as tabulated in Table 1. Data was divided as training (70%), validation (15%), and testing (15%). The regression plots and correlation coefficient for three training algorithms LM, BR, and SCG are shown in Fig. 3 and Table 3, respectively. Decision on the number of hidden layer is solely made considering complexity of the system. Architecture consisted of 2 input variables, 10 hidden layers, and 1 output layer. All the training algorithms used tansig at hidden and logsig at output layer as the transfer function. As discussed, R^2 represents the ability of the model to explain the variance of dependent variable through an independent variable. LM and BR algorithms have shown a similar statistical significance with R^2 value of 0.9561 and 0.9566, respectively, while SCG has shown the highest significance with R^2 value of 1 for training. All three LM, BR, and SCG have shown significant R^2 value of 1 [29] for the testing, denotes the perfect fit for the data in the testing phase. Figure 3a–c represent the input data and the best fit curve regression plot for the architecture trained with LM, BR, and SCG algorithms, respectively. Regression plots highlight the generalization ability of ANN with the above mentioned algorithms.

3.2 Development of an ANFIS Model

Experimental data was divided initially as 70% for training and 30% for validation. Figure 4a illustrates an ANFIS structure having 2 inputs and 1 output. Figure 4b, c depicts the strength of model in terms of training and testing data. Dots and asterisk represent the experimental and predicted data, respectively. Training data graph shows significant overlapping in comparison to the test set data. Reason to this can be attributed to the overfitting of data in comparison to that of the ANN network.

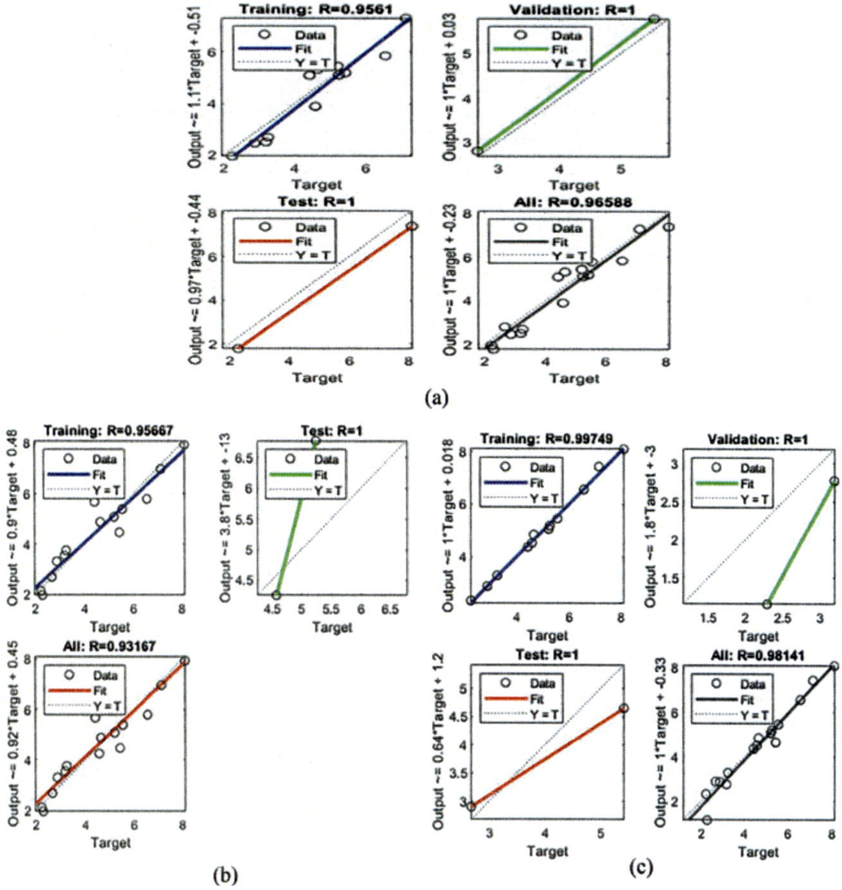

Fig. 3 Regression plots for **a** LM, **b** BR, and **c** SCG

Table 3 ANN training algorithms with R^2 values

Algorithm	Neurons	Activation function (layer)		R^2			
		Hidden	O/P	Training	Testing	Validation	All
Levenberg Marquardt	2-10-1	Tansig	Logsig	0.9561	1	1	0.9658
Bayesian regularization	2-10-1	Tansig	Logsig	0.9566	1	–	0.9316
Scaled conjugate gradient	2-10-1	Tansig	Logsig	0.9974	1	1	0.9814

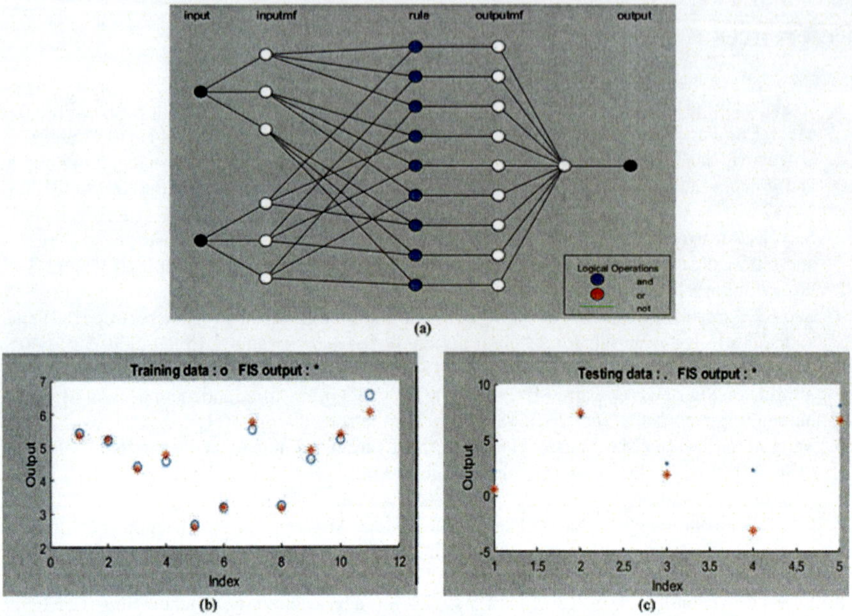

Fig. 4 **a** ANFIS architecture, **b** plot of training data and **c** plot for testing data

4 Conclusion

Prediction of COF at the human finger pad is of vital importance for the manipulative tasks, ergonomics, and motor control activities of an individual and, persons with disabilities. Since non-linear nature contributes to the multi-attribute constraints in the modelling of friction coefficients for biological tissues, the effect of microstructure [30] is crucial in selecting a particular modelling technique. Based on literature normal load and sliding velocity are taken as input variables in the study. Both ANN and ANFIS models have comprehensively been analysed and their accuracy and predictability is evident. Salient results of the study are enumerated as:

- Model is developed based on the experiments designed through Taguchi L16 (4^2) design.
- Both the techniques are suitable enough for the prediction of COF, but ANFIS test data highlights overfitting.
- R^2 value of more than 0.95 in all the three training algorithms, study concludes that ANN is relatively superior technique than ANFIS in the case.

This can very conveniently be attributed to other factors that have not been considered in the model but have contributed to the physical experiments. Developed models are validated through other set of input variables not reported here to keep the study succinct. However, the accuracy of the developed model can be enhanced [31] by increasing the experimental data and the input variables.

References

1. Joodaki H, Panzer MB (2018) Skin mechanical properties and modeling: a review. Proc Inst Mech Eng Part HJ Eng Med 232(4):323–343. https://doi.org/10.1177/0954411918759801
2. Uygur M, de Freitas PB, Jaric S (2010) Frictional properties of different hand skin areas and grasping techniques. Ergonomics 53(6):812–817. https://doi.org/10.1080/00140131003734237
3. Seo NJ, Armstrong TJ, Young JG (2010) Effects of handle orientation, gloves, handle friction and elbow posture on maximum horizontal pull and push forces. Ergonomics 53(1):92–101. https://doi.org/10.1080/00140130903389035
4. Seo NJ, Armstrong TJ (2009) Friction coefficients in a longitudinal direction between the finger pad and selected materials for different normal forces and curvatures. Ergonomics 52(5):609–616. https://doi.org/10.1080/00140130802471595
5. Tomlinson SE, Lewis R, Carré MJ (2009) The effect of normal force and roughness on friction in human finger contact. Wear 267(5–8):1311–1318. https://doi.org/10.1016/j.wear.2008.12.084
6. Veijgen NK, Masen MA, van der Heide E (2013) Variables influencing the frictional behaviour of in vivo human skin. J Mech Behav Biomed Mater 28:448–461. https://doi.org/10.1016/j.jmbbm.2013.02.009
7. Gee MG, Tomlins P, Calver A, Darling RH, Rides M (2005) A new friction measurement system for the frictional component of touch. Wear 259(7–12):1437–1442. https://doi.org/10.1016/j.wear.2005.02.053
8. Lewis R, Menardi C, Yoxall A, Langley J (2007) Finger friction: grip and opening packaging. Wear 263(7–12):1124–1132. https://doi.org/10.1016/j.wear.2006.12.024
9. Sergachev DA, Matthews DTA, van der Heide E (2019) An empirical approach for the determination of skin elasticity: finger pad friction against textured surfaces. Biotribology 18:100097. https://doi.org/10.1016/j.biotri.2019.100097
10. Masen MA (2011) A systems based experimental approach to tactile friction. J Mech Behav Biomed Mater 4(8):1620–1626. https://doi.org/10.1016/j.jmbbm.2011.04.007
11. Zhang S et al (2017) Finger pad friction and tactile perception of laser treated, stamped and cold rolled micro-structured stainless steel sheet surfaces. Friction 5(2):207–218. https://doi.org/10.1007/s40544-017-0147-9
12. Zahouani H, Pailler-Mattei C, Sohm B, Vargiolu R, Cenizo V, Debret R (2009) Characterization of the mechanical properties of a dermal equivalent compared with human skin in vivo by indentation and static friction tests. Ski Res Technol 15(1):68–76. https://doi.org/10.1111/j.1600-0846.2008.00329.x
13. Ramalho A, Szekeres P, Fernandes E (2013) Friction and tactile perception of textile fabrics. Tribol Int 63:29–33. https://doi.org/10.1016/j.triboint.2012.08.018
14. Derler S, Gerhardt LC (2012) Tribology of skin: Review and analysis of experimental results for the friction coefficient of human skin. Tribol Lett 45(1):1–27. https://doi.org/10.1007/s11249-011-9854-y
15. Derler S, Schrade U, Gerhardt LC (2007) Tribology of human skin and mechanical skin equivalents in contact with textiles. Wear 263(7–12):1112–1116. https://doi.org/10.1016/j.wear.2006.11.031
16. Li W, Zhai ZH, Pang Q, Kong L, Zhou ZR (2013) Influence of exfoliating facial cleanser on the bio-tribological properties of human skin. Wear 301(1–2):353–361. https://doi.org/10.1016/j.wear.2012.11.073
17. Barrea A, Bulens DC, Lefevre P, Thonnard JL (2016) Simple and reliable method to estimate the fingertip static coefficient of friction in precision grip. IEEE Trans Haptics 9(4):492–498. https://doi.org/10.1109/TOH.2016.2609921
18. Van Kuilenburg J, Masen MA, Groenendijk MNW, Bana V, Van Der Heide E (2012) An experimental study on the relation between surface texture and tactile friction. Tribol Int 48:15–21. https://doi.org/10.1016/j.triboint.2011.06.003
19. Derler S, Rotaru GM (2013) Stick-slip phenomena in the friction of human skin. Wear 301(1–2):324–329. https://doi.org/10.1016/j.wear.2012.11.030

20. Tang W, Rong Ge S, Zhu H, Chuan Cao X, Li N (2008) The influence of normal load and sliding speed on frictional properties of skin. J Bionic Eng 5(1):33–38. https://doi.org/10.1016/S1672-6529(08)60004-9
21. Klaassen M, Schipper DJ, Masen MA (2016) Influence of the relative humidity and the temperature on the in-vivo friction behaviour of human skin. Biotribology 6:21–28. https://doi.org/10.1016/j.biotri.2016.03.003
22. Klaassen M, de Vries EG, Masen MA (2017) The static friction response of non-glabrous skin as a function of surface energy and environmental conditions. Biotribology 11:124–131. https://doi.org/10.1016/j.biotri.2017.05.004
23. Veijgen NK, van der Heide E, Masen MA (2013) A multivariable model for predicting the frictional behaviour and hydration of the human skin. Ski Res Technol 19(3):330–338. https://doi.org/10.1111/srt.12053
24. Surajudeen-Bakinde NT et al (2018) Path loss predictions for multi-transmitter radio propagation in VHF bands using adaptive neuro-fuzzy inference system. Eng Sci Technol Int J 21(4):679–691. https://doi.org/10.1016/j.jestch.2018.05.013
25. Melin P, Soto J, Castillo O, Soria J (2012) A new approach for time series prediction using ensembles of ANFIS models. Expert Syst Appl 39(3):3494–3506. https://doi.org/10.1016/j.eswa.2011.09.040
26. Gill J, Singh J, Ohunakin OS, Adelekan DS (2018) Artificial neural network approach for irreversibility performance analysis of domestic refrigerator by utilizing LPG with TiO_2–lubricant as replacement of R134a. Int J Refrig 89:159–176. https://doi.org/10.1016/j.ijrefrig.2018.02.025
27. Wen L, Ye X, Gao L (2020) A new automatic machine learning based hyperparameter optimization for workpiece quality prediction. Meas Control (United Kingdom) 53(7–8):1088–1098. https://doi.org/10.1177/0020294020932347
28. Kumar V, Kumar A, Chhabra D, Shukla P (2019) Improved biobleaching of mixed hardwood pulp and process optimization using novel GA-ANN and GA-ANFIS hybrid statistical tools. Bioresour Technol 271:274–282. https://doi.org/10.1016/j.biortech.2018.09.115
29. Ratner B (2009) The correlation coefficient: Its values range between 1/1, or do they. J Target Meas Anal Mark 17(2):139–142. https://doi.org/10.1057/jt.2009.5
30. Pond D, McBride AT, Davids LM, Reddy BD, Limbert G (2018) Microstructurally-based constitutive modelling of the skin—linking intrinsic ageing to microstructural parameters. J Theor Biol 444:108–123. https://doi.org/10.1016/j.jtbi.2018.01.014
31. Benmus TA, Abboud R, Shatter MK (2016) Neural network approach to model the propagation path loss for great tripoli area at 900, 1800, and 2100 MHz bands. In: 16th international conference science tech automation control computer engineering STA 2015, pp 793–798. https://doi.org/10.1109/STA.2015.7505236

Thermal Wear in Disc Brake Friction Pads

Anant Nemade, Arvind Chel, and Rajani Nemade

1 Introduction

In braking, actuation force from master cylinder generates the friction force which opposes to direction of sliding at the interface of disc and pad; hence retardation of the vehicle is achieved. During braking friction pads transform the kinetic energy of vehicle in to heat energy in a very short braking time. This high heat develops thermal stress in braking components [1]. Therefore, it is necessary to design all components of the braking system to sustain this high sudden steep rise in temperature to avoid thermal stresses during braking. The contact forces and rotational stresses on the disc are high, are termed as mechanical loading. Normally, in cars hydraulic type of brake actuation mechanism is used which are designed to generate required pressure [2].

If the heat generation is too high then braking system becomes very hot and braking performance will may reduce [3]. There are so many reasons for reduction in braking performance due to heating of brake such as

Change in the properties of the rotor material at this high temperature.

Reduction in mechanical strength of brake disc.

Friction pad wears very fast due to loosening of bond between composite material.

High temperature affects in reduction of coefficient of friction between rotor and pad called brake fade.

Boiling of hydraulic fluid and deterioration of hydraulic seals in braking system.

A. Nemade (✉)
JNEC Research Centre, MGM's University, Aurangabad 431005, India
e-mail: anantnemade7@gmail.com

A. Chel
Department of Mechanical Engineering, JNEC Research Centre, MGM's University, Aurangabad 431005, India

R. Nemade
Department of Chemistry, Government Polytechnic, Aurangabad 431005, India

A. Kumar et al. (eds.), *Recent Advances in Manufacturing and Thermal Engineering*, Lecture Notes in Mechanical Engineering, https://doi.org/10.1007/978-981-19-8517-1_3

Surface crakes on the disc due to high thermal stresses developed.

Out of these any one or combining all the factors involved during braking [4], then brake becomes inactive. Brake heating situation arises many times during long braking, in this situation driver complain that there was a brake failure at the time of incident but the actual phenomenon was different [5]. There might be one of the reason or combine effect of heating of system during long emergency braking situation.

The brake materials thermal capacity and coefficient of surface heat transfer are the factors responsible for rising the temperature [6]. During vehicle driving, three types of braking situations arise such as

Single Brake application.

Series of repeated brake applications.

Continuous brake application (called Drag Braking).

Brakes are designed to work in all these three conditions [7]. For better performance of brake in all the above situations cooling provision of brake disc plays a very important role. The frictional heat developed during braking must be carried away from the friction surface of disc to avoid high temperature generation due to friction between pad and disc [8]. For this reason, rotors of disc brakes are designed in two solid friction plates/rings, each ring is called cheek, separated by small vanes [9]. These discs are called as air ventilated discs or air-cooled disc (ACD) as shown in Fig. 1. During vehicle movement air enters forcefully on these vanes as well as on the rotor surface and carry away the heat naturally. Vanes are straight or curved depending on the cooling requirements of the vehicle [10].

Fig. 1 Friction disc of brake and different types of vanes on ACD brake disc

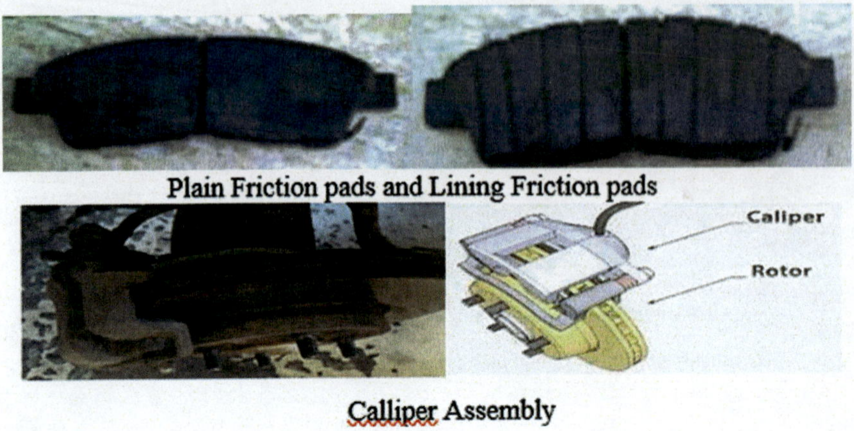

Fig. 2 Friction pads and calliper assembly

In Fig. 2, the assembly of complete braking system with calliper and friction pads is shown. The brakes are applied with the help of actuator operated by mechanical, hydraulic or pneumatic pressure [11]. In single braking velocity of the vehicle reduces with constant braking torque, here deceleration can be assumed to be constant as the braking stroke is single. The braking torque develops gradually with respect to time, when driver applies brakes; braking torque starts developing and at some instance when driver applies full brake it is at maximum value [12]. The time depends upon the response of braking system. When vehicle come to rest condition braking torque becomes zero [13]. This type of braking is observed in most of the cases, here time period of brake application is less as the only objective is to stop the vehicle smoothly. Therefore, heat generation is less; also, this heat dissipates from the surface area and through vanes of disc brakes [14].

In case of repeated braking applications, every time braking force reaches to maximum and friction between disc and pad continue till the driver releases the brake [15]. Every time at per brake, heat is generated and before cooling again the heat is added to next braking stroke. This addition of heat accumulates and brakes become hot. The temperature generated depends upon the number of braking applications and time period of each brake. The rate of cooling of disc depends upon the design and material properties of brake disc [16].

In drag braking, pads are continuously in contact with the disc for a longer time e.g. when vehicle needs continuous braking, there is continuous heat generation by the braking system. Drag braking occurs in situation where vehicle needs to control over a long distance with the application of brakes such as long down slope or vehicle moving with a very high speed and emergency braking situation arises [17]. In this type of braking heat is added continuously from the application of brake to stop the vehicle. Present vehicle braking system is dependent on air flow cooling. When vehicle is at rest the rate of heat transfer to atmosphere via convection is very slow as the vehicle moves forward with the increase of speed the air flow increases and hence

1) Effect of Temp on Pads 2) Effect of Temp on Disc-Cracking 3) Effect of Temp on disc- Conning

Fig. 3 Effect of temperature on friction pads and on disc

force cooling takes place [18]. The heat generation rate and cooling rate are not equal; cooling rate is slow as compared to heat generation. The rate of cooling also depends upon the atmospheric condition, in hot atmospheric regions rate of cooling decreases than the normal [19]. The generated heat has an impact on every component of the braking system. Effect of rise in temperature on rotor (Disc), friction brake pads and other braking components are;

Weakening of Friction pad bonding material.

Drop in the coefficient of friction of pads. (Brake Fade).

Beyond the critical temperature, the failure of different components of braking system may take place., e.g. friction pads, braking fluid, piston, calliper itself, O rings, hose, clips, etc.

Temperature rise also effects on disc material in the form of change in material properties leads in to cracking or conning. Same has been shown in the Fig. 3, it is necessary to dissipate the heat from the local area of disc to avoid such effects on braking system.

These are the hidden by products of heat generation during braking [20]. These cannot be identified during braking operation. In the present study, an attempt has been made to find the effect of temperature on wear rate of friction pads by using brake dynamometer setup [21–23].

2 Research Gap and Objective

Literature review on brake pad wear suggests that there is wear of friction pads due to mechanical friction between disc and pads during braking but only acceptance does not solve the problem, therefore different problem-solving angles have been thought and found that the temperature generation might be the major cause of wear, as in almost all brake researchers have given stress on cooling of disc and this research gap has been identified for the further study. Accordingly, the main objective has been decided that to find the rate of wear of friction pads with respect to temperature generation during the braking process of vehicle.

3 Materials and Methods

To carry out the experimentation off-market friction pads were preferred, these pads were widely used by the local garage mechanics to replace the wear out pads. The common composition of friction pads can be observed in testing report obtained before the experimentations. The results obtained are summarized in Table 1.

Laboratory experimentation methodology is used to predict the wear rate of friction pads. To carry out the experimentations an experimental setup of dynamometer has been designed and the results were obtained as given below.

3.1 *Experimental Setup*

Experimental setup is essential to perform the different experimentations on brake disc [24]. In this research, disc brake has been taken for the study. Experimentations are performed keeping light motor vehicle in to consideration; therefore, car braking system has been installed on brake dynamometer. The basic components required for setup of disc brake are; disc brake assembly with caliper, friction pads, master cylinder, operating paddle, rotating device, i.e. electric motor and inertia wheel [25]. Apart from these basic components, some measuring equipment's are also necessary to record the different readings such as pressure gauge, ABB system for rpm measurement, thermocouple to record the heat generated and fan to create realistic situation for cooling the disc [26].

Inertia brake dynamometers is laboratory-based testing equipment and is designed for brake testing as shown in Fig. 4 [27]. In this type of dynamometer brake disc is attached to the motor driven shaft (AC variable drive electric motor) on which flywheel is fitted to simulate the vehicles inertia and kinetic energy. The brake stator which includes calliper and brake actuation mechanism, mounted on the frame of

Table 1 Testing report of friction pad material (SN heat treatment laboratory)

Lab No. L: 29, 30 Sample Id						Quality		
No.	C %	Si %	Mn %	P %	S %	Cr %	Ni %	Mo %
1	> 1.68	0.105	0.017	0.010	> 0.60	< 0.0015	0.0093	0.0040
2	> 1.68	0.036	0.0034	0.0095	0.246	< 0.0015	0.012	< 0.0020
No	Al%	Cu%	Co%	Ti%	Nb%	V%	W%	Pb%
1	0.0083	< 0.0010	< 0.0015	0.0035	< 0.0030	0.0039	0.010	0.0019
2	0.0020	< 0.0010	0.0017	< 0.0010	< 0.0030	0.0012	< 0.010	< 0.0010
No	B%	Sn%	Zn%	Fe%				
1	< 0.005	0.0013	0.0037	0.430				
2	< 0.005	0.0023	0.0022	0.381				

Fig. 4 Experimental setup of inertia brake dynamometer

dynamometer. The shaft is free to rotate on the bearings and torque arm is installed on tail end to measure the torque [28]. Electric motor drives the complete assembly of flywheel and disc brake mountings; when sufficient speed of the flywheel is achieved brakes are applied to decelerate the flywheel and at the same time motor drive is disconnected [29]. Many simulated situations can be installed on dynamometer to create the realistic situation for brake test. Brake disc is provided with K-type rubbing thermocouple as shown in Fig. 5, which detects the temperature generated on disc surface during braking [30]. This laboratory data then can be validating in actual vehicle testing.

3.2 *Experimentation*

To evaluate experimentally the thermal effect on brake pads by using weighing difference method for brake fade and pad wear [2, 31, 32]. Experiments were carried out

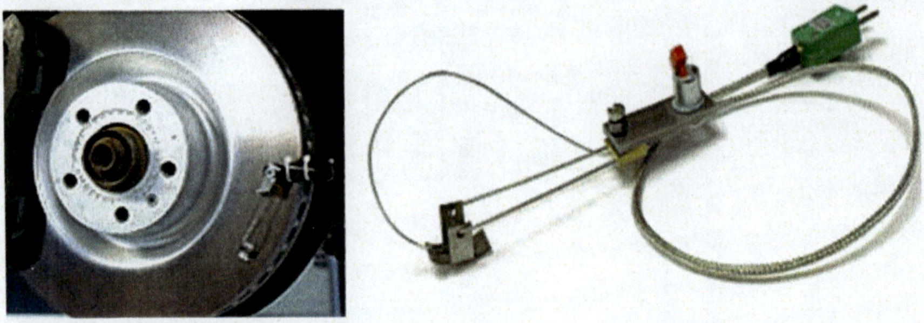

Fig. 5 Rubbing K-type thermocouple

on the inertia brake dynamometer to observe the performance of friction pads against the temperature generated during braking.

3.3 Experimentation for Thermal Brake Fade

Thermal brake fade experimentation was carried out on the inertia dynamometer to observe the temperature of brake fade for ventilated brake disc. Series of brake trials were taken for the observation of brake fade [33]. Before starting actual experimentation, bedding and burnishing of brake pads were done by dry run braking at 180 °C as it is essential for new pads to be used for experimentation [34]. In actual experimentation each time electric motor is rotated at an 1800 rpm to rotate the disc and brakes were applied by keeping all other opera meters constant [35]. Brakes are applied to stop the rotating disc and stopping time is noted for each braking cycle. Braking cycles were repeated with a gap of 1 Min (60 s) at room temperature and without force cooling [36]. The start time of brake and stop time was noted as shown in Table 2. For each braking cycle, temperature generated also noted with the help of rubbing thermocouple. Temperature of disc increases with the number of braking cycles, after nine cycles temperature reaches to 269 °C, and there was no change in disc stopping time. In next three braking cycles, there was slight change in stopping time and also temperature increase observed. After 14 cycles of braking operation temperature reaches to 307 °C and stopping time also increase significantly. In another two braking cycles, temperature reaches to 314 °C and there was significant change in stopping time of disc. This shows that after 300 °C temperature brake fade is observed in disc brake [37]. The obtained reading was plotted on graph as shown in Fig. 6. After experimentation actual pad wear can be observed on friction pads as shown in Fig. 7. The procedure adopted for measuring the wear rate in this experimentation was by change in mass of friction pads. Initial mass readings of both the brake pads (after bedding and burnishing) were noted with the help of accurate weighing balance machine then after experimentation the weight of the friction pads were again noted as shown in Fig. 8 [38].

Table 2 Observations for 1 Min gap braking cycles

Sl. No.	Date	Flywheel weight in Kg	Hydraulic pressure in Mpa	Initial RPM	Final RPM	Braking start time	Braking stop time	Braking time in sec	Initial temp on ventilated disc °C	Max. temp recorded on ventilated disc °C
Col No	01	02	03	04	05	06	07		08	09
01	18/03/2020	42	2.7	1800	0	11:30:00	11:30:02	02	28	78
02		42	2.7	1800	0	11:31:02	11:31:04	02		92
03		42	2.7	1800	0	11:32:04	11:32:06	02		119
04		42	2.7	1800	0	11:33:06	11:33:08	02		142
05		42	2.7	1800	0	11:34:08	11:34:10	02		171
06		42	2.7	1800	0	11:35:10	11:35:12	02		192
07		42	2.7	1800	0	11:36:12	11:36:14	02		218
08		42	2.7	1800	0	11:37:14	11:37:16	02		233
09		42	2.7	1800	0	11:38:16	11:38:19	02		269
10		42	2.7	1800	0	11:39:19	11:39:22	03		278
11		42	2.7	1800	0	11:40:22	11:40:25	03		291
12		42	2.7	1800	0	11:41:25	11:41:28	03		298
13		42	2.7	1800	0	11:42:28	11:42:31	03		302
14		42	2.7	1800	0	11:43:31	11:43:35	04		307
15		42	2.7	1800	0	11:44:35	11:44:39	04		310
16		42	2.7	1800	0	11:45:39	11:45:43	04		314

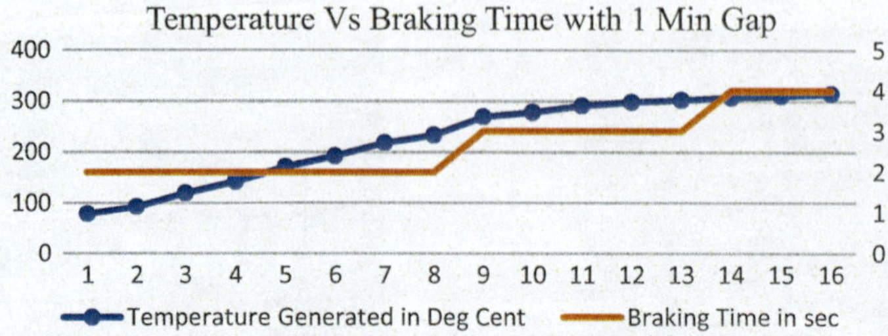

Fig. 6 Graphical representation of temperature rise and braking time

Fig. 7 Brake pads surface after experimentation

3.4 *Observations*

The weight difference after experimentation of friction pads was observed to be 1.2 g. Brake pad wear was on higher side, these wear particles spreads in atmosphere.

3.5 *Experimentation for Thermal Brake Wear*

With the guidelines from standard test procedures for the brake, wear experiments were carried out on the inertia dynamometer [39]. All the operating parameters were kept constant as if it was maintained for above brake fade test. The procedure adopted for measuring the wear rate was by change in mass of pads. Initial mass readings of both the brake pads (after bedding and burnishing) were noted with the help of accurate weighing balance machine. In experimentation pair of brake pad was fitted, and 16 braking cycles were repeated for stopping the disc, from 1800 rpm to stop the disc [40]. The duration between braking cycle time for this test was 12 min, so that the temperature of the disc between two cycles reduces substantially in natural cooling [41]. After each trial start and stop time was noted and at the same time temperature also noted as shown in Table 3. The main aim of this experimentations was to determine the rate of wear during braking without generating temperature on the surface of pad. In this experimentation up to reading number 12

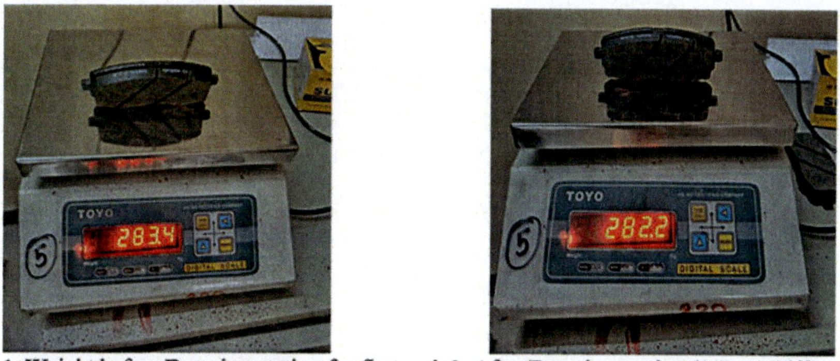

1. Weight before Experimentation for first pad 2. After Experimentation (1.2 gm Difference)

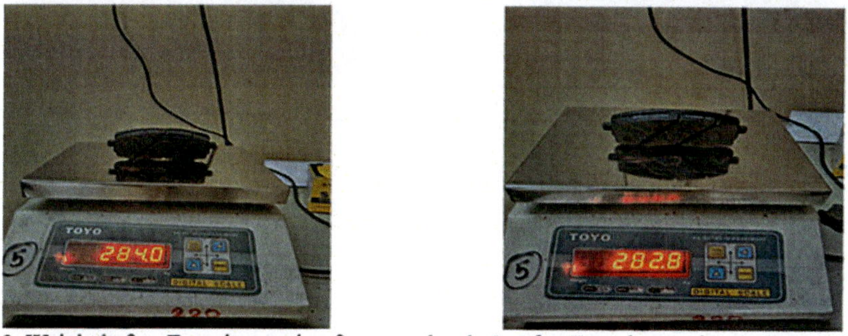

3. Weight before Experimentation for second pad 2. After Experimentation (1.2 gm diff.)

Fig. 8 Brake pads weight before and after wear experimentation

there was no change in braking time, and the temperature was also 112 °C, which is lower as compared to 1 min braking cycle. The temperature generated during this experimentation was plotted on the graph as shown in Fig. 9. After reading number 13, there was a change of 1 s in braking time at the temperature of 116 °C. This implies that the temperature effects on braking time. The wear of friction pads can be observed in Fig. 10.

3.6 Observations

After experimentation, the brake pads were removed from the calliper and weight was noted [42]. The weight of this pads was noted as shown in Fig. 11 and compared with the initial weights and found that there is difference between weight of pads. The weight difference after wear is observed to be 0.2–0.6 g, whereas in 1 min braking cycle after wear the weight difference was 1.2 g. Thus, with the rise of temperature on the surface of the friction pads increases the wear rate and was experimentally verified [43].

Table 3 Observations for 12 Min braking cycle

Sl. No.	Date	Flywheel weight in Kg	Hydraulic pressure in Mpa	Initial RPM	Final RPM	Braking start time	Braking stop time	Braking time in sec	Initial temp on ventilated disc °C	Max. temp recorded on ventilated disc °C
Col No	01	02	03	04	05	06	07		08	09
01	19/03/2020	42	2.7	1800	0	12:30:00	12:30:02	02	30	71
02		42	2.7	1800	0	12:42:02	12:42:04	02		89
03		42	2.7	1800	0	12:54:04	12:54:06	02		85
04		42	2.7	1800	0	1:06:06	1:06:08	02		91
05		42	2.7	1800	0	1:18:08	1:18:10	02		92
06		42	2.7	1800	0	1:30:10	1:30:12	02		97
07		42	2.7	1800	0	1:42:12	1:42:14	02		95
08		42	2.7	1800	0	1:54:14	1:54:16	02		101
09		42	2.7	1800	0	2:06:16	2:06:18	02		99
10		42	2.7	1800	0	2:18:49	2:18:51	02		102
11		42	2.7	1800	0	2:30:51	2:30:53	02		105
12		42	2.7	1800	0	2:30:53	2:30:55	02		112
13		42	2.7	1800	0	2:42:55	2:42:58	03		116
14		42	2.7	1800	0	2:54:58	2:55:01	03		116
15		42	2.7	1800	0	3:07:01	3:07:04	03		120
16		42	2.7	1800	0	3:19:04	3:19:07	03		122

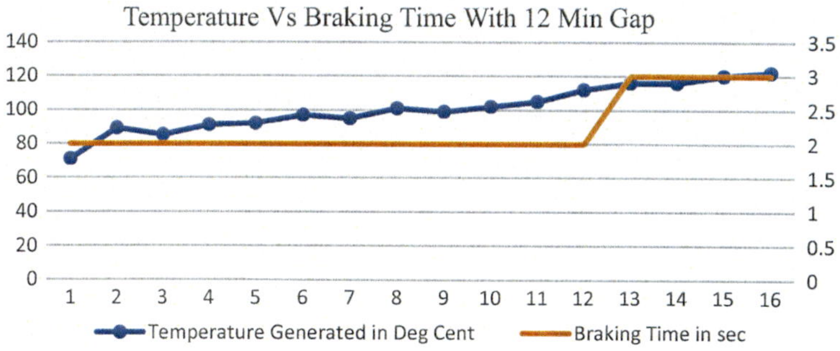

Fig. 9 Graphical representation of temperature rise and braking time

Fig. 10 Brake pads after wear experimentation

4 Results and Discussion

Experimental interpretation can be summarized as; with the increase of temperature friction pad wear also increases. During braking, heat generates and to dissipate the heat in the atmosphere positive cooling is essential. Presently in all most all vehicle's braking system is cooled with the help of natural air flow during vehicle movement. This cooling is termed as natural convection cooling process, from the surface of brake components. The radiation also observed in most of the cases when heat generation is more but researchers in this area neglected it while calculating heat transfer as it is small in quantity. The cooling time provided in second set of experimentation in above experimentations clearly indicates that, if sufficient cooling time is provided to cool the braking components, then the wear rate can be reduced drastically. When the cooling time was 1 min, then the wear of friction pads observed was 1.2 g, whereas when the cooling time was 12 min then the friction pad wear observed was 0.2–0.6 g. Generated heat greatly effects on the bonding properties of

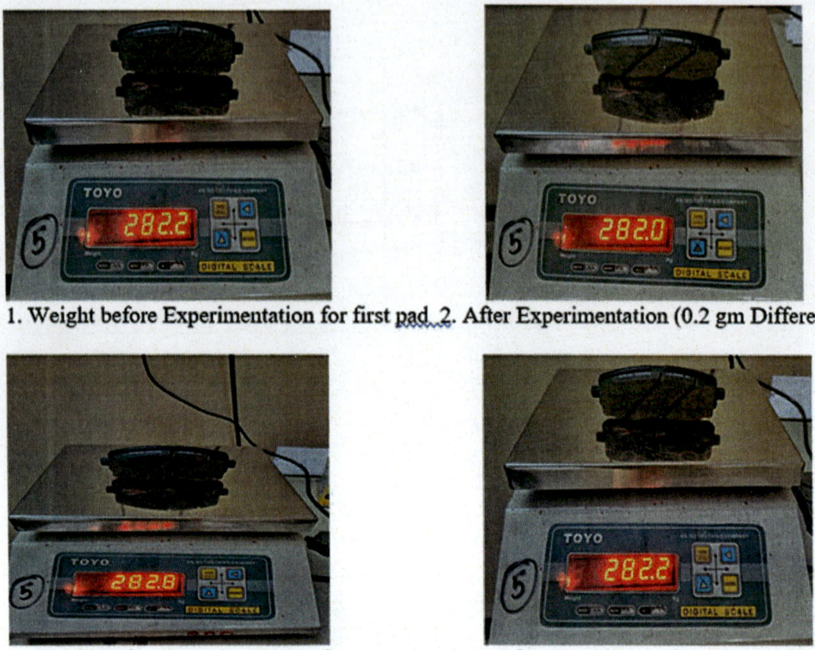

1. Weight before Experimentation for first pad 2. After Experimentation (0.2 gm Difference)

3. Weight before Experimentation for second pad 2. After Experimentation (0.6 gm diff.)

Fig. 11 Brake pads weight before and after wear experimentation

friction pad material, as the temperature increases the friction pad material particles were gets loosed and spreads in to atmosphere causing wear of pad and pollution in the atmosphere. Heating also effects on the braking time of vehicle. In first set of experimentation of 1 min gap braking time increases rapidly as compared to second set of 12 min gap and this leads in to brake fade, the major cause of accidents as shown in graphs 6 and 9.

5 Conclusion

Thermal behaviour at the interface between brake disc and pad is very complicated process to understand. Tribological changes during braking are atmospherically influenced and time dependent. Temperature generation during braking plays a very important role in smooth operation of brakes. During low speed of vehicle air circulation through the ventilated disc dissipates less amount of heat and also, during high speed because of the more air accumulation and opposing pressure there is less dissipation of heat [44]. Overall temperature effects on the operational performance of brakes. Brake fades with the rise in temperature, fading occurs due to loss of coefficient of friction of pads at high temperature and this has been proved

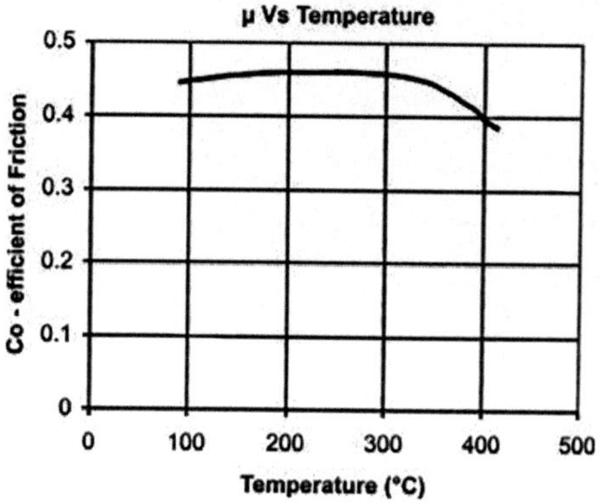

Fig. 12 Brake pads coefficient of friction versus temperature graph (COMPO Advic, Aurangabad, India)

experimentally. Brake pads are manufactured from the fine metallic particles, hence at high operating temperature these particles also lose their bonding from resin and spreads in to air at a higher rate in the form of brake wear; this has been proved from the above experimentations [45]. From the experimentations carried out by different researchers also it has been observed that the critical temperature for disc brake is in between 307 and 314 °C, at this temperature both brake fade and brake wear observed. The results also correlate with the literature made available by different friction pad manufacturers in the form of graph [46].

Graph in Fig. 12 shows off-market friction pad co-efficient of friction reduction with respect to temperature, in this graph after 300 °C coefficient of friction starts reducing and these results also nearer to experimentations carried out. For the modern vehicles, speed and engine power is high therefore, need long braking for stopping which may results in to increase in temperature requires stronger cooling system to avoid the thermal effect. From the experimentation conclusion can be drawn that the temperature increase during braking has effect on the brake pad wear rate and brake fade.

6 Scope for Future Work

Experimentations were carried out on the dynamometer setup have given the correlated results with the off-market friction pad manufacturers results. The same can be experimented on the road vehicles with the same off-market friction pad. The obtained practical results can be compared with setup results, and correction factor can be added to calculate the life and safety of vehicle. This is the need for today's upcoming fast vehicle global market. Experimentation on braking performance will result in reliability and customer safety in automobile market.

References

1. Day AJ, Tirovic M, Newcomb TP (1991) Thermal effects and pressure distribution in brakes. Proc IMechE 205:199–205
2. Belhocine A, Bouchetara M (2013) Temperature and thermal stresses of vehicles gray cast brake. J Appl Res Technol 11(5):674–682. https://doi.org/10.1016/s1665-6423(13)71575
3. Day AJ (2014) Book on braking of road vehicles. Butterworth- Heinemann Publ., Vol-02, pp 25–26 (imprint of Elsevier)
4. Eriksson M, Bergman F, Jacobson S (2002) On the nature of tribological contact in automotive brakes. Wear 252(1–2). Elsevier, pp 26–36. https://doi.org/10.1016/s0043-1648(01)00849-3
5. Bijwe J, Mujumdar NN, Sathapaty BK (2005) Influence of modified phenolic resins on the fade and recovery behavior of friction materials. Wear 259. Elsevier, pp 1068–1078. https://doi.org/10.1016/j.wear.2005.01.011
6. Wagner A, Spelsberg-Korspeter G, Hagedorn P (2012) Structural optimization of an asymmetric automotive brake with cooling channels to avoid sequel. J Sound Vib 333(7):1888–1898. https://doi.org/10.1016/j.jsv.2013.11.035
7. Loizou A, Sheng QH, Day AJ (2013) A fundamental study on the heat partition ratio of vehicle disc brake. J Heat Transf 135(12):121302-1-8. https://doi.org/10.1115/1.4024840
8. Gurunath PV, Bijwe J (2007) Friction and wear studies on brake pad materials based on newly developed resins. Wear. https://doi.org/10.1016/j.wear.2006.12.050
9. Kao TK et al (2000) Brake disc hot spotting and thermal judder: an experimental and finite element study. Int J Veh Des 03(¾):276–296
10. Hwang P, Wu X (2010) Investigation of temperature and thermal stress in ventilated disc brake based on 3D thermos mechanical coupling model. J Mech Sci Techno 24:81–84. https://doi.org/10.1007/s12206-009-1116-7
11. Duzgun M (2012) Investigation of thermo structural behaviours of different ventilation application. J Mech Sci Technol 26(1):235–240. https://doi.org/10.1007/s12206.011.0921-y
12. Palmer E, Mishra R, Fieldhouse J (2009) An optimization study of a multiple row pin vented brake disc to promote brake cooling using computational fluid dynamics. Proc Inst Mech Eng Part D J Automobile Eng 223(7):865–875. https://doi.org/10.1243/09544070jauto1053
13. Zang P et al (2019) Fade behaviour of copper-based brake pad during cyclic emergency braking at high speed and overload condition. Elsevier https://doi.org/10.1016/j.wear.2019.01.126
14. Athanassiou N, Olofson U, Wahlstrom J, Dizdar S (2021) Simulation of mechanical and thermal performance of laser cladded disc brake rotors. J Eng Tribol, pp 1–12. https://doi.org/10.1177/13506501211009102
15. Mathissen M et al (2018) A novel real- world braking cycle for studying brake wear particle. Wear. https://doi.org/10.1016/j.wear.2018.07.020
16. Yevtushenko A, Topczewska K, Kuciej M (2021) Analytical determination of the brake temperature mode during repetitive short-term braking. Materials 14. https://doi.org/10.3390/ma1408 1912
17. Gao F, Chew JW (2021) Evaluating and application of advanced CFD models for rotating disc flows. J Mech Eng Sci. https://doi.org/10.1177/09544062211013850
18. Sing G, Gupta M, Gill HS (2021) Evaluating the cooling performance of light commercial vehicle with water cooled engine systems- an approach beyond regulatory standards. Proc Mater Today. https://doi.org/10.1016/j.matpr.2021.09.126
19. Wei L, Choy YS, Cheung CS (2019) A study of brake contact pairs under different friction conditions with respect to characteristics of brake pad surface. J Tribol Int, pp 99–110. https://doi.org/10.1016/j.triboint.2019.05.016
20. Yan W-T, Feng WS, Xie G (2018) Role of vane configuration on the heat dissipation performance of ventilated brake discs. Appl Therm Eng. https://doi.org/10.1016/j.applthermaleng.2018.03.002
21. Alemani M, Talathi F et al (2009) Analysis of heat conduction in a disk brake system. Springer. https://doi.org/10.1007/s00231-009-0476-y

22. Cho H-J, Cho C-D (2008) A study of thermal and mechanical behaviour for the optimal design of disc brakes. J Automobile Eng Proc 222 (part D). https://doi.org/10.1234/09544070JAUT0722
23. Tirovic M et al (2019) Experimental investigation of the cooling characteristics of a monobloc cast iron brake disc with fingered hub. IME. https://doi.org/10.1177/0954407019838642
24. Cho H-J, Cho C-D (2008) A study of thermal and mechanical behaviour for the optimal design of disc brakes. J Automobile Eng Proc 222 (part D). https://doi.org/10.1234/09544070JAUT0722
25. Timur M et al (2014) Heat transfer of brake pad used in the autos after friction and examination of thermal tension analysis. https://doi.org/10.5755/j01.mech.20.1.6595
26. Agudelo CE, Ferro E (2005) Technical overview of brake performance testing for original equipment and aftermarket industries in the US and European markets. Link technical report FEV205-01. Link Technical Laboratories, Inc. pp 34–39
27. Adamowicz A (2016) Effect of convective cooling on temperature and thermal stresses in disk during repeated intermittent braking. J Frict Wear 37(2):107–112. https://doi.org/10.3103/S1068366616020021
28. Siregar R et al (2018) Experimental analysis in the test rig to detect temperature at the surface disc brake rotor using rubbing thermocouple. East-Eur Enterp Technol. https://doi.org/10.15587/1729-4061.2020.191949
29. Ali B et al (2013) Thermomechanical modelling of disc brake contact phenomena. FME Trans 41:59–65; 60 · VOL. 41, No 1
30. Matějka V et al (2017) On the running-in of brake pads and discs for dyno bench tests. Tribol Int. https://doi.org/10.1016/j.triboint.2017.06.008
31. Neis PD, Ferreira NF, da Silva FP (2014) Comparison between methods for measuring wear in brake friction material. Wear. https://doi.org/10.1016/j.wear.2014.08.004
32. Bijwe J, Mujumdar NN, Sathapaty BK (2005) Influence of modified phenolic resins on the fade and recovery behaviour of friction materials. Wear 259, Elsevier, pp 1068–1078. https://doi.org/10.1016/j.wear.2005.01.011
33. Balotin JG, Jeangb et al (2017) Analysis of the influence of temperature on the friction coefficient of friction materials. In: ABCM symposium series in mechatronics, vol 4, pp 898–906
34. Wei L, Choy YS, Cheung CS (2019) A study of brake contact pairs under different friction conditions with respect to characteristics of brake pad surfaces. Elsevier https://doi.org/10.1016/j.triboint.2019.05.016
35. Tirovic M, Topoaries S, Sherwood G (2021) Experimental Investigation of the cooling characteristics of a monoblock cast iron brake disc with fingerhub. J Automobile Eng 234(1):85–97. https://doi.org/10.1177/0954407019838642
36. Vdovin A, Gustaffson M, Sebben S (2018) A coupled approach for vehicle brake cooling performance simulation. Int J Thermal Sci, pp 257–266. https://doi.org/10.1016/j.ijthermalsci.2018.05.016
37. Talati F, Jalatifar S (2009) Analysis of heat calculation in a disc brake system. Heat Mass Transf. https://doi.org/10.1007/s00231-009-0476-y
38. Qi SH, Day AJ (2007) Investigation of disc/pad interface temperature in friction braking. Wear Elsevier, pp 505–513. 10.1016/j. wear. 2006.08.027
39. Kumbhar BK, Patil SR, Sawant SM (2017) A comparative study on automotive brake testing standards. J Inst Eng Service 98(4):527–531. https://doi.org/10.1007/s40032-016-0289-y
40. Surblys V, Sokolovskij E (2016) Research of the vehicle brake testing efficiency. Procedia Eng 134:452–458. https://doi.org/10.1016/j.proeng.2016.01.067
41. Yan HB et al (2016) Heat transfer enhancement by X-type lattice in ventilated brake disc. Int J Therm Sci 107:39–55
42. Laguna-Camacho JR et al (2015) A study of the wear mechanisms of disk and shoe brake pads. Eng Fail Anal 56:348–359
43. Wei L, Choy YS, Cheung CS (2019) A study of brake contact pairs under different friction conditions with respect to characteristics of brake pad surface. Tribol Int 138:99–110. https://doi.org/10.1016/j.triboint.2019.05.016

44. Filip P, Yun R, Lu Y (2010) Performance and evaluation of eco-friendly brake friction material. Trib Int 43:2010–2019. https://doi.org/10.1016/j.triboint.2010.05.001
45. Mathissen M et al (2018) A novel real-world braking cycle for studying brake wear particle emissions. Wear. https://doi.org/10.1016/j.wear.2018.07.020
46. Compo-Advic India Ltd., Aurangabad, Information Boucher, 2021, (off-market brake pad manufacturer in India)

A Fuzzy Multi-criteria Decision-Making Approach for Finding Energy-Efficient Building Model

Mohd Shahid, Masroof Ahamad, and Munawar Nawab Karimi

Nomenclature

U	Overall heat transfer coefficient
AHP	Analytic hierarchy process
CWENLiP	The criteria weight estimation using nonlinear programming
PIV	Proximity index value
FL	Fuzzy logic

1 Introduction

In recent years, energy problems have become a broad topic among researchers. Buildings are one of the significant energy consumption fields; building contributes approximately 40% of the world's total energy consumption and 33% of total greenhouse gas emissions [1, 2]. Energy consumption in the building can be minimized by giving importance to building retrofitting. Various researchers indicate that the energy efficiency of a building is affected by factors such as building design problems, energy conservation problems, and defective HVAC systems. To improve the energy performance of buildings, it is necessary to introduce several retrofit methods in the existing building. Mills et al. [3, 4] proposed that improving the operation of an existing building in the United States will attain 16% median energy saving. If the

M. Shahid (✉) · M. N. Karimi
Department of Mechanical Engineering, Faculty of Engineering and Technology, Jamia Millia Islamia, New Delhi 110025, India
e-mail: mohammadshahidmech@gmail.com

M. Ahamad
Mechanical Engineering Section, University Polytechnic, Jamia Millia Islamia, New Delhi 110025, India

A. Kumar et al. (eds.), *Recent Advances in Manufacturing and Thermal Engineering*, Lecture Notes in Mechanical Engineering, https://doi.org/10.1007/978-981-19-8517-1_4

retrofitting methods apply to the building envelope like mechanical equipment and lighting systems, there can be substantial energy savings will achieve. According to Lam [5] and Reddy [6], the total energy consumption in the building is affected by the factors, i.e., internal loads, temperature set-point, fenestration, and HVAC equipment efficiency. In India, there are two types of HVAC systems installed in the building, variable air volume (VAV) and conventional HVAC systems. VAV systems are more likely used in office buildings. To identify the influence of inputs on the building model, the parameters are divided into three parts such as internal loads, building envelope, and HVAC systems. For energy-efficient buildings, the selection of parameters plays an important role. Optimization techniques are to be used to find the optimal value of the parameters to reduce the energy consumption in the building. A wide variety of modeling techniques provide possibilities to determine the cooling and heating load of the building and energy consumption in recent years, such as simulation tools [7], ANN [8], and soft computing and fuzzy logic approaches [9]. Soft computing methods like fuzzy logic have become an innovative tool to reduce the total error in the attribute. Fuzzy logic is a well-growing method that describes the knowledge using the linguistic variables in a descriptive human-like manner [10]. The fuzzy systems have some set of rules which make them user-friendly, and this set of rules is derived from the quantitative description. Chibattoni et al. [11] calculate the energy consumption for the Italian residential building using fuzzy logic. They considered the model's domestic habits and occupancy activity, and the mean error was found as 0.52%. Kabak et al. [12] apply the multi-criteria decision-making approach to investigate the energy simulation tool BEP-TR. The multi-criteria decision-making approach is applied in many applications like the electrical vehicle industry [13, 14] and investigates the performance of a dual-fuel engine [15, 16]. Turhan [17] uses the soft computing method ANN and FL to estimate the heat load of the building. Their model used parameters like U-wall, total window area, building area/volume ratio, and total surface area. Few studies categorize alternative buildings based on their energy performance using multi-criteria decision-making, but no one has used the MCDM method based on the proximity index value. Most studies consider the criteria such as geometry shape, mechanical system, location, climate data, building envelope, and cogeneration for the building models, and the effect of each criterion on building energy demand was analyzed. Kajl et al. [18] developed a model based on fuzzy logic to find the correct output by post-processing the results obtained from neural networks.

An accurate determination of criteria weights significantly impacts the ranking of the alternative. Usually, the view regarding the criteria and their relative importance is drawn from groups of experts, and based on their relative importance; values are processed further to calculate criteria weights. Generally, criteria weights are estimated based on subjective and objective approaches [19]. In practice, most MCDM methods used subjective weights in which the analytical hierarchy process [20] was used in wide applications in most decision problems [21]. Many research articles [22–25] are currently based on applying the AHP method in different sectors, like construction. Also, there are several other methods for calculating optimal criteria weights like goal programming [26], logarithmic goal programming [27], fuzzy

programming [28], enhanced goal programming [29], and PCM data utilizing linear programming [30].

The study uses a new method based on the PSO algorithm to estimate the criteria weight. That criteria weight estimation using non-linear programming is used because of their more straight forward computation and gives a more efficient result than AHP. The tool is more reliable for criteria weight calculation which is essential for correct ranking. After that, the multi-criteria decision-making method is considered for ranking the alternative buildings based on their energy performance and identifying the best building model for the energy-efficient building among the alternatives. Multi-criteria decision-making method involves both qualitative and quantitative factors. These approaches are used to choose the probable optimal solution.

In this study, we analyze the different building models and identify the best building model for the energy-efficient buildings among the alternative models using a multi-criteria decision-making method. The study will identify the influence of inputs on the building model. It also provides the weight of each parameter and the optimal values to achieve an energy-efficient building model among the alternative building models. This paper compares two soft computing methods, i.e., AHP and CWENLiP, in determining the weights of each criterion of the building model. And also helps in identifying the best building model with the optimal parameter values.

2 Methodology

The proposed method is based on the proximity of alternatives from the best possible value by applying the Proximity Index value method [31]. The proximity index value is the difference between the normalized value of each alternative and the best available alternative. With due consideration of attribute weights, the proximity index values are linearly added for all the attributes or criteria to give each alternative model's overall proximity index value. After this, it determines the difference between the overall weighted normalized of alternatives from the best alternative.

2.1 AHP

The role of criteria weight in finding the overall proximity index value of alternative is significant. The weight of the parameters has a different consistency ratio. So to find an optimal solution, there must be the least error in the criteria weight. Therefore the weights of the criteria have been calculated using the two methods known as an AHP and criteria weight estimation using non-linear programming. AHP method [32] was developed by Thomas L. Saaty in 1970 and studied worldwide. This method is conveniently applied for deciding problem objectives and evaluating alternatives.

The AHP works by developing priorities for alternatives and the criteria used to judge the alternative.

2.2 Criteria Weight Estimation Using Non-linear Programming

Criteria weight estimation using nonlinear programming [33] is based on a nonlinear constrained optimization model. In this method, the weight of criteria is determined using the PSO algorithm; the PSO algorithm optimizes the weight of the parameters.

2.3 Formulation of the Decision Problem

Step 1: Define the Criterion and alternatives of building models, by defining these criteria and models; create the objective of the decision problem. Assign the different values of criteria for the alternative models.

Step 2: Define the Decision Matrix (DM).

Each row assigns to one alternative source of the decision matrix, and each column is assigned to the criterion.

Step 3: Construction of Normalization matrix of the data.

Since different criteria have different values with different units, hence normalize their values on the same scale. Vector normalization is applied to normalize these values as given in Eq. (1).

$$r_i = \frac{x_i}{\sqrt{\sum_{i=1}^{m} x_i^2}} \tag{1}$$

Step 4: Construction of Weighted normalized Decision Matrix.

The weight of each criterion is calculated by the AHP and CWENLiP methods. With the help of these weights, the weighted normalized decision matrix of alternative models corresponding to the criterion is calculated by Eq. (2) (Fig. 1).

$$v_j = w_j * r_i \tag{2}$$

Step 5: Determination of Weighted Proximity Index (WPI).

Identify the positive attribute and negative attributes. Among the Criterion for the alternative model, evaluate the proximity index of each alternative model. The weighted proximity index matrix is calculated using Eq. (3).

$$\begin{aligned} u_i &= v_{\max} - v_i \quad \text{For benefit attribute} \\ u_i &= v_i - v_{\min} \quad \text{For cost attribute} \end{aligned} \tag{3}$$

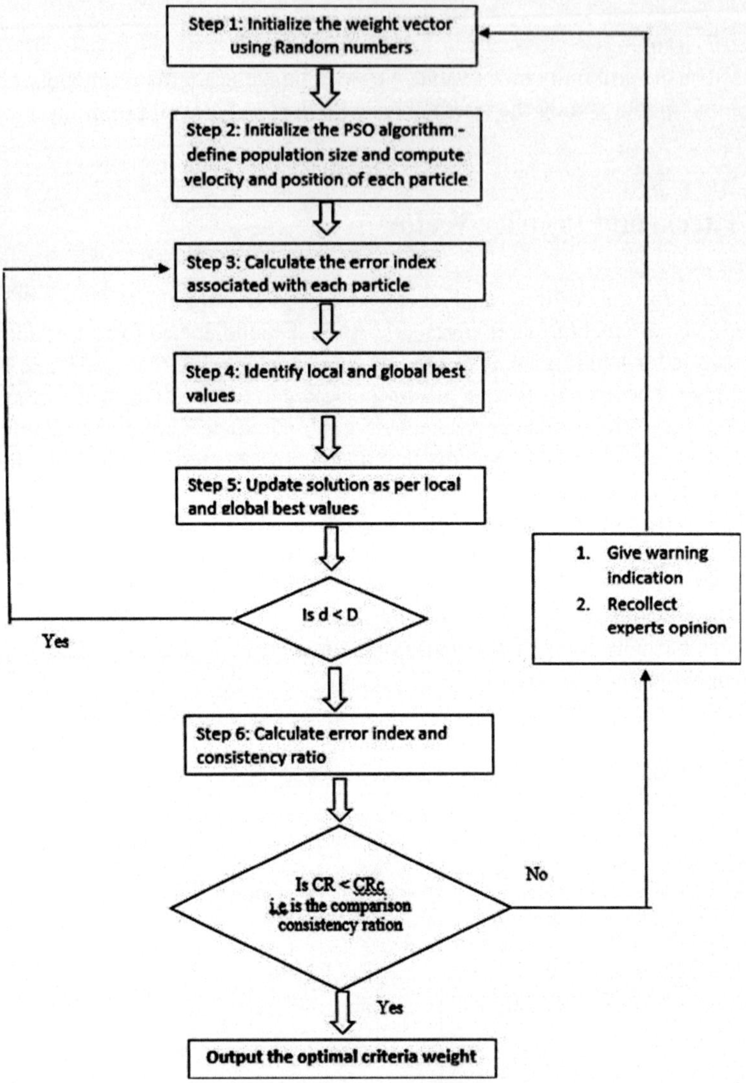

Fig. 1 Flow chart of CWENLiP for determining criteria weight

Step 6: evaluation of Overall Proximity Value.

Using the Eq. (4), the overall proximity value is calculated for each alternative model.

$$d_i = \sum_{j=1}^{n} u_j \tag{4}$$

Step 7: Ranking.

Based on the proximity index value, we are ranking the alternative models. Choose the optimal model among the alternatives with the least overall proximity value.

3 Criteria and Priority Vector

The criteria for the building analysis are determined using the studies and expert opinions like architecture and engineers. After the studies and expert opinion, the main criteria for building analysis are building envelope, internal loads, and HVAC system [34]. These main criteria are further categorized into their sub-parameters. The building envelope is categorized into five sub-parameters, internal load into three sub-parameters, and HVAC systems into five sub-parameters, as shown in Table 1. Finally, there are 13 parameters on which the building models are analyzed. Based on these 13 parameters, the optimum building model for energy-efficient is to be selected using the AHP, CWENLiP, and proximity index method.

The first step is to make the pairwise comparison matrix of the criteria. The consistency ratio (CR) value for these criteria is less than 0.1. Hence, the results are acceptable. The comparison matrix of the main criteria and their sub-criteria are shown in Tables 2, 3, 4, and 5.

Table 1 Criteria of building model

Main criteria	Sub-parameters				
Building envelope ($C1$)	External wall U value ($C11$) (W/m^{2}K)	Roof U value ($C12$) (W/mK)	Window U value ($C13$) (W/m^{2}K)	Shading coefficient ($C14$)	Window wall ratio ($C15$)
Internal loads ($C2$)	Lightning density ($C21$) (W/m^2)	Equipment density ($C22$) (W/m^2)	Occupant density ($C23$)	–	
HVAC system ($C3$)	Cooling set point ($C31$) (°C)	Fresh air ($C32$)	COP ($C33$)	Fan efficiency ($C34$)	Pump efficiency ($C35$)

Table 2 Comparison matrix of main criteria weights

	Building envelope $C1$	Internal loads $C2$	HVAC system $C3$
$C1$	1	5	8
$C2$	0.2	1	4
$C3$	0.125	0.25	1

Table 3 Comparison matrix of building envelope parameters weights

	C11	C12	C13	C14	C15
C11	1	7	3	2	4
C12	0.1428	1	0.5	0.333	0.25
C13	0.333	2	1	0.333	0.5
C14	0.5	3	3	1	0.333
C15	0.25	4	2	3	1

Table 4 Comparison matrix of internal-loads parameters weights

	C21	C22	C23
C21	1	2	4
C22	0.5	1	4
C23	0.25	0.25	1

Table 5 Comparison of HVAC system parameters weights

	C31	C32	C33	C34	C35
C31	1	3	0.1428	2	2
C32	0.333	1	0.125	0.333	0.333
C33	7	8	1	8	7
C34	0.5	3	0.125	1	0.5
C35	0.5	3	0.1428	2	1

4 Results

Energy used in the building is optimized by optimizing the building parameters like building envelope, internal loads, and HVAC system parameters. Building envelope parameters are further divided into five sub-parameters, i.e., External wall U value, Roof (U value), window (U value), shading coefficient, and window wall ratio. By selecting the optimal values of these parameters, energy consumption by the building will minimize. Using the technique AHP and CWENLiP, the weights of each main criteria parameter is calculated as shown in Fig. 2. Figure 2 shows building envelope has a high weightage parameter compared to other prominent factors, in the case of designing a new building, the building envelope plays a significant role.

Building envelope plays a significant role in minimizing the total energy consumption of the building. Most of the heat comes inside the building from the building envelope. Therefore, to minimize the energy consumption, the building envelope and its parameters are more important than internal loads and HVAC systems parameters. As shown in Fig. 2 of criteria weights, engineers and architectures focus more on the design of the building because internal load and HVAC system parameters depending

Fig. 2 Weights of the main criteria for the building

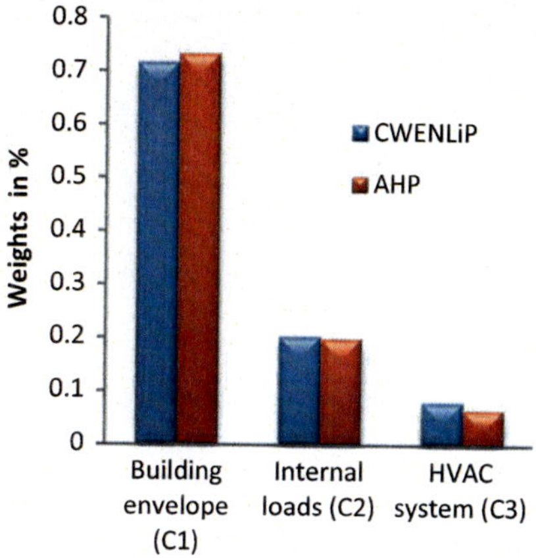

on the building design. When the inside and outside heat transfer reduces from the building, then the internal and HVAC loads reduce significantly. As shown in Fig. 2, after the building envelope parameters, the internal load significantly reduces the energy consumption inside the building.

In the building envelope, the study considers the factors such as U_{wall}, U_{roof}, U_{window}, shading coefficient, and WWR. These factors will focus on reducing energy consumption through the building envelope. From Fig. 3, U_{wall} has the highest criteria weight among the building envelope parameters, which shows U_{wall} plays a significant role in the building envelope. The U_{wall} factor affects the total annual electricity significantly, and engineers and architectures give more weightage than other factors because it has a significant influence on the electricity consumption of the building. After the U_{wall} factor, WWR and shading coefficient are considered significant factors in reducing the heat transfer from the building envelope. WWR affects the energy demand in the building and the daylighting inside the building. With the help of the shading coefficient, a significant amount of energy consumption in the building reduces; therefore, WWR and shading coefficient are considered significant factors in achieving energy-efficient building. As shown in Fig. 3, the weightage of the U_{roof} parameter is significantly less, indicating that U_{roof} is a less significant considered factor in the building envelope than the other building envelope parameters.

There are five sub-parameters among the building envelope; their weights are shown in Fig. 3. As shown in Fig. 3, the external wall U value plays a vital role in the energy-efficient building. After external wall, window wall ratio is a second important factor for the energy-efficient building. The roof U value has the least weightage among these parameters. Similarly, the internal load has three sub-parameters in which lightning density plays a significant role in energy-efficient building. The weights of each sub-parameter are shown in Fig. 4. As shown in Fig. 4, the Lightning density and equipment density is the essential factor in achieving the optimal model of the building. Lightning and equipment density parameters indicate the amount of energy produced within the building. So for an energy-efficient building, there should

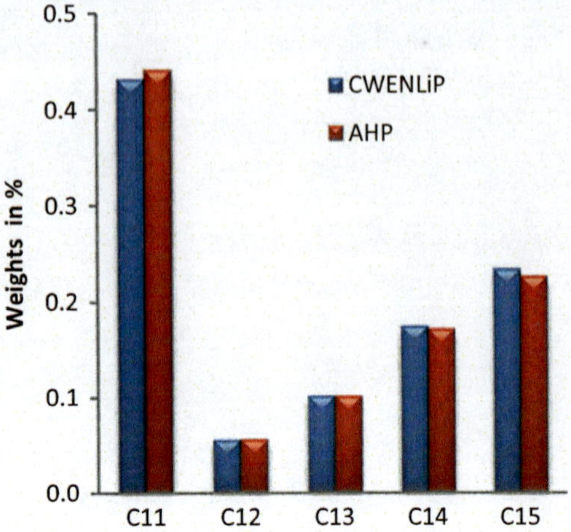

Fig. 3 Weights of building envelope parameters

be a minimum amount of energy produced in the building. As shown in Fig. 4, lighting density has the highest weightage among the other two factors for consideration in achieving energy-efficient buildings. When the parameters of internal loads, i.e., lighting equipment and occupant density, increase, the corresponding electricity used in the building will also increase. Lightning density influences the lighting electricity consumption and can also affect the electricity consumption of air conditioning or HVAC system because the lighting equipment gradually releases heat into the indoor environment of the building. A similar case with the equipment density; therefore, lighting density and equipment density significantly impact the energy demand of the building and are considered a high weightage factor among the internal load parameters.

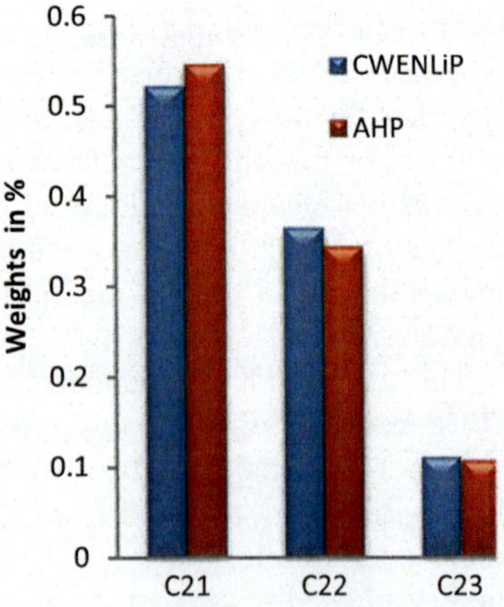

Fig. 4 Weights of the internal load parameters

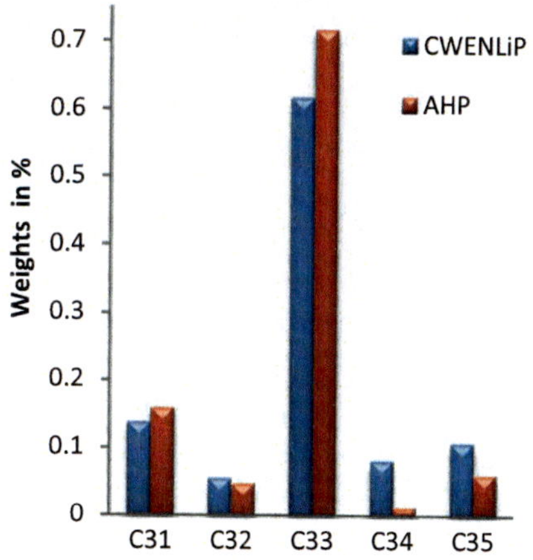

Fig. 5 Weights of the HVAC system parameters

The HVAC system has five sub-parameters; among these parameters, fan efficiency, pump efficiency, and fresh air are the least significant since their weights are the least, as shown in Fig. 5. In comparing other factors, fan efficiency and COP have approximately 7–10% difference in weight using AHP and CWENLiP. As shown in Fig. 5, COP has the highest weightage among the HVAC system parameters; it means COP is the most significant factor in achieving energy-efficient buildings.

In buildings, electricity consumption is affected by HVAC parameters. With the increase of fresh air, the electricity consumption of the air conditioning system increases, and with the increase of remaining HVAC parameters such as chiller COP, pump efficiency, cooling set-point, and Fan efficiency, electricity consumption of the air conditioning system decreases.

As shown in Fig. 5, the COP of the chiller has a higher weightage than other factors, which indicates COP is the most significant parameter for considering in an energy-efficient building. And the weights of fan efficiency and pump efficiency are significantly less compared to other factors, which indicates these factors are least considered in the energy-efficient building among the HVAC system parameters.

Based on the criteria weight analysis, the study analyzes that engineers and architectures considered the building envelope more significant when designing a building and before considering lighting equipment and HVAC system parameters. In addition, among all criteria, COP is the most considered factor in achieving energy-efficient buildings.

The weights of the parameters are calculated using the AHP and CWENLiP. There may be a difference in the weights of the parameters by calculating using these two methods. Therefore, we rank the alternative models using the weightage of the parameters obtained from these two methods. The PIV method is used to rank models to select the best optimal building model. The decision matrix used for the PIV method is shown in Table 6. The thirteen factors have different values for the different building designs. Table 6 shows the values of each parameter for each building model. And units of each parameter are specified in Table 1.

Table 6 Decision matrix

	U_{wall}	U_{roof}	U_{window}	SC	WWR	Lightning density
	*C*1	*C*2	*C*3	*C*4	*C*5	*C*6
MODEL 1	0.3	0.15	1.5	0.3	0.07	5
MODEL 2	0.4	0.2	2	0.35	0.15	7
MODEL 3	0.6	0.25	2.5	0.4	0.2	10
MODEL 4	0.8	0.3	3	0.45	0.25	15
MODEL 5	1.5	0.35	3.5	0.5	0.3	20

	Equipment	Occupant density	Cooling pt	Fresh air	Cop	Fan eff.	Pump eff
	*C*7	*C*8	*C*9	*C*10	*C*11	*C*12	*C*13
MODEL 1	5	5	20	20	3	0.5	0.5
MODEL 2	10	8	21	23	3.5	0.55	0.55
MODEL 3	15	12	22	26	4	0.6	0.6
MODEL 4	20	16	23	29	4.5	0.65	0.65
MODEL 5	25	20	24	32	5	0.7	0.7

The normalized decision Matrix for the alternative models using the Eq. (1) is shown in Table 7.

The weighted normalized decision matrix for the alternative models using Eq. (2) is shown in Tables 8 and 9.

Table 7 Normalized decision matrix

	*C*1	*C*2	*C*3	*C*4	*C*5	*C*6
MODEL 1	0.160356745	0.258199	0.258199	0.330289	0.149274	0.176887284
MODEL 2	0.213808994	0.344265	0.344265	0.385337	0.319874	0.247642198
MODEL 3	0.32071349	0.430331	0.430331	0.440386	0.426498	0.353774569
MODEL 4	0.427617987	0.516398	0.516398	0.495434	0.533123	0.530661853
MODEL 5	0.801783726	0.602464	0.602464	0.550482	0.639748	0.707549138

	C7	C8	C9	C10	C11	C12	C13
MODEL 1	0.13483997	0.167694618	0.40572	0.33952	0.330289	0.370117	0.370117
MODEL 2	0.26967994	0.268311388	0.426006	0.390448	0.385337	0.407128	0.407128
MODEL 3	0.40451992	0.402467083	0.446292	0.441376	0.440386	0.44414	0.44414
MODEL 4	0.53935989	0.536622777	0.466578	0.492304	0.495434	0.481152	0.481152
MODEL 5	0.67419986	0.670778471	0.486864	0.543232	0.550482	0.518163	0.518163

Table 8 Weighted normalized decision matrix using AHP weights

	*C*1	*C*2	*C*3	*C*4	*C*5	*C*6	*C*7
MODEL 1	0.05193955	0.010741	0.019313	0.041616	0.024824	0.019245337	0.00923654
MODEL 2	0.069252733	0.014321	0.025751	0.048553	0.053195	0.026943471	0.01847308
MODEL 3	0.1038791	0.017902	0.032189	0.055489	0.070927	0.038490673	0.02770961
MODEL 4	0.138505466	0.021482	0.038627	0.062425	0.088658	0.05773601	0.03694615
MODEL 5	0.259697749	0.025063	0.045064	0.069361	0.10639	0.076981346	0.04618269

	*C*8	*c*9	*C*10	*C*11	*C*12	*C*13
MODEL 1	0.003620527	0.004301	0.001086	0.015854	0.000329	0.001521
MODEL 2	0.005792843	0.004516	0.001249	0.018496	0.000362	0.001673
MODEL 3	0.008689264	0.004731	0.001412	0.021139	0.000395	0.001825
MODEL 4	0.011585686	0.004946	0.001575	0.023781	0.000428	0.001978
MODEL 5	0.014482107	0.005161	0.001738	0.026423	0.000461	0.00213

Table 9 Weighted normalized decision matrix using CWENLiP weights

WNDM	*C*1	*C*2	*C*3	*C*4	*C*5	C6	*C*7
MODEL 1	0.049726627	0.01038	0.018952	0.041385	0.025048	0.018696986	0.00996467
MODEL 2	0.066302169	0.013839	0.025269	0.048283	0.053675	0.02617578	0.01992935
MODEL 3	0.099453253	0.017299	0.031586	0.05518	0.071566	0.037393972	0.02989402
MODEL 4	0.132604338	0.020759	0.037904	0.062078	0.089458	0.056090958	0.0398587
MODEL 5	0.248633133	0.024219	0.044221	0.068975	0.10735	0.074787944	0.04982337

WNDM	*C*8	*C*9	*C*10	*C*11	*C*12	*C*13
MODEL 1	0.003806668	0.004503	0.001528	0.016448	0.002406	0.003183
MODEL 2	0.006090669	0.004729	0.001757	0.01919	0.002646	0.003501
MODEL 3	0.009136003	0.004954	0.001986	0.021931	0.002887	0.00382
MODEL 4	0.012181337	0.005179	0.002215	0.024673	0.003127	0.004138
MODEL 5	0.015226671	0.005404	0.002445	0.027414	0.003368	0.004456

Table 10 Overall proximity index and rank matrix

Building models	Proximity index (AHP)	Proximity index (CWENLiP)	Rank (AHP)	Rank (CWENLiP)
MODEL 1	0.0613	0.0663	1	1
MODEL 2	0.1157	0.1196	2	2
MODEL 3	0.1812	0.1818	3	3
MODEL 4	0.2544	0.2515	4	4
MODEL 5	0.4142	0.4040	5	5

The overall proximity index value of the five different models is calculated using Eq. (4), and their ranking is shown in Table 10.

The proximity index value is calculated for the five different models of the building. The alternatives were ranked using the overall proximity value method, and the optimum building model was obtained. As shown in Table 10, building model 1 has the least proximity index value among the five-building models. Therefore the optimal model for energy-efficient buildings is model 1.

5 Conclusion

This paper analyzes the building parameters and their significance by calculating their weightage. The weight of parameters is calculated and finds that among the three main building criteria, the building envelope has the highest weightage. The results shows that the weights of building envelope, internal loads and hvac criteria are 0.7169, 0.2024, and 0.0807.Among the sub-parameter of the building envelope, the external wall (U-value) parameter has a high weightage equal to 0.4326 which shows that it is a significant and most considered factor in energy-efficient buildings. Similarly, among the sub-parameters of HVAC systems and internal loads, COP and lightning density parameters have the highest weightage, which indicates their significance in energy-efficient buildings. Results show that in the building envelope most significant factor is the overall heat transfer coefficient of the wall, followed by WWR, shading coefficient, window (U-value), and roof (U-value).

Similarly, based on the results, COP is the most significant factor, and the cooling set point is the second most significant factor for minimizing the energy consumption in the building. Energy consumption in the building is decreased by selecting appropriate lightning density inside the building because lightning density plays a vital role in energy-efficient buildings. Fan efficiency is the most negligible factor in energy-efficient building design. And based on the significance of all parameters, the optimal model is identified by the MCDM technique. The optimal parameters can be applied for the new building under construction as well as the pre-existing buildings. In pre-existing buildings, we can focus more on internal loads and HVAC systems. By providing the optimal values of the parameters of the HVAC system and internal

loads, energy consumption in the building will reduce and achieve energy-efficient building status. But for the new buildings which are to be under construction, engineers and architect can give more importance to building envelope parameters to achieve energy-efficient buildings. And take the parameters based on weights into consideration. This study will help the engineer and architect achieve energy-efficient buildings by focusing more on those parameters which have high weightage and are significant to the energy demand. The study's limitation is that the building's energy consumption is not calculated. In the future studies, more soft computing methods and fuzzy logic techniques can be applied to buildings to achieve energy-efficient buildings.

References

1. Sbci U (2009) Buildings and climate change: a summary for decision-makers. United Nations Environmental Programme, Sustainable Buildings, and Climate Initiative, Paris, pp 1–62
2. Ibn-Mohammed T, Greenough R, Taylor S, Ozawa-Meida L, Acquaye A (2013) Operational vs. embodied emissions in buildings—a review of current trends. Energy Build 66:232–245
3. Mills E, Friedman H, Powell T, Bourassa N, Claridge D, Haasl T, Piette MA (2004) The cost-effectiveness of commercial-buildings commissioning. LBNL-56637
4. Mills E (2011) Building commissioning: a golden opportunity for reducing energy costs and greenhouse gas emissions in the United States. Energ Effi 4(2):145–173
5. Lam JC, Hui SC (1996) Sensitivity analysis of energy performance of office buildings. Build Environ 31:27–39
6. Reddy TA, Maor I, Jian S, Panjapornporn C (2006) Procedures for reconciling computer-calculated results with measured energy data. ASHRAE Research Project
7. Manfren M, Aste N, Moshksar R (2013) Calibration and uncertainty analysis for computer models-A meta-model based approach for integrated building energy simulation. Appl Energy 103:627–641
8. Turhan C, Kazanasmaz T, Uygun IE, Ekmen KE, Akkurt GG (2014) Comparative study of a building energy performance software (KEP-IYTE-ESS) and ANN-based building heat load estimation. Energy Build 85:115–125
9. Kazanasmaz T (2013) Fuzzy logic model to classify effectiveness of daylighting in an office with a movable blind system. Build Environ 69:22–34
10. Tayfur G, Özdemir S, Singh PV (2003) Fuzzy logic algorithm for runoff-induced sediment transport from bare soil surfaces. Adv Water Resourc 26:1249–1256
11. Chiabottoni L, Grisostomi M, Ippoliti G, Longhi S (2014) Fuzzy logic home energy consumption modeling for residential photovoltaic plant sizing in the new Italian scenario. Energy 74:359–367
12. Kabak M, Köse E, Kırılmaz O, Burmaoğlu S (2014) A fuzzy multi-criteria decision-making approach to access building energy performance. Energy Build 72:382–389
13. Pal K, Bahadur Singh L, Kumar S (2021) Selection of a vehicle using multi-attribute decision making. In: Kumar A, Pal A, Kachhwaha SS, Jain PK (eds) Recent advances in mechanical engineering. Lecture notes in mechanical engineering. Springer, Singapore. https://doi.org/10.1007/978-981-15-9678-0_92
14. Kumar S, Pal A (2021) Challenges of battery production: a case study of electrical vehicles in India. In: Kumar A, Pal A, Kachhwaha SS, Jain PK (eds) Recent advances in mechanical engineering. Lecture notes in mechanical engineering. Springer, Singapore. https://doi.org/10.1007/978-981-15-9678-0_94

15. Sharma P, Sahoo BB (2022) Precise prediction of performance and emission of a waste derived Biogas-Biodiesel powered dual-fuel engine using modern ensemble boosted regression tree: a critique to artificial neural network. Fuel 321:124131
16. Said Z, Cakmak NK, Sharma P, Sundar LS, Inayat A, Keklikcioglu O, Li C (2022) Synthesis, stability, density, viscosity of ethylene glycol-based ternary hybrid nanofluids: experimental investigations and model-prediction using modern machine learning techniques. Powder Technol 400:117190
17. Turhan C, Kazanasmaz T, Gökçen Akkurt G (2017) Performance indices of soft computing models to predict the heat load of buildings in terms of architectural indicators. J Therm Eng 3(4):1358–1374. https://doi.org/10.18186/journal-of-thermal-engineering.330180
18. Kajl S, Roberge MA, Lamarche L, Malinovski P (1997) Evaluation of building energy consumption based on fuzzy logic and neural network applications. In: Proceedings of CLIMA 2000 conference, 264–274
19. Vinogradova I, Podvezko V, Zavadskas EK (2018) The recalculation of the weights of criteria in MCDM methods using the Bayes approach. Symmetry 10(6):205
20. Saaty TL (1990) How to make a decision: the analytic hierarchy process. Eur J Oper Res 48:9–26
21. Zavadskas EK et al (2016) Hybrid multiple criteria decision-making methods: a review of applications for sustainability issues. Econ Res-Ekonomska Istraživanja 29(1):857–887
22. Zahedi F (1986) The analytic hierarchy process—a survey of the method and its applications. Interfaces 16(4): 96–108
23. Sipahi S, Timor M (2010) The analytic hierarchy process and analytic network process: an overview of applications. Manag Decis
24. Vaidya OS, Kumar S (2006) Analytic hierarchy process: an overview of applications. Eur J Oper Res 169(1):1–29
25. Subramanian N, Ramanathan R (2012) A review of applications of analytic hierarchy process in operations management. Int J Prod Econ 138(2):215–241
26. Bryson N (1995) A goal programming method for generating priority vectors. J Oper Res Soc 46(5):641–648
27. Wang Y-M, Yang J-B, Xu D-L (2005) A two-stage logarithmic goal programming method for generating weights from interval comparison matrices. Fuzzy Sets Syst 152(3):475–498
28. Mikhailov L (2000) A fuzzy programming method for deriving priorities in the analytic hierarchy process. J Oper Res Soc 51(3):341–349
29. Lin C-C (2006) An enhanced goal programming method for generating priority vectors. J Oper Res Soc 57(12):1491–1496
30. Rezaei J (2016) Best-worst multi-criteria decision-making method: some properties and a linear model. Omega 64:126–130
31. Mufazzal S, Muzakkir SM (2018) A new multi-criterion decision-making (MCDM) method based on proximity indexed value for minimizing rank reversals. Comput Ind Eng 119:427–438
32. Saaty RW (1987) The analytic hierarchy process—what it is and how it is used. Math Modell 9(3–5):161–176
33. Sameera M, Sarfaraz M, Khan NZ, Muzakkir SM, Khan ZA (2021) Towards minimization of overall inconsistency involved in criteria weights for improved decision making. Appl Soft Comput 100:106936
34. Yang T, Pan Y, Mao J, Wang Y, Huang Z (2016) An automated optimization method for calibrating building energy simulation models with measured data: orientation and a case study. Appl Energy 179:1220–1231

A Theoretical Thermodynamic Analysis of R1234yf/CO_2 Cascade Refrigeration System

Ayan Ghosh, Aditya Sharma, Bharat Varshney, Chirag, and Pawan Kumar Kashyap

Nomenclature

$\dot{m}$	Mass flow rate (kg s^{-1})
h	Enthalpy (kJ kg K^{-1})
s	Entropy (kJ kg^{-1} K^{-1})
P	Pressure (MPa)
Q	Heat transfer rate
T	Temperature (K)
W	Work done (Kg m^2 s^{-2})
COP	Coefficient of performance
EL	Exergy Loss (kJ)
EES	Engineering equation solver
GWP	Global warming potential
HFO	Hydrofluoroolefin
HTC	High temperature cycle
LTC	Low temperature cycle
ODP	Ozone depletion potential
TEWI	Total equivalent warming impact

Greek symbols

η	Efficiency

A. Ghosh (✉) · A. Sharma · B. Varshney · Chirag · P. K. Kashyap
JSS Academy of Technical Education, Noida, India
e-mail: ag21072000@gmail.com

P. K. Kashyap
e-mail: pawankashyap@jssaten.ac.in

A. Kumar et al. (eds.), *Recent Advances in Manufacturing and Thermal Engineering*, Lecture Notes in Mechanical Engineering, https://doi.org/10.1007/978-981-19-8517-1_5

$\sum$	Summation

Subscripts

eva	Evaporator
cond	Condenser
H	High temperature circuit
L	Low temperature circuit
F	Cooling space
ex	Exergetic
amb	Ambient
casc	Cascade condensor
EVH	Expansion valve at HTC
EVL	Expansion valve at LTC

1 Introduction

Many industrial aspects such as food storage, liquefaction vapor from petroleum and natural gasses, etc., require a low temperature refrigerant at room temperature from 243 to 193 K. Single performance stage systems are good as long as evaporator with condenser temperature difference is small. So, for low temperature range the cascade system can be a suitable option when using the right refrigerants on HTC and LTC.

Choosing refrigerants with the cascade system is highly dependent on the temperature of what refrigerant is required in a particular system. Besides the features of heat transfer, safety, material and to dry the cooler coil of the refrigerator should be considered [1]. In a cascade system, each region is different in the refrigerator, low temperature units are used gradually refrigerators for low-boiling point, i.e., a circuit with a low temperature uses low-boiling point refrigerants like CO_2 while high temperature circuit uses a relatively high boiling point refrigerant such as R1234yf. For the past three decades, the R134a refrigerant has been used as an effective liquid in many areas such as space heating and cooling, food refrigerator, and hot water production. Research has shown that R134a is relatively common in refrigeration systems (with cooler temperatures of 273–288 K). Although R134a doesn't have the characteristic to deplete ODP, but has the GWP of 1430 which advances the greenhouse effect. In 2006, Act No. 842/2006 the use of low-cost refrigerants GWP was sanctioned [2]. Therefore, the R134a refrigerator (and other liquids with higher GWP as R410A and R407C) is deliberated to phased out in Europe and replaced with R1234yf refrigerators. Following is a list of refrigeration systems research done by other authors previously with combinations of CO_2 and other refrigerants to find an optimum combination (Table 1).

Table 1 Other cascade refrigeration systems combinations researched on

Author	Refrigerants	T_{eva} (K)	T_{cond} (K)	COP
Yilmaz [3]	CO_2/R404A	248	298	2.146
Tripathy [4]	CO_2/NH_3	223	303	2.01
Bellos [5]	CO_2/CO_2	223	313	1.05
Mançuhana [6]	CO_2/R134A	253	318	2.03
Mançuhana [6]	CO_2/R152A	253	318	2.15

Table 2 Refrigerant properties from [8]

Refrigerants	ODP	GWP	Critical pressure (MPa)	Critical temperature (K)
CO_2	0	1	7.1	304.45
R1234yf	0	4	3.3	368

Thermo-physical properties of R744 and R1234yf are calculated using engineering equation solver (EES). Chief trait is high thermodynamic correctness in conjunction with a freight forwarding website provided for many items in a way that makes it a good software to solve equations. The cycle is modeled on application of maximum mass and energy balance equation for each individual cycle process [7] (Table 2).

In this study, thermodynamic analysis was conducted for a cascade refrigeration system below the sub-critical category where CO2 and R1234yf are refrigerants at low and high temperature cycles, respectively. The best working conditions to increase COP and reduce system damage are being tested.

2 Cascade Refrigeration System

The system we use comprises of low as well as high temperature cycles, in which CO2 and R1234yf are used as refrigerants, respectively. Figure 1 shows a detailed diagram of a two-phase refrigeration system made with EES software that we used to describe our model. The HTC consists of a R1234yf compressor, a cascade condenser equipped with a heat exchanger evaporator, a condenser and an expansion valve. On the contrary, the low temperature cycle comprises of the same components as in the HTC but the compressor and the cascade condenser are replaced by an evaporator and a CO2 compressor. Heat exchange between cycles occur with the help of a cascade heat exchanger. In our cascade refrigeration system, CO2 condenses and R1234yf evaporates to provide heat transfer from the CO2 side to the R1234yf side.

In a cascade refrigeration system, there are various temperature zones, namely, the evaporator temperature T_e, the condenser temperature T_c and the cascade condenser temperature T_{casc}. The differentiation between cooling space and evaporating temperature is kept to be at 5 K. Variation of these factors above will result in variation of

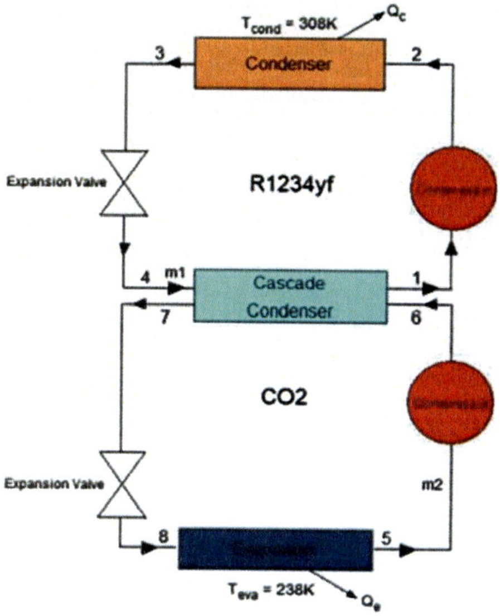

Fig. 1 Schematic diagram of R1234yf-CO_2 cascade refrigeration system

various performance factors such as COP, exergetic efficiency, and total exergy losses which shall be discussed further in this paper.

3 Thermodynamic Equations

Thermodynamic Model

A thermodynamic model of our cascade system is created that go after Ist and IInd laws of thermodynamics. Mass, energy, and exergy balance equations are written for low and high temperature cycles. The COP and exergetic efficiency are then calculated under different operating conditions [3].

Subsequent assumptions have been considered while defining the two-stage cascade refrigeration to conduct the thermodynamic analysis of our system.

- All components are presumed to operate in a steady form. Therefore, there will be negligible changes in potential as well as kinetic energy in the system.
- There will be insignificant pressure as well as heat losses in all components.
- Isenthalpic expansion of refrigerants will take place in expansion valves in both circuits.
- The cascade heat exchanger efficiency is presumed to be 90%.
- The compressors of the high and low temperature circuit are considered to be adiabatic but non-isentropic, and their isentropic efficiency can be expressed as a pressure ratio role.
- Condenser and cascade condenser outlet conditions are in subcooled condition and evaporator outlet condition is in superheated condition [4].

Via above presumptions, a thermodynamic analysis is performed by creating energy and exergy equations to determine the work input of the compressor of both circuits, the heat transfer rate of the evaporator, condenser, cascade condenser, also exergy loss on all components.

3.1 Energy Analysis

Mass Balance

$$\sum_{\text{in}} \dot{m} = \sum_{\text{out}} \dot{m} \tag{1}$$

Energy Balance

$$\dot{Q} - \dot{W} + \sum_{\text{in}} \dot{m}h - \sum_{\text{out}} \dot{m}h = 0 \tag{2}$$

Work done by compressor of High Temperature Circuit,

$$W_H = m_H(h_2 - h_1) \tag{3}$$

Work done by compressor of Low Temperature Circuit,

$$W_L = m_L(h_6 - h_5) \tag{4}$$

Heat Transfer in Evaporator,

$$Q_{\text{eva}} = m_L(h_5 - h_8) \tag{5}$$

Heat Transfer in Condenser,

$$Q_{\text{cond}} = m_H(h_3 - h_2) \tag{6}$$

High Temperature Circuit Throttling,

$$h_4 = h_3 \tag{7}$$

Low Temperature Circuit Throttling,

$$h_8 = h_7 \tag{8}$$

Heat Transfer in Cascade Condenser,

$$Q_{\text{he}} = m_L(h_6 - h_7) \tag{9}$$

$$Q_{\text{he}} = m_H(h_1 - h_4)\eta_{\text{he}} \tag{10}$$

COP of High Temperature Circuit,

$$\text{COP}_H = m_H(h_1 - h_4)/W_H \tag{11}$$

COP of Low Temperature Circuit,

$$\text{COP}_L = m_L(h_5 - h_8)/W_L \tag{12}$$

COP of system,

$$\text{COP} = m_L(h_5 - h_8)/(W_L + W_H) \tag{13}$$

3.2 Exergy Analysis

Exergy Lost at compressor of High Temperature Circuit,

$$\text{EL}_H = W_H - m_H(h_2 - h_1 - T_{\text{amb}}(s_2 - s_1)) \tag{14}$$

Exergy Lost at compressor of Low Temperature Circuit,

$$\text{EL}_L = W_L - m_L(h_6 - h_5 - T_{\text{amb}}(s_6 - s_5)) \tag{15}$$

Exergy Lost at evaporator,

$$\text{EL}_{\text{eva}} = Q_{\text{eva}}(1 - (T_{\text{amb}}/TF)) + m_L(h_8 - h_5 - T_{\text{amb}}(s_8 - s_5)) \tag{16}$$

Exergy Lost at condenser,

$$\text{EL}_{\text{cond}} = m_H(h_2 - h_3 - T_{\text{amb}}(s_2 - s_3)) \tag{17}$$

Exergy Lost at Expansion Valve of High Temperature Circuit,

$$\text{EL}_{\text{EVH}} = m_H(h_3 - h_4 - T_{\text{amb}}(s_3 - s_4)) \tag{18}$$

Exergy Lost at Expansion Valve of Low Temperature Circuit,

$$E_{\text{LEVL}} = m_L(h_7 - h_8 - T_{\text{amb}}(s_7 - s_8)) \quad (19)$$

Exergy Lost at Cascade Condenser,

$$\text{EL}_{\text{casc}} = m_H((h_4 - h_1) - T_{\text{amb}}(s_4 - s_1)) - m_L((h_7 - h_6) - T_{\text{amb}}(s_7 - s_6)) \quad (20)$$

Total Exergy Lost,

$$\text{EL}_{\text{total}} = \text{EL}_H + \text{EL}_L + \text{EL}_{\text{eva}} + \text{EL}_{\text{cond}} + \text{EL}_{\text{EVH}} + \text{EL}_{\text{EVL}} + \text{EL}_{\text{casc}} \quad (21)$$

Exergetic Efficiency,

$$\eta_{\text{ex}} = Q_{\text{eva}}((T_{\text{amb}}/T_{\text{eva}}) - 1)/(W_H + W_L) \quad (22)$$

4 Result

The thermodynamic model established above is defined using the software EES, which is used to conduct the energy analysis to find out the COP of system and exergy analysis as well find exergy losses of components. These performance and efficiency characteristics are computed using EES and then plotted for several conditions. Different operating conditions are taken into account to plot different case studies on graph and then study their characteristics. The evaporator temperature is varied from 253 to 223 K, the condensing temperature is varied from 293 to 323 K. The intermediate temperature at the cascade condenser is set to be at 268 K. The difference in temperature between the cooling space temperature (T_f) and the evaporator temperature is set to be 2 K. Several curves are plotted down below which are obtained after the variation of these factors in EES.

4.1 The Influence of Evaporator Temperature on COP of System

This case is demonstrated in Fig. 2. In each of the curved lines, evaporator temperature is varied at a fixed condenser temperature which gives different COP values for the system. Figure 2 shows that the COP of the system increases with rise in evaporator temperature This is because the pressure ratio in each phase reduces with evaporator temperature and this results in an increment in COP of the system [9].

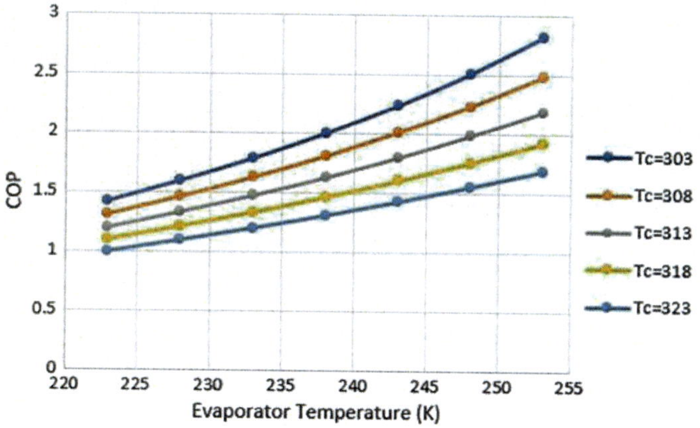

Fig. 2 Variation of COP with evaporator temperature

4.2 *The Influence of Condenser Temperature on COP of the System*

The demonstration is highlighted in Fig. 3. In each of the curved lines condenser temperature is varied at a fixed evaporator temperature which gives different COP values for the system. The figure shows that with increasing condenser temperature, the COP of the system decreases [6].

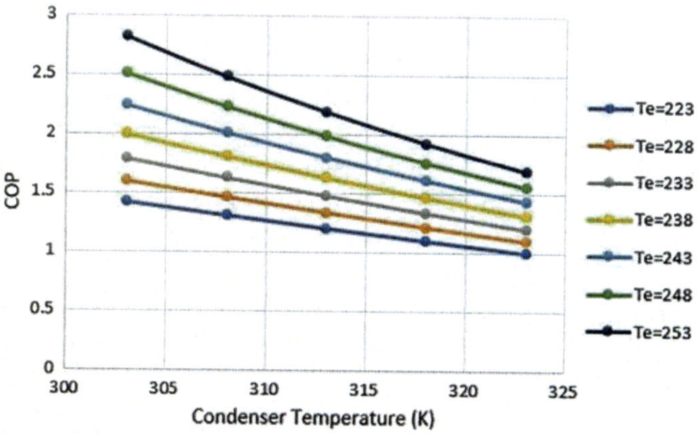

Fig. 3 Variation of COP with condenser temperature

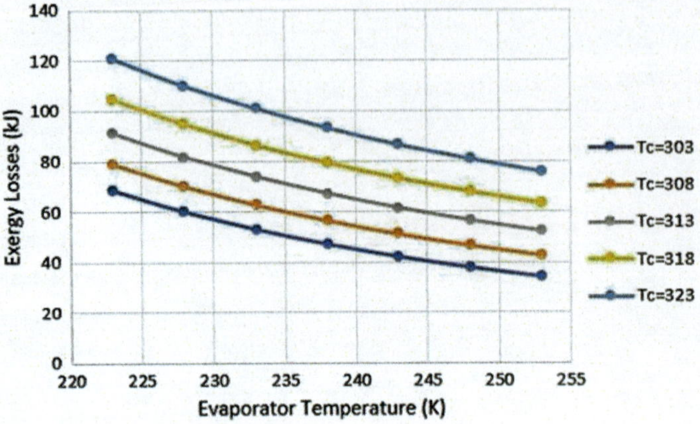

Fig. 4 Variation of exergy losses with evaporator temperature

4.3 *The Influence of Evaporator Temperature on Total Exergy Losses*

The facts have been demonstrated in Fig. 4. In each of the curved lines evaporator temperature is varied at a fixed condenser temperature which gives different exergy loss values for the system. It is inferred that as evaporator temperature increases, total exergy destruction decreases. This is because with increment of evaporator temperature the difference in temperature between the evaporator and the cooling space reduces which results in reduction of exergy losses in the evaporator and thus the total exergy losses as well [9].

4.4 *The Influence of Condenser Temperature on Total Exergy Losses*

Illustration of the following has been demonstrated in Fig. 5. In each of the curved lines condenser temperature is varied at a fixed evaporator temperature which gives different exergy loss values for the system. It depicts that with the increase in condenser temperature, the total exergy losses increase [9].

4.5 *The Influence of Evaporator Temperature on Exergetic Efficiency*

Representation of the following is showcased in Fig. 6. In each of the curved lines evaporator temperature is varied at a fixed condenser temperature which provides us

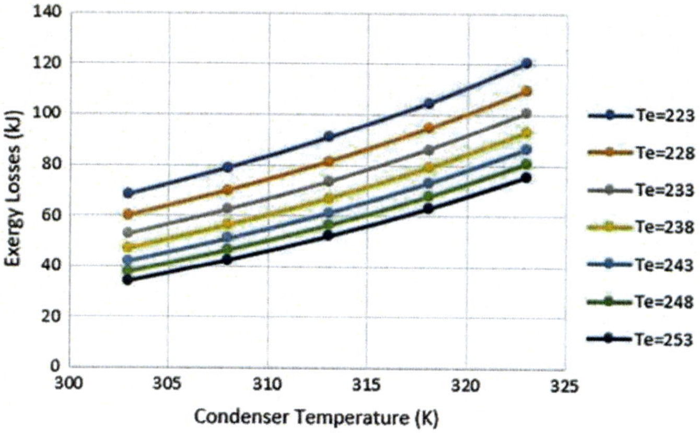

Fig. 5 Variation of total exergy losses with condenser temperature

with different exergetic efficiency for the system. Exergetic efficiency was observed to increase with increasing evaporation temperature but after a certain point it starts decreasing. The decrement occurs at different points for different condenser temperatures. The increment in exergetic efficiency occurs as the evaporator temperature is increased, the compressor work is reduced which reduces the exergy losses in the compressor and the mass flow rate of the refrigerants decrease thus increasing the exergetic efficiency while it starts decreasing after a peak as we approach the ambient temperature as the exergy losses start to increase [9].

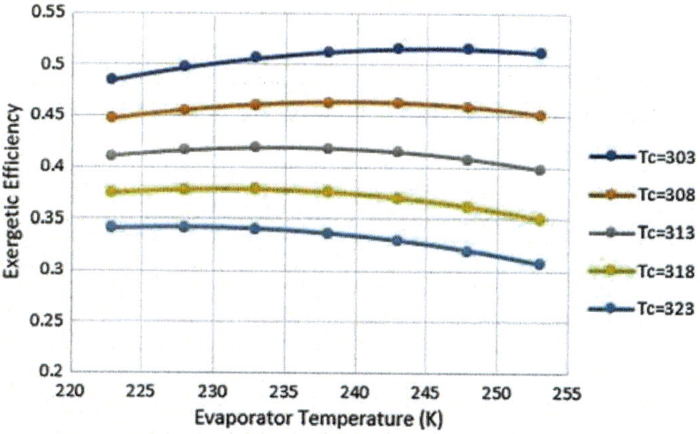

Fig. 6 Variation of exergetic efficiency with condenser temperature

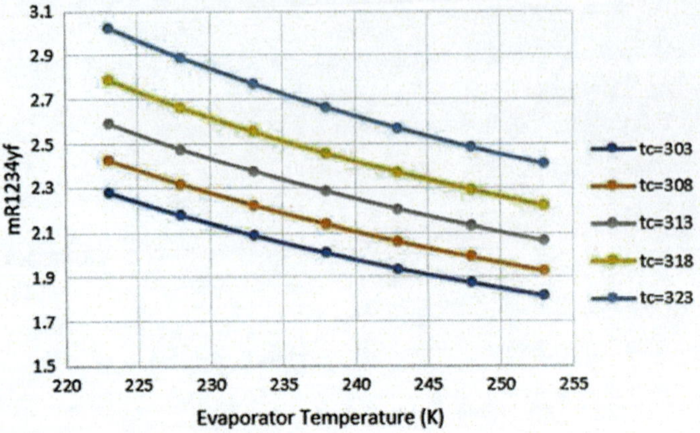

Fig. 7 Variation of mass flow rate of R1234yf with evaporator temperature

4.6 *The Influence of Evaporator Temperature on Mass Flow Rate of R1234yf*

Examination of the context is demonstrated in Fig. 7. In each of the curved lines evaporator temperature is varied at a fixed condenser temperature which gives different mass flow rate values for the system. It is evident that the mass flow rate of R1234yf decreases as we increase the evaporator temperature. This happens as we escalate temperature of evaporator, the compressor work of the high temperature circuit also reduces which in turn results to a lesser mass flow rate value.

4.7 *The Influence of Evaporator Temperature on Mass Flow Rate of CO2*

The inspection of this case is demonstrated in Fig. 8. The evaporator temperature is varied at a fixed condenser temperature which gives different mass flow rate values for the system. It shows that the mass flow value of CO2 decreases as we increase the evaporator temperature but after 248 K it starts increasing again. Reason is, as we rise evaporator temperature the compressor work of the low temperature circuit also reduces which in turn results to a lesser mass flow rate value but starts increasing back up again after 248 K.

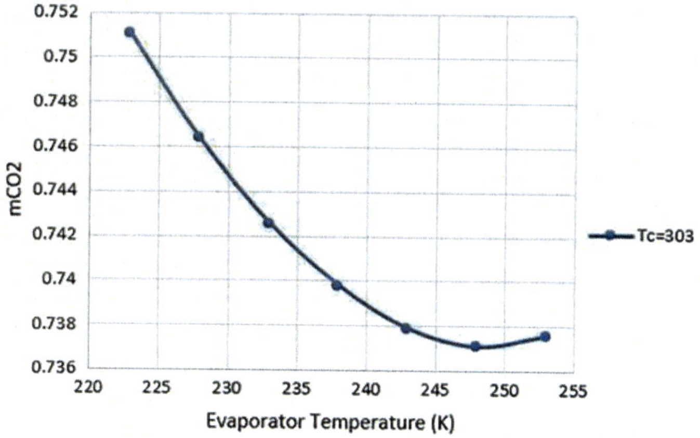

Fig. 8 Variation of mass flow rate of CO_2 with evaporator temperature

4.8 Exergy Losses in Different Components of System

The exergy losses per components of the two-stage sub-critical cascade refrigeration system using CO2-R1234yf as refrigerants for fixed condenser as well as evaporator temperature, i.e., 303 K and 253 K, respectively, are illustrated under in a pie chart. It is noticed that the maximum exergy losses occur at cascade condenser which is 22% and minimum exergy losses occur at expansion valve of low temperature circuit which is 8% [9].

Fig. 9 Distribution of exergy losses per component

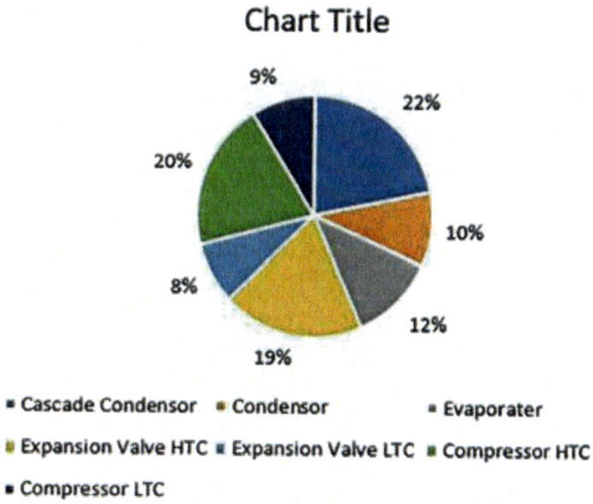

Acknowledgements All the authors recognize the assistance provided by the Department of Mechanical Engineering, JSSATE, Noida for carrying out the research work.

References

1. Nissin Refrigeration & Engineering Ltd. (Online). Available: https://www.nissin-ref.co.jp/english/product_blog/1-2.html
2. Umwelt Bundesamt (Online). Available: https://www.umweltbundesamt.de/en/topics/climate-energy/fluorinated-greenhouse-gases-fully-halogenated-cfcs/statutes-regulations/eu-regulation-concerning-fluorinated-greenhouse#eu-regulation-no-5172014-on-fluorinated-greenhouse-gases
3. Yilmaz B, Erdonmez N, Sevindir MK, Mancuhan E (2014) Thermodynamic analysis and optimization of cascade condensing temperature of a CO_2(R744)/R404A cascade refrigeration system. In: International refrigeration and air conditioning conference, West Lafayette
4. Tripathy S, Jena J, Padhiary DK, Roul MK (2014) Thermodynamic analysis of a cascade refrigeration system based on carbon dioxide and ammonia. Int J Eng Res Appl, pp 1–6
5. Bellos E, Tzivanidis C (2019) A theoretical comparative study of CO_2 cascade refrigeration systems. MDPI pp 1–19
6. Mançuhan E, Tunç B, Yetkin K, Çelik C (2019) Comparative analysis of cascade refrigeration systems performance and environmental impacts. JOTCSB, pp 1–12
7. F-chart Software (Online). Available: https://fchartsoftware.com/ees/
8. Singh S, Dasgupta MS (2016) Thermodynamic analysis of a low tewi (R1234YF-R744) cascade system. In: National conference on recent trends in mechanical engineering, Pilani
9. Chowdhury S, Roy R, Mandal BK (2019) A review on energy and exergy analysis of two-stage vapour compression refrigeration system. Int J Air- Conditioning Refrig, pp 1–9

Analyzing the Factors Influencing the Electric Vehicle Selection Using Fuzzy AHP and TOPSIS-SAW-COPRAS-ELECTRE Framework

Saumya Diwan, Shristi Mehrotra, Saumya Singh, and Pravin Kumar

1 Introduction

Vehicle emissions have long been the most significant source of pollution in the environment. These contaminants have both acute and chronic environmental consequences in addition to global warming, acid rain, and human health issues. The automobile sector is one of the major contributors to air pollution and ozone depletion, which in turn lead to climate change due to greenhouse gas emissions such as CO, CH_4, N_2O, and CO_2 [1]. Such predicaments have led to an increasing adoption of alternative fuel vehicles (AFVs) not only worldwide, but also across the country. At a global level, reducing the effects of global warming and promoting the energy transition are two of the most important concerns that drive the transition away from fossil fuels and toward renewable energy sources [2]. These vehicles utilize green modes of energy rather than fossil fuels.

Electric vehicle (EV) is one type of AFV, which has become a major theme in the global automotive industry. Many developed countries, in order to support sustainable development that ensures environmental, social, and economic development in the long term, have shifted toward electric vehicles (EVs) [3]. EVs can make a major contribution to reducing global carbon emissions at a reasonable cost [4]. There are three major types of EVs available: battery electric vehicle (BEV), hybrid EV (HEV),

S. Diwan (✉) · S. Mehrotra · S. Singh · P. Kumar
Delhi Technological University, Delhi 110042, India
e-mail: diwansaumya@gmail.com

S. Mehrotra
e-mail: shristi.mehrotra.dtu@gmail.com

S. Singh
e-mail: saumyasingh30aug@gmail.com

P. Kumar
e-mail: pravin.dce@gmail.com

A. Kumar et al. (eds.), *Recent Advances in Manufacturing and Thermal Engineering*,
Lecture Notes in Mechanical Engineering, https://doi.org/10.1007/978-981-19-8517-1_6

and plug-in hybrid EV (PHEV). Of these three, only BEVs run solely on electricity, and are zero-emission vehicles, thus being a cleaner AFV [5].

After dominating the EV market in 2020, the BEV segment is expected to be worth USD 116.80 billion by 2030 due to consumers' growing preference for EV over gasoline-powered vehicles, as well as vehicle CO_2 emission restrictions. Leading EV manufacturers have started development processes in India as a result of the growing popularity of EVs, which is predicted to boost the country's market growth [6]. The EV market is expected to sharply increase by 2030. As India is a major contributor of global greenhouse gas and global warming levels, and at the same time harbors one the biggest markets in the world, increasing BEV manufacturing in the country is a major stepping stone towards a sustainable future [7]. Thus, there is a necessity for a comprehensive decision framework to assess the available BEVs in the market and select the most appropriate one. Catering to such a necessity, this study proposes a multi-criteria decision-making (MCDM) framework in which selecting a BEV is dependent on a variety of conflicting factors/criteria.

MCDM approaches have been successfully used in the field of vehicle selection [8] and ranking of electric vehicle battery technologies [4]. However, on reviewing contemporary research works, it was found that most of the approaches are focused on specifically the internal combustion engine vehicles and factors to be considered while their selection and very few studies on BEV purchasing. A significant amount of work can be seen on the electric vehicle battery supply chain (EVSC) and factors influencing the same in addition to its technologies and lack of availability of raw materials. There seems to be an inadequacy of studies analyzing the factors influencing the purchasing decision of BEVs. This research article aims to provide a comprehensive outlook to the factors most affecting the consumers decisions while considering to buy a BEV. Several factors were identified, validated and prioritized on the basis of data collected from domain experts and consumers. The following are the contributions of this research paper:

- Identification and grouping of 10 major factors and the BEV purchasing decision
- Formulation of framework to prioritize the factors from the industry and consumer perspective based on F-AHP
- Ranking the BEV alternatives utilizing a combination of 4 MCDM techniques
- Providing a consolidated ranking by Rank Aggregation for the BEV alternatives in the Indian context
- Providing consumers with a reliable and data-driven purchasing decision support tool.

This article is further organized in the following manner: Sect. 2 consists of the extensive Literature Review for identifying the factors influencing the BEV purchasing. In Sect. 3, the proposed research methodology consists of data collection, MCDM techniques used: Fuzzy-AHP (for weight calculation) and a combination of TOPSIS, SAW, ELECTRE and COPRAS model. Also, the robust rank aggregation (RRA) algorithm is discussed in this section. The results are presented in Sect. 4 followed by the implications and discussions in Sect. 5. Section 6, finally gives the Conclusion, Limitations, and Future Scope of the study.

2 Literature Review

Fuzzy MCDM techniques have long been used in industrial applications of ranking of particular alternatives based on well-defined attributes to aid prioritization. As the research analyzing the EV industry develops with time and infusion of resources, research works covering the applications of such methods in the EV industry are increasing as well. However, being a growing field, substantial research for the same is lacking, especially in the case of EV selection. The following are the factors identified through the literature review performed.

2.1 Rapid Charging Time (*Via* DC Charging)

Serradilla et al. [9] discussed the impact of a well-built rapid charging infrastructure, has over EV adoption among customers. They mentioned that while the rapid charging time is dependent on various external factors such as external temperatures and state of charge (SOC), it is highly dependent upon the input charge. The research addresses the cost–benefit payoff for public and private EV chargers and suggests an economically feasible model that can help spurt EV adoption. Zhang et al. [10] studied various factors that can drive EV sales and consequently, adoption among the general. The paper mentions that it is necessary to improve fast charging technologies in order to considerable increase the market potential of BEVs.

2.2 Acceleration

In contrast to a gas-powered car, EVs offer "immediate torque," which allows for rapid acceleration. Electric cars are notorious for providing swift, smooth, and noise-free acceleration. Grunditz et al. [11] elaborated on the role and importance of acceleration performance of an EV and state that on average, small BEVs can reach an acceleration of 0–100 km/h in 13 s while large BEVs can reach the same in 10 s. The works of Skippon et al. [12], and Zhang et al. [10] show that acceleration is a weighty factor for customers of EVs. Skippon et al. [12] showed that this correlation remains positive among the markets of England while Zhang et al. [10] showed the same for Japanese markets.

2.3 Full Charge Time (*Via* Normal AC Charging)

EVs require DC current, but because electricity distribution systems supply AC current, a rectifier is required. At-home chargers also use AC current and so, this is

the most commonly used charging type [13]. Yang et al. [14], Nazari et al. [15], and Egbue and Long [16] have pointed out that longer charging time is one of the major factors that can negatively impact EV adoption. Nazari et al. [15] studied this factor with respect to the customers' willingness to pay for an EV. Tortós et al. [13] studied the EV charging behavior in the UK, which is known to be one of the largest EV markets in the world.

2.4 Initial (Purchasing) Price

Neves et al. [17] explained that the cost of buying an EV is, obviously, higher than that of an ICE vehicle, pertaining to the new technologies deployed and so, higher costs are a barrier to EV adoption in any geography. This statement has been deduced and explained by Egbue and Long [16], and Zhang et al. [10]. Steinweg [18] suggested that the relatively high cost of electric cars can be explained by the high production costs that are caused by the lack of economies of scale. The total cost for the production of electric car batteries is expected to decrease with increased manufacturing and further developments. Lin et al. [19] have studied and enlisted the various reasons for high EV adoption in China, and have mentioned decreasing initial price as one of the main aspects. Hagman et al. [20] have explained the total cost of ownership of BEVs and stresses on the importance of policy making to reduce the costs and increase the EV adoption among people.

2.5 Battery Capacity/Performance

Vilchez et al. [21] explained that battery capacity is measured in kilowatt-hour (kWh), and a larger battery capacity is anticipated to lead to higher sales. BEVs typically have a larger battery capacity than PHEVs. Larger vehicles have larger batteries for each of these two powertrains. Steinweg [18] and Vilchez et al. [21] discussed that current technological advancements are mostly aimed at extending the life of Li-ion batteries that is measured by the number of charge-and-discharge cycles as well as overall battery age. Steinweg [18] also suggested that as a part of improving the battery, increasing battery performance is an essential feature. Recent research includes producing peak power at low temperatures, state-of-charge measurement, and thermal management.

2.6 Range

Numerous researchers have elaborated that range is a major driving factor of EV purchase decisions among customers, a few of them Rezvani et al. [22] and Egbue

and Long [16]. Bonges et al. [23] indicated that without certainty that they will be able to simply recharge their vehicle, customers find the risk to be too big for the vehicle's expense. They also showed that show negative correlation between BEV sales and range anxiety. Egbue and Long [16] and Rezvani et al. [22] showed that restricted range is one of the main customers' concerns. Similarly, Sovacool et al. [24], Skippon et al. [12], and Nazari et al. [15] elucidated that range distance is a pivotal barrier to EV penetration.

2.7 *Top Speed*

While EVs are quicker, i.e., take less time to reach one point from another, than ICE vehicles, their top speeds are relatively lesser than those of ICE vehicles. Sovacool et al. [24] studied the effect of top speed on customer buying preferences over different demographics and shows that men give more importance to speed. Zhang et al. [10] used top speed as an input parameter and reported that it was effective in enhancing BEV adoption. Burgess et al. [25] showed that while most non-EV users feel that fossil fuel driven cars have a much better top speed, those who have had EV drivers feel quite the opposite, thus suggesting that exposure to EVs is essential.

2.8 *Maximum Power*

Egbue et al. [16] explained that the driving force necessary for the BEV to accelerate is generated by the battery power. Sovacool et al. [24] stated that the performance of EVs is limited due to the weight of the battery. Nazari et al. [15] stated that maximum power is not a noteworthy factor for both, BEV and non-BEV customers. Gerssen-Gondelach et al. [26] discussed the performance of EV batteries over short as well as long term usage.

2.9 *Number of Airbags*

Basu et al. [27] explained the importance of proper EV sensors for airbags in preventing injuries. They explained that airbags are extensively used in all kinds of vehicles, not just EVs, in order to protect the passengers in case of a very fast change in acceleration of the vehicle. This association between the number of airbags and the perceived safety of the vehicle, and subsequently, the purchasing preferences of the customers, is explained by Kim et al. [28].

2.10 Ground Clearance

Nagapan et al. [29] explained the importance and function of ground clearance for vehicles on Indian roads. This work suggests that a significant proportion of vehicle customers give adequate importance to ground clearance. Balakrishnan et al. [30] have shown that better design factors such as ground clearance make it easier for the automobile companies to market the vehicles, thereby increasing the sales. Table 1 summarizes the identified factors.

In addition to the above literature survey, 4 key EV Manufacturers were identified in the Indian EV market. The five selected EV models have been briefly introduced as follows:

Tata Nexon EV: The information below has been sourced from the official brochure published on the official Tata Nexon EV website [31, 32]. The Tata Nexon EV is the electric variant of the Tata Nexon compact SUV. It is the first EV launched by Tata Motors and was the best-selling EV in India in 2020. Tata Nexon EV was the winner of the Green Card Award of the Indian Car of the Year Award 2021. Nexon is the first Indian-made car to achieve a 5-star rating in the Global NCAP crash tests. The vehicle is powered with a 30.2 kW Li-ion battery with an IP67 certification. It comes with two charging options: rapid charging and regular 15A charging.

Tata Tigor EV: The information below has been sourced from the official brochure published on the official Tata Tigor EV website [33, 34]. The Tata Tigor EV is the all-electric version of the Tata Tigor. It is the one of the most affordable EVs in the March 2022 lineup. The Tigor EV was earlier available only for commercial use, but the company launched an updated version that is now available to private customers as well. The vehicle has an IP67 certified impact resistance battery pack that adds to

Table 1 Factors identified through literature review along with their source

Factors	Source
Quick charging time, F_1 (time in minutes to charge an EV from 0 to 80% via DC charging)	[9, 10]
Acceleration, F_2 (Time in seconds to accelerate from 0 to 100 kmph)	[10–12]
Full charging time in hours, F_3 (time in minutes to fully charge an EV via normal AC charging)	[12–16]
Initial purchasing price, F_4 (selling price of the vehicle, in ₹ lacs)	[10, 16–20]
Battery capacity/performance, F_6 (energy stored in the battery in kWh)	[18, 21]
Range, F_7 (distance, in km, that can be covered by the EV in a single charge	[12, 15, 16, 22–24]
Top speed, F_8 (maximum achievable speed, in km/h)	[10, 24, 25]
Maximum power, F_9 (Electric drive unit in bhp)	[15, 16, 24, 26]
Number of airbags, F_{11} (Number of airbags installed in the vehicle)	[27, 28]
Ground clearance, F_{12} (Least distance, in mm, between the lower end of the vehicle and the ground)	[29, 30]

the safety features. It received the 4-star safety rating by the Global NCAP in 2021, making it the safest (as of March 2022) electric sedan in India.

MG ZS EV: The information below has been sourced from the official brochure published on the official MG ZS EV website [35, 36]. The MG ZS EV was launched in India in March 2022 and is the first EV launched by MG Motor India. As of March 2022, it is available in only one variant, MG ZS EV Exclusive, but the Excite variant is expected to launch in July 2022. The MG ZS EV European spec variant was awarded 5-star safety rating by the Euro NCAP on 18 Dec, 2019. The car has three driving modes: ECO, Normal, and Sport. The vehicle has an ASIL-D, IP69K safety rated and IP67, UL2580 certified 50.3 kW battery and has the following charging options: AC Fast Chargers, Portable Charger with the car, and DC Super-Fast Charging.

Mahindra e-Verito: The information below has been sourced from the official brochure published on the official Mahindra e-Verito website [37, 38]. The Mahindra e-Verito is the all-electric version of the Mahindra Verito sedan. The vehicle has been solely used for commercial purposes. With upcoming developments, it is now being used as government vehicles. The vehicle is the first one in India to deploy the regenerative braking technology. It also has the Mahindra Electric patented REVive Feature that enables the car to run 8 additional kilometers in case it runs out of charge. The company claims that its running cost is ₹1.15/km, which is lesser than that of all other cars.

Hyundai Kona EV: The information below has been sourced from the official brochure published on the official Hyundai Kona EV website [39, 40]. Launched in July 2019, the Hyundai Kona Electric SUV is the first EV launched in the country by Hyundai Motor India. Kona Electric is also the first fully electric long-range SUV launched in the country. Although globally, it is available in two battery options—39.2 and 64 kW, only the first (39.2 kW) model is available in India. The car has three driving modes: ECO/ECO+ , Comfort, Sport. The vehicle supports 3 forms of charging: DC Quick Charger CCS Type 2, AC Wall Box Charger, and Portable Charger.

Based on the literature review, we have identified the following research gaps:

- A large proportion of research works are focused specifically on the EV battery supply chain and technologies and lack of availability of raw materials.
- A lack of studies analyzing the factors affecting purchase of BEVs in the market. Most of such studies revolve around ICE vehicles (combustion vehicles), machine parts etc.
- The comprehensive prioritization of factors in the domain BEV purchasing in terms of experts' opinions was relatively absent.

3 Research Methodology

In this study, the proposed research methodology consists of three phases. In the first phase i.e., the extensive literature review, focused on the contemporary studies and reports related to the of EV Selection domain. Here, a set of 10 broad factors,

Table 2 7-point fuzzy linguistic scale

Linguistic variables	Very low	Low	Fairly low	Medium	Fairly high	High	Very high
Triangular fuzzy numbers	(0, 0.05, 0.15)	(0.1, 0.2, 0.3)	(0.2, 0.35, 0.5)	(0.3, 0.5, 0.7)	(0.5, 0.65, 0.8)	(0.7, 0.8, 0.9)	(0.85, 0.95, 1)

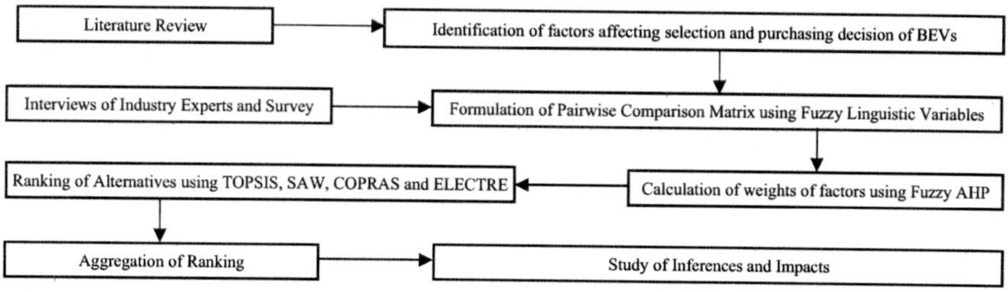

Fig. 1 Overview of proposed methodology

F_i (Table 1), were formulated that affect most, the purchasing decisions of BEVs. In the second phase, the factors identified were validated and assessments in terms of pairwise factor importance were obtained from industry experts and customers. Five key BEVs in India were taken as alternatives. Data regarding each of the factors (here attributes) for the respective alternatives, was collected through the company's published brochures and news articles. These were summarized into a decision matrix. Following this, in the third phase a F-AHP was used to determine the weights of these attributes and a combination of MCDM Techniques: TOPSIS, SAW, COPRAS, and ELECTRE were used to combine these importance weights and ratings in order to provide an outlook towards the EVSC and rank the manufacturers. These were aggregated using the robust rank aggregation (RRA) algorithm. Finally, the resultant ranking was corroborated through a general survey where the respondents were made to give their ratings for each of them based on the identified 10 factors. The overview of the methodology is provided in Fig. 1.

3.1 Data Collection

The data for this research has been collected via company's published brochures, reports, news articles and a set of extensive surveys and questionnaires.

The opinions of various industry experts and seasoned professionals have been accounted to maintain the credibility and reliability of the data collected and hence the model further used. A detailed set of pairwise comparison values were collected from a set of 13 EV customers. Of these, 5 people were regular users of EVs, 5 are occasional EV users, i.e., have EV as secondary vehicle, and 3 are just about

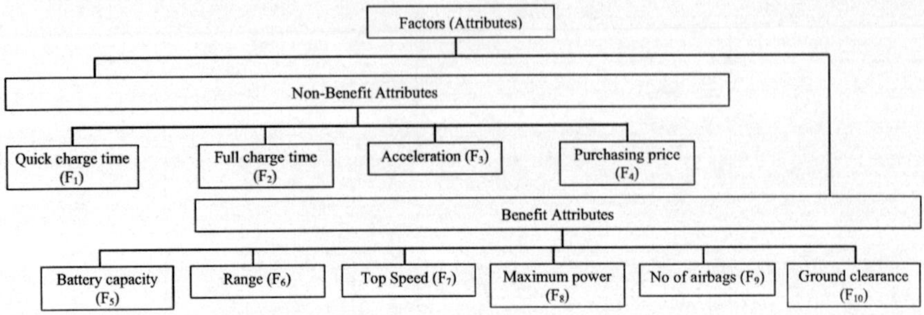

Fig. 2 Factors considered for weight determination and MCDM

to switch to an EV. The data was collected was collected, through a questionnaire, in linguistic variables on a 7-point Linguistic Scale (Table 2) so as to capture the subjectivity in each of their assessment. This also helped maintain uniformity in scale over the collected data (Table 3). These data points were used to construct a pairwise comparison matrix for the set of 10 factors that would be considered during EV selection and purchasing decision. A pool of industry and academia members in the domain of and EVs and EVSC was also created for the study. To validate the factors, 13 people were contacted, but five respondents gave their consent for the process. All the five experts are currently involved in Indian EV companies and hence, their insights encompass the Indian market. Three of them belong to Pune, one from Halol, Gujrat and one from Chennai. For wider scope of opinions, the respondents belonging to the different companies considered and different levels and roles in their respective organizations were taken.

Initially, the respondents were made to go through the identified factors in order to provide validation. Further, the respondents were made to provide pairwise comparisons amongst the identified 10 factors. Using the obtained data, a pairwise comparison matrix was formed for the F-AHP model to determine the weights (Table 5).

Standardization of data: the data for the case study was obtained from respective BEV company brochures and news reports. It was noted that different measurement units for full charge time and acceleration were used. These were linearly approximated under the assumption of uniform variation over the standard measuring intervals. Full charge time values were interpolated for Tata Nexon EV (A_1) and Tata Tigor EV (A_2) to the interval of 0–100% and the Acceleration values were interpolated for Tata Tigor EV (A_2) and Mahindra e-Verito (A_4) to the interval of 0–100 kmph time. The standardized data points have been summarized in Table 4.

The weights corresponding to the identified seven factors obtained by F-AHP were used further for four MCDM techniques namely: TOPSIS, SAW, COPRAS and ELECTRE. Multiple techniques were considered in order to provide a contrasting rankings. The factors were classified into six benefit factors and four non benefit factors as shown in Fig. 2.

Table 3 Collected dataset for case study of five alternatives of BEV in Indian market

	Quick charge time (0–80%)	Full charge time (0–100%)	Acceleration (0–100 kmph time)	Purchasing price (lakhs)	Battery capacity (kWh)	Range (KMFC)	Top speed (kmph)	Max power (BHP)	No. of airbags	Ground clearance (mm)
Tata Nexon EV [31, 32]	60	8.5 (10–90%)	9.9	15.845	30.2	312 [41]	120 [41]	127.235	2	205
Tata Tigor EV [33, 34]	65	8.75 (0–80%)	5.7 (0–60 kmph)	12.815	26	306	80 [42]	55	2	172
MG ZS EV [35, 36]	60	8.5	8.5	23.935	50.3	461	173.809 [43]	174.332	6	177 [44]
Mahindra e-Verito [37, 38]	90	11.5	18.666 (0–60 kmph)	9.29	14.4	189	86	41.572	1	172
Hyundai Kona EV [39, 40]	57	6.166	9.7	24.365	39.2	452	167 [45]	134.102	6	158 [45]

Table 4 Standardized data

	Quick charge time (0–80%)	Full charge time (0–100%)	Acceleration (0–100 kmph time)	Purchasing price (lakhs)	Battery capacity (kWh)	Range (KMFC)	Top speed (kmph)	Max power (BHP)	No. of airbags	Ground clearance (mm)
Tata Nexon EV	60	10.625	9.9	15.845	30.2	312	120	127.235	2	205
Tata Tigor EV	65	10.9375	9.5	12.815	26	306	80	55	2	172
MG ZS EV	60	8.5	8.5	23.935	50.3	461	173.809	174.332	6	177
Mahindra e-Verito	90	11.5	31.11	9.29	14.4	189	86	41.572	1	172
Hyundai Kona EV	57	6.166	9.7	24.365	39.2	452	167	134.102	6	158

3.2 Fuzzy-Analytical Hierarchy Process (F-AHP)

The Analytical Hierarchy Process (AHP) was introduced as MCDM problem solving approach by Saaty [46]. It employs a pairwise comparison matrix for the selected attributes in order to provide weights for the same. The following are the steps for the weight calculation by F-AHP:

Step 1: Formulation of Pairwise Comparison Matrix: the pairwise importance ratings collected in the data collection process (Section) are concisely expressed in an $(n \times n)$ matrix, D of triangular fuzzy numbers $\widetilde{x_{ij}}$, ,whose element indicate the importance rating of attribute (i.e., factor) i, with respect to the attribute j

$$D = \begin{matrix} & \begin{matrix} X_1 & \cdots & X_n \end{matrix} \\ \begin{matrix} X_1 \\ \vdots \\ X_n \end{matrix} & \begin{pmatrix} \widetilde{x_{11}} & \ldots & \widetilde{x_{1n}} \\ \vdots & \ddots & \vdots \\ \widetilde{x_{n1}} & \cdots & \widetilde{x_{nn}} \end{pmatrix} \end{matrix} \tag{1}$$

where

$$\widetilde{x_{ij}} = \left(a_{ij},\ b_{ij},\ c_{ij}\right)$$

Step 2: Calculate the fuzzy geometric mean

$$\widetilde{w_i} = \widetilde{r_i} \times (\widetilde{r_1} + \widetilde{r_2} + \widetilde{r_3} + \cdots + \widetilde{r_n})^{-1} \tag{2}$$

Step 3: Weight normalization and Defuzzification of weights by center of area (COA) method

$$w_i = \frac{(l_i + m_i + n_i)}{3} \tag{3}$$

where $\widetilde{w_{ij}} = (l_i, m_i, n_i)$

The 7 factors identified in the literature review were formulated into a pairwise comparison matrix which was filled by the entries the respondents provided on the 7-point fuzzy linguistic scale in table.

3.3 Technique for Order Preference by Similarity to Ideal Solution (TOPSIS)

TOPSIS is a method of solving MCDM problems, found by Hwang and Yoon [47]. It works on the basic principle that the chosen alternative should have the least distance

from the ideal solution and the most distance from the negative ideal solution. The procedure for the same is explained in a step wise algorithmic manner below-

Step 1: Assign the aggregated weights obtained from F-AHP along with the Decision Matrix

$$R = [r_{ij}] \text{ where } r_{ij} \text{ refers to the collected datapoint for alternative } i \text{ and attribute } j. \tag{4}$$

Step 2: normalize the datapoints decision matrix W

$$r_{nij} = \left(\frac{r_{ij}}{\sqrt{\sum_{i=1}^{m} r_{ij}^2}} \right) \tag{5}$$

$j = 1, 2, \ldots n$ attributes

Step 3: Calculate the weighted normalized fuzzy decision matrix, which is given as follows

$$R_{wn} = (r_{wnij}) \tag{6}$$

where $r_{wnij} = r_{nij} \times w_j, w_j = w_i'$

Step 4: Calculate the Positive Ideal Solution, PIS and Negative Ideal Solution, NIS, which are given as-

$$A^* = (r_{wnij}^+), A^- = (r_{wnij}^-) \tag{7}$$

where $r_{wnij}{}^+ = \max_i \{r_{wnij}\}, r_{wnij}{}^- = \min_i \{r_{wnij}\}$

Step 5: Compute the Euclidean Distance from each alternative to the PIS and to the NIS, respectively

$$d_i^* = \sum_{j=1}^{n} d(r_{wnij}, r_{wnij}^+), d_i^- = \sum_{j=1}^{n} d(r_{wnij}, r_{wnij}^-) \tag{8}$$

$$d(x, y) = \sqrt{\frac{1}{2}[(x - y)^2]} \tag{9}$$

where x, y are two numeric values

Step 6: Compute the Closeness Coefficient CC_i

$$CC_i = \frac{d_i^-}{d_i^* + d_i^-} \tag{10}$$

Step 7: Rank the alternatives.

3.4 Simple Additive Weighting (SAW)

Simple additive weighting (SAW) method. SAW was introduced by Fishburn in 1967 and MacCrimmon in 1968 to be used as a method in solving multi-criteria problems [48]. The simple additive weighting (SAW) method is known as the weighted addition method. In SAW the performance of each alternative is calculated as follows:

Step 1: Assign the aggregated weights obtained from F-AHP along with the Decision Matrix using Eq. (4)

Step 2: Normalize the datapoints decision matrix W using Eq. (5)

Step 3: Calculate the weighted normalized decision matrix using Eq. (6)

Step 4: Calculate the Performance Value P_{wn} of each Alternative

$$P_{wn} = \sum (r_{nij}) \tag{11}$$

3.5 COmplex PRoportional ASsessment (COPRAS)

The COmplex PRoportional ASsessment (COPRAS) method was introduced by Zavadskas, Kaklauskas, and Sarka in 1994 [49]. This method is used to assess the maximizing and minimizing index values, and the effect of maximizing and minimizing indexes of attributes on the results assessment is considered separately.

Step 1: Assign the aggregated weights obtained from F-AHP along with the Decision Matrix using Eq. (4)

Step 2: Normalize the datapoints decision matrix W

$$r_{nij} = \left(\frac{r_{ij}}{\sqrt{\sum_{i=1}^{m} r_{ij}}} \right) \tag{12}$$

$j = 1, 2, \ldots n$ attributes

Step 3: Calculate the weighted normalized decision matrix using Eq. (6)

Step 4: Calculate the maximizing and minimizing indices by summation of r_{wnij}

$$S_{+i} = \sum_{j=1}^{g} r_{wnij}, \; S_{-i} = \sum_{j=g+1}^{n} r_{wnij} \tag{13}$$

where $i = 1, \ldots, m$

Step 5: Calculate the relative significance values using

$$Q_i = S_{+i} + \frac{\sum_{i=1}^{m} S_{-i}}{S_{-i} \sum_{i=1}^{m} \frac{1}{S_{-i}}} \tag{14}$$

3.6 ELimination Et Choix Traduisant La REalite (ELECTRE)

The ELECTRE (ELimination Et Choix Traduisant la REalite) method was first introduced by Roy [50] that evaluates all alternatives using outranking comparisons, and ineffective and eliminates low-attractive alternatives.

Step 1: Assign the aggregated weights obtained from F-AHP along with the Decision Matrix using Eq. (4)

Step 2: Normalize the datapoints decision matrix W using Eq. (5):

Step 3: Calculate the weighted normalized decision matrix using Eq. (6).

Step 4: Calculate Dominant Matrix, Dominated Matrix, Concordance Matrix and Discordance Matrix.

Step 5: Formulate the Aggregated Dominant Matrix.

3.7 Rank Aggregation

In this study, the Robust Rank Aggregation [51] was used. In this algorithm, first the rankings are formulated in form a probabilistic representation using binomial probability. Finally, a minimization function is used to determine the final p values and subsequent rankings.

$$B_{k,n}(x) := \sum_{i=k}^{n} {}^{n}C_{k} x^{l} (1-x)^{n-l} \tag{15}$$

$$p(r) = \min B_{k,n}(r) \tag{16}$$

where $k = 1 \ldots n$

4 Results

In Sect. 2, 10 broad factors were identified as shown in Table 1. These factors were considered for the questionnaire in the data collection (Sect. 3.1) through which a pairwise comparison matrix was formulated. The weights of each of these attributes were determined through F-AHP following which a case of 5 eV models were taken which were ranked using the mentioned techniques utilizing the determined weights. The following are the description of the models.

4.1 Weight Determination

Based on the responses collected from the experts, a pairwise comparison matrix was formulated. The data was collected in the form of linguistic variables using the 7-point scale in Table 2 and is summarized in Table 3. The geometric mean of the fuzzy values was taken. Finally, following equations the weights were calculated for each of the 10 Factors and normalized further (Table 5).

The final normalized weights in crisp form obtained from F-AHP are shown in Table 6.

4.2 Ranking of Alternatives

The weights obtained, were summarized with the final decision matrix for the ranking model in Table 7.

The rankings for each of the MCDM techniques were calculated. These were then aggregated using the robust rank aggregation algorithm to give a final consolidated ranking (Table 8) for the consumer data for the identified alternatives collected through the literature survey and expert validation.

5 Discussions and Implications

Based on the F-AHP, the most important weighted factors were noted to be the purchasing price and full charge time which were very closely ranked (Fig. 3). Being a developing technology in the market, the price of the electric 4-wheel vehicles is still high. EV Lithium is a limited resource its supply isn't abundant and certain protocols for recycling of such batteries need to be put in place. The demand of EV while rising is still, however, constrained due to several social and economic aspects as well which drive the EVSC and sales. In a developing nation such as India, the Full charge time is also a major factor since the cost of charging, i.e., electricity consumption is an additional expense. Also, there might be cases where a long uninterrupted power supply may not be available. These were followed by the quick time charge factor. Its position lower than full time charge seems justified as in the present scenario there is a lack of quick charging stations for BEVs with limited accessibility. Range and ground clearance were subsequently ranked in this order based on the obtained weights. Range plays a huge factor since there is very high amount of inter-city and inter-state travel and its importance is furthered by the lack of quick charging stations as well. The presence of well-established charging facilities could help reduce the dependence of range being a highly weighted factor. With the exception of the metropolitan areas and certain National Highways, the roads of most of the Indian states are friendly to low ground clearance vehicles which

Table 5 Pairwise comparison matrix

		Quick charge time	Full charge time	Acceleration	Purchasing price	Battery capacity	Range	Top speed	Max power	No. of airbags	Ground clearance
		F_1	F_2	F_3	F_4	F_5	F_6	F_7	F_8	F_9	F_{10}
Quick charge time	F_1	1	FL	H	FL	FH	FL	FH	FH	FL	L
Full charge time	F_2		1	VH	M	H	M	VH	H	M	FL
Acceleration	F_3			1	L	L	L	H	FL	L	L
Purchasing price	F_4				1	VH	FL	VH	H	H	FL
Battery capacity	F_5					1	L	H	FH	FL	L
Range	F_6						1	VH	VH	H	M
Top speed	F_7							1	L	L	L
Max power	F_8								1	M	L
No. of airbags	F_9									1	L
Ground clearance	F_{10}										1

Table 6 Normalized crisp weights for the considered factors

Factors		Quick charge time	Full charge time	Acceleration	Purchasing price	Battery capacity	Range	Top speed	Max power	No. of airbags	Ground clearance
		F_1	F_2	F_3	F_4	F_5	F_6	F_7	F_8	F_9	F_{10}
Weights	w_i	0.14539	0.25315	0.06498	0.25126	0.02562	0.09261	0.03484	0.02240	0.02530	0.0844

Table 7 Final decision matrix with weights

Weights		0.145	0.253	0.065	0.251	0.026	0.093	0.035	0.022	0.025	0.084
Factors		Quick charge time	Full charge time	Acceleration	Purchasing price	Battery capacity	Range	Top speed	Max power	No. of airbags	Ground clearance
		F_1	F_2	F_3	F_4	F_5	F_6	F_7	F_8	F_9	F_{10}
Tata Nexon EV	A_1	60	10.625	9.9	15.845	30.2	312	120	127.23	2	205
Tata Tigor EV	A_2	65	10.937	9.5	12.815	26	306	80	55	2	172
MG ZS EV	A_3	60	8.5	8.5	23.935	50.3	461	173.80	174.33	6	177
Mahindra e-Verito	A_4	90	11.5	31.11	9.29	14.4	189	86	41.571	1	172
Hyundai Kona EV	A_5	57	6.166	9.7	24.365	39.2	452	167	134.10	6	158

Table 8 Ranking Aggregation

		MCDM techniques				Robust rank aggregation	
		TOPSIS	SAW	COPRAS	ELECTRE	*P* value	Aggregated ranking
Alternatives	A_1	2	3	3	3	0.924147	3
	A_2	1	5	4	5	0.971853	4
	A_3	5	1	2	1	0.104441	1
	A_4	3	4	5	2	0.999718	5
	A_5	4	2	1	4	0.878497	2

makes it another important factor a consumer considers while making a purchasing difference.

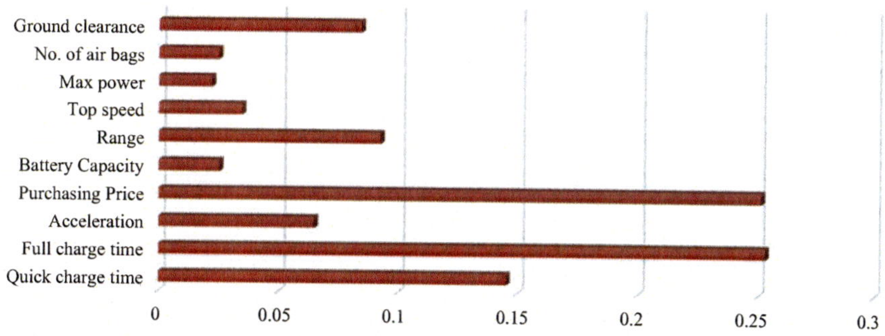

Fig. 3 Weights of factors obtained from fuzzy AHP

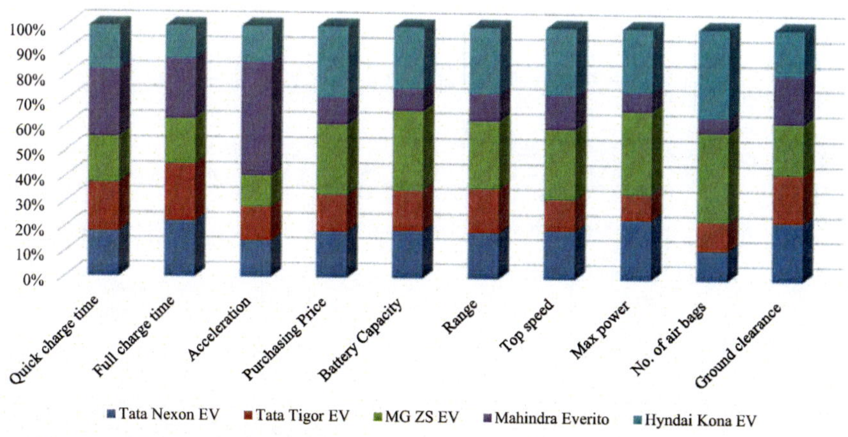

Fig. 4 Data Distribution amongst the alternatives

Based on the ranking model, MG ZS EV was the highest ranked alternative followed by Hyundai Kona EV, Tata Nexon EV, Tata Tigor EV, and lastly by Mahindra e-Verito in that order.

Figure 4 shows the comparison of alternatives by a percentage based data distribution for each feature. The MG ZS EV and Hyundai Kona Electric both stand out in having the lowest full charge times as 8.5 and 6.166 h. As this is the factor having the most importance, these two vehicles are highly ranked in the resulting hierarchy. These two even have very low quick charging times of 60 min and 57 min, respectively. These two vehicles even have the highest values for the range while Mahindra e-Verito has the lowest range. As these two vehicles fair very highly over the top three benefit attributes, they are ranked highly in the resultant hierarchy. These two vehicles are priced higher than the rest of the alternatives considered while the Mahindra e-Verito is priced very low as its sales began in the commercial segment. Even with the purchasing price being a cost attribute, their specification in other highly weighted parameters outweigh it.

In benefit attributes (factors) like range, ground clearance and acceleration, MG ZS EV stands out. Owing to its lower price range than the top two alternatives and balanced quality specifications, Tata Nexon EV outweighs the rest of the alternatives. It has the highest ground clearance of 205 mm which is more than 30 units above all other alternatives. It also has the highest sales and customer demand in the present Indian scenario. The Tata Nexon EV and Tata Tigor EV have moderate values over most of the attributes, they are ranked moderately in the hierarchy. The Mahindra e-Verito is has lower values for most benefit attributes and higher values for most cost attributes. Thus, it is ranked the lowest in the hierarchy.

6 Conclusions, Limitations, and Future Scope

The rapidly growing EV technology and markets have made the study of such supply chains studies essential. In this study, through an extensive literature review and comprehensive distribution of factors that affect the EV supply chain and have been explored in some way in contemporary research works. A total of 10 broad factors have been procured which a consumer considers or should consider while making a purchasing or BEV selection decision. Further, these were validated through a pool of experts and a set of consumers who also linguistically scored them as attributes based on their importance on a 7-point linguistic scale. Fuzzy Analytical Hierarchy Process (F-AHP) model was used to translate collected data to attribute weights. The case of five major 4-wheel EV Models by the key Manufactures in the Indian context were considered and they were taken as alternatives and were subsequently ranked in terms of their supply chain using a combination of four MCDM techniques. The data collected for the ranking model was collected through company's published brochures literature reports, news articles and public reports. The rankings were consolidated using the robust rank algorithm to provide a final aggregated ranking. The results show that the purchasing price, full charge time were the most highly

weighted factors. These were followed by the quick charge time, range, ground clearance, and acceleration in that order.

Based on the ranking model, MG ZS EV was the highest ranked alternative followed by Hyundai Kona EV, Tata Nexon EV, Tata Tigor EV, and lastly by Mahindra e-Verito in that order. Thus, the study provided a data driven overview of the EVSC and EV selection. Since manufacturers release data in different formats and scales, the data collected have some inconsistencies, as discussed earlier. To counter these discrepancies, certain assumptions and calculation shave been made, which do not necessarily denote real life results. For example, Tata has depicted its full charge time for the Nexon EV and the Tigor EV as the time taken to charge the battery from 10 to 90%, while all other manufacturers have released this data from 0 to 100%. Moreover, not all official sources consist of all the required data points and so, data from other trustable, but outside sources have been considered. To illustrate, the values of the range and top speed for the Tata Nexon EV have not been mentioned in the official website or brochure, and the values considered for the analysis have been taken from news articles.

Since the findings of this study are in the Indian context where the EV market is at an initial stage, these may be applicable to other nations where the EV market is developing and is in the nascent stages. These might help industry and academia to gain insights as to the potential avenues that need to be to be looked into. Similar study could be done in other developing countries to draw similarities and in developed nations to draw comparative differences and recommendations. Furthermore, this study focusses on the economy segment of vehicles. Luxury automotive brands are expanding their EV segments, and thus, this study can be further extended to cover the luxury EV models as well.

Acknowledgements The authors acknowledge the support from Department of Mechanical Engineering, Delhi Technological University, New Delhi, India.

References

1. Kumar RR, Alok K (2020) Adoption of electric vehicle: a literature review and prospects for sustainability. J Clean Prod 253:119911
2. Sun X, Hao H, Zhao F, Liu Z (2019) The dynamic equilibrium mechanism of regional lithium flow for transportation electrification. Environ Sci Technol 53:743–751
3. Samaie F, Meyar-Naimi H, Javadi S, Feshki-Farahani H (2020) Comparison of sustainability models in development of electric vehicles in Tehran using fuzzy TOPSIS method. Sustain Cities Soc 53:101912
4. Kumar S, Pal A (2021) Challenges of battery production: a case study of electrical vehicles in India. In: Kumar A, Pal A, Kachhwaha SS, Jain PK (eds) Recent advances in mechanical engineering. ICRAME 2020. Lecture notes in mechanical engineering. Springer, Singapore. https://doi.org/10.1007/978-981-15-9678-0_94
5. Wilberforce T, El-Hassan Z, Khatib FN, Al Makky A, Baroutaji A, Carton JG, Olabi AG (2017) Developments of electric cars and fuel cell hydrogen electric cars. Int J Hydrogen Energy 42(40):25695–25734

6. Wang W, Zhang Q, Peng Z, Shao Z, Li X (2020) An empirical evaluation of different usage pattern between car-sharing battery electric vehicles and private ones. Transp Res Part A Policy Pract 135:115–129
7. India electric vehicle market size, share & trends analysis report by product (BEV, PHEV), by vehicle type (Passenger cars, commercial vehicles), and segment forecasts, 2021–2030
8. Pal K, Bahadur Singh L, Kumar S (2021) Selection of a vehicle using multi-attribute decision making. In: Kumar A, Pal A, Kachhwaha SS, Jain PK (eds) Recent advances in mechanical engineering . ICRAME 2020. Lecture notes in mechanical engineering. Springer, Singapore. https://doi.org/10.1007/978-981-15-9678-0_92
9. Serradilla J, Wardle J, Blythe P, Gibbon J (2017) An evidence-based approach for investment in rapid-charging infrastructure. Energy Policy 106:514–524
10. Zhang H, Song X, Xia T, Yuan M, Fan Z, Shibasaki R, Liang Y (2018) Battery electric vehicles in Japan: human mobile behavior based adoption potential analysis and policy target response. Appl Energy 220:527–535
11. Grunditz EA, Thiringer T (2018) Electric vehicle acceleration performance and motor drive cycle energy efficiency trade-off. In: 2018 XIII international conference on electrical machines (ICEM). IEEE, pp 717–723
12. Skippon S, Garwood M (2011) Responses to battery electric vehicles: UK consumer attitudes and attributions of symbolic meaning following direct experience to reduce psychological distance. Transp Res Part D Transp Environ 16(7):525–531
13. Quirós-Tortós J, Ochoa LF, Lees B (2015) A statistical analysis of EV charging behavior in the UK. In: 2015 IEEE PES innovative smart grid technologies Latin America (ISGT LATAM). IEEE, pp 445–449
14. Yang S, Deng C, Tang T, Qian Y (2013) Electric vehicle's energy consumption of car-following models. Nonlinear Dyn 71(1):323–329
15. Nazari F, Rahimi E, Mohammadian AK (2019) Simultaneous estimation of battery electric vehicle adoption with endogenous willingness to pay. ETransportation 1:100008
16. Egbue O, Long S (2012) Barriers to widespread adoption of electric vehicles: an analysis of consumer attitudes and perceptions. Energy Policy 48:717–729
17. Neves SA, Marques AC, Fuinhas JA (2019) Technological progress and other factors behind the adoption of electric vehicles: empirical evidence for EU countries. Res Transp Econ 74:28–39
18. Steinweg T (2011) The electric car battery: sustainability in the supply chain
19. Lin B, Wu W (2018) Why people want to buy electric vehicle: an empirical study in first-tier cities of China. Energy Policy 112:233–241
20. Hagman J, Ritzén S, Stier JJ, Susilo Y (2016) Total cost of ownership and its potential implications for battery electric vehicle diffusion. Res Transp Bus Manag 18:11–17
21. Vilchez JG, Exploring the battery market for electric cars
22. Rezvani Z, Jansson J, Bodin J (2015) Advances in consumer electric vehicle adoption research: a review and research agenda. Transp Res Part D Transp Environ 34:122–136
23. Bonges HA III, Lusk AC (2016) Addressing electric vehicle (EV) sales and range anxiety through parking layout, policy and regulation. Transp Res Part A Policy Pract 83:63–73
24. Sovacool BK, Kester J, Noel L, de Rubens GZ (2018) The demographics of decarbonizing transport: the influence of gender, education, occupation, age, and household size on electric mobility preferences in the Nordic region. Glob Environ Chang 52:86–100
25. Burgess M, King N, Harris M, Lewis E (2013) Electric vehicle drivers' reported interactions with the public: driving stereotype change? Transport Res F: Traffic Psychol Behav 17:33–44
26. Gerssen-Gondelach SJ, Faaij AP (2012) Performance of batteries for electric vehicles on short and longer term. J Power Sourc 212:111–129. Hsieh MH, Lindridge A (2005) Universal appeals with local specifications. J Product Brand Manag
27. Basu AK, Tatiya S, Bhattacharya S (2019) Overview of electric vehicles (EVs) and EV sensors. In: Sensors for automotive and aerospace applications. Springer, Singapore, pp 107–122
28. Kim HS, Kim HJ, Son B (2006) Factors associated with automobile accidents and survival. Accid Anal Prev 38(5):981–987

29. Nagappan M, Vinoth Kanna I (2020) A novel technique and detailed analysis of cars in Indian roads to adopt low ground clearance. Int J Ambient Energy 41(10):1089–1095
30. Balakrishnan N, Chakravarty AK, Ghose S (1997) Role of design-philosophies in interfacing manufacturing with marketing. Eur J Oper Res 103(3):453–469
31. https://nexonev.tatamotors.com/. Accessed 25th Mar 2022
32. https://nexonev.tatamotors.com/wp-content/themes/tata-nexon/images/brochure/Tata-Motors-Nexon-EV-Brochure.pdf. Accessed 3/25/22
33. https://tigorev.tatamotors.com/. Accessed 3/25/22
34. https://tigorev.tatamotors.com/wp-content/themes/limberev/Tata-Tigor-EV-Brochure.pdf. Accessed 3/25/22
35. https://www.mgmotor.co.in/vehicles/mgzsev. Accessed 3/25/22
36. https://s7ap1.scene7.com/is/content/mgmotor/mgmotor/documents/mg-dc-pdf-0186.pdf. Accessed 3/25/22
37. https://www.mahindraelectric.com/vehicles/everito/. Accessed 3/25/22
38. https://www.mahindraelectric.com/pdfs/eVERITO-Brochure.pdf. Accessed 3/25/22
39. https://www.hyundai.com/in/en/find-a-car/kona-electric/highlights. Accessed 3/25/22
40. https://www.hyundai.com/content/dam/hyundai/in/en/data/brochure/Hyundai_KONA_SUV_brochure.pdf. Accessed 3/25/22
41. https://autos.maxabout.com/cars/tata/nexon-ev. Accessed 3/25/22
42. https://autos.maxabout.com/cars/tata/tigor-ev/tigor-ev-zx-plus. Accessed 3/25/22
43. https://www.mg.co.uk/new-cars/new-mg-zs-ev. Accessed 3/25/22
44. https://www.v3cars.com/mg-cars/zs-ev/dimensions. Accessed 3/25/22
45. https://autos.maxabout.com/cars/hyundai/kona/kona-electric. Accessed 3/25/22
46. Saaty TL (1980) The analytic hierarchy process, vol 324. McGraw-Hill. New York
47. Hwang CL, Masud ASM (2012) Multiple objective decision making—methods and applications: a state-of-the-art survey, vol 164. Springer Science & Business Media
48. Kaliszewski I, Podkopaev D (2016) Simple additive weighting—a metamodel for multiple criteria decision analysis methods. Expert Syst Appl 54:155–161
49. Podvezko V (2011) The comparative analysis of MCDA methods SAW and COPRAS. Eng Econ 22(2):134–146
50. Figueira JR, Greco S, Roy B, Słowiński R (2010) ELECTRE methods: main features and recent developments. In: Handbook of multicriteria analysis. Springer, Berlin, Heidelberg, pp 51–89
51. Kolde R, Laur S, Adler P, Vilo J (2012) Robust rank aggregation for gene list integration and meta-analysis. Bioinformatics 28(4):573–580

Simulation Modeling of a Greenhouse Integrated with Earth-Air Heat Exchanger System

Tarun Kumar, Utkarsh Jha, Yashaswi Raj, and Anil Kumar

Abbreviations

$T_{d,t}$	Soil temperature at time t(s) at depth d(m)
α_s	Ground thermal diffusivity (m^2/s)
α_t	Time since the start of calendar year (day)
T_a	Average soil surface temperature (°C)
m_a	Mass flow rate
t_o	Phase constant of ground surface (s; days)
C_a	Specific heat
h_c	Convective heat transfer coefficient
T_{fi}	Inlet temperature of fluid
T_o	Outlet temperature of air
L	Length of the pipe
Q_u	Heat transfer between pipe and air
A_{ex}	Experimental value
A_p	Predicted value
N	No of observations

1 Introduction

Air conditioning devices nowadays are commonly used to achieve thermal comfort in any three-dimensional space like buildings, residential spaces, workplaces, etc. Utilizing earth-air heat exchangers is one of the most effective methods. Among the

T. Kumar · U. Jha (✉) · Y. Raj · A. Kumar
Department of Mechanical Engineering, Delhi Technological University, Bawana Rd., Shahbad Daulatpur Village, Rohini, Delhi 110042, India
e-mail: utkarshj722@gmail.com

A. Kumar et al. (eds.), *Recent Advances in Manufacturing and Thermal Engineering*, Lecture Notes in Mechanical Engineering, https://doi.org/10.1007/978-981-19-8517-1_7

several actions undertaken for protected agriculture, to bring any change in a greenhouse temperature is supposed the most energy-intensive. The air inside greenhouse increases considerably high during summer especially in Northern India. Tiwari et al. [1] recorded that for Northern India, temperatures inside the greenhouse can reach 45 °C of high temperature in the summer and as low as 6 °C in the winters. Hence concluding in summer and winter, a substantial difference in temperature between day and night hours damages the health of the crop in the greenhouse. Therefore, it is very important to maintain favorable condition for the effective growth of crops [2]. As a result, numerous researchers have proposed employing an EAHE infrastructure to keep greenhouse temperatures consistent.

When the network of tubes are positioned under the earth's surface a constant depth of 1.5–2 m, it helps the tubes in contact with the soil to remain at a constant temperatures throughout the year. An EAHE heating and cooling system is a thermodynamic system where heat transfer takes place between the network of tubes ground and the flowing air [3]. For the air which passes in the tube, it gets heated up in the winter and cools down in the summer resulting in air condition within the greenhouse. As a result, an EAHE's heating/cooling system can be utilized all year. Also, the constant temperature when measured at a depth of 1.5–2 m is equivalent to an area's annual average temperature [4, 5]. As illustrated in Fig. 1, during operations, air is extracted from the greenhouse using a blower which is positioned at the EAHE buried pipe entry.

Various studies conducted throughout the years to verify the effectiveness of EAHE's. Short-term studies, lasting around 24 h, were conducted by Goswami and Dhaliwal [6] who evaluated the change in temperature of the air when it was supplied to an EAHE. The earth-air heat exchanger was 25-m-long and 0.3-m in diameter which was setup roughly at a depth of 2.4 m. For an airflow of around 390 m^3/h (1.5 m/s), roughly 8 °C of temperature reduction in temperature was noted. According to the study's findings, long-term trials should be done to evaluate the practicality of such systems, the performance of an EAHE setup in the winter season which was built in Bhopal, Central India [7]. The heating potential as observed from an experimental configuration of EAHE system was between 0.59 and 1.23 (MJ h) for velocities of

Fig. 1 Earth-air heat exchanger schematic

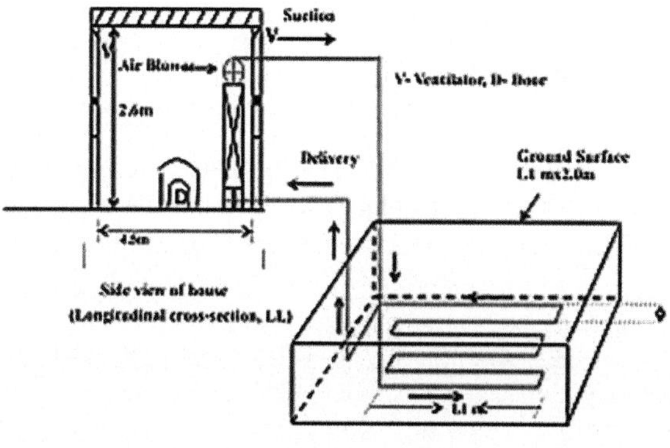

air between 2 and 5 m/s. Energy indicators for the EAHE system, such as seasonal energy efficiency ratio (SEER) and energy payback time (EPBT), were computed in another of their studies [8]. Boundary conditions of Bhopal (Central India) were used to develop the simulation model; it was also reported that heating and cooling capacities were increased by 45% on average when airflow speed increased from 2 to 3.5 and 3.5 to 5 m/s.

Kumar et al. [9] used an artificial neural network to investigate the thermal potential of an EAHE. The performance of EAHEs for household structures was investigated by Ajmi et al. [10] and discovered that, due to their cooling capacity during the summer, they can nearly reduce energy consumption in a normal house by 30%. In order to lower the expense of achieving comfort conditions utilizing subsurface thermal energy, Hamada et al. [11] devised, analyzed and improved EAHE employing a no-dig approach. In hot temperatures in Tehran, Iran, Khalajzadeh et al. [12] tested the heating potential of an evaporative cooler hybrid system. They determined that the hybrid system may effectively replace a regular ACs and that its cooling efficiency is more than one. In an experiment conducted by Li et al., it was shown that the performance of an EAHE improved when the moisture content of the surrounding soil increased [13]. Pakari et al. [14] conducted two experiments in two instalments in which earth-to-air heat exchanger ran continuously for 20 days in the first and the second run intermittently, with the on-cycle lasting for about 5 h every day. It was shown that there is no considerable difference in the performance under continuous and intermittent operation.

In all previous studies, significant problems arise for further development as physical methods require time and lots of monetary resources.

The purpose of this research is to utilize the CFD simulation approach to evaluate the heating and cooling performance of the EAHE system integrated with greenhouse in the summer and winter climates of Delhi, India. The research entails creating a quasi-steady state 3D simulation model and running CFD simulations on it using ANSYS Fluent 2021-R2 to identify hourly variations in greenhouse air temperature equipped with EAHE.

2 Design and Mechanism

The greenhouse paired with the EAHE model is based on an actual experimental setup at IIT Delhi [15] which has a floor size of 6 × 4 m and is oriented in east to west direction where the perpendicular height of the slanting roof is 2 m meters tall and all the walls are 3 m tall. On the west side of the greenhouse, a copper heat exchanger with a 54-m total length and a 0.06-m diameter is placed 1 m below the ground. They are laid out in a serpentine pattern of 35 m in length buried under the ground and 0.5 m spacing, with eight turns. The EAHE's delivery end sits below the greenhouse, with four vents arranged in a plus sign to provide consistent air mixing within. Similarly, the suction end has two openings placed 1 m above the ground

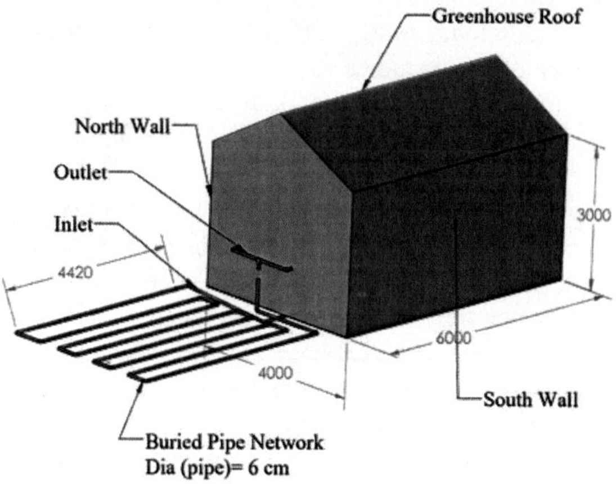

Fig. 2 Greenhouse integrated EAHE arrangement

attached to one of the sidewall. All the modeling and modification were done in SolidWorks2020 precisely to the scale with the existing model, as shown in Fig. 2.

3 CFD Modeling and Experimental Validation

3.1 CFD Modeling and Simulation

Computational fluid dynamics (CFD) is currently highly popular among researchers to study the performance of various EAHE systems [16]. CFD simulations are generally used to determine that how the flow of the ambient air takes place in an EAHE system when the input and output variables are specified under the given condition. These conditions can be defined by imposing suitable boundary conditions. CFD modeling is then used to determine the values of airflow parameters at a large number of EAHE locations. In most cases, these points are linked using a numerical grid or mesh. CFD codes are built on numerical techniques for dealing with fluid flow problems. CFD modeling provides the quantitative solution of partial differential equations in a discretized form that govern the processes involving air flow and heat transfer [17].

3.2 Thermal Modeling and Analysis

It should be mentioned that during the formulation of the CFD model, various numerical approximations and assumptions were made. To ensure that CFD tools are used correctly, a thorough understanding of their applicability range and limitations is

required. The following assumptions were taken into consideration when using CFD to simulate the EAHE system.

- The ground's surface temperature is the same as temperature of the ambient which is also the temperature at which air is entering the greenhouse.
- The yearly average temperature of the site is used to estimate the Earth's undisturbed temperature (Delhi).
- EAHE uses copper pipe with a consistent cross section.
- Because the pipe thickness utilized in EAHE is so thin, the pipe material's thermal resistance is negligible.
- As it is assumed that the soil temperature around the pipe would remain constant, therefore, the pipe's surface temperature in the axial direction is uniform.

3.3 Earth's Undistributed Temperature

The thermal diffusivity of the soil and the temperature of it at a given depth z and time t can be calculated using orthotropic soil parameters

$$T_{\mathrm{d,t}}, = T_{\mathrm{a}} - A_{\mathrm{s}} \exp\exp\left[-d\left(\frac{\pi}{365\pi_s}\right)^{\frac{1}{2}}\right] \times \cos\left\{\frac{2\pi}{365}[t - t_{\mathrm{o}} - \frac{d}{2}\left(\frac{365}{\pi\alpha_s}\right)^{\frac{1}{2}}\right\} \tag{1}$$

The exact value for the temperature of the earth using $t(d, t)$ cannot be calculated because it is computed for average soil parameters. Therefore, a hypothetical value is assumed equal to earth's average ground temperature of a location where temperature of the ground equals the surrounding ambient air temperature. The temperature does not change significantly and can be assumed to be constant at a depth d equal to 0.8 m.

3.4 Energy Analysis

Figure 3 shows the heat transfer that takes place between the pipe of constant wall temperature and the flowing fluid. From the Fig. 3, energy balance equation for a finitely small element of length dx can be written by:

$$\dot{m}_{\mathrm{a}} C_{\mathrm{a}} \frac{\mathrm{d}T(x)}{\mathrm{d}x} \mathrm{d}x = 2\pi r h_{\mathrm{c}} (T_{\mathrm{o}} - T(x)) \mathrm{d}x \tag{2}$$

Solving the above differential equation using boundary conditions at $x = 0$, $T(x) = T_{\mathrm{fi}}$ and at $x = L$, $T(x) = T_{\mathrm{o}}$ we get:

$$T_{\mathrm{fo}} = T_{\mathrm{o}}\left(1 - \mathrm{e}^{-\frac{2\pi r h_{\mathrm{c}} L}{\dot{m}_{\mathrm{a}} C_{\mathrm{a}}}}\right) + T_{fi} \mathrm{e}^{-\frac{2\pi r h_{\mathrm{c}} L}{\dot{m}_{\mathrm{a}} C_{\mathrm{a}}}} \tag{3}$$

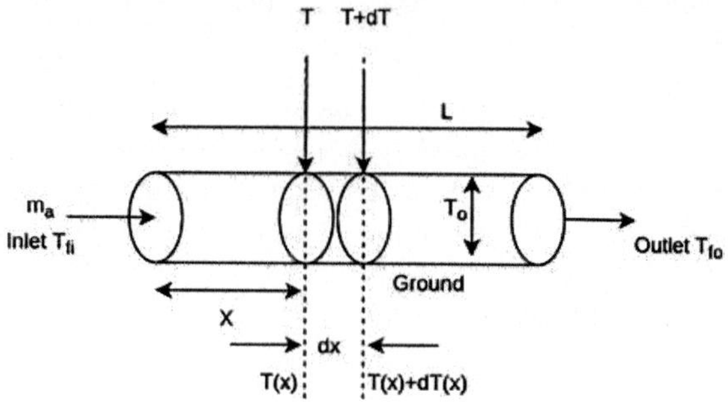

Fig. 3 Heat transfer between flowing air and ground

The thermal gain by the air from the EAHE is given by:

$$\dot{Q}_{\mathrm{u}} = \dot{m}_{\mathrm{a}} C_{\mathrm{a}} (T_{\mathrm{fo}} - T_{\mathrm{fi}}) \tag{4}$$

$$\dot{Q}_{\mathrm{u}} = \dot{m}_{\mathrm{a}} C_{\mathrm{a}} (T_{\mathrm{o}} - T_{\mathrm{fi}}) \left[1 - \mathrm{e}^{-\frac{2\pi r h_{\mathrm{C}} L}{\dot{m}_{\mathrm{a}} c_{\mathrm{a}}}} \right] \tag{5}$$

3.5 *Boundary Conditions*

Inlet Boundary Conditions

A turbulent flow is measured at the EAHE pipe's intake with subsonic flow characteristics. For a mass flow rate of 1000 kg/h, an average airflow speed of 8.2 m/s was determined. For the summer season, the static air temperature at the entrance was set to Delhi's average summer temperature in the month of May and similarly in winter for a typical day of the month, January. The ambient temperature values from Ghosal's study [15] were taken at hourly instant 24-h day cycle for both the days and are depicted in Table 1. At inlet air temperature, thermophysical properties such as thermal conductivity and dynamic viscosity and material properties of air, namely specific heat, density were specified.

Outlet Boundary Conditions

The relative pressure at the EAHE pipe's outlet was set to 0 atm in the subsonic flow regime.

Wall

The locations along the EAHE pipe where air temperature had to be measured, such as where the air entered the room from the EAHE pipe, were determined. In the axial direction, the temperature on the pipe's surface (wall) was kept constant and set to Delhi's ambient temperature (27 °C). A no-slip wall boundary condition with smooth wall conditions was applied on the pipe's interior surface. Also, the room-wall temperature was taken as ambient temperature at that instant, and convective

Table 1 Inlet ambient temperatures of Delhi for a typical summer and winter day

Time of the day	Inlet (ambient) temperature for summer	Inlet (ambient) temperature for winter
1	27	5
3	30	6
5	31	5
7	32	7
9	32.7	10
11	36	13
13	39	16
15	42	16.5
17	42.2	13.5
19	41	10
21	35	7.5
23	27	6

Table 2 Thermophysical properties of materials used in simulation

Material	Density (kg/m^3)	Specific heat capacity (J/kg K)	Thermal conductivity (W/m K)
Air at 40 °C	1.126	1006.9	0.020
Air at 10 °C	1.246	1005	0.025
Copper pipe	8690	385	386

heat transfer coefficient having value 10 w/m^2k was taken for the convection heat transfer between room walls and the air.

The study was carried out utilizing the basic mathematical equations involving the fluid flow coupled with heat transfer. CFD serves as a tool to investigate the impact of operating factors such as pipe size, airflow velocity and ambient temperature on the thermodynamic performance of the EAHE system. Table 2 gives the thermophysical characteristics of copper pipe and air at typical day times of summer and winter at 40 °C and 10 °C, respectively, and in CFD simulations, these variables are utilized as input variables.

3.6 *Mesh Generation*

To establish an efficient mesh for the simulation, a grid convergence study is carried out. This is to verify that additional mesh refining does not have a substantial impact on the results. Mesh independence is studied using three distinct mesh models: coarse, medium and fine. To achieve appropriate mesh quality, the refinement is based on

Table 3 Mesh sizing comparison

	Coarse	Refinement medium	Fine
Number of elements	83,572	139,863	300,977
Number of nodes	84,377	140,859	302,340

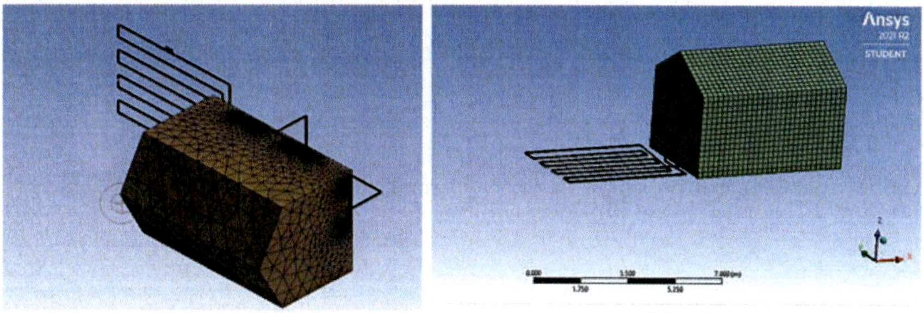

(a) Coarse Mesh (b) Fine Mesh

Fig. 4 CFD mesh generation for greenhouse model

size reduction of the components on the buried pipes and at the pipe-to-greenhouse contact. For comparison, the RANS-based k-turbulence model with the standard wall function is used in all three scenarios. Table 3 gives the features of the three grids tested in the mesh independence study (Fig. 4).

The simulations were run hourly to estimate fluctuation in greenhouse temperature during the day for the three mesh sizes, with the intake air temperature set to the ambient temperature at the time (Table 1) and the temperature of the EAHE pipes set to 300 K. Figure 5 shows that when the mesh is refined, the values of greenhouse temperature do not vary considerably. The average difference in R^2 score between the medium and fine mesh products in reference to the experimental model [15] is just 1.32%. In conclusion, the medium mesh sizing is chosen for all simulations to produce acceptable results with the advantage of reduced computation time.

4 Results and Discussions

4.1 Turbulence Modeling

ANSYS FLUENT R2020 is used to model and simulate the governing equations for unstable the Newtonian incompressible turbulent flow using the Reynolds-averaged Navier–Stokes (RANS) model.

For pressure–velocity coupling, SIMPLE algorithm matrix technique is utilized. According to previous research, a turbulence intensity value of less than or equal to 1% is regarded as mild, and a number more than 10% is typically deemed high. The default values of 5% and 10% for turbulence intensity and viscosity ratio are used at

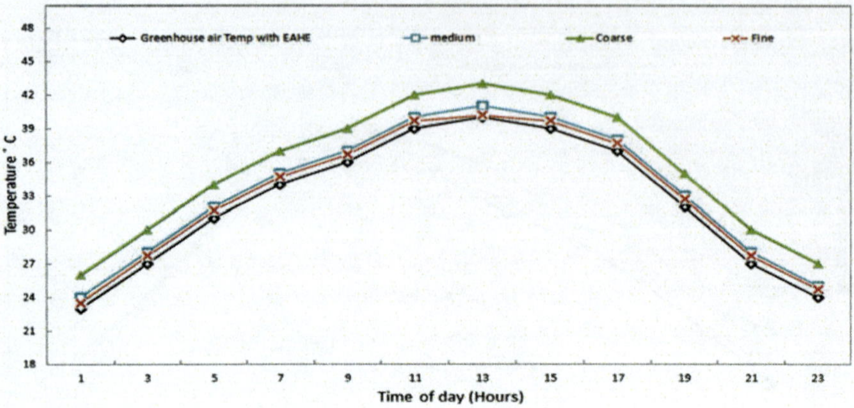

Fig. 5 Coarse versus medium versus fine mesh for CAD model

the inlet and outflow boundary conditions, respectively. The convergence conditions for the residual equations of continuity, momentum and turbulence are set to 10^5.

4.2 *Turbulence Model and Simulation Validation*

The obtained findings are compared to current experimental data to evaluate both the turbulence and simulation models, as well as to justify the best possible turbulence models for the research. The turbulence models employed are the standard k–ω and the standard k–ε, with the former k-epsilon model being most suited for flow away from the wall, such as free-surface flow, and the latter k-omega model is best suited for flow near the wall, where an adverse pressure gradient develops (Fig. 6).

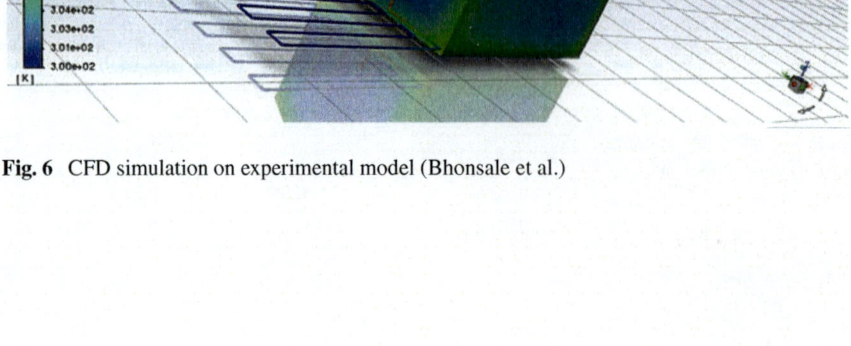

Fig. 6 CFD simulation on experimental model (Bhonsale et al.)

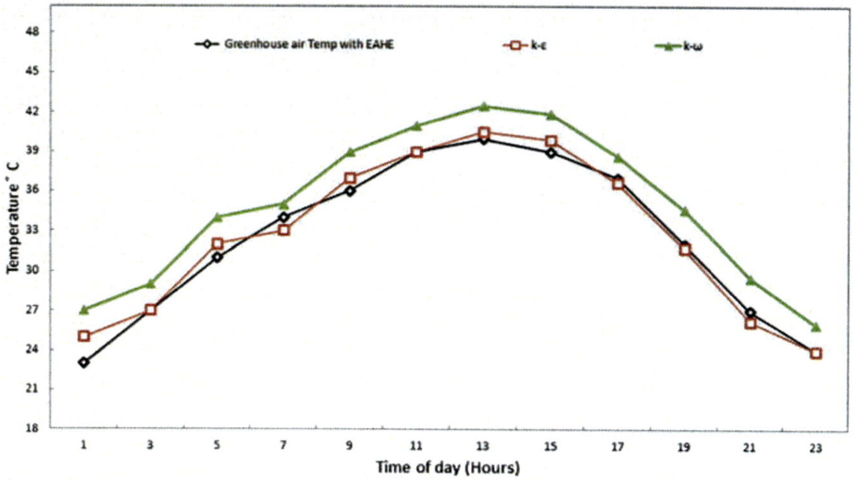

Fig. 7 k-epsilon versus k-omega comparison

The simulation and experimental results of the hourly fluctuations of the greenhouse temperature are compared for a typical summer day with all the same boundary conditions. The experimental data from Ghosal et al. model [15] and simulated results for the two turbulence models are shown in Fig. 7.

In contrast to the experimental findings, the simulated findings are consistent with those of experimental values [14], the coefficient of determination (R^2) and root mean square error (RMSE) [16] are to see which of the turbulence model is more accurate:

$$R^2 = 1 - \frac{\sum_{i=1}^{n} \left(A_{\text{p},i} - A_{\text{ex},i}\right)^2}{\sum_{i=1}^{n} \left(A_{\text{ex},i}\right)^2} \tag{6}$$

$$\text{RMSE} = \sqrt{\frac{\sum_{i=1}^{n} \left(A_{\text{p},i} - A_{\text{ex},i}\right)^2}{N}} \tag{7}$$

Table 4 gives the R^2 value and RMSE from the experimental data for the two turbulence models. The typical k–ε turbulence model produces the best results among alternative models, with an R^2 of 0.989 and RMSE of 0.862. Thus, we went on with the standard k–ε turbulence model for our analysis. Furthermore, with a percentage error of 2.17% for the standard k–ε turbulence model between experimental results and simulation results, the steps performed in the simulation were hence verified.

Table 4 Statistical validation of turbulence models used

Turbulence model	R^2	RMSE
Standard k–ε	0.989	0.862
Standard k-ω	0.975	2.536

4.3 Performance Testing for Typical Summer and Winter Day

The temperature distribution inside the greenhouse at steady state conditions for an average daytime in summer and winter is illustrated in Figs. 8 and 9, respectively. Also, the variation of the average temperature of the air inside greenhouse throughout the day is plotted in Fig. 10 in comparison with temperature variation in the experimental model.

For a typical summer day, when the ambient temperature in the summer varied from 27 to 44 °C, the greenhouse temperature varied from 21 to 40 °C, and on a typical daytime, when the temperature of ambient air was high nearly to 40 °C, the temperature of greenhouse dropped by nearly 7–8 °C on an average.

Similarly, for an average winter daytime, when the ambient temperature varied from 5 to 20 °C, the greenhouse temperature varied from 13 to 24 °C. When the ambient air temperature was as low as 10 °C on a normal summer day, the temperature

Fig. 8 CFD Simulation for summer air conditioning

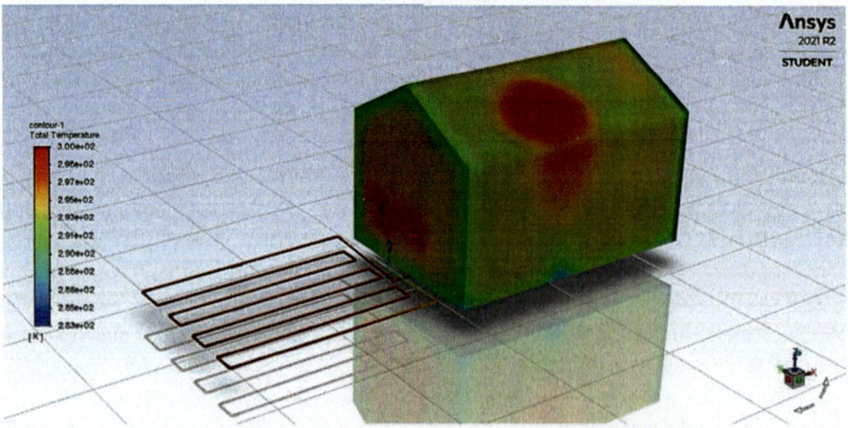

Fig. 9 CFD simulation for winter air conditioning

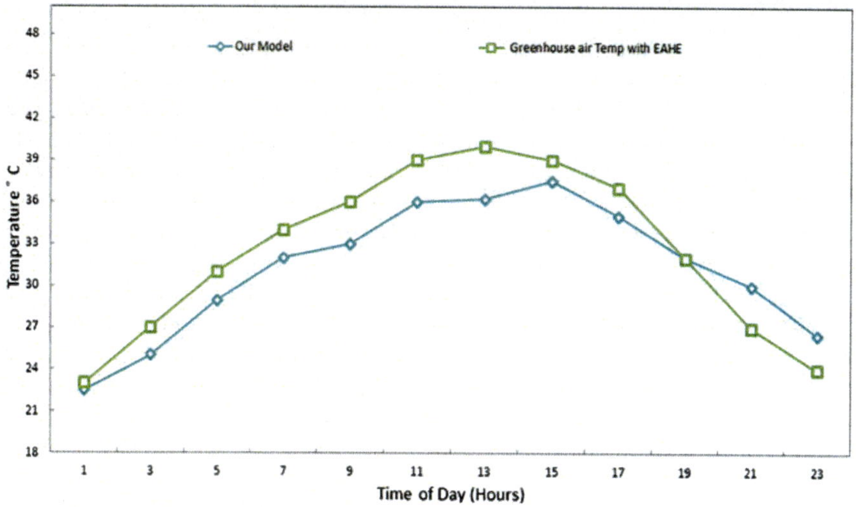

Fig. 10 Model comparison with Ghosal et al. experimental data [14]

within the greenhouse soared by about 8–9 °C on average. An overall decrease in average temp of 6.7% was observed compared to the experimental model for a typical summer day and an average increase of 4.1% for a typical winter day with the same input conditions. This could be due to a more uniform mixing of the air from EAHE since the model had multiple uniformly spaced openings.

5 Conclusions

The greenhouse temperature at different instants of time throughout the day was computed by performing the CFD simulation on the overall system by modeling all the aspects in our thermal analysis related to heat transfer between air and integrated heat exchanger and evaporative cooling taking place inside the greenhouse due to cool air. The predicted temperatures inside the modified greenhouse air in the modified model had values of coefficient of determination of 0.989, and a root mean squared error of 0.928, exhibiting fair agreement with the experimental results of experimental model EAHE. The introduction of the EAHE resulted in a 7–8 °C rise for winter period and a 5–6 °C for the summer period in greenhouse air temperatures.

The results of the modified designed model showed better performance by an overall average of 5.4% as due to multiple air inlets to the greenhouse caused a more uniform temperature distribution when compared to the original model. Hence, our modified model is feasible and, as a result, can also be implemented on a practical scale. An overall decrease in average temperature of 6.7% was observed compared to the experimental model for a typical summer day and an average increase of 4.1%

for a typical winter day with the same input conditions. This could be due to a more uniform mixing of the air from EAHE since the model had multiple uniformly spaced openings. Similarly, results from the parametric study were evaluated for different mass flow rates and pipe diameters.

Acknowledgements We also thank our alma mater Delhi Technological University for providing us with a platform to showcase our research as a part of the curriculum. Different researchers' experience has aided us in numerous ways in improving our study and arriving at the final results. All errors that may inevitably remain are entirely our responsibility.

References

1. Tiwari GN, Sharma PK, Goyal RK, Sutar RF (1998) Estimation of an efficiency factor for a greenhouse: a numerical and experimental study. Energy Build 28(3):241–250
2. Ghosal MK, Tiwari GN (2006) Modeling and parametric studies for thermal performance of an Earth to air heat exchanger integrated with a greenhouse. Energy Convers Manag 47(13–14):1779–1798
3. Puri V (1987) Earth tube heat exchanger performance correlation using boundary element method. Trans ASAE 30(2):514–520
4. Bharadwaj SS, Bansal NK (1981) Temperature distribution inside ground for various surface conditions. Build Environ 16(3):183–192
5. Ghosal MK, Tiwari GN (2004) Mathematical modeling for greenhouse heating by using thermal curtain and geothermal energy. Solar Energy 76(5):603–613
6. Goswami DY, Dhaliwal AS (1985) Heat transfer analysis in environmental control using an underground air tunnel. J Solar Energy Eng 107(2):141–145
7. Bisoniya TS, Kumar A, Baredar P (2014) Heating potential evaluation of Earth–air heat exchanger system for winter season. J Build Phys 39(3):242–260
8. Bisoniya TS, Kumar A, Baredar P (2015) Energy metrics of earth–air heat exchanger system for hot and dry climatic conditions of India. Energy Build 86:214–221
9. Kumar R, Kaushik SC, Garg SN (2006) Heating and cooling potential of an Earth-to-air heat exchanger using artificial neural network. Renew Energy 31(8):1139–1155
10. Al-Ajmi F, Loveday DL, Hanby VI (2006) The cooling potential of earth–air heat exchangers for domestic buildings in a desert climate. Build Environ 41(3):235–244
11. Hamada Y, Nakamura M, Saitoh H, Kubota H, Ochifuji K (2007) Improved underground heat exchanger by using no-dig method for space heating and cooling. Renew Energy 32(3):480–495
12. Khalajzadeh V, Farmahini-Farahani M, Heidarinejad G (2012) A novel integrated system of ground heat exchanger and indirect evaporative cooler. Energy Build 49:604–610
13. Li H, Ni L, Yao Y, Sun C (2019) Experimental investigation on the cooling performance of an Earth to air heat exchanger (EAHE) equipped with an irrigation system to adjust soil moisture. Energy Build 196:280–292
14. Pakari A, Ghani S (2019) Performance evaluation of a near-surface earth-to-air heat exchanger with short-grass ground cover: An experimental study. Energy Convers Manag 201:112163
15. Bisoniya TS, Kumar A, Baredar P (2014) Parametric analysis of earth–air heat exchanger system based on CFD modeling. Int J Power Renew Energy Syst 1:36–46
16. Ghosal MK, Tiwari GN, Srivastava NSL (2004) Thermal modeling of a greenhouse with an integrated Earth to air heat exchanger: An experimental validation. Energy Build 36(3):219–227
17. Dejchanchaiwong R, Arkasuwan A, Kumar A, Tekasakul P (2016) Mathematical modeling and performance investigation of mixed-mode and indirect solar dryers for natural rubber sheet drying. Energy Sustain Dev 34:44–53

Comparative Study of Ethanol-Blended Fuels Using a Stirling Engine Experimental Model

Sparsh Sharma and Yash Sharma

1 Introduction

1.1 Ethanol as a Fuel

Ethanol is a transparent liquid with a unique, characteristic odor and belongs to the –OH group. It has a low melting point of −114 °C, a boiling point of 78.37 °C, and at 20 °C, a density of 0.8 g/ml. Ethanol has a high octane rating of 113 which enables it to operate smoothly on high-compression engines. The fuel gets combusted completely due to the presence of oxygen molecules in ethanol which results in fewer emissions and lesser residue deposits making it a more sustainable and eco-friendly alternative for gasoline.

Ethanol can be blended with any concentration of petrol to create different possible mixtures. In the study, two variants of ethanol-blended fuels: E-85 (maximum of 85% anhydrous ethanol blended with a minimum of 15% gasoline) and E-100 (around 95 vol % ethanol mixed with ~5 vol % water) were used. E-N fuel, where N is the number attached to it, implies that the fuel mixture comprises N% of ethanol along with (100-N) % of gasoline.

The usage of ethanol as an eco-friendly fuel [1] has been now around for centuries. However, the usage of ethanol-blended fuels as a transport fuel (or as a possible bio-fuel) on roads has been growing at a rapid pace recently. Ethanol-blend as a fuel was first used in Samuel Morey's ethanol turpentine I.C. engine. A lot of automobiles, in the twentieth century beginning, started operating on ethanol but they were not as efficient. The major source of organic ethanol production is through the milling process of sugarcane or by refining corn into a fuel which are readily available and economically sound crops available in India. Depending on the volume concentration of ethanol in the blend, ethanol fuels contain lesser energy per liter compared to

S. Sharma (✉) · Y. Sharma
Delhi Technological University, New Delhi 11042, India
e-mail: sparsh2209@gmail.com

A. Kumar et al. (eds.), *Recent Advances in Manufacturing and Thermal Engineering*,
Lecture Notes in Mechanical Engineering, https://doi.org/10.1007/978-981-19-8517-1_8

gasoline. Denatured ethanol (98% ethanol) contains around 30% lesser energy per liter than the former. Another big advantage of using ethanol-blended fuels is that it produces usable by-products that are DDGs (Dried Distiller Grains; used as a high-protein animal feed) and carbon dioxide [2]. Carbon dioxide can be captured from the ethanol production process which can be further used in the production of dry ice and for cryogenic freezing. Also, one metric ton of DDGs could act as a possible replacement for 1.2 metric tons of soybean and corn being used as food products. Ethanol was first properly used as fuel in a car when Henry Ford designed his 1908 model to operate on alcohol. Our prototype is a real-life working model which uses the basic principle of a general thermodynamic cycle—the Stirling cycle for external combustion.

The energy demand [3] in the world is increasing rapidly as a result of the growing economy, expanding population, increasing urbanization, evolving lifestyles, and rising spending power. Around 98% of the fuel requirement in the road transportation industry is currently met by fossil fuels and the remaining 2% is managed by bio-fuels. The vehicular population is estimated to increase as a result of this which in turn will further increase the demand for transportation fuels.

1.2 Stirling Engine and Cycle

The Stirling engine, invented by R. Stirling in 1816, is a heat engine that incorporates external combustion and is very different from the internal-combustion engine that is usually present in our cars. It is a type of an external combustion engine. Over the years, manufacturers have realized that the Stirling engine can be proven to be more efficient than a diesel or gasoline engine. For the experiment, a portable working model of a Stirling engine was built using cheap, non-complex, and easily accessible parts to measure the extent to which we can increase the efficacy of the Stirling engine. Further, it was used to demonstrate and compare the effectiveness, economic, and environmental results for different fuels working on the same cycle.

A Stirling thermodynamic cycle consists of two constant volume isochoric processes and two constant temperature isothermal expansion and compression processes. Stirling engine is principled on the Stirling cycle in which air inside the engine is provided heat from a source outside the cylinder, instead of the fuel combustion inside the cylinder as in the case of normal I.C. engines. The gases that are used inside a Stirling engine are trapped inside the engine's chambers. It contains no exhaust valves that allow high-pressure gases to escape as is in the case of petrol or a diesel engine. Despite these high-pressure entrapments, no explosions take place inside the cylinders. The major principle of a Stirling cycle engine is that a fixed quantity/volume of gas get entrapped inside the confines of the engine and it facilitates a chain of events that alter the pressure of the gas within the engine, causing it to do and produce work output.

The approach in which this research was different from the ones listed below is that this study has used the Stirling engine as the basic framework of operation for

facilitating combustion. Most of the research done in the past has used normal IC engines that are in use for diesel and petrol as a working fuel. The advantage of using the Stirling engine as a base is that it is external combustion which allows us to use different heat sources and combustion chambers according to the space requirements. Stirling engines produce lower vibrations and noise and showcase good consistency with a linear electric machine.

2 Literature Review

Several varied research papers were studied on what methodologies were proposed to obtain measurements on ethanol blending in the past.

In the paper "Effect of lower ethanol-gasoline blends on performance and emission characteristics of the single-cylinder SI engine" [4], the observations on low-concentration ethanol-gasoline blends (20% by vol) were measured for output power and emission characteristics. It was performed on the Bajaj Kawasaki bike, K TEC model which is a single-cylinder four-stroke SI engine. Tests were performed for different blends on a concentration volume basis, specifically 5, 10, and 20%. The torque, fuel consumption, power output, and BMEP were measured along with exhaust emissions for the concentration of CO, CO_2, and HC. The bike was made to run at a variable engine speed from 4000 to 8000 rpm.

In the study "The Effect of Using Ethanol-Gasoline Blends on the Mechanical, Energy, and Environmental Performance of In-Use Vehicles" [5], the effect of using blends of ethanol and gasoline on the torque and power produced along with the fuel consumption and type of gases hydrocarbon emissions was measured. The research also tried to measure the changes in volatile organic compounds such as formaldehyde and acetaldehyde which were insignificantly varying.

A flex-fuel vehicle [6] is typically a vehicle that can run on more than one type of fuel. FFVs have a single fuel tank, fuel system, and engine like conventional gasoline vehicles. Stirling cycles are fascinating examples of the work that can be done by simple thermodynamic processes on mechanical devices. It can be used for taking energy in the form of heat and turning it into mechanical energy. A Stirling engine avoids these problems. The working fluid present continues to remain inside the arrangement throughout the cycle. This, in turn, provides higher operating efficiency and versatility along with lower complexity.

Taking into consideration the paper, "Effects of Ethanol–Diesel on the Combustion and Emissions from a Diesel Engine at a Low Idle Speed" [7], there were four types of ethanol-blended test fuels that were used. The ethanol blending ratios were 0% for pure diesel, and 3, 5, and 10% for 3, 5, and 10% ethanol mixtures (by vol.%). The maximum heat release rate increased by 13.5%, the brake specific fuel consumption (BSFC) increased, up to 5.9%, and the brake thermal efficiency (BTE) for diesel-ethanol-blended fuels was maintained at 23.8%. The coefficient of variation (COV) of the indicated mean effective pressure (IMEP) was consistently lower than 1% when ethanol was blended.

In another paper, “Study of Impact of Ethanol Blends on SI engine performance and Emission” [8], the experiments were carried out on the four-stroke spark-ignition engine with pure petrol, pure ethanol, and 10, 20, 40, 60 and 80% ethanol-gasoline blend. The operation was carried out at a constant speed of 3000 rpm for various loads of one-fourth, half, three-fourth, and full load and conclusions were made that ethanol blending results in lower exhaust emissions under part load and full load operations.

Recent results from Jegan and Chitra et al. [9] were carried out on an existing SI engine for E-10, E-20, and E-25 ethanol blends to diagnose aspects such as air–fuel ratio, operating cylinder pressure ignition timing and compression ratio related to the performance parameters. It was noted how these ethanol blends outperformed by providing certain modifications in the engine.

Another similar study by Pal [10] tested lower percentage ethanol blends on a Maruti Suzuki Wagon R engine for brake power, thermal efficiency, and emission parameters like CO, HC, NO_x, and CO_2 to display that there was a marginal increase in BHP, SPFC, and thermal efficiency by using these blends.

The major gaps that we realized in the papers and which we tried to furnish through ours were (a) most of the studies tested only for a few ethanol blends and that too in the low ethanol percentage range; (b) none of the papers used Stirling engine as their operational framework. Stirling engines work better than IC engines in low temperature zones.

3 Materials and Methods

3.1 Experimental Setup

The schematic diagram (refer Fig. 1) shows how the model was created for the study to facilitate and simulate a combustion process inside the engine by using simple, readily available components.

The major components include:

- Heat Source—A small spirit lamp was made using a transparent vial (which would contain the 10 ml of fuel) and inserted a cotton taper through a hole on the cap which will be used to burn the fuel by lighting through a match stick. This was thought to be a convenient and economical solution for achieving the design of a burner/spirit lamp.
- Design of the Cylinders—A 10 ml glass test tube was used to create the cylinder where the two chambers for combustion could be demarcated effectively. The tube was divided into two sections that are, the hot chamber and cold chamber with the help of a wooden cork so that components do not get mixed between the two. The hot chamber contained steel wool which is used as an economizer to pre-heat the gas in the tube, while the cold chamber contained the graphite piston for its to and fro motion.

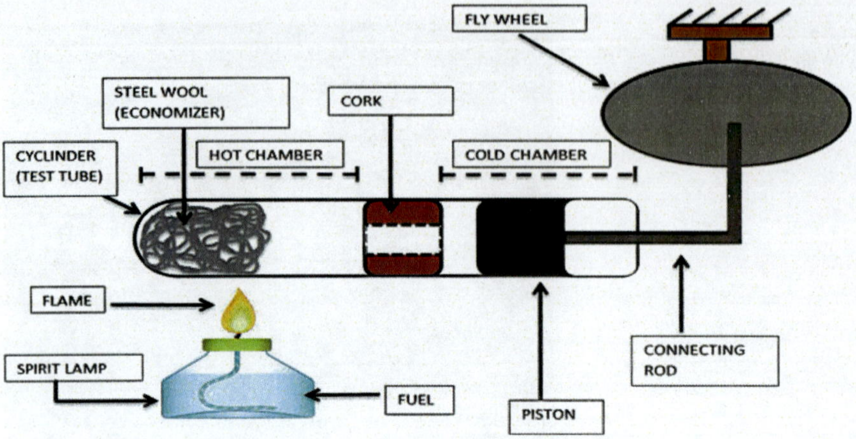

Fig. 1 Schematic diagram of the model design for a working Stirling engine

- Design of the Piston—A graphite electrode (15 mm diameter and 150 mm length) was used as a suitable substitute for a real piston. The reason for choosing graphite as the material is because of its porous nature, its heat resistance, and lubricating properties which facilitate us in not using any form of coolant/lubrication for our prototype taking into consideration the size constraints. Graphite has an extremely high melting point of ~3600 degrees Celsius.
- Design of the Connecting Rod—A simple GI (galvanized steel) wire of length 6 cm was used and shaped it as our connecting rod by twirling its ends and using it as a linkage between the piston and the flywheel.
- Design of the Flywheel—A thick wooden circular structure with rims and holes in between was used as the frame of the flywheel. Wood was used to increase the mass of the flywheel which resulted in a higher moment of inertia and rims were punched to increase the airflow and output revolutions per minute by decreasing the air drag. A ball bearing was used at the center which was fixed to the wooden framework from one side and the flywheel from the other facilitating smooth rotation and helping us create maximum output rpm (refer Fig. 2).

3.2 *Working of the Model*

The fuel present inside the spirit lamp is burnt by lighting the candle lamp, and the heat generation increases the temperature within the left section (hot chamber) containing the steel wool. This sudden increase in the temperature of the sealed hot chamber creates excess pressure inside it, as a result of which, the hot air is forced to move toward the second half of the test tube (cold chamber). The Stirling cycle creates power only during the first part of the cycle and this is that part of the Stirling cycle that does the work.

Fig. 2 Experimental set up with all the components

As a result of this isothermal expansion, the hot air is pushed into the cold chamber [cooled cylinder, i.e., on a lower temperature scale thermally], which quickly cools the gas and reaches thermal equilibrium with the cooling source (i.e., the surroundings here), in turn lowering its pressure (isochoric pressure decrease). This sudden decrease in pressure makes it easier to facilitate compression of the gas in the next part of the cycle.

Now, the cold chamber starts to compress the gas due to a pressure difference. The heat that is generated by this compression of the piston (graphite rod) is removed from the cooling source (surroundings). This makes the piston move up (right, in this case) to decrease the pressure inside.

Due to this sudden decrease in pressure inside the chamber, the gas is again forced back into the heated cylinder (hot chamber), where it quickly heats up, builds up pressure, and the cycle starts repeating itself in the same manner until the entirety of fuel is burnt.

3.3 *Observation*

For the study, 10 ml each of diesel, petrol, E-10, E-20, E-30, E-40, E-50, E-60, E-75, E-85, and E-100 fuels was tested to observe their combustion cycles. The calorific values of these different fuels were mapped on a line graph; see Fig. 3a, to understand how much energy is released in reality by burning 1 kg of that fuel. The fuels were tested only once per day to avoid excessive load on the prototype and to avoid errors from heating and frictional losses.

Observations were made to measure the output of the flywheel in revolutions per minute using a digital tachometer. The output time (duration in which the entire 10 ml of fuel was combusted) was also noted down in seconds. Lastly, the type of smoke and residue left after the combustion of each fuel was also carefully noticed as given

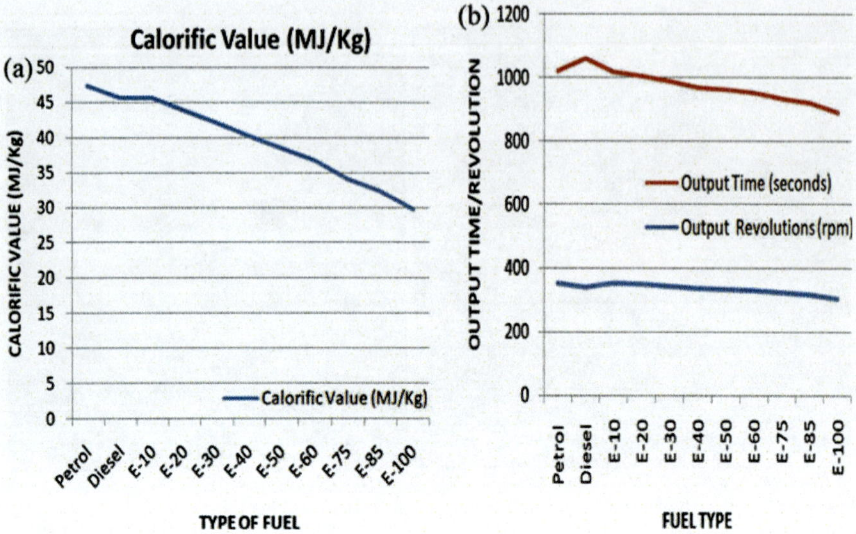

Fig. 3 **a** Calorific values of the different fuels; **b** comparing the output time (in seconds) and the output rpm for different fuel

in Table 1. A line graph was plotted to compare the output time (in seconds) and the output revolutions (in rpm) for the different concentration blends of ethanol, see Fig. 3b. The residue deposit on the combustion of petrol and E-10 fuels is shown in Fig. 4a, b to demonstrate the difference in carbon footprint left by the two fuels.

Table 1 Observation table for noting down output rpm, cycle time, and type of smoke produced

Type of fuel	Fuel quantity (ml)	Calorific value (MJ/kg)	Output (rpm) revolutions	Output time (s)	Type of smoke
Petrol	10	47.30	352	670	Black
Diesel	10	45.50	340	721	Black
E-10	10	45.54	351	667	Black
E-20	10	43.78	349	655	Black
E-30	10	42.02	342	647	Black
E-40	10	40.26	336	633	Light black
E-50	10	38.50	332	631	Light black
E-60	10	36.74	327	624	Light black
E-75	10	34.10	321	612	Very little in qty
E-85	10	32.34	316	602	Very little in qty
E-100	10	29.74	302	588	Clear(negligible)

Fig. 4 **a** Black carbon deposit on combustion of petrol; **b** no residue left on combustion of E-100 fuel

3.4 Methodology

The following are the different formulas used in the calculation for different physical properties that we referred from a paper related to rotational motion dynamics [11]. These calculations can be given in Table 2.

$$\omega = (2\pi N)/60 \tag{1}$$

$$F = M\omega^2 r \tag{2}$$

$$\tau = F * r \tag{3}$$

$$P = \tau * \omega \tag{4}$$

$$E = P * t \tag{5}$$

R = Radius of flywheel = 4 cm = 0.04 m; M = Mass of flywheel = 100gm = 0.1 kg; N = Output rev (rpm).

ω = Angular Velocity (rad/s); F = Force (N); τ = Torque (Nm); P = Power (W); E = Energy (J); t = Time(s).

Table 2 Calculations for the different physical properties of each fuel using the formulas given above

Type of fuel	Angular velocity (m/s)	Force (N)	Torque produced (Nm)	Power output (W)	Energy output (J)
Petrol	36.86	5.43	0.2172	8.01	5366.70
Diesel	35.60	5.07	0.2028	7.22	5205.62
E-10	36.76	5.40	0.2160	7.94	5295.98
E-20	36.55	5.34	0.2136	7.81	5115.55
E-30	35.81	5.13	0.2052	7.35	4755.45
E-40	35.19	4.95	0.1980	6.97	4412.01
E-50	34.77	4.84	0.1936	6.73	4246.63
E-60	34.24	4.69	0.1876	6.42	4006.08
E-75	33.62	4.52	0.1808	6.08	3720.96
E-85	33.09	4.38	0.1752	5.80	3491.60
E-100	31.63	4.00	0.1601	5.06	2975.28

4 Results and Discussion

The calculations in Table 2 were focused on how much torque can be produced and how much power output we are able to obtain from each of the fuels. A bar graph was plotted for comparing the energy output of the various blends; see Fig. 5a and for the power output, as shown in Fig. 5b.

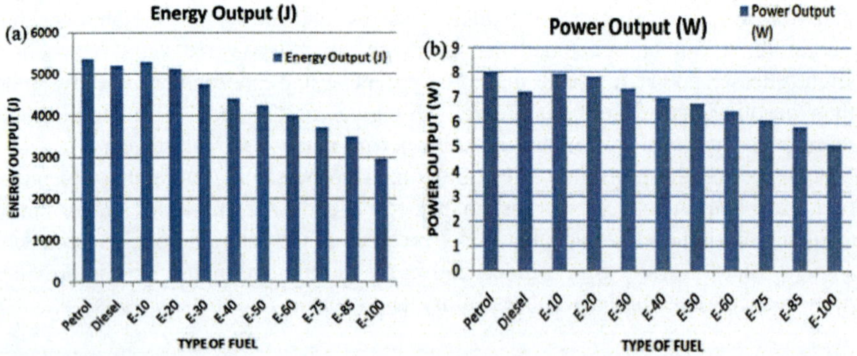

Fig. 5 **a** Energy produced (in Joules) by the different fuels on combustion; **b** power output (in Watts) by the combustion of fuels

5 Conclusions

Analyzing the above results, it can be noted that ethanol-blended fuels produce slightly lesser power and energy output as compared to diesel and petrol for the same quantity of fuel. The trends show a slight decrease in power and energy output of ethanol-blended fuels as compared to diesel and petrol for the same quantity of fuel as we move from E0 to E100. But as the percentage of ethanol was increased in the mixture, the smoke produced got cleaner and lesser polluted with carbon monoxide and nitrogen dioxide effluents.

Ethanol has the property of vaporizing in high temperate so we can see that the run time started decreasing as the % of ethanol in the mixture started increasing. However, the torque produced by each of the fuels on the flywheel is almost equal.

There is a significant difference in the energy output of E-100 fuel (2.97 kJ) as compared to diesel (5.20 kJ) and the gap would further increase when these fuels are used in the engine of an entire automobile raising concerns on whether these fuels are transport-worthy or not. The research further tried to study the behavior of fuels like E30, E40, E50, E60, and E75 which was not present in other research papers and the pattern was similarly decreasing for the different physical quantities.

In terms of cost efficiency, E-100 fuel is 25–30% more cost-efficient than petrol and 23–28% more cost-effective than diesel for producing the same amount of energy, while E-85 is 20–25% more cost-efficient than petrol and diesel for producing the same amount of energy.

Using the results from the experimental model analysis, it was concluded that ethanol-blended gasoline fuels (specifically in the range of 10–25% in the concentration of ethanol) are much more efficient as they produce almost the same energy output with lesser emissions of oxides of carbon and nitrogen in the atmosphere. Therefore, it can be concluded that although ethanol-powered engines produce slightly lesser power it is way more cost-effective and environment-friendly than their gasoline and diesel counterparts. This gives us the scope of extending our research to find means for improving their performance in the future years.

Considering these insights, it can be concluded by saying that ethanol-blended fuels are the future in creating a sustainable, eco-friendly alternative for our currently available non-renewable fuels but major research in the future is required to reach a stage where it easily gets blended within the automotive industry by improving performance, range, mileage, accessibility, and comfort for the daily users.

References

1. Bielaczyc P, Woodburn J, Gandyk M, Szczotka A (2016) Ethanol as an automotive fuel—a review. Combust Engines 166(3):39–45. https://doi.org/10.19206/CE-2016-338
2. Flugge M, Lewandrowski J, Rosenfeld J, Boland C, Hendrickson T, Jaglo K, Kolansky S (2017) A life-cycle analysis of the greenhouse gas emissions of corn-based ethanol. USDA Contract No: AG-3142-D-16–0243

3. Sarwal R, Kumar S, Mehta A, Varadan A, Singh SK, Ramakumar SSV, Mathai R (2021) Roadmap for ethanol blending in India 2020–25. Publishing Agency: NITI Aayog ISBN: 978–81–949510–9–4
4. Kumbhar VS, Mali DG, Pandhare PH, Mane RM (2012) Effect of lower ethanol gasoline blends on performance and emission characteristics of the single cylinder SI engine. Int J Instrum Control Autom (IJICA) ISSN: 2231–1890 1(3,4)
5. Tibaquirá JE, Huertas JI, Ospina S, Quirama LF, Niño JE (2018) The effect of using ethanol-gasoline blends on the mechanical, energy and environmental performance of in-use vehicles. Energies 221. https://doi.org/10.3390/en11010221
6. Zhai H, Frey HC, Rouphail NM, Gonçalves GA, Farias TL Comparison of flexible fuel vehicle and life-cycle fuel consumption and emissions of selected pollutants and greenhouse gases for ethanol 85 versus gasoline. J Air Waste Manage Assoc 59(8):912–924
7. Kim HY, Ge JC, Choi NJ (2020) Effects of ethanol–diesel on the combustion and emissions from a diesel engine at a low idle speed. Appl Sci 10:4153. https://doi.org/10.3390/app10124153
8. Pai S, Tasneem HRA, Rao A, Shivaraju N, Sreeprakash B (2013) Study of impact of ethanol blends on SI engine performance and emission. In: National conference on challenges in research and technology in the coming decades (CRT 2013), pp 1–7
9. Jegan TMC, Chitra R, Glivin G (2021) Testing of ethanol as an alternative fuel for IC engine. Energy Res J 12(1):13–21. https://doi.org/10.3844/erjsp.2021.13.21
10. Pal A (2011) Blending of ethanol in gasoline Impact on SI engine performance and emissions. Int J Thermal Technol 4(1):1–5. https://doi.org/10.14741/IJTT/MAR.2014.01
11. Pyrhönen J, Hrabovcová V, Semken R (2016) Torque and force production and power. https://doi.org/10.1002/9781119260479.ch5

Execution of CNG on Two-Wheeler Vehicles

Ankit Saurabh, Anjani Kumar Jha, Aditya Tiwari, and Amit Pal

1 Introduction

Transportation is the backbone of any economy in the modern times. Humans need to travel for their work. Their food comes from a sophisticated supply chain, which involves transportation of goods, and transportation is a key means even for their leisure activities like tour and travel. The human mobility got wings after the discovery of automobiles and two-wheelers. They provide easy and fast commute for our daily needs, though this comes at a heavy price for our Mother Nature and planet. Currently, the majority of the automobiles run on non-renewable fossil fuels such as petrol and diesel, and they are a major source of air pollution and causing Global Warming [1, 2]. There is an urgent need for eco-friendly fuels today, which would allow better environment and clean air to breathe and live. Compressed natural gas (CNG) is basically natural gas—which is primarily methane (CH_4) along with some other constituents—compressed at a very high pressure. CNG is an attractive low emission alternative for the fossil fuels, and it delivers similar horsepower ratings. The Supreme Court of India had directed an obligatory use of CNG for the taxis by 31 March 2016 to curb the deadly levels of pollution in the city [3]. Still the usage of CNG in India is limited to mostly commercial purposes heavy vehicles. Two-wheeler vehicles are a major source of transportation in India and they also account for almost

A. Saurabh (✉) · A. K. Jha · A. Tiwari · A. Pal
Department of Mechanical Engineering, Delhi Technological University, Shahbad Daulatpur, Delhi 110042, India
e-mail: ankit.saurabh.tech@gmail.com

A. K. Jha
e-mail: anjani.jha.1552@gmail.com

A. Tiwari
e-mail: imtaditya15@gmail.com

A. Pal
e-mail: amitpal@dce.ac.in

A. Kumar et al. (eds.), *Recent Advances in Manufacturing and Thermal Engineering*, Lecture Notes in Mechanical Engineering, https://doi.org/10.1007/978-981-19-8517-1_9

1/3rd of the air pollutants in India [4]. So, it is imperative to minimize the pollution caused by two-wheelers to solve the growing pollution problem. CNG could be the perfect alternative to the fossil fuels in two-wheelers. However, the implementation of CNG as a fuel for motorcycles vehicles is negligible due to certain factors which include safety, compactness, ergonomics, and also the availability of gas stations. CNG and the benefits it provides over traditional fossil fuels are discussed in this paper. The working of CNG engine is very similar to that of petrol. This means that the vehicles running on petrol can be made to run also on CNG by using some equipment. This paper provides a brief about the implementation issues and cost benefit analysis followed by emission analysis.

1.1 CNG and Its Advantages

CNG is made by compressing the natural gas to less than 1% of the volume at atmospheric pressure [5]. Natural gas is a naturally occurring colourless and odourless mixture of gases which is majorly composed of methane (70–90%) and some other hydrocarbons [6]. It is stored and distributed in hard cylindrical or spherical shaped containers at a pressure of 20–25 MPa. The composition of CNG sold in India is given in Table 1 [7]. It is not only cheaper and economical than other fuels, but also emits significantly less amount of harmful pollutants in the environment. CNG is 35–75% less expensive compared to diesel and petrol, approximately 20% compared to electric, and around 60% compared to the petrol-electric hybrid [8]. It is also called green fuel because of its lead- and sulphur-free character. Its non-corrosive nature and absence of any lead or benzene content enhances the longevity of the spark plugs by eliminating lead fouling which looks like yellowish-brown deposits on the spark plug's insulator nose. It can cause the engine to misfire at high rpm and under hard acceleration. The dry nature of CNG increases engine life as it leaves no residues behind in the piston. CNG also increases the life of lubricating oil as it does not contaminate and dilute the crankcase oil [9]. Overall, vehicles running on CNG require low maintenance cost than running on other fuels. CNG is safer than other fuels. It is lighter than air and it rises and disperses quickly in the air in case of accidental spill, thus minimizing the risk of fatality. Self-ignition is also prevented here and it is less likely to ignite on hot surfaces due to a high auto-ignition temperature which is approximately 540 °C and a narrow range (5–15%) of flammability. For comparison, auto-ignition temperatures for petrol and diesel are, 280 °C and 210 °C, respectively [10]. CNG has high octane number of approximately 130 compared to 90 for premium gasolines, which allows for the increased engine compression and combustion efficiency of CNG [9].

Table 1 Typical composition of natural gas [7]

Chemical name	Chemical formula	Percentage composition (%)
Methane	CH_4	80–90
Ethane	C_2H_6	7–8
Propane	C_3H_8	2–3
Butane	C_4H_{10}	0–1
Carbon dioxide	CO_2	0–8
Oxygen	O_2	0–0.2
Nitrogen	N_2	0–5
Hydrogen sulphide	H_2S	0–5
Other rare gases	Ar, He, Ne, Xe	Trace

1.2 Basic Parts and Working of a CNG Engine

The basic working principles of CNG engine is similar to that of petrol that is why, it is easy to convert engines running on petrol to CNG by using some additional equipment. It is important to understand the main parts of CNG engine, as shown in Fig. 1 and explained further to appreciate the parts of the conversion kits described later.

- **CNG Fuel Tank**: It stores the CNG in the vehicle until the engine needs it. A typical cylinder with a 50-L water-carrying capacity is capable of carrying approximately 9 kg of CNG. This is equivalent to 12.5 L of petrol.

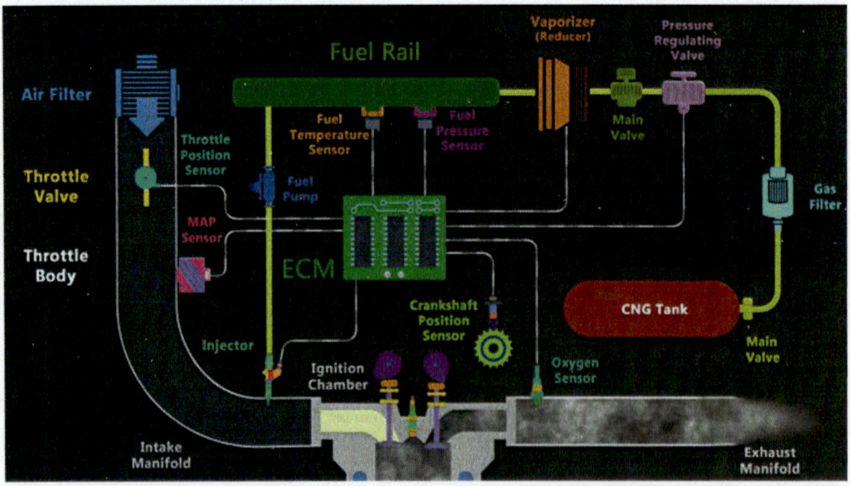

Fig. 1 Schematic design of CNG engine [11]

- **Gas Filter**: The gas filter separates any particulate matter in the CNG which could affect the engine. It is generally installed in CNG run vehicles to forestall contaminants in the fuel tank from getting into the engine, securing basic engine parts.
- **Main Valve (MV) and Pressure Regulating valve (PRV)**: The output can be turned off and on manually by the main valve. The PRV regulates the high pressure of gas coming from the tank. It also operates when the gas pressure in tank exceeds the limiting pressure.
- **Vaporizer/Reducer**: A vaporizer is necessary in CNG conversion systems to vaporize CNG to a low pressure so that it can be later mixed with air for combustion.
- **Fuel Pump**: It transfers fuel from CNG tank to the engine fuel injection system through the fuel rail. The pressure is then decreased to a lower level, and the fuel is brought into the intake manifold or ignition chamber, where it is blended in with air.
- **Electronic Control Module (ECM)**: It is used to control engine performance factors such as fuel efficiency, gasoline flow rate, emission, and ignition time. Furthermore, it monitors the operation of the vehicle.
- **CNG/Gasoline Switch**: The switch is wired into the vehicle's electrical framework. It is programmed to check that fuel is properly flowing either from CNG or from petrol, and that they both are not open or close at the same time.
- **Fuel Injection System**: It is used to inject the compressed natural gas into the engine's combustion chambers for ignition process.
- **Sensors**: Sensors are used to convey the information such as pressure and temperature to the main electronic control module.

1.3 Components of CNG Prototype

As discussed previously, the working and design of CNG engine is similar to that of petrol engine. So, a two-wheeler running on petrol can be made to run also on CNG by using certain parts including reducer, selector switch, high- and low-pressure pipes, manometer, air gas, filling and fitting valves. The different parts of the CNG prototype (Fig. 2) are discussed ahead.

- **Reducer**: A device which is used to reduce the high pressure of CNG which is approximately equal to 200 bar to required pressure for engine to work of about 5–7 bar. The reduction happens in two stages, first being from 200 to 100 bar and second being from 100 to 7 bar. It is the next bulkiest part of the conversion after the CNG tanks and adds to the major portion of weight of the whole system. Size is about 15 cm in diameter and 8 cm depth.
- **Air Gas Valve/Mixer**: It mixes the air with CNG according to the requirement of the engine and it is also used to regulate the flow of CNG from reducer to carburettor at reduced pressure (5–7 bar). Size is 6.5 cm and Dia. is 4.5 cm.

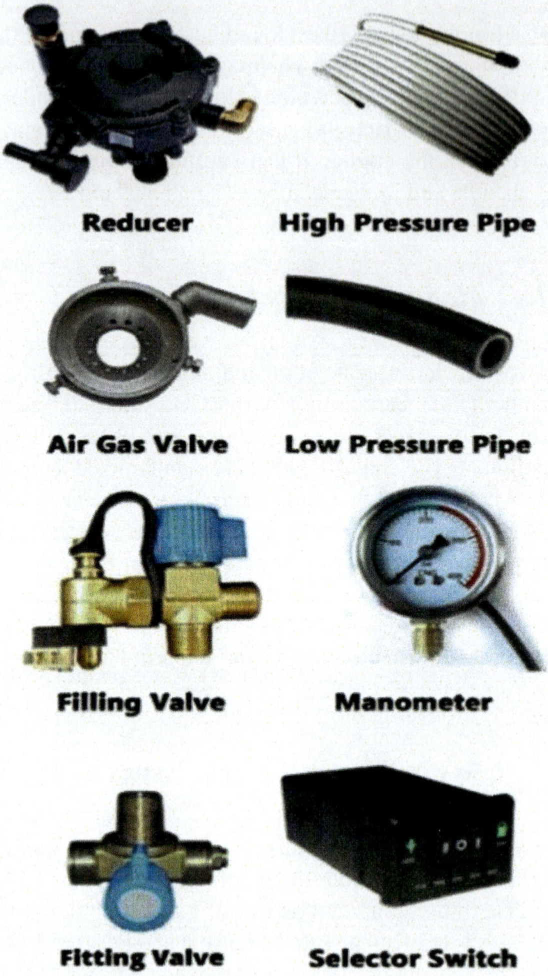

Fig. 2 Components of CNG prototype [12]

- **Filling Valve**: It is used to add and refill CNG into the tanks which are installed on the body of the vehicle. Length is typically 10 cm.
- **Fitting Valve**: It is used to regulate the flow of CNG from cylinder to the filling valve. It is installed at the inlet of cylinder. Size is typically 8 cm in length.
- **High-Pressure Pipe**: Connecting pipes which are used to connect between various stages of conversion where the pressure is high. The diameter of this pipe is 0.5 cm. This is used in three stages: Filling valve to reducer; cylinder to filling valve; and in between multiple cylinders if they exist.
- **Low-Pressure Pipes**: As the name suggests, these are the pipes which are employed to work under low pressure (5–7 bar). The diameter is 2.5 cm. It is used to transfer the CNG from the reducer outlet to engine inlet.

- **Manometer**: It is used to indicate the pressure of the gas in the CNG tank cylinder at the time of filling. The manometer is also connected with the switch to indicate the amount of gas which is left inside the cylinder.
- **Selector Switch**: It is used to switch between the fuel (CNG or petrol) which is used in the engine. It can be either digital or mechanical. It is installed on the handle and its size can be as small as 2 cm^2.

1.4 Major Pollutants from Exhaust

Two-wheelers are one of the major sources of multiple pollutants such as particulate matter (PM), carbon monoxide (CO), many different types of hydrocarbons (HC), etc. In this paper, pollution under control test (PUC) is performed on the 2 major pollutants HC and CO which are standardized and widely accepted.

Above are the two major emissions which are measured in an emission test by an AVL *Exhaust Gas Analyser*. These emission tests are based on lambda coefficient (λ) which is a measure of percentage of oxygen in the exhaust gases. In case the mixture is lean, NOx will be produced at elevated temperatures along with CO_2. In case the mixture is rich, it will be impossible to convert all fuel into CO_2 and it will lead to the formation of CO and unburnt HC.

2 CNG in Two-Wheelers: Issues and Insights

The whole idea behind this paper is to find a way in which CNG can be employed in motorcycles by introducing minimal changes in the already existing design.

The most important aspect of having anything new and easily accepted by the masses is to blend it with the current and existing technologies; that is, an innovation is effective and practical not only when it is new and out of the box, but also feasible if it comes into the practice easily and is acceptable and compatible for the masses without critical changes in the traditional structure or the blueprint.

This thing is taken care into this paper. Hundreds of thousands of two-wheelers are already used across the globe. Rather than wiping out all of them completely and making a new design, attachments and accessories can be added to them which will make petrol running vehicles compatible with the new green and abundant fuel CNG.

Certain equipment can be used as described in the previous section to make the vehicle CNG compatible. However, design issues such as bulkiness and safety are critical problems to solve before these equipments can be used in the two-wheelers.

Design issues: CNG vehicles require greater amount of space for fuel storage than conventional gasoline powered vehicles since it is a gas, rather than a liquid like gasoline. CNG tank can be installed easily over the front storage in scooters,

which has been tried on a very small scale. Similar principle can be applied in other two-wheelers like bike, but not exactly the same idea because:

1. **Space constraint**: There is no storage space present either below the handle or under the seat.
2. **Safety consideration**: In scooters, the tank is shielded from the front, avoiding direct damage in case of collision and is far from the direct heat from the engine, which is not easily possible in the bike.

The tanks are conveniently stored in the front of the scooters. It is shielded from the front and is far from the engine, which makes it an ideal location.

3 Methodology

To decide the optimal position of CNG tanks in the two-wheelers, brainstorming was done for various possible designs, then eliminating among them by analytical and practical methods. Not only the location but also various other factors such as safety, stability, and space considerations were also discussed in the design. For this purpose, a *Hero Honda Splendor Plus* bike properly functioning on petrol was used. This was taken as the subject on which various factors were tested. Two dummy loads were taken (equivalent to the weight and size of two CNG cylinders used in scooters). These loads were attached to various locations on the bike and the stability of the bike was tested while driving it. Apart from this, safety was also taken into consideration by mounting the dummy loads to the least exposed locations. Many locations were discussed and/or tried, but later in the Results section, only those locations are shown which are easily feasible and does not alter the stability of the bike.

3.1 Online Survey

For estimating various key factors such as the maximum cost users are willing to pay for CNG prototype, design and environmental preferences of users, availability of CNG filling stations, etc. an online survey was conducted in which various questions were asked using an online google form. The lists of all questions and their responses have been provided in the appendix.

A total of 46 responses were received (age range: 18–44 years, Fig. 3), majority of them being young subjects. The results of the survey are taken in the upcoming sections, wherever it is necessary. One of the interesting results from the survey is that almost 69% (31/45) of the users indicated that the motivation to use CNG for them is environment compared to lesser cost, even if cost is a major deciding factor for India buyers in general. Some other results of user survey are that 60% (27/40) travel alone, 88% (35/40) travel 50 km or less daily. Also, 36% (15/42) users indicated that

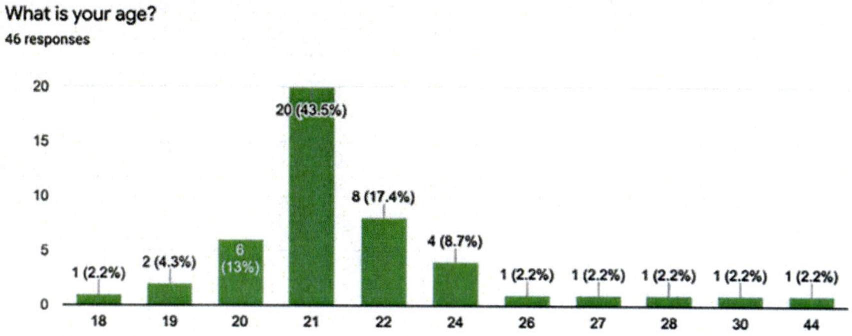

Fig. 3 Age distribution of respondents in online survey

there are ≥5 CNG pumps in 25 km zone from their home. Also 86% (37/43) of the users would like to have hybrid options (CNG + Petrol) rather than only CNG.

3.2 Possible Location of Tanks

- **Location 1: Replacing Side Box with CNG Tanks**

In this position, the CNG tanks take place of the side box which is installed majorly in the Indian subcontinent. This is replaced by a permanent cavity which contains two CNG cylinders. It is an easy location to install CNG tanks, just a carriage for CNG tanks is needed. This is most suitable for single person bikes, or delivery bikes such as for pizzas, grocery, etc.; that is, where the pillion is not there. This does not mean that no pillion can sit in this case. In India, people sit along with *side box* also, so it can be taken into consideration for those who are ready for the benefits of CNG sacrificing a little bit of comfort.

- **Location 2: Back of the Bike**

This is the second position in which the cylinder is installed in the back. In this case, there is no problem for the pillion, and the *side box* space is free to use. The tank is far from the exhaust, which is a positive factor but it is more exposed to impact by collision, which can often occur from the back. To protect from that, the CNG cylinders should be placed inside a sturdy case. One of the drawbacks of this design is that the bike will be a little bit longer than what it was, and it reduces the stability by shifting the centre of gravity towards the rear end.

- **Location 3: Both Side on Foot-rest** (Fig. 4)

This is the third and final position (Fig. 4) proposed for the prototype**.** The cylinders are arranged along the length of motorcycle over the foot-rest one on each side. The foot-rest is providing sufficient auxiliary support to the cylinders along with the

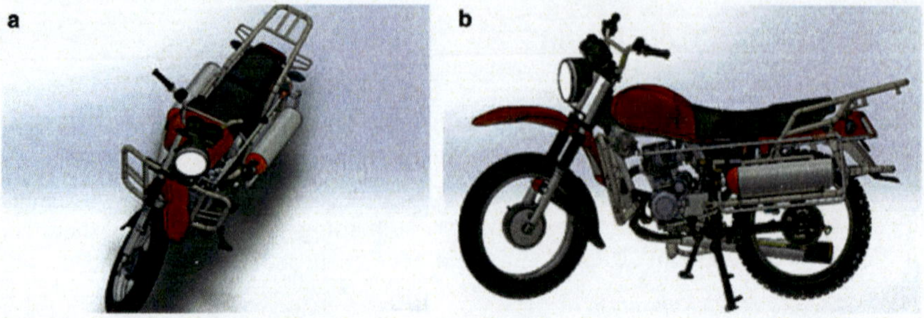

Fig. 4 **a** Isometric view of location (3) of CNG tanks; **b** side view of possible location (3) of CNG tank

welded brackets which contributes to the safety of the tanks. The overall dimension of the cylinder is such that it lies within the enclosure created by the leg guard. Hence, it is also protected from the front side, which adds on extra safety to it. The effective length of the motorcycle also remains same, and the centre of gravity is not affected along the transverse axis. However, it is shifted downward along the height of the prototype which in turn leads to better stability. Along with the safety and stability, this location is also more comfortable to the pillion as it provides some extra leg space when compared to "location 1".

3.3 Possible Location of Reducer

The Reducer is the heaviest and bulkiest part of the prototype after the CNG tanks. But it can be easily fit into the probable locations as shown in Fig. 5a. The other parts are small and can be fit easily in the overall design.

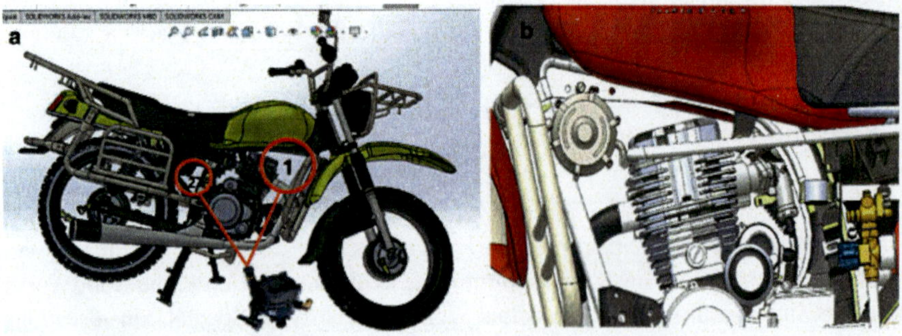

Fig. 5 **a** Possible location of reducer; **b** solidWorks model of reducer

4 Results and Conclusion

The petrol bikes can be made to work with CNG by switching the fuel inlet from petrol to CNG with addition of certain equipment like Reducer, various Valves, and Storage Tank as discussed before. Alternative choices and more than one options are always preferred over a single if in case something goes wrong. So, there is a provision to switch from CNG to petrol, and vice versa according to the needs.

4.1 Final Location of Tank

Out of all, the suggested locations as discussed in the methodology part, the location (3) is taken as the final location due to the following reasons:

- Safest location: It is protected from external damage as it is shielded by the leg guard.
- Sturdy: It is supported by the foot-rest as well as the welded bracket.
- Most Stable: The centre of gravity is shifted towards the ground which increases the stability of the motorcycle.
- Comfortable: It provides ample leg space for the pillion and is the most comfortable location practically possible.

4.2 Final Location of Reducer

Out of all, the suggested locations as discussed in the methodology part, the location (1) is taken as the final location due to the following reasons:

- Safest location: It is shielded from external damage as it lies within the enclosure formed by the leg guard.
- Feasibility: Keeping the space constraint in mind, this is the only practical solution.

4.3 Final Prototype

Taking all factors into considerations, which include optimal location of tank, reducer, high- and low-pressure pipes, pressure gauge, selector switch, air gas, fitting and filling valve, the following working prototype is proposed. Firstly, the SolidWorks model is shown followed by the actual working prototype. Special considerations have been given to the piping system, so that it does not interfere with the moving parts of the motorcycle such as pedals of gears, suspensions, handle, and wheels. Practical assessment was also done throughout the brainstorming and final prototype assembly which ensures the model is stable and safe to function.

4.4 Cost Benefit Analysis

The biggest factor which decides whether a project is sustainable or not is how economical it is. Cost is the deciding factor of many things which is made to practice. In the survey, 69% (31/45) users preferred cost over comfort (24%; 11/45) or design and aesthetics (7%; 3/45).

To estimate the overall cost of constituent parts available in the market, an online marketplace [12] was used, which is a major retailer for India. The costs of the items are as given in Table 2.

Based on the current cost given in Table 2, the total cost of the prototype is ₹9800. Adding some miscellaneous charges like cost of installation, the total cost could be around ₹12,000. It is worth noting here that these prices are an upper limit as the prices will be usually lower if the items are purchased in bulk, and still further lower if they are manufacture in-house on a large scale. In the survey, most of the users responded that they are willing to pay a maximum of ₹10,000 (56%, 24/43), followed by ₹7000 (40%; 17/43), and ₹15,000 (4%; 2/43). None of the users opted for the highest price ₹20,000.

Market researches [13, 14] say that the CNG kit equipped scooters offer a running cost of around ₹0.6 per km. This includes nearly ₹0.48 per km for the fuel and ₹0.12 per km for maintenance. However, running cost of a standard scooter on petrol is ₹1.75 per km, indicating that running cost of CNG per km **is about one third as that of petrol**.

Hence, total cost benefit (C_b) achieved per km = 1.756−0.596 = ₹1.16 per km (Table 3).

Break-even point: Kilometres required to recover the cost.

Table 2 Cost of components of CNG prototype from [12]

Item	Specifications	Final cost (in ₹)
Reducer	200/50–7/5 bar	3500
Tank	1–1.2 kg and 200 bar	3100
High-pressure pipes	0.5 cm diameter × 8 feet	350
Low-pressure pipes	2.5 cm diameter × 4 feet	200
Air gas valve	4.5 cm diameter × 6.5 cm	300
Selector switch	2 cm^2	100
Filling valve	10 cm	900
Fitting valve	8 cm	550
Pressure gauge	5–200 bar	800
Miscellaneous	N/A	2200

Table 3 Running cost calculations

For CNG	For petrol
Current prevailing cost (C_p) of CNG is ₹71.61 per kg [15]	Current prevailing cost (C_p) of petrol is ₹105.41 per ltr [15]
Mileage (M) of the prototype is 120 km/kg	Mileage (M) of the prototype is 60 km/ltr
Running cost $C_p/M =$ 71.61/120 = ₹0.596 per km	Running cost = $C_p/M =$ 105.41/60 = ₹1.756 per km

Total price of prototype/Cost Benefit per km (C_b) = ₹12,000/(₹1.16/km) = 10,344 km.

So, the two-wheeler can recover all the cost in just 10,344 km of the journey. To be noted here that the average life of two-wheeler bikes in India is more than 10 years and it covers more than 100,000 km during that time.

This same can also be found by equating the total cost incurred in CNG operated motorcycle and its petrol counterpart. The total cost involved in CNG prototype as a function of distance travelled in kilometre (x) is given as 12,000 + 0.596x, and the cost estimated for petrol counterpart is given as 1.756x. Thus, 12,000 + 0.596x = 1.756x → x = 10,344 km. This is shown in Fig. 9.

The survey indicated that almost 70% (32/46) of the respondents travel 5 days or more a week and almost 73% (29/40) travel 20 km or more daily. So, in one year, the approximate distance they travel = (20 km/day) * (5 days/week) * 52 week = 5200 km. The cost of CNG prototype could be recovered in roughly two years of travel.

4.5 *Emission Analysis*

After the cost benefit analysis, the next major factor which decides the feasibility of the prototype is whether it is sustainable for the environment or not. CNG is known to be a green fuel for four-wheelers and heavy vehicles owing to less emission of pollutants when compared to their petrol counterparts.

Similar results have been obtained for the prototype by performing pollution testing on it. Quantity of CO and HC (in % volume and ppm, respectively) at idling RPM for both CNG and petrol is measured. A side by side comparison of the two fuel type is given in Table 4.

4.6 *Conclusion*

This paper reviewed and investigated the state-of-the-art usage of CNG in two-wheeler vehicles, their design, and the problems which prevent their mass adoption.

Table 4 Various emission and standards

Fuel	Petrol	CNG
Prescribed standard CO [16]	3.0	0.3
Measured level CO	**1.27**	**0.11**
Prescribed standard HC [17]	3000	200
Measured level HC	**553**	**33**

The bold text signifies that it is the result obtained

The motivation behind this paper is to re-purpose the already existing vehicles by using some attachments to allow them to run on CNG. This solves and eliminates the problem of redesigning the whole prototype of a two-wheeler, so it is more economical and feasible as the origin and basic working remains the same.

This paper proposes possible solutions to complex design problems arising due to safety and space consideration while replacing petrol with CNG for two-wheelers. The pros and cons of proposed locations of both CNG tank and reducer are discussed in Figs. 4 and 5, respectively. In the result section, the final location of tank in Fig. 6 followed by final location of reducer in Fig. 7 is shown. Lastly, in Fig. 8a, b, final model of motorcycle is shown. Finally, the economic feasibility of the solution through a break-even analysis is done.

Fig. 6 Final location of tank

Fig. 7 Final location of reducer

Fig. 8 **a**, **b** Final prototype with labelled constituent parts

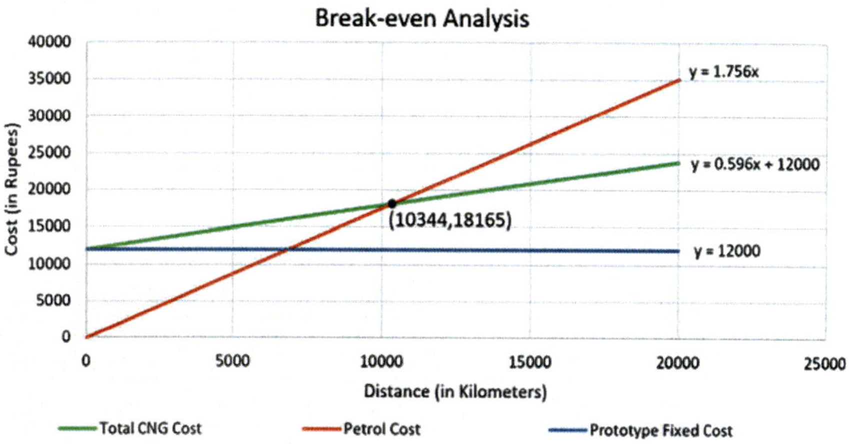

Fig. 9 Break-even graph

1. The calculations show that the cost of the prototype could be recovered in 10,344 km of journey. It is worth noting that on an average, a two-wheeler in India covers more than 100,000 km of journey during its lifetime.
2. It is also concluded that the cost of the equipment is compensated in just 2 years, this means that other than recovering cost of the prototype, there will be substantial economic benefits as well.

CNG in two-wheeler vehicles with the employment of various attachments is a viable, innovative and feasible option. It is an economical option for an environment-friendly future.

Acknowledgements We thank Dr. Vijay Gautam of Delhi Technological University, Shahbad Daulatpur, Main Bawana Road, Delhi-110042, India, for providing his valuable insights.

We thank Dr. Santosh Anand, Department of Informatics, Systems and Communications (DISCo), University of Milano-Bicocca, Piazza dell'Ateneo Nuovo, 1, Milan 20126, Italy, for his constant support while writing this paper.

We thank Shikha Patel of SandStorm Infotech Pvt. Ltd., New Delhi, India (http://www.sand-storm.com) for helping us with the illustrations. Lastly, we thank Mr. Pawan, Mechanic at Rahul Bike Services, Shahbad Dairy, Delhi-110042 for his help in assembly of the constituent parts.

Appendix A: Survey for execution of CNG in two-wheeler vehicles

We are designing a prototype that will help us to employ CNG as a fuel in two-wheeler vehicles. There are certain equipments for motorcycles which allows CNG to be used as a fuel in petrol engines with some modifications. Our paper aims to make two-wheelers currently running on petrol to use CNG as an alternative fuel. This survey will help us to better design and plan the prototype.

In the following bar charts, y-axis represents the number of people, whereas x-axis represents the respective quantities (Figs. 10, 11, 12, 13, 14, 15, 16, 17, 18, 19, 20, 21, 22, 23 and 24).

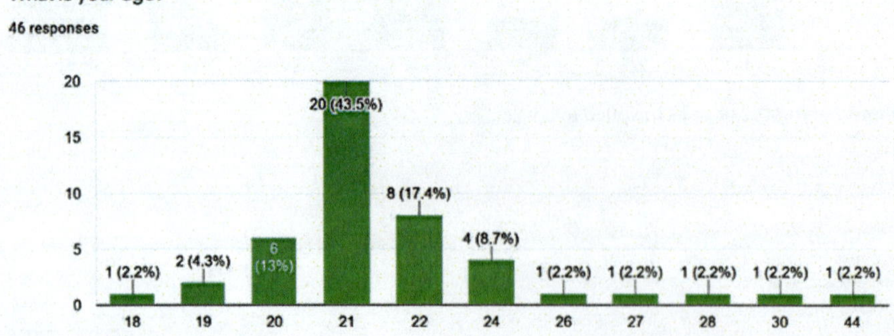

Fig. 10 Age distribution

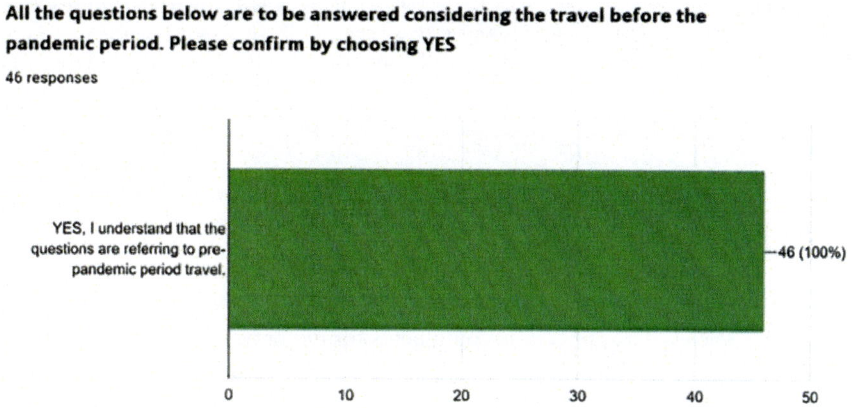

Fig. 11 Pre-COVID-19 declaration

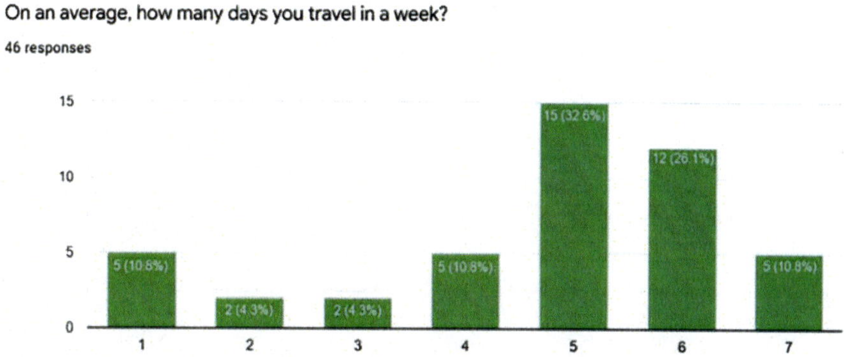

Fig. 12 Number of days travelled in a week

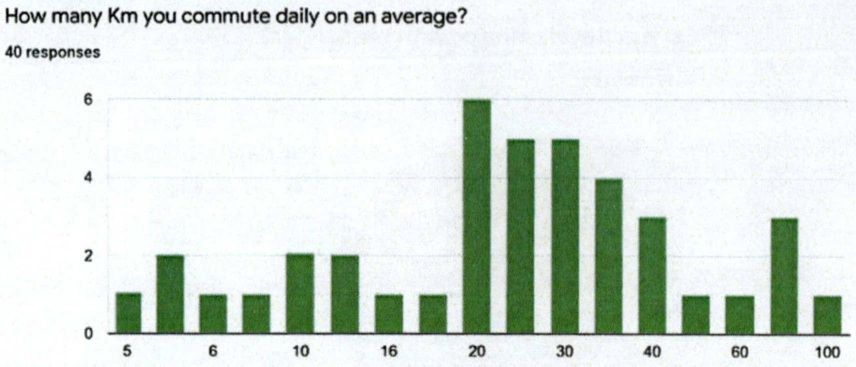

Fig. 13 Distance travelled in a week

Fig. 14 Travelling long distance

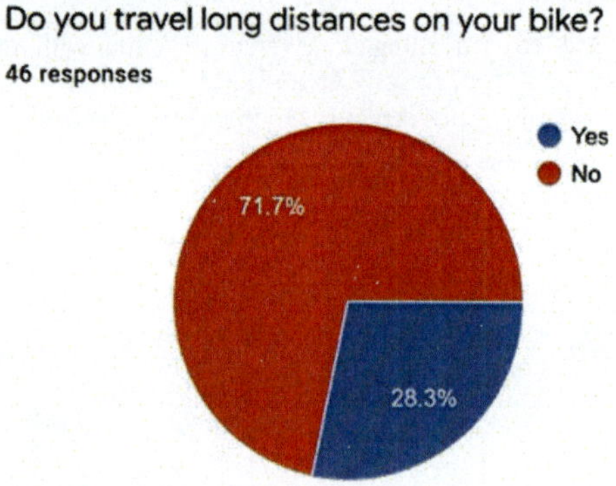

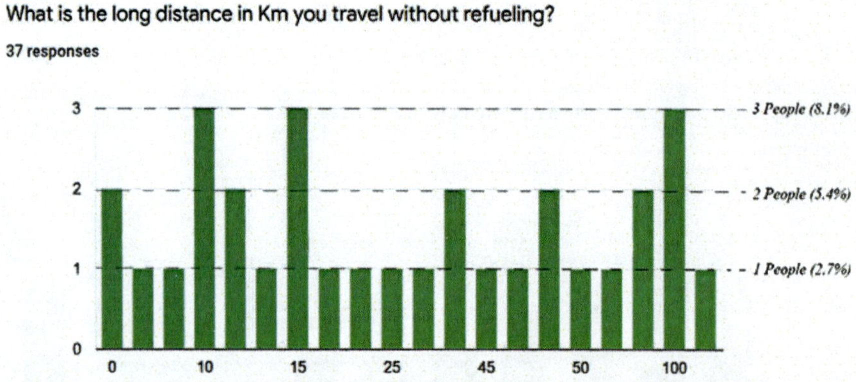

Fig. 15 Longest distance travelled without refuelling

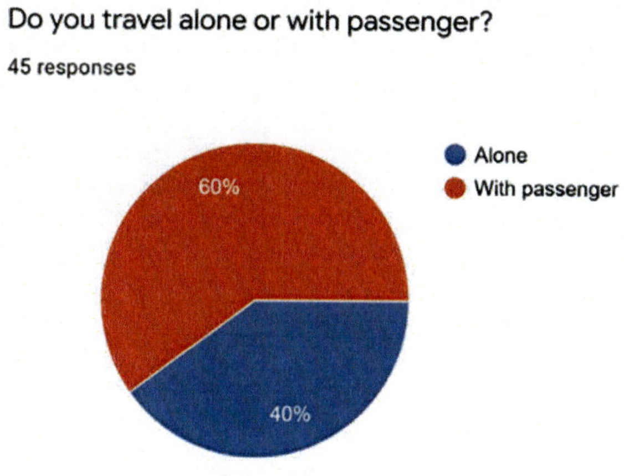

Fig. 16 Travelling alone versus travelling with pillion

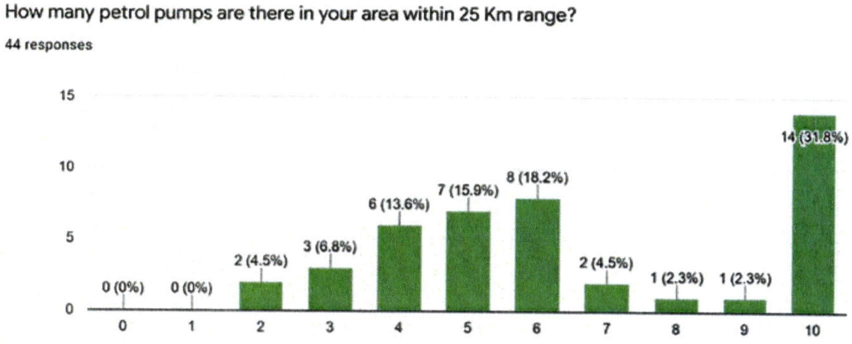

Fig. 17 Number of petrol pumps within 25 km radius

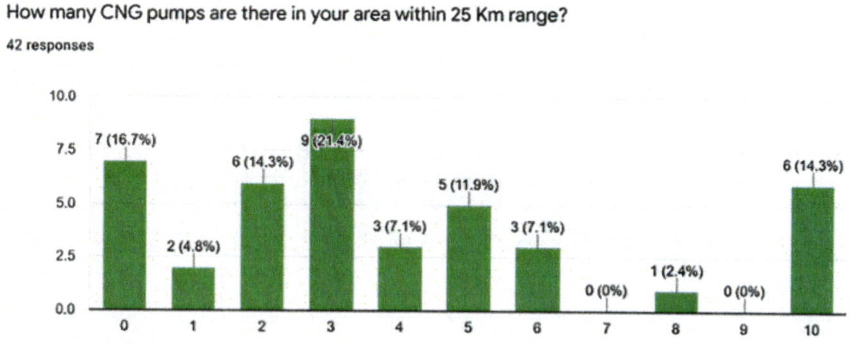

Fig. 18 Number of CNG pumps within km radius. 25

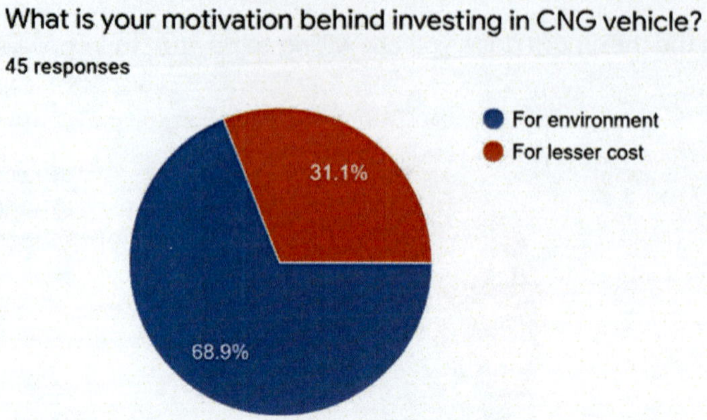

Fig. 19 Motivation behind CNG vehicle

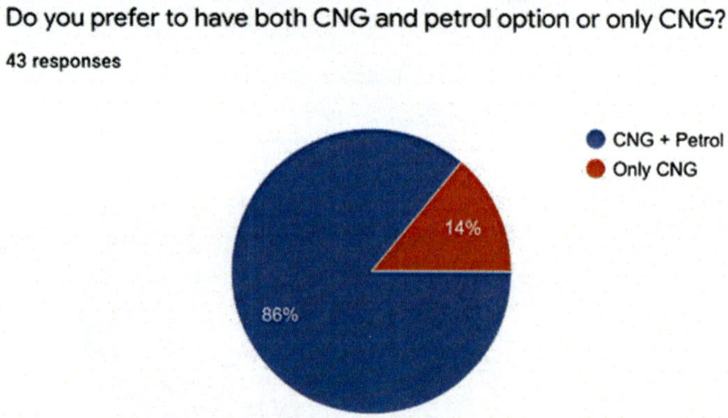

Fig. 20 Preference of solo CNG versus hybrid

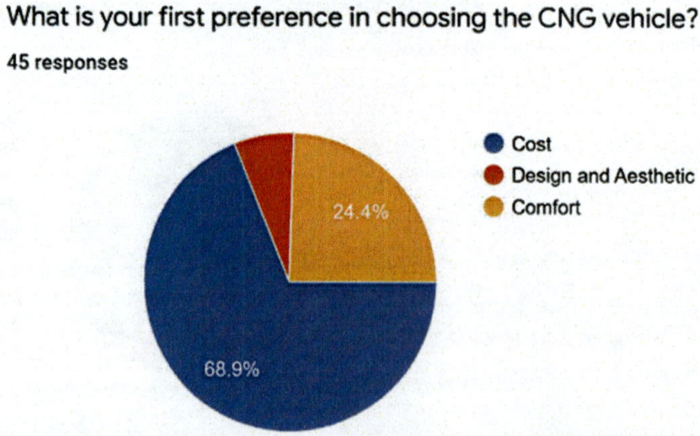

Fig. 21 Preference of cost versus design versus comfort

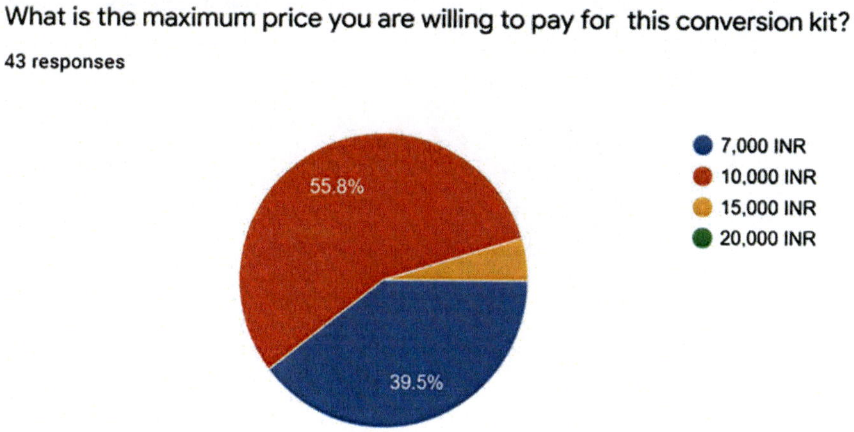

Fig. 22 Maximum price willing to pay for equipment

Fig. 23 Life of vehicles in years

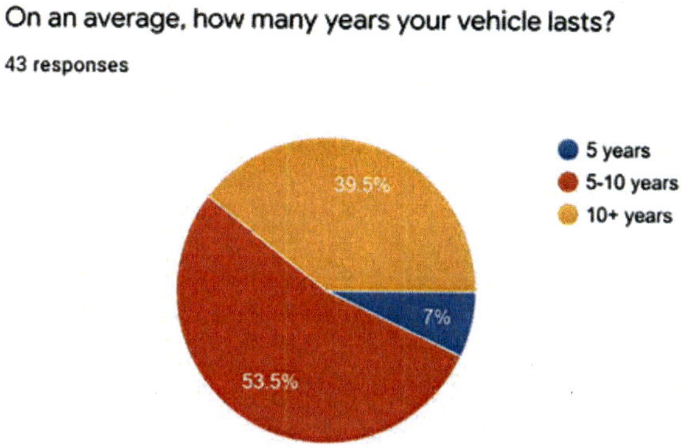

Fig. 24 Maintenance cost of two-wheeler per year

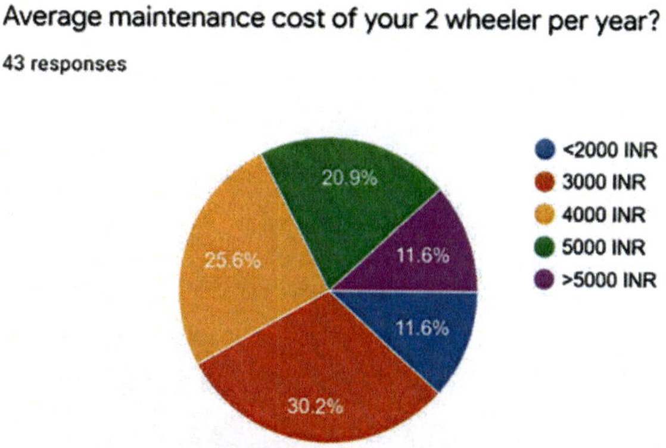

References

1. Mayer A, Czerwinski J, Kasper M, Ulrich A, Mooney J J (2012) Metal oxide particle emissions from diesel and petrol engines. SAE Technical Paper 2012; 2012–01–0841. https://doi.org/10.4271/2012-01-0841
2. Bhandarkar S (2013) Vehicular pollution, their effect on human health and mitigation measures. Vehicle Eng 1(2):33–40
3. Supreme Court Orders Taxis to Run on CNG in Delhi. [Online]. Available: https://www.businessstandard.com/article/news-cd/supreme-court-orders-taxisto-run-on-cng-in-delhi-115121601347_1.html. Accessed: 28 April 2022. Archived at http://archive.is/3zb8n
4. Two-wheelers are causing maximum air pollution. [Online]. Available: https://www.indiatoday.in/mailtoday/story/two-wheelers-are-causing-maximum-airpollution-1377531-2018-10-29. Accessed: 28 Apr 2022. Archived at http://archive.is/8G2lC
5. Khan MI, Yasmin T, Shakoor I (2015) Technical overview of compressed natural gas (CNG) as a transportation fuel. Renew Sustain Energy Rev 51:785–797
6. Natural Gas: Typical Composition of Natural Gas. [Online]. Available: http://naturalgas.org/overview/background. Accessed: 28 Apr 2022. Archived at http://archive.is/VDCo4
7. Sonthalia A, Rameshkumar C, Sharma U, Punganur A, Abbas S (2015) Combustion and performance characteristics of a small spark ignition engine fuelled with HCNG. J Eng Sci Technol 10(4):404–419
8. CREG: Commission de Régulation de I'Électricité et du Gaz. Study on the cost-effectiveness of natural gas (CNG or compressed natural gas) used as fuel in cars, 29 March 2018. https://www.creg.be/sites/default/files/assets/Publications/Studies/F1736EN.pdf. Accessed on 28 Apr 2022
9. Singh R, Maji S (2012) Performance and exhaust gas emissions analysis of direct injection CNG-Diesel dual fuel engine. Int J Eng Sci Technol 4(3):833–846
10. Fuels and Chemicals—Autoignition Temperatures. [Online]. https://www.engineeringtoolbox.com/fuels-ignition-temperatures-d_171.html. Accessed: 28 Apr 2022. Archived at https://archive.is/pH1oS.
11. Al-Saadi AAA, Aris IB (2015) CNG-diesel dual fuel engine: a review on emissions and alternative fuels. In: IEEE 2015 10th Asian control conference. https://doi.org/10.1109/ASCC.2015.7244858
12. Indiamart.com, 'Buyers and Sellers website.' [Online]. Available: https://www.indiamart.com/. Accessed: 28 Apr 2022
13. Mahanagargas.com, 'Launch of CNG-fueled Two Wheelers in Mumbai.' [Online]. Available: https://www.mahanagargas.com/News.aspx?lid=92. Accessed: 28 Apr 2022. Archived at http://archive.is/a1yva
14. Auto.ndtv.com, 'Honda Activa 3G With CNG Kit Review.' [Online]. Available: https://auto.ndtv.com/reviews/honda-activa-3g-with-cngkit-review-1647696. Accessed: 28 Apr 2022. Archived at http://archive.is/mmlPa
15. Goodsreturns.in, 'Fuel Prices Today.' [Online]. Available: https://www.goodreturns.in/fuel-price.html. Accessed: May 1 2022. Archived at https://archive.is/BrvDn
16. Heromotocorp.com, 'PUC Certification.' [Online]. Available: https://www.heromotocorp.com/en-in/rider-zone/biking-tips/puc-certification.html. Accessed: Apr 28 2022. Archived at https://archive.is/CGZI1
17. Transport.delhi.gov.in, 'Know the exhaust emission standards.' [Online]. Available: https://transport.delhi.gov.in/content/know-exhaust-emission-standards. Accessed: Apr 28 2022. Archived at https://archive.is/SDDXM

CFD Analysis of Two-Phase Ejector Impacts of C-D Nozzle Gometry

Ruen Farzan, Pema Wangdi, Rishabh Kumar Chauhan, and M. Zunaid

Nomenclature

T-Dia	Throat Diameter
Div-L	Divergent Length
P-in	Primary Inlet pressure
Prim-mf	Primary inlet mass flowrate
Sec-mf	Secondary Inlet mass flow rate
Outlet-mf	Ejector outlet mass flow rate
COP	Coefficient of Performance
h_g	Enthalpy of saturated Vapor
h_f	Enthalpy of saturated liquid
Re	Entrainment Ratio
CFD-Re	Entrainment Ratio by CFD approach
Exp-Re	Entrainment Ration by xperimental approach

1 Introduction

There are several ways and solutions to save energy in the vapor compression refrigeration cycle by superheating, using heat exchangers, or controlling the rotation of

R. Farzan (✉) · P. Wangdi · R. K. Chauhan · M. Zunaid
Department of Mechanical Engineering, Delhi Technological University, Delhi 110042, India
e-mail: 11.ruenfarzan@gmail.com

P. Wangdi
e-mail: pemawangdi_2k18me152@dtu.ac.in

R. K. Chauhan
e-mail: rishabhchauhan_2k18me172@dtu.ac.in

A. Kumar et al. (eds.), *Recent Advances in Manufacturing and Thermal Engineering*,
Lecture Notes in Mechanical Engineering, https://doi.org/10.1007/978-981-19-8517-1_10

the pump [1] but another way used by researchers in the last decades is using a two-phase ejector instead of expansion valve in refrigeration cycle which proved to enhance the Coefficient of Performance(COP) by 17% and by 22%. The efficiency of the system which is considerable in terms of saving energy and characterized better than expansion valve [2]. Different features play a significant role in the performance of an ejector; however, many experimental and numerical investigations have been done to identify and improve the impactful feature in ejector design to have the maximum energy efficiency and performance [3, 4].

Mainly ejectors contain four parts Fig. 1. (a) first is the motive nozzle, which inlet fluid from the heater Fig. 1. (B) flow through it with high temperature, high pressure, and high velocity in the nozzle outlet, and this study concentration is also in the geometry and dimensions aspects of this part, then fluid enters to the suction chamber, in the suction chamber, suction fluid inlet exists for the entrance of secondary vapor fluid, the two-fluid from two different inlets which is motive inlet and suction inlet mixes and enters to the mixing tunnel, and the mixing tunnel is a constant area of mixing which two phases from two inlets mix and leave to the diffuser. Diffuser is the last part of the ejector in which the fluid leaves the ejector to the condenser. The high saturated fluid from the heater through the primary nozzle and dry vapor from the evaporator through the suction inlet mixes and creates wet vapor to enter the condenser Fig. 1. (b) after the condensation process liquid enters to a tank for separation into two different cycles, one the high saturated liquid enters the evaporator by passing an electric valve which is to control pressure between the tank and evaporator. Another stream from the tank by the use of a pump move to the heater [5–8].

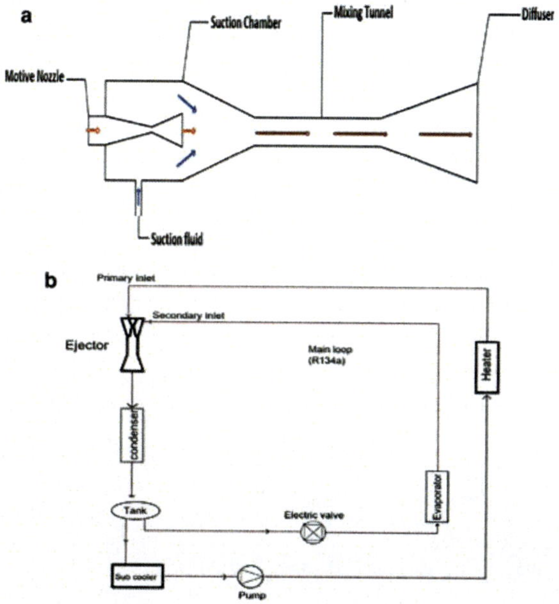

Fig. 1 **a** Schematic diagram of ejector **b** Refrigeration system diagram [7]

Attempts to modify the ejector design take time and huge financial investment which is hard to conduct expensive experimentation, while the fluid inside the ejector is not visible to study the geometric aspects of the ejector in a better way; Computational Fluid Dynamic is used as an effective tool to study the behavior of fluid inside the two-phase ejector. CFD is used to predict the design and improve the performance of an ejector [9, 10] minimizing friction on the boundary and between fluid particles will lead to the high flow of secondary fluid as the secondary fluid moves near or parallel to the supersonic fluid flow the entrainment ratio increases without any further increase in pressure [9, 11–14].

In a refrigeration system, high pressure and temperature in saturation state come from the heater, it is the motive fluid that causes the secondary fluid which is vapor refrigerant with low pressure to flow along with it, and the mixture move to the condenser a portion of that goes back to the evaporator and remaining goes to the heater [15, 16]. To evaluate the efficient performance of the system, we use a term called Coefficient of Performance (COP)

$$\text{COP} = \text{Re}\frac{h_{g-\text{evap}} - h_{f-\text{con}}}{h_{g-\text{boil}} - h_{f-\text{con}}} \tag{1}$$

The main goal for the design and development of a two-phase ejector is to improve the Entrainment Ratio (Re) of the ejector is:

$$\text{Re} = \frac{\text{Secondary inlet flow}}{\text{Primary inlet flow}} \tag{2}$$

To increase the entrainment ratio (Re) many investigations and researches have been conducted to study the thermal aspects, testing different types of refrigerants, geometry, and many more, but in this paper, we study the performance of an ejector in different primary nozzle dimensions, 1,1,12-Tetrafluoroethane(R134a) was used as a refrigerant by using CFD technique the analysis was done [17–23].

2 Literature Review

The impact of geometry of an ejector was investigated by Garcia del Valle et al. [9]; in this experimental study, they considered an ejector having the same constant area of mixing with three different profiles and the result showed that the nozzle position has no impact on flow rate for standard design condition, but for two other geometric situations, the nozzle position caused change on flowrate.

Hu et al. [10] changed the throat diameter (Dt = 0.91.2 mm) of a two-phase ejector in an air conditioning system in an experimental setting. The authors employed R410A as the refrigerant, and the findings revealed that when Dt = 1 mm, the ejector efficiency and system performance were the best. The authors also experimented with the use of a needle to modify the throat diameter.

Kim et al. [25] investigated the effects of different Dt (1.04–1.21 mm) on the performance of an air conditioner that used R410A and an ejector as an expander. Experiments carried out under various environmental settings yielded findings indicating that the smallest diameter had the highest COP gain.

Ameur et al. [7] conducted an experiment to see how the effect of nozzle shape affects performance under various conditions. Three different nozzles were tested separately in the same ejector to look at the impacts of the nozzle's divergence and throat diameter for varied NXP positions and a wide range of primary subcooling levels. The results showed that nozzle shape had little or no effect on nozzle position relative to the mixing section and that the three tested nozzles all had about the same optimal NXP.

Sierra Pallares et al. [10] studied the flow physics of an ejector based on an experimental and computational study which found out that SST K-Ɛ and SKE k-Шboth were fine for studying the two-phase flow turbulence model, after comparison with experimental data, the error was 15% and 4.68%, respectively, for each turbulence model. For the design of a two-phase ejector, fluid or refrigerant should be chosen carefully it impacts the performance of the ejector as much as correct measurement and geometry.

3 Computational Approach

In this study, we used Computational Fluid Dynamic (CFD) to analyze the performance of a two-phase ejector by testing different dimensions (i.e., nozzle throat diameter, divergent length, and divergent angles) of the nozzle in the primary inlet of the ejector to describe the effects of geometry and dimensions on the performance of an ejector, and on the other hand, three different pressure inlets was used to find the suitable geometry for a specific pressure input. A set of governing equations was used, i.e., Navier–Stokes equation for compressibility of the model, mass, energy, and momentum conservation equations used to evaluate fluid behavior inside the ejector [17]. Solidworks used for geometry modeling and ANSYS fluent Analysis tool.

4 Geometry, Discretization, and Boundary Conditions

In this study, we used an ejector geometry which is based on Ameur [7] experimental studies, and all aspects and dimensions of the geometry including suction chamber, mixing tunnel, and diffuser were the same but the primary nozzle dimensions were variable for scenario each we have evaluated. A convergent-divergent nozzle been used instead of a convergent nozzle because convergent nozzle is having a high speed and mass flowrate which brings a need for the use of expansion valve but C-D nozzle operates fine, therefore, by using C-D nozzle, evaporator efficiency increases.

Table 1 Dimensions of the ejector for different geometry condition

Details		Dimensions
C-D nozzle	Inlet diameter (mm)	10
	Throat diameter (mm)	1.25, 1.45, 1.75
	Divergent length (mm)	45, 50, 55
	Divergent angle (°)	0.7, 2.6
Secondary inlet	Width (mm)	7
NXT	Distance (mm)	32
Mixing chamber	Diameter (mm)	7
	Length (mm)	164.3
Diffuser	Divergent angle (°)	8.14
	Divergent length (mm)	174.1

The 2D model is based on Khaled Ameur's [7] experimental data Table 1., geometry of the ejector has been modified each time; however, different throat diameter, divergent length, and divergent angle have been examined but the remaining dimensions of ejector remained the same, i.e., suction inlet, mixing tube and diffuser details mentioned in Table 1.

For discretization of the 2D model 93,700 number cells have been generated which is satisfying for fluid flow analysis inside an ejector[baek], the refinement has been done and high concentration of cells was locally defined near the C-D nozzle, Secondary inlet and the mixing walls because thermodynamic behavior and high velocity in the area, therefore, the mesh was fine and guaranteed the convergence of the computational results (Fig. 2).

A set of boundary conditions were defined for validation of the model which contains the primary inlet, secondary inlet, outlet, and standard walls, for the primary inlet, an inlet pressure was defined in the range of 2.5–6.5bars and temperature 328 °F

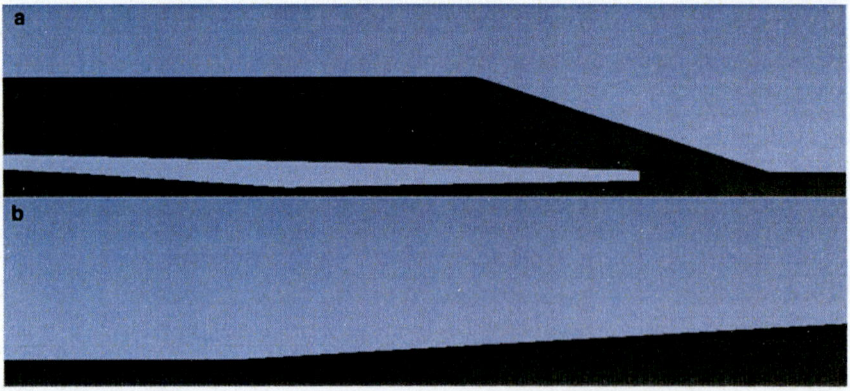

Fig. 2 **a** Generated mesh locally defined in c-d nozzle and suction chamber. **b** Generated mesh in mixing tube and diffuser. *Sources* This figure is generated by Ansys software

the secondary inlet had pressure between 0.42 and 1.35bars and 278 °F temperature besides these boundary conditions a set of functions also have been defined like, no sleep condition to describe a realistic flow inside the ejector.

5 Model Setup and Verification

We utilized ANSYS Fluent for CFD analysis of compressible, supersonic, transient, and two-phase flow of fluid inside the ejector, pressure-based solver was used for solving the governing equations at high speed and low-speed regions [17]. The physical models that were defined are, Energy model, the Realizable gas model, SST K-Ɛ model applied to describe the turbulence behavior of fluid inside the ejector. R134a was used as an environmentally friendly refrigerant, and the properties of vapor and fluid phases taken from REFPROP [16] data were imported by the user defined function Table 2. There are two phases of flow with different velocities and homogenous; therefore, mixture model is used to model with the flow having strong coupling.

After applying the liquid phase as motive fluid in the primary nozzle and vapor phase in the second inlet, the walls were defined with a no-slip condition which had the lowest velocity near the wall and highest in the center. In solution, SIMPLIC Scheme was used for pressure and velocity coupling followed by PRISTO! For pressure and second-order differential equation for solving the equation to get the most realistic results. Reports were defined for features, for instance, mass flow rate, total pressure, static pressure, and velocities for all inlets and outlets.

For method verification of computational analysis, two types of geometry were used and the results were compared with the analysis of similar study have been done by Baek [24]. The original geometry has been modified by filet command and some reduction in the dimensions to be similar to the shape of Baek [24] geometry, then the ANSYS code implemented with the same procedure and methods to obtain the solution as shown in Fig. 3. The blue line shows the static pressure along the ejector length given by Baek [24] and the orange line shows the recent studies.

Table 2 Properties of the refrigerant from REFPROP

Properties	Saturated liquid	Vapor
Density(ƥ)	1078.6[Kg/m^3]	15.679[Kg/m^3]
Specific Heat (Cp)	1608.33[J/Kg K]	0.9051[J/Kg K]
Thermal cond	0.00683 [W/m k]	0.0132 [W/m k]
Viscosity	0.003264[Kg/m s]	0.0001154[Kg/m s]
Specific weight	102.03[Kg/Kmol]	102.03[Kg/Kmol]
Steady state enthalpy	13398001[KJ/Kgmol]	27356780[KJ/Kgmol]
Reference temperature	328°K	294°K

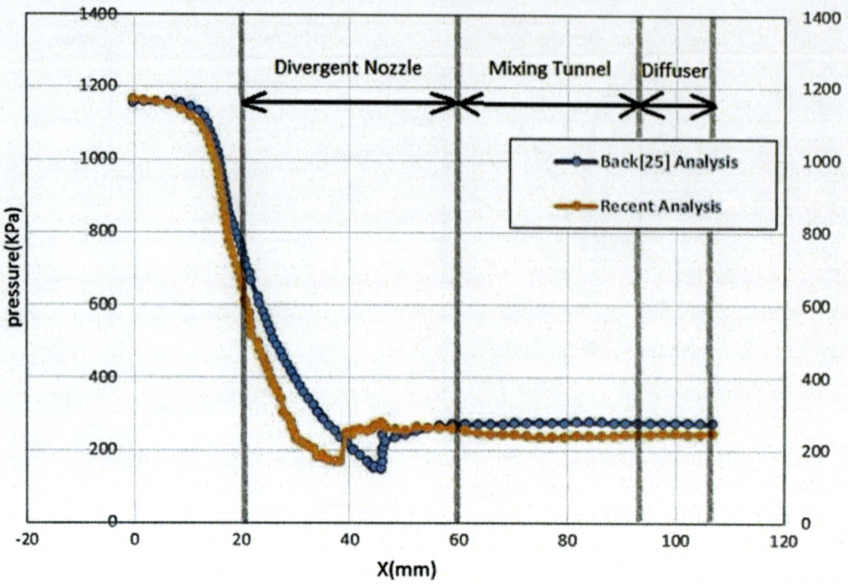

Fig. 3 Comparison of Baek [24] static pressure behavior and the recent study

The deviation appeared in the line chart is because of higher pressure defined in the primary inlet which results differences in the nozzle throats pressures but even by having this amount of deviation which is very less we can tell that the method has been used for analysis is the correct method which was proven by Baek [24] paper in terms of static pressure and velocity data which same behavior occurred for more detail of the verification please refer to [11, 24].

6 Results and Validation

Pressure and Fluid Behavior Inside the Ejector

In this paper, we used a two-phase ejector which is having a convergent-divergent nozzle with variable dimensions' Table 1. To investigate the fluid behavior inside the ejector and impacts of primary nozzle dimensions on performance of the ejector to extract the best dimensions which performs well. In Fig. 4., nozzles with 3 different diameters have been shown, however, ejector with 1.75(mm) T-Dia (green line) of nozzle is having the highest expansion rate which is flowed by ejector having nozzle 1.45 T-Dia(blue line) and the ejector having nozzle 1.25 T-Dia is having lowest pressure in the inlet but the less efficient in terms of expansion compare to the other two C-D nozzle diameter. And ejector with largest throat diameter (green line) Fig. 4. is having the highest expansion rate.

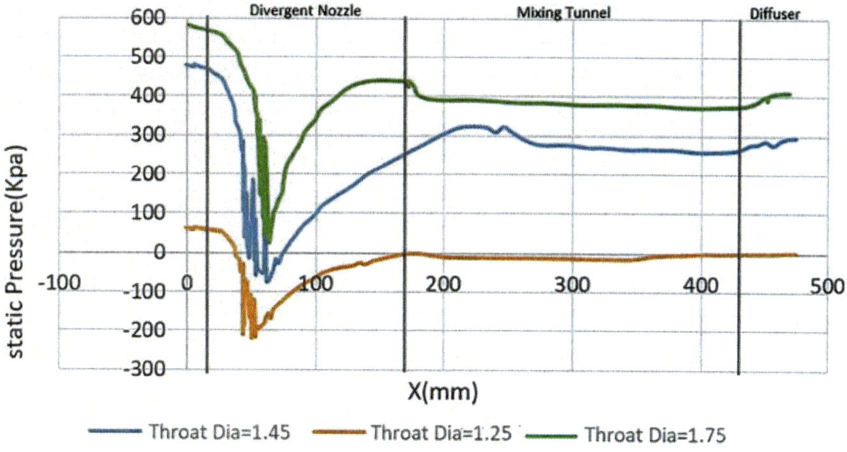

Fig. 4 Static pressure along ejector length(X-axis) for different throat diameters

The same concept was proven by contours as well, which is shown in Fig. 5. For the three cases mentioned on Fig. 4, the contours shown Fig. 5 are having the largest T-Dia and pressure ratio all over the ejector is equal rate except the C-D part of the nozzle which is considered to have the highest velocity. For bigger T-Dia, the outlet pressure is also high which causes back pressure and pressure distribution across ejector.

Entrainment Ratio and Model Validation

The ejector having the largest nozzle T-Dia is having the highest amount of flowrate in the outlet and the nozzle with smallest T-Dia having lowest flowrate in the outlet but the case for entrainment ration is the opposite when the T-Dia is smaller, rate of secondary flow increases Table 3 and velocity is higher in the nozzle outlet. As the nozzle outlet pressure reduces, the ejector outlet pressure increases [7].

In the Case 1–6, the primary flow is more the secondary flow Fig. 6 because of outlet pressure and high back flow but from 7 to 9, secondary flow rate is high or equal to the primary mass flow rate. But in terms of quantity, the nozzle with higher

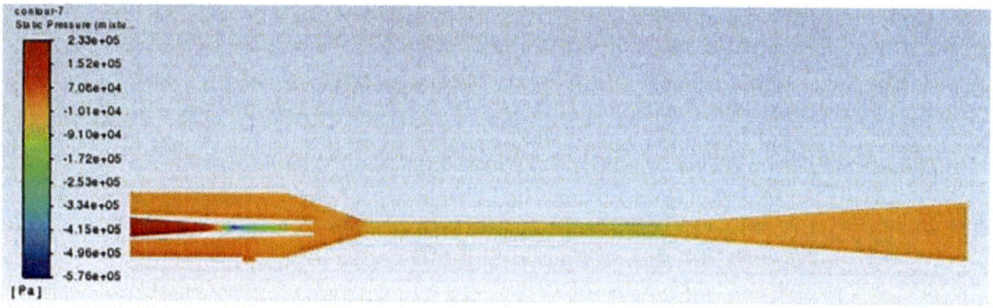

Fig. 5 Pressure contours for Throat Dia = 1.75

Table 3 Primary, secondary and outlet mass flow rates in nine cases with differences in geometry

Case No.	T-Dia (mm)	Div-L (mm)	P-in (Mpa)	Prim-mf (g/s)	Sec-mf (g/s)	Outlet-mf (g/s)
1	1.75	55	6.5	79,412.25	73,636.1	121,090.8
2	1.75	55	4.5	65,179.9	60,471.9	169,973.8
3	1.75	55	2.5	47,419.8	42,863.3	208,182
4	1.75	45	6.5	79,201.4	73,559.1	207,716
5	1.75	45	4.5	65,037	60,408.2	118,581.4
6	1.75	45	2.5	47,121.4	40,763.4	118,581.4
7	1.25	55	6.5	67,515	70,937.5	185,710.9
8	1.25	55	4.5	39,800	39,623.5	103,111.5
9	1.25	55	2.5	55,257.5	58,255.7	150,747.8

throat diameter is having higher total flow. Therefore, smaller throat diameter has higher entrainment ratio which is proven by experimental investigation as well.

Mass flow rate was measured for analysis of each geometry condition to validate the CFD analysis with experimental data Table 4. The entrainment ratio which is the ratio of secondary fluid flow and primary fluid flow was taken as reference and compared with experimental investigation of Amuer [7]; after calculation, the maximum error of 12.26% was found which is favorable, as the CFD analysis results cannot be 100% similar with experimental outcomes 12.26% is very close as compared to the other CFD studies[24] which is 15% and more errors were seen.

Velocity and Thermodynamic Behavior

The refrigerant(R134a) leaving the boiler with high pressure and temperature to the primary nozzle the concept of Venturi effects comes into the picture, when the high pressure–temperature passes from nick shape area the velocity increases and

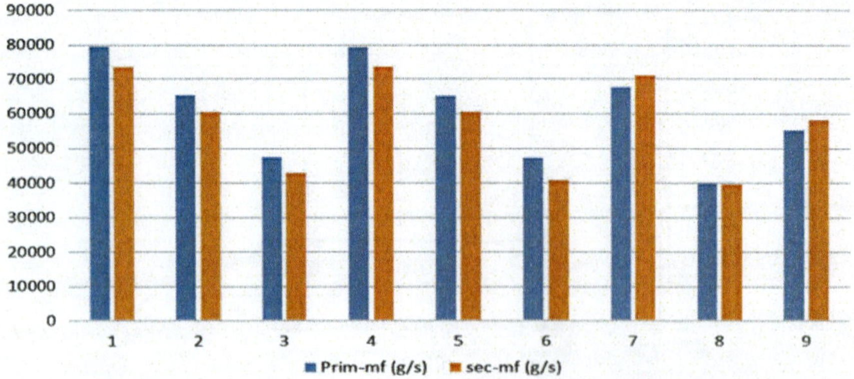

Fig. 6 Primary mass flow rate (blue graph) and secondary mass flow rate(orange graph)

Table 4 Validation of CFD analysis with respect to experimental studies [7]

Case No.	CFD Re	Exp Re	Error (%)
1	0.5243	−0.7057	12.3
2	0.43682	−0.78428	12.211
3	0.22909	−0.9799	12.0899
4	0.52473	−0.68268	12.0741
5	0.4368	−0.7781	12.149
6	0.18597	−1.03628	12.2225
7	0.57672	−0.6502	12.2692
8	0.19155	−1.0166	12.0815
9	0.47515	−0.741	12.1615

pressure drops will be seen. In experimental investigations, it is not possible to see the velocity raise and drop areas inside the ejector but by CFD method and generating contours are visible and high-velocity points are colored red and the lighter velocity locations have blue colors from the contour it is visible in Fig. 8. It shows that nozzles having three minimum velocity magnitude with high pressure input in the inlets. In the nozzle, velocity rise is happening as the diameter reduces Fig. 7 and it has a reduction in the suction chamber which is recovered by entering to the mixing tunnel, however, as much as velocity increases shock waves are seen in the nozzle outlet and mixing tunnel. In the diffuser, area increases and velocity drop occurring which is not similar in all cases.

In Fig. 7, velocity behavior is plotted, the nozzle having smallest T-Dia(orange line) is having the highest velocity across ejector, and nozzle with medium throat diameter has lesser velocity as compare to the two other cases.

According to line graph Fig. 7, the inlet velocity starts from 100 m/s ejector having (T-Dia = 1.75) for inlet nozzle maximum velocity in the nozzle outlet is

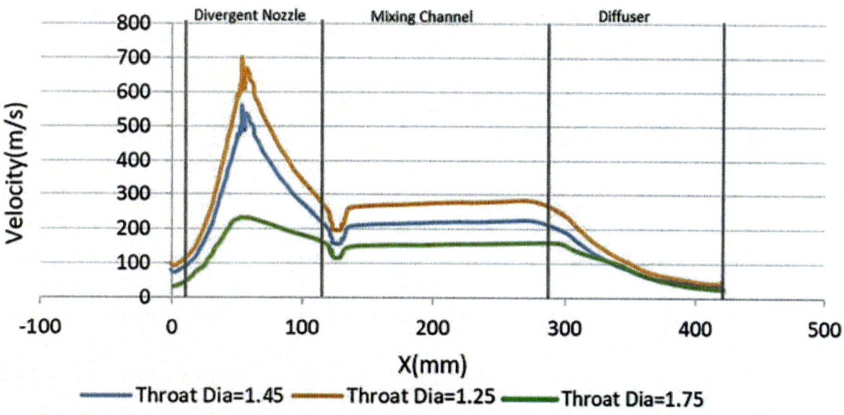

Fig.7 Velocity along ejector length(X-axis) for different throat diameters

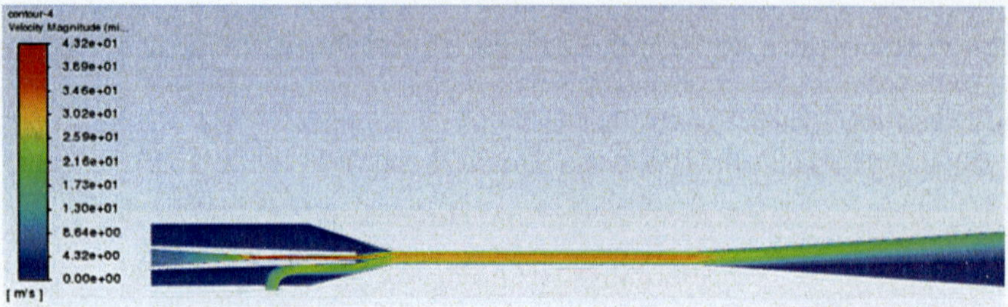

Fig. 8 Velocity contours for Throat Dia = 1.75

about (250 m/s) on the other hand ejector with (T-1.45) reaches up to 560 m/s the maximum nozzle outlet velocity but the highest velocity rate observed in the ejector having smallest (T-Dia = 1.25) have the highest velocity (700 m/s) the flow is having the highest shock wave as well because of ejector outlet back pressure supersonic fluid flow changes to subsonic.

Velocity contour illustrated in Fig. 8 shows the two stream, one from primary and other from secondary inlet which is flowing through mixing tube and diffuser but the stream form primary inlet touched with inlet of mixing tunnel, therefore, for there is a need to change mixing tube diameter as well with respect to nozzle geometry.

7 Conclusion

A 2D analysis of two-phase ejector has been done by using CFD method. R134a was used as refrigerant and properties were entered by user defined function, three divergent lengths and three throat diameters were tested by three pressure inputs for 27 different cases. Our aim was to find the high entrainment ratio, which case have the high expansion rate, vapor quality in the outlet of ejector, and utilizing different physical models to validate the CFD results with experimental investigations. Nozzle having (1.25 mm) throat diameter with longest divergent length (55 mm) had the high speed flow, however, the shock wave has been generated as a result was higher than the other cases, less amount of wall friction was observed, and low pressure have been generated in the ejector outlet. But the phase transition and expansion rate were less. For ejector having large throat diameter and short divergent length, expansion rate was high but flowrate decreased because of high pressure in the ejector outlet even the even the velocity vector shown that there are some back flows around the secondary inlet. By modifying the geometry, the verification of models and physical principles were applied have been done. Energy equation, realizable gas model, and K-Ɛ model used to calculate numerical for the above cases, and the similar result was reviled in the validation about 12.6% error was observed and the reason is friction and perpendicular direction of secondary inlet with flow stream, in case of CFD that was sensitive. The second flow stream was impacting the direction of first high

speed stream of flow in CFD which will not occur in the experimental process and not mention in case of experimental and CFD analysis before, therefore, using outlet boundary condition in ejector outlet and primary mass flowrate was tested to be the right inputs for boundary conditions. The ejector having nozzle throat diameter of (1.45 mm) was observed to perform well by having (55 mm) divergent length.

Acknowledgements The paper is a group work for B.Tech major project, we could never imagine of doing this project without the support of our professors. It was because of their effort and support that we could do this project. We want to express our gratitude and special thanks to our mentor **Dr. Mohammad Zunaid** Assistant prof. in ME Department, it was all because of his guidance, support, and motivation that could complete this paper. Last but not least, we want to express our deep sense of gratitude to Ms Sushila Rani Assistant prof in Mechanical Department of Delhi Technological University.

References

1. Liu Y, Yu M, Yu J (2022) An improved 1-D thermodynamic modeling of small two-phase ejector for performance prediction and design. Appl Therm Eng 204:118006
2. Rahamathullah MR, Palani K, Aridass T, Venkatakrishnan P, Palani S (2012) A review on historical and present developments in ejector systems. Int J Eng Res Appl 3(2):10–34
3. Li Y, Deng J, Ma L, Zhang Y (2018) Visualization of two-phase flow in primary nozzle of a transcritical CO_2 ejector. Energy Convers Manage 171:729–741
4. Haida M, Smolka J, Hafner A, Ostrowski Z, Palacz M, Nowak AJ, Banasiak K (2018) System model derivation of the CO_2 two-phase ejector based on the CFD-based reduced-order model. Energy 144:941–956
5. Sumeru K, N H (2012) A review on two-phase ejector as an expansion device in vapor compression refrigeraton cycle. Renew Sustain Energy Rev
6. Sopian K, E BS (2017) Effect of the nozzle exit position on the efficiency of ejector cooling system using R134A. ARPN J Eng Appl Sci
7. Ameur K, A Z (2020) Experiment performance of a two-phase ejector: nozzle geometry and subcooling effects. Inventions
8. Banasiak K, H A (2012) Experimental and numerical investigation of the influence of the two-phase ejector geometry on the performance of the R744 heat pump. Int J Refrig
9. del Valle JG, S J (2014) An experimental investigation of a R-134a ejector refrigeration system. Int J Refrig
10. Hu J, S J (2014) Numerical and experimental investigation on nozzle parameters for R410A ejector air conditioning system. Int J Refrig
11. Lucas C, K J (2012) Experimental investigation of the COP improvement of a refrigeration cycle by use of an ejector. Int J Refrig
12. Lawrence N, Elbel S (2015) Analysis of two-phase ejector performance metrics and comparison of R134a and CO_2 ejector performance. Sci Technol Built Environ 21(5):515–525
13. Chen W, H C (2020) Experimental and numerical investigation of two phase ejector performance with the water injected into the induced flow. Int J Adv Nucl Reactor Des Technol
14. Smolka J, B Z (2012) A computational model of a transcritical R744 ejector based on a homogeneous real fluid approach. Appl Math Model
15. Bansiak K, H A (2012) Mathematical modelling of supersonic two phase R744 flows through converging-diverging nozzles: the effects of phase transition models. Appl Therm Eng
16. Chen W, S C (2014) Numerical and experimental analysis of two phase flow in ejector. Scie Direct Energy Proc

17. Hakkaki-Fard A, A Z (2015) A computational methodology for ejector design and performance maximisation. Energy Convers Manage
18. Mazzelli F, M A (2014) Performance analysis of a supersonic ejector cycle working with R245fa. Int J Refrig
19. Natthawut R, T T (2013) CFD simulation on the effect of primary nozzle geometries for a steam ejector in refrigeration cycle
20. Tillner-Roth R, Baehr HD (1994) An international standard formulation of the thermodynamic properties of 1,1,1,2-tetrafluoroethane (HFC-134a) covering temperatures from 170 K to 455 K at pressures up to 70 MPa. J Phys Chem Ref Data 23:657–729
21. Bilir N, Ersoy HK (2009) Performance improvement of the vapour compression refrigeration cycle by a two-phase constant area ejector. Int J Energy Res 33(5):469–480
22. Sunghoon B, S K, S S, S R (2018) Numerical study of high-speed two-phase ejector performance with R134a refrigerant
23. Sarkar J (2017) Performance analyses of novel two-phase ejector enhanced multi-evaporator refrigeration systems. Appl Therm Eng 110:1635–1642
24. Baek S, K S (2018) Numerical study of high speed two-phase ejector performance with R134a refrigerant. Int J Heat Mass Transfer.lishing Inc; 1999:281–304
25. Kim D, Jeon Y, Jang DS, Kim Y (2018) Performance comparison among two-phase, liquid, and vapor injection heat pumps with a scroll compressor using R410A. Appl Therm Eng 137:193–202

Optimum Location Selection for Smog Tower Installation in Delhi

Suraj Kumar Jha, Shivang Dutt, Sheshank Pandey, Tarun Phore, and Anil Kumar

1 Introduction

Air pollution is a major health risk globally, related to many health disorders, including respiratory diseases, cardiovascular illness, stroke and cancer, resulting in about 7 million deaths worldwide in 2016 [1]. Delhi is the world's most polluted capital city [2]. Delhi is located in one of India's key industrial sectors, the Gurgaon-Delhi-Meerut industrial region, which is one of the nation's most important economic regions, and industrial areas cover 51.81 km^2 of land in the National Capital Territory of Delhi [3]. Despite the relocation of major polluting enterprises, Delhi's lung patient incidence rate is twelve times higher than the national average, and over 30% of the city's population suffers from breathing problems due by air pollution [4].

Industrial emission, domestic pollution, vehicular emissions, road dust and construction activities are largely responsible for air pollution in Delhi. Increase in pollution levels in recent years is also due to exponential growth of registered industries. Pollution controlling devices in industries are either missing or are not in a working condition, which results in release of unfiltered pollutants directly into the atmosphere [5]. Also, using short chimneys in industries restricts escape of pollutants in upper layers of atmosphere and increases pollution load.

Authorities in the city have introduced several strategies to mitigate air pollution. Some of the measures taken or proposed are banning biomass burning, emission standards for industries, restricting ash content in coal for thermal power plants, banning 10 years old diesel vehicles and improving public transport network. These measures also include constructing public infrastructure like Wind Augmentation and purifying Unit (WAYU) by CSIR-National environmental engineering research institute and installation of smog towers by government of NCT of Delhi to remove particulate matter from ambient air. Smog towers are large-scale air purifiers used

S. Kumar Jha (✉) · S. Dutt · S. Pandey · T. Phore · A. Kumar
Department of Mechanical Engineering, Delhi Technological University, Delhi, India
e-mail: 19suraj11kumar@gmail.com

A. Kumar et al. (eds.), *Recent Advances in Manufacturing and Thermal Engineering*,
Lecture Notes in Mechanical Engineering, https://doi.org/10.1007/978-981-19-8517-1_11

to remove pollutants in the atmosphere. The limitation of open-air purifiers is that it has no boundaries and the air is dynamic, which results in the homogenization of air after a certain time [6]. Smog towers treat particulate matter that is dangerous to human health [7] and can increase mortality rate [8].

Installation of smog tower should be such that there is a maximum utilization of capacity of the smog tower. Therefore, it is extremely important to find the optimum locations for installing smog towers. A location study to find the optimum location for installing smog towers has been performed by Janani Bharatraj for the capital city of Tamil Nadu, Chennai [9]. Currently, no studies have been conducted so far to find optimum location for smog towers in Delhi.

This study aims to find optimum location for installing smog towers in Delhi, using the AHP and TOPSIS decision-making methodologies. The alternatives are compared based on various factors which affect location of smog towers. The presence of industrial or commercial centers is considered an initial criterion for finding locations for this study. Due to COVID-19 induced lockdowns and other restrictions, the air pollution levels have improved by 50% [6]. Hence, by reducing emissions at source, air pollution can be controlled.

The following factors have been considered for this analysis.

A. *Particulate Matter 2.5 (PM 2.5)*

PM 2.5 is suspended particulates with a diameter of less than 2.5 microns. It is taken as an average value of PM levels in 1 year. Its unit is taken as $\mu g/m^3$.

B. *Particulate Matter 10 (PM 10)*

PM 10 is particulate matter dispersed in the air with a diameter of less than 10 microns. It is taken as an average value of PM levels in 1 year. Its unit is taken as $\mu g/m^3$.

C. *Number of Hospitals nearby*

Number of hospitals situated within a 1 km radius of the location.

D. *Green Area*

Natural vegetation absorbs pollution and acts as a pollution damper [10], and it takes into account the effective green area in km^2 within a 1km radius of the location.

E. *Population Density*

High population density contributes to high pollution levels due to increased human activity in a region. It is taken as persons per km^2.

2 Methodology

This study is conducted in the National Capital Territory of Delhi; industrial location is taken as the initial criteria for identifying locations. Historical particulate matter data for the year 2019 is considered for this study, and COVID-19 induced lockdowns started in Delhi in March 2020 and continued in some form or the other till 2021. Due to this, pollution data during this period is not reliable and does not represent normal conditions.

2.1 *Analytical Hierarchy Process (AHP)*

Saaty's analytic hierarchy process (AHP) is a multi-criteria decision-making method that has been used in economics, politics and engineering. It is an approach for making decisions that entails breaking process down into a framework that ranks aspects by comparing them subjectively. It pertains to the study of both material and immaterial factors, and the comparison of each pair of elements is done on a scale of 1–9 as given in Table 1. If there are N factors, an $N \times N$ matrix is constructed, where C_{ij} is the relative weight of I over j, and Cji is the relative importance of j over i, also $C_{\text{ij}} = 1/C_{\text{ji}}$. The normalized Eigenvector called the priority vector is calculated for each criterion; this priority vector is nothing but the relative weightage of the criteria. The sum of the products of every component of the Eigenvector as well as the total of the columns of the inverse matrix yields the primary Eigenvalue, known as max.

The consistency ratio (CR) is calculated by dividing the consistency index (CI) by the random index (RI) and is used to ensure that the author has made an informed decision on the issue at hand. The consistency ratio of less than 0.1 is considered satisfactory. The CI is calculated using Eq. (1) and CR using Eq. (2)

$$\text{CI} = \frac{\lambda\max - N}{N - 1}, \tag{1}$$

where N is number of attributes

$$\text{CR} = \frac{\text{CI}}{\text{RI}} \tag{2}$$

Table 1 Saaty's pairwise comparison scale [11]

No.	Importance
1	Equal importance
3	Moderate importance
5	Strong importance
7	Very strong importance
9	Absolute importance

Table 2 RI values for associated N

N	3	4	5	6	7	8	9	10
RI	0.52	0.89	1.11	1.25	1.35	1.4	1.45	1.49

Random index (RI) values are given in Table 2 for corresponding values of N

2.2 *Technique for Order Preference by Similarity to Ideal Solution (TOPSIS)*

According to the principle, the best option has to have a minimal Euclidean distance from the best optimal solution and a greatest possible Euclidean distance from the worst optimal solution. Hwang and Yoon [12] devised it. The matrix produced after vector normalization provides standardized units that are comparable. A decision matrix of $M \times N$, where M denotes the number of options and N denotes the set of characteristics.

The matrix is normalized using vector normalization. The weight of each criterion is multiplied by the standardized decision matrix to make a weighted normalized matrix. The best and the worst alternative are selected for respective criteria, and then the Euclidean distance for each alternative is calculated from the best and worst points. Then, the performance score for each alternative is calculated. Based on the performance score, the alternatives are ranked.

3 Results and Discussion

The locations selected for this study have been given in Table 3. The locations were marked using Google Earth [13] and represented in Fig. 1. The locations chosen for this study are major industrial or commercial hubs, where the air pollution level is generally high.

The relative weightage of the variables is represented by the AHP matrix, which is calculated by comparing the qualities together. After normalization of the matrix using linear min–max normalization, the weightage of attributes, CR and CI are evaluated. The weightage obtained is given in Table 4.

Table 5 and 6 provide top 5 locations after analyzing alternatives using AHP and TOPSIS methods, respectively.

It is observed that 4 locations from top 5 using the two MADM methods are essentially the same. The rankings are not consistently own to varied approaches of the methods employed for analysis. The smog towers should be installed at these locations after a detailed survey and finding suitable land for the same.

Table 3 List of locations for the analysis

S. No.	Area	S. No.	Area
A1	Narela	A10	Kirti Nagar
A2	Bawana	A11	Uttam Nagar
A3	Badli	A12	Karol Bagh
A4	Jahangirpuri	A13	Chandni Chowk
A5	Wazirpur	A14	Connanught Place
A6	Mangolpuri	A15	Dhaula Kuan
A7	Trinagar	A16	Lajpat Nagar
A8	Mundka	A17	Okhla
A9	Anand Parbat	A18	Anand Vihar

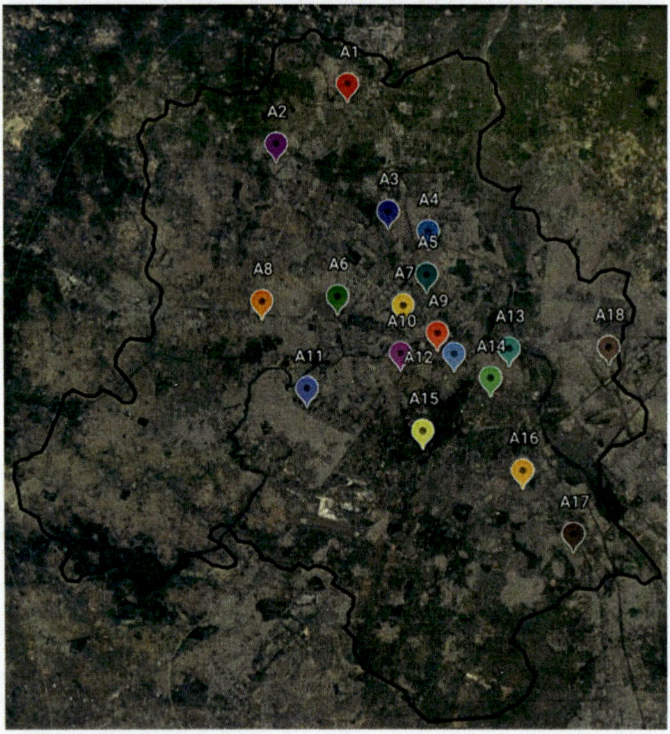

Fig. 1 Location for analysis on Delhi map

4 Conclusion

In this study, location analysis to find the optimum location for installation of smog towers in Delhi is attempted, where the decision-making process is based on different criteria affecting the location. Anand Vihar tops the priority list with a score of 0.729

Table 4 Final weightage calculated for each factor using the AHP method

Attributes	Weightage
PM 2.5	0.393
PM 10	0.393
Hospital	0.079
Green area	0.034
Population density	0.101

Table 5 Top 5 locations as per the AHP method

Rank	Code	Area	Score
1	A18	Anand Vihar	0.729
2	A5	Wazirpur	0.671
3	A8	Mundka	0.649
4	A2	Bawana	0.504
5	A15	Dhaula Kuan	0.499

Table 6 Top 5 locations as per the TOPSIS method

Rank	Code	Area	Score
1	A18	Anand Vihar	0.738
2	A13	Chandni Chowk	0.546
3	A8	Mundka	0.544
4	A5	Wazirpur	0.493
5	A2	Bawana	0.482

through AHP method and 0.738 through TOPSIS method also 4 locations among the top 5 ranks are same. However, the priority score is not very close to 1, as this is a preliminary study and lays foundation for further research in this direction. A comprehensive study can be conducted by considering other factors of pollution like vehicular emission, construction sites, etc.

References

1. World Health Organization (2021) World Health Statistics 2021: monitoring health for the SDGs, sustainable development goals. In: Industry and higher education, vol 3(1)
2. IQAir (2020) World air quality report. 2020 World Air Quality Report, August
3. Parveen N, Siddiqui L, Sarif MN, Islam MS, Khanam N, Mohibul S (2021) Industries in Delhi: air pollution versus respiratory morbidities. Process Saf Environ Prot 152:495–512. https://doi.org/10.1016/j.psep.2021.06.027
4. Pandey JS, Kumar R, Devotta S (2005) Health risks of NO2, SPM and SO2 in Delhi (India). Atmos Environ 39(36). https://doi.org/10.1016/j.atmosenv.2005.08.004

5. ENVIS Center C (2016) Air pollution of Delhi: an analysis. In: ENVIS centre on control of pollution water, air, and noise. http://www.cpcbenvis.nic.in/envis_newsletter/Air%20pollution%20in%20Delhi.pdf
6. Guttikunda S, Jawahar P (2020) Can we vacuum our air pollution problem using smog towers? Atmosphere 11(9):922. https://doi.org/10.3390/atmos11090922
7. Kappos AD, Bruckmann P, Eikmann T, Englert N, Heinrich U, Höppe P, Koch E, Krause GHM, Kreyling WG, Rauchfuss K, Rombout P, Schulz-Klemp V, Thiel WR, Wichmann HE (2004) Health effects of particles in ambient air. Int J Hyg Environ Health 207(4):399–407. https://doi.org/10.1078/1438-4639-00306
8. Pascal M, Falq G, Wagner V, Chatignoux E, Corso M, Blanchard M, Host S, Pascal L, Larrieu S (2014) Short-term impacts of particulate matter (PM10, PM10-2.5, PM2.5) on mortality in nine French cities. Atmos Environ 95:175–184. https://doi.org/10.1016/j.atmosenv.2014.06.030
9. Bharatraj J (2022) Location selection for smog towers using Zadeh's Z-numbers integrated with WASPAS. In: Fuzzy systems—theory and applications, 219. https://doi.org/10.5772/intechopen.95906
10. Kończak B, Cempa M, Pierzchała, Deska M (2021) Assessment of the ability of roadside vegetation to remove particulate matter from the urban air. Environmental Pollution, 268, 115465 https://doi.org/10.1016/j.envpol.2020.115465
11. Saaty TL (1977) A scaling method for priorities in hierarchical structures. J Math Psychol 15(3):234–281. https://doi.org/10.1016/0022-2496(77)90033-5
12. Yoon KP, Hwang CL (1995) Multiple attribute decision making: an introduction. Sage publications
13. https://earth.google.com/web/%4028.64668961,77.09293244,212.49022657a,91155.63985732d,30y,0h,0t,0r/data=OgMKATA?authuser=0 [Accessed: 07th May 2022]

Application of Machine Learning Approach in Internal Combustion Engine: A Comprehensive Review

Sanjeev Kumar, Prabhakar Sharma, and Kiran Pal

1 Introduction

The United Nations proposed and globally accepted sustainable development goals (SDGs), and more particularly SDG 7 emphasizes upon universal availability of clean and cheap energy to one and all. Among the various energy-demanding sectors, the transportation sector is one of the largest consumers of energy. Finding the greatest option for vehicle systems, transportation fuel, energy sources, and improved resource use has prompted researchers and manufacturers to improve efficiency [1]. The demand for sustainable assets and vehicle systems is swelling over the globe, owing to a rise in oil costs, security of energy, and global climate change. CO_2 remains the most common gas component of worldwide GHG emissions, accounting for up to 64.9% of total global GHG [2]. As a result, there would be a significant shift in global climate change, urban air pollution, and the depletion of nonrenewable energy sources. All of these issues have led researchers and manufacturers to develop innovative technology for the improved engine for vehicles [3, 4].

The industrial revolution and exploding population initiated unsustainable worldwide energy consumption and demand. Despite several studies examining the correlation between energy utilization and economic growth, the world is witnessing alarming rates of energy consumption and demand. Internal combustion engines (ICEs) are the heart of a wide array of aviation, terrestrial, industrial, and marine transportation, making transportation a major consumer. However, given the detrimental effects of using fossil fuels, exertions have been undertaken to mitigate

S. Kumar (✉) · P. Sharma
Department of Mechanical Engineering, Delhi Skill and Entrepreneurship University, Delhi 110089, India
e-mail: ersanjeevkumar124@gamil.com

K. Pal
Department of Mathematics, DITE, Okhla, New Delhi 110042, India

A. Kumar et al. (eds.), *Recent Advances in Manufacturing and Thermal Engineering*, Lecture Notes in Mechanical Engineering, https://doi.org/10.1007/978-981-19-8517-1_12

the imminent fuel crunch and emission-related environmental problems. Alternative energy sources that can integrate renewables and sustainability into the ever-increasing energy demand are urgently required [5]. Fatty oil methyl ester also known as biodiesel for potential application as blended fuel in compression ignited (CI) engines is one of the most promising alternative fuels, along with hydrogen, alcohol, ether, liquefied petroleum gas, producer gas, and compressed natural gas [6, 7]. Biodiesel is an extensive array of fuels derived from animal or vegetable oil fatty acids that have remained reasonably prominent due to its environmental benefits. Besides, being devoid of sulfur, it emits fewer pollutants as it is oxygenated and burns in the improved way [8]. It contains less energy density and emits more oxides of nitrogen, though. It possesses promising physio-chemical characteristics, such as cetane index, viscosity, flash point, lubrication, and density, but its maintenance costs are higher. It economically increases energy sufficiency but has a significant installation expense [5, 9].

Biofuel research has seen a significant surge in recent years, owing to the many benefits it offers. Another benefit to consider is the reduction of emissions and the simplicity of manufacturing. Several tactics are used to increase the efficiency of biofuels while also achieving the ideal mix ratio between diesel and gasoline. Additionally, while determining the best fuel mix, predictive modeling of engine emissions is taken into consideration as well [10]. It is the process of developing outcomes that are based on statistical strategies and probability assumptions that are known as predictive modeling. To deliver exact results, this approach requires certain input data [11]. Increased amounts of experimental data fed into the model improve the accuracy of the forecast. The precision of the predictions is vital component in instituting the accuracy of the model. In this article, we have also discussed current technological developments in the field of Machine Learning, which employs both Artificial Intelligence and Model Prediction to generate a highly accurate and compact model that is both accurate and compact [12].

In general, although different methodologies may yield a range of answers, experimental research can provide a straightforward examination of a practical system and accurate data. However, they are generally resource intensive and incur significant engineering expenditures, in terms of both capital and time expenditure [13]. Alternatively, although analytical procedures by their very nature give proof of concept, they also rely on conventions and approximations that are possibly unsuitable for real-world employment in certain cases. Since computational techniques fail to consider geometrical complexity and account for physical and in the majority of cases, they do provide alternate workable solutions through the use of geographically constrained approximations [14].

The present review has several important objectives, one being to consider the feasibility of applying frequently known statistical and machine learning tools into ICE research. Alternative approaches to predicting and optimizing the behavior of ICEs would be presented to readers. The main goal is to offer the necessary mathematical/statistical basis to link the procedures from the perspective of ICE research and therefore how such techniques may be applied in ICE studies. Even though large literature for each one of the methodologies under discussion is readily accessible,

this analysis has provided a thorough perspective on the multiple possibilities that are readily offered for use in ICE investigation.

2 Review of Important Machine Learning Techniques in the ICE Domain

In recent times, several machine learning techniques have been employed specifically for model forecasting and optimization of IC engine-operating parameters and outputs. An IC engine has primarily fuel injection parameters, engine loading, ratio of compression, and speed as independent input factors. Engine performance and efficiency parameters like brake thermal efficiency (BTE), specific fuel consumption (SFC), maximum/peak pressure in the cylinder (P_{max}), and emission characteristics such as oxides of nitrogen (NO_x), carbon monoxide (CO), di-oxide of carbon (CO_2), unburnt hydrocarbon (UHC), etc., are studied by most of the researchers as response variables. The important AI- and ML-based prediction and optimization are discussed in the following paragraphs.

2.1 Artificial Neural Network

Artificial neural network (ANN) has been the most employed AI tool for model prediction of ICE in both single fuel and dual fuel domains. The ANN is a useful technology in the domain of artificial intelligence for model forecasting since it saves time while generating more accurate results than other methodologies. Over the last decade, the usage of the ANN approach has gained in popularity in the energy and engineering science. Artificial neural networks (ANN) are easy to use, simple, and fast computer approaches capable of addressing challenging nonlinear problems. The only downside is that careful network design and a larger dataset for simulation are required, both of which are prohibitively costly.

The most basic NN architecture is the multi-layer perceptron (MLP) design. This kind of network, also termed as a feed-forward NN, is differentiated by the fact that all neurons in every layer have forward connections to all neurons in the layer above them. The MLP-ANN technique is shown by three layers in Fig. 1: the initial layer supplied with incoming data; the inter layers termed as concealed layers; and last layer, also known as the output layer, that shows ANN's predicted results.

Several studies have been published on the employment of ANN in the ICE domain for output prediction. Gene expression programming (GEP) and ANN were examined by Sharma [13], for their ability to forecast emission characteristics and engine performance. Biodiesel/diesel blends derived from linseed oil were utilized to power a stationary diesel engine. The engine input control factors, especially fuel injection settings, the ratio of biodiesel/diesel blends, and engine loadings were modified over

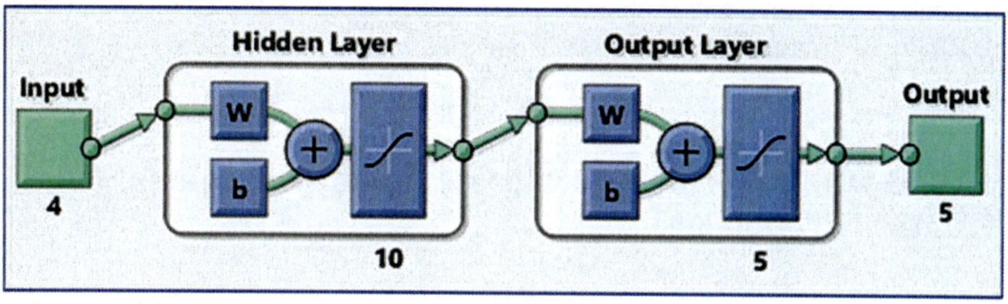

Fig. 1 A typical ANN architecture

a total of sixty laboratory tests. Based on statistical data like absolute percentage of variance 0.9667–0.997 and 0.9923–0.9998 for GEP and ANN, individually, the correlation coefficient was 0.98–0.99 and 0.99–0.99 for GEP and ANN, and it was determined that these two models are effective prognostic tools.

Uslu [15] also employed the ANN for predictive modeling. It was determined that the ANN is capable of accurately modeling exhaust output and performance characteristics with coefficients of determination (R^2) ranging from 0.8665 to 0.9857. In comparison to the experimental data, the highest mean relative error (MRE) is less than 10 percent. On a more realistic note, Salam and Verma [16] combined ANN-based empirical modeling with the limited responses obtained from direct numerical simulation to reliably estimate engine behavior while using little computing resources. They were able to properly estimate 17 combustion, performance, and emission parameters using injection pressure, fuel blends ratio, and loading. Similarly, Uslu [17] used ANN to approximate a prognostic model for nine engine outputs. All the above-mentioned studies discovered empirical idleness among engine parameter variables, implying that empirically reduced models are practical.

It can be concluded that, even with very simple designs, ANN has revealed its predictive ability for employment in the ICE domain. The creation of user-friendly interfaces across several platforms, such as the Rapidminer and nntool toolkit, has allowed the employment of ANN for the quick and easy characterization of ICE behavior.

2.2 *Gene Expression Programming*

GEP is a well-known method for leveraging computer algorithm evolution to solve nonlinear and complex engineering issues. Typically, GEP computer programs are encoded by gene expression strings of a defined length that are generated using nature-inspired operators such as crossover and mutation [10]. It has been demonstrated that GEP facilitates the creation of an accurate and concise computer-based model for output prediction. Time-series forecasting, classification difficulties, regression issues, data mining, and knowledge discovery are a few of the effective

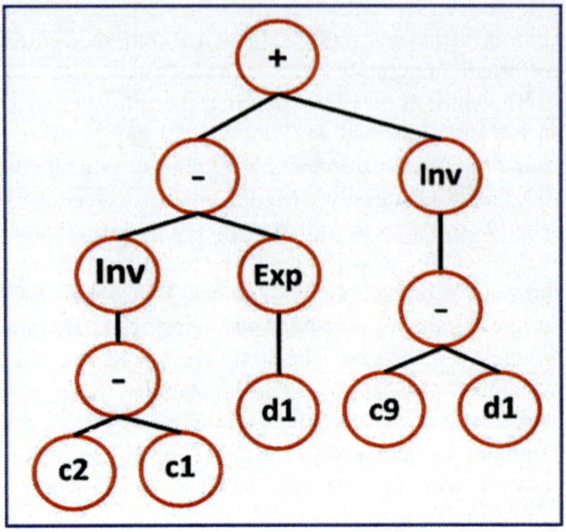

Fig. 2 A typical GEP model architecture

practical uses where it has been used [18]. It utilizes the problem's historical dataset to provide an Expression Tree solution (ET).

This approach was invented by Ferreira [19] in 1999 and made public in 2001. The GEP algorithm make use of basic elements involved in the formation of two popular algorithms i.e., genetic algorithm and programming, denoted as GA and GP, respectively. In GEP, chromosomal genotype is equal to that of a GA; however, the chromosomal phenotype is a tree type structure of data with variable size and length, similar to genetic programming. Of course, logical association among parameters should be investigated. During the model generation process, three ETs are typically created, and the total of these becomes the entire model. Figure 2 depicts a typical ET scenario [20, 21]. Applying the GEP approach, a set of undeviating natured chromosomes is primarily constructed to investigate relationship between parameters '*a*', '*b*', and '*y*'.

GEP, which is based on machine learning, was utilized by Sharma [22] to construct a prognostic model for the efficiency and emission features. The test CI engine was powered with linseed biodiesel mixes. Adjusting the engine loadings, the ratio of biodiesel/ diesel blends, injection pressure, engine load, and fuel injection duration yielded the experimental results for GEP. To predict BTE, SFC, NOx, and HC emissions, a model based on GEP was developed. Seventy percent of the gathered data was employed for learning, while thirty percent has been employed for model validating. The suggested GEP model was sufficiently efficient to accurately predict engine emission and performance physiognomies. For the output prediction, the correlation coefficient (R) was between 0.993 and 0.999, and the determination coefficient (R^2) was between 0.985 and 0.9998. The GEP model predicted outcomes with a root mean square error (RMSE) in the range of 0.0048–2.597 and a mean absolute error (MSE) ranging from 0.00369 to 4386, respectively. In another similar study by Sharma and

Sharma [10], it was reported that GEP combined with response surface methodology performed marginally better than GEP alone.

Bhowmik et al. [23] examined the influence of contaminated oxygenated fuel on the emissions and working of a CI engine. The study observed diesel adulteration dramatically decreases NO_X emissions while decreasing BTE, and increasing UHC, and CO emissions. Mixing ethanol with contaminated diesel decreases engine exhaust emissions devoid of changing performance characteristics. Based on experimental findings, GEP models for modeling the input (share of kerosene, loading, and share of ethanol) and outcomes (BTE, NO_x, HC, and also CO)' connection as the diesel-ethanol paradigm was developed. The study reported GEP as a robust prognostic technique. The study by Kakati and Banerjee [24] analyzed the GEP technique, with its particular adaptability qualities in model construction, for its trustworthiness as an expert modeling system in such an innovative combustion paradigm. To demonstrate the applicability and proficiency of the GEP approach, its performance was compared to that of the RSM that remains the gold standard for engine output mapping goals. Several classic and enhanced statistical curve fitting criteria, including relative error and absolute metrics, were used to investigate and evaluate model integrity.

GEP was used by Roy et al. [25] to anticipate the emission and performance parameters of a CRDI engine working in dual fuel settings powered with a CNG/diesel combination. Based on testing data, a GEP-based prognostic framework was created to forecast BTE, BSFC, NO_x, HC, and PM. The control factors' parameters considered for the model were energy share of CNG, load, and injection settings. The built GEP model accurately predicted performance and emission characteristics, as shown by correlation values in the band of 0.999–0.9999. While mean absolute % error for the prognostic model was 0.0356–1.089%, in conjunction with very low root mean squared errors, which provided a sufficient assessment of the sturdiness of the forecasted accuracy, the collected findings were equated to an ANN-based framework created using the same parameters, and the GEP model was observed superior in forecasting the response output.

Roy et al. [20] employed GEP, to establish a link between the inputs and outputs of a CRDI engine with EGR. GEP was used to simulate the emission and performance parameters (BTE, BSFC, CO_2, PM, and NO_x) with injection pressure, load, and EGR % serving as inputs. According to the results, the GEP was capable of accurately simulating observed engine performance and emission indices under a variety of CRDI operating modes with EGR. In addition, the outcomes of the GEP were compared to an ANN-based prognostic model constructed using the same parameter ranges. Comparing the results, the GEP model fared superior than the ANN model in forecasting the expected response variables.

2.3 Adaptive Neuro-Fuzzy Inference System

ANFIS is a hybrid method of AI that is mix of fuzzy logic with ANN. ANFIS blends the training capabilities of ANN and its interpersonal structure combined with decision-making of fuzzy logic. ANFIS similar to ANN can be trained employing vast quantities of data. Thus, ideal ANFIS architecture for addressing a linked problem is revealed. The finished ANFIS design is evaluated for its responsiveness to new samples. The network architecture of the ANFIS is divided into two sections: the premise and effect sections ANFIS training entails identifying the parameters associated with these components using an optimization technique. During training, ANFIS utilizes existing pairs of data for control and response variables. Then, IF–THEN fuzzy standards are developed which show that such subsystems are interrelated.

In the scientific literature, ANFIS are also termed as fuzzy type models, fuzzy controllers, fuzzy-rule frameworks, and associative fuzzy memory. An ANFIS consists of five working elements: a rule base containing a database describing the membership functions and several fuzzy if–then rules; a processor unit which does the inference; and a inference performing unit. Figure 3 depicts that ANFIS consists of five layers. This diagram represents an ANFIS structure with double inputs having a mono output. The first layer termed as the fuzzification layer. Using MF, the layer of fuzzification creates fuzzy clusters from input data.

Dirik [26] employed the ANFIS to develop a prognostic model for a natural gas-powered combined cycle power plant. The data gathered from the pollution monitoring system were employed to model ANFIS-GA and forecast emissions of NOx. First, ANFIS frameworks were established employing fuzzy C-Means, and to decrease the inaccuracy, the parameters were then improved using a GA. The

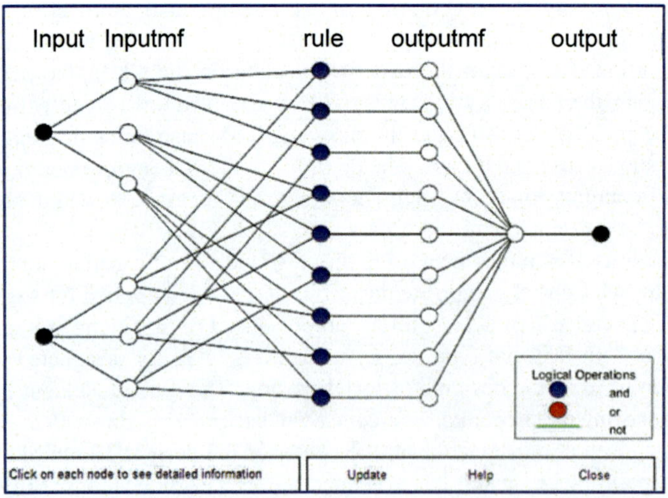

Fig. 3 A typical ANFIS architecture with 5 layers

obtained findings reveal that R^2 ranges between 0.8 and 0.904 for test and training data divided at various speeds. Sharma and Sahoo [27] used a hybrid ANFIS-RSM approach for prognostic modeling of dual fuel engine powered syngas-diesel combinations. In four distinct combinations, the syngas performance ($CO + H_2$), a unique imitation gaseous type fuel, had been investigated. Engine emission and performance data gathered over the whole load range were used to create RSM and ANFIS-based prognostic models. The ANFIS outperformed RSM in terms of model forecast, the RSM proved beneficial in creating algebraic connections among engine control factors and response. The constructed ANFIS framework has a high correlation between R (0.997–0.999) and R^2 (0.9918–0.997). Singh el al. [28] employed a different combination of ANIS and optimization techniques to model-optimize the efficiency and emission characteristics of a jojoba-fueled engine. The input factors include fuel injection pressure and timing, biodiesel mixes %, and engine loading, and the analysis takes into account corresponding output factors such as BTE, NOx, and UHC. The determination coefficient, on the other hand, shows that the ANFIS-PSO models with R^2 (0.9825, 0.9877, and 0.9895 for BTE, UHC, and NOx) provide a fair enhancement in steadiness, especially when compared to the projected ANFIS model. Several other authors [18, 29–31] reported the robust prognostic efficiency of the ANFIS technique.

2.4 Response Surface Methodology

Response surface methodology (RSM) is a computational and statistics-based technique for identifying elements that influence response. Because traditional procedures for improving the functioning of CI engines take more effort and time consuming, the RSM method is quick and inexpensive. The experiment design is a strategy utilized to design the experiment done on the CI engine performance and emissions, the necessary variables to be investigated utilizing affecting factors. Among the input parameters and the answers, RSM also helps in creating a 3-D surface to show cause and effects. The input factors for CI engine operation, such as the ratio of compression (CR), engine loadings, and fuel mix %, and the responses, such as performance and emissions parameters [32, 33].

The fractions of 3 m may be used to diminish the cost of carrying out the experimental analysis. Central composite design (CCD) is often applied for second-order designs. CCD consists of at least three components. The first element is a complete or fractional 2 m factorial type design with levels of factor designated −1 and 1 and is referred to as the design's factorial portion. The second element is an axial section consisting of 2 m points, where the twin data points in line with every control factor are chosen at a distance from the center of the design. The third element is the center point piece, which has a defined number of replications. Box-Behnken design is independent of factorial and fractional factorial designs. Box-Behnken design (BBD) positions dealings at the medians of the process space's boundaries

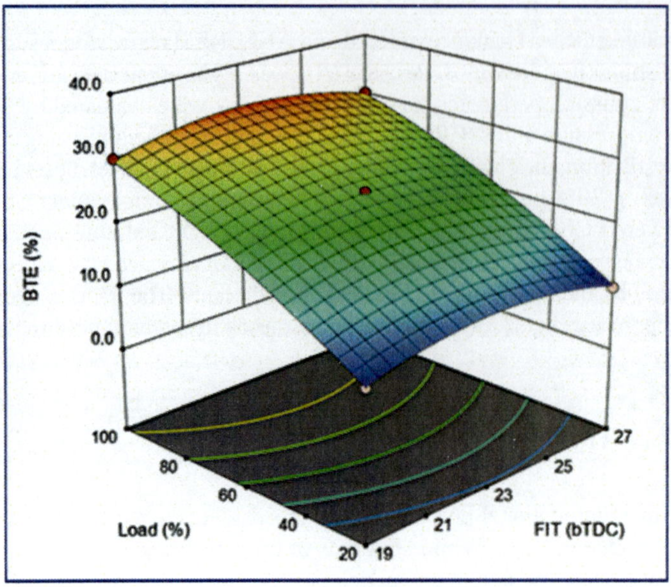

Fig. 4 A typical surface diagram

and its center. In comparison to CCD, the BBD is rotatable, needs three layers for each component, and has limited orthogonal blocking capability [34, 35].

For decades, RSM has been used as a strategy for enhancing and obtaining information on the combustion of alternative fuels coupled with other fuels in CI engines. Numerous studies have been reported in the recent past that employ RSM to improve diesel engine performance and emission physiognomies by employing diverse design methodologies, single or multiple input factors, and distinct results. Besides the prognostic ability, the RSM helps in creating 3D surface and 2D contour diagrams to illustrate the two factors at a time approach [36–39] (Fig. 4).

The RSM has been employed by numerous investigators to develop correlation, modeling, and optimization of IC engine parameters in the recent past. The influences of alumina nanoparticle addition in biodiesel/diesel blends and speed of engine on the performance and emission parameters of a six-cylinder, CI engine were examined [40]. A biodiesel/diesel blended fuel was supplemented with nanoparticles of alumina at concentrations of 160, 120, 80, and 40 ppm. These blends were used to fuel a diesel engine that was operated at varying engine speeds (800–1000 rpm). Using response surface methodology, the interplay of control factors on the emission as well as performance of CI engine was then investigated. The greatest values of braking power and torque were 42.82 kW and 402.8 Nm, respectively, at the engine speed of 1000 rpm. With the emergence of computer software, it has been easier to evaluate the aptness of biodiesel for employment in diesel engines by conducting fewer tests. Simsek et al. [41] investigated the ideal ratio of animal waste fat biodiesel and the associated engine responses employing ANN and RSM. In addition, test data were

used to examine the performance of RSM and ANN. RSM regression results indicate that the absolute percentage of variance (R^2) was higher than 0.95 for each output model. The emission parameters of a VCR CI engine were evaluated by [42] under different compression ratios (CR), injection pressure, and other operating conditions. In this investigation, the engine was powered by Jatropha biodiesel diesel blends B30 (70% diesel + 30% biodiesel). Compared to diesel, the employment of biodiesel blends powered CI engines has dramatically raised CO_2 and emissions by 13.3% and 2.12%, respectively. The environmental impact of elevated CO_2 emissions will be countered by the cultivation of biodiesel based plants. The RSM was utilized by several other researchers [33, 43–46] in the biodiesel diesel/dual fuel engine domain.

2.5 *Other AI and ML Methods*

The growing computational power is helping researchers to explore newer AI techniques for modeling ICEs. While hundreds of these techniques have been reported in the recent past, a few significant techniques used for engine parameters modeling are discussed here. Liu et al. [46] employed machine learning to model and predict a heavy-duty engine. The objective was to evaluate four distinct ML techniques—support vector machine, ANN, random forest, and XGBoost—concerning a analytical one dimensional CFD model. Model inputs included the timing of spark, engine speed, equivalency ratio, and speed of engine. The ANN approach was the most suitable; however, it needed the effort to tune its hyper-parameters. Overall, the findings demonstrated that well-learned ML models may complement more sophisticated physical models and aid in improving engine performance, emissions, and longevity. Wang et al. [47] examined ML models, including linear regression (LR), regression tree (TR), tree ensembles (TE), support vector machine (SVM) [48], and Gaussian process regression (GPR) [49] to forecast the combustion data as well as other engineering problems. Variations of the different primary fuel (n-butanol + gasoline), loads (low to part load), the timing of ignition, hydrogen vol. %, and excess air ratio were tested. For data processing and model optimization, the test data was partitioned into learning and test set of data. The method of normalization, K-fold type cross-validation, and the Bayesian approach based optimization were utilized. The LR model had the shortest training time among the five ML models, but the TR model had the lowest generalization ability. The TR model's minimum leaf size has a significant impact on generalization and regression. Based on these assumption, the TR framework enhanced regression capacity but needed more learning time.

3 Conclusion

Biodiesel does have the ability to profoundly contribute to the long-term reliability of transportation fuels. Because biodiesel production and consumption processes are

complicated and nonlinear, rapid and precise modeling tools are necessary for their monitoring, design, control, and optimization. ML techniques have outperformed traditional approaches for modeling such enormously complex systems. Furthermore, the benefits and drawbacks of using ML technology for biodiesel powered engine studies are presented to inspire future R&D efforts in this subject. This article examines the applications of ML approaches in combustion. The intersection of these two disciplines has engrossed considerable consideration in contemporary time due to the expansion of machine learning and continuing combustion research. This paper also explored the uses of machine learning in combustion and emission modeling. ML provides a huge advantage when it comes to detecting patterns hidden behind enormous data, discovering and envisaging high-dimensional operating control spaces, extracting complex modeling from inputs and outputs, and reducing computing cost and memory use. The ANN approach is the most used machine learning (ML) methodology. ANN is a soft computing learning approach that simulates the neurological processing power of the human brain to map input–output links of uncertain systems. Due to its huge generalization capacity, ANN has grown approval for handling complex nonlinear scientific and engineering problems in the real world. Other important approaches such as neuro-fuzzy, RSM, gene expression programming, random forest, and boosted regression tree are also frequently used. This article examines and evaluates several ML technology applications used in alternative fuel-powered engines.

References

1. Shyu C-W (2021) A framework for ‘right to energy’ to meet UN SDG7: Policy implications to meet basic human energy needs, eradicate energy poverty, enhance energy justice, and uphold energy democracy. Energy Res Soc Sci 79:102199. https://doi.org/10.1016/J.ERSS.2021.102199
2. Su W, Ye Y, Zhang C, Baležentis T, Štreimikienė D (2020) Sustainable energy development in the major power-generating countries of the European Union: the pinch analysis. J Clean Prod 256:120696. https://doi.org/10.1016/J.JCLEPRO.2020.120696
3. Sharma P, Sharma AK (2021) Combustion and thermal performance of dual fuel engine: influence of controlled producer gas substitution with pilot B20 (WCOME biodiesel—diesel) blending. Lect Notes Mech Eng 20:341–353
4. Veza I, Karaoglan AD, Ileri E, Afzal A, Hoang AT, Tamaldin N, Herawan SG, Abbas MM, Farid M, Said M (2022) Multi-objective optimization of diesel engine performance and emission using grasshopper optimization algorithm. Fuel 323:124303. https://doi.org/10.1016/j.fuel.2022.124303
5. Salam S, Choudhary T, Pugazhendhi A, Verma TN, Sharma A (2020) A review on recent progress in computational and empirical studies of compression ignition internal combustion engine. Fuel 279:118469. https://doi.org/10.1016/J.FUEL.2020.118469
6. Bora BJ, Tran TD, Shadangi KP, Sharma P, Said Z, Kalita P, Buradi A, Nguyen VN, Niyas H, Pham MT, Le CTN, Tran VD, Nguyen XP (2022) Improving combustion and emission characteristics of a biogas/biodiesel-powered dual-fuel diesel engine through trade-off analysis of operation parameters using response surface methodology. Sustain Energy Technol Assessments 53:102455. https://doi.org/10.1016/J.SETA.2022.102455

7. Said Z, Le DTN, Sharma P, Dang VH, Le HS, Nguyen DT, Bui TAE, Nguyen VG (2022) Optimization of combustion, performance, and emission characteristics of a dual-fuel diesel engine powered with microalgae-based biodiesel/diesel blends and oxyhydrogen. Fuel 326:124987. https://doi.org/10.1016/j.fuel.2022.124987
8. Sharma P, Sharma AK (2021) Statistical and continuous wavelet transformation-based analysis of combustion instabilities in a biodiesel-fueled compression ignition engine. J Energy Resour Technol 144. https://doi.org/10.1115/1.4051340
9. Bae C, Kim J (2017) Alternative fuels for internal combustion engines. Proc Combust Inst 36:3389–3413. https://doi.org/10.1016/j.proci.2016.09.009
10. Sharma P, Sharma AK (2021) AI-based prognostic modeling and performance optimization of CI engine using biodiesel-diesel blends. Int. J. Renew. Energy Resour. 11:701–708
11. Said Z, Nguyen TH, Sharma P, Li C, Ali HM, Nguyen VN, Pham VV, Ahmed SF, Van DN, Truong TH (2022) Multi-attribute optimization of sustainable aviation fuel production-process from microalgae source. Fuel 324:124759. https://doi.org/10.1016/j.fuel.2022.124759
12. Said Z, Sharma P, Sundar LS, Afzal A, Li C (2021) Synthesis, stability, thermophysical properties and AI approach for predictive modelling of Fe3O4 coated MWCNT hybrid nanofluids. J Mol Liq 117291. https://doi.org/10.1016/J.MOLLIQ.2021.117291
13. Sharma P (2021) Artificial intelligence-based model prediction of biodiesel-fueled engine performance and emission characteristics: a comparative evaluation of gene expression programming and artificial neural network. Heat Transf. https://doi.org/10.1002/htj.22138
14. Said Z, Cakmak NK, Sharma P, Sundar LS, Inayat A, Keklikcioglu O, Li C (2022) Synthesis, stability, density, viscosity of ethylene glycol-based ternary hybrid nanofluids: experimental investigations and model -prediction using modern machine learning techniques. Powder Technol 117190. https://doi.org/10.1016/J.POWTEC.2022.117190
15. Uslu S, Celik MB (2018) Prediction of engine emissions and performance with artificial neural networks in a single cylinder diesel engine using diethyl ether. Eng. Sci. Technol. an Int. J. 21:1194–1201. https://doi.org/10.1016/j.jestch.2018.08.017
16. Salam S, Verma TN (2019) Appending empirical modelling to numerical solution for behaviour characterisation of microalgae biodiesel. Energy Convers Manag 180:496–510. https://doi.org/10.1016/J.ENCONMAN.2018.11.014
17. Uslu S (2020) Optimization of diesel engine operating parameters fueled with palm oil-diesel blend: comparative evaluation between response surface methodology (RSM) and artificial neural network (ANN). Fuel 276:117990. https://doi.org/10.1016/j.fuel.2020.117990
18. Sharma P, Said Z, Memon S, Elavarasan RM, Khalid M, Nguyen XP, Arıcı M, Hoang AT, Nguyen LH (n.d.) Comparative evaluation of AI-based intelligent GEP and ANFIS models in prediction of thermophysical properties of Fe3O4-coated MWCNT hybrid nanofluids for potential application in energy systems. Int J Energy Res. https://doi.org/10.1002/er.8010
19. Ferreira C (2001) A new adaptive algorithm for solving problems. Complex Syst
20. Roy S, Ghosh A, Das AK, Banerjee R (2015) Development and validation of a GEP model to predict the performance and exhaust emission parameters of a CRDI assisted single cylinder diesel engine coupled with EGR. Appl Energy 140:52–64. https://doi.org/10.1016/j.apenergy.2014.11.065
21. Siddiki SYA, Mofijur M, Kumar PS, Ahmed SF, Inayat A, Kusumo F, Badruddin IA, Khan TMY, Nghiem LD, Ong HC, Mahlia TMI (2022) Microalgae biomass as a sustainable source for biofuel, biochemical and biobased value-added products: an integrated biorefinery concept. Fuel 307:121782. https://doi.org/10.1016/J.FUEL.2021.121782
22. Sharma P (2020) Gene expression programming-based model prediction of performance and emission characteristics of a diesel engine fueled with linseed oil biodiesel/diesel blends: an artificial intelligence approach. Energy Sources Part A Recover Util Environ Eff. https://doi.org/10.1080/15567036.2020.1829204
23. Bhowmik S, Paul A, Panua R, Ghosh SK, Debroy D (2019) Artificial intelligence based gene expression programming (GEP) model prediction of Diesel engine performances and exhaust emissions under Diesosenol fuel strategies. Fuel 235:317–325. https://doi.org/10.1016/j.fuel.2018.07.116

24. Kakati D, Banerjee R (2021) Assessing the competency of a semi-parametric expert system in the realms of response characterization uncertainty in premixed methanol dual fuel diesel combustion strategies: in critique to RSM. Expert Syst Appl 185:115516. https://doi.org/10.1016/J.ESWA.2021.115516
25. Roy S, Ghosh A, Das AK, Banerjee R (2014) A comparative study of GEP and an ANN strategy to model engine performance and emission characteristics of a CRDI assisted single cylinder diesel engine under CNG dual-fuel operation. J. Nat. Gas Sci. Eng. 21:814–828. https://doi.org/10.1016/j.jngse.2014.10.024
26. Dirik M (2022) Prediction of NOx emissions from gas turbines of a combined cycle power plant using an ANFIS model optimized by GA. Fuel 321:124037. https://doi.org/10.1016/J.FUEL.2022.124037
27. Sharma P, Sahoo BB (2022) An ANFIS-RSM based modeling and multi-objective optimization of syngas powered dual-fuel engine. Int J Hydrogen Energy. https://doi.org/10.1016/j.ijhydene.2022.04.093
28. Singh NK, Singh Y, Sharma A, Rahim EA (2020) Prediction of performance and emission parameters of Kusum biodiesel based diesel engine using neuro-fuzzy techniques combined with genetic algorithm. Fuel 280:118629. https://doi.org/10.1016/J.FUEL.2020.118629
29. Aghbashlo M, Peng W, Tabatabaei M (2021) Machine learning technology in biodiesel research: a review. Prog Energy Combust Sci 85:100904. https://doi.org/10.1016/j.pecs.2021.100904
30. Saravanakumar L, Prakash R (2020) Validation of performance and emissions of a CI engine fueled with calophyllum inophyllum methyl esters using soft computing technique. Fuel 266:117070. https://doi.org/10.1016/J.FUEL.2020.117070
31. Razavi R, Sabaghmoghadam A, Bemani A, Baghban A, Chau KW, Salwana E (2019) Application of ANFIS and LSSVM strategies for estimating thermal conductivity enhancement of metal and metal oxide based nanofluids. Eng Appl Comput Fluid Mech 13:560–578. https://doi.org/10.1080/19942060.2019.1620130
32. Elkelawy M, El Shenawy EA, Alm-Eldin Bastawissi H, Shams MM, Panchal H (2022) A comprehensive review on the effects of diesel/biofuel blends with nanofluid additives on compression ignition engine by response surface methodology. Energy Convers Manag X 14:100177. https://doi.org/10.1016/J.ECMX.2021.100177
33. Sharma P (2021) Prediction-optimization of the effects of di-tert butyl peroxide-biodiesel blends on engine performance and emissions using multi-objective response surface methodology (MORSM). J Energy Resour Technol 1–26. https://doi.org/10.1115/1.4052237
34. Sharma P, Chhillar A, Said Z, Huang Z, Nguyen VN, Quy P, Nguyen P, Nguyen XP (2022) Energy sources, part a: recovery, utilization, and environmental effects experimental investigations on efficiency and instability of combustion process in a diesel engine fueled with ternary blends of hydrogen peroxide additive/biodiesel/diesel. https://doi.org/10.1080/15567036.2022.2091692
35. Elkelawy M, Bastawissi HAE, Esmaeil KK, Radwan AM, Panchal H, Sadasivuni KK, Suresh M, Israr M (2020) Maximization of biodiesel production from sunflower and soybean oils and prediction of diesel engine performance and emission characteristics through response surface methodology. Fuel 266:117072. https://doi.org/10.1016/J.FUEL.2020.117072
36. Sharma P, Sharma AK (2021) Application of response surface methodology for optimization of fuel injection parameters of a dual fuel engine fuelled with producer gas- biodiesel blends, energy sources. Part A Recover Util Environ Eff 00:1–18. https://doi.org/10.1080/15567036.2021.1892883
37. Thodda G, Madhavan VR, Thangavelu L (2020) Predictive modelling and optimization of performance and emissions of acetylene fuelled ci engine using ann and rsm, energy sources. Part A Recover Util Environ Eff 00:1–19. https://doi.org/10.1080/15567036.2020.1829191
38. Ravindra K, Kaur-Sidhu M, Mor S, Chakma J, Pillarisetti A (2021) Impact of the COVID-19 pandemic on clean fuel programmes in India and ensuring sustainability for household energy needs. Environ Int 147:106335. https://doi.org/10.1016/J.ENVINT.2020.106335

39. Sharma A, Ansari NA, Pal A, Singh Y, Lalhriatpuia S (2019) Effect of biogas on the performance and emissions of diesel engine fuelled with biodiesel-ethanol blends through response surface methodology approach. Renew Energy 141:657–668. https://doi.org/10.1016/j.renene.2019.04.031
40. Ghanbari M, Mozafari-Vanani L, Dehghani-Soufi M, Jahanbakhshi A (2021) Effect of alumina nanoparticles as additive with diesel–biodiesel blends on performance and emission characteristic of a six-cylinder diesel engine using response surface methodology (RSM). Energy Convers. Manag. X. 11:100091. https://doi.org/10.1016/j.ecmx.2021.100091
41. Simsek S, Uslu S, Simsek H (2022) Proportional impact prediction model of animal waste fat-derived biodiesel by ANN and RSM technique for diesel engine. Energy 239:122389. https://doi.org/10.1016/J.ENERGY.2021.122389
42. Singh A, Sinha S, Choudhary AK, Panchal H, Elkelawy M, Sadasivuni KK (2020) Optimization of performance and emission characteristics of CI engine fueled with Jatropha biodiesel produced using a heterogeneous catalyst (CaO). Fuel 280:118611. https://doi.org/10.1016/J.FUEL.2020.118611
43. Kashyap D, Das S, Kalita P (2021) Exploring the efficiency and pollutant emission of a dual fuel CI engine using biodiesel and producer gas: An optimization approach using response surface methodology. Sci Total Environ 773:145633. https://doi.org/10.1016/j.scitotenv.2021.145633
44. Elkelawy M, Bastawissi HAE, Esmaeil KK, Radwan AM, Panchal H, Sadasivuni KK, Suresh M, IsrarM (2020) Maximization of biodiesel production from sunflower and soybean oils and prediction of diesel engine performance and emission characteristics through response surface methodology. Fuel. https://doi.org/10.1016/j.fuel.2020.117072
45. Boodaghi H, Etghani MM, Sedighi K (2021) Performance analysis of a dual-loop bottoming organic Rankine cycle (ORC) for waste heat recovery of a heavy-duty diesel engine, Part I: Thermodynamic analysis. Energy Convers Manag 241:113830. https://doi.org/10.1016/J.ENCONMAN.2021.113830
46. Mahla SK, Ardebili SMS, Sharma H, Dhir A, Goga G, Solmaz H (2021) Determination and utilization of optimal diesel/n-butanol/biogas derivation for small utility dual fuel diesel engine. Fuel 289:119913. https://doi.org/10.1016/j.fuel.2020.119913
47. Liu J, Huang Q, Ulishney C, Dumitrescu CE (2021) Machine learning assisted prediction of exhaust gas temperature of a heavy-duty natural gas spark ignition engine. Appl Energy 300:117413. https://doi.org/10.1016/J.APENERGY.2021.117413
48. Wang H, Ji C, Su T, Shi C, Ge Y, Yang J, Wang S (2022) Comparison and implementation of machine learning models for predicting the combustion phases of hydrogen-enriched Wankel rotary engines. Fuel 310:122371. https://doi.org/10.1016/J.FUEL.2021.122371
49. Said Z, Sharma P, Elavarasan RM, Tiwari AK, Rathod MK (2022) Exploring the specific heat capacity of water-based hybrid nanofluids for solar energy applications: a comparative evaluation of modern ensemble machine learning techniques. J. Energy Storage. 54:105230. https://doi.org/10.1016/J.EST.2022.105230
50. Said Z, Sharma P, Tiwari AK, Le VV, Huang Z, Bui VG, Hoang AT (2022) Application of novel framework based on ensemble boosted regression trees and Gaussian process regression in modelling thermal performance of small-scale Organic Rankine Cycle (ORC) using hybrid nanofluid. J Clean Prod 360:132194. https://doi.org/10.1016/j.jclepro.2022.132194

Experimental Validation of Damage Detection in Concrete Beams Under Impact Load Using Piezo Sensors

Arya Sajith and Shilpa Pal

1 Introduction

Structural health monitoring is a widely accepted technique in the field of civil engineering. It gives a real-time analysis of the structure [5]. With passage of time, some sort of flaws or localized damages are bound to happen to a structure. Inspection and monitoring of structures are necessary to avoid failure of the structure that might even lead to collapse. Visual inspection and some non-destructive techniques (NDT) that uses low-frequency response of structure might not be suitable for detecting insipient damages [8]. This is where smart sensors such as lead zirconate titanate (PZT) play an important role.

Structures are subjected to different types of loads during their life time. One among these loads is the impact load. Impact load arises in various situation such as a structure when is hit by a vehicle, a marine structure subjected to ice load as an impact, and terrorism attacks [2]. In this study, lead zirconate titanate is used as the smart sensor which is embedded inside the concrete beam to analyze the damage under impact load by employing electro-mechanical impedance technique. The EMI technique uses both the direct and converse piezoelectric effect which allows the PZT sensor to act as both the sensor and the actuator [1]. It basically creates a relation between structures' mechanical property and sensor's electrical property.

Bhalla [5] in his study on evaluation of smart piezoceramic (PZT) transducers for the structural health monitoring of concrete found that location of damage or failure can be localized using an array of PZT transducers. Aabid [1] focuses on the various opportunities, applications, and challenges of structural health monitoring with piezo electric material. Park G [9] in his study investigated the efficiency of implementing the impedance based technique for structural health monitoring

A. Sajith (✉) · S. Pal
Delhi Technological University, Shahabad Daulatpur, Main Bawana Road, Delhi, India
e-mail: aryasajith@gmail.com

A. Kumar et al. (eds.), *Recent Advances in Manufacturing and Thermal Engineering*,
Lecture Notes in Mechanical Engineering, https://doi.org/10.1007/978-981-19-8517-1_13

on composite-reinforced concrete walls and concluded that this technique has the capability to monitor the conditions of real life civil structures.

1.1 Electro-Mechanical Impedance Technique of PZT Patches

In EMI based technique, the PZT patches can be bonded with structure in three ways—non bonded, surface bonded, and embedded [4]. It works on both direct and converse piezoelectric effect. LCR meter is a device used to record the vibration response. This LCR meter excites the piezo patches attached on the structure at high frequency varying in a range from 30 to 500 kHz [8]. This causes the PZT patches to produce mechanical strain or vibrations i.e., due to the excitation, a strain will be developed in the structure. Hence the PZT patch acts as an actuator in this case. Now acting as a sensor, the patch receives a signal back which consists of structural properties and this is called an admittance signature. These admittance signatures are very unique to each structure. It consists of a real part called conductance and an imaginary part called inductance. Any change or damage in structure will alter the admittance signature thereby indicating damage in the structure. The admittance signature is altered because when the structure has a damage, and it affects the mechanical impedance which is essentially a function of mass, stiffness, and damping of the structure [3].

Ling et al. [6] found out a relation between the electrical and structural impedance for one dimensional PZT patches. This was later extended for 2 D by Bhalla and Song [3] and expressed the admittance $\overline{Y}$ in ohms as Eq. (1).

$$\overline{Y} = G + Bj = 4\omega\frac{l^2}{h}\left(\overline{\varepsilon_{33}^T} - \frac{2d_{31}^2\overline{Y^E}}{1-\nu}\right) + \frac{8\omega d_{31}^2\overline{Y^E}l^2}{h(1-\nu)}\left(\frac{Z_{a,\text{eff}}}{Z_{s,\text{eff}} + Z_{a,\text{eff}}}\right)\overline{T}j \tag{1}$$

where, conductance is denoted by G, susceptance by B, x is the angular frequency, the mechanical impedance of PZT patch and host structure are represented by $Z_{a,\text{eff}}$ and $Z_{s,\text{eff}}$, respectively. $\overline{Y^E}$ is the Young's modulus and $\overline{\varepsilon_{33}^T}$ is the electric permittivity of the patch, l represents the length and h denotes the thickness of PZT patch, piezoelectric strain coefficient is represented by d_{31}, Poisson's ratio of the patch is denoted by ν, and tangent ratio by $\overline{T}$ and it is given as Eq. (2).

$$\overline{T} = \frac{1}{2}\left(\frac{\tan C_1kl}{C_1kl} + \frac{\tan C_2kl}{C_2kl}\right);$$
$$k = \omega\sqrt{\frac{\rho(1-\nu^2)}{\overline{Y^E}}} \tag{2}$$

where k is the wave number and ω is the density of the PZT patch.

In this study, PZT patch has been embedded in a concrete beam of size 700 mm × 150 mm × 150 mm to determine the damage detection capacity of the piezo sensors under impact load. To quantify the damage or deviation, RMSD or root mean square index has been used [7]. The aim of the study is also to understand how the sensors respond to impact load and to what extend this technique can be used in real-life structural health monitoring.

2 Materials and Methods

PZT patches are very thin and brittle with a size of around 0.2–0.3 mm. The PZT patches have been initially soldered with the soldering wire by fixing the positive and negative side, so that the same can be followed throughout the work. The soldering should be very carefully done as the patches are very sensitive and brittle as shown in Fig. 1. In order to protect the sensors while using it as embedded sensors, epoxy adhesives have been used. Araldite has been used as the epoxy adhesive to cover the soldered PZT patch. For this, a plastic mold has been taken and it is filled halfway with araldite. The patch was placed inside it once the epoxy starts to dry and then it has been completely filled. After it has been completely dried, the plastic mold is taken out and the PZT patch is now completely protected to use while casting the concrete as shown in Fig. 2.

Fig. 1 Soldered PZT patch

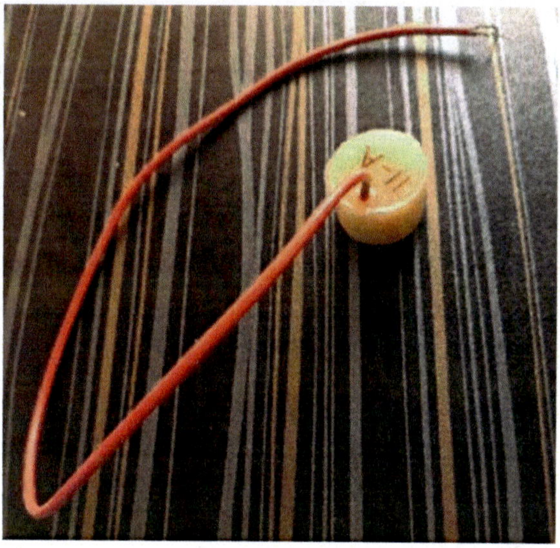

Fig. 2 PZT patch covered with epoxy

2.1 Casting the Beams

Mix design for M30 grade of concrete has been done, and compressive strength test was carried out to confirm the ratio obtained. The center portion of the mold was marked, and the concrete was poured into the mold till halfway. The epoxy-covered PZT patch has been placed in the center, and the mold is then completely filled with the concrete as shown in Figs. 3 and 4. After 24 h, the mold is removed and the beam is placed in a curing tank for 28 days.

Fig. 3 Placing the PZT patch while casting

Fig. 4 Casting the beam in the mold

2.2 *Impact Test*

A spherical ball of 5 kg in weight has been used as the impact ball and is dropped from a height of 3 m. A vertical pipe was designed to make sure that the point of impact after dropping the ball is center. The voltage has been monitored at the time of the impact with the help of an oscilloscope. Before and after each impact, the altered admittance signature from the PZT patch has been recorded with the help of an LCR meter.

2.3 *Experimental Setup*

Experimental setup is shown in Fig. 5. The vertical pipe has been used to guide the ball that is being dropped from a height of 3 m. The impact ball is a steel ball of 13 cm diameter and weighs 5 kg and is tied with the help of a rope for dropping as shown in Fig. 6. One end of the LCR meter is connected to the laptop to record the data and save it for future references, and the other end is connected to the PZT patch inside the sample to obtain the admittance signature as shown in Fig. 7.

3 Results and Discussion

Impact load is an impulsive dynamic load which occurs in a short time. Impact resistance of a structure is the ability to absorb and dissipate energy without causing damage to the structure. In this study, the ball was dropped from a height of 3 m through the vertical pipe. The beam cracked exactly at the center after the second impact. The crack formation is shown in Figs. 8 and 9.

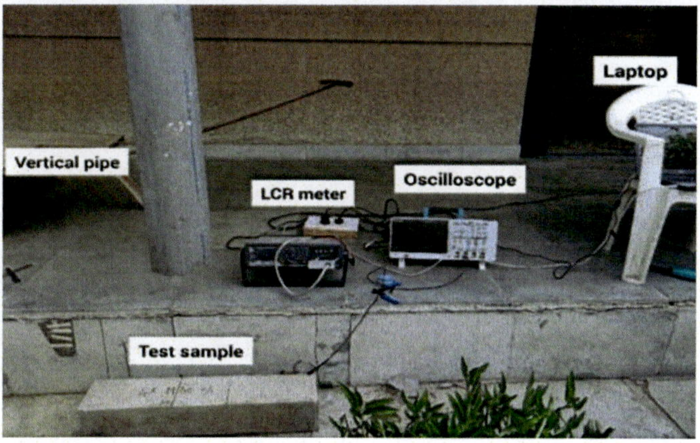

Fig. 5 Experimental setup

Fig. 6 Impact ball

3.1 Admittance Signature Analysis

The admittance signature is the unique signature of the structure with all the properties. After the impact, the structural properties undergo some changes and hence the admittance signature get altered. This alteration indicates damage. In this study, admittance signature was measured before and after impact with the help of a LCR meter. The other end of wire that was soldered into the PZT sensors are connected with the LCR meter by noting the positive and negative sides. Measurements were taken for a frequency range of 30–600 kHz, and the admittance signature was recorded. The obtained result is shown in the form of a graph in Figs. 10, 11 and 12.

Fig. 7 Connection point

Fig. 8 Initial crack after 1st impact

Fig. 9 Crack at failure

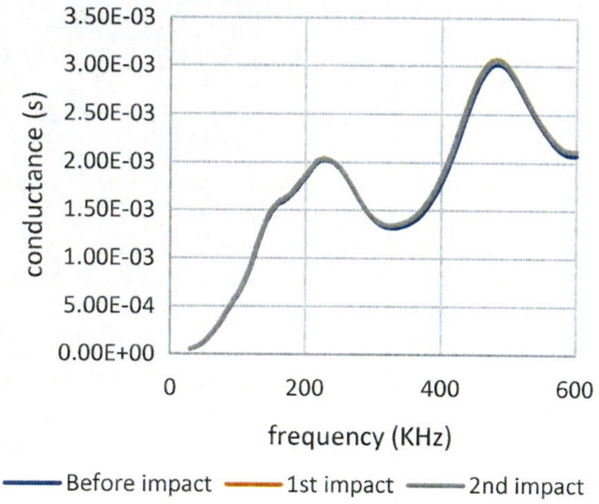

Fig. 10 Frequency versus conductane graph

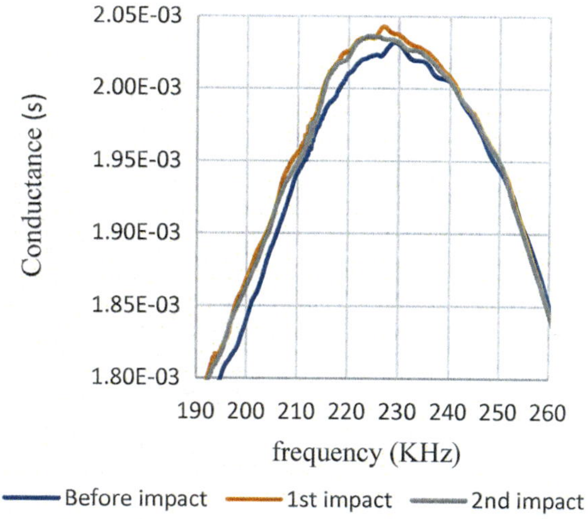

Fig. 11 Enhanced frequency versus conductane graph

From the obtained admittance signature, it can be seen that with each impact the admittance signature is shifting, this is because after the damage the structural parameters like damping, stiffness, and mass distribution are changing and this change alters the signature as well, which thereby indicates that the sensor is able to recongnise the damage. The peak values obtained from the admittance signatures are shown in Table 1. It can be concluded that with each impact, the maximum value of the admittance signature is increasing, which indicates that the extend of damage is also increasing.

The almost straight line in the frequency versus susceptance graph as per Fig. 12 indicates that the PZT patches were intact in the beam and were working in proper conditions (Fig. 13).

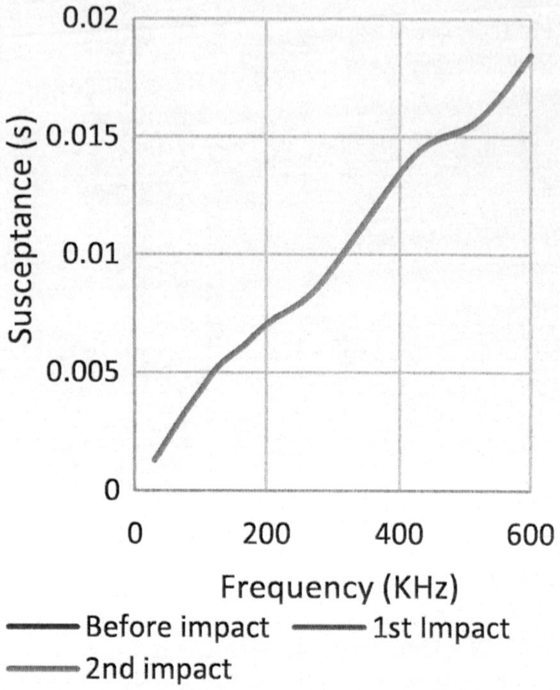

Fig. 12 Frequency versus Susceptance graph

Table 1 Peak value of conductance

Impact number	Peak value (sec)
1	3.02×10^{-3}
2	3.06×10^{-3}
3	3.07×10^{-3}

3.2 RMSD Index Analysis

RMSD index or root mean square index is a method used to quantify the damage. It is a technique used to calculate the amount of deviation of the admittance signature after impact i.e., when the damage has occurred in the structure with the baseline admittance signature.

$$\text{RMSD}\% = \sqrt{\frac{\sum_{i=1}^{i=N} (b_i - a_i)^2}{\sum_{i=1}^{i=N} a_i^2}} \times 100 \tag{3}$$

where a_i is the real part of the impedance signal (conductance) of the baseline signature i.e., before impact and b_i is the real part of the impedance signal after the damage has occurred. N indicates the total number of frequencies.

Fig. 13 Enhanced frequency versus Susceptance graph

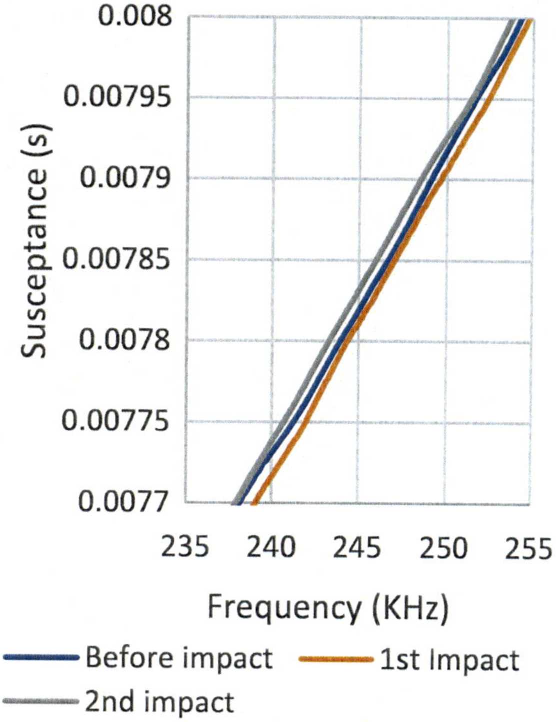

Applying Eq. (3), RMSD percentage for first impact and second impact has been found out and graph has been plotted as per Fig. 14. It can be concluded that RMSD value is increasing with each impact, which shows the deviation of the admittance signature value from the baseline indicating the damage.

Fig. 14 RMSD % comparison

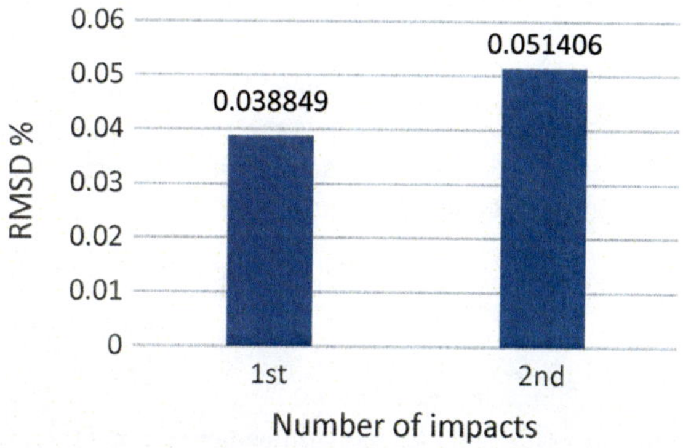

4 Conclusion

The present work has explored the damage detection capability of PZT transducers by using EMI technique. It can be inferred from the study that PZT patches are highly sensitive and have the capability to sense even the slightest change in the structure in terms of damage. From the admittance signature obtained, it can be observed that the peak value for conductance which is the real part of impedance signal increases with number of impact which indicates that the deviation in the signature has been recorded with the help of PZT patches. RMSD index, which was used to quantify the damage, also showed that the percentage RMSD value is increasing with increase in the number of impact, which shows that the extent of damage is also increasing.

EMI technique has proved to be one of the best techniques for structural health monitoring. It is very effective in detecting local damage by using the high-frequency range. This helps in an early detection of even the incipient damages and hence can be applied for real-time analysis. Thus, it can be successfully used in structural health monitoring for maintenance and prevention of failure of structures.

References

1. Aabid A, Parveez B, Raheman M, Ibrahim Y, Anjum A, Hrairi M, Parveen N, Zayan J (2021) A review of piezoelectric material-based structural control and health monitoring techniques for engineering structures: challenges and opportunities. Actuators 10(5):101
2. Anil O, Kantar E, Yilmaz MC (2015) Low velocity impact behaviour of RC slabs with different support types. Constr Build Mater 93:1078–1088
3. Bhalla S, Soh CK (2004) Structural health monitoring by piezo-impedance transducers II: application. J Aerosp Eng 17:166–175
4. Divsholi B, Yang Y (2011) Comparison of embedded, surface bonded and reusable piezoelectric transducers for monitoring of concrete structures. In: Proceedings of SPIE sensors and smart structures technologies for civil, mechanical, and aerospace systems, vol, 798151–1–10
5. Dixit A, Bhalla S (2018) Prognosis of fatigue and impact induced damage in concrete using embedded piezo-transducers. Sens Actuators A 274:116–131
6. Liang C, Sun FP, Rogers CA (1994) Coupled electro-mechanical analysis of adaptive material systems-determination of the actuator power consumption and system energy transfer. J Intell Mater Syst Struct 5(1):12–20
7. Singh I, Pal S, Dev N Impedance based damage assessment of concrete under the combined effect of impact and temperature using different piezo configurations
8. Naidu A, Bhalla S (2003) Damage detection in concrete structures with Smart Piezoceramic transducers. In: SPIE Proceedings SPIE smart materials, structures, and systems, vol 5062. Bangalore, India, 684–690
9. Park G, Cudney H, Inman D (2000) Impedance-based health monitoring of civil structural components. J Infrastruct Syst 6:153–160

Experimental Investigation for Enhancing Surface Quality in 3D Printing Process Using Non-planar Layer Method

Mriganka Maity, Somnath Das, Ranjan Kumar, and Joydip Kumar Mondal

1 Introduction

Printing of 3D model into a physical/solid object, the process is called three-dimensional (3D) printing process and it is possible to print an object with complex geometry that cannot be created by any other artificial resources. Non-planar 3D printing is a different type of printing process that uses all axes at the same time and is a unique method of Fused Deposit Modeling (FDM). In the case of regular FDM 3D printers, the layers are locked at the same height, but in the case of non-planar 3D printing, this moderation is removed and printing of smooth, curved layers is possible, resulting in smoother surfaces [1]. Non-planar 3D printing processes help eliminate the famous "stair effect" usually found on surfaces created by regular FDM printers. The idea of non-planar 3D printing has already demonstrated its potential in a variety of applications where curved surfaces are required. The team there came up with the idea of non-planar 3D printing, but the non-planar 3D printing process is still in its infancy [2]. It is true that layer-based additions are the production process instinctively the influence of prudence greatly known and this method has long been addressed by many more researchers. The non-planar layer 3d printer is designed to improve the surface quality of printed object and to determine how it gradually becomes dominant in areas with an object surface tendency. Several research teams have already started efforts to mitigate the effects of stair-steps and recover surface quality [3]. In the case of normal 3d printing, stair-stepping artifacts can be suggestively reduced by increasing the Z-direction, but in this case the printing time is greatly increased [4]. Adapted slicing process is a very unique process that basically balances the transaction between 3D printed parts quality and printing speed where

M. Maity (✉) · R. Kumar · J. K. Mondal
Swami Vivekananda University, Barrackpore, Kolkata 700121, India
e-mail: mriganka14maity@gmail.com

S. Das
Swami Vivekananda Institute of Science and Technology, Sonarpur, Kolkata 700145, India

A. Kumar et al. (eds.), *Recent Advances in Manufacturing and Thermal Engineering*,
Lecture Notes in Mechanical Engineering, https://doi.org/10.1007/978-981-19-8517-1_14

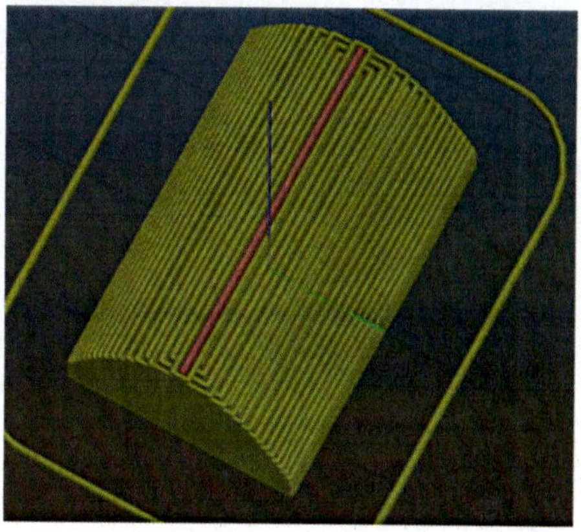

Fig. 1 Slicing view of FDM 3D-printing object using planar layer method

the geometry of the surface reduces the thickness of the layer in places where higher resolution is required. With non-planar layers, the extrusion planner follows exactly the contour of the real surface of the 3D printed object in its place of slicing into the layer [5]. Non-planar layers are used to smooth out almost perfect surfaces as shown in Fig. 2, and Fig. 1 shows the slicing view of FDM 3D-printing object using planar layer method which is not smooth. In the case of non-planar layer, formatting G-code is not a very simple task like planar layer, it is a partially complex task. Only surfaces with 1D-curvature can be printed with linear extrusion without distortion and in general cases 5-axis equipment is required [6]. But self-collision and displacement between different non-planar surfaces is somehow the main problem. In general, non-planar layers are strongly affected by stair-stepping for flat surfaces [7, 8]. This research work discusses experimental studies on product surface quality that were not possible for the planner tool path. To enhance the surface quality of 3D printed object, non-planar layer method has been employed and its efficacy has been compared with respect to slicing time, printing time, and printing cost over planer layer method.

2 Toolpath Generation for Non-planar Layer

Non-planar 3D printing is another type of FDM printing style that allows 3D printing on the curvature of any part. Non-planar printing is the process of printing a part with curved layers where the curvature extends across the X-, Y-, and Z-axes. This printing style usually helps to print curved parts with a smooth exterior finish regardless of other factors like filament type, hot finish, post-processing, etc. The 3D printing process is capable of creating different objects without creating specific tooling and without using different tools. Planer slicing should be increased to create non planner toolpath so that the non-planar layers replace the top of the object separately [9]. For this reason, for all surface improvements, regular planer slicing is implemented and

Fig. 2 Slicing view of FDM 3D-printing object using non-planar layer method

then the non-planar printable area surfaces are moved to the highest level which is able to distort their intended position to create toolpaths.

2.1 *Process of Generation Layer*

It is cut into horizontal slices to create layers from the object model. The layers of a printed object are always evenly distributed along the z-axis. To represent its outline, multiple polygons are formed in each level and points are formed from these polygons whose horizontal layer intersects with any of the sides. Planer layers have some areas where the surface becomes smoother when printed with non-planner layers. To exchange these, all areas above the surface and below the non-planner surface should be marked in advance. Potential NPL layers are obtained using polygon operations of PNL, where Ln means layer area for non-planar layer and NS means projection of the non-planar layer [10]. In the case of non-planners, the stair-step occurs at the highest planning level as shown in Fig. 3.

$$\mathrm{PNL} = (\mathrm{Ln} \setminus \mathrm{Ln} + 1) \cap \mathrm{NS} \tag{1}$$

2.2 *Process of Generation Toolpath*

Toolpath generation is the most important part for the non-panel layer. In this research work, the non-planar toolpath were created using surface finish algorithms and planar perimeters. Perimeter toolpaths are formed by relocating internal outline polygons by half the extrusion width, and this process will continue until the number of the perimeters is found. The collider cone shows the evolution of the collision as shown in

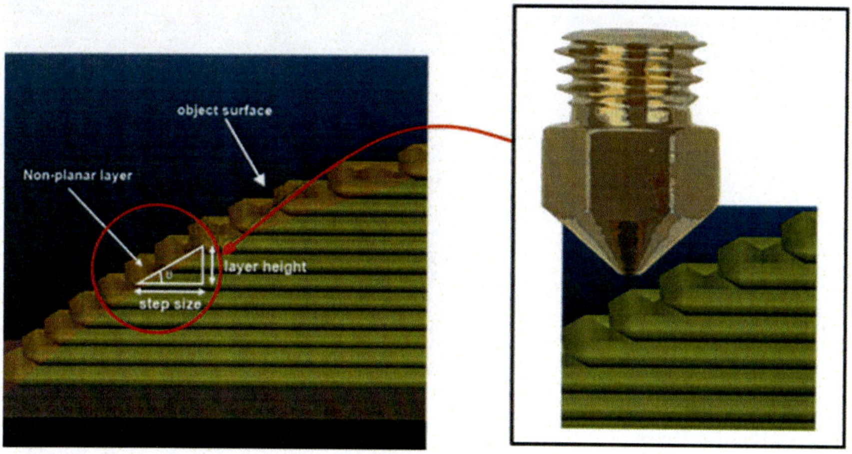

Fig. 3 For smaller angles, non-planner extrusions over planar layers have a stair-stepping effect

Fig. 4. The surface layers are then divided into several sections. Non-planar surfaces have already been categorized so there is no need to re-classify them. The three-dimensional tool path is formed by projecting down any extrusion route located on a level with a connected non-planar surface from a previously created two-dimensional tool path [11].

In the case of a three-dimensional object, a path is formed from multiple two-dimensional points, and a z-dimension is added to each point of this path so that it is converted into a three-dimensional extrusion path. Individually, the point directly above the non-planar layer surface is projected vertically downward at the height of the mesh to create an extrusion path to follow the real surface geometry of the printed object. Every point in the extrusion path is tested for each point so that all aspects of the main surface mesh are able to find the equivalent part for each point. The connection points help to create planar points in the extrusion path so that each point is always predictable downstairs from its own direction. In the case of 3d objects, the x- and y-directions remain the same, and only the z-direction point has to be added so that the process is repeated with an additional z-direction of height for

Fig. 4 Layer-based collision detection

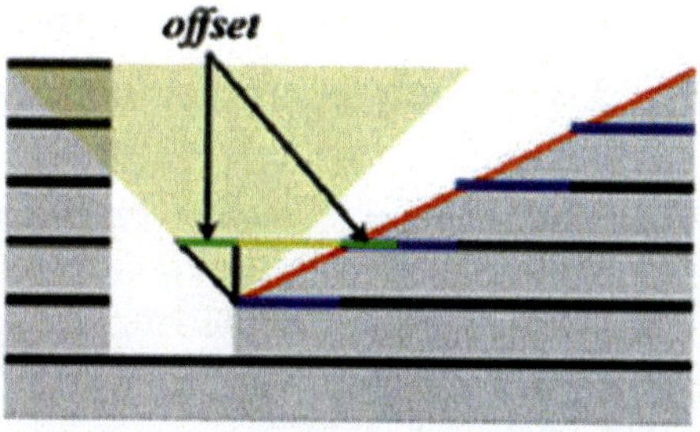

each shell layer. In the case of each extrusion line, the end points coincide with the surface of the object, but the lines do not follow curvature properly. To solve this problem, the intersections of an extrusion line with two-dimensional planar space are injected into the calculated point and each connection is expected in the direction of correct z-height. The G-code of the non-planner toolpath was first created to print solid objects using the non-planner toolpath on an FDM 3D printer. The G-code of the non-planner is the same as the G-code of the Planner toolpath, but the slicing software is used separately. The slicer software used to create the non-planar tool path is Slic3r.

3 Process of Collision Avoidance for Non-planar Layer

In the case of FDM 3d printers, the printed layers do not collide with the planer slicing and, in this case, the printhead always travels upwards and never returns to the bottom layer. However, when printing non-planar layers, there is a possibility of collision, if somehow the print head moves closer to the already printed layers. These should be avoided during toolpath generation because the printer cannot actively detect these collisions. In the case of non-planar areas, collisions are impossible because non-planar surfaces have an angle on each side which is as small or equal to θnp as described in seconds which can only cause collisions in the case of high-level flat areas [12]. To avoid this, the non-planner tool thoroughly observes the path before the collision.

Collision checking begins with the lowest level of a particular non-planner layer so that collisions between non-planar surface areas are permanently ignored because they are not true. A collider polygon connection is made for each layer and the present layer polygon. If these intersections collide due to irrationality, the intersection of the current layer and the non-planar surface are joint using a collider. The model of this collision is then offset by how much width will be created within the height of a level. The entire extrinsic path of the non-planar layer remains collision free between the newly formed objects while all layers are collision free. Figure 5 shows how collisions can occur when traveling from a point below the maximum printed layer. The simple method used in this research work to avoid these collisions is to at all times move the printhead to its current maximum print height, then move the printhead down again to the desired position as shown in Fig. 6. This makes it less likely that the printhead will collide while moving from one place to another.

The G-code program for printing is built from a toolpath that does not collide although traveling. Non-planar surfaces are not very good for normal printing because the layers collide with the printhead in any complex object so a special printhead is designed for non-planar printing, so that the layers do not collide with the printhead in any way while printing non-planar surface and construct a 3d object with a smooth surface. So, creating a new FDM 3D printer nozzle is a good way to avoid collisions of non-planar layers.

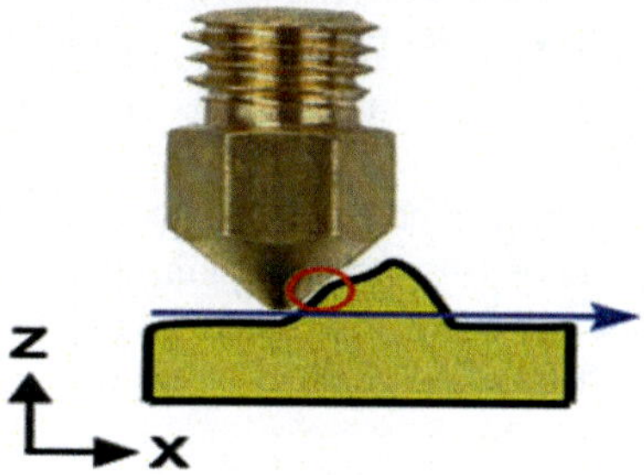

Fig. 5 Collisions can occur when traveling from a point below the maximum printed layer

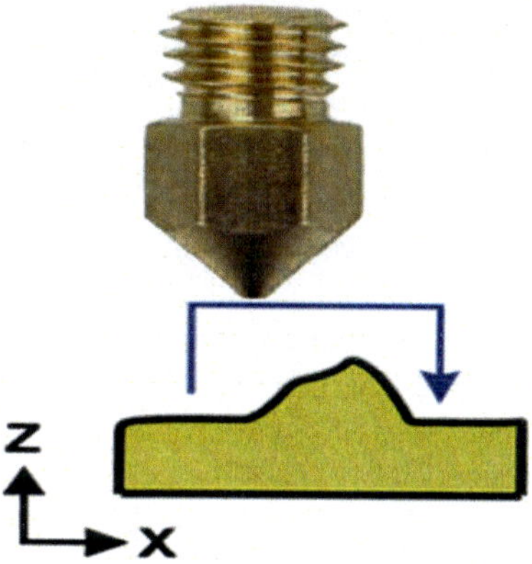

Fig. 6 Move the printhead to its current maximum print height, then move the printhead down again to the desired position

The nozzle of an FDM 3D printer is designed for this research work and is designed to avoid collisions in the case of a non-planar layer which is shown in Fig. 7. Figure 8 shows the create final object using normal printing method, and Fig. 9 shows the create final object using non-planar layer printing method. After comparing Figs. 8 and 9, it clearly shows that non-planar layer method object provides better surface finish than the normal printing method object.

Fig. 7 FDM 3D printer nozzle

Fig. 8 Create solid object using normal printing method

Fig. 9 Create solid object using non-planar layer printing method

4 Result and Discussion

In the case of 3D printing, most of the time is spent on printing but slicing takes very little time to adjust the settings. In this study, the time required to complete the slicing process and then the printing time for printing from the 3d model is shown in Table 1. The slicing time is comparatively higher than the non-planner layer method but not too much compared to the planar layer method, so it is very easy to optimize and speed up the process. However, in the case of non-planar layers, slicing will never be as fast as planer slicing because many extra steps have been added to the non-planar layer printing process. Figure 7 shows that slicing time of non-planner object, printing time, cost of printing, and required raw material are same as planar printing, but surface quality of non-planner object is much improved.

See Figs. 10, 11, 12 and 13.

Table 1 Calculate printing time, slicing speed, and printing cost

	Slicing time (s)	Printing time (s)	Printing cost (Rs.)	Required material (gm.)
Planar layer	7	290	2	1
Non-planar layer	11	300	2	1

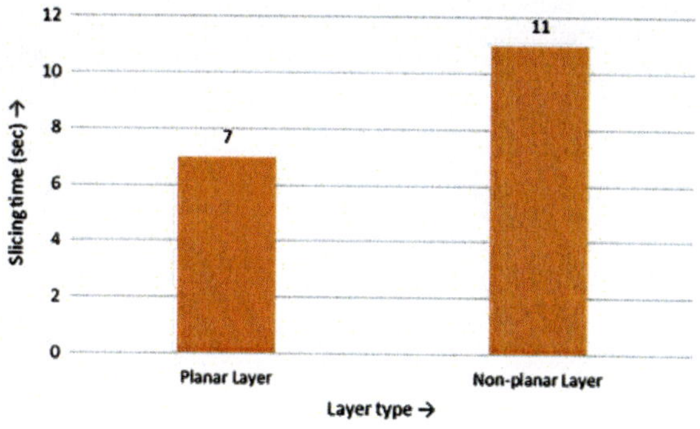

Fig. 10 Slicing time versus layer type bar chart for planar and non-planar layer

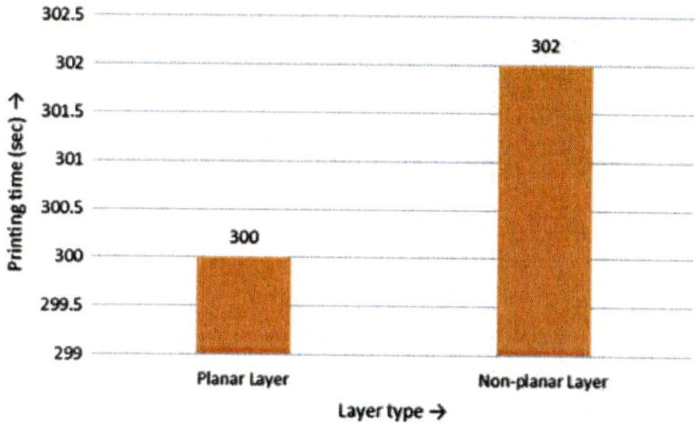

Fig. 11 Printing time versus layer type bar chart for planar and non-planar layer

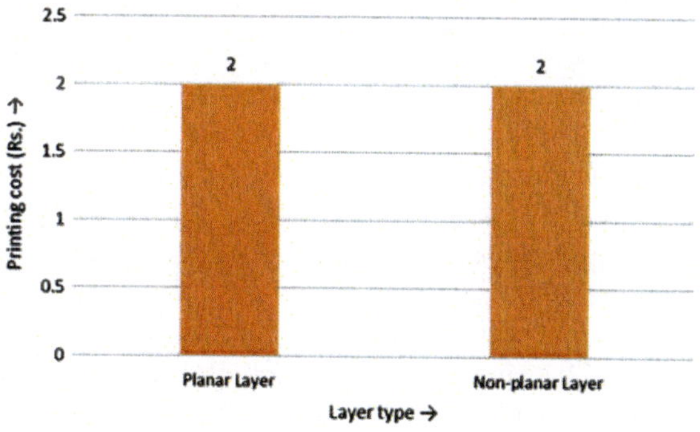

Fig. 12 Printing cost versus layer type bar chart for planar and non-planar layer

4.1 Approximation Error

The approximate error is the difference between the planar layer and the non-planar layer printed object. Estimated errors are common when the object is larger than a model, but the design of these regions is smaller with a negative estimation error. The side profiles of each object have been enlarged so that each region at the top of

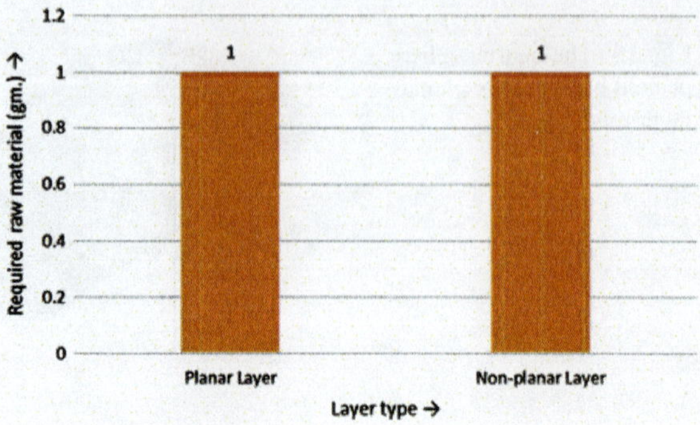

Fig. 13 Raw material versus layer type bar chart for planar and non-planar layer

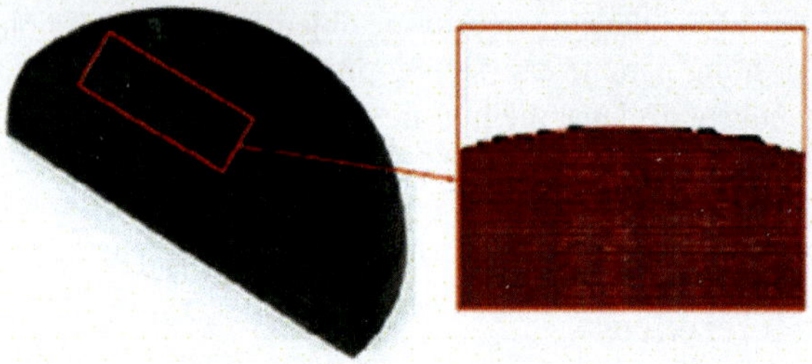

Fig. 14 The quarter sphere printed object using planar layer method

the sphere is able to show a positive guess error and each region below is able to show a negative guess error.

Approximate error of planar object usually the non-planar is much larger than the printed object which the toolpath follows the surface geometry due to the approximate error theory of the non-planar object at the non-planar level. This study found that the surface of the surface somehow produces roughness piercing extruders that produce an approximate error from the actual object, and this error increases with the angle of the surface itself. Also, the change among the non-planner and normal printing object layer is somewhat rough which shows from the side view that the surface of the non-planar object is smoother than the surface of the normal printing object as shown in Figs. 13 and 14. Thus, the work of this study proves that the approximate error of a non-planar layer 3d-printed object is less than that of a planar layer 3d-printed object because it depends on the geometry of the object and the possibility of creating a collision-free non-planner surface (Fig. 15).

5 Conclusions

Experimental study has been presented to enhance the surface quality of 3D printed object using non-planar layer method. The G-code program for printing is built from a toolpath that avoids collisions while printhead traveling. Non-planar layer method

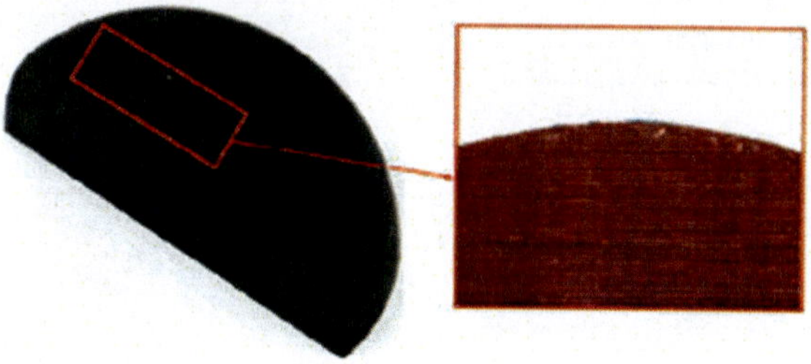

Fig. 15 The quarter sphere printed object using planar layer method

has shown better surface quality that were not possible for the planar toolpath method. Further, it is observed that the surface of the non-planar object is smoother than the surface of the normal printing object which is desirable. It has been identified that 3D printing parameters (slicing time, printing time) and printing cost of printed object using non-planar method are similar to planar layer method with an improved surface quality.

References

1. Chakraborty D, Aneesh Reddy B, Choudhury A (2008) Extruder path generation for curved layer fused deposition modeling. Comput Aided Des 235:243–340
2. Douglas DH, Peucker TK (1973) Algorithms for the reduction of the number of points required to represent a digitized line or its caricature. Cartographica: Int J Geographic Inf Geovisual 112:122–10.
3. Huang B, Singamneni S (2014) Curve layer fused deposition modeling with varying raster orientations. Appl Mech Mater 446(447–263):269
4. Wasserfall F (2015) Embedding of SMD populated circuits into FDM printed objects. Solid Freeform Fabric Symp 180:189–226
5. Llewellyn-Jones T, Allen R, Trask R (2016) Curved layer fused filament fabrication using automated toolpath generation. 3D Print Additive Manuf 236:243–3
6. Lim S, Buswell RA, Valentine PJ, Piker D, Austin SA, De Kestelier X (2016) Modelling curved-layered printing paths for fabricating large-scale construction components. Addit Manuf 216:230–312
7. Kubalak JR, Wicks AL, Williams CB (2018) Using multi-axis material extrusion to improve mechanical properties through surface reinforcement. Virtual Phys Prototyping 13:32–38
8. Jin Y, Du J, He Y, Fu G (2017) Modeling and process planning for curved layer fused deposition. Int J Adv Manuf Technol 273:285–291
9. Falaudi J, Hu Z, Alrashed S, Braunholz C, Kaul S, Kassaye L (2015) Does material choice drive sustainability of 3D printing? Int J Mech Mechatron Eng 216:223–229
10. Chen L, Chung MF, Tian Y, Joneja A, Tang K (2019) Variable-depth curved layer fused deposition modeling of thin-shells. Robot Comp Integr Manuf 422:434–457
11. Ahlers D, Wasserfall F, Hendrich N, Zhang J (2019) 3D printing of nonplanar layers for smooth surface generation. In: Proceedings of the IEEE 15th International Conference on Automation Science and Engineering (CASE) vol 1737, p 1743
12. Micalizzi S, Lantada AD, De Maria C (2019) Shape-memory actuators manufactured by dual extrusion multimaterial 3D printing of conductive and non-conductive filaments. Smart Mater Struct 1050:1025–1028

Smart System for Monitoring the Toxic Gases in Sewerage System via Wireless Network

Puneet Dhankad, Amaan Ahmad, Anil Kumar Shukla, and Sanmukh Kaur

1 Introduction

Effluence has been designed around an Internet of Things system and network that detects hazardous gases as a safety safeguard for sanitation workers who put their lives on the line to ensure reduced health dangers. Sanitation workers have died as a result of these harmful substances. Over the last few years, the rates have risen considerably. When you get there, levels of danger and a lack of effective sewage decontamination are the most threatening factors. As an aggregate of these injuries, sewage cleanup workers have died. Influenza and dysentery are two examples of specific ailments induced by exposure to a toxic gas not only for a short stretch of time but also for a long duration of time. Septic pipes are pipelines that progress toward the septic tank. Constructions can be perched in a variety of places, ranging from offering treatment for waste material detection from household regions to largely industrialized urban locations, a company that treats waste materials and diagnoses them. Sewage gases are usually produced when organic waste decomposes. Compost products and combinations result in the assemblage of poisonous trash that emits harmful gases such as sequence of alkanes, carbonic acid gas, additional sulfur dioxide elements, ammonia, hydrogen sulfide, methane, and carbon which are all present in trace amounts. Monoxide gas traces are most commonly seen in septic tanks. The gas in the tank sensor [1] is used to determine the amount of toxins present in the environment. Once a hazardous gas has been discovered in the waste product environment, a signal will be sent to the linked mobile gadget

P. Dhankad (✉) · A. Ahmad · A. K. Shukla · S. Kaur
Amity School of Engineering and Technology, Amity University Uttar Pradesh, Noida, India
e-mail: dhankadpuneet6@gmail.com

A. K. Shukla
e-mail: akshukla2@amity.edu

S. Kaur
e-mail: skaur2@amity.edu

A. Kumar et al. (eds.), *Recent Advances in Manufacturing and Thermal Engineering*,
Lecture Notes in Mechanical Engineering, https://doi.org/10.1007/978-981-19-8517-1_15

of the approved personnel who are remotely placed in the job. Gas sensors detect harmful toxins such as hydrogen sulfide, alkane series, and carbon monoxide gas generated by the waste product at each moment and modify them until they meet the requirements. The goal of this study is to demonstrate how to monitor the level of gas accumulation at a waste product power station or a septic system in real time. The suggested system would be deployed in various areas for waste products, both rural and urban. Aside from having industrial plants, the system may be customized to fit into any home with dynamic small style parameters. The ThingSpeak [1] IoT platform would be available on the internet from anywhere in the globe, allowing for remote monitoring of contamination or toxicity levels. The growth of the Internet of Things framework as a fundamental part of the gadgets for controlling the installation was also identified in this study. This method provides a more complex sewer [2] setting solution than previous methods, which neglected real-time monitoring and changes in gas and moisture concentrations in the air. This is due to the fact that the flux of waste product water fluctuates substantially over time and is dependent on a variety of factors such as filter quality, energy accumulation, and plant disruption. Due to electronic monitoring, the above approach tracks variables in such a way that every minute evaluation is visible via isolated areas, allowing for a firm interpretation of CH4, CO, and alternative sewage pollutants, as well as the release through gutters, and helps to track and efficiently sanitize municipal management, a feature that previous solutions lacked. In the event that breathed over an extended period of time, these gases can be perilous, and in the event that huge concentrations are ingested into the circulation, they can cause genuine infections within the working environment. Seepage frameworks identify gases such as sulfur dioxide, methane, alkali, nitrogen dioxide, carbon dioxide, and carbon monoxide. As a result, these noxious vapors are perilous and can possibly murder people, especially sewage laborers and cleaners.

2 Literature Survey

The system for sewage inspection was implemented to keep safe the existence of workers in dangerous environments. It conveys a message to the offices that engage these people when the ppm levels of certain gases exceed the allowed levels. The survey equipment meets all of the specifications for a system that keeps an observation on the working of sewage.

Cloud-based IoT device control for optimal air quality monitoring. Cloud computing with air quality monitoring, ground-level ozone, as well as air defilement that cause asthma, anaphylaxis, lung cancer, pulmonary eosinophilic disorders, anaphylaxis, and other respiratory ailments were monitored by the previous systems. It uses a platform that is based on cloud to monitor the ppm degree of pollutants and save data from various sensors.

Sewerage gas monitoring and notifying system sewage regions methane gas sensor from the IoT. This has resulted in a method for measuring gas ppm levels. Temperature and humidity have a significant impact on the well-being of sewerage workers on the job, and our technology recognizes this. The framework plan incorporates a sensor that recognizes sewage level, a controller that orders, and a communication arrangement that records complaints around persistent sewerage level rises and, on the off chance that any, blockages. To preserve track of the data, a database must be kept. Earlier to flood, the framework sends out caution signals within the frame of complaints to the fitting offices by means of mail and SMS. For modern cities' underground drainage and manhole monitoring systems, automated [3] Internet of Things, the manhole monitoring system, and the underground drainage are adapted and designed in this work for IoT [4] applications. The proposed model includes a framework for checking the water level, air temperature, and pressure inside a manhole, as well as ensuring that the lid is open. From afar, the system can monitor the condition of the manholes instantaneously. Appraisal of nursery gas emanations from sewage treatment offices, assessment of greenhouse gas emissions from multiple municipal water treatment plants around the city, as well as assurance of their sewage treatment systems' long-term viability. Although the paper emphasizes indirect greenhouse gas emissions from the aforementioned sewerage water [5] treatment plants, the proposed mitigation strategies are vague and ineffective. It fails to make clear and practical recommendations for reducing indirect emissions. An IoT-based underground sewerage monitoring system should be developed and deployed that regulates sewerage conditions and provides a mechanism to maintain and control the subsurface infrastructure using methodological techniques. The model contains a significant number of complex components that must be regularly maintained and examined. There is currently no specialized network for managing numerous types of sensors at the same time in the existing system. In such a situation, failure is a possibility. For using wireless network in a web-based real-time underground seepage or sewage checking framework, sensor systems are utilized: contains a complex arrangement of low-cost, long-lasting components that empower metropolitan specialists to persistently screen the sewage environment and water levels, guaranteeing the security and well-being of sewage laborers. Light sensors are a cost-effective arrangement to guarantee that the sewer vent is continuously fixed, as well as that the framework and all of its components are in great working order and not helpless to theft.

3 Hardware Design

Distinct gas sensors, such as MQ9, MQ2, MQ135, and MQ136 for observing any sum of destructive substances inside squander, as well as dampness and temperature sensors, are utilized within the range. One collector conclusion is provided for the control of the current setup at totally diverse nodal positions. From the sewage to the control unit, the gadget gives a rich amount of information. Sensors shown in Figs. 1

and 2 have been quantified in order to support the concept of sensor connections, allowing them to be used in both commercial and residential settings. The throughput is transferred to the cloud via the GSM [6] module according to a predetermined scenario. The ThingSpeak IoT framework is used in this paper.

Figure 3 shows the assessing essential sites, in what manner the archetype is employed. The archetype is utilized at strategic sites by analyzing each drainage map and establishing a field, such as an assessment criterion, that searches for the best position. These are commonly seen around the sewer's [7] start. The delivery timetable can often be altered depending on the task at hand. Operators can use the ThingSpeak IoT platform's system built graph, for illustration, to get its changes within the level of in-sewer gas, stickiness, or temperature over time. Blockages can moreover be identified in development using the live video spilling capability.

Fig. 1 Smart sewerage system view 1

Fig. 2 Smart sewerage system view 2

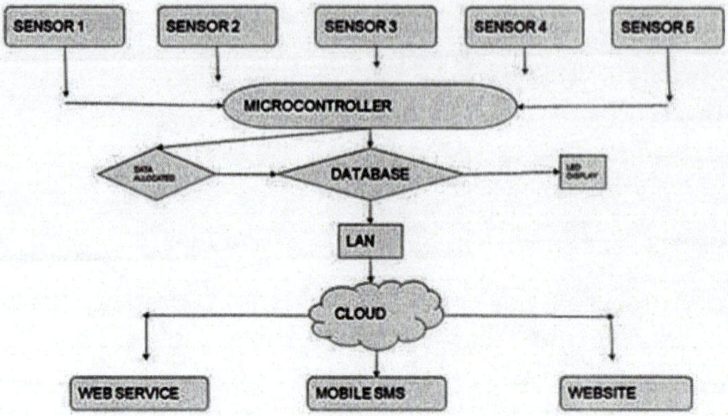

Fig. 3 Flow diagram 1

4 Hardware Description

This paper's architecture relies heavily on the Arduino and GSM modules. The Arduino [8] can display sensor data such as concentration values from the (MQ2) and (MQ9) of various waste substances, such as sewage, as well as humidity and temperature sensors as shown in Fig. 4. In practice, the Thingspeak IoT platform is used to reliably convey these real-time ppm measurements to the database. Pictographic representations of the ppm values of various gases are displayed using an inquisitive tool. Finally, when the measurements are above the GSM module's threshold, the status is sent to the user. The client registers and tracks sensor ppm data rates in order to avoid accidents when working in biodegradable pollution tanks and to protect themselves from diseases caused by toxic and harmful gases.

Fig. 4 Smart sewerage system view 3

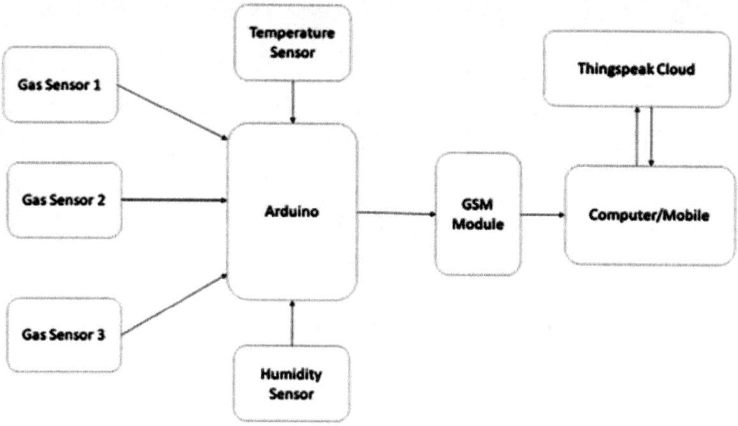

Fig. 5 Flow diagram 2

5 Methodology

The device architecture, which includes the Arduino microprocessor, is depicted in the figure. This microcontroller handles sensor simulation, software-based SMS production, and phone calls. Before sensors can be utilized, they must be calibrated. The sensor produces an analog voltage that can be converted using an ADC. The transformed data [9] is used to compute the expected ppm gas value for design elements. As a result, sensors for humidity and temperature have been installed to keep track of [10] the levels of these components in the sewage environment, making it safer for people working there. ThingSpeak is an Online of Things (IoT) application that stores information obtained from gadgets through channels. The information is at that point transported to and received from this channel, and it may be recouped too by changing the channel setup parameters. The open has get to know about knowledge-sharing network. It is a Web of Things (IoT) application that employs channels to store information obtained from gadgets. This channel gets and transmits information [11], and it may too be recovered by changing the channel's setup parameters. Everybody is welcome to connect knowledge-sharing networks (Fig. 5).

6 Conclusion

The objective of this paper is to offer a strategy for identifying perilous vaporous fabric releases in seepage frameworks, social lodging, and mechanical offices. Sewage is additionally included within the normal creation of harmful gases. An IoT-based monitoring system is being introduced to decrease exposure to such industrial risks. Previous programs suggested manual sampling for sewer gas analysis at

predefined time intervals. Humidity, temperature, and the production were all aspects that were disregarded. The problems in the present system will be communicated for utilizing the most recent design. Humidity and temperature sensors, in addition to gas sensors, can be utilized to analyze the sewage's overall environment. In future, the camera can be connected which will then feed live footage, allowing sewage workers to check for bottlenecks.

References

1. Novo O, Beijar N, Ocak M, Kjallman J, Komu M, Kauppinen T (2015) Capillary networks—bridging the cellular and IoT worlds. In: Proceedings of the 2015 IEEE 2nd World Forum on Internet of Things (WF-IoT), Milan, Italy, 14–16 December 2015; Institute of Electrical and Electronics Engineers (IEEE): Piscataway, NJ, USA, pp 571–578
2. Talukder S, Sakib II, Talukder ZR, Das U, Saha A, Bayev NSN (2017) USenSewer: ultrasonic sensor and GSM-Arduino based automated sewerage management. In: Proceedings of the 2017 international conference on current trends in computer, electrical, electronics and communication (CTCEEC), Mysore, India, 8–9 September 2017; Institute of Electrical and Electronics Engineers (IEEE): Piscataway, NJ, USA, 12–17
3. Talukder ZR, Das U, Saha A, Bayev NSN (2017) USenSewer: ultrasonic sensor and GSM-arduino based automated sewerage management. IN: Proceedings of the 2017 international conference on current trends in computer, electrical, electronics and communication (CTCEEC), Mysore, India, 8–9 September 2017; Institute of Electrical and Electronics Engineers (IEEE): Piscataway, NJ, USA, 12–17.
4. Li L, Xiaoguang H, Ke C, Ketai H (2011) The applications of WiFi-based wireless sensor network in internet of things and smart grid. In: Proceedings of the 2011 6th IEEE conference on industrial electronics and applications, Beijing, China, pp 789–793
5. Cao L (2009) Wireless mesh monitoring system for sewage treatment plant. In: Proceedings of the 2009 ISECS international colloquium on computing, communication, control, and management, Sanya, China, vol 4, pp 350–353
6. Nasution TH, Muchtar MA, Siregar I, Andayani U, Christian E, Sinulingga EP (2017) Electrical appliances control prototype by using GSM module and Arduino. In: Proceedings of the 2017 4th international conference on industrial engineering and applications (ICIEA), Nagoya, Japan, 21–23 April 2017; Institute of Electrical and Electronics Engineers (IEEE): Piscataway, NJ, USA, 355–358
7. Edmondson V, Cerny M, Lim M, Gledson B, Lockley S, Woodward J (2018) A smart sewer asset information model to enable an 'Internet of Things' for operational wastewater management. Autom Constr 91:193–205
8. Haswani NG, Deore PJ (2018) Web-Based realtime underground drainage or sewage monitoring system using wireless sensor networks. In: 2018 fourth international conference on computing communication control and automation (ICCUBEA), pp 1–5. https://doi.org/10.1109/ICCUBEA.2018.8697512
9. Islam M, Uddin J, Kashem MA, Rabbi F, Hasnat W (2020) Design and implementation of an IoT system for predicting aqua fisheries using arduino and KNN. In: Proceedings of the intelligent human computer interaction, Daegu, Korea; Singhm M, Kang D-K, Lee J-H, Tiwary US, Singh D, Chung WY (eds) (2021) Springer International Publishing, Cham, Switzerland, pp 108–118.
10. Pasha S (2016) Thingspeak based sensing and monitoring system for IoT with Matlab analysis. Int J New Technol Res 2:5

11. Latif SL, Afzaal HA, Zafar NA (2017) Modeling of sewerage system using internet of things for smart city. In: Proceedings of the 2017 international conference on frontiers of information technology (FIT), Islamabad, Pakistan, 18–20 December 2017; Institute of Electrical and Electronics Engineers (IEEE): Piscataway, NJ, USA, 46–51

Scrutinizing the Enablers of Flexible Manufacturing Competence of Organizations Using DEMATEL Approach

Asmit Karadbhajane, Inayat Ullah, Sourabh Shukla, and Anand Babu Kotta

1 Introduction

Nowadays, customer demand is likely the single most powerful motivator for changes in corporate strategy. Businesses have traditionally competed by concentrating on essential consumer demands or production criteria such as pricing, quality, service, or delivery. However, increasing globalization in the twentieth century enabled businesses to reach a far larger consumer sector. Because each individual consumer base had different product requirements, the diversity of client marketplaces produced a demand for suppliers to enable customization. In response to the increased unpredictability of the environment, enterprises are concentrating on improving their flexible manufacturing competency (FMC), which is broadly regarded as a foundation of competitive advantage [1]. FMC improves a company's business strategy from a strategic viewpoint by decreasing uncertainty and ensuring a smooth manufacturing flow [2]. Customers appreciate flexible production capabilities, such as volume and mix flexibility, as an extrinsic component of competition. In reaction to changing socioeconomic conditions, volume flexibility is defined as a company's ability to change volume levels lucratively and with low disruption. The capacity of a production system to cope with increased product variety while preserving system performance is referred to as mix flexibility [3]. Customers place a high value on these outward qualities since they immediately benefit them. FMC is a significant internal component of competitiveness that customers are unaware of [4]. It includes machine, labor, material handling, capacity, and route flexibility.

Businesses are under tremendous pressure to please customers who want personally designed items delivered at a fair cost and in the shortest time possible [5]. As a result, businesses are refocusing their efforts on embracing consumers in order to

A. Karadbhajane · I. Ullah · S. Shukla · A. B. Kotta (✉)
Department of Mechanical Engineering, G H Raisoni College of Engineering, Nagpur 440016, India
e-mail: anand7850@gmail.com

A. Kumar et al. (eds.), *Recent Advances in Manufacturing and Thermal Engineering*, Lecture Notes in Mechanical Engineering, https://doi.org/10.1007/978-981-19-8517-1_16

better understand and anticipate their demands, as well as using FMC as a strategic technique to meet the aforementioned objectives and gain a competitive advantage [6]. The implementation of FMC is growing rapidly since it improves an organization's capacity to respond to a wide range of client requests while delivering greater results [7]. For an industry, there are many enablers which affect its FMC. These enablers have their own way of influencing the industries. Each enabler affects an industry in its own way. The influence of an enabler may vary depending on the type of industry. For instance, some enablers may have a huge effect on the development of an industry, while some enablers may have less effect on the growth of an industry. In order to know how these enablers influence the industries, we can use several Multi Criteria Decision Making (MCDM) methods. In our case, we are using DEMATEL methodology. Not much research work is available on the internet about division of the enabler affecting FMC into cause-and-effect groups. The objective of this research is to categorize the enablers into cause-and-effect groups, so that industries can know which enablers they need to focus more on.

2 Literature Review

This literature provides a brief description of the enablers shortlisted. These enablers help the economy by cutting logistical costs (such as labor expenses), boosting productivity and efficiency, enhancing customer satisfaction through shorter lead times, and improving data quality and responsiveness. These enablers are now divided into three categories—strategic, planning, and operational. Strategic enablers are those enablers which have a long-term effect on the industry, i.e., these enablers will show their influence for nearly 5–10 years, or even more. In other words, they affect the industry in the long run. Planning enablers are the enablers which have their effect on the industry for around a year. Operational enablers are those which affect the industry on daily basis. The enablers are displayed in Table 1.

AMT is a set of technologies that automate and combine several design, manufacturing, planning, and control processes [8]. Absorptive capacity is influenced by an organization's creative behaviors in reaction to external disruptions, both proactive and reactively [9]. Organizational culture is the beliefs, expectations, and practices that govern and impact the activities of all team members. Consider it a set of characteristics that define our organization [10]. Top management commitment is viewed as crucial to accomplishing corporate goals. Top management takes strategic decisions, and the consequences of such actions affect the company's performance [11]. Organizational structure is a system that defines how specific tasks are directed in order to fulfill an organization's goals [12]. Internal learning is the circumstance in which people acquire new skills and information while doing ordinary and demanding jobs as a team or individually. External learning is defined as employees gaining skills, information, and experience while interacting with suppliers, clients, and other third-party service providers [13]. Adaptive human resources can deal with

Table 1 Categorization of enablers with their description

Sr. No.	Category	Enabler	References
1	Strategic enablers	Advanced manufacturing technologies (F1), absorptive capacity (F7), organizational culture (F12), top management commitment (F15), organizational structure (F16)	[8–12]
2	Planning enablers	Internal and external learning (F4), adaptive human resources (F5), modularity in products and processes (F8), capacity flexibility (F10), postponement implementation (F11), integrated management systems (F13), value chain management (F14)	[13–22]
3	Operational enablers	Operational improvement practices (F2), supplier flexibility (F3), sourcing flexibility (F6), time-based manufacturing (F9)	[3, 8, 11, 18]

a variety of confusing and complicated work environments and make good judgments in a variety of situations [14]. Product modularity is frequently listed as a goal of good design practice, while receiving less research than other current design practices such as design-for-assembly. As corporations try to simplify their product lines and provide even more variation at lower prices, modularity is becoming a main area of emphasis [15, 16]. Process modularity is a method of standardizing manufacturing sub-processes so that they may be easily reordered or new modules can be quickly added in response to changing product requirements. Capacity flexibility is defined as the capacity to change total production volume and variety while utilizing both conditional and fixed resources [17, 18]. Hiring temporary workers, using overtime/flexi-time, installing new machinery, and using subcontractors are all examples of operations that boost capacity flexibility. Postponement implementation allows manufacturers to include a wide range of client requests into their goods without disrupting established production schedules and operational routines, allowing them to respond to demand changes [19]. Integrated management systems (IMS) are smart systems that combine all parts of a firm's procedures and standards [20, 21]. A good IMS reduces the need for several management systems, saving time and effort. In response to the turbulent business climate, value chain management is a method that gives both supplier and sourcing flexibility, allowing for fast changes in production levels, raw material procurement, and transportation capacity and schedules [22]. OIPs are a collection of just-in-time (JIT) manufacturing methods that focus on continuous improvement in manufacturing, such as set-up time reduction, preventative maintenance, cellular layout, pull production, overall quality management, and more, and help organizations increase their volume and mix flexibility [8]. SFL refers to a supplier's ability to respond to changes needed by purchasing organizations in terms of volume, variety, and innovation. Suppliers have a direct and considerable impact on the cost, quality, technology, and time-to-market of new products [11]. Businesses must determine whether or not to pursue source flexibility. Based on the link between flexible sourcing and delivery performance, firms should

seek for either high or low degrees of flexibility, with middle levels avoided [18]. Time-based manufacturing is a production method that focuses on meeting changing client requests fast. Time-based manufacturing depends on time reduction measures to boost responsiveness and other competitive characteristics such as FMC [3].

3 Methodology

The DEMATEL technique has been used in operations to highlight the interrelationships among criteria and to identify the primary criteria that indicate the efficacy of enablers/aspects. It has also been used in a variety of contexts, including marketing tactics, control systems, safety issues, global management competency development, and group decision-making. Hybrid models that combine the DEMATEL and other methodologies have also been widely employed in a variety of industries, including e-learning assessment and aviation safety monitoring [23, 24].

The operations performed in DEMATEL are: [25]

Step 1: Identification of enablers

The DEMATEL method begins with the identification of relevant enablers through literature review. A sum total of 16 enablers are identified and their description is provided in the literature review section.

Step 2: Creation of the direct relation matrix

For assessing the link between different criteria, we utilize four scales: 4 (very high influence), 3 (high influence), 2 (moderate influence), 1 (low influence), and 0 (no influence). Following that, decision-makers create sets of pair-wise comparisons in terms of impacts and directionality between criteria. The initial data may then be represented by the direct relation matrix, which is a (n × n) matrix A with each member denoting the degree to which the criteria i impacts the criterion j.

Step 3: Normalizing the direct relation matrix

Normalization is performed using the following:

$$X = k.A \tag{1}$$

$$\text{Where}\quad k = \frac{1}{\max_{1 \le i \le n} \sum_{j=1}^{n} a_{ij}} \tag{2}$$

Step 4: Attaining the total relation matrix

The total relation matrix T may be constructed using the equation below once the normalized direct relation matrix X has been obtained. I represents the identity matrix here.

$$T = X(1 - X)^{-1}. \tag{3}$$

Step 5: Producing a causal diagram

Through the equations below, the sum of rows and the sum of columns are designated as vector D and vector R, respectively. Then, by adding D to R, the horizontal axis vector (D + *R*) is created, revealing the relative significance of each criterion. Similarly, the 'Relation' vertical axis (D−R) is created by subtracting D from *R*, which may be used to separate criteria into cause-and-effect categories. When (D−R) is positive, the criteria indicate the cause group, but when (D−R) is negative, it represents the effect group. As a result, by mapping the dataset of (D + R, D−R), the causal diagram can be constructed, offering some insight for decision-making [25].

$$T = [t_{ij}]_{n\times n}, i, j = 1, 2, \ldots, n \tag{4}$$

$$D = \left[\sum_{j=1}^{n} t_{ij}\right]_{n\times 1} = [t_{i.}]_{n\times 1} \tag{5}$$

$$R = \left[\sum_{i=1}^{n} t_{ij}\right]_{1\times n} = \left[t_{.j}\right]_{1\times n} \tag{6}$$

where vector D and vector R reflect the sum of rows and columns from the total relation matrix $T = \left[t_{ij}\right]_{n\times n}$, respectively [25].

4 DEMATEL (Decision-Making Trail and Evaluation Laboratory) Approach for Modelling the Enablers

Step 1: Identification of enablers

First of all, 16 key enablers affecting Flexible Manufacturing Competence (FMC) were identified. The brief description of these enablers is mentioned in the literature review section. The enablers are represented by the letter **F.**

Step 2: Development of direct relation matrix (**A**)

A direct relation matrix is computed as the average of all the matrices taken into consideration. These matrices are created with the help of industry experts. In our case, we have three matrices. These three matrices represent three different opinions of experts on how certain enablers affect FMC. For generating the matrix, we take the average of the three matrices provided by industry experts.

Step 3: Normalization of the direct relation matrix

There are 3 steps for normalizing a matrix:

i. Addition of all the rows.
ii. Identification of the highest value in the addition column.
iii. Dividing the entire matrix by the highest value identified in the previous step.

From Table 2, the highest value in the last column is 41.21767. After dividing the direct relation matrix with this value, the normalized direct relation matrix (**X**) is generated, which is shown (Table 3):

Step 4: Generating the total relation matrix (**T**)

For calculating the total relation matrix, the following formula is used:

$$\boldsymbol{T} = \boldsymbol{X}(\boldsymbol{I} - \boldsymbol{X})^{-1}$$

where X = Normalized direct relation matrix
$\boldsymbol{I}$ = Identity matrix

Now, subtraction of normalized direct relation matrix from the identity matrix $(I{-}X)$ is done. Upon subtraction, Table 4 is generated:

Now, we have to take the inverse of the above matrix. Upon taking the inverse, we get Table 5$[(I - X)^{-1}]$:

Now, we multiply the above inverse matrix with the normalized direct relation matrix, as stated in the formula for calculating the total relation matrix. After multiplying, we get Table 6:

In the table given below, D represents the sum of the rows and R represents the sum of columns. The total relation matrix (Table 6) along with its D and R values is as follows:

From here on, (D + R) has been represented by the alphabet 'M' and (D−R) has been represented by the alphabet 'N'. Now, another table representing M and N has been constructed below. This table will indeed assist in creating a causal diagram. This causal diagram will ultimately help in understanding the relationship between the different enablers identified. This table (Table 7) also shows the cause-and-effect groups. The table of M and N is as follows:

Step 5: Plotting the causal diagram

5 Discussion and Conclusions

Because the development and enhancement of FMC involves considerable resources and efforts, businesses must be able to identify the most essential aspects, and which approach should be used to obtain the greatest FMC advantages. The classification of the enablers given in this study will aid organizations in acquiring critical insights

Table 2 Direct relation matrix with an additional column indicating the sum of all the rows in the matrix

	F1	F2	F3	F4	F5	F6	F7	F8	F9	F10	F11	F12	F13	F14	F15	F16
F1	0	0.25	0.55	0.83	0.44	0.94	0.55	0.28	0.25	0.3	0.32	0.5	0.55	0.5	2.17	1
F2	4	0	2.11	0.42	3.33	0.55	3	4	0.28	0.28	0.25	2.67	0.61	1.17	2.44	3
F3	2.33	1.22	0	0.75	0.42	0.25	1	0.3	0.25	0.39	0.33	0.67	0.39	0.33	2.67	0.67
F4	1.33	2.67	2	0	1.5	0.5	0.25	0.3	0.36	0.5	0.33	0.36	0.5	0.33	2.67	3.33
F5	2.33	0.3	2.67	2.75	0	0.39	1.33	0.33	0.25	0.3	0.3	0.25	0.55	0.39	2.17	0.33
F6	1.83	2.33	4	2	2.67	0	0.89	0.25	0.25	0.3	0.3	0.67	0.39	0.42	2.67	0.67
F7	2.33	0.33	1	4	1.44	2.17	0	0.33	0.25	0.33	0.33	0.44	1.55	0.25	3.33	2.67
F8	3.67	0.25	3.33	3.33	3	4	3	0	0.3	0.25	0.28	1.22	1.5	1.55	2.5	2.33
F9	4	3.67	4	3	4	4	4	3.33	0	2.67	3.33	2.67	3.67	1.42	2.44	2.5
F10	3.33	3.67	2.67	2	3.33	3.33	3	4	0.42	0	0.33	1.33	2.5	2.83	1.61	3.33
F11	3	4	3	3	3.33	3.33	3	3.67	0.3	3	0	0.55	1.22	2.78	2.44	0.5
F12	2	0.39	1.67	3	4	1.67	2.33	2.11	0.42	0.83	2.33	0	1.22	0.28	4	3.33
F13	2.33	2	2.67	2	2.33	2.67	2.08	2.75	0.28	0.86	2.11	2.11	0	0.28	2.44	0.3
F14	2	2.78	3	3	2.67	2.67	4	2.08	1.67	0.83	1.17	3.67	3.67	0	4	2.78
F15	0.92	1.19	0.42	0.42	0.92	0.42	0.3	0.86	0.3	1.75	1.19	0.25	1.19	0.25	0	0.28
F16	1	0.33	1.67	0.3	3	1.67	0.42	0.55	1.42	0.3	3	0.3	3.33	1.17	3.67	0
Total	36.42	25.37	34.75	30.8	36.39	28.55	29.16	25.15	6.99	12.9	15.91	17.66	22.84	13.93	41.22	27.02

Table 3 Normalized direct relation matrix (**X**)

	F1	F2	F3	F4	F5	F6	F7	F8	F9	F10	F11	F12	F13	F14	F15	F16
F1	0	0.01	0.01	0.02	0.01	0.02	0.01	0.01	0.01	0.01	0.01	0.01	0.01	0.01	0.05	0.02
F2	0.1	0	0.05	0.01	0.08	0.01	0.07	0.1	0.01	0.01	0.01	0.06	0.01	0.03	0.06	0.07
F3	0.06	0.03	0	0.02	0.01	0.01	0.02	0.01	0.01	0.01	0.01	0.02	0.01	0.01	0.06	0.02
F4	0.03	0.06	0.05	0	0.04	0.01	0.01	0.01	0.01	0.01	0.01	0.01	0.01	0.01	0.06	0.08
F5	0.06	0.01	0.06	0.07	0	0.01	0.03	0.01	0.01	0.01	0.01	0.01	0.01	0.01	0.05	0.01
F6	0.04	0.06	0.1	0.05	0.06	0	0.02	0.01	0.01	0.01	0.01	0.02	0.01	0.01	0.06	0.02
F7	0.06	0.01	0.02	0.1	0.04	0.05	0	0.01	0.01	0.01	0.01	0.01	0.04	0.01	0.08	0.06
F8	0.09	0.01	0.08	0.08	0.07	0.1	0.07	0	0.01	0.01	0.01	0.03	0.04	0.04	0.06	0.06
F9	0.1	0.09	0.1	0.07	0.1	0.1	0.1	0.08	0	0.06	0.08	0.06	0.09	0.03	0.06	0.06
F10	0.08	0.09	0.06	0.05	0.08	0.08	0.07	0.1	0.01	0	0.01	0.03	0.06	0.07	0.04	0.08
F11	0.07	0.1	0.07	0.07	0.08	0.08	0.07	0.09	0.01	0.07	0	0.01	0.03	0.07	0.06	0.01
F12	0.05	0.01	0.04	0.07	0.1	0.04	0.06	0.05	0.01	0.02	0.06	0	0.03	0.01	0.1	0.08
F13	0.06	0.05	0.06	0.05	0.06	0.06	0.05	0.07	0.01	0.02	0.05	0.05	0	0.01	0.06	0.01
F14	0.05	0.07	0.07	0.07	0.06	0.06	0.1	0.05	0.04	0.02	0.03	0.09	0.09	0	0.1	0.07
F15	0.02	0.03	0.01	0.01	0.02	0.01	0.01	0.02	0.01	0.04	0.03	0.01	0.03	0.01	0	0.01
F16	0.02	0.01	0.04	0.01	0.07	0.04	0.01	0.01	0.03	0.01	0.07	0.01	0.08	0.03	0.09	0

Table 4 Table representing ($I—X$)

	F1	F2	F3	F4	F5	F6	F7	F8	F9	F10	F11	F12	F13	F14	F15	F16
F1	1	0.01	0.01	0.02	0.01	0.02	0.01	0.01	0.01	0.01	0.01	0.01	0.01	0.01	0.05	0.02
F2	0.1	1	0.05	0.01	0.08	0.01	0.07	0.1	0.01	0.01	0.01	0.06	0.01	0.03	0.06	0.07
F3	0.06	0.03	1	0.02	0.01	0.01	0.02	0.01	0.01	0.01	0.01	0.02	0.01	0.01	0.06	0.02
F4	0.03	0.06	0.05	1	0.04	0.01	0.01	0.01	0.01	0.01	0.01	0.01	0.01	0.01	0.06	0.08
F5	0.06	0.01	0.06	0.07	1	0.01	0.03	0.01	0.01	0.01	0.01	0.01	0.01	0.01	0.05	0.01
F6	0.04	0.06	0.1	0.05	0.06	1	0.02	0.01	0.01	0.01	0.01	0.02	0.01	0.01	0.06	0.02
F7	0.06	0.01	0.02	0.1	0.04	0.05	1	0.01	0.01	0.01	0.01	0.01	0.04	0.01	0.08	0.06
F8	0.09	0.01	0.08	0.08	0.07	0.1	0.07	1	0.01	0.01	0.01	0.03	0.04	0.04	0.06	0.06
F9	0.1	0.09	0.1	0.07	0.1	0.1	0.1	0.08	1	0.06	0.08	0.06	0.09	0.03	0.06	0.06
F10	0.08	0.09	0.06	0.05	0.08	0.08	0.07	0.1	0.01	1	0.01	0.03	0.06	0.07	0.04	0.08
F11	0.07	0.1	0.07	0.07	0.08	0.08	0.07	0.09	0.01	0.07	1	0.01	0.03	0.07	0.06	0.01
F12	0.05	0.01	0.04	0.07	0.1	0.04	0.06	0.05	0.01	0.02	0.06	1	0.03	0.01	0.1	0.08
F13	0.06	0.05	0.06	0.05	0.06	0.06	0.05	0.07	0.01	0.02	0.05	0.05	1	0.01	0.06	0.01
F14	0.05	0.07	0.07	0.07	0.06	0.06	0.1	0.05	0.04	0.02	0.03	0.09	0.09	1	0.1	0.07
F15	0.02	0.03	0.01	0.01	0.02	0.01	0.01	0.02	0.01	0.04	0.03	0.01	0.03	0.01	1	0.01
F16	0.02	0.01	0.04	0.01	0.07	0.04	0.01	0.01	0.03	0.01	0.07	0.01	0.08	0.03	0.09	1

Table 5 Representation of the inverse matrix $[(I - X)]^{-1}$

	F1	F2	F3	F4	F5	F6	F7	F8	F9	F10	F11	F12	F13	F14	F15	F16
F1	1.01	0	−0.01	-0.01	0	−0.02	-0.01	0	0	0	0	−0.01	−0.01	−0.01	−0.04	−0.02
F2	−0.08	1	−0.03	0.01	−0.06	0.01	-0.06	-0.09	0	0	0	−0.06	0	−0.02	−0.03	−0.06
F3	−0.05	−0.03	1.01	−0.01	0	0	-0.02	0	0	-0.01	0	-0.01	0	-0.01	−0.06	−0.01
F4	−0.02	−0.06	-0.04	1.01	−0.02	−0.01	0	0	0	-0.01	0	0	0	0	−0.05	−0.07
F5	−0.05	0	−0.06	−0.06	1.01	0	-0.03	0	0	0	0	0	−0.01	−0.01	−0.04	0
F6	−0.03	-0.05	−0.09	−0.04	−0.05	1.01	-0.01	0	0	0	0	−0.01	0	−0.01	−0.05	0
F7	−0.05	0.01	−0.01	−0.09	−0.02	−0.05	1.01	0	0	0	0	−0.01	−0.03	0	−0.06	−0.05
F8	−0.07	0.01	−0.06	−0.06	−0.05	−0.09	-0.06	1.01	0	0	0	−0.02	−0.02	−0.03	−0.03	−0.04
F9	−0.05	−0.06	−0.06	−0.04	−0.06	−0.07	−0.07	−0.05	1.01	−0.05	−0.07	−0.05	−0.07	−0.02	−0.01	−0.03
F10	−0.05	−0.08	−0.03	−0.02	−0.05	−0.06	−0.05	−0.08	0	1.01	0	−0.01	−0.04	−0.06	0	−0.06
F11	−0.04	−0.08	−0.04	−0.05	−0.05	−0.06	−0.05	−0.07	0	−0.07	1.01	0	−0.01	−0.06	−0.03	0.02
F12	−0.03	0.01	−0.02	−0.05	−0.08	−0.02	−0.04	−0.04	−0.01	−0.01	−0.05	1.01	−0.01	0	−0.07	−0.07
F13	−0.03	−0.03	−0.04	−0.03	−0.03	−0.05	−0.03	−0.05	0	−0.01	−0.05	−0.04	1.01	0	−0.03	0.01
F14	−0.01	−0.05	−0.04	−0.04	−0.03	−0.04	−0.08	−0.03	−0.04	−0.01	−0.01	−0.08	−0.07	1.01	−0.06	−0.04
F15	−0.01	−0.02	0	0	−0.01	0	0	−0.01	−0.01	−0.04	−0.03	0	−0.02	0	1.01	0
F16	−0.01	0.01	−0.02	0.01	−0.06	−0.03	0.01	0	−0.03	0	−0.06	0	−0.07	−0.02	−0.07	1

Table 6 Total relation matrix

	F1	F2	F3	F4	F5	F6	F7	F8	F9	F10	F11	F12	F13	F14	F15	F16	R
F1	−0.01	0	0.01	0.01	0	0.02	0.01	0	0	0	0	0.01	0.01	0.01	0.04	0.02	0.151
F2	0.08	0	0.03	−0.01	0.06	−0.01	0.06	0.09	0	0	0	0.06	0	0.02	0.03	0.06	0.452
F3	0.05	0.03	−0.01	0.01	0	0	0.02	0	0	0.01	0	0.01	0	0.01	0.06	0.01	0.2
F4	0.02	0.06	0.04	−0.01	0.02	0.01	0	0	0	0.01	0	0	0	0	0.05	0.07	0.278
F5	0.05	0	0.06	0.06	−0.01	0	0.03	0	0	0	0	0	0.01	0.01	0.04	0	0.255
F6	0.03	0.05	0.09	0.04	0.05	−0.01	0.01	0	0	0	0	0.01	0	0.01	0.05	0	0.338
F7	0.05	−0.01	0.01	0.09	0.02	0.05	−0.01	0	0	0	0	0.01	0.03	0	0.06	0.05	0.356
F8	0.07	−0.01	0.06	0.06	0.05	0.09	0.06	−0.01	0	0	0	0.02	0.02	0.03	0.03	0.04	0.514
F9	0.05	0.06	0.06	0.04	0.06	0.07	0.07	0.05	−0.01	0.05	0.07	0.05	0.07	0.02	0.01	0.03	0.75
F10	0.05	0.08	0.03	0.02	0.05	0.06	0.05	0.08	0	−0.01	0	0.01	0.04	0.06	0	0.06	0.582
F11	0.04	0.08	0.04	0.05	0.05	0.06	0.05	0.07	0	0.07	−0.01	0	0.01	0.06	0.03	−0.02	0.57
F12	0.03	−0.01	0.02	0.05	0.08	0.02	0.04	0.04	0.01	0.01	0.05	−0.01	0.01	0	0.07	0.07	0.483
F13	0.03	0.03	0.04	0.03	0.03	0.05	0.03	0.05	0	0.01	0.05	0.04	−0.01	0	0.03	−0.01	0.428
F14	0.01	0.05	0.04	0.04	0.03	0.04	0.08	0.03	0.04	0.01	0.01	0.08	0.07	−0.01	0.06	0.04	0.622
F15	0.01	0.02	0	0	0.01	0	0	0.01	0.01	0.04	0.03	0	0.02	0	−0.01	0	0.147
F16	0.01	−0.01	0.02	−0.01	0.06	0.03	−0.01	0	0.03	0	0.06	0	0.07	0.02	0.07	0	0.336
D	0.5616	0.4178	0.5431	0.4772	0.5855	0.4713	0.4833	0.4258	0.1058	0.2081	0.2556	0.2895	0.3732	0.2266	0.614	0.4246	

Table 7 Division of enablers into cause-and-effect groups

	D	R	M (D + R)	N (D−R)	Cause/Effect
	0.561632	0.151149	0.71278	0.410483	Cause
F2	0.417813	0.452358	0.870171	−0.03455	Effect
F3	0.543103	0.199707	0.74281	0.343396	Cause
F4	0.477206	0.278396	0.755602	0.198809	Cause
F5	0.585506	0.255214	0.84072	0.330292	Cause
F6	0.471255	0.337664	0.808919	0.133592	Cause
F7	0.483314	0.356322	0.839637	0.126992	Cause
F8	0.425769	0.514195	0.939964	−0.08843	Effect
F9	0.105751	0.749551	0.855302	−0.6438	Effect
F10	0.208061	0.582313	0.790375	−0.37425	Effect
F11	0.25563	0.569765	0.825395	−0.31413	Effect
F12	0.289493	0.482848	0.772341	−0.19335	Effect
F13	0.373186	0.428296	0.801482	−0.05511	Effect
F14	0.226573	0.622039	0.848612	−0.39547	Effect
F15	0.61399	0.14672	0.76071	0.46727	Cause
F16	0.424565	0.336311	0.760876	0.088255	Cause

into the aspects that require immediate attention in order to improve their ability to cope with demand volatility and fierce competition.

DEMATEL study recommends evaluating each enabler's relevance and categorizing the elements into two groups: 'influential' and 'influenced.' Because important group variables have such a large influence on FMC's goal, they have substantial research implications [25]. It is important to note that strengthening just one or two components would not improve the overall system because the enablers are intertwined [26]. The enablers must be classified into cause-and-effect groups in order to make successful judgments [27]. The influential group is upgraded first, followed by the impact group. The adoption of FMCs is the subject of this research, which focuses on the enablers [28]. Initially, the enablers of FMC adoption are determined using an integrated strategy that includes a literature review and expert feedback in the manufacturing area [29]. The DEMATEL approach is used to create a causal relationship when the enablers are finalized. The impact these enablers have on the industries can be understood from the causal diagram (Fig. 1). The value of M indicates how much important a certain enabler is. The greater the value of M, the greater is its importance. Whereas N, which represents (D−R), has a different meaning. If the value of N of a specific enabler is positive, it indicates that the enabler has its influence over other enablers. If the value of N of a certain enabler is negative, it means that other enablers have an influence on that enabler.

It can be observed from Table 7 that the enabler 'modularity in products and processes' has the highest value of M, i.e., 0.93. It denotes that this enabler is the most crucial enabler for firms seeking to improve their FMC. The finding seems

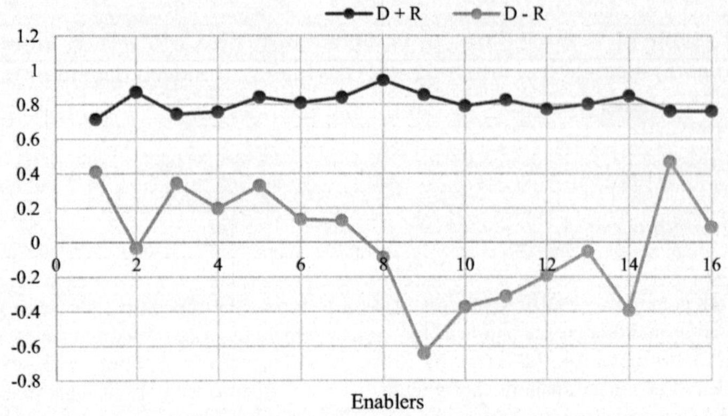

Fig. 1 The causal diagram

logical, given that this enabler allows many manufacturing operations to run concurrently. Independent components can thus be developed and tested individually before being merged into a modular product design. This contributes to the overall quality of the product. This enabler can be referred to as a primary enabler in the quest of improving FMC. Following enabler F8, the next two enablers with the highest value of M are 'operational improvement practices' and 'time-based manufacturing', having the value of M as 0.87 and 0.85, respectively. Operational improvement practices are helpful to service firms in improving operational efficiency as well as developing employee creativity by inducing intrinsic motivation. Since time-based manufacturing is an externally focused production system that emphasizes quick response to changing customer needs, its primary purpose is to reduce end-to-end time in manufacturing, which indeed assists in improving the FMC of an organization. Since these three enablers have the highest values of M, we can conclude that they have the most influence over the organizations, as compared to other enablers. One noteworthy observation is that all of these enablers have a negative value of N, which means other enablers can impact these three enablers, and that they don't act have complete autonomy of their own. This shows us that organizations need to focus on other enablers as well, in order to improve their FMCs. Apart from the most important enablers, focusing on others enablers will indirectly lead to the improvement of the important enablers as well.

This research adds to the existing amount of knowledge and wisdom on the formation and enhancement of FMC in manufacturing companies by identifying essential elements and categorizing them into cause-and-effect categories, which is the main objective of this study. This may also help managers establish and implement plans, in order to develop flexible operational capabilities and convert existing systems into adaptive systems. With environmental turbulence developing at an alarming rate and predicted to intensify, firms must come up with new ideas and build unique tactics to magnify their FMC while grasping opportunities to thrive and addressing the present business climate. In this sense, the current study will be beneficial to firms looking

to gain a competitive edge through the creation of FMC. In future, more research can be done on the enablers affecting FMC, which will indeed benefit organizations by increasing their FMC.

References

1. Ullah I, Narain R (2020) Achieving mass customization capability: the roles of flexible manufacturing competence and workforce management practices. J Adv Manag Res
2. Zhang Q, Vonderembse MA, Cao M (2006) Achieving flexible manufacturing competence: the roles of advanced manufacturing technology and operations improvement practices. Int J Oper Prod Manag 26:580–599
3. Jin Y, Norbis M, Awudu I (2022) Supplier-dedicated resources and flexibility's roles in a manufacturer's superior performance. J Gen Manag 47:86–96
4. Ullah I, Narain R (2020) Analyzing the barriers to implementation of mass customization in Indian SMEs using integrated ISM-MICMAC and SEM. J Adv Manag Res
5. Jost PJ, Süsser T (2020) Company-customer interaction in mass customization. Int J Prod Econ
6. Ullah I, Narain R (2018) Analysis of interactions among the enablers of mass customization: an interpretive structural modelling approach. J Model Manag
7. Ullah I, Narain R (2020) The impact of customer relationship management and organizational culture on mass customization capability and firm performance. Int J Customer Relat Mark Manag (IJCRMM). 11:60–81
8. Cagliano R, Spina G (2000) Advanced manufacturing technologies and strategically flexible production. J Oper Manag 18:169–190
9. Wang Y, Guo B, Yin Y (2017) Open innovation search in manufacturing firms: the role of organizational slack and absorptive capacity. J Knowl Manag 21:656–674
10. Isensee C, Teuteberg F, Griese KM, Topi C (2020) The relationship between organizational culture, sustainability, and digitalization in SMEs: a systematic review. J Cleaner Prod
11. Tarigan ZJ, Siagian H, Jie F (2020) The role of top management commitment to enhancing the competitive advantage through ERP integration and purchasing strategy. Int J Enterp Inf Syst (IJEIS). 16:53–68
12. Koçyiğit Y, Akkaya B (2020) The role of organizational flexibility in organizational agility: A research on SMEs. Bus Manag Strategy 11:110–123
13. Okoro CS, Nkambule M, Kruger A (2020) The state of restroom facilities as a measure of cleaning service quality in an educational institution. J Corp Real Estate 23:55–68
14. Do BR, Yeh PW, Madsen J (2016) Exploring the relationship among human resource flexibility, organizational innovation and adaptability culture. Chin Manag Stud 10:657–674
15. Wang Z, Zhang M (2020) Linking product modularity to supply chain integration and flexibility. Prod Plan Control 31:1149–1163
16. Ravi V, Shankar R, Tiwari MK (2005) Productivity improvement of a computer hardware supply chain. Int J Product Perform Manag 54:239–255
17. Alp O, Tan T (2008) Tactical capacity management under capacity flexibility. IIE Transactions J; 40:221–37
18. Holtewert P, Bauernhansl T (2016) Increase of capacity flexibility in manufacturing systems by substitution of product functions. Procedia CIRP. 57:92–97
19. Ferreira KA, Flávio LA, Rodrigues LF (2018) Postponement: bibliometric analysis and systematic review of the literature. Int J Logistics Syst Manag 30:69–94
20. Vasiliev VA, Velmakina YV, Mayborodin AB, Aleksandrova SV (2020) Use of information technologies for the integration of an enterprise quality management system with the requirements of the related standards. Russ Metall 2020:1644–1648

21. Purwanto A, Asbari M, Santoso PB (2020) Effect of integrated management system of ISO 9001: 2015 and ISO 22000: 2018 implementation to packaging industries quality performance at Banten Indonesia. Jurnal Ilmiah MEA (Manajemen, Ekonomi, & Akuntansi). 4:17–29
22. Soon QH, Udin ZM (2011) Supply chain management from the perspective of value chain flexibility: an exploratory study. J Manuf Technol Manag 22:506–526
23. Shukla S, Patil AP, Kawale AP, Singh SK, Thombre MA (2021) Effect of thermal ageing and deformation on microstructural evolution of 304 and 202 grade steel. Mater Today Proc 38:3238–3245
24. Yadav S, Luthra S, Garg D (2020) Internet of things (IoT) based coordination system in Agri-food supply chain: development of an efficient framework using DEMATEL-ISM. Oper Manag Res
25. Amiri M, Sadaghiyani J, Payani N, Shafieezadeh M (2011) Developing a DEMATEL method to prioritize distribution centers in supply chain. Manag Sci Lett 1:279–288
26. Dhakne A, Jaju S, Shukla S (2022) Review on analysis of enhancing wear properties through thermo-mechanical treatment and grain size. Mater Today Proc
27. Gaidhane J, Ullah I, Khalatkar A (2022) Tyre remanufacturing: A brief review. Mater Today Proc
28. Ullah I, Narain R (2020) Linking supplier selection and management strategies with mass customization capability. J Bus Ind Mark 36:1213–1228
29. Singh C, Singh D, Khamba JS (2020) Analyzing barriers of Green Lean practices in manufacturing industries by DEMATEL approach. J Manuf Technol Manag

Design, Material, and Performance Study of Modified Solar Cooker

Bhupendra Koshti, Rahul Dev, and Priyank Srivastava

Nomenclature

T_{w1}	initial temperature of the water, ℃
T_p	Absorber plate temperature, ℃
I_{av}	average solar radiation
T	time required, s
P_c	cooking power, W
P_s	Standard cooking power, W
I_d	Diffused solar insulations W/m^2
R_T	thermal resistance W/m·K
L_{gw}	length of glass wool mm
K_{gi}	Thermal conductivity of G.I W/m·K
α	Absorptivity
ε_g	Hemispherical total emittance of the polycarbonate cover plate.
T_{w2}	Final temperature of the water, ℃
T_d	Temperature difference of initial and final temperature of the water, ℃
m	mass of water taken, kg
U_L	Overall heat loss
F'	solar collector efficiency factor
MBSC	modified box type solar cooker
T_f	Final water temperature ℃

B. Koshti (✉) · R. Dev · P. Srivastava
Department of Mechanical Engineering, MNNIT Allahabad, Prayagraj, Uttar Pradesh 211004, India
e-mail: bhupendra@mnnit.ac.in

R. Dev
e-mail: rahuldsurya@mnnit.ac.in

P. Srivastava
e-mail: priyank.srivastava25@gmail.com

A. Kumar et al. (eds.), *Recent Advances in Manufacturing and Thermal Engineering*,
Lecture Notes in Mechanical Engineering, https://doi.org/10.1007/978-981-19-8517-1_17

L_{gi} length of G.I sheet mm
L_{pc} length of polycarbonate sheet mm
K_{gw} Thermal conductivity of glass wool W/m·K
T Transmissivity
β Collector tilt angle
T_a ambient temperature, °C
I_s Solar insulations, W/m^2
C_p specific heat capacity of water, (J/Kg°C)
F_1 First Figure of merit
F_2 Second Figure of merit
ρ Humidity %
V Wind Velocity m/s
K_{air} Thermal conductivity of air W/m·K
L_{air} length of air gap mm
K_{pc} Thermal conductivity of polycarbonate sheet W/m·K
ε_p Hemispherical total emittance of the absorber plate
L Air gap spacing between the absorber and the glass cover plate

1 Introduction

Solar energy is a sustainable energy source, present in unlimited quantities among all the energy sources that are present in the world compared to other energy sources like the business-as-usual approach, nuclear energy, big Hydro, thermal power plant, etc. The energy generated by solar energy is clean and pollution-free, and valuable for all living organisms present on the earth. Solar energy does not create global warming and limits the percentages of carbon and Ozone depletion; due to this, the earth's temperature is maintained.

Food is the basic need of human beings that provides essential nutrients and calories for the development of metabolic activities. When food is prepared in a solar cooker, its essential nutrients and calories retain in the food due to proper heating, and thermal energy is maintained inside the solar cooker. The solar cooker does not require much maintenance, and there is no probability of accidents. No harmful gases are evolved and occur while cooking food as compared to food cooked in LPG, kerosene stoves, induction cookers, microwave ovens, electric heaters, conventional Chula's, etc.

In South Asian countries like India, Pakistan, Nepal, and Bangladesh, about 80% population relies on the burning of conventional biomass, which was created indoor. In South Asia, solid pollutants cause health issues [1]. Therefore, solar energy is a convenient option for cooking at home.

A modified solar cooker is made from a transparent sheet, and a 10 mm air gap is maintained between the inside and outside sides of the solar cooker that increases the

heat transfer augmentation in the cooking area by improving the materials, design of solar cookers, and cooking time [2].

1.1 Literature Review

There are various types of designs designed for solar cookers. Mahavar investigated using transparent acrylic (PMMA) material that increases the solar radiation falling on the cooking pots, and thermal performance is increased by a considerable amount [2].

Mullick designed and fabricated a box type of solar cooker, whose outer body is made from teakwood, inner body from aluminum sheet, and the cooking pots from aluminum materials and they performed the thermal test on solar cooker under different climate conditions, different food and formed the Figure of Merit (F_1 and F_2) [3]. Grupp et al. presented modified cooking pots fixed in a simple box solar-powered cooker and good heat transfer with the absorber plate fixed under shine. The experimental result showed that thermal performance is increased [4]. Rathore et al. experimentally investigated two types of solar cooker: a simple box solar cooker constructed from steel and another concentrating type consisting of PPMA having a reflectivity of 0.75. The author concluded that energy produced by concentrating solar cooker was found in the range of 0.65–39.3 W and for solar box type cooker 7.44–33.49 W, exergy produced between 0.92 and 2.58 W [5]. Nahar et al. experimentally investigated with transparent insulating material (TIM), and a 158 °C stagnation temperature is attained [6]. Sharma et al. concentrated on a PCM storage unit and compared it to a traditional solar cooker, experimentally finding that evening cooking is possible with PCM storage materials; a standard solar cooker is not compatible with a PCM storage unit [7]. Ali designed and fabricated a solar cooker made of wooden pieces and outer covered by plywood based on Sudanese conditions and compared it with the Indian solar cooker. Results from the experiments showed the cooker's acceptability and were successful [8]. Ozturk evaluated the energy and exergy analysis of box-type cookers made of hard plastic and concentrating solar cookers. The efficiency range is found to be 3.05–35.2%, 0.58–3.52%, and 2.79–15.65%, 0.4–1.25% [9]. Kumar selected two parameters for the experiment: optical efficiency and heat capacity for double glazing solar cooker with an aperture area of $0.245m^2$ made of fiber for design and forecast Figures of merit F_1 and F_2. They considered linear regression analysis of F_2 on the water at different loads and concluded that F_2 is the function of water [10]. Reddy and Rao compared the efficiency of a solar oven with an annular gap filled with thermic fluid [11]. According to Kimambo, they analyzed, fabricated, and evaluated the realization of six cookers, mainly Sun stoves, simple box type cookers, wooden cookers, panel type cookers, reflector type cookers with unvarnished aluminum, and varnished reflectors made of glass, and found that most of the solar cooker is capable of cook food in house with specific location and types of materials used with mid and maximum insulations [12]. Harmim and Boukar developed a solar cooker with a finned vessel.

Compared with a finned vessel, this cooker required less time than food cooking [13]. According to Senger, solar cookers made from packing cardboard are more efficient in terms of temperature profile as well in Figure of merit (F_1 and F_2) and less cost and weight than fiber body cookers [14]. Senger et al. constructed a masonry and cement plaster solar cooker having a cylindrical shape, and comparative thermal performance tested results show that the building-material-housing solar cooker (BHMC) is slightly better than the commercial solar cooker (CSC) [15]. Folaranmi experimentally investigated the double-glazed simple box solar cooker made of sawdust that has been compressed with binder and painted aluminum absorber plate with painted black and double-glazed lid [16]. Akoy et al.'s three different types of solar cookers were devised and built. They are mainly concerned with calculating the thermal performance of simple cookers, panel-type solar cookers, and concentrating solar cookers. They are made of domestically available materials like wood and cardboard [17]. Sagade et al. focused on a new design technique in the solar cooker by modifying cooking pots and increasing heat transfer in the solar cooker, which reduces cooking time and efficiency [18]. Poonia et al. calculated the performance of the stagnation test and water boiling test of solar cooker made of galvanized sheet outer body and aluminum sheet inner body and parameter came out and showed that this hot insulations comes under the category of A-class solar cooker. This cooker reduces carbon dioxide by 815.30 annual bases [19]. Khallaf et al. designed a dome shape solar cooker with an internal reflector used [20]. Engoor et al. developed a solar cooker with a Fresnel lens integrated to predict the thermal performance and energy exergy analysis [21].

1.2 The Objective of the Modified Solar Cooker

Based on previous research and a review of the literature, a lot of works were carried out on solar cookers based on different designs, materials, insulations, and thermal performance of solar cookers. However, some research also focused on different designs of cooking utensils/pots used in solar cookers. Additionally, many more researches are supervised to enhance the thermal efficiency of solar cookers. However, no researcher is concentrated on the concept of equivalent thickness for the same heat flux and temperature difference and maintaining an air gap between all sides of the cooker.

2 Materials and Methods

See Table 1.

See Figs. 1 and 2.

Table 1 Experimental details of Modified box type solar cooker

S. No.	Parameters	Specifications
1	Orientation	East–West
2	Placed	MNNIT Allahabad, Prayagraj, India
3	Coordinates	25.4358° N, 81.8463° E
4	Meteorologic conditions	Warm and Humid
5	Outer body materials of solar cooker	North- FRP (Fibre-Reinforced Plastic) (18 × 48 cm) South- Polycarbonate (18 cm × 48 cm) East- Polycarbonate (18 cm × 48 cm) West- Polycarbonate (18 cm × 48 cm) Top cover- FRP (Fibre-Reinforced Plastic) (48.5 cm × 48.5 cm)
6	Inner body Materials of solar cooker	North- FRP (Fibre-Reinforced Plastic) (14 cm × 46 cm) South- Polycarbonate (14 cm × 46 cm) East- Polycarbonate (14 cm × 46 cm) West- Polycarbonate (14 cm × 46 cm) Top glass – Polycarbonate (46.5 cm × 46.5 cm) Base- FRP (Fibre-Reinforced Plastic) (45 cm × 45 cm)
7	Reflector	Simple plane (46 cm × 46 cm)
8	FRP(Thickness)	5 mm
10	Polycarbonate (Thickness)	5 mm

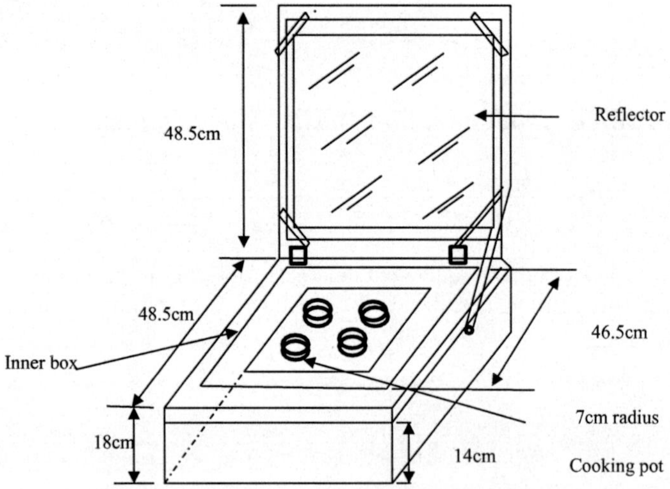

Fig. 1 Schematic diagram of modified box-type solar cooker (MBSC)

Fig. 2 Photograph of modified box-type solar cooker (MBSC)

2.1 Equivalent Thickness and Design Specifications

$$R_T = \frac{L_{gi}}{K_{gi}} + \frac{L_{gw}}{K_{gw}} + \frac{L_{gi}}{K_{gi}} = \frac{L_{pc}}{K_{pc}} + \frac{L_{air}}{K_{air}} + \frac{L_{pc}}{K_{pc}} \tag{1}$$

The heat transfer rate and equivalent thickness for two G.I. sheets with glass wool replacing two polycarbonate sheets with air gap obtained is 10 mm (Fig. 3).

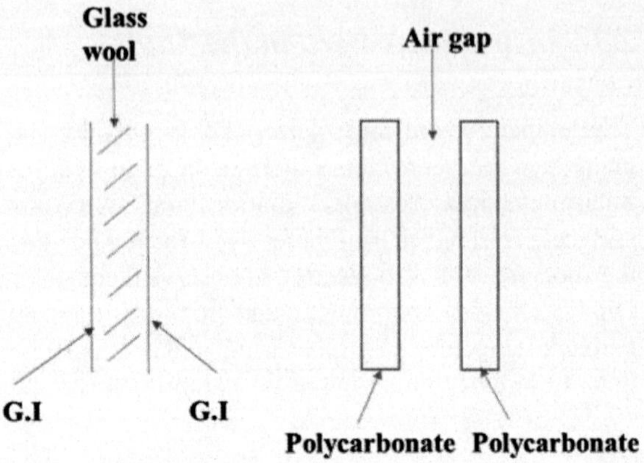

Fig. 3 Schematic sketch of equivalent thickness for same heat flux

2.2 Design Equations for Solar Cooker Are as Follows

Absorber plate-$A_p = \frac{mc_p(T_{w1}-T_{w2})}{t[\tau_p\alpha_p I_{av}-U_L(T_p-T_a)]}$ (2)

The top heat loss coefficient-

$$U_t^{-1} = \left[\frac{12.75\{(T_p - T_{i)} \cos\beta\}^{0.264}}{(T_p + T_i)^{0.46} L^{0.21}} + \frac{\sigma\left(T_p^2 + T_i^2\right)(T_p + T_i)}{\frac{1}{\varepsilon_p} + \frac{1}{\varepsilon_g} - 1}\right]^{-1} + \left[h_w + \frac{\sigma\varepsilon_g(T_i^4 - T_a^4)}{(T_i + T_a)}\right]^{-1} + \frac{T_g}{K_g} \quad (4)$$

Samdarshi and Mullick developed top heat loss equations [23]

Overall heat loss-
$U_{total} = U_{top} + U_{bot} + U_{sides}$ (3)

Bottom heat loss coefficient
$\frac{1}{U_B} = \frac{t}{K}$ (5)

Side heat loss coefficient
$\frac{1}{U_s} = \frac{1}{4L_{air\ gap}}\left[\frac{t_{inner}}{K_{inner}} + \frac{t_{out}}{K_{outer}}\right]$ (6)

Bottom and side heat loss equations proposed by Kumar et al. [24]

Energy balance of solar cooker-$(mc_w)\frac{dT_w}{dt} = F'[(\tau\alpha)I_s(t) - U_L(T_w - T_a)]$ (7)
This energy balance proposed by Schwarzer et. al [25]

F' = solar collector efficiency factor $F' = 0.85$, after integrating above equation with initial conditions $t = 0$ and $T_w = T_{w0} \cdot t_0 = \frac{(mc_w)}{F' A_p U_L}$ (8)

See Table 2.

Table 2 Calculated value of modified box-type solar cooker

S. No.	Specifications	Values
1	Top heat loss coefficient, U_T. (W/m^2 K)	2.31
2	Side heat loss coefficient, U_s (W/m^2 K)	0.056
3	Bottom heat loss coefficient, U_B. (W/m^{2}K)	0.8
4	Overall heat loss coefficient, U_L. (W/K)	3.166
5	Absorber plate area, A_p (m^2)	0.25

2.3 Instruments Used in the Study

The temperature of the solar cooker is outside, and inside surfaces are measured using a copper–constantan thermocouple (T type) attached to the display scanner. Solarimeter measures solar radiation intensity, top cover, side surfaces, and horizontal surfaces. Infrared thermometer used for outside surfaces temperature. A Mercury thermometer was also inserted to observe the inside temperature of the solar cooker. Digital-Thermo-Anemometer and Humidity Temperature Meter are used to measure ambient temperature. A measuring beaker is used for the quantity of water taken in pots. Before beginning the test's, calibrated instruments were utilized (Table 3).

Table 3 Instruments used for an experiment

S. No.	Instruments	Measurement characteristics	Specifications	Variability
1	Solarimeter	Solar insulations	Company: AMPROBE SOLAR-100, Make: Amprobe, Everett, WA accuracy: ± 5%, range: 0–1999 W/m^2 , resolution: 0.1 W/m^2	± 1.4
2	I.R. thermometer	Outside surface temperatures	Model: FLUKE 62 MAX, Make: Fluke Corporation, Everett, WA range: −30–500 °C, accuracy: ± 1.5 °C, resolution: 0.1 °C	± 0.14
3	Thermocouples	Inside and outside surface temperatures and cooking pots	Copper–constantan T-type (New Delhi, India) accuracy: ± 1 °C, range: −40–350 °C	± 0.22
4	Digital Thermo-Anemometer	Wind velocity and ambient temperatures	Range 0–30 m/s, Resolution 0.1 m/s, accuracy ± (5%rdg + 0.5), Temperature range -10 - 50 °C, Resolution 0.1 °C, accuracy ± 2 °C, Sampling rate	0.5
5	Thermometer	Ambient temperature	Range −10–110 °C	

2.4 Thermal Performance Tests of Solar Cooker

No-load test (First Figure of merit)—Solar cookers were placed outside in the sunshine without cooking pots. The thermocouples were fixed to the cooker's base plate and top cover. The thermocouple was sealed with silicone, and the output of the thermocouple was connected with a digital temperature scanner. The reflector is not considered in the test. The test started at 9:00 h, and during the test, solar insulations, base plate temperature, and surrounding temperature were measured. At 12 PM, the maximum temperature reached the base plate, and stagnation temperature was attained in the solar cooker. At this stagnation point, the temperature of the base plate, surrounding temperature, and solar insulations were computed and calculated the value of F_1. This test can also be carried out in the solar simulator in an indoor environment.

The figure of merit $F1$ is defined as the ratio of optical efficiency to absorber plate heat loss factor, as well as the measurement of the differential temperature obtained by the absorber plate at a given amount of solar insulations.

$$F_1 = \frac{(T_P - T_a)}{I_S} \tag{9}$$

First figure of merit proposed by Mullick et al. [3] (Table 4).

Full-Load Test (Second Figure of Merit)—In this test, cooking pots are filled with water. To calculate the value of F_2, solar cooker's reflector is removed. The initial temperature of water was taken at 60–95°C, and the final temperature of water at 90–95°C was taken respectively. Solar radiation, ambient air temperature, and water temperature are taken in 10 min interval of time [22].

The second figure of merit concludes that the boiling temperature intervals of water in cooking pots should be small as possible.

$$F_2 = \frac{F1\left(mC_p\right)}{At} \ln \frac{1 - (1/F1)((Tw1 - Ta)/I_s}{1 - (1/F1)((Tw2 - Ta)/I_s)} \tag{10}$$

Second figure of merit proposed by Mullick et al. [3] (Table 5).

Cooking Power—The solar cooker was placed in the sun, and cooking pots were filled with 2 kg of water. Change in water temperature was measured at 10 min time interval. The product of mass, water-specific heat, and water temperature gradient is divided by a 10 min gap (600 s).

$$P_c = \frac{\mathrm{m}c_p(T_{w1} - T_{w2})}{600} \tag{11}$$

Cooking power by Folaranmi [16] (Table 6).

Table 4 No-load test data for modified box-type solar cooker 11 December 2021

Time(h)	Solar insulations Is (Global) (W/m^2)	Diffused solar insulations Id (W/m^2)	Walls solar insulations (W/m2)	Humidity ρ (%)	Wind Velocity v (m/s)	Ambient temperature T_a (ºC)	Absorber plate T_p (ºC)	Temperature difference $(T_p - T_a)$ ºC
10:00	649.0	242.0	1211.0	68.3	1.9	19.3	62.0	42.7
10:30	700.2	237.0	1160.0	66.1	2.3	20.2	67.0	46.8
11:00	752.0	233.0	1110.0	64.0	2.1	21.2	72.0	50.8
11:30	785.0	226.0	1186.0	61.8	1.7	21.9	79.5	57.6
12:00	818.0	220.0	1263.0	59.7	1.3	22.7	87.0	64.3
12:30	766.0	198.5	1124.0	57.5	1.1	22.9	96.0	73.1
13:00	714.0	226.0	1185.0	55.3	1.0	23.2	105	81.8
13:30	678.4	201.0	1256.0	52.2	0.8	23.1	98.0	74.9
14:00	643.0	177.0	1327.0	49.2	0.4	23.1	91.0	67.1
14:30	539.2	161.5	1223.5	49.3	0.4	23.0	84.0	61
15:00	436.0	145.0	1120.0	49.4	0.8	22.9	77.0	54.1

Table 5 Full-load test data for a modified box-type solar cooker on 14 December 2021

Time(h)	Solar insulations I_s (W/m^2)	Diffused solar insulations (I_d) (W/m^2)	Walls solar insulations (W/m^2)	Humidity ρ (%)	Wind Velocity V (m/s)	Ambient temperature T_a (°C)	Absorber plate T_p (°C)	Inside pot water temperature T_w (°C)	Temperature difference $(T_p - T_a)$ °C
10:00	627.0	204.0	1204.0	61.5	0.4	16.8	57.0	44.4	40.2
10:30	659.0	218.0	1248.0	60.1	0.6	17.4	60.6	52.8	43.3
11:00	692.0	233.0	1293.0	58.7	0.8	18.0	64.2	61.3	46.2
11:30	706.0	236.0	1299.0	56.8	1.25	18.5	71.8	67.2	53.3
12:00	720.0	239.0	1306.0	55.3	1.7	19.0	79.5	73.2	60.5
12:30	685.0	218.5	1296.0	52.9	1.25	19.8	68.9	75.7	64.9
13:00	650.0	198.0	1287.0	50.5	0.8	20.6	89.4	78.2	69.4
13:30	589.0	174.0	1230.0	49.3	0.6	19.9	90.7	71.1	71.1
14:00	528.0	150.0	1173.0	48.2	0.4	19.2	92.0	64.0	72.8
14:30	446.0	145.0	932.0	49.9	0.75	19.0	86.4	57.0	67.4
15:00	364.0	140.0	691.0	51.6	1.1	18.9	80.9	52.0	62.0

Table 6 Cooking power test data for a modified box-type solar cooker on 06 April 2022

Time(h)	Solar insulations I_s (W/m^2)	Water temperature Tw_1 (°C)	Final water temperature T_f (°C)	Cooking power (P_c)	Temperature difference (°C)	Ambient temperature T_a (°C)
10:00	847.0	60.0	63.3	31.82	3.3	35.5
10:10	868.0	63.3	66.6	30.86	6.6	37.2
10:20	889.0	66.6	69.2	30.13	9.2	38.2
10:30	910.0	69.2	71.3	28.52	11.3	41.0
10:40	931.5	71.3	74.9	26.96	14.9	42.7
10:50	952.1	74.9	77.0	27.25	17.0	44.5
11:00	973.0	77.0	80.1	26.67	20.1	45.5
11:10	977.2	80.1	83.8	24.84	23.4	47.1
11:20	996.0	83.8	86.6	24.26	26.8	47.2
11:30	999.5	86.6	89.6	25.12	29.8	47.3
11:40	1002.7	89.6	92.1	25.04	32.3	47.6
11:50	1012.1	92.1	95.8	24.87	36.0	47.8
12:00	1026.0	95.8	98.3	24.47	38.5	48.1
12:10	1028.0	98.3	101.2	24.43	41.4	47.6
12:20	1030.0	101.2	105.1	24.38	45.3	47.4
12:30	1032.0	105.1	102.2	23.52	48.4	47.2
12:40	1034.0	102.2	99.8	19.41	51.4	46.9
12:50	1036.0	99.8	97.7	16.97	53.5	46.7
13:00	1038.0	97.7	95.7	16.38	55.5	46.4
13:10	989.5	95.7	93.8	16.07	57.4	46.1
13:20	1009.0	93.8	92.1	14.10	59.1	46.0
13:30	961.0	92.1	90.5	13.93	60.7	45.8
13:40	912.5	90.5	89.0	13.76	62.2	45.6
13:50	932.1	89.0	87.5	13.47	63.7	45.4
14:00	884.0	87.5	85.9	13.25	65.1	45.2

Standard cooking power

For calculating standard cooking power, the standard solar insulations was multiplied by cooking power and dividing the average insulations measured during this interval.

$$P_s = P_c \frac{700}{I_{av}} \tag{12}$$

Standard cooking power by Folaranmi [16] (Table 7).

Table 7 Standard cooking power test data for a modified box-type solar cooker on 06 April 2022

Time(h)	Solar insulations I_s (W/m^2)	Water temperature Tw_I (°C)	Final water temperature T_f (°C)	Cooking power (P_c)	Temperature difference (°C)	Standard cooking power (Ps)	Ambient temperature T_a (°C)
10:00	847.0	60.0	63.3	31.82	3.3	22.94	35.5
10:10	868.0	63.3	66.6	30.86	6.6	22.25	37.2
10:20	889.0	66.6	69.2	30.13	9.2	21.725	38.2
10:30	910.0	69.2	71.3	28.52	11.3	20.56	41.0
10:40	931.5	71.3	74.9	26.96	14.9	19.43	42.7
10:50	952.1	74.9	77.0	27.25	17.0	19.64	44.5
11:00	973.0	77.0	80.1	26.67	20.1	19.23	45.5
11:10	977.2	80.1	83.8	24.84	23.4	17.91	47.1
11:20	996.0	83.8	86.6	24.26	26.8	17.49	47.2
11:30	999.5	86.6	89.6	25.12	29.8	18.11	47.3
11:40	1002.7	89.6	92.1	25.04	32.3	18.05	47.6
11:50	1012.1	92.1	95.8	24.87	36.0	17.93	47.8
12:00	1026.0	95.8	98.3	24.47	38.5	17.64	48.1
12:10	1028.0	98.3	101.2	24.43	41.4	17.61	47.6
12:20	1030.0	101.2	105.1	24.38	45.3	17.57	47.4
12:30	1032.0	105.1	102.2	23.52	48.4	16.95	47.2
12:40	1034.0	102.2	99.8	19.41	51.4	13.99	46.9
12:50	1036.0	99.8	97.7	16.97	53.5	12.23	46.7
13:00	1038.0	97.7	95.7	16.38	55.5	11.81	46.4
13:10	989.5	95.7	93.8	16.07	57.4	12.04	46.1
13:20	1009.0	93.8	92.1	14.10	59.1	10.16	46.0
13:30	961.0	92.1	90.5	13.93	60.7	10.40	45.8
13:40	912.5	90.5	89.0	13.76	62.2	9.92	45.6
13:50	932.1	89.0	87.5	13.47	63.7	9.71	45.4
14:00	884.0	87.5	85.9	13.25	65.1	9.55	45.2

2.5 *Methodology*

The solar data is taken for the winter and summer seasons at Prayagraj (U.P.) from the experimental observations. The modified box-type solar cooker's methodology approach is as follows:

- Solar intensity on top of the polycarbonate cover and side walls, and the temperature at different locations of the solar cooker are taken experimentally.
- With the use of experimental data, the thermal performance and solar data analyses of the solar cooker were done.

- Plot different graphs on the bases of experimental data of solar intensity and temperature.

3 Results and Discussion

Figure 4 shows global solar radiation (W/m^2) plotted against time (h) for a sunny day in December 2021 and April 2022. In the early morning, the quantity of solar radiation incident on solar cooker was small after noon time progressively rises to a maximum value and at evening decreases at the sunshine. Based on Prayagraj (U.P.) latitude (25.4358° N, 81.8463° E), the solar radiation intensity in summer is maximum compared to winter due to the attitude angle in the sun path diagram. From the graph, the maximum solar radiation in April 2022 at 11:00–12:00 h and diffused solar radiation at 11 AM. In December 2021, the maximum solar radiation was at 12:00–13:00 h and diffused solar radiation at 13:00 PM because of the period of daylight in the northern hemisphere, where Allahabad (U.P.) is situated.

No-load test (stagnation test for First Figure of Merit F_1)

The test was conducted on 11 December 2021 under a clear sky at Prayagraj (25.4358° N, 81.8463° E), Uttar Pradesh, India. This experiment test was carried out from 10:00 AM to 14:00 PM till the maximum absorber plate temperature attained was 105°C at 13:00 PM. The following data were recorded during the test: solar radiation, ambient temperature, and absorber plate temperature. The calculated value

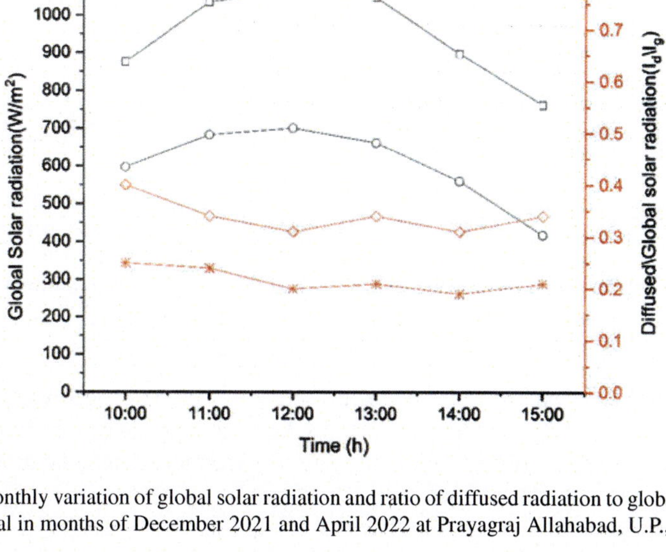

Fig. 4 Monthly variation of global solar radiation and ratio of diffused radiation to global radiation for a typical in months of December 2021 and April 2022 at Prayagraj Allahabad, U.P., India

of $F_1 = 0.114$ is obtained from $T_a = 23.2$°C, $T_p = 105$°C and $I_s = 714$ W/m^2. This value resembled the modified solar cooker as A class cooker (Greater than $F_1 = 0.111$).

Full-Load Test (Second Figure of Merit)

This test was conducted on 14 December 2021 in the clear sky. The following data were used to calculate the full-load test. Initial temperature was taken (T_{w1}) = 61.1 °C, final temperature of water (T_{w2}) = 78.2 °C, Average solar radiation (I_s) = 606.0 W/m^2, First figure of merit (F_1) = 0.11456, Mass of water(m) = 2 kg, specific heat of water (C_p) = 4186 J/kg °C, Time required (t) = 7200 s, Average ambient temperature (T_a) = 18.8 °C. F_2 value is obtained at 0.20 (Fig. 5).

A cooking power test was conducted on a modified box type solar cooker on 06 April 2022 and followed the international standard method. The load is taken with 2 kg of water in the cooking pot. The solar cooker was placed to the sun between 10:00 h and 14:00 h. The ambient temperature, initial water temperature, final water temperature, and solar radiation were measured for every 10 min. of the interval, as shown in the table.

The standard cooking power of the modified box-type solar cooker is plotted against the time difference. The maximum temperature was 43°C and the minimum was 28°C. Standard cooking power graph is plotted between standard cooking power and temperature difference in Fig. 9. $P_s = 23.727 - 0.2019\, T_d$ equation of regression was obtained from the Figure. The standard recommended value of coefficient R^2 is 0.925>0.75. The cooking power $P_{50} = 13.63$ W at a 50°C temperature difference was

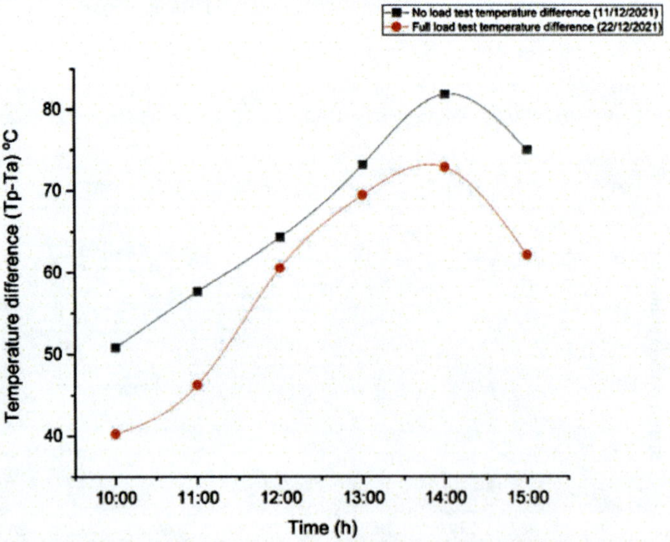

Fig. 5 Temperature difference $(T_p - T_a)$ of modified box-type solar cooker with respect to time (h) for typical days in the month on 11th and 22nd December 2021

calculated on the basis of the regression relationship which is given above (Figs. 6 and 7).

Figure 8 shows plotted total incident radiation(W/m^2) on a modified box-type solar cooker (MBSC) with respect to time (h). On 11 December 2021 and 06 April 2022, the maximum incident solar radiation on the solar cooker at 10:00 h was 2890 W/m^2 and a minimum of 1556 W/m^2 at 15:00 h.

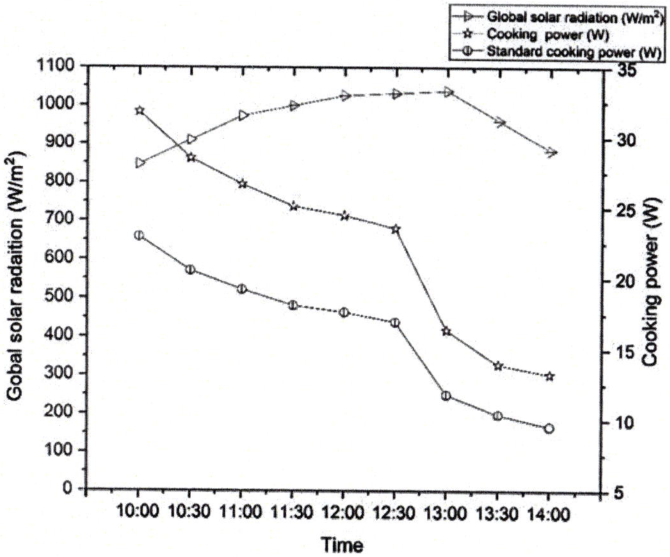

Fig. 6 Comparison of cooking power and standard cooking power modified box-type solar cooker with respect to time (h) for typical days in the month on 11th and 22nd December 2021

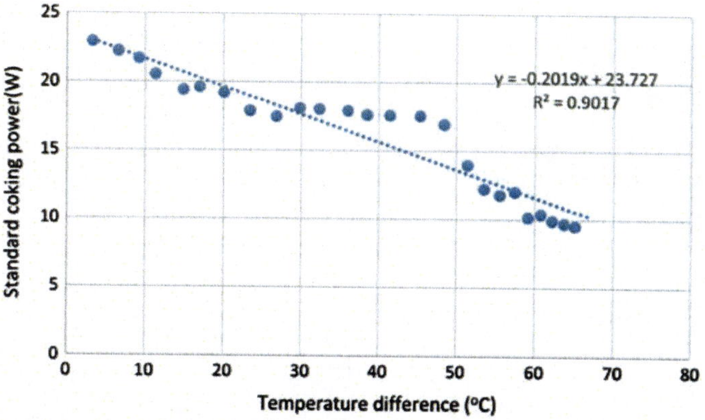

Fig. 7 Ambient temperature of the modified box-type solar cooker for a typical day in the month on 6th April 2022

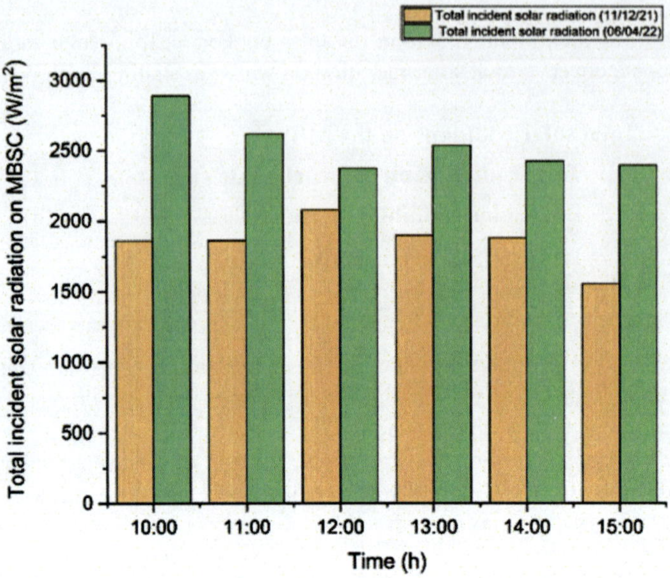

Fig. 8 Total incident solar radiation (instantaneous) for the month on 11th December 2021 and 6th April 2022

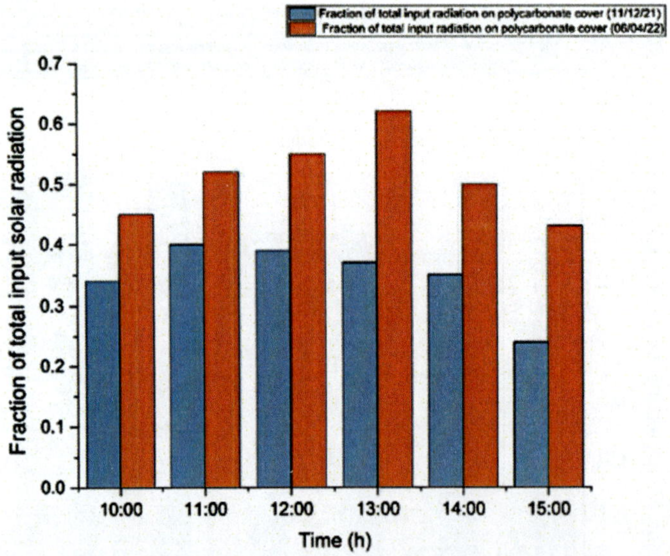

Fig. 9 Fraction of total input solar radiation on polycarbonate cover (instantaneous) for a typical day in the month on 11th December 2021 and 6th April 2021

The total incident solar radiation on solar cooker = Total solar radiation into polycarbonate cover + total solar radiation on walls (excluding north wall).

$$\begin{aligned}
&\textbf{Total solar raditions on the MBSC} \\
&\quad = \textbf{Total radiation on polycarbonate cover} \\
&\quad + \textbf{Total solar radiation on walls(East + west + south)} \\
&\quad = 16601\left(\mathrm{W/m^2}\right)(\text{06th April 2022}) \\
&\quad = 11228\left(\mathrm{W/m^2}\right)(\text{11th December 2021})
\end{aligned}$$

Figure 9 shows the fraction of total incident radiation on Polycarbonate ($\mathrm{W/m^2}$) with respect to time (h) on the month of 11 December 2021 and 06 April 2022. The maximum ratio is 0.62 at 13:00 h and the minimum is 0.24 at 15:00 h.

Figure 10 graphs are plotted between the fraction of total input solar radiation on the side walls for the month of 11 December 2021 and 06 April 2022 and with respect to time (h). The maximum fraction value is 0.71, and the minimum fraction value is 0.4.

Figure 11 inner air temperature of the modified box-type solar cooker is plotted against the time; the maximum temperature attained was 111.5°C at 12:00 h, and the minimum was 62°C at 15:00 h.

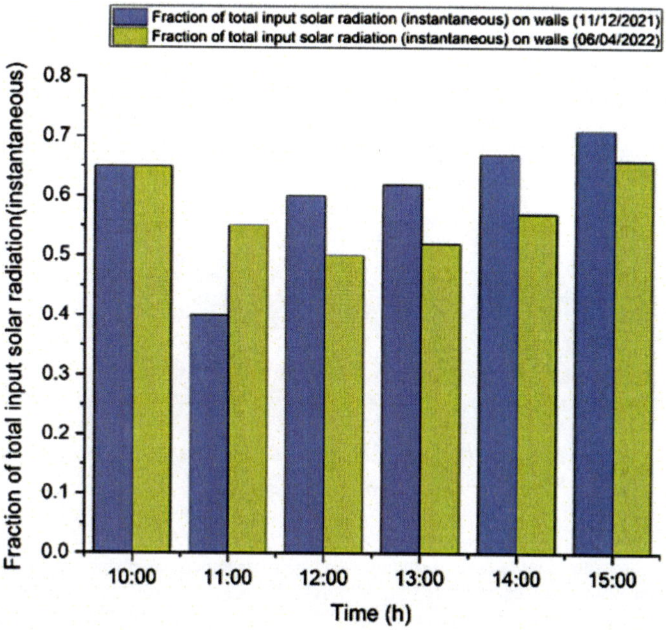

Fig. 10 Fraction of total input solar radiation on side walls (instantaneous) for a typical day in the month on 11th December 2021 and 6th April 2022

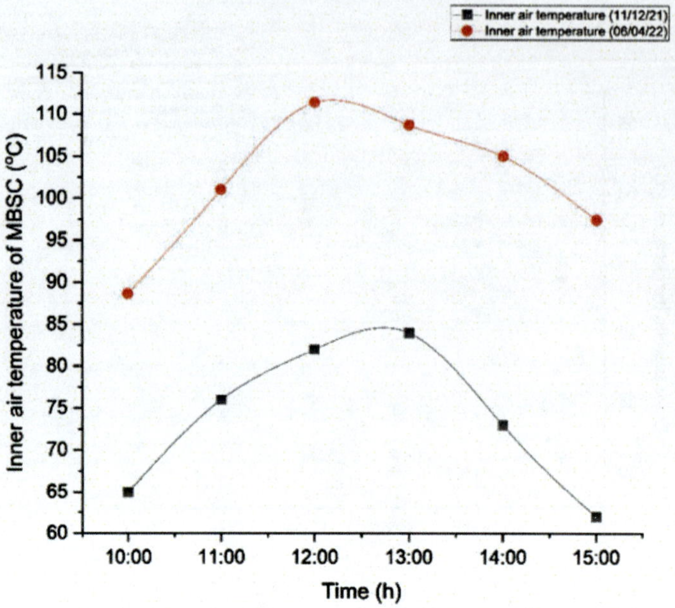

Fig. 11 Inner air temperature of the modified box-type solar cooker for typical days in the month on 11th December 2021 and 6th April 2022

Figure 12 outside the wall's temperature of the modified box-type solar cooker is plotted against the time. The maximum temperature of the outside north wall was 55.9°C and the minimum was 33.8°C. The maximum outside the south wall temperature was 49.1°C, the minimum was 31.4°C, the maximum east wall outside temperature was 48.9°C, the minimum was 32°C, and the maximum west wall outside temperature was 54.9°C, and the minimum was 28.7°C, respectively.

Figure 13 inside the wall's temperature of the modified box-type solar cooker is plotted against the time. The maximum temperature of the inside north wall was 76°C and the minimum was 30.9°C. The maximum inside south wall temperature was 68°C, the minimum was 42°C, the maximum inside east wall temperature was 71°C, and the minimum was 28.9°C, and the maximum inside west wall temperature was 56.5 °C and the minimum was 28.8°C, respectively.

Figure 14 ambient temperature of the modified box-type solar cooker is plotted against the time. The maximum ambient temperature was 46.4°C at 13:00 h, the minimum was 19.3°C at 10:00 h, and wind velocity varied from 0.4 to 2.9 m/s.

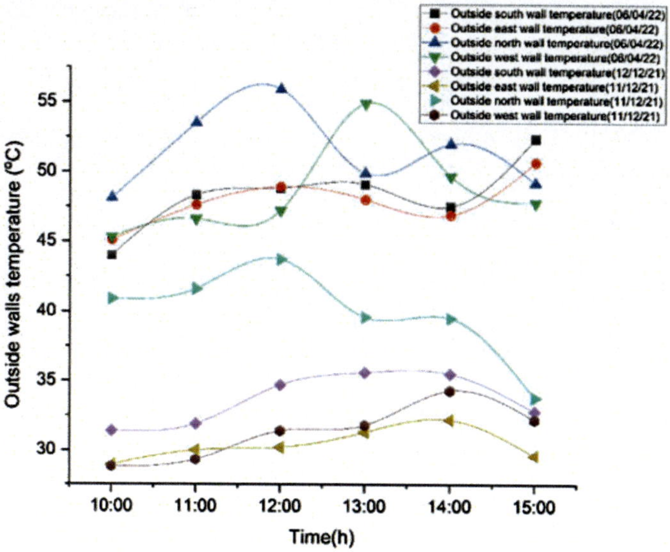

Fig. 12 Outside wall's temperature of the modified box-type solar cooker for a typical day in the month on 6th April 2022

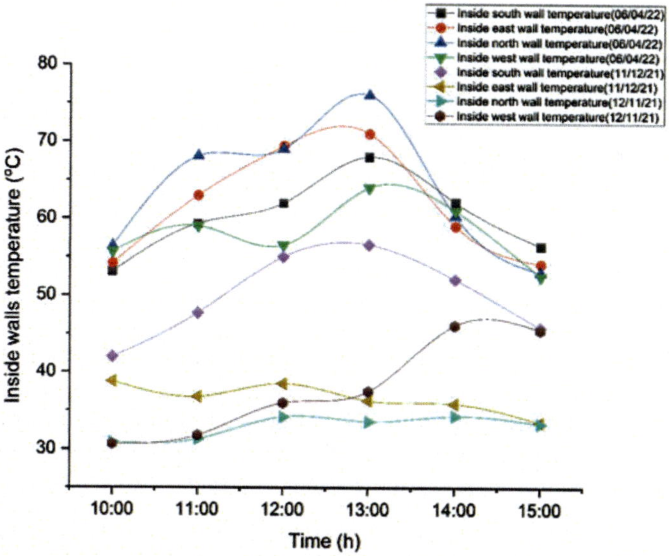

Fig. 13 Inside wall's temperature of the modified box-type solar cooker for a typical day in the month on 11th December 2021 and 6th April 2022

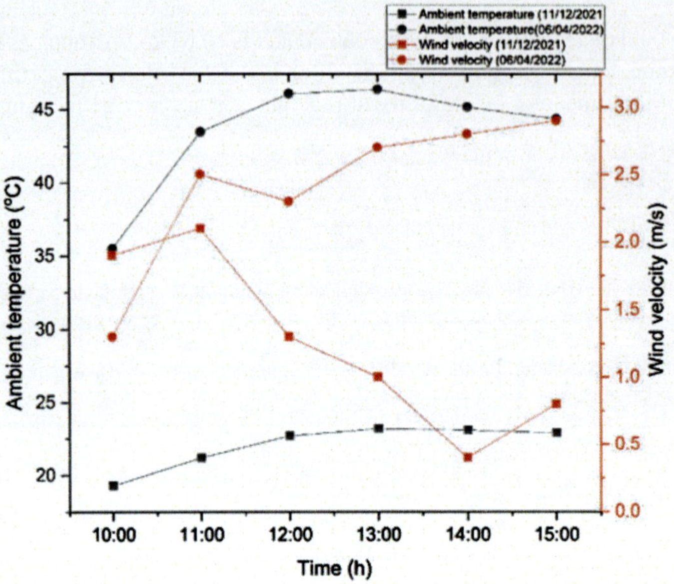

Fig. 14 Ambient temperature of the modified box-type solar cooker for a typical day in the month on 11th December 2021 and 6th April 2022

4 Conclusion

In this study, a modified box-type solar cooker was designed, developed, and its results were evaluated in the conditions of Prayagraj, Uttar Pradesh, India. Based on this research, the following are the findings:

- In this modified box-type solar cooker, the polycarbonate and FRP sheets were used for fabrication. The outer and inner walls, east, west, and south, are transparent to increase the input solar radiation, while the north walls were non-transparent for better heat entrapment. The percentage increase of solar radiation compared to the conventional solar cookers is 36%.
- In April 2022, the peak solar radiation measured was 1038 W/m^2. The average amount of fraction solar radiation of 0.47 falls on the side walls and 0.44 on the polycarbonate cover of the solar cooker.
- In the proposed setup, the maximum temperature of water achieved 105°C, the figure of merits F_1 is 0.11456, $F_2 = 0.26$, and cooking power is 13.63 W and found satisfactory for box-type solar cooker.
- The total fabrication cost of a modified box-type solar cooker was 4130 Rs /53$ (1US = 78.08 current price), and a conventional solar cooker cost was 4750 Rs/60$ on 04/10/2019.

The weight of a modified box-type solar cooker is 10.67 kg. Advantages like small cooking time, longer life, and ease of handling make this setup a competitive option for the current time, and saving of fossil fuel makes it sustainable for future use.

References

1. Junaid M, Syed HJ et al (2018) Status of indoor air pollution (IAP) through particulate matter (PM) emissions and associated health concerns in South Asia. Chemosphere 191:651–663
2. Mahavar S, Verma S et al (2012) Fabrication and experimental findings of a solar rice cooker. In: European association for the development of renewable energies, environment and power quality (EA4EPQ)
3. Mullick SC, Kandpal TC et al (1987) Thermal test procedure for box-type solar cookers. Sol Energy 39(4):353–360
4. Grupp M, Montagn P et al (1991) A novel advanced box-type solar cooker. Solar Energy 47(2):107–113
5. Rathore N, Shukla SK (2009) Experimental investigations and comparison of energy and exergy efficiencies of the box type and Solar Parabolic Cooker. Int J Energy Technol Policy 7(2):201–212
6. Nahar NM, Marshall RH et al (1994) Studies on a hot box solar cooker with transparent insulation materials. Energy Convers Manage 35(9):787–791
7. Buddhi D, Sharma SD et al (1999) Design, development and performance evaluation of a latent heat storage unit for evening cooking in a solar cooker. Energy Convers Manage 41:1497–1508
8. Ali Mohamed BS (2000) Design and testing of Sudanese solar box cooker. Renew Energy 21:573–581
9. Ozturk HH (2004) Comparison of energy and exergy efficiency for solar box and parabolic cookers. J Energy Eng 133(1):53–62
10. Subodh K (2005) Estimation of design parameters for thermal performance evaluation of box-type solar cooker. Renew Energy 30:1117–1126
11. Reddy AR, Rao AN (2007) Prediction and experimental verification of performance of box-type solar cooker—part I. Cooking vessel with central cylindrical cavity. Energy Convers Manage 48(7):2034–2043
12. Kimambo CZM (2007) Development and performance testing of solar cookers. J Energy Southern Africa 18(3):41–51
13. Harmim A, Boukar M (2008) Experimental study of a double exposure solar cooker with finned cooking vessel. Sol Energy 82(4):287–289
14. Namrata S, Prabha D et al (2010) Low-cost solar cooker: promising solution towards reducing indoor air pollution from solid fuel use. Indian J Sci Technol 3:1038–1042
15. Namrata S, Prabha D et al (2011) Experimental studies, energy savings and payback periods of a cylindrical building-material-housing solar cooker. Int J Energy Inf Commun 2:3
16. Folaranmi J (2013) Performance evaluation of a double-glazed box-type solar oven with reflector. Hindawi Publishing Corporation. J Renew Energy Article ID 184352, p 8
17. Akoy OME, Ahmed IAA (2015) Design, construction and performance evaluation of solar cookers. J Agricult Sci Eng 1:75–82
18. Sagade AA, Samdarshi SK et al (2019) Experimental determination of the thermal performance of a solar box cooker with a modified cooking pot. Renew Energy 150:1001–1009
19. Poonia S, Singh AK (2019) Development and performance evaluation of high insulation box type solar cooker. J Agricult Eng 43(1)
20. Khallaf AM, Tawfik MA, El-Sebaii AA, Sagade AA (2022) Mathematical modeling and experimental validation of the thermal performance of a novel design solar cooker. Sol Energy 207:40–50

21. Engoor GG, Shanmugam S, Veerappan AR (2022) Energy and exergy-based study on a box type solar cooker coupled with a Fresnel lens magnifier. Int J Green Energy. https://doi.org/10.1080/15435075.2022.2043868
22. Buddhi D, SD Sharma, Sawhney RL (1999) Performance test of a box-type solar cooker: effect of load on the second Figure of merit. Int J Energy Res 23:827–830
23. Samdarshi SK, Mullick SC (1991) Analytical equation for the top heat loss factor of a flat-plate collector with double glazing. J Solar Energy Eng 113–117. https://doi.org/10.1115/1.2929955
24. Kumar ABC, Maniyeri R, Anish S (2021) Design, fabrication and performance assessment of a solar cooker with an optimum composition of heat storage materials. Environ Sci Pollut Res 28:63629–63637. https://doi.org/10.1007/s11356-020-11024-3
25. Schwarzer K, Silva MEVD (2008) Characterization and design methods of solar cookers. Solar Energy 82:157–163

On Identifying the Suitable Substrate Medium for Induction Heating-Based Metal Wire Additive Manufacturing

Rahul Kumar Choubey, Gourav Kumar Sharma, and Prashant Kumar Jain

Abbreviations

AM	Additive Manufacturing
FDM	Fused Deposition Modeling
EMAM	Extrusion based Metal Additive Manufacturing
FFF	Fused Filament Fabrication
IH	Induction Heating

1 Introduction

Additive manufacturing is a type of manufacturing technology that includes the deposition of multiple layers of material which create a part using digital computer data. The procedure involves printing successful layers of materials that are formed on top of each other [1, 2]. In 1986, Charles Hull firstly developed stereolithography (SLA) technology, which was followed by development such as powder bed fusion, fused deposition modeling (FDM) and inkjet printing. American society for testing and materials (ASTM) floated a standard definition of AM as "processing of joining materials to make object from 3D model data, usually layer upon layer, as opposed to subtractive manufacturing methodologies" [3]. However, until 2015, the AM process classification and standardization were unknown. Therefore, based on the part development approach, the AM process has been categorized into seven types in 2015, i.e., VAT-photopolymerization, Material extrusion, Material jetting, Binder jetting, Direct energy deposition, Powder bed fusion, and Sheet lamination

R. K. Choubey · G. K. Sharma · P. K. Jain (✉)
FFF Lab, Mechanical Engineering Discipline, PDPM Indian Institute of Information Technology, Design and Manufacturing, Jabalpur 482005, India
e-mail: pkjain@iiitdmj.ac.in

A. Kumar et al. (eds.), *Recent Advances in Manufacturing and Thermal Engineering*,
Lecture Notes in Mechanical Engineering, https://doi.org/10.1007/978-981-19-8517-1_18

[4, 5]. Since then, AM has been hailed for its ability to quickly convert a conceptual idea into a unique shape at a lower cost than traditional production methods [6]. The research into diverse applications and material domains has helped AM technology gain popularity in the last decade. According to a report, the share of AM in the global industrial outlook is expected to grow exponentially to USD 41.5 billion by 2027, up from USD 13.9 billion in 2020 (Metal AM report 2020). When compared to traditional methods, AM has the potential to improve materials efficiency, reduce life cycle impacts, and enable greater engineering functionality by requiring less special tooling in part fabrication.

Recently, the extrusion-based metal additive manufacturing (EMAM) technique has been used for the rapid production of metals. The metal filament/wire input form of material is extruded in this method. These technologies rely on increased design freedom and cost effectiveness for producing parts. EMAM is not yet widely explored, although scientific and technical literature on the subject is quickly expanding. However, the literature on fused deposition modeling (FDM) for polymers or polymer metal matrix is extensively available but is still lacking for direct use of bare metal wire as a feedstock material [7]. This study seeks to fill that need of FDM and it has been highlighted as a technology that comes before or enables. EMAM process requires high energy source to melt the metal and this high temperature requirement has been fulfilled by using induction heating as a primary heating [8].

Induction heating (IH) is a non-contact heating process that has been used to heat and melt ferrous and non-ferrous metals. Its inherent advantages include shorter heating times, efficient energy consumption, localized heating, low maintenance, safe working conditions, and adjustable and repeatable heating [8–10]. The IH technique starts when a coil carrying a medium or high frequency alternating current generates an electromagnetic field in its vicinity (according to Ampere's Law). When a metal conductor is placed in this magnetic field, eddy currents are generated. These eddy currents increase internal energy because of the metal's resistance to the flow of the eddy current. As a result of this increase in internal energy, metal temperature and thermal energy dissipation (as per Joule's Law) rise. Furthermore, energy dissipation occurred as a result of hysteresis heating caused by changing the magnetic field [11].

In fused filament fabrication process material deposition takes place through extrusion method. In this method, material is extruded through the nozzle and deposited in the form of layer upon layer onto the substrate, where it congeals into a shape consistent with the digital spatial model. The extrusion-based additive manufacturing process has a challenge that the initial layer of the deposited material does not stick to the substrate, because of which next layer which was going to be deposited on the initial layer will slide away and the desired sample will not be deposited in the defined shape as given in the CAD model. Therefore, in polymer-based FFF process, Kapton/blue tape has been used to stick the extruded polymer filament on the substrate [12]. Like FFF, in extrusion-based metal wire additive manufacturing (EMWAM), substrate material and its adhesion with the printed sample are very important to obtain the desired dimensional sample. Therefore, some of the

researchers have explored initial layer adhesion on the substrate in metal extrusion-based additive manufacturing. Jabbari et al. achieved good adhesion of tin–lead alloy on a galvanized iron substrate [13]. Du et al. had achieved good metallurgical bonding between tin–lead alloy and copper clad substrate [14]. Mireles et al. utilized ABS as a substrate material for tin–bismuth solder alloys [15]. Deposition of solder material on the substrate has been achieved due to low oxidation, but when aluminum is taken as filament material, the deposition of aluminum material on the substrate is a critical task due to its rapid oxidation. Thirumangalath et al. has deposited the liquid aluminum on a high-temperature stainless steel substrate using the drop-on-demand method. During this condition, the force of the fully melted liquid metal drop interacts with the substrate and metallurgical bonding happens [16].

In this manuscript, through extrusion-based metal additive manufacturing, the adhesion of semi-solid aluminum from the nozzle of the extruder onto the substrate has been investigated using three different types of substrates. This investigation will be helpful for depositing aluminum material in open or close air environments.

2 Methodology

2.1 Principle of Extrusion-Based Metal Wire Additive Manufacturing

Metal extrusion using induction heating is a relatively new technology in the additive manufacturing domain. In this process, solid filament is pushed inside the induction heated extruder through a pinch roller, where it gets converted from solid into semi-solid form. Solid filament acts as a piston to push the semi-solid metal from the nozzle of the extruder onto the built platform, and material is deposited in layer-by-layer fashion on the built platform, like the wildly popular polymer-based FDM process. In this process, the extruder mounting has been fixed, and the built platform, along with the substrate, moves over the part in the x and y axes for each layer. After the deposition of a layer, the built platform then lowers for the addition of a new layer on top of the previous layer. Extrusion-based additive manufacturing has long been used for polymers, but it has only lately become popular for metal parts. In Fig. 1, the principle of extrusion-based metal wire additive manufacturing using induction heating source has been shown.

2.2 Feedstock and Substrate Material

Aluminum magnesium alloy (i.e., 5356 grade) has been chosen in a wire form as a feedstock material, which is 1.6 mm in diameter and has been used for sample deposition. This material has been widely used in a variety of fabrication sectors,

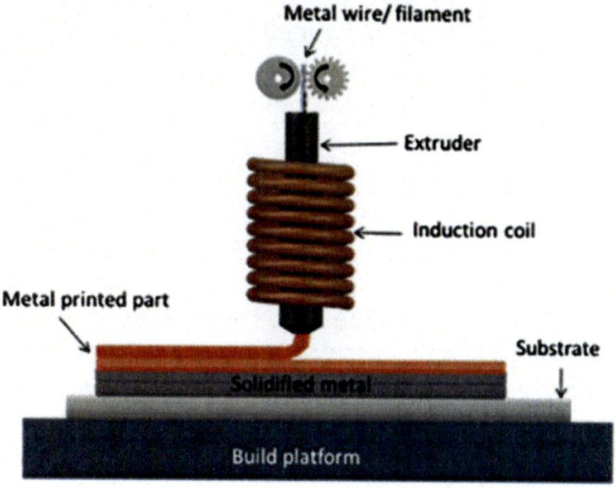

Fig. 1 Principle of extrusion-based metal wire additive manufacturing using induction heating

Table 1 Chemical composition of filament (Al-5356) and substrate (Al-5083) material

Element	Mg	Fe	Mn	Cr	Si	Ti	Cu	Zn	Al
5356	5.59	0.27	0.16	0.13	0.11	0.09	0.004	0.003	Bal
5083	4.9	0.4	1.0	0.25	0.4	0.15	0.1	0.25	Bal

including marine, automotive, and aviation [17–19]. Now, this material has to stick or bond with the base material made of the same grade alloy or aluminum 5083 grade, which is widely used in welding processes [19]. Therefore, in this case, the substrate material has been selected as aluminum 5083 grade with a dimension of $200 \times 200\ mm^2$ and a thickness of 3.0 mm. Table 1 shows the chemical composition of wire/filament and substrate materials.

2.3 *Experimental Setup Description*

Figure 1 depicts the operation of an extrusion-based metal wire additive manufacturing using induction heating. The experimental setup consists of various key components, which according to work are: filament feeding system to feed the wire into the extruder; extruder head for melting the metal and deposition, induction heater with coil arrangement for providing heat to the extruder for filament melting, positioning unit that translates the built platform with substrate to deposit the track or layer for directional contours. Another setup controlling component is the Proportional Integral Derivative (PID) based temperature controller for maintaining the extruder temperature in the semi-solid range of filament material and a user control interface to input the contour coordinates for movement of the positioning system, as shown in Fig. 2. An induction heater with a 1.0 kW capacity and a customized helical coil configuration has been employed for heating the cast iron material for melting

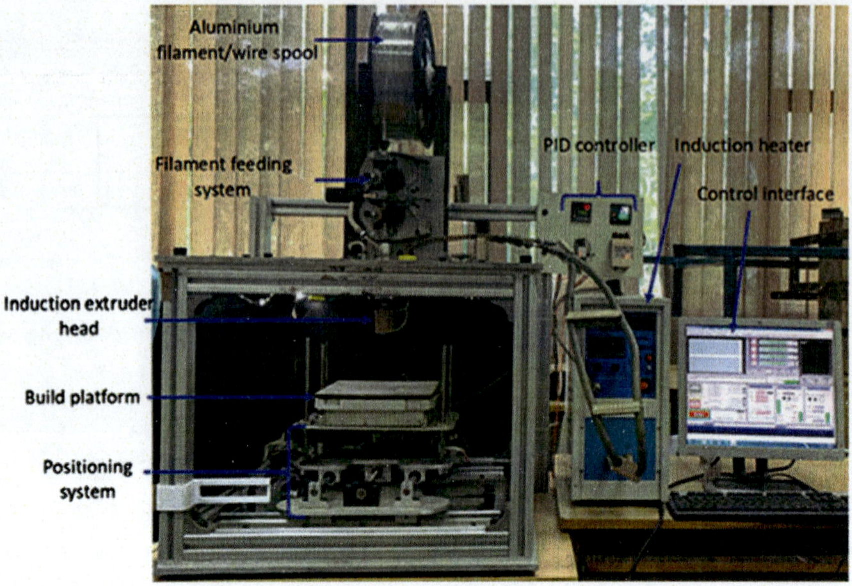

Fig. 2 Extrusion-based metal wire additive manufacturing using induction heating

the aluminum filament. The cast iron extruder was chosen because of its strong electrical resistivity, high magnetic permeability, and high melting temperature properties. The filament must be extruded in a semi-solid state for an extrusion-based process. The semi-solid temperature range of aluminum 5356 is 575–635 °C. Therefore, the extruder temperature has been maintained within the range with the help of a controller. The customized coil has nine coil turns with a tube diameter of 5.0 mm with a coil radius and length of 15 and 50 mm, respectively. The coil arrangement around the extruder dimension is 12 mm in diameter and 55 mm in length. These coil parameter selections have nine number of turns, which provide an efficient electromagnetic field for quick heating and providing temperature range in which metal can be melted inside the extruder [11]. The envelope size of an extrusion-based metal wire additive manufacturing system with an induction heating setup is 600 × 600 × 450 mm^3, with a built volume of 200 × 200 × 120 mm^3.

2.4 Technological Adhesion Gap

In additive manufacturing, a product's permanent sticking to the bed during manufacturing and easy peeling off afterward are both required for the product's desired form, one of the major technological challenges in achieving this condition [20]. The adhesive force holds the element generated on the bed to its surface (Fig. 3a). It is commonly observed in additive manufacturing that the internal adhesion forces of

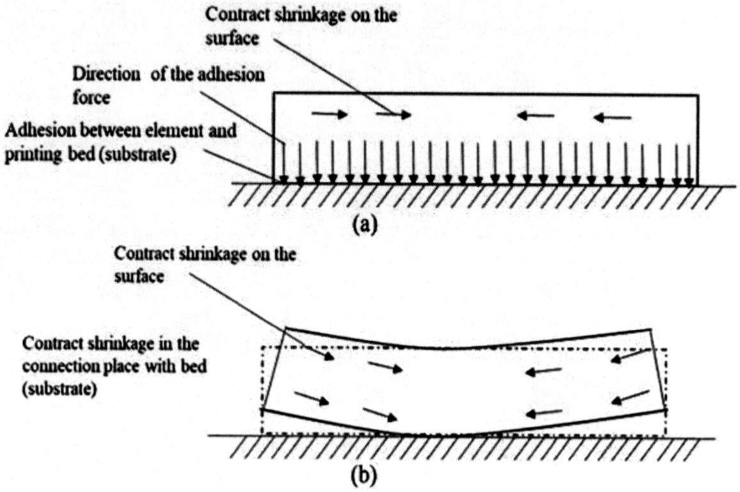

Fig. 3 Printed sample on substrate: **a** Adhesion point between the printed element and the substrate **b** Shrinkage, oxide development, and insufficient adhesive force cause the edges of the element to peel away

the elements and the bed are usually smaller than the cohesion strength, because of which the edge of the generated part breaks off due to processing shrinkage (Fig. 3b). High adhesive force is required to prevent processing shrinkage as well as the element and the 3D printer bed from moving against each other [21]. Due to the oxide layer on the surface of the printed sample, the adhesive force between the printed sample and the substrate gets reduced and creates insufficient adhesion force, which results in detachment or deformation of the manufactured element from the printing bed.

2.5 *Methods for Increasing Adhesion Between Printing Substrate and Produced Sample*

In order to achieve successful adhesion between the deposited sample and the extrusion-based wire additive manufacturing (EMWAM) substrate, three different substrates are employed and examined through experimentation. Figure 4 depicts the three different substrate surfaces that were used in EMWAM.

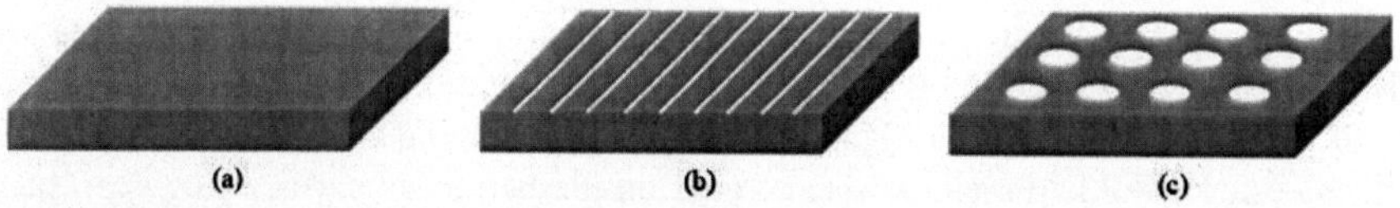

Fig. 4 Priting bed substrate surface: **a** Smooth surface, **b** Rough surface, **c** Performated surface

The deposition of aluminum material on the three different substrates of the printing bed has been investigated using the same operational process parameters (extruder temperature, heating bed temperature, extruder nozzle diameter, printing speed, stand-off distance, and filament feed rate). Experiments were carried out to stick the initial layer and uniformly deposit layer upon layer of extruded semi-solid material onto a substrate for an aluminum filament with a diameter of 1.6 mm. The following deposition parameters are taken for this experiment: extruder temperature = 600 °C, bed temperature = 350 °C, extruder nozzle diameter = 1.5 mm, printing speed = 120 mm/min, filament/wire feed rate = 120 mm/min, and standoff distance = 1.8 mm.

To create a smooth surface, a 5083-grade aluminum substrate is cleaned with 100-grit sandpaper to remove the oxide layer, followed by ethanol to remove oil and debris. In the first experiment, deposition of a multiple layer square-shaped contour sample of 45 × 45 mm^2 on the smooth surface substrate has been done as shown in Fig. 5. As seen in Fig. 5, there is very little adhesion between the printed sample and the substrate due to oxide formation on the surface of the printed sample. The initial layer does not attach to the substrate, separating the manufactured sample from the printing bed. This sample separation problem on the smooth substrate has been addressed with the modification of the substrate. In the second experiment, a smooth surface has been converted into a rough surface by scratching it at regular intervals, creating a pattern of scratches, and sample adhesion to the substrate has been checked, as shown in Fig. 6. Multiple layers have been deposited on the rough surface. It has been observed that the first semi-solid layer is exposed to a rough surface with scratches, which provides a larger surface area and provides partial adhesion between the substrate and the printed sample.

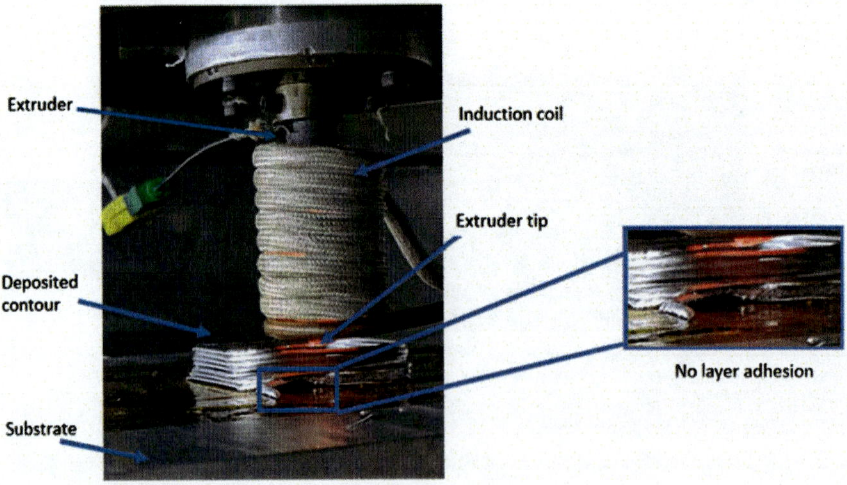

Fig. 5 Printing of multiple layers square contour sample on smooth substrate

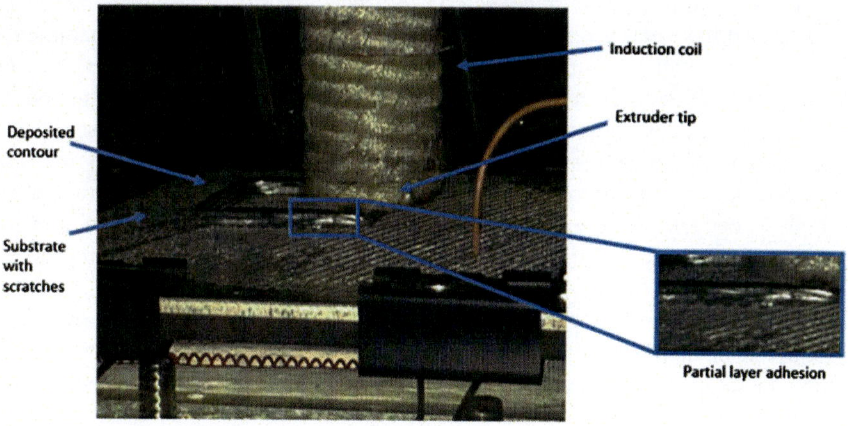

Fig. 6 Printing of multiple layers square contour sample on rough surface

Furthermore, in the final experiment, the substrate has been modified into a perforated pattern by creating a pattern hole of 1.8 mm in diameter with equal spacing for sufficient adhesion of the initial layer on it. The size of the holes was governed by the nozzle diameter and the thermal expansion of the deposited material. A multiple layer has been deposited on the perforated substrate. It is observed that the initial layer has been shown to have sufficient adhesion to the substrate, as shown in Fig. 7. Because of the accessible holes in the perforated substrate, a portion of an initial semi-solid layer gets within the pores of the perforated substrate and provides sufficient adhesion between each other. The perforated substrate adheres better to the initially deposited layer in comparison to smooth and rough substrates.

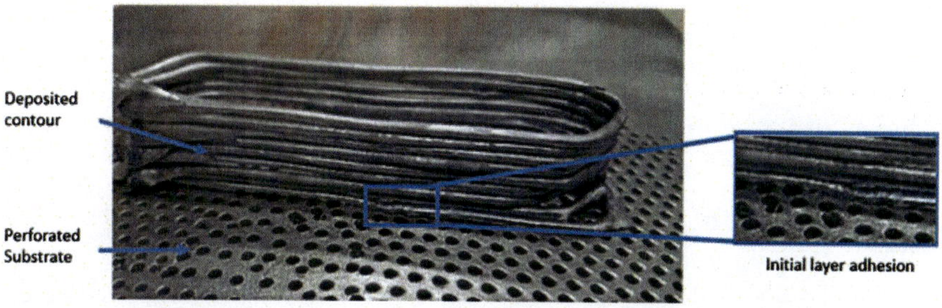

Fig. 7 Printing of multiple layers square contour sample on perforated substrate

3 Results and Discussion

In additive manufacturing, sticking the initial layer ensures the correct dimension of the printed sample. Through experimentation, the difficulties of adhering an initial semi-solid layer of Al 5356 to the substrate using extrusion-based metal wire additive manufacturing with induction heating have been checked. Three different substrates have been utilized in the experiment. In the first experiment, when semi-solid aluminum material is deposited on a smooth surface, the initially deposited layer does not cling to the substrate because an oxide layer forms on the printed sample. Due to this, the initially deposited layer does not adhere to the substrate, and a distorted sample that differs from the CAD model is produced. In the next experiment, the deposited initial layer adhesion has been checked on a rough-surface substrate with regular scratches. It has been observed that the initially deposited layer on the substrate gets removed from the substrate after the deposition of multiple layers on top of the initial layer. Partial adhesion has been observed between the rough substrate and the initial layer. Furthermore, in the final experiment, multiple layers have been stacking on the perforated pattern substrate, and it has been observed that the initial layer and the substrate show sufficient adhesion between each other. Because the perforated substrate has a cavity on its surface for the deposition of the initial printing layer, and adhesion has developed between them due to the rapid oxidation of the printing aluminium material in ambient conditions.

4 Conclusion and Future Scope

In the experimentation work, aluminum has been deposited on various substrates using induction heating as an energy source. The difficulties of adhering an initial semi-solid layer of Al 5356, extruded through a cast iron extruder on the substrate have been identified through experimentation. Three different substrates made of Al 5083 grade material have been utilized in three different experiments. The deposition parameters have been considered for the experiment were: 350 °C bed temperature, 600 °C extruder temperature, 1.5 mm extruder nozzle diameter, 120 mm/min filament feed rate, 120 mm/min substrate speed, and 1.8 mm standoff distance. The following point has been observed in the experiments:

- The smooth surface substrate suffers from oxidation problems, resulting in the deposited material (Al 5356) remaining isolated from the substrate due to insufficient adhesion.
- A scratched rough-surfaced substrate shows partial adhesion with the deposited material.
- The perforated substrate shows complete adhesion with the deposited material because the portion of deposited material slides inside the holes and provides

additional grip to printed sample. For a filament size of 1.6 mm diameter, perforated substrate with 1.8 mm hole size would be suitable for printing aluminum material in open or close air environment conditions.

As a result of the experiments, it can be inferred that the perforated substrate would be suitable for printing aluminum in extrusion-based metal additive manufacturing methods involving induction heating.

Future Scope Metal additive manufacturing technologies that rely on extrusion could be benefitted from the use of a perforated substrate. Perforated substrate can also be explored while printing other highly reactive material like copper, titanium, and silver through extrusion-based additive manufacturing using induction heating.

References

1. Taufik M, Jain PK (2020) Part surface quality improvement studies in fused deposition modelling process: a review. Aust J Mech Eng 00:1–25. https://doi.org/10.1080/14484846.2020.1723342
2. Francis V, Jain PK (2016) Experimental investigations on fused deposition modelling of polymer-layered silicate nanocomposite. Virtual Phys Prototyp 11:109–121. https://doi.org/10.1080/17452759.2016.1172431
3. ASTM F2792-12a, Standard Terminology for Additive Manufacturing Technologies, vol. 10.04
4. Gawali SK, Kumar N, Jain PK (2022) Additive manufacturing of large size parts through retrofitment of three-axes CNC machining centre. In: Proceedings of the international conference on industrial and manufacturing systems (CIMS-2020). pp 421–437
5. Gao W, Zhang Y, Ramanujan D, Ramani K, Chen Y, Williams CB, Wang CCL, Shin YC, Zhang S, Zavattieri PD (2015) The status, challenges, and future of additive manufacturing in engineering. CAD Comput Aided Des 69:65–89. https://doi.org/10.1016/j.cad.2015.04.001
6. Thompson SM, Bian L, Shamsaei N, Yadollahi A (2015) An overview of direct laser deposition for additive manufacturing; Part I: Transport phenomena, modeling and diagnostics. Addit Manuf 8:36–62. https://doi.org/10.1016/j.addma.2015.07.001
7. Rane K, Strano M (2019) A comprehensive review of extrusion-based additive manufacturing processes for rapid production of metallic and ceramic parts. Adv Manuf 7:155–173. https://doi.org/10.1007/s40436-019-00253-6
8. Sharma GK, Pant P, Jain PK, Kankar PK, Tandon P (2021) On the suitability of induction heating system for metal additive manufacturing. Proc Inst Mech Eng Part B J Eng Manuf 235:219–229. https://doi.org/10.1177/0954405420937854
9. Hascoët JY, Parrot J, Mognol P, Willmann E (2018) Induction heating in a wire additive manufacturing approach. Weld World 62:249–257. https://doi.org/10.1007/s40194-017-0533-y
10. Lucia O, Maussion P, Dede EJ, Burdio JM (2014) Induction heating technology and its applications: past developments, current technology, and future challenges. IEEE Trans Ind Electron 61:2509–2520. https://doi.org/10.1109/TIE.2013.2281162
11. Sharma GK, Pant P, Jain PK, Kankar PK, Tandon P (2021) Numerical and experimental analysis of heat transfer in inductive conduction based wire metal deposition process. Proc Inst Mech Eng Part C J Mech Eng Sci. https://doi.org/10.1177/09544062211028267
12. Gawali SK, Kumar N, Jain PK (2020) Investigations on the development of heated build platform for additive manufacturing of large-size parts. In: Manufacturing engineering. pp 1–17. Springer

13. Jabbari A, Abrinia K (2018) Developing thixo-extrusion process for additive manufacturing of metals in semi-solid state. J Manuf Process 35:664–671. https://doi.org/10.1016/j.jmapro.2018.08.031
14. Fang X, Du J, Wei Z, He P, Bai H, Wang X, Lu B (2017) An investigation on effects of process parameters in fused-coating based metal additive manufacturing. J Manuf Process 28:383–389. https://doi.org/10.1016/j.jmapro.2017.01.008
15. Mireles J, Espalin D, Roberson D, Zinniel B, Medina F, Wicker R (2012) Fused deposition modeling of metals. 23rd Annu Int Solid Free Fabr Symp—An Addit Manuf Conf SFF 2012:836–845
16. Thirumangalath SC, Vader S, Vader Z (2014) Liquid metal 3D printing: a magnetohydrodynamic approach. Transplantation 97:1200
17. Mills RJ, Lattimer BY, Case SW, Mouritz AP (2018) The influence of sensitization and corrosion on creep of 5083–H116. Corros Sci 143:1–9. https://doi.org/10.1016/J.CORSCI.2018.07.036
18. Holroyd NJH, Burnett TL, Sei M, Lewandowski JJ (2017) Materials science & engineering a improved understanding of environment-induced cracking (EIC) of sensitized 5XXX series aluminium alloys. 682:613–621. https://doi.org/10.1016/j.msea.2016.11.088
19. Yi AG, Poplawsky JD, David A, Wang Z, Free ML (2017) Characterizing and modeling the precipitation of Mg-rich Phases in Al 5xxx alloys aged at low temperatures. J Mater Sci Technol. https://doi.org/10.1016/j.jmst.2017.02.001
20. Manufacturing EA, Spoerk M, Id JG, Lichal C, Cajner H (2018) Optimisation of the adhesion of polypropylene-based materials during. https://doi.org/10.3390/polym10050490
21. Płaczek D (2019) Adhesion between the bed and component manufactured in FDM technology using selected types of intermediary materials. 2:1–9

Kinetic Analysis of *Phoenix Dactylifera* and *Phyllanthus Emblica* Seeds Through Thermogravimetric Analyser: Determination of Activation Energy

Indra Mohan, Satya Prakash Pandey, and Sachin Kumar

1 Introduction Including Literature Review

A rapid decrement in the reserves of fossil fuel along with the negative effects on the environment has put mankind under enormous pressure for the production of energy through sources which are renewable in nature. Considering all the available renewable sources, biomass resources could certainly yield a permanent solution to the incessant usage of traditional fuels. Sources of biomass are available in various categories as oilseeds, forestry and agricultural biomass. Also, of all the available biomass oilseeds, *Phoenix Dactylifera* (*PD*) and *Phyllanthus Emblica* (*PE*) also known as Dry dates and Amla respectively, are available widely in the Indian subcontinent available for almost all parts of the year as it is also widely used for consumption purpose. Additionally, the bio-oil yield through plants can also solve multiple problems at once, such as the production of renewable fuel, and higher production through agriculture as well as an income source for the rural sector. Renewable fuels and other valuable chemicals can be obtained through biomass with the help of different thermochemical and biochemical conversion processes. Some of the thermochemical methods for conversion include gasification, liquefaction, and pyrolysis, out of which, the most practical and extensively used method is pyrolysis. Analysis of the kinetic parameters with the help of a thermogravimetric analyser is important to assess the energy potential of any organic material. The kinetic triplets are defined through different model-fitting and model-free approaches obtained by the thermogravimetric analyser [1].

The activation energy at various conversion rates is used to assess the model-free approaches [2] however, if a single value of activation energy is required, then model-fitting methods are used. It is however also tough to determine the mechanism of reaction with an increase in the change in conversions, thus providing non-real

I. Mohan · S. P. Pandey · S. Kumar (✉)
Central University of Jharkhand, Ranchi, India
e-mail: sachin.kumar.01@cuj.ac.in

A. Kumar et al. (eds.), *Recent Advances in Manufacturing and Thermal Engineering*, Lecture Notes in Mechanical Engineering, https://doi.org/10.1007/978-981-19-8517-1_19

values of kinetic parameters [3]. There have only been a few studies regarding the study of the kinetics of various oilseeds in the past. The kinetic, as well as the thermodynamic parameters of acai seeds, were assessed through two isoconversional models, Kissinger–Akahira–Sunose (KAS) and Flynn–Wall–Ozawa (FWO). A high heating value was found in the acai seeds at 121.1 M/kJ which proved that the acai seeds could be used as a renewable feedstock to produce alternate fuel [4]. Different Isoconversional models like KAS, Friedman, Starink, FWO, Avrami and DAE (Distributed Activation Energy) were used to find out the kinetic triplet of a non-edible oilseed named *Mesua ferrea L.* The result proved the suitability of the selected seed as an alternative biofuel resource [5]. In a recent study of Argemone Mexicana seeds, different isoconversional kinetic models namely the STR, KAS, VYZ, FWO and FRM were applied. The average values of activation energy found through the FWO, KAS, FRM, STR and VYZ kinetic models respectively were 185.08, 174, 212.86, 175.11 and 174.36 kJ/mol, proving its worth to obtain renewable biofuel [6].

From the past successful attempts, it could certainly be established that if there are multiple biomass feedstocks available for producing biofuels, especially the non-edible wastes, then it could save the waste seeds from getting dumped besides contributing to the generation of biofuels at a large scale. It would also be helpful in the expansion of small and medium sectors, particularly the rural economy. Thus, to contribute further to this aspect, two non-edible seeds were selected to study for different properties so as to compare the results with past works for their suitability to generate biofuel. Considering all the available literature, the present work is novel in nature as the kinetic modelling of the selected seeds, *Phoenix Dactylifera* (*PD*) and *Phyllanthus Emblica* (*PE*), have never been carried out or studied in the past.

2 Methodology

2.1 Feedstock (Raw Seeds)

The PD and PE seeds were accumulated after purchasing the fruit. The separation of seeds from the fruit was done manually after which the fruits were used for consumption purposes. The collected seeds were then dried in the sunlight for 24 h. After this, the sun-dried seeds were placed inside the oven for 2 h at 60 °C temperature to extract any moisture left. The dried seeds were then crushed to a fine powder to be used for thermogravimetric analysis. Proximate analyses of both seeds were done as per ASTM test standards.

2.2 TGA (Thermogravimetric Analysis) of Raw Seeds

The TGA of Phoenix Dactylifera and Phyllanthus Emblica seeds was carried out with the help of the Thermogravimetric analyser-DTA-DSC thermal analyser of the PerkinElmer (lab system STA 6000). Around 10 mg of powdered seeds were placed in an aluminium oxide crucible. The crushed seeds were then heated to 900 °C (maximum temperature) with three varying heating rates at 10, 20 and 30 °C/min. Inert gas (nitrogen) was utilized at 40 ml/min flow rate. The OriginPro programme was applied to extract TGA parameters for both seeds through the thermogravimetric data.

2.3 Kinetic Theory

The analysis of the kinetics of pyrolysis is prominent for a deep investigation of the mechanism of the process. There is a complex arrangement in the biomass feedstock which displays hindrances while assessing the mechanism of the reaction. Thus, an insightful understanding of the mechanism of the reaction is important to optimize an efficient design of the reactor for large-scale commercial usage. The pyrolysis mechanism of *Phoenix Dactylifera* and *Phyllanthus Emblica* seeds can be generally described as:

$$Phoenix\ Dactylifera\ \text{seeds} \rightarrow \text{Volatile matter} + \text{residue char}$$
$$Phyllanthus\ Emblica\ \text{seeds} \rightarrow \text{Volatile matter} + \text{residue char}$$

The kinetic process rate can be expressed as the following equation:

$$\frac{\mathrm{d}\alpha}{\mathrm{d}t} = kf(\alpha)\, or\, \frac{\mathrm{d}\alpha}{\mathrm{d}t} = \frac{A}{\beta}\exp\left(-\frac{E}{\mathrm{RT}}\right)f(\alpha) \tag{1}$$

where dα/dt being the conversion rate of the reaction, k-rate constant, $f(\alpha)$-model of the reaction, A-Pre-exponential factor (s^{-1}), E-energy of activation (KJ/mol), β-rate of heating (°C/min) and R has a value of 8.314 J mol^{-1} K^{-1} being the universal gas constant. To calculate the activation energy, the kinetic parameters have been obtained with the help of KAS method.

The following equation shows the conversion during the pyrolysis process

$$\alpha = (W_i - W)/\left(W_i - W_f\right) \tag{2}$$

where, W_i, W and W_f are the initial, total and final weight, respectively.

The equation of rate could be expressed as

$$\frac{\mathrm{d}\alpha}{\mathrm{d}t} = k(T)f(a) \tag{3}$$

where, $k(T)$ is the rate constant and $f(a)$ is the reaction model.

Arrhenius's equation is expressed as

$$k(T) = A\exp\left(-\frac{E}{\mathrm{RT}}\right) \tag{4}$$

where, T is the absolute temperature.

Now, equating the Eqs. (3) and (4), we get

$$\frac{\mathrm{d}\alpha}{\mathrm{d}t} = Ae^{\left(-\frac{E}{\mathrm{RT}}\right)}f(\alpha) \tag{5}$$

β can also be expressed as a temperature function as the following equation

$$\beta = \frac{\mathrm{d}T}{\mathrm{d}t} = \frac{\mathrm{d}T}{\mathrm{d}\alpha}\cdot\frac{\mathrm{d}\alpha}{\mathrm{d}t} \tag{6}$$

Now, a combination of Eqs. (5) and (6) provides an equation, expressed as

$$g(a) = \int_0^{\alpha}\frac{d\alpha}{f(\alpha)} = \int_0^{T}\frac{A}{\beta}e^{\left(-\frac{E}{\mathrm{RT}}\right)}\mathrm{d}T$$
$$= \frac{AE}{\beta R}\int_x^{\infty}u^{-2}e^{-u}\mathrm{d}u = \frac{AE}{\beta R}p(x) \tag{7}$$

where, $x = E/\mathrm{RT}$, and $g(a)$ is the integration of different models of reaction, while the RHS of the above equation defines the temperature integration as having no exact integral solution.

However, the numerical approximations can be applied for solving Eq. (7).

2.4 *Kinetic Modelling of Pyrolysis of* Phoenix Dactylifera *and* Phyllanthus Emblica *Seeds*

The activation energy of *Phoenix Dactylifera* and *Phyllanthus Emblica* seeds was calculated through a model-free, KAS (Kissinger–Akahira–Sunose) method.

The equations used in the KAS method can be written as:

$$\frac{\mathrm{d}T}{\beta} = \mathrm{d}t \tag{8}$$

$$\frac{d\alpha}{f(\alpha)} = \frac{A}{\beta}\exp\left(-\frac{E}{RT}\right)dT \quad (9)$$

$$\int_0^{\alpha}\frac{d\alpha}{f(\alpha)} = \int_0^{T}\frac{A}{\beta}e^{\left(\frac{-E}{RT}\right)}dT \quad (10)$$

where β is the TGA rate of heating.

$$g(a) = \int_0^{\alpha}\frac{d\alpha}{f(\alpha)} = \int_0^{T}\frac{A}{\beta}e^{\left(\frac{-E}{RT}\right)}dT \quad (11)$$

$$g(a) = \frac{AE}{\beta R}p(x) \quad (12)$$

where $x = E/RT$.

KAS model also applies an approximation which can be expressed as

$$p(x) = x^2e^{-x} \quad (13)$$

$$\ln\frac{\beta}{T^2} = \ln\left[\frac{AE}{Rg(a)}\right] - \frac{E}{RT} \quad (14)$$

In the above equation, $ln\frac{\beta}{T^2}$ and 1/T are considered as the dependent and independent variables. 'E' (activation energy) was calculated through the linear equation slope.

3 Results and Discussion

3.1 *Raw Seeds Characterization*

Proximate analyses for *Phoenix Dactylifera* and *Phyllanthus Emblica* seeds are shown in Table 1. There is a major influence of the content of moisture of biomass which impacts the behavior of the pyrolysis process besides the quality and physical properties of the product obtained through pyrolysis. At higher temperatures, dry biomass feed might produce bio-oil with high viscosity [7]. If there is a high content of moisture, thermal pre-treatment must be performed for the removal of excess moisture which would result in an improved quality of the obtained product along with a better thermal efficiency during the process [8]. The content of ash provides an evaluation of non-combustible compounds and non-volatile matter in the biomass [9]. The amount of ash in PD and PE seeds was 1.08% and 2.78% respectively which

Table 1 Proximate analyses of *PD* and *PE* seeds

Characteristics (wt%)	*Phoenix Dactylifera* seeds	*Phyllanthus Emblica* seeds
Amount of moisture	1.08	2.78
Amount of ash	5.28	4.83
Amount of volatile matter	84.47	51.46
Amount of fixed carbon	9.17	10.93

was found to be lower than that of castor seeds [10], cherry seeds [11], neem seeds [12], sal seeds [13] and karanja seeds [14]. Additionally, an important parameter, the content of volatile matter in the PD and PE seeds was calculated to be 84.47 wt% and 81.46 wt% respectively.

Likewise, PD and PE seeds have a fixed carbon content of 9.17 and 10.93 wt% respectively. There is an increment in the content of fixed carbon of the biomass when the temperature of pyrolysis is increased which results in a better heating value of the biomass [15]. Table 1 depicts the proximate analyses of PD and PE seeds.

3.2 *Thermogravimetric-DTG Analyses at Selected Rates of Heating*

The profile for thermal degradation of the PD and PE seeds has been acquired with the help of a thermogravimetric analyser. Figures 1 and 2 show that during the thermal conversion of both seeds, there is a weight loss (%) when there is an increment in the temperature at different rates of heating fixed at 10, 20 and 30 °C/min. Hemicellulose generally gets decomposed at a 220–315 °C temperature range while that for cellulose is 315–400 °C. Lignin decomposition occurs at around 900 °C [4]. The TGA profiles of the three primary components when convoluted provide the TGA patterns for lignocellulosic biomass, while the degree of evolution is determined by the content of individual components [16]. There are three stages of thermal degradation of PD and PE seed which are stage I, stage II and stage III. The initial process of loss of biomass weight in pyrolysis is the dehydration or drying stage which occurs till the temperature has reached 160 °C from the ambient [17].

There was a rapid decrement in the content of moisture in the PD and PE seeds in the 50–240 °C temperature range because a huge amount of moisture evaporated as the temperature increased. In both seeds, stage I represents the dehydration stage which takes place in a 25–240 °C temperature range which is additional to the evolution of moisture and extractive compounds. If there is a presence of water in the biomass feedstock, it results in a reduction of the flame temperature hence a decrement in the heating value which again provides a longer ignition delay with a decrement in the rate of combustion [18].

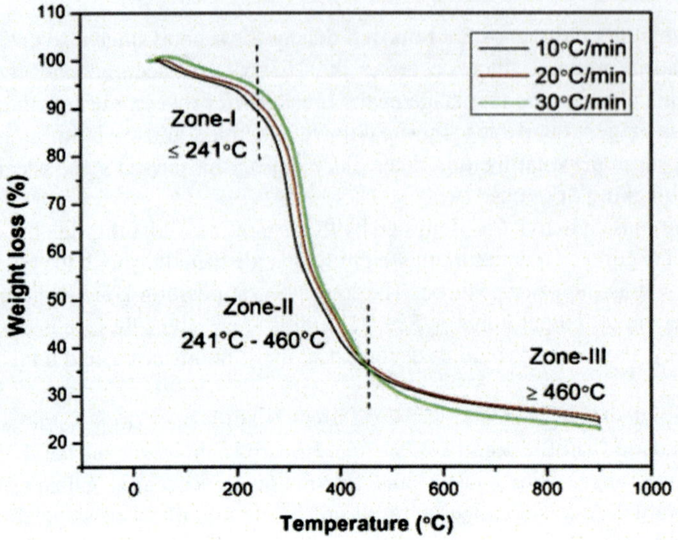

Fig. 1 TGA plot of PD seeds

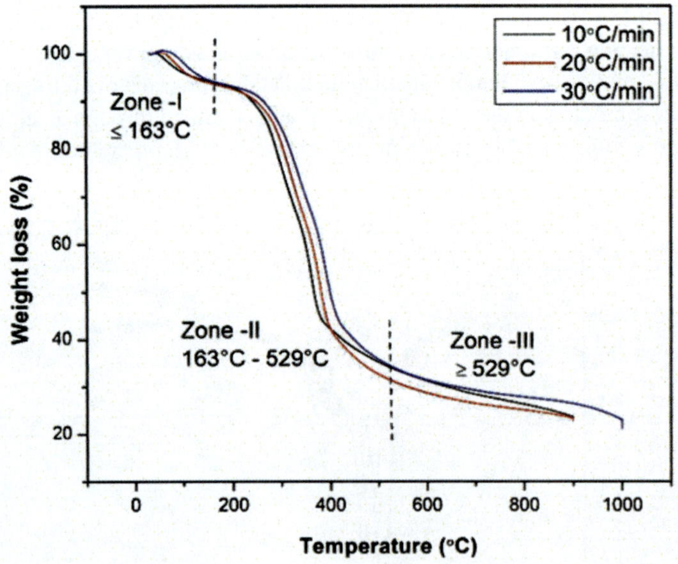

Fig. 2 TGA plot for PE seeds

The volatile compounds are removed during the second stage also called the de-volatilization process which occurs in the 163–529 °C accompanied temperature range. This phase is the first stage of the breakdown process, during which volatile chemicals are generated due to the breakdown of hemicelluloses, lignin and cellulose. The maximum de-volatilization takes place during the second stage because of the thermal breaking of bonds [19].

The biomass reactivity is identified by the rate of mass loss through the derivative DTG or TG curve. The maximum weight loss is identified through the peak temperature shown as the peak position. The lower decomposition zone is the third stage which occurs at temperatures $\leq$ 529 °C. In this stage, the cellulose is decomposed completely, the heavier volatiles decompose, the C bonds break and finally the char forms.

As depicted in Figs. 3 and 4, the different heating rates do not primarily affect the degradation profile trend of *PD* and *PE* seeds, however, at the 175–500 °C temperature range, there is just a small fetch in the upper region, called as the active pyrolysis zone. At lower temperatures during the de-volatilization stage, the shoulder peaks were caused because of the pyrolysis of hemicellulose. The removal of moisture and lighter volatile matter indicates the primary peak in the DTG curve till 175 °C. In the de-volatilization zone, the breakdown of cellulose and hemicellulose forms the second peak in the curve. DTG peaks moved to a higher temperature zone at increased heating rates, but this had no effect on biomass conversion.

Figures 3 and 4 also demonstrate that the DTG thermograph of both seeds at the 3 selected heating rates in the nitrogen atmosphere is gained through an endothermic process providing unstable curves at less than 175 °C temperature. Additionally,

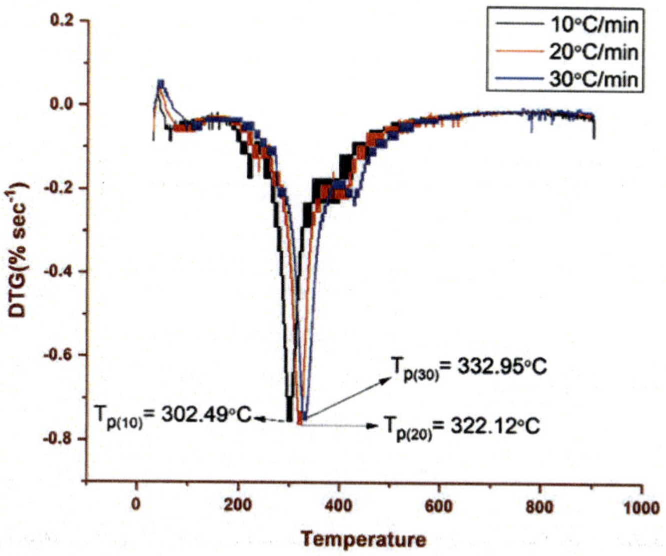

Fig. 3 DTG plot of PD seeds

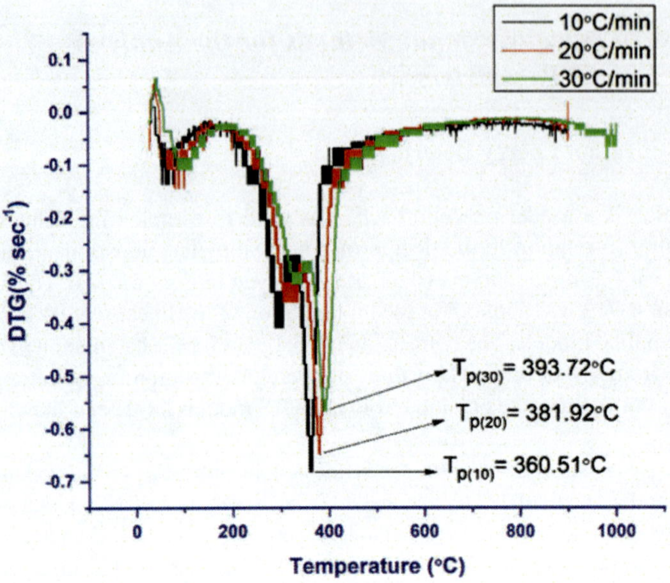

Fig. 4 DTG plot of PE seeds

the two formed peaks are because of the breakdown of celluloses and hemicelluloses in the second zone, the decomposition of lignin takes place at a slower rate at temperatures greater than 575 °C.

The highest temperatures where the optimum loss of weight takes place are respectively at 302.49 °C, 322.12 °C and 332.95 °C with heating rates of 10, 20 and 30 °C/min for *PD* seeds while for *PE* seeds, the peak temperatures observed were 360.51 °C, 381.92 °C and 393.72 °C at the same heating rates respectively. DTG peaks are discovered to be substantially nearer together. This could be because of the catalytic behavior of the mineral elements present in biomass [20]. Because cellulose breakdown is shifted towards reduced temperatures by inorganic salts, the DTG peaks of cellulose and hemicelluloses are extensively overlapping [21]. As indicated in Figs. 3 and 4, the extractives are represented as a downward slope in degradations. The first small peak in the DTG plot demonstrates the moisture removal of the seeds, however, the later peaks in the 230–900 °C temperature range depict the pyrolysis of seeds.

The primary decomposition occurs around a temperature of 250 °C. At roughly 270 °C, a curve at lower temperature signalizes multiple heating rates of *PD* and *PE* seeds, showing hemicellulose breakdown. However, at higher temperatures around 490 °C for all heating rates, shoulder characterizes for both samples due to the devolatilization of cellulose and de-polymerization reaction [22]. The DTG curve was nearly stationary above 700 °C, indicating that the pyrolysis reaction with the residual carbonaceous solid components was complete.

3.3 *Use of Isoconversional Methods for the Analysis of Kinetic Parameters*

3.3.1 Activation Energy Determination

KAS, which is a model-free approach, was used to estimate the value of energy of activation in relation to the degree of conversion. The above-mentioned model produced various linear curves, which are displayed in Figs. 5 and 6. The coefficient of regression (R^2) was applied for the investigation of the linear regressions obtained from the applied models. The conversion range opted for the calculation of activation energy is from 0.1 to 0.50. The values of average activation energy along with an enhancing conversion rate obtained from the KAS models have been shown in Figs. 5 and 6.

During pyrolysis of *PD* and *PE* seeds, the activation energy calculated through the KAS method was calculated to be constant in between the conversion rate of 0.3–0.4. Alterations in the values of energy of activation were observed when the conversion was changed owing to the cellulose, hemicellulose and lignin decomposition taking place at different rates of heating during the three different zones of pyrolysis. The increment in the energy of activation along with an increment in conversion is because of the occurrence of complicated processes at higher temperatures (endothermic and decarboxylation of carboxylic acid salts) [23].

The average values of energy of activation for *PD* and *PE* seeds as found through the KAS model are 111.935 and 157.272 kJ/mol. The error in activation energy

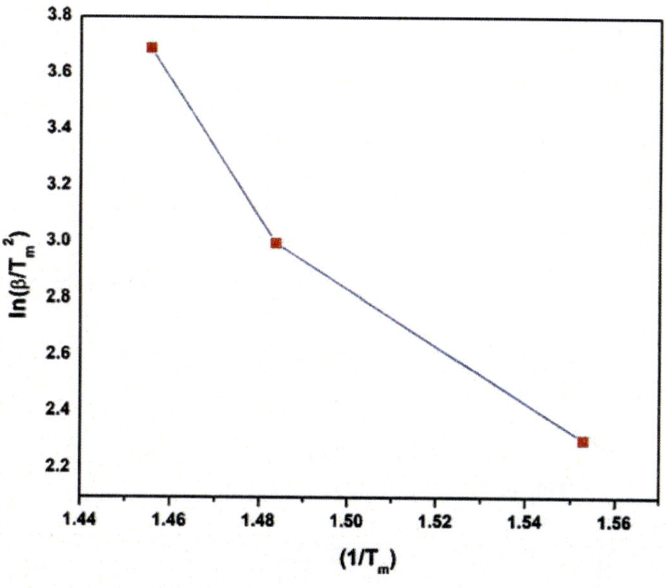

Fig. 5 Kinetic study plot of PD seeds at conversion rate from 0.05 to 0.50 via KAS method

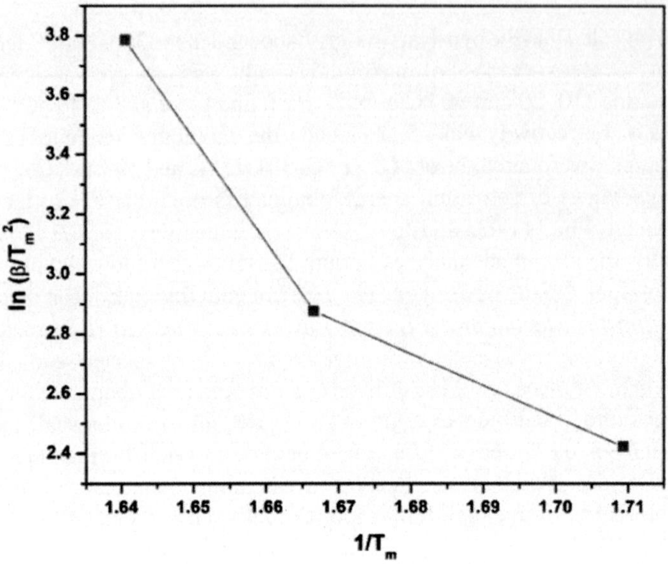

Fig. 6 Kinetic study plot of PE seeds at conversion rate from 0.05 to 0.50 via KAS method

(standard deviation) is calculated respectively to be 23.155 and 34.66 kJ/mol. In the applied KAS model, the value of activation energy depicted an enhancing trend with an increment in the conversion rate. Similarly, for the pyrolysis of *Cassia fistula L.* and *Syzygium cumini* seeds, the average activation energies were calculated as 201.19 and 223.49 kJ/mol via the KAS model.

In the present research study, the average value of energy of activation calculated through the FWO model was calculated to be 222.66 kJ/mol and 200.82 kJ/mol [24]. In addition to this, the value of activation energy for the pyrolysis of acai seed via FWO method was calculated to be 159.12 kJ/mol, while it was 157.62 kJ/mol as found through the KAS method [25].

4 Summary and Conclusion

Kinetic analysis of the pyrolysis of biomass is considered to be quite complicated because of the various coinciding reactions. However, isoconversional methods are helpfully able to define the multiple reactions taking place inside the feedstock. The current study aimed to calculate the pyrolysis kinetic parameters of *Phoenix Dactylifera* (*PD*) and *Phyllanthus Emblica* (*PE*) seeds at different rates of heating of 10, 20 and 30 °C/min with the help of a thermogravimetric analyser. The proximate analyses of both the seeds were also carried out to check for the basic suitability of seeds for consideration in the generation of biofuels. The TGA showed that the thermal decomposition of *PD* and *PE* seeds took place in 3 different zones, however,

it was also revealed that the pyrolysis mainly happened in the 350–800 °C temperature range. For *PD* seeds, the loss of maximum weight was observed at three different rates of heating (10, 20 and 30 °C/min), were found to be at 302.49 °C, 322.12 °C and 332.95 °C respectively while for *PE* seeds, the maximum weight loss at the same 3 heating rates was found to be at 360.51 °C, 381.92 °C and 393.72 °C respectively. The average values of activation energy obtained through the KAS kinetic model were evaluated to be 111.935 and 157.272 kJ/mol respectively for *PD* and *PE* seeds. Additionally, the proximate analyses of both the seeds also showed their worth as a source of biofuel. On the basis of current experimental findings, it can be concluded that both *Phoenix Dactylifera* (*PD*) and *Phyllanthus Emblica* (*PE*) seeds possess some similar properties as that demonstrated in previously selected seeds. It can also be derived that both seeds could be utilized as a raw feedstock source for the pyrolysis process for biofuel and bioenergy production. The future works shall include the pyrolysis of *PD* and *PE* seeds at different temperatures to yield the maximum quantity of bio-oil. Besides, it shall also include the investigation to upgrade the quality of the obtained biofuel for usage in different applications such as CI engines or to abstract useful chemicals.

Acknowledgements We are grateful to our supervisor and the Central University of Jharkhand, Ranchi for providing the facilities for research and analyses.

References

1. White JE, Catallo WJ, Legendre BL (2011) Biomass pyrolysis kinetics: a comparative critical review with relevant agricultural residue case studies. J Anal Appl Pyrol 91(1):1–33. https://doi.org/10.1016/j.jaap.2011.01.004
2. Slopieck K, Bartocci P, Fantozzi F (2012) Thermogravimetric analysis and kinetic study of poplar wood pyrolysis. Appl Ener 97:491–497. https://doi.org/10.1016/j.apenergy.2011.12.056
3. Hu M, Chen Z, Guo D, Liu C, Xiao B, Hu Z, Liu S (2015) Thermogravimetric study on pyrolysis kinetics of Chlorella pyrenoidosa and bloom-forming cyanobacteria. Biores Technol 177:41–50. https://doi.org/10.1016/j.biortech.2014.11.061
4. Sahoo A, Kumar S, Mohanty K (2020) Comprehensive characterization of non-edible lignocellulosic biomass to elucidate their biofuel production potential. Biomass Conv Bioref. https://doi.org/10.1007/s13399-020-00924-6
5. Komandur J, Vinu R, Mohanty K (2022) Pyrolysis kinetics and pyrolysate composition analysis of Mesua ferrea L: A non-edible oilseed towards the production of sustainable renewable fuel. Biores Technol 351:126987. https://doi.org/10.1016/j.biortech.2022.126987
6. Pandey SP, Kumar S (2020) Valorisation of argemone mexicana seeds to renewable fuels by thermochemical conversion process. J Environ Chem Eng 8:104271. https://doi.org/10.1016/j.jece.2020.104271
7. Demirbas A (2004) Effect of initial moisture content on the yields of oily products from pyrolysis of biomass. J Anal Appl Pyrol 71(2):803–805. https://doi.org/10.1016/j.jaap.2003.10.008
8. Kan T, Strezov V, Evans TJ (2016) Lignocellulosic biomass pyrolysis: A review of product properties and effects of pyrolysis parameters. Renew Sust Energy Rev 57:1126–1140. https://doi.org/10.1016/j.rser.2015.12.185

9. Angın D (2013) Effect of pyrolysis temperature and heating rate on biochar obtained from pyrolysis of safflower seed press cake. Biores Technol 128:593–597. https://doi.org/10.1016/j.biortech.2012.10.150
10. Singh RK, Shadangi KP (2011) Liquid fuel from castor seeds by pyrolysis. Fuel 90(7):2538–2544. https://doi.org/10.1016/j.fuel.2011.03.015
11. Duman G, Okutucu C, Ucar S, Stahl R, Yanik J (2011) The slow and fast pyrolysis of cherry seed. Biores Technol 102(2):1869–1878. https://doi.org/10.1016/j.biortech.2010.07.051
12. Nayan NK, Kumar S, Singh RK (2013) Production of the liquid fuel by thermal pyrolysis of neem seed. Fuel 103:437–443. https://doi.org/10.1016/j.fuel.2012.08.058
13. Singh VK, Soni AB, Kumar S, Singh RK (2014) Pyrolysis of sal seed to liquid product. Biores Technol 151:432–435. https://doi.org/10.1016/j.biortech.2013.10.087
14. Nayan NK, Kumar S, Singh RK (2012) Characterization of the liquid product obtained by pyrolysis of karanja seed. Biores Technol 124:186–189. https://doi.org/10.1016/j.biortech.2012.08.004
15. Demirbas A (2006) Effect of temperature on pyrolysis products from four nut shells. J Anal Appl Pyrol 76(1–2):285–289. https://doi.org/10.1016/j.jaap.2005.12.012
16. Dhyani V, Bhaskar T (2018) A comprehensive review on the pyrolysis of lignocellulosic biomass. Renewable Energy 129:695–716. https://doi.org/10.1016/j.renene.2017.04.035
17. Cai J, Liu R (2007) Research on water evaporation in the process of biomass pyrolysis. Energy Fuel 21(6):3695–3697. https://doi.org/10.1021/ef700442n
18. Cai J, Chen S (2008) Determination of drying kinetics for biomass by thermogravimetric analysis under nonisothermal condition. Dry Technol 26(12):1464–1468. https://doi.org/10.1080/07373930802412116
19. Roque-Diaz P, Villas L, Shemet CVZ, Lavrenko VA, Khristich VA (1985) Studies on thermal decomposition and combustion mechanism of bagasse under non-isothermal conditions. Thermo Acta 93:349–352. https://doi.org/10.1016/0040-6031(85)85088-7
20. Antal MJJ, Varhegyi G (1995) Cellulose pyrolysis kinetics: the current state of knowledge. Indus Eng Chem Res 34(3):703–717. https://doi.org/10.1021/ie00042a001
21. Varhegyi G, Antal MJ Jr, Jakab E, Szabó P (1997) Kinetic modeling of biomass pyrolysis. J Anal App Pyrol 42(1):73–87. https://doi.org/10.1016/S0165-2370(96)00971-0
22. Sahoo A, Kumar S, Mohanty K (2021) Kinetic and thermodynamic analysis of Putranjiva roxburghii (putranjiva) and Cassia fistula (amaltas) non-edible oilseeds using thermogravimetric analyzer. Renewable Energy 165:261–277. https://doi.org/10.1016/j.renene.2020.11.011
23. Sahoo A, Gautam R, Kumar S, Mohanty K (2021) Energy optimization from a binary mixture of non-edible oilseeds pyrolysis: kinetic triplets analysis using thermogravimetric analyser and prediction modeling by artificial neural network. J Environ Manage 297:113253. https://doi.org/10.1016/j.jenvman.2021.113253
24. Pal DB, Tiwari AK, Srivastava N, Hashem A, Allah EFA (2021) Thermal studies of biomass obtained from the seeds of Syzygium cumini and Cassia fistula L. and peel of Cassia fistula L. fruit. Biomass Conv Bioref. https://doi.org/10.1007/s13399-021-01492-z
25. Santos VO, Queiroz LS, Araujo RO, Ribeiro RCP, Guimaraes MN, Da Costa CEF, Chaar JS, De Souza LKC (2020) Pyrolysis of acai seed biomass: kinetics and thermodynamic parameters using thermogravimetric analysis. Bioresour Technol Rep 12:100553. https://doi.org/10.1016/j.biteb.2020.100553

Revisiting the Recent Advancements in the Design and Performance of Solar Greenhouse Dryers

Anil Singh Yadav, Abhay Agrawal, Amit Jain, Rajiv Saxena, Manoj Kumar, Abhishek Sharma, and Sonali Singh

1 Introduction

Conventional energy sources such as oil, gas, coal, and nuclear power are currently dominating the current energy scenario. Most developing countries cannot economically benefit from these sources because of their finite nature. Rather than relying on fossil fuels, renewable energy sources such as solar, wind, and biomass are now being used. The sun, wind, and rain aren't the only renewable energy sources. Tides,

A. S. Yadav (✉)
Mechanical Engineering Department, IES College of Technology, Bhopal, Madhya Pradesh 462044, India
e-mail: anilsinghyadav@gmail.com

A. Agrawal
Mechanical Engineering Department, Rewa Engineering College, Rewa, Madhya Pradesh 486002, India

A. Jain
Department of Chemical Engineering, Guru Ghasidas Vishwavidyalaya, Bilaspur, Chhattisgarh 495009, India

R. Saxena
Infinity Management and Engineering College, Sagar, Madhya Pradesh 470001, India

M. Kumar
Mechanical Engineering Department, Guru Gobind Singh Educational Society's Technical Campus, Bokaro, Jharkhand 827013, India

National Institute of Foundry and Forge Technology, Ranchi, Jharkhand 834003, India

A. Sharma
Department of Mechanical Engineering, Birsa Institute of Technology Sindri, Dhanbad, Jharkhand 828123, India

S. Singh
State Water Data Centre, Hydrometeorology, Bureau of Design and Drawings of Hydraulic and Irrigation Structures, Bhopal, Madhya Pradesh 462042, India

A. Kumar et al. (eds.), *Recent Advances in Manufacturing and Thermal Engineering*,
Lecture Notes in Mechanical Engineering, https://doi.org/10.1007/978-981-19-8517-1_20

waves, and sunlight may also be used to generate electricity. The concept of solar power has been around since the industrial revolution. It has the potential to produce large-scale power and can be widely distributed [1]. The increase in the number of people living in the world has become a major issue for the world's food supply. The demand for food has to increase in order to meet this demand. This is why it is important that the food is stored in order to keep it fresh. In order to preserve the nutrients in the food, various methods are used to reduce the moisture content of the food. One of these is solar drying, which is an olden technique that involves using the energy of the sun to remove harmful organisms from the food. After the harvest, the crop can be stored for a longer period of time to ensure that it's not damaged. This process can also improve the quality of the food and prevent it from going through any problems after the harvest. The use of heat transfer techniques such as convection, thermal, and radiation can also help in the drying process. A natural sun drying method involves the laying of the crops on the floor or a mat in the full sun for an entire day. They become contaminated with various pests and diseases as a result of exposure to the sun. Researchers have developed various methods to remove these harmful organisms. Compared to other methods, greenhouse solar dryer is more eco-friendly. It uses less energy and is able to provide the best possible quality of dried products. Also, this type of dryer does not use fossil fuels, which means that it is more sustainable to use. In addition, this type of dryer also occupies less space, which means that it can produce better quality products with minimal energy consumption. The idea of the greenhouse effect serves as the conceptual underpinning for the process of greenhouse sun drying. This technique enables the greenhouse's long-wave radiations to absorb the sun's incoming short-wave radiations so that the greenhouse may continue to be heated by the sun. This process can be used for various agricultural activities such as crop cultivation and aquaculture. The type of greenhouse solar dryer that's used for this process is classified as either active or passive. In the passive mode, a ventilation system is provided at the bottom of the greenhouse to allow the air to enter the facility [2–11].

A greenhouse, also referred to as a glasshouse or a hothouse, is a structure that's made of glass and designed to grow plants that require certain environmental conditions. These types of structures are typically smaller compared to industrial-sized buildings. A cold frame, also referred to as a mini greenhouse, is a structure that's designed to keep its contents warm inside. The cultivation of fruits and vegetables is one of the most typical purposes served by this kind of greenhouse. It is well known that commercial glass greenhouses are high-tech facilities that are able to care for a wide range of plant species. The greenhouse is filled with various equipment such as lights, screens, and cooling systems, which can be controlled by a computer. These kinds of facilities can also be used to improve the conditions of the plants by analyzing the air temperature, relative humidity, and vapor-pressure deficit. This process can help to reduce the risk of crop production. The idea of growing plants in controlled environments has been around since the Roman era. During the time of the Roman emperor Tiberius, he would eat a type of vegetable known as a cucumber every day. The vegetables were grown using various artificial methods. They were placed inside wheeled carts and kept inside to keep them warm at night. According

to Pliny the Elder, the cucumbers were stored in frames or houses that were covered with thick sheets of selenite or oil-based materials known as specularia. The first documented mention of a heated greenhouse was in the Sanga Yorok, a book written by a physician during the 1450s. In his treatise, Tiberius explains how to construct a greenhouse that can be used to cultivate vegetables and produce flowers, as well as ripen fruit inside an artificially heated environment. Early attempts at building greenhouses were often very challenging due to their design. They required a huge amount of work to maintain a balanced and adequate heat supply. The first heated greenhouse was constructed in the UK in 1681. During the seventeenth century, the construction of greenhouses in Europe continued to improve as technology improved. One of the most prominent examples of this type of structure was the Palace of Versailles, which was built over 150 m long. Charles Lucien Bonaparte, a French botanist, is known for building the first modern greenhouse in the Netherlands, during the 1800s. The growing popularity of botany eventually led to the establishment of greenhouses at universities. The French first introduced the concept of greenhouses to the world by building orangeries, which were designed to protect fruit trees from freezing. As the popularity of pineapples grew, more pineapple pits were built. After then, large glasshouses were also being erected in world. These types of structures, which had sufficient height for large trees, were referred to as palm houses. Public gardens and parks were also known to feature greenhouses. During the nineteenth century, the use of glass and iron architecture led to the development of palm houses and other similar structures. These types of structures were commonly used in large buildings, such as railway stations and exhibition halls. During the nineteenth century, several large greenhouses were built. In Japan, the first greenhouse was constructed in 1880 by Samuel Cocking. During the twentieth century, the use of geodesic domes was also introduced to the growing number of greenhouses. In the 1960s, the use of wider sheets of polyethylene film allowed for the construction of new types of greenhouses. These structures, which are known as hoop houses, were made by a variety of companies. Construction costs were also greatly reduced due to the use of various materials, such as aluminum extrusions and steel tubing. Due to the increasing number of greenhouses, smaller farms and garden centers were also being established. In the 1970s, the use of UV-inhibitors was also introduced to improve the film's durability. This resulted in the film's usable life being increased from one to three or even four years. In the 1980s and 1990s, the use of gutter-connected greenhouses started to increase. These types of structures have two or more bays that are connected by a row of support posts or a common wall. Due to the increased floor area, the heating inputs of these types of greenhouses have been reduced. These kinds of greenhouses are often seen at manufacturing facilities as well as other contexts in which plants are made available to the general public for purchase. A gutter-connected greenhouse is commonly covered with plastic materials that are designed to provide a higher heating efficiency. This type of structure is also known to absorb solar radiation. The temperature inside a greenhouse can increase due to the presence of solar radiation. Unlike other greenhouse constructions, the temperature inside a gutter-connected structure can rise due to the lack of natural convection. This phenomenon is different from the greenhouse effect, which is a theory that states that the temperature inside

a greenhouse increases due to the presence of warm air. Some studies claim that the effects of infrared radiative cooling can have economic implications. An analysis of the effects of near-infrared radiation on a greenhouse's heat demand was carried out. It concluded that the use of screens with high coefficient of reflection could reduce the greenhouse's heat demand by around 8%. Other methods such as the use of dye-based transparent surfaces could also help to decrease the greenhouse's heat consumption. Ventilation is one of the most essential elements necessary for the operation of a profitable greenhouse. This is a process that involves regulating the humidity and temperature levels in the greenhouse to prevent the development of plant pathogens. This is done to prevent the plants from developing diseases that prefer to live in still air conditions. One of the most important factors that a greenhouse can consider when it comes to becoming energy-efficient is the use of ventilation. The utilization of vents and recirculation fans are two methods that may be used to complete this procedure. Aside from reducing the greenhouse's energy consumption, ventilation can also help to boost the plant's respiration and photosynthesis. Unlike other buildings, which have solid opaque walls, a greenhouse's heat loss through its covering is one of the main factors that prevents it from functioning properly. This issue can be solved by using plastic coverings that are designed to allow light to enter the structure. When supplemental heat is needed, most greenhouse owners turn to natural gas or electric heaters. However, passive heating can be used to capture the greenhouse's heat and use it to boost the temperature. This process involves capturing solar energy during times of relative abundance. Aside from keeping the greenhouse cool, livestock waste can also be used to heat it. For instance, if a chicken coop is placed inside a greenhouse, the waste heat from the animals can be used to cool the greenhouse. Another process that can be used to cool a greenhouse is by opening the windows when the temperature inside the structure gets too hot. This process can be performed manually or through an automated system. For instance, window actuators can be used to open and close the windows depending on the weather conditions. Electronic controllers can also be used to monitor the greenhouse's temperature. This type of system can be very simple, but it can be very complex in larger operations. A shade house can also be used to provide cooling during hot weather. This type of structure can be placed over the windows to allow light to enter the greenhouse. In addition, some greenhouses have grown lights that are designed to increase the amount of light that the plants receive. This method can help to boost the plant's yield and reduce the greenhouse's energy consumption. For over a hundred years, greenhouse cultivation has been known to increase the plant's growth rate by around 1100 parts per million through the use of carbon dioxide enrichment. This process was first introduced in the Netherlands following the invention of many pieces of apparatus that allows the accumulation of carbon dioxide in successive layers. High-quality secondary metabolites, such as cardiac glycosides, can be produced in a greenhouse by increasing the greenhouse's temperature and carbon dioxide concentration. This process can also help to decrease the greenhouse's water usage. By reducing the greenhouse's air flow, carbon dioxide enrichment can help to meet the plant's needs for carbon. Due to the location of commercial greenhouses near industrial facilities, they can be beneficial to both the plant and the environment. The process can also

help to reduce greenhouse emissions. For instance, by using carbon dioxide enrichment, a refinery can reduce its carbon dioxide emissions by about 70%. On the other hand, a nursery can boost its tomato production by about 70% through the process. This method only works if the greenhouse has a limit on carbon dioxide. Glass with a thickness of 3 mm, which is of high quality and does not have air bubbles, is often used in the construction of residential greenhouses. Plastics are commonly used in these types of structures. Commercial greenhouses, which are often high-tech facilities used for the production of flowers and vegetables, are also commonly made of glass. These structures are equipped with various equipment such as cooling and heating systems, and they can be controlled by a computer. Compared to the smaller panes used in modern domestic designs, the 730 × 1422 mm size of the glass provides a larger area of glass for the greenhouse. Additionally, for a given greenhouse size, this kind of construction needs a greater amount of support. The Dutch Light design refers to a form of greenhouse that has sloping sides and employs the usage of glass that has not been cut. This type of greenhouse also has a cold frame with either a half or full pane [12].

In addition to being useful for drying crops, solar energy can also be utilized in various other applications, such as energy storage. Due to the abundance of solar radiation in different regions, such as tropical countries, solar drying can be a promising technology. Unfortunately, traditional agricultural dryers are not able to meet the energy requirements of these regions, which are caused by greenhouse gas emissions. With the proper technical specifications, a greenhouse can provide a variety of control over the plant's growing environment. Some of these factors include the temperature, humidity, and levels of light. These types of structures can also be used to improve the quality of the land's growing season. In addition, they can help to boost the food production in marginal areas by overcoming shortcomings in the area's growing season. A shade house is also commonly used in greenhouses to provide shade during hot weather. These structures are becoming more important as countries with high-latitude climates look to increase their food supply. Commercial greenhouses are commonly used for the production of various crops, such as flowers, vegetables, and fruits. Some of these are also used for the development of special greenhouse varieties. These may be started inside in the latter part of winter or the early part of spring, and then moved outdoors once the weather is warm enough. Aside from these, greenhouse equipment can also be used to improve the efficiency of the plant's growing process. For instance, by using hydroponic systems, they can allow the plant to grow in the interior space of the greenhouse. In addition, some types of bees are also known to perform pollination services in the greenhouse. Unlike outdoor production facilities, which are usually equipped with various equipment and methods, a greenhouse's management requirements are unique. These include controlling pests and diseases, as well as providing adequate water supply. In addition, greenhouse plants often require significant amounts of light and heat to produce their best results. These types of structures can also be used for other applications, such as oil recovery. Although greenhouse dryers are commonly used for large-scale drying of various materials, their thermal performance is not always ideal. Since the main power requirement of greenhouse dryers is thermal energy, the use of PV

modules can eliminate the need for electricity. Due to the varying factors that affect their performance, such as ambient temperature, wind speed, and solar radiation, the most recent methods to improve the thermal performance of these structures are presented [13–39]. The techniques presented in this study involve the integration of various solar energy sources, such as solar thermal collectors and photovoltaic modules, into greenhouse dryers.

2 Advanced Solar Greenhouse Dryers

Janjai et al. [40] constructed a PV-mounted greenhouse solar dryer that has a black concrete floor (Fig. 1). The greenhouse was equipped with a 44-meter^2 floor area. The outer surface of the greenhouse was covered with a layer of polycarbonate sheets. A 53 W rating solar panel was used to power the fan. To investigate its performance, the researchers loaded 150 kg of fresh chillies. The results of the study revealed that the dryer dried the chillies from 80 to 10% moisture content. The average drying period for these types of plants is approximately 2 to 3.5 days.

Kumar and Tiwari [41] researched the effects of the greenhouse's design and the various factors that affect the mass transfer coefficient of onion flakes in the dryer (Fig. 2). They found that the weight of the onions varied continuously while they were being dried. The experiment was conducted in three different conditions: natural, forced convection, and open sun drying. The results indicated that the varying coefficient of mass transfer during convection of the onions resulted in a significant increase in the overall mass transfer coefficient. The researchers also noted that the greenhouse's floor area and its UV film contribute to the overall performance of the dryer.

In order to achieve a drying capacity of 100 kg under forced mode, Barnwal and Tiwari [14] constructed a hybrid PV/T greenhouse dryer that has a 30-degree inclined

Fig. 1 PV-ventilated greenhouse solar dryer [40]

Fig. 2 Pictorial view of the greenhouse solar dryer for onion drying [41]

roof type (Fig. 3). The setup was made up of a DC fan and two 75 W modules. The air coming out of the top and passing through the mesh tray system is then collected and distributed through the three tiers of leaves and wire mesh. The structural frame of the greenhouse was also coated with a UV-stabilized polyethylene sheet to prevent infrared radiation from entering the structure. The researchers then compared the performance of the hybrid greenhouse solar dryer to a conventional one.

An analysis of the energy and exergy properties of a greenhouse was performed by Nayak and Tiwari [42] to predict its behavior when integrated with a PV/T collector. The green house was constructed at IIT Delhi's solar energy park (Fig. 4). The 8 PV modules were mounted on a wooden structure. One of the fans of the facility has a capacity of 12 W. The solar module's power is stored in 12 DC batteries, which are rated at 6 V each. The inverter can be used to convert DC to AC. The greenhouse's exergy efficiency was also improved by about 4%.

Fig. 3 Pictorial view of the hybrid PV/T greenhouse solar dryer [14]

Fig. 4 Pictorial view of the greenhouse solar dryer [42]

Janjai et al. [43] presented simulation and experimental performance on the drying of bananas and longan inside a greenhouse dryer equipped with a PV-integrated roof type structure (Fig. 5). The concrete-floored roof type dryer has a capacity of 8 × 5.5 × 3.5 m^3, and is covered with a layer of polycarbonates plates. The 50 W PV module is used to run three fans, which are required to maintain the facility's ventilation. During the operation of the greenhouse, the temperature difference between longan and banana was varying from 21 °C to 58 °C and 30 °C to 60 °C. The same process can also be performed within the greenhouse, as the temperature gap between the two plants is only 4 and 3 days. On the other hand, the same process can be performed under open sun, which takes 6 and 5 days. Compared to other dried products, greenhouse dried product has a better color and taste.

Sethi and Arora [44] used a reflecting north wall that has an inclination angle that is optimized for the width of the tray in greenhouse (Fig. 6). The researchers tested the greenhouse's performance under different conditions. In addition, they were able

Fig. 5 Pictorial view of the greenhouse solar dryer [43]

to perform the experiment with bitter gourd slices. The modified greenhouse was located in India's Punjab state. The improved greenhouse dryer was equipped with a 6 m × 4 m floor area and a reflective north wall. The north wall was also equipped with a 12 mm wooden ply board. The interior portion of the facility was also covered with a UV-stabilized polyethylene sheet. The results of the experiment revealed that the modified greenhouse's performance was significantly improved when the north wall was used in forced convection or natural convection mode.

Ganguly et al. [45] were able to demonstrate a model of a greenhouse that was equipped with a storage system and a power generation facility (Fig. 7). The integrated system allowed them to power the facility in a sustainable manner. The researchers were able to generate electricity from 51 solar PV modules. The researchers were able to restore power to the facility using two 480 W PEM fuel cell systems and an electrolyzer with a combined output of 3.3 kV. Some of the power generated by the solar panels is used by the greenhouse's appliances, while the remaining energy is utilized by the electrolyzer to generate hydrogen.

Sevda and Rathore [46] were able to successfully achieve a capacity of 1500 papers per batch by using a greenhouse dryer. The facility was equipped with five chimneys and an exhaust fan to maintain the required temperature and humidity inside the dryer. The researchers built a tunnel-shaped solar dryer that was designed to dry handmade papers. The facility was constructed at the Vidya Bhawan Society's cellulosic waste recycling education project in Rajasthan, India. The design of the tunnel-shaped solar dryer was made possible through the use of UV-stabilized polyethylene sheet.

Rathore and Panwar [47] were able to demonstrate the performance of a tunnel solar dryer by successfully drying seedless grapes at a temperature of 10 to 28 degree Celsius. The design of the solar tunnel dryer was based on a hemi-cylindrical metallic frame with a UV-stabilized polyethylene sheet (Fig. 8). The exhaust fan and the other components of the facility are insulated to prevent heat loss.

Almuhanna [48] constructed a greenhouse dryer at the King Faisal University in Saudi Arabia. The facility was developed so that its thermal performance could be evaluated, along with its feasibility. The structure was constructed using an aluminum frame and has a roof that has an inclination of 30°. Two plywood boxes containing

Fig. 6 Pictorial view of the improved greenhouse solar dryer [44]

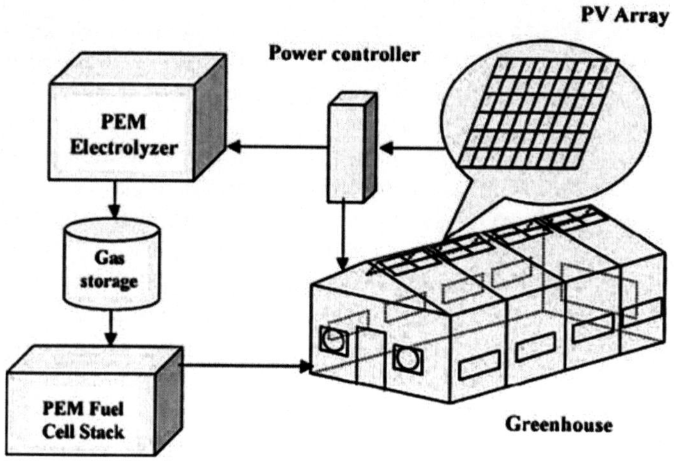

Fig. 7 Model of the integrated system [45]

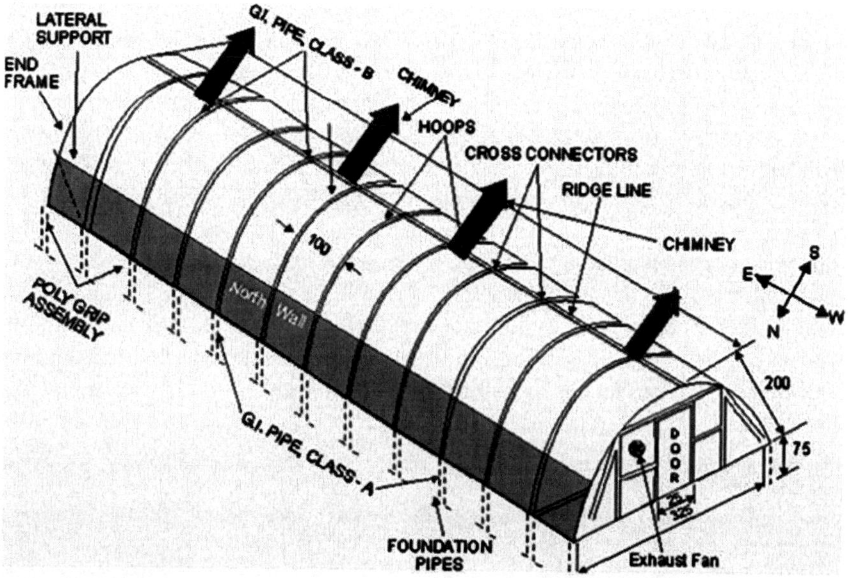

Fig. 8 Solar tunnel dryer [47]

a total of 1 square meter were placed inside the drying area. The box was equipped with an axial fan to maintain the air flow rate. During the experiment, the overall thermal efficiency of the facility was 60.11%.

In Thailand, Janjai [49] developed a greenhouse dryer that was equipped with a gas-fired cooking gas (LPG) burner. The facility was designed for small-scale dry

food processing plants. It has a capacity of 560 m^3 and can load up to 1000 kg. The gas-fired cooking gas can be used during off-sunshine periods. A 15-W direct current fan and three 50-W photovoltaic modules provide the electricity for the dryer. During the process of drying tomatoes, the temperature varied from 35 °C to 65 °C. The drying period that was observed was two to three days less than what would have been expected from natural sun drying.

Adu et al. [50] built a tent-type solar dryer that was equipped with a transparent roof at Nigeria. The facility's drying platform and two long side walls were covered with black cloth. The remaining two walls of the solar dryer are half covered with black cloth. The roof of greenhouse is covered with a transparent plastic cover. The performance of the dryer was obtained by drying okra, which was subjected to different conditions. After 23 h, the initial moisture content of the okra had decreased to 3.43%.

In order to study the effects of solar energy on the exergy and energy requirements of a hybrid solar drying system, Fudholi et al. [51] constructed a system that was consisted of a V-groove solar air collector, a rotating rack, and a PV array. The system dried salt-cured silver jewfish in 8 h at a moisture content of 64 to 10%. The average exergy efficiency of the system was 31%.

In order to evaluate the performance on the different floor conditions, Prakash et al. [18] develop a modified greenhouse solar dryer (Fig. 9). The evaluation took place in three different conditions: barren floor, floor with black painted concrete floor, and the enclosed black PVC sheet. The modified greenhouse solar dryer was tested in both load and no-load conditions. Various vegetables such as potato chips and tomatoes were used as crop in the active mode. The modified greenhouse solar dryer was equipped with an exhaust fan that was powered by a solar panel. The results of the study revealed that the floor with the enclosed black PVC sheet had better performance when compared to the other conditions. The researchers also found that the modified greenhouse solar dryer had better drying performance when compared to the passive version.

Jitjack et al. [52] constructed two greenhouse structures using the area-enhanced panels (Fig. 10). They tested the system under different conditions, such as an empty greenhouse and rubber sheets drying. The two greenhouse structures, which are referred to as the Parabolic and additional area-enhanced panels. The black painted floor and the transparent polycarbonate sheets enclosed the greenhouse. The results of the study revealed that the cost of the greenhouse with the additional panels was only 7% higher than that of the one without them. The efficiency of the system also increased by 15%.

A mixed-mode solar greenhouse dryer was constructed by the Tiwari et al. [53] at IIT Delhi (Fig. 11). The thermal modeling was carried out so that the system's energy and exergy parameters could be analyzed and evaluated. The experimental setup was enclosed with 3 mm thick glass. The floor area of the dryer was 1.066 m^2. The system was designed to maintain the required air circulation through two DC fans and three PV modules. The researchers also utilized the roof as an enclosed PV panel to prevent the crop from decolorating. The results of the study revealed that

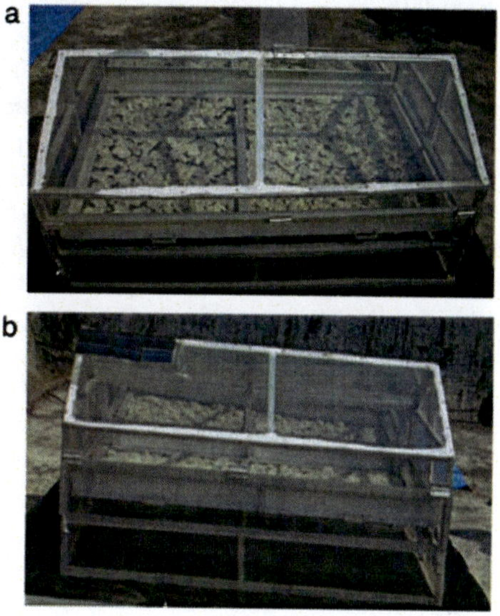

Fig. 9 Photographs for modified greenhouse dryer [18]

Fig. 10 Pictorial view of the improved greenhouse solar dryer [52]

the system's efficiency and room temperature increase significantly which increased the system's performance.

In an experiment conducted in a greenhouse pilot in Morocco, Belloulid, Hamdi, Mandi, and Ouazzani [54] dewatered sludge samples and molded them into 30 cylindrical cakes. The models depicted in this study show the various features of the experimental plant. The dimensions of an open greenhouse solar dryer were 160 × 60 cm. It was equipped with a transparent polycarbonate sheet of 1 cm thick. In cold and hot seasons, the moisture content decreased significantly. The volume reduction achieved in these seasons was 80%.

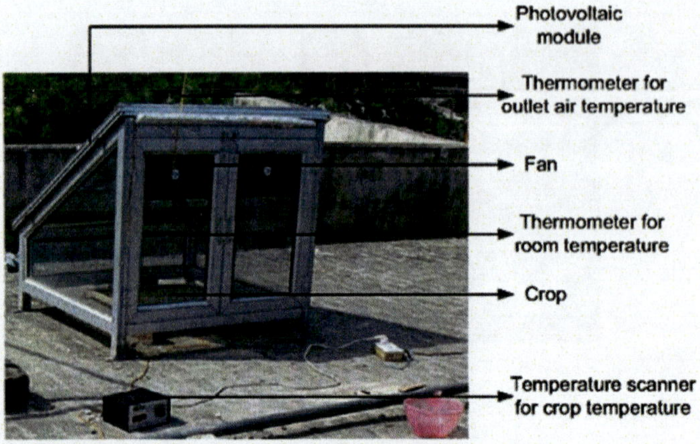

Fig. 11 Photograph of photovoltaic-integrated greenhouse dryer [53]

Morad, El-Shazly, Wasfy, and El-Maghawry [55] tested the greenhouse's temperature control system by controlling the air flow rate. They also tested the system's various operating conditions. These comprised two distinct plant conditions as well as three unique peppermint load configurations. The system also had two fan operating systems (Fig. 12). To dry peppermint plants, the researchers constructed three solar tunnel dryer units. These are equipped with a transparent plastic film that has a dimension of $2 \times 1 \times 0.8$ m, and they are powered by an electric motor. The results of the study revealed that the continuous fan operating condition and the load of 4 kg/m^2 having air flow of 2.10 m^3/min allowed the plant to get a good drying condition.

Azaizia et al. [20] attached a solar air collector and a chapel-shaped greenhouse solar dryer to a research facility in Borj Cedria, Tunisia (Fig. 13). The performance of the dryer was studied in terms of its drying efficiency. A mathematical model was developed to predict the various factors that affect the product's drying. The increase in the area of the product that was molded resulted in a decrease in the water evaporation rate. This effect also increased the plant's drying time. The floor area of the system was approximately 14.8 m^2. The simulation results indicated that the collector area of 2 m^2 can compensate for the losses caused by the fans.

The concept of greenhouse has been studied by Tiwari [56], who states that it is a system that provides a controlled environment for the plant growth. The main objective of this type of system is to provide a good environment for the plants to grow well. This is usually done by using greenhouse technology for the production of fruits, vegetables, and flowers. The agriculture sector is the backbone of India's economy and has shown the strong correlation between economic growth and agricultural development. Despite the various achievements that have been made in the field, the country's agricultural productivity is still below that of other countries. If India is to become an economic power, it needs to improve its agricultural productivity. There is a need for a new and effective technology that can improve the efficiency

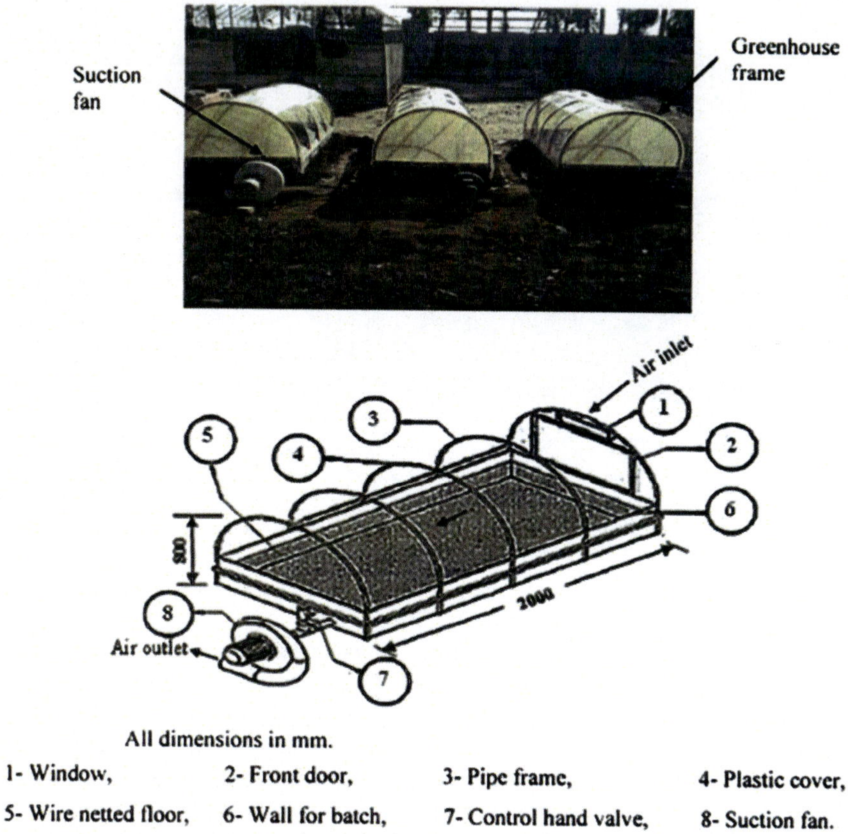

Fig. 12 Solar tunnel greenhouse dryer [55]

Fig. 13 Pictorial view of the improved greenhouse solar dryer [20]

and profitability of our farming systems. One of the most common technologies is the greenhouse. Although it is centuries old, this technology is still being used in India. The potential of greenhouse cultivation is immense due to the semi-arid climate in India. Since the government does not provide subsidy for the construction of greenhouses, the farmers are forced to build these structures on their own. This study aims to review various design of greenhouse structure that can be utilized by the farmers.

3 Conclusions

For farming in harsh and challenging environments, such as those in extreme terrains, utilizing greenhouses is a great alternative. They can help to grow plants in these conditions. However, it is not always easy to create a habitable environment for the organisms living in it. Through the use of technology and sensors, these hurdles can be easily ignored, and greenhouses can be an effective alternative to traditional farming. In areas where there is a shortage of resources, greenhouse farming has turned the tide. As a result of the review, the following conclusions can be drawn:

1. The active mode of a greenhouse solar dryer is also better than the passive one.
2. Natural convection is ideal for low-moisture content crops.
3. The nutrients, taste, and color of the dried product are better in a greenhouse solar dryer than in an open sun.
4. For remote locations, an integrated greenhouse solar dryer is the best choice. It can be used with thermal storage material to maintain the greenhouse's temperature. This helps in reducing the drying period.
5. The insulated north wall of a greenhouse solar dryer helps in preventing heat loss to the surrounding area.
6. It can also be used to pre-heat the air to increase its efficiency. A solar collector or LPG burner can also be used to improve the efficiency of the dryer.
7. For instance, if a crop is being dried in a certain location, a simulated model can be used to determine which type of solar dryer is ideal for the job.
8. Aside from agricultural produce, solar greenhouse dryers can also be used to process marine and poultry products. These products can be dried more efficiently and with better quality.
9. Since solar greenhouse dryers can be installed in remote areas, they can be more affordable compared to other methods. Also, due to the lack of grid electricity, the integration of solar systems and thermal energy storage systems can be more feasible.

The rapid emergence and evolution of sustainable structures such as greenhouse dryers and solar-based systems are expected to greatly improve their efficiency and profitability in the future. The ultimate goal of these innovations is to improve the returns that farmers can expect from their agricultural operations.

References

1. Yadav AS, Bhagoria JL (2013) Renewable energy sources-an application guide: energy for future. Int J Energy Sci 3:70–90
2. Singh P, Shrivastava V, Kumar A (2018) Recent developments in greenhouse solar drying: a review. Renew Sustain Energy Rev 82:3250–3262
3. Prasad R, Yadav AS, Singh NK, Johari D (2019) Heat transfer and friction characteristics of an artificially roughened solar air heater. Lecture notes in mechanical engineering, pp 613–626
4. Yadav AS, Shrivastava V, Ravi Kiran T, Dwivedi MK (2021) CFD-based correlation development for artificially roughened solar air heater. Lecture notes in mechanical engineering, pp 217–226
5. Yadav AS, Dwivedi MK, Sharma A, Chouksey VK (2022) CFD based heat transfer correlation for ribbed solar air heater. Mater Today: Proc 62:1402–1407
6. Shrivastava V, Yadav A, Shrivastava N (2021) Comparative study of the performance of double-pass and single-pass solar air heater with thermal storage. Lecture notes in mechanical engineering, pp 227–237
7. Yadav AS, Prakash Shukla O, Singh Bhadoria R (2022) Recent advances in modeling and simulation techniques used in analysis of solar air heater having ribs. Mater Today Proc 62:1375–1382
8. Yadav AS, Gattani A (2022) Revisiting the influence of artificial roughness shapes on heat transfer enhancement. Mater Today Proc 62:1383–1391
9. Yadav AS, Gattani A (2022) Solar thermal air heater for sustainable development. Mater Today Proc 60:80–86
10. Yadav AS, Agrawal A, Sharma A, Sharma S, Maithani R, Kumar A (2022) Augmented artificially roughened solar air heaters. Mater Today Proc 63:226–239
11. Yadav AS, Sharma SK (2021) Numerical simulation of ribbed solar air heater. In: Sikarwar BS, Sundén B, Wang Q (eds) Advances in fluid and thermal engineering. Lecture notes in mechanical engineering. Springer, Singapore, pp 549–558
12. Wikipedia Contributors. Greenhouse, Wikipedia, The Free Encyclopedia. https://en.wikipedia.org/w/index.php?title=Greenhouse&oldid=1088943458. Accessed 21 May 2022
13. Shrivastava V, Yadav AS, Shrivastava N (2022) Thermal performance assessment of greenhouse solar dryer operated under active mode. Lecture notes in mechanical engineering, pp 75–82
14. Barnwal P, Tiwari GN (2008) Grape drying by using hybrid photovoltaic-thermal (PV/T) greenhouse dryer: an experimental study. Sol Energy 82:1131–1144
15. Prakash O, Kumar A (2014) Solar greenhouse drying: a review. Renew Sustain Energy Rev 29:905–910
16. Chauhan PS, Kumar A (2016) Performance analysis of greenhouse dryer by using insulated north-wall under natural convection mode. Energy Rep 2:107–116
17. Patil R, Gawande R (2016) A review on solar tunnel greenhouse drying system. Renew Sustain Energy Rev 56:196–214
18. Prakash O, Kumar A, Laguri V (2016) Performance of modified greenhouse dryer with thermal energy storage. Energy Rep 2:155–162
19. Tiwari S, Tiwari GN, Al-Helal IM (2016) Development and recent trends in greenhouse dryer: a review. Renew Sustain Energy Rev 65:1048–1064
20. Azaizia Z, Kooli S, Elkhadraoui A, Hamdi I, Guizani A (2017) Investigation of a new solar greenhouse drying system for peppers. Int J Hydrogen Energy 42:8818–8826
21. Chauhan PS, Kumar A, Gupta B (2017) A review on thermal models for greenhouse dryers. Renew Sustain Energy Rev 75:548–558
22. Chauhan PS, Kumar A (2018) Thermal modeling and drying kinetics of gooseberry drying inside north wall insulated greenhouse dryer. Appl Therm Eng 130:587–597
23. Tiwari S, Agrawal S, Tiwari GN (2018) PVT air collector integrated greenhouse dryers. Renew Sustain Energy Rev 90:142–159

24. Choab N, Allouhi A, El Maakoul A, Kousksou T, Saadeddine S, Jamil A (2019) Review on greenhouse microclimate and application: design parameters, thermal modeling and simulation, climate controlling technologies. Sol Energy 191:109–137
25. Azaizia Z, Kooli S, Hamdi I, Elkhal W, Guizani AA (2020) Experimental study of a new mixed mode solar greenhouse drying system with and without thermal energy storage for pepper. Renewable Energy 145:1972–1984
26. Azam MM, Eltawil MA, Amer BMA (2020) Thermal analysis of PV system and solar collector integrated with greenhouse dryer for drying tomatoes. Energy 212:118764
27. Kiburi FG, Kanali CL, Kituu GM, Ajwang PO, Ronoh EK (2020) Performance evaluation and economic feasibility of a solar-biomass hybrid greenhouse dryer for drying Banana slices. Renew Energy Focus 34:60–68
28. Nimnuan P, Nabnean S (2020) Experimental and simulated investigations of the performance of the solar greenhouse dryer for drying cassumunar ginger (Zingiber cassumunar Roxb.). Case Stud Therm Eng 22:100745
29. Gupta V, Sabharwal Gupta K, Khare R (2021) Experimental analysis for drying of potato slices on detachable solar greenhouse dryer. Mater Today Proc 47:6269–6273
30. Lingayat A, Balijepalli R, Chandramohan VP (2021) Applications of solar energy based drying technologies in various industries—a review. Sol Energy 229:52–68
31. Mhd Safri NA, Zainuddin Z, Mohd Azmi MS, Zulkifle I, Fudholi A, Ruslan MH, Sopian K (2021) Current status of solar-assisted greenhouse drying systems for drying industry (food materials and agricultural crops). Trends Food Sci Technol 114:633–657
32. Mishra L, Sinha A, Gupta R (2021) Energy, exergy, economic and environmental (4E) analysis of greenhouse dryer in no-load condition. Sustain Energy Technol Assess 45:101186
33. Mishra S, Verma S, Chowdhury S, Dwivedi G (2021) Analysis of recent developments in greenhouse dryer on various parameters—a review. Mater Today Proc 38:371–377
34. Singh P, Gaur MK (2021) Sustainability assessment of hybrid active greenhouse solar dryer integrated with evacuated solar collector. Curr Res Food Sci 4:684–691
35. Srinivasan G, Muthukumar P (2021) A review on solar greenhouse dryer: design, thermal modelling, energy, economic and environmental aspects. Sol Energy 229:3–21
36. Colorado A, Morales O, Ossa D, Amell A, Chica E (2022) Modeling the optimal condition for drying rumen contents using a solar greenhouse dryer. Case Stud Therm Eng 30:101678
37. El-Mesery HS, El-Seesy AI, Hu Z, Li Y (2022) Recent developments in solar drying technology of food and agricultural products: a review. Renew Sustain Energy Rev 157:112070
38. Selimefendigil F, Şirin C, Ghachem K, Kolsi L, Alqahtani T, Algarni S (2022) Enhancing the performance of a greenhouse drying system by using triple-flow solar air collector with nano-enhanced absorber coating. Case Stud Therm Eng 34:102011
39. Singh S, Gill RS, Hans VS, Mittal TC (2022) Experimental performance and economic viability of evacuated tube solar collector assisted greenhouse dryer for sustainable development. Energy 241:122794
40. Janjai S, Khamvongsa V, Bala BK (2007) Development, design, and performance of a PV-Ventilated greenhouse dryer. Int Energy J 8:249–258
41. Kumar A, Tiwari GN (2007) Effect of mass on convective mass transfer coefficient during open sun and greenhouse drying of onion flakes. J Food Eng 79:1337–1350
42. Nayak S, Tiwari GN (2008) Energy and exergy analysis of photovoltaic/thermal integrated with a solar greenhouse. Energy Build 40:2015–2021
43. Janjai S, Lamlert N, Intawee P, Mahayothee B, Bala BK, Nagle M, Müller J (2009) Experimental and simulated performance of a PV-ventilated solar greenhouse dryer for drying of peeled longan and banana. Sol Energy 83:1550–1565
44. Sethi VP, Arora S (2009) Improvement in greenhouse solar drying using inclined north wall reflection. Sol Energy 83:1472–1484
45. Ganguly A, Misra D, Ghosh S (2010) Modeling and analysis of solar photovoltaic-electrolyzer-fuel cell hybrid power system integrated with a floriculture greenhouse. Energy Build 42:2036–2043

46. Sevda MS, Rathore NS (2010) Performance evaluation of the semicylindrical solar tunnel dryer for drying handmade paper. J Renew Sustain Energy 2:013107
47. Rathore NS, Panwar NL (2010) Experimental studies on hemi cylindrical walk-in type solar tunnel dryer for grape drying. Appl Energy 87:2764–2767
48. Almuhanna EA (2012) Utilization of a solar greenhouse as a solar dryer for drying dates under the climatic conditions of the eastern province of Saudi Arabia: Part I: thermal performance analysis of a solar dryer. J Agric Sci 4:237
49. Janjai S (2012) A greenhouse type solar dryer for small-scale dried food industries: development and dissemination. Int J Energy Environ 3:383–398
50. Adu EA, Bodunde AA, Awagu EF, Olayemi FF (2012) Design, construction and performance evaluation of a solar agricultural drying tent. Int J Eng Res Technol 1:1–11
51. Fudholi A, Yendra R, Basri DF, Ruslan MH, Sopian K (2016) Energy and exergy analysis of hybrid solar drying system. Contemp Eng Sci 9:215–223
52. Jitjack K, Thepa S, Sudaprasert K, Namprakai P (2016) Improvement of a rubber drying greenhouse with a parabolic cover and enhanced panels. Energy Build 124:178–193
53. Tiwari S, Tiwari GN, Al-Helal IM (2016) Performance analysis of photovoltaic–thermal (PVT) mixed mode greenhouse solar dryer. Sol Energy 133:421–428
54. Belloulid MO, Hamdi H, Mandi L, Ouazzani N (2017) Solar greenhouse drying of wastewater sludges under arid climate. Waste Biomass Valorization 8:193–202
55. Morad MM, El-Shazly MA, Wasfy KI, El-Maghawry HAM (2017) Thermal analysis and performance evaluation of a solar tunnel greenhouse dryer for drying peppermint plants. Renewable Energy 101:992–1004
56. Tiwari GN (2003) Greenhouse technology for controlled environment. Alpha Science International Ltd., Oxford

Application of Multi-Criteria Decision-Making Tool for Choosing Right Biogas Plants: Process Controllability, Suitability, and Cost Perspectives in the Indian Context

Haris Jamal, M. K. Loganathan, P. G. Ramesh, Mandeep Singh, and Girish Kumar

1 Introduction

Biogas is formed naturally from organic matter under anaerobic conditions. Biomass digesters were built in New Zealand and India in the mid-nineteenth century in an effort to utilize anaerobic digestion (AD) of biomass. When rising oil costs drove research into alternate energy sources in the 1970s, biogas technology gained popularity [1]. Depending on the feedstock, it is mostly made up of CH_4, with tiny quantities of H_2S, moisture, and other trace impurities [2]. Biogas production has the potential to reduce greenhouse gas emissions, provide a renewable source of energy, converts waste into valuable fertilizer, and lead to lower pollution impacts from waste disposal [3]. The use of anaerobic bacteria to handle animal manure in biogas plants greatly reduces pollution and diseases in the environment [4]. Animal feces and slurry, agriculture debris and by-products, fermentable organic scraps from food and agribusiness, organic fraction of municipal waste and food service, sewage sludge, and exclusive energy crops are the most prevalent biomass classifications used in European biogas production [5].

A biogas plant is a mechanism that permits fermentation to occur in the absence of oxygen. Through the intake line or input tank, fresh waste matter (typically in

H. Jamal · M. K. Loganathan (✉)
Department of Mechanical Engineering, The Assam Kaziranga University, Jorhat 785006, India
e-mail: loganathanmk123@gmail.com

P. G. Ramesh
Triveni Engineering and Industries Ltd., Mysuru 570016, India

M. Singh
School of Mechanical and Mechatronic Engineering, University of Technology Sydney, Ultimo, NSW 2007, Australia

G. Kumar
Department of Mechanical Engineering, Delhi Technological University, Delhi 110042, India

A. Kumar et al. (eds.), *Recent Advances in Manufacturing and Thermal Engineering*, Lecture Notes in Mechanical Engineering, https://doi.org/10.1007/978-981-19-8517-1_21

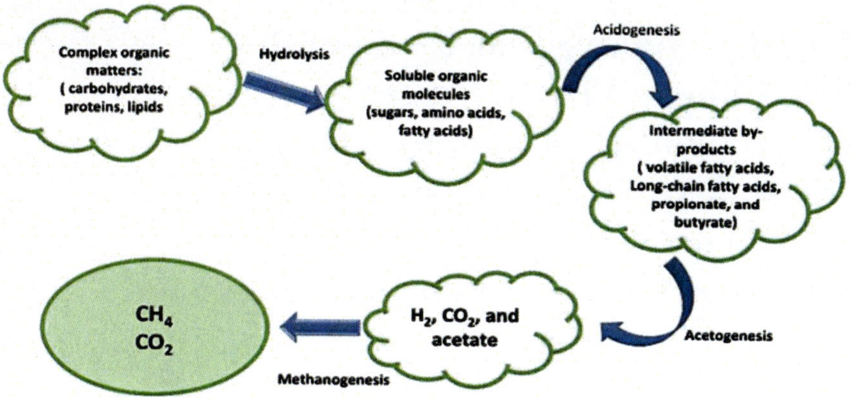

Fig. 1 The anaerobic digestion

the form of a homogenous slurry) is delivered into the plant's digester. Microbial activity causes fermentation inside the digester, producing biogas and rich in humus and other nutrients organic manure. Slurry is discharged into the slurry storage pit through the outflow pipe [6]. The process of anaerobic digestion is divided into four stages: hydrolysis, acidogenesis, acetogenesis, and methanogenesis, each of which is carried out by a distinct species of bacteria. The first phase of Anaerobic Digestion is hydrolysis, in which complex organic materials (polymers) are reduced into smaller components. The acidogenesis step produces volatile fatty acids in addition to ammonia, carbon dioxide, hydrogen sulfide, and other metabolic ends. Anaerobic digestion continues with acetogenesis, the third phase. Acetogens digest the plain molecules formed during the acidogenesis phase, giving acetic acids, hydrogen, and carbon dioxide. Methanogens are responsible for the last step of anaerobic digestion. Methanogens create methane, carbon dioxide, and water by using in-between products from earlier phases [6]. Figure 1 presents the four digestion steps in a simplified flow chart.

It is critical to understand how various process factors (temperature, pH, carbon-to-nitrogen ratio, organic loading rate, hydraulic retention duration, and feedstock size) impact the operation of a biogas plant in order to assure constant biogas generation. The parameters, if controlled properly, will help to improve the efficiency of the plant. The parameters are enumerated in the following line.

Temperature: Depending on the type of microbe used, biogas plants may function at a variety of temperatures. When the temperature falls below 15 °C, digesters without insulators are rendered ineffective [7–9]. The higher the temperature, the more gas is produced.

pH: The pH of the methanogens is best between 6.7 and 7.5. In a double-stage biogas plant, the pH in stage 2 is greater than stage 1 [7, 8]. The concentration of volatile fatty acids (VFAs), ammonium, and alkalinity all influence the pH level.

Ratio of Carbon to Nitrogen: If the ratio of carbon-to-nitrogen ratio is greater than ideal, the breakdown rate will be slower. If the ratio of Carbon to Nitrogen is lower than the ideal one, i.e., 20 to 30, the buildup of ammonia might occur, suppressing bacterial action [10].

Rate of Organic Loading Rate: The type of waste put into the digester determines the actual loading rate [11]. A high organic loading rate (OLR) causes fatty acid accumulation during acetogenesis, which reduces methane production [10].

Hydraulic Retention Time: It is the mean number of days for a unit amount of substrate to remain in the digester. [11]. It is highly dependent on the substrate and typically ranges from 30 to 60 days. In general, the hydraulic retention time(HRTs) for non-stirring digesters is 30 days. However, digester with decomposition rates can be reduced to 10 to 20 days [10]. Short hydraulic retention times allow for greater efficiency of process and reduced costs of capital, but longer hydraulic retention times are necessary for lignocellulosic waste digestion. [12].

Feedstock Size: Large materials might block the digester and create difficulties for bacteria to break them down. Reduction in size of feed leads to increase surface area which facilitates faster decomposition. Fibrous substrates are tough for enzymes that transform carbohydrates into simple sugars to break down [10]. Anaerobic digestion of food waste/crop residue straw in the optimum ratio of 5:1 gives the highest gas production when the particle size range is 0.45 to 0.60 mm [13].

The suitability and expense of a small-scale biogas plant have been a source of contention for decades [6, 14]. Suitability means how well a particular type biogas plant is suitable to meet the various location and multiple feedstock requirement. The biogas communities all over the world need to choose the biogas plant that suits to their location and feedstock requirement. Such biogas plants are not the standardized, and customizing the plant and process introduce a significant amount of production, quality, and maintenance issues during installation and operation. The plants cannot be operated sustainably in a long run and this will increase capital and operational costs [15–17], thus lowering payback period. Additionally, if the process is not controlled properly, then the gas production would be reduced with slow rate of return on investment. The plant with versatility of suiting to various climatic locations, feedstock requirement, and socio-economic conditions would be a standard choice for the biogas community.

Small-scale biogas digester programs are frequently promoted in developing countries by governments or non-profit organizations with financial subsidies, institutional support, and technical knowledge and skills. However, as mentioned earlier, the selection of appropriate plants must be done based on factors such as process controllability, suitability, and cost, in order to ensure sustainable operation of the plant. Over the last few decades, multiple research have been undertaken to enhance the AD process as well as analyze the efficacy of various systems. There are some efforts directed toward application of MCDM methods to select the biogas plants but based on customized community requirement, but not generic model based on technical, process, and financial requirements like process controllability, suitability,

and cost. The practical issue is the lack of a tool to assist in the selection of an appropriate biogas plant model. In this paper, an MCDM-based methodology is proposed to help in selecting a suitable biogas plant model.

MCDM analysis for sustainable energy decision-making simply gives the means to reduce the challenge, and it has piqued the interest of decision-makers for quite some time. MCDA is a well-known method for assessing long-term sustainability [18]. The use of MCDM to pick the ideally best Li-Ion battery for EV applications and related parametric studies have been proposed recently [19–21]. There are three steps to using any numerical analysis of alternatives decision-making technique:

(a) Identify the relevant criteria and options. (b) Assign numerical values to the relative relevance of the criteria and the effects of the options on them. (c) Sort the numerical data to provide a rating to each option [22]. Multi-criteria analysis has been shown in several studies to be useful method for evaluating, designing, and selecting sustainable energy programs, avoiding failures during implementation and management. A multi-criteria decision support tool was created and validated for assessing residential biogas digester programs in rural parts of Latin America [23].

2 Overview of Biogas Plants

This section presents an overview of various biogas digesters. In general, there are five types of biogas digester plants normally employed for domestic and industrial applications, as described in the following lines.

2.1 *Fixed-Dome Biogas Plant*

A digester and a permanent, stationary gas holder lie atop the digestion in a fixed-dome biogas system. Gas production pushes the slurry into the outflow tank. The increase in gas pressure is directly proportional to the gas stored and the difference between the digester and the outflow tank slurry level heights. The plant is constructed below ground, which protects it from physical harm while also saving the area. During the winter night, the underground digester is protected from the cold. In the summer, though, buried digesters warm very slowly. It is reasonably priced. There are no moving components, so it's straightforward. The lack of steel parts extends the plant's life to 20 years or more. All concrete structures are long-lasting investments. Fixed-dome plants require a lot of work, which means more jobs are created in the area. Plants with permanent domes should be built under the direction of qualified technicians. Gas usage becomes less effective if gas pressure varies dramatically. Gas burners and other basic equipment cannot be set to their best settings. In India, the maximum storage capacity of a home biogas plant is 33% of the rated gas output per day. Animal manure is the main source of feedstock. The internal pressure of up to 0.15 bar is mitigated by soil covering the dome up to the top. Because of

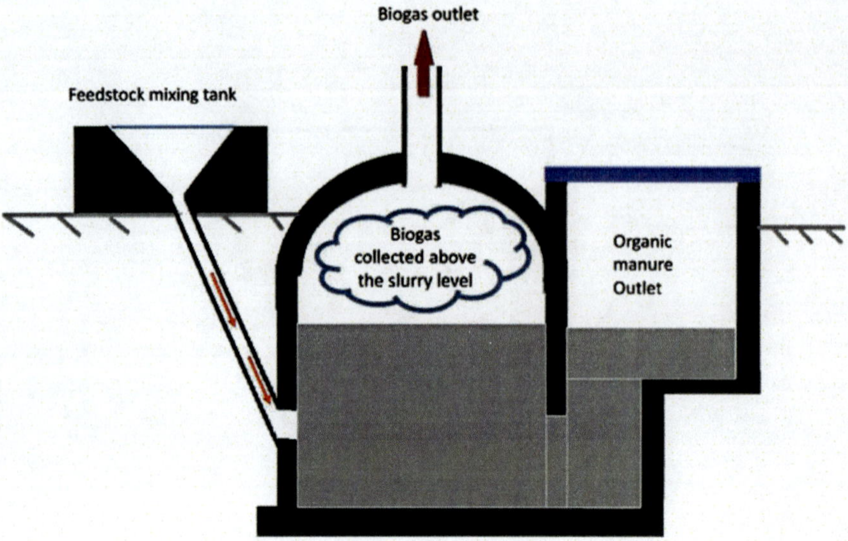

Fig. 2 Fixed-dome biogas plant

cost concerns, the suggested minimum plant size is 5 m^3. Digester volumes of up to 20 m^3 have been reported and are technically feasible. Gas leaks are a major problem with fixed-dome BGPs. The amount of gas generated is not easily noticeable to the operator, and operation is unclear; a stationary dome necessitated careful level planning, and bedrock excavation may be difficult and expensive. Locally sourced construction materials are utilized [6]. From winter to summer, the rate of decrease in CH_4 emissions fluctuates between 11.5 and 15% [15]. Figure 2 shows a sketch of the fixed-dome type biogas plant.

2.2 *Floating Drum Biogas Plant*

In this concept, a storage tank is created by a floating upside-down drum that rests on top of a digester. The tank's up-and-down motion is caused by fluctuations in the collected gas near the top. The inverted drum's weight produces the required pressure for the gas to flow down the pipeline and be utilized. The surface scum is broken by rotating the drum, which contains welded bracing. In the range of 8 to 10 cm of the water column, the pressure in the gas holder remains constant. In the gas holder, only half of the gas production per day is held. In the case of non-consumption of biogas, excess gas escapes from the rim in the form of bubbles. Due to their greater construction costs, floating drums constructed of novel materials such as high-density polyethylene and plastics reinforced with glass fibers are not popular.

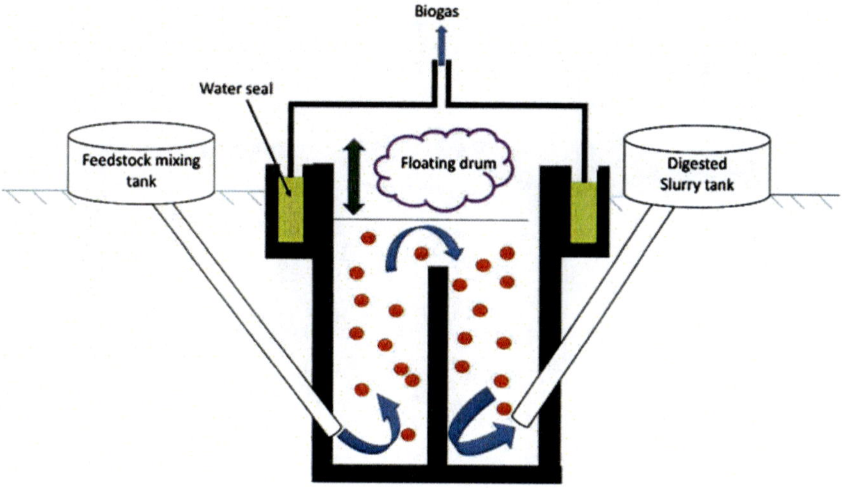

Fig. 3 Floating drum biogas plant

Concrete floating drums reinforced with wire mesh are porous and prone to paper-thin cracking. They necessitate a gastight and elastic interior covering. Due to their lack of UV resistance, polyvinyl chloride (PVC) barrels are not appropriate (UV). Floating drums are easy to use and comprehend. The volume of gas held is plainly observable by the position of the drum, which supplies gas at constant pressure. As long as the gasholder is de-rusted and painted on a regular basis, gas tightness is not an issue. Due to corrosive wear of many steel elements, a floating drum-type biogas plant can survive up to 15 years. Steel drums have a maximum life of 5 years in coastal locations of tropic zone. Painting requires frequent upkeep, which adds to the expensive building cost. [6]. A single person cannot revolve a floating drum with a diameter of 5 m. The unbroken scum will become increasingly solid over time which will decline gas production. In facilities with digester volumes more than 50 m^3, poking no longer delivers enough agitation. Stirring or agitation facilities are necessary. A 5 m diameter floating drum requires a more accurate guiding structure; otherwise, the drum would tilt to the point of jamming. In this way, water-jacket plants are particularly sensitive [17]. Figure 3 shows a sketch of a floating drum biogas plant.

2.3 *Flexible Bag Biogas Plant (Balloon-type)*

A balloon-type biogas plant consists of a flexible bag for biogas storage. Available bag is built of plastic or rubber. By putting weight over the balloon, the required pressure for the gas to flow down the pipeline to the consumption point may be accomplished.

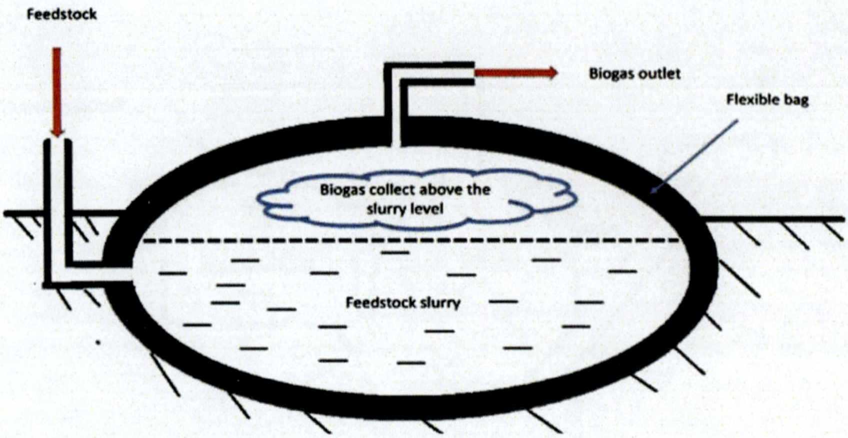

Fig. 4 Flexible bag biogas plant

Balloon skin may be harmed if the gas pressure exceeds the balloon's capacity. Agitation of the slurry inside the balloon proves advantageous to the digestion process. This sort of plant utilizes water hyacinths which is considered as a difficult feed. It is necessary to use UV-resistant balloon material. When compared to other Indian modest domestic biogas plants, this plant has a substantially shorter practical operating life. The discharge is still contagious because of the poor effectiveness of treatment. Due to sullage intrusion, it is not suited when the groundwater table is high. [6]. Biogas output in the balloon-type digester dropped by 70% over the winter, opposed to just 17% in the hemispherical fixed-dome type (Deenbandhu). Because the rubber balloon type is impacted by ambient temperatures, it is not recommended for usage in mountainous places. In steep terrain, construction material transportation expenses are significant. On the other hand, this type of digester is simple to transport, lowering the digester's overall expense. It is also difficult to excavate a large hole under the earth to install digesters at high elevations [24]. Figure 4 shows a flexible bag biogas plant.

2.4 Fixed Dome with Expansion Chamber

This type is similar to the fixed one except that the additional component, which is known as displacement tank or expansion chamber, is provided to help to adjust the pressure inside the main chamber by displacing the slurry. When gas production starts, the pressure inside the reactor increases that helps the slurry moves gradually to the expansion chamber, thereby causing the balanced pressure between reactor and displacement tank. The gas pressure goes up when volume of the gas increases, i.e., due to the difference in heights between two slurry level. The gas pressure will

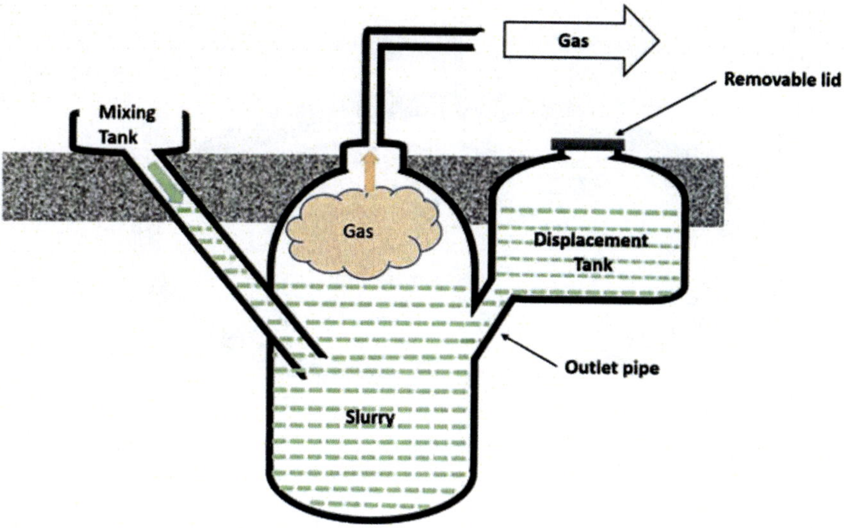

Fig. 5 Fixed dome with expansion chamber

be low if the gas volume is less in the reactor [25, 26]. The fixed dome with expansion chamber is represented in Fig. 5.

2.5 *Double-Stage Biogas Plant*

To maximize operating conditions, anaerobic digestion can be done in two tanks, which is known as a double-stage biogas plant. The retention period is shorter in the double-stage plant. These plants have the ability to produce more biogas, but they come with a greater cost of capital. It is important to install agitation equipment in the digester with a capacity greater than 100 m^3. There are different digester designs available. The types and designs of digesters are chosen based on the sorts of feedstock to be digested [27]. The double-stage method could be used in wastewater treatment, food waste AD plants, or high solid substrates like lignocellulose. In terms of material, structure, degree of mechanism, and efficiency, the options for technological modification are nearly limitless. Hygienically secure digestion, increased gas production, and higher expenses are all advantages of a double-stage system with integrated gas holding [17]. According to a techno-economic study, double-stage AD might be roughly 3% more expensive than single-stage AD. Depending on the feedstock, the operation and purpose of a double stage may differ. Depending on the operating temperatures, most single-stage systems have a retention duration of 18–30 days. The majority of double-stage systems have a cumulative retention duration of 10–18 days. The digester's size is reduced by 25–45% as a result of this. Other features of the double-stage AD include lower digestion capacity, improved

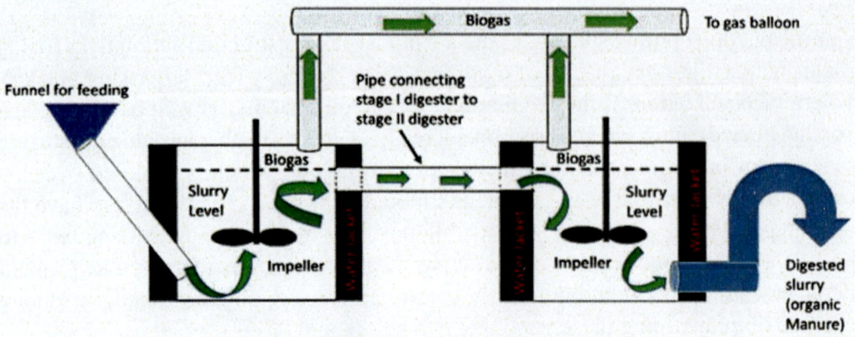

Fig. 6 Biogas plant with two stages, a hot water jacket, and a gas holder

odor management, increased vs destruction, a design to reduce foam throughout the process, and optimization of the retention duration in each stage through solids control [28]. Continuously stirred tank reactors to facilitate adaptability to a wider range of feedstock, reduced treatment time, compact size, standard designs, and suitability for various climates [29]. Figure 6 shows a biogas plant with two stages, a hot water jacket, and a gas holder.

3 Key Factors as Selection Criteria

The plant's long-term viability is hampered by certain design constraints. The operation and maintenance of biogas plants are crucial to their long-term viability. Correct training and quality control, as well as a steady supply of feedstock and utilization of all anaerobic digestion end- and by-products, are all necessary conditions for long-term biogas systems that are suitable for the community and climate. Aside from energy generation and by-products, other benefits include the creation of high-skilled employment in biogas design, engineering, operation, and maintenance, particularly in remote locations [29, 30].

3.1 Process Controllability Factor

Controlling the digestion process will help to ensure the sustenance of AD. This is important to obtain good yield of biogas. By churning the slurry, the bacteria are dispersed throughout. The generation of gas is accelerated by regular stirring manually or automatically [17]. In many times, agitation helps to prevent internal fibrous scum from forming on top of the digesting liquid. Excessive gas pressure may force the substrate out of the outlet openings if the scum is not broken. The froth on the surface of the digester causes it to jam [11]. The retention period is reduced

significantly by optimal stirring. There are numerous construction alternatives for big plants with a digester capacity of more than 20 m^3 (= daily feed surpassing 500 kg), where mechanization of the feeding and discharge systems, as well as mixing and stirring devices, must be properly considered [31]. It is difficult to provide the stirring mechanism in the most of the reactors except the one which is installed over the earth surface. The double-stage reactors mounted above the earth surface have the advantage of having provision for stirring mechanism. Though the stirring helps to ensure homogeneous mixture of the slurry, the process parameters discussed should be taken care of. Most autonomously controlled reactors built across the world are capable of maintaining these parameters at an optimal level.

3.2 *Suitability Factor*

Here the suitability can be categorized into two types, which are suitability based on location and suitability based on feedstock. The suitability of biogas plants to the location is the one of the major criteria for the selecting the biogas plant. Also, a significant effect of feedstock quality on the life of reactor is observed. The next sub-sections go through these in detail.

3.2.1 Suitability Based on Location (SL)

Location decides which type of biogas plant can be employed. Low ambient temperatures have an influence on the biogas fermentation process because they increase the viscosity of the liquids involved, limiting microbial activity, and methane generation [4]. When the temperature drops below 15 °C in biogas digesters without insulation, the process becomes unstable. Only the mesophilic range (temperature up to 35 °C) is of importance in basic plants [31]. Gas production is reduced by 23–37% during the lower winter months (13–14 °C ambient temperature) compared to the rest of the year (between 16 and 24 °C ambient temperature), according to research done in Kashmir State [32].

Supplies for construction and maintenance are crucial aspects to consider when selecting a biogas digester plant. An engineer must consider the local availability of digester parts while choosing and/or changing an existing digester design. Spares are a problem for automatically managed plants. Parts are difficult to come by in rural areas, especially those without access to the highway. Parts availability and hence the selection of biogas plant will differ from nation to country. The diameters of floating drums in the model might be depending on the size of floating drums accessible in developing countries [33].

Operator expertise, professional employees, and well-trained workers are all important factors in biogas plant production. There is a lack of technical competence or services necessary for biogas in rural regions, as well as a shortage of experienced people for building and repairs. Apart from that, there are issues such as a lack of

competent human resources for resolving technical issues that arise when operating the biogas plant [4, 6]. Again, the plant which is having complex controlling mechanism shall have to be maintained and operated by technically competent people. Low-cost fixed or floating dome type reactor may.

Many things must be examined before the digester is erected at a certain area. These considerations include the amount of feedstock and water accessible on a daily basis, as well as the available area, soil type on the site, and groundwater level. Water logging is an issue in low-lying locations, thus subsurface digesters are not recommended. Plants that are placed in an open environment receive plenty of sunshine throughout the day, all year. Collecting water will be difficult if it is not readily accessible. As a result, installing biogas is not suggested if the water source is more than 20 m away [6]. Sufficient space should be available for effluent disposal and usage [11].

3.2.2 Suitability Based on Feedstock (SF)

Most of the biogas plants can digest all type of biomass if the appropriate retention time, and other parameters are maintained optimally. However, the biomass containing more H_2S, sulfur, H_2O will accelerate the corrosion on the walls or the gas holders of the biogas plants. The steel material of floating dome type plants is more susceptible to corrosion if the biomass is generally more corrosive in nature. In general, the balloon type does not have this issue.

Organic soluble, which may be easily transformed to VFAs, are abundant in food waste. Since early on, rapid synthesis of volatile fatty acids has a detrimental impact on the anaerobic digestion process, anaerobic digestion of food waste is significantly more vulnerable than anaerobic digestion of agricultural leftovers. Excessive VFA conversion early in the digestive process might produce a pH decrease and hinder the methanogenesis process. Sedimentation and floating layers might be generated based on the feedstock quality and amount. Sedimentation occurs as a result of the excessive number of pollutants in the substrates. The feedstock parameters, such as dry matter, organic dry matter, biogas production potentials, and carbon to nitrogen ration, identify the anaerobic digestion process, as well as the dimensions and form of the digester [7, 34].

3.3 Cost Factor

The costs associated with biogas plants can be categorized into two groups: capital cost and operational cost. These costs play an important role in deciding the better biogas plant.

3.3.1 Capital Cost

The floating dome model from the Indian Khadi & Village Industry Commission (KVIC), with capacities ranging from 1 to 6 m^3, has the highest installation cost, followed by the Janta and lastly the Deenbandhu type biogas plants. The same trend is seen for the annual operational cost. Steel drum cost accounts for around 40% of the overall cost of the installation for the KVIC model [35]. The initial capital expenses of larger, more sophisticated digesters are significantly greater than those of smaller, less complicated ones. In this type of biogas plant, the handling of organic feedstock per unit digester volume is improved, resulting in more generation of biogas for the same volume of feedstock handled. [11, 36]. In mountainous terrain, transportation expenses for the materials needed to build the digester are significant, resulting in a high capital cost [24].

3.3.2 Operational Cost

The operating expenses of biogas generation vary depending on the source [24, 31]. For an agriculturally based biogas plant, estimates vary from 10 to 16% of capital [24]. From discussions with industry the cost of maintenance and overheads for an agricultural biogas plant are in the region of €5/t feedstock. The digesters which require more high maintenance are the ones that has corrosion issues. Normally, floating-type drums, which are vulnerable to corrosion failures, need to be maintained more frequently, that leads to high operational cost. If the reactor which does not have the provision for increasing or maintaining temperature, ambient temperature will require pre-heated feedstock. Fixed-type plants require minimal maintenance due to their relative simplicity in terms of design and lack of moving components. The reactors with heating and stirring facility that consume more energy do incur high operating cost.

4 MCDM Model

The MCDM methods are best—suited for addressing selection problem based on several criteria. The simple but effective method; WPM has been applied to select the best biogas plant. This is described below.

The chosen performance factors are rated appropriately as in Table 1. The factors' performance may differ from excellent (E) to poor (P). The codes "average" (A) and "good" (G) are used in between. On a scale of 0–10, each rating is assigned a performance code [25]. The performance code will be standardized based on the sort of factor that may be favorable or unfavorable, using a modified version of the scale described in [25]. Each kind of biogas plant has a different performance code, which varies based on the degree of performance of each element. A performance code is assigned between 0 and 4 if a factor's performance is low, and between 8 and 10 if

Table 1 Performance rating

Type of biogas plant	PC	SL	SF	CC	OC
Fixed dome	P	G	P	A	G
Fixed dome with expansion chamber	A	G	P	A	G
Floating drum	A	A	P	P	P
Balloon type	P	P	A	A	G
Double stage with hot water jacket and stirrer	E	P	E	P	P

Table 2 Performance code for each biogas plant's factors

Type of biogas plant	PC	SL	SF	CC	OC
Fixed dome	4	8	4	7	8
Fixed dome with expansion chamber	7	8	4	7	8
Floating drum	7	7	4	4	4
Balloon type	4	4	7	7	8
Double stage with hot water jacket and stirrer	10	4	10	4	4

the factor's performance is outstanding. "Average" and "good" performance codes are allocated to the ranges 4–7 and 7–8, respectively.

Where there is a consistent energy supply and access to components, a double stage with hot water jacket and stirrer will be ideal [33]. The floating drum biogas plant, followed by the fixed-dome type of biogas plant, has the highest installation and hence yearly running cost for all capacities ranging from 1 to 6 m^3 [35]. A study undertaken to see if weed could be utilized as a biogas feedstock in a tubular digester came to the conclusion that weed could be used as a biogas feedstock [26]. Table 2 shows the performance code for each biogas plant's factors.

Referring to Table 2, let 'A_{ij}' be the performance code of the 'ith' biogas plant, using the 'jth' factor. As previously indicated, we may categorize the factors as favorable or unfavorable, and then normalize the quantitative value of the unfavorable factor using Min $(A_{ij})/A_{ij}$ and the favorable factor using $A_{ij}/\text{Max}\ (A_{ij})$ to create a normalized performance code, as shown in Table 3. All variables, save the final two, CC and OC, were determined to be favorable, implying that the greater the value of the factor, the better the biogas plant. Table 3 shows the normalized performance code.

According to WPM, suitable weights are used to assess the performance codes for each component depending on their relevance, and the product of the normalized weighted performance code of all factors is used to evaluate the performance codes.

$$A_i^{\text{WPM}} = \prod_{j=1}^{n} x_{ij}^{wi}$$

where ith type of biogas plant is having, $j = 1, 2, \ldots, n$ factors.

Table 3 Normalized performance code

Type of biogas plant	PC	SL	SF	CC	OC
Fixed dome	0.4	1	0.4	0.57	0.5
Fixed dome with expansion chamber	0.7	1	0.4	0.57	0.5
Floating drum	0.7	0.87	0.4	1	1
Balloon type	0.4	0.5	0.7	0.57	0.5
Double stage with hot water jacket and stirrer	1	0.5	1	1	1

Table 4 Performance score and ranking

Type of biogas plant	Performance score	Ranking
Fixed dome	0.46	4
Fixed dome with expansion chamber	0.53	3
Floating drum	0.70	2
Balloon type	0.44	5
Double stage with hot water jacket and stirrer	0.84	1

All of the parameters are given equal weights (say, $w = 0.25$), as they are all equally significant. The performance codes are then rated to help with the biogas plant selection. Table 4 analyzes and rates the performance of different types of biogas plants. Table 4 shows the performance score and ranking.

5 Conclusion

In this paper, a simple yet powerful MCDM method; WPM (Weighted Product Method) has been employed to choose the best biogas plant. Five different biogas plants have been considered in this work. From the study, it has been observed that the double-stage biogas with hot water jacket and stirrer mechanism has scored well in overall performance, with the performance code of **0.84**. The selected plant consumes more power that leads to increased operational cost but with excellent process controllability and suitability to feed stocks places it in first place. The gas productivity is also good as the process is well controlled. The floating drum and fixed dome with expansion chamber ranked as 2 and 3, respectively, and the fixed and balloon-type models are least preferred. A further study is required to perform the life cycle assessment in order to check whether the process is sustainable and cost effective by analyzing the energy consumption and environmental impact.

References

1. Bond T, Templeton MR (2011) History and future of domestic biogas plants in the developing world. Energy Sustain Dev 15(4):347–354
2. Kapoor R, Ghosh P, Tyagi B, Vijay VK, Vijay V, Thakur IS, Kamyab H, Nguyen DD, Kumar A (2020) Advances in biogas valorization and utilization systems: a comprehensive review. J Clean Prod 273:123052
3. Wellinger A, Murphy JD, Baxter D (eds) (2013) The biogas handbook: science, production and applications. Elsevier
4. Nevzorova T, Kutcherov V (2019) Barriers to the wider implementation of biogas as a source of energy: a state-of-the-art review. Energ Strat Rev 26:100414
5. Al Seadi T (2008) Biogas handbook. 2008 ed. Syddansk Universitet, Esbjerg
6. Baredar P, Khare V, Nema S (2020) Design and optimization of biogas energy systems. Academic Press
7. Nsair A, Onen Cinar S, Alassali A, Abu Qdais H, Kuchta K (2020) Operational parameters of biogas plants: a review and evaluation study. Energies 13(15):3761
8. Deublein D, Steinhauser A (2011) Biogas from waste and renewable resources: an introduction, 2nd edn. Wiley, Germany
9. Patinvoh RJ, Taherzadeh MJ (2019) Challenges of biogas implementation in developing countries. Curr Opin Environ Sci Health 12:30–37
10. Gummert M, Hung NV, Chivenge P, Douthwaite B (2020) Sustainable rice straw management. Springer Nature, p 192
11. Mattocks R (1984) Understanding biogas generation. Volunteers in Technical Assistance, Virginia
12. Meegoda JN, Li B, Patel K, Wang LB (2018) A review of the processes, parameters, and optimization of anaerobic digestion. Int J Environ Res Public Health 15(10):2224
13. Yong Z, Dong Y, Zhang X, Tan T (2015) Anaerobic co-digestion of food waste and straw for biogas production. Renewable Energy 78:527–530
14. Arya A, Badgujar S, Kumari A, Badgujar K, Singh RK (2020) State-of-the-art, challenges, and issues of biogas production technology in India: a review. Microbiol Res Int 8(4):57–75
15. Khoiyangbam RS, Kumar S, Jain MC, Gupta N, Kumar A, Kumar V (2004) Methane emission from fixed dome biogas plants in hilly and plain regions of northern India. Biores Technol 95(1):35–39
16. Budiman I (2020) The role of fixed-dome and floating drum biogas digester for energy security in Indonesia. Indonesian J Energy 3(2):83–93
17. Sasse L (1988) Biogas plants. Deutsches Zentrum für Entwicklungstechnologien, Wiesbaden. Germany
18. Wang JJ, Jing YY, Zhang CF, Zhao JH (2009) Review on multi-criteria decision analysis aid in sustainable energy decision-making. Renew Sustain Energy Rev 13(9):2263–2278
19. Loganathan MK, Mishra B, Tan CM, Kongsvik T, Rai RN (2021) Multi-Criteria decision making (MCDM) for the selection of Li-Ion batteries used in electric vehicles (EVs). Mater Today Proc 41:1073–1077
20. Loganathan MK, Ming Tan C, Mishra B, Msagati TAM, Snyman LW (2019) Review and selection of advanced battery technologies for post 2020 era electric vehicles. In: 2019 IEEE transportation electrification conference (ITEC-India). Bengaluru, India, pp 1–5. https://doi.org/10.1109/ITEC-India48457.2019.ITECINDIA2019-254
21. Loganathan MK, Tan CM, Sultana S, Yun Lisa Hsieh I, Kumaraswamidhas LA, Rai RN (2021) Parametric performance analysis of battery operated electric vehicles. In: Proceedings of International conference on sustainable energy and future electric transportation (SEFET), 21–23 Jan 2021. Hyderabad, India, pp 1–6. https://doi.org/10.1109/SeFet48154.2021.9375788
22. Triantaphyllou E (2000) Multi-criteria decision making methods. In: Multi-criteria decision making methods: a comparative study. Springer, Boston, MA, pp 5–21

23. Ferrer-Martí L, Ferrer I, Sánchez E, Garfí M (2018) A multi-criteria decision support tool for the assessment of household biogas digester programmes in rural areas. A case study in Peru. Renew Sustain Energy Rev 95:74–83
24. Rajendran K, Aslanzadeh S, Taherzadeh MJ (2012) Household biogas digesters—a review. Energies 5(8):2911–2942
25. Saleh A (2015) Comparison among different models of biogas plants. Department of Chemical Engineering, COMSATS Institute of Information Technology, Lahore
26. Kabeyi MJB, Olanrewaju OA (2020) Development of a biogas plant with electricity generation, heating and fertilizer recovery systems. In: Proceedings of the International conference on industrial engineering and operations management (2020)
27. Samer M (2012) Biogas plant constructions. In: Kumar S (ed) Biogas, pp 343–368
28. Rajendran K, Mahapatra D, Venkatraman AV, Muthuswamy S, Pugazhendhi A (2020) Advancing anaerobic digestion through two-stage processes: current developments and future trends. Renew Sustain Energy Rev 123:109746
29. Task IB (2018) Integrated Biogas Systems: IEA Bioenergy
30. Afridi ZU, Qammar NW (2020) Technical challenges and optimization of biogas plants. ChemBioEng Rev 7(4):119–129
31. Eggeling G (1981) Biogas: manual for the realisation of biogas programmes. Ubersee-Museum Bremen, Borda
32. Breitenmoser L, Gross T, Huesch R, Rau J, Dhar H, Kumar S, Hugi C, Wintgens T (2019) Anaerobic digestion of biowastes in India: opportunities, challenges and research needs. J Environ Manage 236:396–412
33. Rowse LE (2011) Design of small scale anaerobic digesters for application in rural developing countries. University of South Florida
34. Li Y, Park SY, Zhu J (2011) Solid-state anaerobic digestion for methane production from organic waste. Renew Sustain Energy Rev 15(1):821–826
35. Singh KJ, Sooch SS (2004) Comparative study of economics of different models of family size biogas plants for state of Punjab, India. Energy Convers Manage 45(9–10):1329–1341
36. Theuerl S, Herrmann C, Heiermann M, Grundmann P, Landwehr N, Kreidenweis U, Prochnow A (2019) The future agricultural biogas plant in Germany: a vision. Energies 12(3):396

Mathematical Modeling and Simulation of Dual Fuel Cycle Using Natural Gas and Diesel/Biodiesel

Brijesh Dager, Ajay Kumar, R. S. Sharma, Ajay Chhillar, and Prabhakar Sharma

1 Introduction

The rising cost and imperfect supply of crude petroleum fuels carry researchers toward the investigation of performance, emission, and combustion characteristics of the engine on a cheap and easily available alternative source of energy for surface transportation [1, 2]. Moreover, rapid change in the pollution standards leads toward clean and green alternative fuels instead of using conventional petroleum-based fuels. The broad cause of the above problem is created by vehicular emissions using gasoline and diesel-fuelled internal combustion engine [3, 4]. Therefore researchers should investigate their results by using a cheaper and cleaner alternative fuel. Compressed natural gas seems to be an aspiring fuel for the above problem as its main constituent is methane. It is abundantly available and cheaper than gasoline and conventional diesel fuel. Gasoline fuelled engines gives quiet performance using CNG but diesel-fuelled engine requires more research work to use CNG as the main fuel with petroleum-based pilot diesel in dual fuel mode [5, 6].

To optimize a process or operation experimentally is a complex task as it is associated with various problems related to design and optimization. To eliminate this problem, mathematical modeling and computer simulation have become influential and economical tools, which reduce the setup cost of experiments and time [4,

B. Dager (✉) · A. Chhillar · P. Sharma
Mechanical Engineering Department, Delhi Skill and Entrepreneurship University, Delhi 110089, India
e-mail: brijesh_dagar1297@yahoo.com

A. Kumar
Mechanical Engineering Department, D.C.R. University of Science and Technology, Sonepat, Haryana 131039, India

R. S. Sharma
Automobile Engineering Department, G.B. Pant Institute of Technology, Okhla, New Delhi 110020, India

A. Kumar et al. (eds.), *Recent Advances in Manufacturing and Thermal Engineering*, Lecture Notes in Mechanical Engineering, https://doi.org/10.1007/978-981-19-8517-1_22

7]. The combustion process in an internal combustion engine is a highly nonlinear and complex phenomenon. While designing the engine in earlier days, extensive testing of the prototype was a very hard and complex task. Furthermore, selecting the optimized condition of design from such testing was very difficult too [8]. Therefore, optimization analysis utilizing computer simulation has become quite popular in recent years [9, 10]. The various processes (inlet, compression, expansion, and exhaust strokes) of a direct injection engine can easily be simulated by using a mathematical modeling tool [11, 12]. Depending upon the equation which gives its predominant structure based on energy conservation, the model is treated as thermodynamic (zero-dimensional) [13]. The performance characteristics can be analyzed using this model as it is based on the first law of thermodynamics. Whereas when based on full analysis of the fluid motion is treated as a fluid dynamic (multidimensional) model and gives a spatial detail inside the engine at the cost of a large investment in computational time and equipment [14, 15]. A reasonable prediction of pressure and temperature concerning time can be determined by using these models.

The main objective of the present research work is to develop computer codes to simulate the dual fuel cycle processes using conventional and alternate fuels of the form $C_XH_YO_Z$. This model delineates the process that occurred in the working fluid of the CNG-diesel/biodiesel dual fuel compression ignition engine and also gives an idea about the performance restrictions and trends as a function of system variables. A step-by-step simulation approach of analysis is applied in the present study. Firstly, Ideal Cycle Simulation (ICS) approach is applied which is based on the ideal air standard diesel cycle and it is assumed that the working fluid of the engine is air. Secondly, the Fuel–air Cycle Simulation approach (FCS) is applied by introducing a mixture of air and petroleum diesel as a working fluid by incorporating adiabatic flame temperature for combustion analysis. It is also assumed in the second step that combustion is taking place under adiabatic conditions and at constant pressure. In the third step, the Dual fuel Cycle Simulation approach (DFCS) is applied first by introducing CNG-air mixture and pilot diesel as working fluid, and later on, DFCS is applied first by introducing CNG-air mixture and pilot biodiesel as the working fluid. Hence the values of all the parameters like temperature, volume, pressure, power output, network output, thermal efficiency, mean effective pressure, etc. at all the state points of ICS, FCS, and DFCS during adiabatic combustion of a chemically correct air–fuel mixture are calculated. The main constituent of CNG available in the National Capital region is 84.5% of methane, 7.7% of ethane, and 2.4% of propane on a mole basis. In the present study, the effect of all the constituents of CNG is incorporated into the performance parameters of the dual fuel engine. The predicted results are computed and presented graphically.

Table 1 Pipeline CNG Composition in India (Mole %)

Constituent	Delhi	IPCL	Mumbai
Methane (CH_4)	84.5	88.42	82.55
Ethane (C_2H_6)	7.7	8.79	7.67
Propane (C_3H_8)	2.4	1.59	3.85
I-Butane (I- C_4H_{10})	0.26	0.29	0.64
N-Butane (n-C_4H_{10})	0.32	0.28	0.78
I-Pantene (I-C_5H_{12})	0.18	0.05	0.13
N-Pentane (n-C_5H_{12})	0.19	0.05	0.13
Hexane (C_6H_{14})	0.17	0.04	0.09
Nitrogen (N_2)	0.12	0.20	0.07
Carbon dioxide (CO_2)	4.23	0.27	0.07

2 Materials and Methods

2.1 *Fuel Compositions and Their Properties*

The composition of natural gas depends upon the source of the gas whether it is from an oil condensate field or exists by itself. The oil condensate gas is referred to as associated gas, whereas the other is non-associated gas. The earlier one contains heavier hydrocarbons like butane, pentane, ethane, pentane, and hexane in significant amounts along with methane up to 50%, whereas the non-associated gas contains a much higher percentage of methane up to 85% [16]. The largest component of CNG is methane as compared to other gases; therefore, it has the same characteristics as that methane. Therefore, it becomes essential here to know about the various components of CNG in this study and their effects on the combustion phenomena in a diesel engine. CNG is odorless, tasteless, colorless, non-volatile, toxic, and lighter than air having a relative vapor density of 0.68 [17]. The various compositions of pipeline CNG supplied in India are shown in the Table 1 on a mole basis.

Mustard oil was extracted from raw mustard in the oil mill and then with the help of the transesterification process, it was converted to methyl ester, a very most attractive process of production of biodiesel [18–20]. The temperature during the preparation of MOME such as reaction temperature was kept 60 °C. Diesel was purchased from a local filling station, and the various properties of the fuels are listed in Table 2.

2.2 *Numerical Simulation of CI Engine Using Air as Working Fluid*

It is advantageous to devise closed cycles that approximate open diesel cycle for analysis of compression ignition processes. One such approach is ideal cycle simulation

Table 2 Properties of CNG, diesel, and MOME

Properties	CNG	Diesel	MOME
Boiling point (K at 1 atm)	147	433–655	N/A
State	Gas	Liquid	Liquid
Auto ignition temperature (K)	905	477–533	N/A
Density (Kg/m^3)	128	785–881	880
Research octane number	130	N/A	N/A
Net energy content (MJ/Kg)	49.5	43.9	39.85
Combustion energy (KJ/m^3)	24.6	36	N/A
Cetane index	N/A	55	50
Flash point (K)	124	325	313

(ICS) with air as the working medium. This simulation is based on the following assumptions [21]:

- The working medium is assumed to be an ideal gas.
- There are no intake and exhaust processes.
- Working fluid throughout the cycle is a fixed mass of air.
- Process of heat addition from an external source is incorporated in place of combustion process is replaced.
- Heat addition takes place at constant pressure.
- The process of heat rejection to the surrounding takes place at constant volume.
- All processes are internally reversible.
- No heat transfer to the surrounding.
- Friction is neglected.
- The working medium has constant specific heats (C_P and C_V).

Based on these assumptions, computer codes were developed in EES software for the simulation of ideal diesel cycle, the values of various engine parameters are computed, and the predicted results for ideal cycle simulation are represented graphically.

2.3 *Mathematical Modeling of CNG-Diesel Dual Fuel–Air Diesel Cycle (DFCS)*

Based on the ideal cycle simulation, for diesel-air cycle processes (FCS), computer codes were developed in EES software for chemically correct ($Y = Y_{CC}$), chemically lean ($Y > Y_{CC}$), and chemically rich ($Y < Y_{CC}$) fuel–air mixtures. In FCS simulation, fuel–air mixture is considered as the working medium and the intake process with simplified assumptions is also included. To simulate the combustion process, constant pressure adiabatic combustion calculations are incorporated and this simulation was termed fuel–air cycle simulation (FCS). The main aim of the

Table 3 Engine specifications used in the simulation

Parameters	Values
Stroke length (S)	110.0 mm
Cylinder bore (B)	80.0 mm
Connecting rod length (L)	230.0 mm
Compression ratio (r)	16.5
Displacement volume (V_{disp})	553 (cc)
Volume at TDC (v_{tdc})	35.6 (cc)
Volume at BDC (v_{bdc})	588.6 (cc)

simulation using adiabatic combustion calculations is to evaluate the power output and thermal efficiency for a given engine speed, ambient air temperature, and fuel and fuel–air ratio assuming naturally aspirated conditions. The predicted results of the diesel-air cycle (FCS) and dual fuel cycle (DFCS) simulation are presented graphically. Numerical simulation of CNG-diesel dual fuel cycle processes is achieved by suitably modifying the FCS equations by assuming the CNG-air mixture behaves as a perfect gas and does not undergo any chemical change during the compression process. The dual fuel cycle remains based on the diesel cycle. The processes of the diesel cycle i.e., compression, constant pressure heat addition, expansion, exhaust, and intake are modified to simulate the dual fuel cycle processes assuming constant pressure adiabatic combustion. Once again computer codes were developed in EES software for chemically correct ($Y = Y_{CC}$), chemically lean ($Y > Y_{CC}$), and chemically rich ($Y < Y_{CC}$) air and fuel mixtures for evaluating the performance parameters of the dual fuel cycle engine with 15% diesel fuel and 85% CNG fuel with 100% methane and with 84.5% methane, 7.7% ethane, and 2.4% propane content in it. The predicted results of dual fuel thermodynamic cycles simulation are tabulated and presented graphically for 15% diesel and 85% CNG. The details of the engine under consideration for simulation are given in Table 3.

For estimation of pressure, volume, and temperature at all the state points of the CNG-diesel dual fuel cycle and for computing the thermal efficiency, power output, mean effective pressure, and network output computer codes were developed in EES software. The initial conditions at state point 1 in the dual fuel cycle are assumed as $P_1 = P_5 = P_a = P_{ag} =$ ambient pressure $= 101.325$ kPa. $V_1 = V_{bdc}$, $T_a = T_m = T_{ag} = 298$ K. Following equations were formulated for compression, combustion, expansion, and exhaust and intake processes to compute various CNG-diesel dual fuel engine cycle parameters:

2.4 Isentropic Compression Process of Dual Fuel Cycle

Initially, let us assume the value of N_{Xg} and T_1 (Fig. 1) both depending upon the previous cycle, which in turn depends on the cycle before that, and so on [22]. To

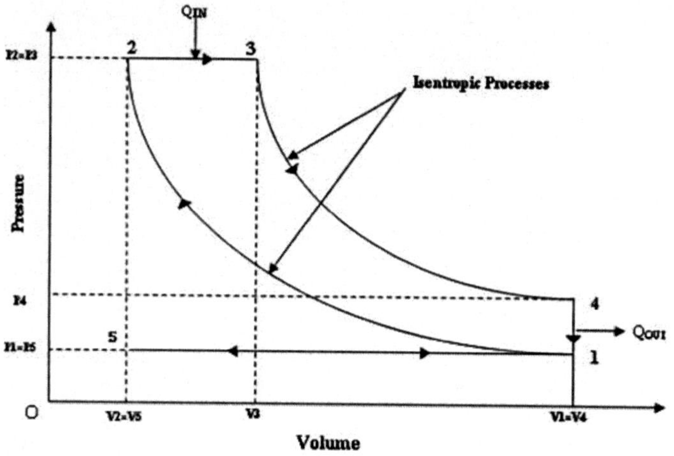

Fig. 1 P–V diagram of CNG-Diesel dual fuel cycle for naturally aspirated conditions

start with, let us assume, $N_x = 0$, $T_1 = T_a = T_{ag} = 298\,\text{K}$.

$$N_{ag} = \frac{P_1 \times V_{bdc}}{R_U \times T_{ag}} \tag{1}$$

where, N_{ag} is kilo mole of air and CNG taken in during suction,

P_1 = Pressure of air + CNG at state point 1,

R_U = Universal gas constant = 8.314 kJ/Kg mol. K,

$T_1 = T_{ag}$ = Absolute temperature at state point1.

$$\text{Kr}_{\text{dual}} = \frac{\text{Cpr}_{\text{dual}}}{\text{Cpr}_{\text{dual}} - R_U} \tag{2}$$

where, Kr_{dual} is polytropic index of compression for the reactant's mixture of dual fuels.

$$\text{Cpr}_{\text{dual}} = \frac{\text{Cpf}_{\text{dual}} + Y_{cc} \times \text{Cp}_{\text{O2}} + 3.76Y_{cc} \times \text{Cp}_{\text{N2}}}{N_{mo}} \tag{3}$$

where Cpr_{dual} stands for constant pressure heat capacity of reactant mixture of air and CNG.

$$\text{Cpf}_{\text{dual}} = N_{\text{biodiesel}} \times \text{Cpf}_{\text{biodiesel}} + N_{\text{diesel}} \times \text{Cpf}_{\text{diesel}} + N_{\text{cng}} \times \text{Cpf}_{\text{cng}} \tag{4}$$

$$\text{Cpf}_{\text{cng}} = N_{\text{methane}} \times \text{Cpf}_{\text{CH}_4} + N_{\text{ethane}} \times \text{Cpf}_{\text{C}_2\text{H}_6} + N_{\text{propane}} \times \text{Cpf}_{\text{C}_3\text{H}_8} \tag{5}$$

$$N_{mo} = 1 + 4.76 \times Y_{cc} \tag{6}$$

where N_{mo} represents kilo moles of air + biodiesel + diesel + CNG vapors in a mixture containing 01 kmol of fuel.

$$T_2 = T_1 \times r^{(\mathrm{Kr}_{\mathrm{dual}}-1)} \tag{7}$$

where T_2 is the temperature at state point 2 in the CNG-diesel dual fuel cycle and $\mathrm{Kr}_{\mathrm{dual}}$ is ratio of specific heats of reactant mixture of dual fuels.

$$P_2 = P_1 \times r^{\mathrm{Kr}_{\mathrm{dual}}} \tag{8}$$

where, P_2 is the pressure at state point 2 in the dual cycle,

$$V_2 = V_{\mathrm{tdc}} \tag{9}$$

2.5 Adiabatic Combustion Process

Combustion process is assumed to take place adiabatically and under constant pressure. For no heat transfer, the energy equation can be written as [21]:

$$\mathrm{Hr}(T_2) + \mathrm{Hf}\big(T_f\big) = \mathrm{Hp}(T_3) \tag{10}$$

where the suffixes *r, f,* and *p* denote the $N_a + N_g + N_X$ moles of reactants, fuel (Diesel + CNG), and the products of combustion. Solution of Eq. 10 will give the value of T_3. Since the combustion process is at constant pressure, $P_3 = P_2$. Knowing P_3 and T_3, volume at point3 is given by Eq. 11.

$$V_3 = N_p \times R_U \times \left(\frac{T_3}{P_3}\right) \tag{11}$$

For a constant pressure adiabatic combustion, the energy released, when the products are cooled to initial temperature T_1 of the reactant, when a unit quantity of fuel burn at constant pressure, Q_p is given by Eq. 12.

$$Q_p = \sum\{N_i \times [h_i(T_3) - h_i(T_2)]\} \tag{12}$$

This equation was solved using Newton–Raphson iteration technique by knowing the product gas mole numbers, Ni in Eq. 12.

If Y_{cc} represents moles of O_2 per mole of ($C_{10}H_{20}O_2 + C_{10}H_{22}$ + CNG) fuel for chemically correct mixture then the reactants fuel–air mixture for dual fuel combustion is

$$N_{\text{biodiesel}} \times C_{10}H_{20}O_2 + N_{\text{diesel}} \times C_{10}H_{22}$$
$$+ N_{\text{cng}} \times \left(N_{\text{methane}} \times CH_4 + N_{\text{ethane}} \times C_2H_6 + N_{\text{propane}} \times C_3H_8\right)$$
$$+ Y_{cc} \times O_2 + 3.76 \times Y_{cc} \times N_2 \tag{13}$$

where $N_{\text{biodiesel}}$, N_{diesel}, N_{CNG}, N_{methane}, and N_{propane} denote percentage number of moles of diesel, CNG, methane, ethane and propane, respectively

$$Y = Y_{cc} \times \text{RAF} \tag{14}$$

where, Y represents actual number of moles of O_2 per mole of ($C_{10}H_{20}O_2 + C_{10}H_{22}$ + CNG) fuels. *RAF* stands for relative air–fuel ratio.

$$Y_{cc} = N_{mc} + 0.25 \times N_{mh2} - 0.5 \times N_{mo2} \tag{15}$$

$$N_{mc} = N_{\text{biodiesel}} \times 10 + N_{\text{diesel}} \times 10 + N_{\text{cng}} \times \left(N_{CH_4} + N_{C_2H_6} \times 2 + N_{C_3H_8} \times 3\right) \tag{16}$$

$$N_{mh2} = N_{\text{biodiesel}} \times 20 + N_{\text{diesel}} \times 22 + N_{\text{cng}}$$
$$\times \left(N_{CH_4} \times 4 + N_{C_2H_6} \times 6 + N_{C_3H_8} \times 8\right) \tag{17}$$

$$N_{mo2} = N_{\text{biodiesel}} \times 2 + N_{o2} \times 2 \tag{18}$$

where N_{mc}, N_{mh2}, and N_{mo2} represent the number of moles of carbon, hydrogen, and oxygen atoms, respectively, in the fuel. For CI engines converted to dual fuel, a mixture of $C_{10}H_{20}O_2 + C_{10}H_{22}$ + CNG is used as fuel, the various product mole numbers are set as:

$$N_{co} = N1 \text{ and } N1 = 0, \tag{19}$$

$$N_{CO_2} = N2 \text{ and } N2 = 2.51, \tag{20}$$

$$N_{H_2O} = N3 \text{ and } N3 = 3.49, \tag{21}$$

$$N_{N_2} = N4 \text{ and } N4 = 3.76 \times Y_{CC}, \tag{22}$$

$$N_{O_2} = N5 \text{ and } N5 = 0, \tag{23}$$

where N_{CO}, N_{CO_2}, N_{H_2O}, N_{N_2}, and N_{O_2} represent the product mole numbers of carbon monoxide, carbon dioxide, water, nitrogen, and oxygen present in the combustion products, respectively. With the known values of product mole numbers N1, N2, N3 N4, and N5 in the combustion products, the value of the adiabatic flame temperature

of all the three cases described above can be calculated by using the following equations:

$$\text{Qp}_{\text{dual}} = \left(\text{Hrp}_{\text{dual}} - N_{\text{CO}} \times 282800\right) \times \frac{N_m}{N_{\text{mo}}} \tag{24}$$

where Qp_{dual} represents the heat energy released at constant pressure when the products are cooled to initial temperature $T_1 = 298$ K, of the reactants consisting of CNG and biodiesel when a unit quantity of these fuels burns at constant pressure.

$$H_{\text{rp}_{\text{dual}}} = N_{\text{biodiesel}} \times H_{\text{rp}_{\text{biodiesel}}} + N_{\text{diesel}} \times H_{\text{rp}_{\text{diesel}}} + N_{\text{CNG}} \times H_{\text{rp}_{\text{CNG}}} \tag{25}$$

$$H_{\text{rp}_{\text{CNG}}} = N_{\text{CNG}} \times \left(0.845 \times H_{\text{rp}_{\text{CH}_4}} + 0.125 \times H_{\text{rp}_{\text{C}_2\text{H}_6}} + 0.03 \times H_{\text{rp}_{\text{C}_3\text{H}_8}}\right) \tag{26}$$

where $H_{\text{rp}_{\text{dual}}}$ and $H_{\text{rp}_{\text{CNG}}}$ represent heat of reaction for dual and CNG at constant pressure, respectively, and is defined as the heat energy added to bring the products of combustion to the initial temperature $T_1 = 298$ K when a unit quantity of the fuel and chemically correct oxygen burn at constant pressure. The adiabatic flame temperature for constant pressure combustion process is given by Eq. 27.

$$\text{Tnew}_{\text{dual}} = T_{\text{dual}} - \frac{\left(\text{HpT}_{\text{dual}} - \text{HpTr}_{\text{dual}} - \text{Qp}_{\text{dual}}\right)}{\text{Cp}_{\text{dual}}} \times \frac{N_p}{N_{\text{po}}} \tag{27}$$

$$T_3 = \text{Tnew}_{\text{dual}} \tag{28}$$

where T_3 is the absolute temperature at state point3, T_{dual} is the assumed value of adiabatic flame temperature of the products of combustion for a dual fuel combustion process, and $\text{Tnew}_{\text{dual}}$ is value of AFT obtained after first iteration. The values of Hp_{Tdual}, Qp_{dual}, and Cp_{Tdual} have been calculated using standard JANAF tables.

2.6 *Isentropic Expansion Process*

After computing the values of T_3, P_3, and V_3, corresponding values at point 4 in CNG-diesel dual fuel cycle can be calculated by Eq. 29 to Eq. 31.

$$T_4 = T_3 \times (V_3/V_{\text{bdc}})^{(\text{Kp}_{\text{dual}}-1)} \tag{29}$$

$$P_4 = P_3 \times \left(\frac{V_3}{V_{\text{bdc}}}\right)^{\text{Kp}_{\text{dual}}} \tag{30}$$

$$V_4 = V_{\text{bdc}} \tag{31}$$

where T_4, P_4, and V_4 are the values of temperature, pressure, and volume at state point4 of dual fuel cycle, respectively.

$$\mathrm{Kp}_{\mathrm{dual}} = \frac{\mathrm{Cpp}_{\mathrm{dual}}}{\mathrm{Cpp}_{\mathrm{dual}} - R_U} \tag{32}$$

where $\mathrm{Kp}_{\mathrm{dual}}$ is specific heat ratio of the products of combustion of dual fuels.

$$\mathrm{Cpp}_{\mathrm{dual}} = \frac{\mathrm{Cp}(T3_{\mathrm{dual}})}{N_{\mathrm{po}}} \tag{33}$$

where $\mathrm{Cp}_{\mathrm{dual}}$ represents constant pressure specific capacity of products of dual fuel combustion and $\mathrm{CpT}_{3\mathrm{dual}}$ is the constant pressure specific heat capacity of products at temperature $\mathrm{T}_{3\mathrm{dual}}$.

2.7 Exhaust and Intake Processes

At state point 5, pressure drops to ambient pressure and the temperature T_5 can be calculated using Eq. 34.

$$T_5 = T_4 \times \left(\frac{P_1}{P_4}\right)^{(\mathrm{Kp}_{\mathrm{dual}}-1)/\mathrm{Kp}_{\mathrm{dual}}} \tag{34}$$

$$N_{\mathrm{xg1}} = \frac{P_1 \times V_{\mathrm{bdc}}}{R_U \times T_5} \tag{35}$$

where $\mathrm{N}_{\mathrm{Xg1}}$ is the number of moles of exhaust gas fraction left in the combustion chamber after the first cycle. The absolute temperature of air, CNG, and residual gas fraction at state point1 i.e., at the commencement of next cycle was calculated using Eq. 36. This temperature of air in the combustion chamber in beginning of next cycle will be little higher than ambient air temperature due to the hot gases left from the previous cycle.

$$T_{\mathrm{ag1}} = \frac{r \times T_{\mathrm{ag}}}{r - 1 + \frac{T_{\mathrm{ag}}}{T_5}} \tag{36}$$

where r stands for compression ratio, T_{ag} and T_5 represent ambient air-CNG temperature and exhaust gas temperature, respectively. Number of kmoles of air and CNG mixture taken in during intake process of the next cycle is given by Eq. 37.

$$N_{ag1} = \frac{P_1 \times V_{tdc}}{R_U \times T_{ag}} - N_{xg} \tag{37}$$

Table 4 Predicted results of diesel fuel cycle simulation (FCS)

Parameters	Cycle					
	1	2	3	4	5	6
T_1 (K)	298	313	328.6	344.7	361.4	378.6
T_2 (K)	800.4	833.2	866.8	900.9	935.7	970.9
T_3 (K)	2998	2999	2999	2999	2999	2999
T_4 (K)	2095	2075	2055	2036	2017	1999
T_5 (K)	1433	1435	1438	1440	1443	1446
P_2 (kPa)	4491	4451	4410	4369	4329	4288
P_3 (kPa)	4491	4451	4410	4369	4329	4288
P_4 (kPa)	712.2	671.6	633.6	598.4	565.6	535.1
Rc	3.746	3.599	3.459	3.329	3.205	3.089
Power (kW)	24.42	23.19	22.00	20.88	19.80	18.77
Pmep (bar)	17.43	16.55	15.70	14.90	14.13	13.39
η thermal	48.44	48.31	48.13	47.90	47.63	47.30
Wnet	976.7	927.4	880.1	835.0	791.8	750.8

With new values of T_{ag1}, N_{ag1}, and N_{xg1}, new value of Cpr_{dual} is calculated and the computations proceeded through a second cycle commencing with the compression stroke using Eq. 1. This way the third cycle can follow the second cycle and so on. The predicted results for diesel cycle (FCS), dual fuel cycle (DFCS) using pilot diesel and dual fuel cycle (DFCS) using pilot biodiesel computations of six consecutive cycle parameters are shown in Tables 4, 5, and 6, respectively.

2.8 Indicated Thermal Efficiency, Work Output, Mean Effective Pressure, and Power Output Computations for Dual Fuel Cycle

Network output during one cycle was computed using Eq. 38.

$$\text{Wnet}_{\text{dual}} = \text{Wexp}_{\text{dual}} + \text{Wcomb}_{\text{dual}} - \text{Wcomp}_{\text{dual}} \tag{38}$$

$$\text{Wexp}_{\text{dual}} = \text{Up}(T_{3\text{dual}}) - \text{Up}(T_{4\text{dual}}) \tag{39}$$

where, $\text{Wnet}_{\text{dual}}$, $\text{Wexp}_{\text{dual}}$, $\text{Wcomb}_{\text{dual}}$, and $\text{Wcomp}_{\text{dual}}$ are the net, expansion, combustion, and compression work, respectively, for dual fuel cycle.

$$\text{Up}(T_{3\text{dual}}) = \text{Sf} \times \text{Hp}(T_{3\text{dual}}) - R_U \times N_p \times T_{3\text{dual}} \tag{40}$$

Table 5 Predicted results of dual fuel cycle simulation (DFCS) using diesel as pilot fuel

Parameters	Cycle					
	1	2	3	4	5	6
T_1 (K)	298	312	326.4	341.2	356.4	372
T_2 (K)	515.1	533.5	552.2	571.1	590.4	609.9
T_3 (K)	2993	2993	2993	2993	2994	2994
T_4 (K)	2327	2307	2289	2170	2252	2235
T_5 (K)	1140	1146	1151	1156	1161	1167
P_2 (kPa)	2890	2859	2828	2798	2769	2741
P_3 (kPa)	2890	2859	2828	2798	2769	2741
P_4 (kPa)	791.2	749.4	710.4	674.2	640.3	608.7
Rc	5.813	5.613	5.424	5.244	5.074	4.911
Power (kW)	22.74	21.82	20.94	20.09	19.27	18.49
Pmep (bar)	16.23	15.57	14.94	14.34	13.75	13.20
η thermal	50.29	50.52	50.71	50.86	50.97	51.04
Wnet	909.6	872.8	837.4	803.4	770.8	739.6

Table 6 Predicted results of dual fuel cycle simulation (DFCS) using biodiesel as pilot fuel

Parameters	Cycle					
	1	2	3	4	5	6
T_1 (K)	298	312.6	327.7	343.3	359.4	375.9
T_2 (K)	1030	1017	1113	1156	1199	1243
T_3 (K)	2994	2995	2995	2995	2995	2995
T_4 (K)	1909	1890	1872	1854	1837	1820
T_5 (K)	1303	1306	1308	1311	1313	1316
P_2 (kPa)	5778	5729	5679	5629	5578	5528
P_3 (kPa)	5778	5729	5679	5629	5578	5528
P_4 (kPa)	649.2	612.7	578.8	547.2	517.8	490.5
Rc	2.906	2.794	2.689	2.590	2.497	2.409
Power (kW)	22.46	21.21	20.02	18.89	17.81	16.79
Pmep (bar)	16.03	15.14	14.29	13.48	12.71	11.98
η thermal	51.3	51.6	51.68	51.82	51.84	51.67
Wnet	898.5	805.7	781.2	755.6	734.2	711.9

$$\text{Up}(T_{4\text{dual}}) = \text{Sf} \times \text{Hp}(T_{4\text{dual}}) - R_U \times N_p \times T_{4\text{dual}} \quad (41)$$

where Up(T) is the internal energy of products of dual fuel combustion at temperature $T_{3\text{dual}}$ and $T_{4\text{dual}}$, respectively, N_p is the number of moles of gaseous products and Hp(T) stands for enthalpy of products at temperature $T_{3\text{dual}}$ and $T_{4\text{dual}}$.

$$\mathrm{Sf} = \frac{N_m}{N_{\mathrm{mo}}} + \frac{N_{\mathrm{xg}}}{N_{\mathrm{po}}} \tag{42}$$

Sf denotes the scale factor which reduces the mole numbers to a proper size to fit the engine, once N_{ag} and N_{xg} are known. N_{m} and N_{po} represent moles of fuel vapor and air in engine during the compression stroke, kilomoles of air–fuel vapor in a mixture containing 1 kmol of diesel, and CNG fuels and kilomoles of products formed from the combustion of N_{mo}, respectively.

$$\mathrm{Wcomb_{dual}} = N_p \times R_U \times (T_3 - T_2) \tag{43}$$

$$\mathrm{Wcomp_{dual}} = \left(N_{\mathrm{ag}} + N_{\mathrm{xg}}\right) \times \left(\mathrm{Cpr_{dual}} - R_U\right) \times (T_2 - T_1) \tag{44}$$

Thermal efficiency of the dual fuel cycle was computed using formulated Eq. 45, mean effective pressure using Eq. 46, and power output using Eq. 47 given below:

$$\eta_{\mathrm{dual}} = \frac{\mathrm{Wnet_{dual}} \times N_{\mathrm{mo}}}{-\mathrm{Hrp_{dual.}} \times N_{\mathrm{ag}}} \tag{45}$$

$$\mathrm{Pmep_{dual}} = \frac{\mathrm{Wnet_{dual}} \times 4}{\pi \times B^2 \times S \times 101325} \tag{46}$$

$$\mathrm{Poutput_{dual}} = \frac{\mathrm{Wnet_{dual}} \times RPM}{60 \times 1000} \tag{47}$$

By using Eq. 27, the adiabatic flame temperature for diesel fuel is 2999 K and 2995 K for CNG-diesel dual fuel, i.e., diesel adiabatic temperature is more than CNG adiabatic temperature. Therefore, conversion of an existing compression ignition engine to CNG-diesel dual made can be carried out without any change in engine design and basic engine structure as there will be no effect of heat socks due to the replacement of diesel fuel with CNG [23].

3 Results and Discussion

The predicted results of six consecutive simulation cycles for all the combinations of fuel are computed and shown in Tables 4, 5, and 6 for a chemically correct mixture. In the first case, conventional diesel ($C_{10}H_{22}$) is taking as a main fuel. Whereas in the second one, 85% CNG (a mixture of 84.5% CH_4 + 12.5% C_2H_6 + 3% C_3H_8) with 15% of pilot diesel ($C_{10}H_{22}$). And the third one is 85% CNG (a mixture of 84.5% CH_4 + 12.5% C_2H_6 + 3% C_3H_8) with 15% of B20 (blend of 20% $C_{10}H_{20}O_2$ + 80%$C_{10}H_{22}$) as pilot fuel. The compression ratio of the engine is assumed to be 16.5 with initial temperature and pressure conditions as 298 K and 101.325 kPa at 1500 rpm under full load conditions. The disparity of thermal efficiency of 6

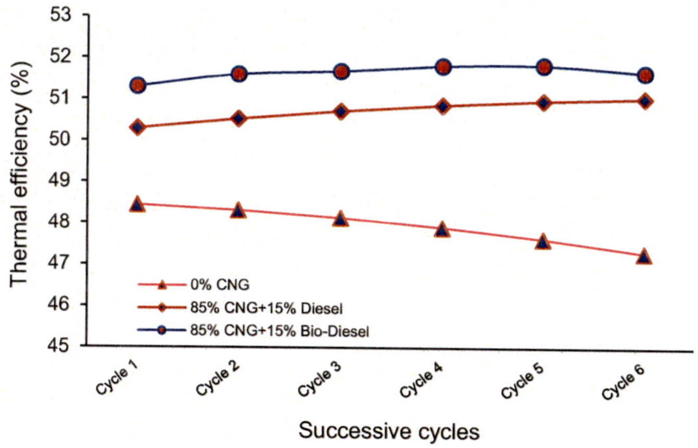

Fig. 2 Variation of thermal efficiency with successive cycles ($r = 16.5$)

consecutive cycles in the aforesaid three cases is shown in Fig. 2. It is very clear from the figure that the thermal efficiency of dual fuel cycle using B20 is more than dual fuel with diesel and pure diesel cycles. This may be because of the availability of oxygen molecule bonded in the biodiesel itself that improves the combustion quality and hence indicates the combustion of dual fuel cycle with B20 as pilot fuel is complete and smooth.

The variations of predicted results of mean effective pressure of pure diesel and dual fuel simulation for six consecutive cycles are shown in Fig. 3. The mean effective pressure of the pure diesel cycle is more than both the other modes and as the successive cycles proceed one after the other, its values decrease smoothly. Figure 4 depicts the variances in expected power output values for pure diesel and dual fuel simulations across six consecutive cycles. The power output of the pure diesel cycle is greater than that of the other two modes, and its values fall steadily as the subsequent cycles progress one after the other. Figure 5 displays the variations in projected work output values for pure diesel and dual fuel simulations throughout a six-cycle period. The pure diesel cycle exhibits more network than the other two modes. Work output follows the same trend as followed by the power output.

3.1 Validation Test

The developed model's estimated engine performance and combustion characteristics were validated using lab-based experiments. The engine was operated over the entire operating range, and the performance and combustion outputs were recorded. The Tables 7, 8, and 9 illustrate the results of an experiment conducted under ideal operating conditions, as well as the percent deviation from the model's predicted

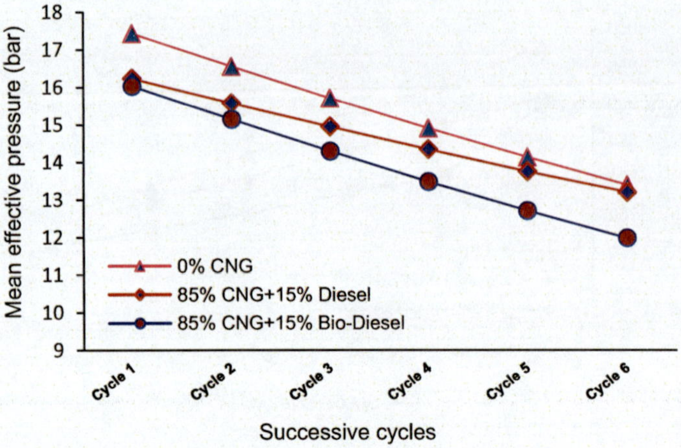

Fig. 3 Variation of mean effective pressure with successive cycles ($r = 16.5$)

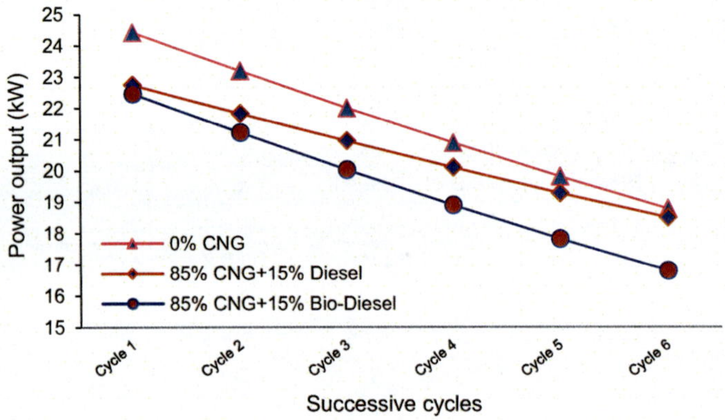

Fig. 4 Variation of power output with successive cycles ($r = 16.5$)

outputs. At ideal operating circumstances, all experimental output was within 8% of the model-predicted values.

4 Conclusion

The projected simulation results show that there are negligible cyclic variations in the different performance characteristics of the FCS and DFCS modes. This means that the conventional diesel engine may be used in dual fuel mode without any changes to the engine design, such as the compression ratio. However, since natural gas's self-ignition temperature is fairly high, it may also be used at higher compression ratios.

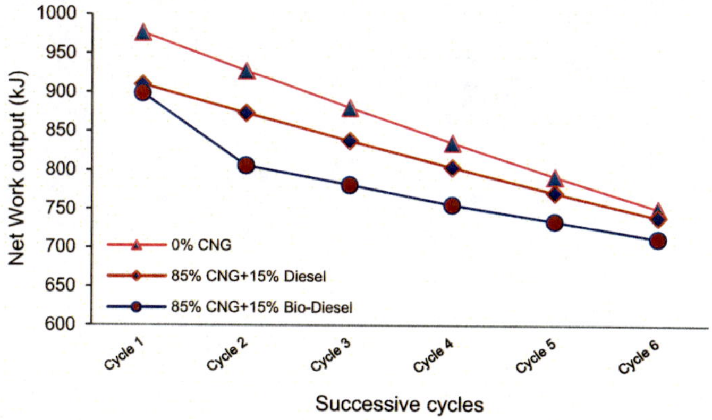

Fig. 5 Variation of network output with successive cycles ($r = 16.5$)

Table 7 Validation test results of diesel fuel cycle simulation (FCS)

Parameters	Cycle					
	1	2	3	4	5	6
T_1 (K)	298.0	314.0	330.6	346.0	363.0	380.0
T_2 (K)	776.4	808.2	840.8	873.9	907.6	941.8
T_3 (K)	2968.0	2969.0	2969.0	2969.0	2969.0	2969.0
T_4 (K)	2116.0	2095.8	2075.6	2056.4	2037.2	2019.0
T_5 (K)	1461.7	1463.7	1466.8	1468.8	1471.9	1474.9
P_2 (kPa)	4401.2	4362.0	4321.8	4281.6	4242.4	4202.2
P_3 (kPa)	4446.1	4406.5	4365.9	4325.3	4285.7	4245.1
P_4 (kPa)	690.8	651.5	614.6	580.4	548.6	519.0
Rc	3.6	3.5	3.3	3.2	3.1	3.0
Power (kW)	24.7	23.4	22.2	21.1	20.0	19.0
Pmep (bar)	17.3	16.4	15.5	14.8	14.0	13.3
η thermal	46.0	45.9	45.7	45.5	45.2	44.9
Wnet	966.9	918.1	871.3	826.7	783.9	743.3

To convert the engine to dual fuel mode, a few hardware adjustments are necessary, including CNG storage, a CNG conversion kit, and electrical controllers to operate the solenoid valves. Furthermore, an air gas mixer is needed to generate a homogenous and chemically correct mixture to improve the combustion performance of the engine. It was revealed that the adiabatic flame temperature had no significant effect in all three test cases. Furthermore, the model was validated with the experimental parameters and was found within 8% of the model-predicted values under optimal operating conditions.

Table 8 Validation test results of dual fuel cycle simulation (DFCS) using diesel as pilot fuel

Parameters	Cycle					
	1	2	3	4	5	6
T_1 (K)	298	313	328	344	358	376
T_2 (K)	504.8	522.9	541.2	559.7	578.7	597.8
T_3 (K)	3052.9	3052.9	3052.9	3052.9	3053.9	3053.9
T_4 (K)	2373.5	2353.1	2334.8	2213.4	2297.0	2279.7
T_5 (K)	1162.8	1168.9	1174.0	1179.1	1184.2	1190.3
P_2 (kPa)	3034.5	3002.0	2969.4	2937.9	2907.5	2878.1
P_3 (kPa)	2832.2	2801.8	2771.4	2742.0	2713.6	2686.2
P_4 (kPa)	767.5	726.9	689.1	654.0	621.1	590.4
Rc	5.6	5.4	5.2	5.0	4.9	4.7
Power (kW)	24.1	23.1	22.2	21.3	20.4	19.6
Pmep (bar)	15.9	15.3	14.6	14.1	13.5	12.9
η thermal	49.8	50.0	50.2	50.4	50.5	50.5
Wnet	927.8	890.3	854.1	819.5	786.2	754.4

Table 9 Validation test results of dual fuel cycle simulation (DFCS) using bio-diesel as pilot fuel

Parameters	Cycle					
	1	2	3	4	5	6
T_1 (K)	298.0	315.5	329.4	346.7	364.3	378.7
T_2 (K)	1019.7	1006.8	1101.9	1144.4	1187.0	1230.6
T_3 (K)	2964.1	2965.1	2965.1	2965.1	2965.1	2965.1
T_4 (K)	1851.7	1833.3	1815.8	1798.4	1781.9	1765.4
T_5 (K)	1276.9	1279.9	1281.8	1284.8	1286.7	1289.7
P_2 (kPa)	5893.6	5843.6	5792.6	5741.6	5689.6	5638.6
P_3 (kPa)	5662.4	5614.4	5565.4	5516.4	5466.4	5417.4
P_4 (kPa)	642.7	606.6	573.0	541.7	512.6	485.6
Rc	2.8	2.7	2.6	2.5	2.4	2.3
Power (kW)	27.0	25.5	24.0	22.7	21.4	20.1
Pmep (bar)	15.9	15.0	14.1	13.3	12.6	11.9
η thermal	48.7	49.0	49.1	49.2	49.2	49.1
Wnet	1078.2	966.8	937.4	906.7	881.0	854.3

References

1. Sharma P, Sahoo BB, Said Z, Hadiyanto H, Nguyen XP, Nižetić S, Huang Z, Hoang AT, Li C (2022) Application of machine learning and Box-Behnken design in optimizing engine characteristics operated with a dual-fuel mode of algal biodiesel and waste-derived biogas. Int J Hydrogen Energy. https://doi.org/10.1016/J.IJHYDENE.2022.04.152
2. Siddiki SYA, Mofijur M, Kumar PS, Ahmed SF, Inayat A, Kusumo F, Badruddin IA, Khan TMY, Nghiem LD, Ong HC, Mahlia TMI (2022) Microalgae biomass as a sustainable source for biofuel, biochemical and biobased value-added products: an integrated biorefinery concept. Fuel 307:121782. https://doi.org/10.1016/J.FUEL.2021.121782
3. Othman MF, Adam A, Najafi G, Mamat R (2017) Green fuel as alternative fuel for diesel engine: a review. https://doi.org/10.1016/j.rser.2017.05.140
4. Bora BJ, Dai Tran T, Prasad Shadangi K, Sharma P, Said Z, Kalita P, Buradi A, Nhanh Nguyen V, Niyas H, Tuan Pham M, Thanh Nguyen Le C, Dung Tran V, Phuong Nguyen X (2022) Improving combustion and emission characteristics of a biogas/biodiesel-powered dual-fuel diesel engine through trade-off analysis of operation parameters using response surface methodology. Sustain Energy Technol Assessments 53:102455. https://doi.org/10.1016/J.SETA.2022.102455
5. Verma S, Das LM, Bhatti SS, Kaushik SC (2017) A comparative exergetic performance and emission analysis of pilot diesel dual-fuel engine with biogas, CNG and hydrogen as main fuels. Energy Convers Manage 151:764–777. https://doi.org/10.1016/j.enconman.2017.09.035
6. Dager B, Kumar A, Singh Sharma R (2022) Exploring the effects of pilot injection timing and natural gas flow rates on the performance of twin-cylinder compression ignition engine. Energy Sources Part A 44:2730–2747. https://doi.org/10.1080/15567036.2022.2059598
7. Said Z, Nguyen TH, Sharma P, Li C, Ali HM, Nguyen VN, Pham VV, Ahmed SF, Van DN, Truong TH (2022) Multi-attribute optimization of sustainable aviation fuel production-process from microalgae source. Fuel 324:124759. https://doi.org/10.1016/j.fuel.2022.124759
8. Sharma P, Chhillar A, Said Z, Huang Z, Nguyen VN, Quy P, Nguyen P, Nguyen XP (2022) Experimental investigations on efficiency and instability of combustion process in a diesel engine fueled with ternary blends of hydrogen peroxide additive/biodiesel/diesel. Energy Sources Part A. https://doi.org/10.1080/15567036.2022.2091692
9. Sharma P. Prediction-optimization of the effects of di-tert butyl peroxide-biodiesel blends on engine performance and emissions using multi-objective response surface methodology (MORSM). J Energy Resour Technol 1–26. https://doi.org/10.1115/1.4052237
10. Sharma P, Sahoo BB (2022) Precise prediction of performance and emission of a waste derived biogas-biodiesel powered dual-fuel engine using modern ensemble boosted regression tree: a critique to artificial neural network. Fuel 321:124131. https://doi.org/10.1016/j.fuel.2022.124131
11. Sharma P, Sahoo BB (2022) An ANFIS-RSM based modeling and multi-objective optimization of syngas powered dual-fuel engine. Int J Hydrogen Energy. https://doi.org/10.1016/j.ijhydene.2022.04.093
12. Sharma P (2021) Prediction-optimization of the effects of di-tert butyl peroxide-biodiesel blends on engine performance and emissions using multi-objective response surface methodology (MORSM). J Energy Resour Technol 1–26. https://doi.org/10.1115/1.4052237
13. Sridhar G, Babu R (2010) Facts about producer gas engine. Paths Sustain Energy. https://doi.org/10.5772/13030
14. Sharma P, Sharma AK (2021) Combustion and thermal performance of dual fuel engine: influence of controlled producer gas substitution with pilot B20 (WCOME biodiesel–diesel) blending. Lect Notes Mech Eng 20:341–353
15. Sharma P, Le MP, Chhillar A, Said Z, Deepanraj B, Cao DN, Bandh SA, Hoang AT (2022) Using response surface methodology approach for optimizing performance and emission parameters of diesel engine powered with ternary blend of Solketal-biodiesel-diesel. Sustain Energy Technol Assess 52:102343. https://doi.org/10.1016/j.seta.2022.102343

16. Shahir VK, Jawahar CP, Vinod V, Suresh PR (2020) Experimental investigations on the performance and emission characteristics of a common rail direct injection engine using tyre pyrolytic biofuel. J King Saud Univ Eng Sci 32:78–84. https://doi.org/10.1016/j.jksues.2018.05.004
17. Ramachander J, Gugulothu SK, Sastry GR, Surya MS (2021) Statistical and experimental investigation of the influence of fuel injection strategies on CRDI engine assisted CNG dual fuel diesel engine. Int J Hydrogen Energy 46:22149–22164. https://doi.org/10.1016/J.IJHYDENE.2021.04.010
18. Justin Abraham Baby S, Suresh Babu S, Devarajan Y (2018) Performance study of neat biodiesel-gas fuelled diesel engine. Int J Ambient Energy. https://doi.org/10.1080/01430750.2018.1542625
19. Sharma P, Sharma AK (2020) Experimental evaluation of thermal and combustion performance of a di diesel engine using waste cooking oil methyl ester and diesel fuel blends. In: Smart innovation, systems and technologies. https://doi.org/10.1007/978-981-15-2647-3_50
20. Sharma P, Sharma AK (2021) Statistical and continuous wavelet transformation based analysis of combustion instabilities in a biodiesel fuelled compression ignition engine. J Energy Resour Technol 1–26. https://doi.org/10.1115/1.4051340
21. Heywood JB (1988) Internal Combustion Engine Fundementals. https://doi.org/10987654
22. Abagnale C, Cameretti MC, De Simio L, Gambino M, Iannaccone S, Tuccillo R (2014) Numerical simulation and experimental test of dual fuel operated diesel engines. Appl Therm Eng 65:403–417. https://doi.org/10.1016/j.applthermaleng.2014.01.040
23. Ravi K, Mathew S, Pradeep Bhasker J, Porpatham E (2016) Gaseous alternative fuels for CI engines—a technical review. Int J Pharm Technol 8:5257–5268

Laser-Induced Spark Ignition of Methane-Air Mixtures in Constant Volume Combustion Chamber

Prashant Patane, Vishal Kolapte, Milankumar Nandgaonkar, and Subhash Lahane

1 Introduction

The major global issues in the energy sector are depletion of fossil fuels and depreciation in the environment due to their burning. Thus, there is a need to use fossil fuels more efficiently by reducing losses in energy extraction processes. Considering the environmental aspect and the rate of depletion of fossil fuel, forces us to find renewable alternative fuels which are environment friendly like methane, compressed natural gas (CNG), hydrogen, or hythane. The evidence from the literature suggests using a lean mixture, for better efficiency and lower emissions. Nitrogen oxide (NOx) emissions reduce significantly with a leaner air–fuel ratio due to a decrease in overall flame temperature. However, the use of a lean mixture results in a significant decrease in power density, which can be overcome by improved inlet pressure. Turbocharging or supercharging can be used for increasing inlet pressure [1–4]. The traditional spark ignition system has limitations in igniting such high-pressured fuel, as inlet pressure inside the chamber has a directly proportionate with the supply voltage. High voltage is required to burn mixture at extreme lean mixtures which result in electrode erosion and thus reduces the spark plug life [5].

The above limitations of the traditional ignition system encourage researchers to find alternative ignition methods. The laser ignition (LI) method has come up with the most viable alternative, because of the advantages it offers [6–9]. LI is an electrode-less ignition method thus there is not any issue with erosion of electrode. The absence of the electrode quenching effect reduces NOx emission, which is being the motivation for developing large-sized laser ignition gas engines. Better

P. Patane (✉) · V. Kolapte · M. Nandgaonkar · S. Lahane
Mechanical Engineering Department, COEP, Pune, Maharashtra 411005, India
e-mail: prashant.patane007@gmail.com

P. Patane
School of Mechanical Engineering, Dr. Vishwanath Karad, MIT World Peace University, Pune, India

A. Kumar et al. (eds.), *Recent Advances in Manufacturing and Thermal Engineering*, Lecture Notes in Mechanical Engineering, https://doi.org/10.1007/978-981-19-8517-1_23

control and regulation over the traditional ignition method is possible; hence, it gives precise ignition time. Laser ignition gives freedom to select the focal position inside the combustion chamber. To speed up the flame propagation in lean combustion, a multipoint ignition system can be easily developed [10–12].

Laser beam transfers energy to the combustible mixture or medium with four laser ignition mechanisms, namely resonant ignition [13, 14], photochemical ignition [15–17], thermal ignition [5, 18, 19], and non-resonant ignition [20]. Out of four LI mechanisms, the non-resonant laser ignition is usually used to start the combustion process as the mechanism is not dependent on laser wavelength, and it resembles the traditional electric spark plug ignition mechanism [21]. The non-resonant ignition mechanism is called laser-induced spark ignition. In LISI, the nanosecond laser pulses can be focused into the chamber by using plano-convex lenses [22, 23]. If the focal spot intensity becomes greater than the breakdown threshold intensity, the breakdown of the mixture takes place and it leads to generation of plasma [22, 23]. If the energy at the spark is sufficient, then the ignition takes place [23]. However, there is significant difference in duration of energy transfer with an electric spark plug and laser ignition, which is in the order of microseconds and nanoseconds, respectively [7]. The process of non-resonant breakdown begins with the ionization of some of gas molecules, which will liberate electrons that gain more photons, with an increase in kinetic energy. These energized electrons hit the other molecules in the gas and ionize them, which leads to electron avalanche. This process of non-resonant breakdown requires impurities in the combustible mixture to generate the initial electrons. These initial electrons absorb laser radiation and increase temperature in local region, which starts the avalanche process [24]. Ionization potential for most of the gases are in the order of 10 eV but the available photon energy employed in this work is about 1 eV; hence, multiphotons processes are essential in initial stages of breakdown [25].

Many of the researchers have carried out laser ignition of different fuels at different chamber conditions and at different air–fuel ratios in CVCC. Weinrotter et al. [26] has analyzed laser ignition of H_2-air mixture in CVCC. It has been summarized that as initial pressure of chamber increases, peak pressure increases linearly; however, time to reach peak pressure also increases at all equivalence ratios. The minimum laser energy (MPE) needed for ignition reduces as initial chamber pressure increases. Increase in initial chamber temperature results in decrease in MPE for ignition for all air–fuel ratios. Increase in laser energy above the MPE does not show any effect on combustion duration and peak pressure. LI ignition at the center position of the chamber results in the reduction of total combustion duration. It has been also seen that, increase in the plasma size results in reduction in combustion duration but MPE required goes on increasing. Srivastava et al. [27] studied and measured the laser plasma size and its propagation in atmospheric air and seen that, the plasma approaching toward the laser beam grows at faster rate as compared to plasma moving in the laser beam direction. Srivastav et al. [22] further analyzed the LI of CNG-air mixtures and observed that increase in chamber pressure; the breakdown threshold energy needed for ignition goes on decreasing. The pressure–time history at different air–fuel mixtures shows prolonged combustion duration and reduction in peak pressure as mixtures becomes leaner. Phuoc et al. [28] have experimentally calculated

the breakdown threshold intensities of CH_4, H_2, N_2, and O_2 for the pressure range of 150–3040 torr and observed that as the pressure increases the breakdown threshold intensity goes on decreasing. The result shows that the breakdown threshold intensity is inversely varied with pressure (Ithr $\propto p\text{-}n$). Kopecek et al. [29] investigated LI of CH4-air mixtures with 355, 532, and 1064 nm laser wavelength. It has been summarized that, the use of laser beam nearer Gaussian beam and optimum optics reduces the MPE required for ignition. However, no any notable wavelength dependence on MPE for ignition was observed for gas mixtures. The lean mixture of equivalence ratio of 0.56 can be ignited by slightly increasing the laser pulse energy. Prasad et al. [30] carried out LI of HCNG-air mixture ($\lambda = 1.1$, 1.3 and 1.5) and analyzed the results of different initial chamber pressure, λ and H_2 percentage on peak pressure, combustion duration and flame propagation. It has been observed that, at any initial chamber pressure, increase in λ results in decrease in the peak pressure. The maximum pressure and minimum combustion time were observed for $\lambda =$ 1.1. The longer combustion time was observed as mixture becomes leaner. As initial pressure of chamber increases, the combustion time and maximum combustion pressure increases. Increase in hydrogen percentage improves the maximum combustion pressure but the combustion time reduces. Srivastav et al. [23] carried out the LI of natural gas–air mixtures in a CVCC and studied the impact of quality of beam and focal length on the air–fuel mixture combustion performance. The minimum energy needed for ignition increases as the focal length of lens increases for all λ values. The peak pressure was same for all the focal lengths, but the minimum combustion duration was observed for focal length 150 mm. The minimum energy required for ignition reduces as beam quality factor ($M2$) value approaches toward 1 for all values of λ. However, marginal change in peak pressure and combustion time was observed with change in beam quality. Dharamshi et al. [31] carried out laser ignition of H_2-air mixture in a CVCC and summarized that as λ increases from 2.0 to 5.0, the pressure rise rate, maximum combustion pressure, and flame speed decrease. Increase in laser energy slightly decreases the time to attain peak pressure; however, the peak pressure remains almost same. The minimum pulse energy needed for ignition increases with increase in λ. Dharamshi et al. [32] further observed that, increase in initial chamber pressure increases the maximum pressure and flame kernel observed more wrinkled. The increase in initial chamber temperature results in reduction in MPE to ignite the mixture. The plasma position at the center of the chamber results in reduction in combustion duration; however, peak pressure remains unchanged. The MPE needed for ignition reduces as focal length of lens decreases.

The aim of this chapter is to study and analyze the LI behavior of CH_4-air mixture in a CVCC. The minimum pulse energy and breakdown threshold intensity required for breakdown of pure air and methane at various initial chamber pressures and at 298 K initial temperature was measured. The methane-air mixture is ignited at 2.5, 5.0, and 7.5 bar chamber pressure and at 298 K temperature for ϕ from 0.6 to 1.4. The influence of chamber pressure on peak pressure and combustion duration was studied. The minimum pulse energy necessary for burning of methane-air mixtures for all initial pressure and at various equivalence ratios was measured. Further the impact

of initial pressure on pressure rise rate, peak pressure, and combustion duration for all ϕ is analyzed.

2 Experimental Setup

The schematic layout for the experimentation is as seen in Fig. 1. LI of methane-air mixture is carried out in a CVCC having 90 mm internal diameter and 107 mm height. The chamber was designed and developed for 100 bar maximum pressure and 1000 temperature. Four optical windows are mounted diametrically opposite on chamber out of which two windows are for laser beam entry and exit and two optical windows to visualize the combustion. The chamber is made up of mild steel; however, optical windows are made up of sapphire material. A Litron makes Q-switched (Nano LG 225–10) Nd:YAG laser was used to initiate the ignition of methane-air mixture which can deliver maximum 225 mJ pulse energy and 4–6 ns pulse duration at 1064 nm wavelength. The diameter of laser beam is 5 mm. Laser energy needed for ignition is measured by using a Coherent make (FieldMax II–TOP) laser energy meter. Thorlabs makes plano-convex lens of 125 mm focal length is used to focus the laser beam in to the chamber. All the experiments are carried out by focusing the laser beam at the center of the chamber. A static pressure sensor was used to prepare the initial pressure of methane-air mixture. A piezoelectric pressure sensor was mounted on the top of the chamber for acquire the P–t history after the combustion. Before conducting the experiments, the minimum energy necessary for the plasma formation in the air and methane for different initial chamber pressure is measured. The minimum laser energy for all experimental conditions is selected so that the LI was achieved for all equivalence ratios.

All the experiments are carried out with compressed air (moisture free) and methane was used for combustion of methane-air mixture. To achieve required ϕ, it is important to maintain the partial pressure of air and methane using static pressure sensor. Air compressor and methane cylinder are fitted with pressure regulator. The methane and air flow from cylinder and compressor to the combustion chamber via high pressure pipe. The required ϕ inside the chamber is maintained by using Dalton's law of partial pressure. Initially, methane is filled into the chamber as the partial pressure of methane is low and then air is passed with high partial pressure. The high pressure of air ensures the high turbulence and thus homogeneity of the mixture. After filling the compressed air, the mixture is then kept for 1 min for thermal stabilization and homogeneity. Laser is then triggered for combustion of methane-air mixture which then taken out through exhaust valve. The residual gases remaining inside the chamber are flushed out using compressed air. Thus, the chamber is now ready for the preparation of next equivalence ratio. A piezoelectric pressure sensor was used to log the pressure–time history after combustion. The signals from the pressure sensor were sent to computer for data analysis via data acquisition system.

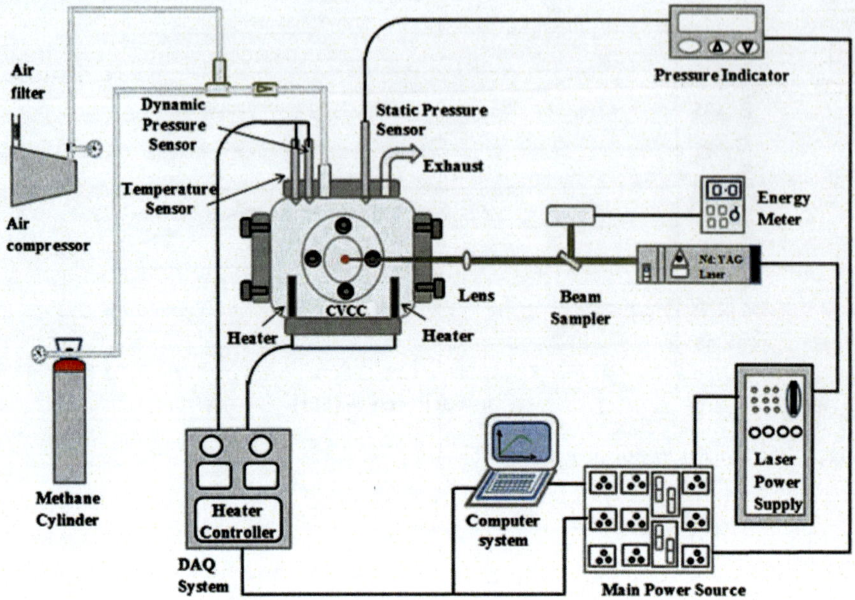

Fig. 1 Schematic layout of experimental setup

3 Results

3.1 Minimum Breakdown Energy

The breakdown threshold is the energy at which any medium or air would breakdown for greater than 50% of the laser shots [22, 28]. The breakdown was simply observed as it was always appeared as bright light at the focal point associated with cracking noise and quick laser pulse absorption passed through focal area [28]. The breakdown takes place via electron cascade process. The electron cascading needs the availability of initial electrons. The electrons further gain more photons through the inverse bremsstrahlung process. If the electrons gain adequate energy, they ionize other gas molecules on impact, resulting to a cascading of electron and the gas breakdown. The minimum breakdown threshold energy required for air and methane at various initial pressures inside chamber and at 298 K initial temperature is measured and it is depicted in Figs. 2 and 3, respectively.

It has been observed that as initial pressure inside the chamber for air and methane increases, the energy required for breakdown goes on decreasing. This is mainly due to large number of air and methane molecules will be available as the initial chamber pressure increases. However, the minimum breakdown energy required for methane is less than air for at all the pressure conditions. This may be due to the presence of

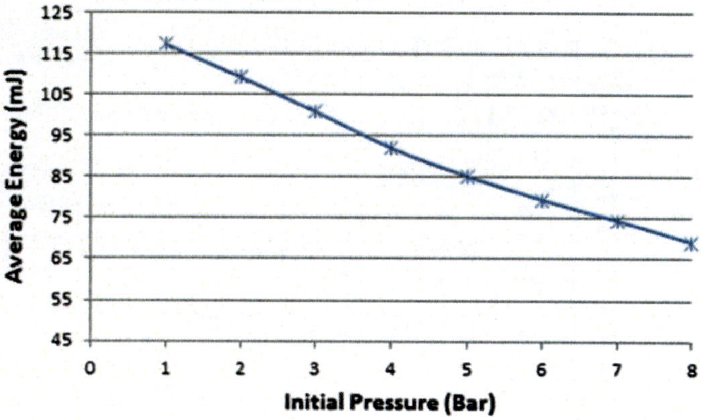

Fig. 2 Threshold energy of air at various chamber pressures and at 298 K chamber temperature

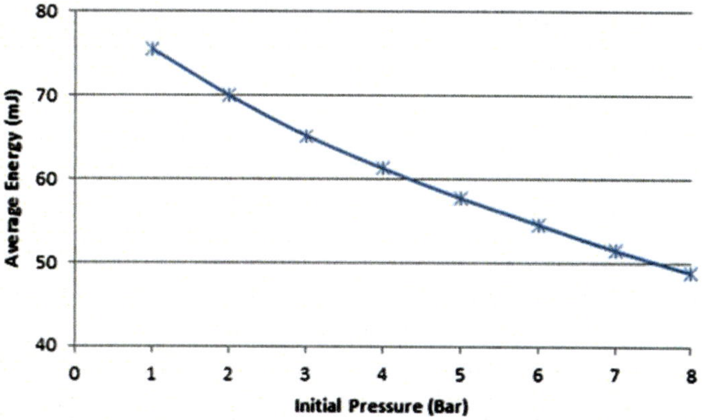

Fig. 3 Threshold energy of methane at various chamber pressures and at 298 K chamber temperature

more impurities in the methane compared to air which can facilitate the formation of the initial electrons.

3.2 *Breakdown Threshold Intensity* ($I_{threshold}$)

The breakdown threshold intensity (I) is the intensity at which breakdown of the mixture take place resulting to generation of plasma spark [33]. If the spark energy at focal spot is high enough, then mixture ignites. The breakdown threshold intensity (I) is calculated by using Eq. (1).

$$I_{\text{threshold}} = \frac{\text{Minimum pulse energy (MPE)}}{\text{laser pulse width} \times \text{focal spot area}} \tag{1}$$

$$I_{\text{threshold}} = \frac{\text{Minimum pulse energy (MPE)}}{\text{laser pulse width} \times \left(\pi/_{4}d_{\text{f}}^{2}\right)}$$

where,

d_f = diameter at focal spot.

The diameter at the focal spot is calculated by Eq. (2) [23],

$$d_f = \frac{4 \times f \times \lambda \times M^2}{\pi \times d_b} \tag{2}$$

where,

f = focal length of lens = 125 mm, λ = Laser wavelength = 1064 nm.

M^2 = beam quality = 1.8, d_b = Laser beam diameter = 5 mm.

The breakdown threshold intensity (I) for air and methane at different initial pressure is depicted in Figs. 4 and 5, respectively. The breakdown threshold intensity (I) at all initial pressure lies in the range of 10^{10} to 10^{12} W/cm^2. It has been observed that the threshold intensity reduces quickly as the pressure increases. As the pressure increases from 1 to 8 bar, the threshold intensity for air decreases from 2.2560×10^{11} to 1.3302×10^{11} W/cm^2 and for methane it decreases from 1.4525×10^{11} to 9.3877×10^{10} W/cm^2. The breakdown threshold for methane found to be lower than air for all initial pressure conditions.

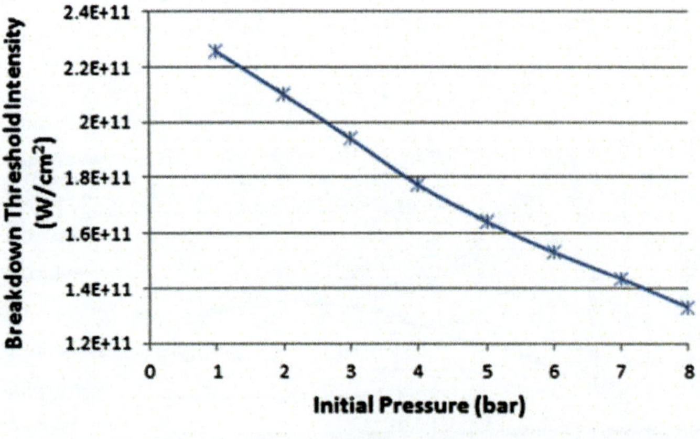

Fig. 4 Threshold intensity of air at various chamber pressures and at 298 K chamber temperature

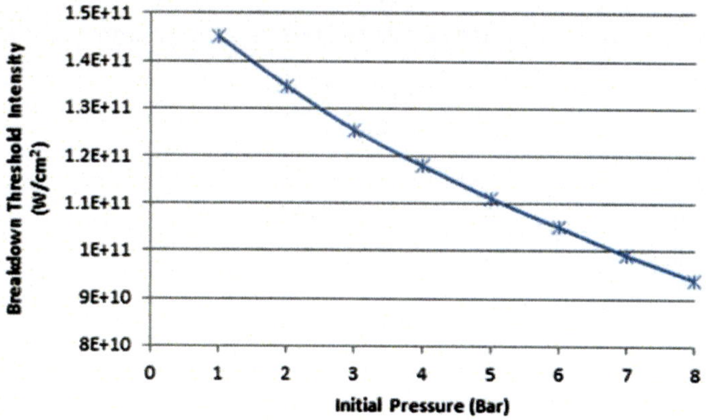

Fig. 5 Threshold intensity of methane at various chamber pressures and at 298 K chamber temperature

3.3 *P–t History*

The P–t history for methane-air mixture at 2.5, 5.0, and 7.5 bar chamber pressure and 298 K temperature is plotted for ϕ of 0.6 to 1.4 and are depicted in Figs. 6, 7, and 8 respectively. It has been seen that for all initial pressure of 2.5, 5.0, and 7.5 bar initial pressure, the peak pressure goes increasing from ϕ of 0.6 to 1.0 and further as mixture becomes rich the peak pressure starts decreasing. The peak pressure for all initial pressures is seen at ϕ of 1.0 due to correct chemical air–fuel mixture. The peak pressure was observed to be 17.31, 36.03, and 53.39 bar for initial pressure of 2.5 bar, 5.0 bar, and 7.5 bar, respectively.

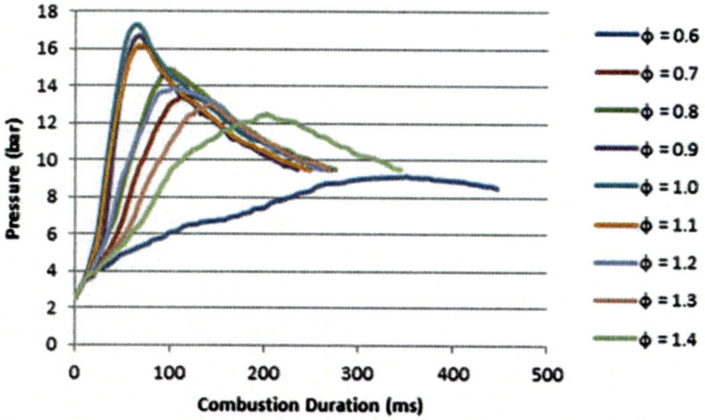

Fig. 6 P–t history for methane-air mixture at 2.5 bar chamber pressure and 298 K chamber temperature for various ϕ

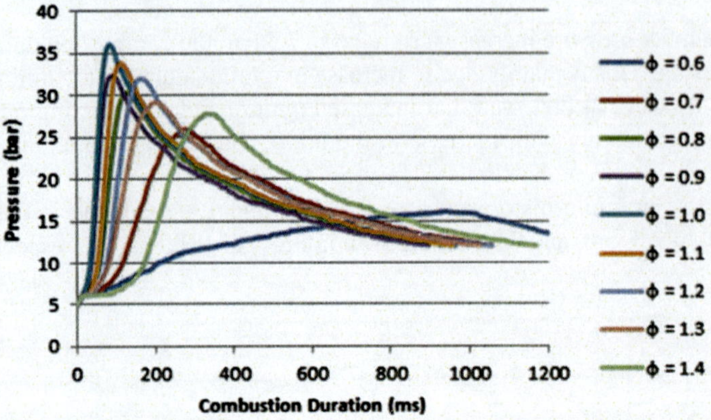

Fig. 7 P–t history for methane-air mixture at 5.0 bar chamber pressure and 298 K chamber temperature for various ϕ

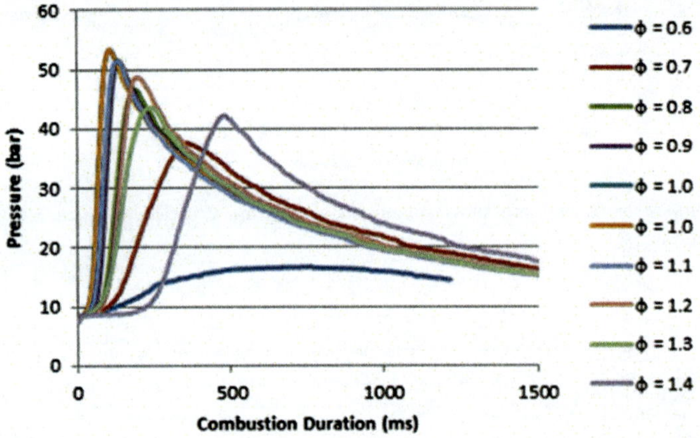

Fig. 8 P–t history for methane-air mixture at 7.5 bar chamber pressure and 298 K chamber temperature for various ϕ

The combustion duration is the minimum time required to achieve peak pressure. The minimum combustion duration for all initial pressure condition was observed for equivalence ratio of 1.0 due to higher laminar burning velocity of methane at ϕ of 1.0. The lowest combustion duration was seen to be 64.1, 84.08, and 106.7 ms for initial pressure of 2.5 bar, 5.0 bar, and 7.5 bar, respectively.

The comparison of peak pressure as well as duration of combustion for various equivalence ratios of methane-air mixture at 2.5, 5.0, and 7.5 bar initial pressure is depicted in Figs. 9 and 10, respectively. It has been observed that with increase in the chamber pressure from 2.5 to 7.5 bar, the peak pressure shows increasing trend. Increase in chamber pressure results in increase in energy content of mixture, thus maximum chamber pressure is seen after ignition. Similarly, it has been seen that

as the chamber pressure increases from 2.5 to 7.5 bar, the combustion duration goes on increasing. This is mainly due to increase in ignition delay in the initial phase of ignition. It has been observed that increased ignition delay at starting phase of the pressure–time history with increase in the initial chamber pressure, and therefore, overall duration of combustion also increased.

The pressure–time history at $\phi = 0.7$, 1.0, and 1.4 at all the initial pressure condition of 2.5, 5.0, and 7.5 bar is shown in Figs. 11, 12, and 13, respectively.

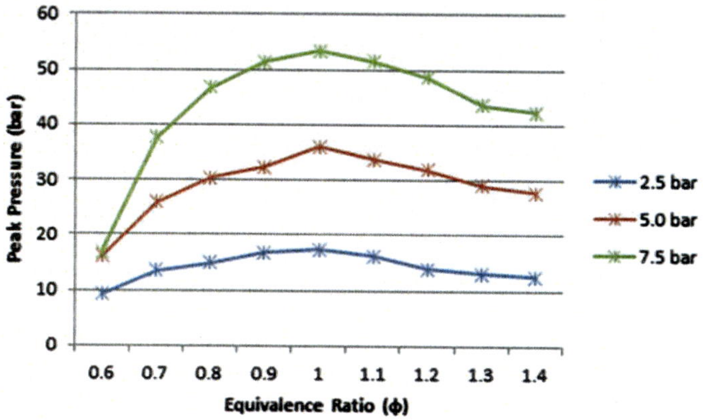

Fig. 9 Comparison of peak pressures of methane-air mixture at various ϕ at 2.5, 5.0, and 7.5 bar chamber pressures

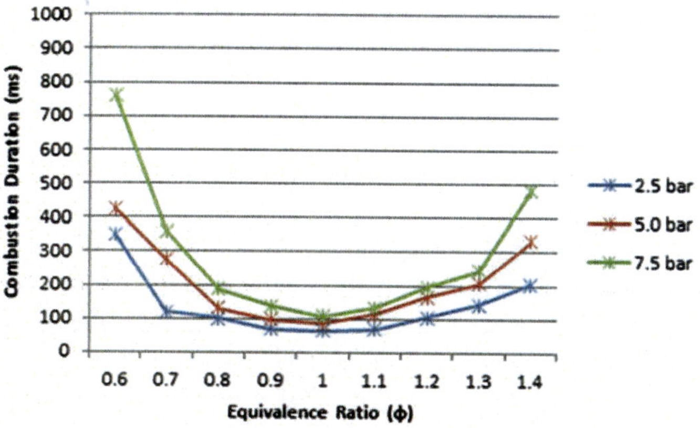

Fig. 10 Comparison of combustion duration of methane-air mixture at various ϕ at 2.5, 5.0, and 7.5 bar chamber pressures

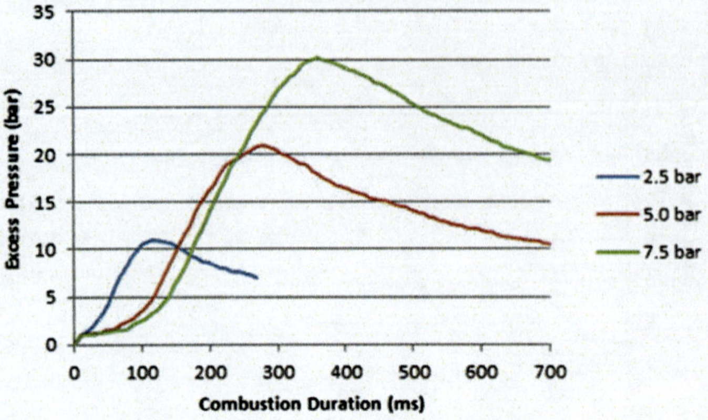

Fig. 11 Comparison of pressure–time curve of CH_4-air mixture at ϕ of 0.7 for 2.5, 5.0, and 7.5 bar chamber pressure

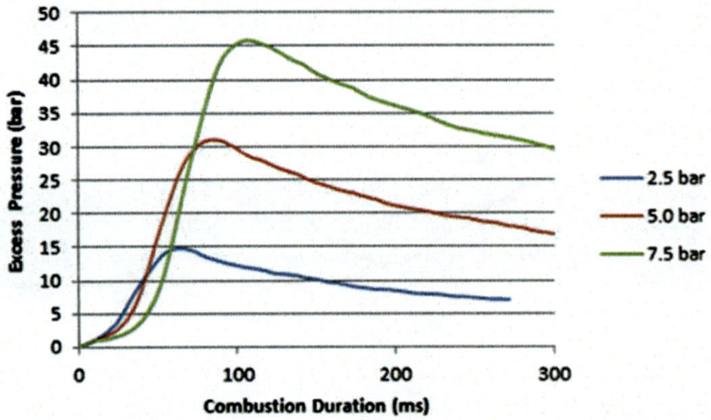

Fig. 12 Comparison of pressure–time curve of CH_4-air mixture at ϕ of 1.0 for 2.5, 5.0, and 7.5 bar chamber pressure

3.4 Minimum Pulse Energy (MPE)

The minimum laser pulse energy (MPE) is the least laser energy necessary to ignite the combustible mixture and is measured by using laser energy meter. The MPE for combustion to take place is seen at $\phi = 1.0$ for all the initial pressure conditions due to correct chemical air–fuel mixture at ϕ of 1.0. At stoichiometric mixture, the minimum pulse energy required at 2.5, 5.0, and 7.5 bar initial pressure was observed to be 94, 83, and 71 mJ, respectively. However, it has been observed that, as initial pressure increases the minimum pulse energy required decreases. This is mainly due to increase in the methane-air mixture density inside the CVCC which will reduce

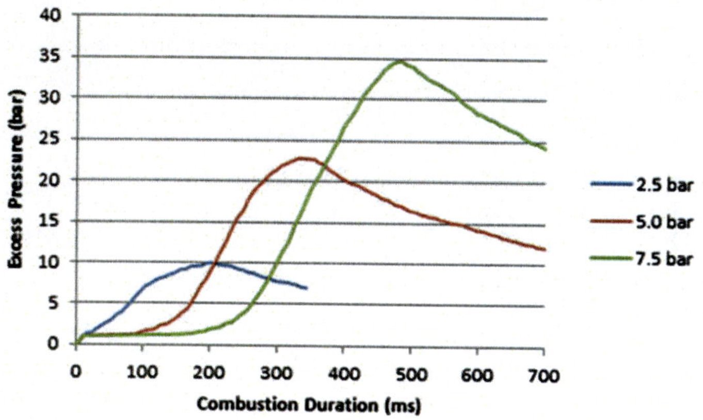

Fig. 13 Comparison of pressure–time curve of CH_4-air mixture at ϕ of 1.4 at 2.5, 5.0, and 7.5 bar chamber pressure

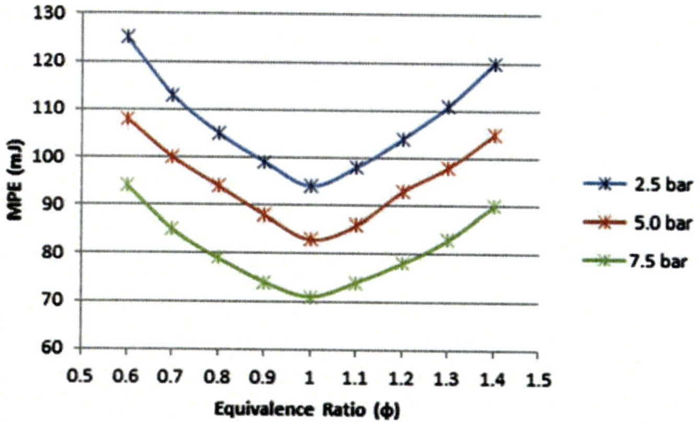

Fig. 14 Minimum pulse energy (MPE) for various ϕ of methane-air mixture at 2.5, 5.0, and 7.5 bar initial chamber pressure

the amount of energy required. The minimum pulse energy required at different equivalence ratios at different initial pressure is shown in Fig. 14.

4 Conclusions

The combustion characteristics like pressure–time history, minimum pulse energy (MPE) breakdown threshold energy, and breakdown threshold intensity (I) of CH_4-air mixture ignited using single-point LISI were discussed. The major outcomes can be concluded in the following points.

i. The minimum breakdown threshold energy and breakdown threshold intensity for both air and methane go on decreasing as initial chamber pressure goes on increasing.
ii. The maximum pressure for methane-air mixture was seen at ϕ of 1.0 for all initial pressure conditions and the peak pressure was observed to be 17.31, 36.03, and 53.39 bar for 2.5, 5.0, and 7.5 bar initial pressures, respectively. The peak pressure increases by almost 7.0 times the initial pressures for all the initial pressure maintained inside the chamber.
iii. The minimum combustion duration for methane-air mixture was seen at ϕ of 1.0 for all initial pressure conditions. The minimum combustion duration was observed to be 64.1, 84.08, and 106.7 ms for 2.5, 5.0, and 7.5 bar initial pressures, respectively.
iv. Increase in initial chamber pressure results in increase in the peak pressure and the combustion duration, however, the pressure rise rate decreases in the initial phase.
v. As the initial chamber pressure increases from 2.5 to 7.5 bar, the minimum pulse energy needed for combustion of methane-air mixture decreases. The minimum pulse energy required was observed at ϕ of 1.0 for all initial pressure conditions. The minimum pulse energy required at an equivalence ratio of 1.0 was observed to be 94, 83, and 71 mJ for 2.5, 5.0, and 7.5 bar initial pressure, respectively.

Acknowledgements I would like to thank AICTE and Technical Education Quality Improvement Programme (TEQIP) for the funding for this project.

References

1. Srivastava DK, Agarwal AK (2014) Comparative experimental evaluation of performance, combustion and emissions of laser ignition with conventional spark plug in a compressed natural gas fuelled single cylinder engine. Fuel 123:113–122
2. Karim GA (2003) Hydrogen as a spark ignition engine fuel. Int J Hydrogen Energy 28:569–577
3. Cho HM, He BQ (2007) Spark ignition natural gas—a review. Energy Convers Manage 48(2):608–618
4. Das A, Watson HC (1997) Development of a natural gas spark ignition engine for optimum performance. Proc Inst Mech Eng D J Automobile Eng 211:361–378
5. Hill RA, Laguna GA (1980) Laser initiated combustion of CH_4 and O_2 mixtures. Opt Commun 32(3):435–439
6. Phuoc T (2006) X, Laser-induced spark ignition fundamental and applications. Opt Laser Eng 44:351–397
7. Bradley D, Sheppard CGW, Suardjaia IM, Woolley R (2004) Fundamentals of high-energy spark ignition with lasers. Combust Flame 138:55–77
8. Patane P, Nandgaonkar M (2020) Review: multipoint laser ignition system and its applications to IC engines. Opt Laser Technol 130:106305
9. Patane P, Nandgaonkar M (2021) Numerical simulation of combustion characteristics and emission predictions of methane-air and hydrogen-air mixtures in a constant volume combustion chamber using multi-point laser-induced spark ignition. Energy Sources Part A: Recovery, Utilization Environmental Effects. https://doi.org/10.1080/15567036.2021.1910383

10. Phouc TX (2000) Single-point versus multi-point laser ignition: experimental measurements of combustion time and pressures. Combust Flame 122:508–510
11. Morsy MH, Ko YS, Cho P (2001) Laser induced twopoint ignition of premixture with a single-shot laser. Combust Flame 125:724–727
12. Morsy MH, Chung SH (2003) Laser-induced multi-point ignition with a single-shot laser using two conical cavities for hydrogen/air mixture. Exp Thermal Fluid Sci 27:491–497
13. Forch BE, Miziolek AW (1991) Laser-based ignition of H_2/O_2 and D_2/O_2 premixed gases through resonant multiphoton excitation of H and D atoms near 243 nm. Combust Flame 85:254–262
14. Forch BE, Miziolek AW (1986) Oxygen-atom two-photon resonance effects in multiphoton photochemical ignition of premixed H_2/O_2 flows. Opt Lett 11(3):129–131
15. Lavid M, Stevens JG (1985) Photochemical ignition of premixed hydrogen/oxidizer mixtures with excimer lasers. Combust Flame 60:195–202
16. Chou M-S, Zukowski TJ (1991) Ignition of $H_2/O_2/NH_3$, H_2/Air/NH_3, and $CH_4/O_2/NH_3$ mixtures by excimer-laser photolysis of NH_3. Combust Flame 87:191–202
17. Lucas D, Dunn-Rankin D, Hom K, Brown NJ (1987) Ignition by excimer laser photolysis of ozone. Combust Flame 69:171–184
18. Raffel B, Warnatz J, Wolfrum J (1985) Experimental study of laser-induced thermal ignition in O_2/O_3 mixtures. Appl Phys B 37:189–195
19. Hill RA (1981) Ignition-delay times in laser initiated combustion. Appl Opt 20(13):2239–2256
20. Morgan CG (1975) Laser-induced breakdown of gases. Rep Prog Phys 38:621–665
21. Kopecek H, Maier H, Reider G, Winter F, Wintner E (2003) Laser ignition of methane–air mixtures at high pressures. Exp Thermal Fluid Sci 27:499–503
22. Srivastav DK, Wintner E, Agarwal AK (2011) Flame kernel characterization of laser ignition of natural gas-air mixture in a constant volume combustion chamber. Opt Lasers Eng 49:1201–1209
23. Srivastav DK, Wintner E, Agarwal AK (2014) Effect of focal size on the laser ignition of compressed natural gas-air mixture. Opt Laser Eng 58:67–79
24. Weinberg FJ, Wilson JR (1971) A preliminary investigations of the use of focused laser beams for minimum ignition energy studies. Proc Roy Soc Lond A 321(1544):41–52
25. Radziemski LJ, Cremers DA (1989) Laser-induced plasmas and applications. Marcel Dekker, New York, Basel
26. Weinrotter M, Kopecek H, Winter E, Lackner M, Winter F (2005) Application of laser ignition to hydrogen-air mixtures at high pressures. Int J Hydrogen Energy 30:319–326
27. Srivastav DK, Weinrotter M, Iskrac K, Agarwal AK, Wintner E (2009) Characterization of laser ignition in hydrogen-air mixtures in a combustion bomb. Int J Hydrogen Energy 34:2475–2482
28. Phuoc TX (2000) Laser spark ignition; experimental determination of Laser-induced breakdown thresholds of combustion gases. Opt Commun 175:419–423
29. Kopecek H, Maier H, Reider G, Winter F, Winter E (2003) Laser ignition of methane-air mixtures at high pressures. Exp Thermal Fluid Sci 27:450–499
30. Prasad RK, Jain S, Verma G, Agarwal AK (2017) Laser ignition and flame kernel characterization of HCNG in a constant volume combustion chamber. Fuel 190:318–327
31. Dharamshi K, Pal A, Agarwal AK (2013) Comparative investigations of flame kernel development in a laser ignited hydrogen-air mixture and methane-air mixture. Int J Hydrogen Energy 38(25):10648–10653
32. Dharamshi K, Agarwal AK (2014) Parametric study of a laser ignited hydrogen-air mixture in a constant volume combustion chamber. Int J Hydrogen Energy 39:20207–20215
33. Morsy M (2012) Review and recent developments of laser ignition for internal combustion engines applications. Renew Sustain Energy Rev 16:4849–4875

Application of Industrial High-Performance Waste and Cigarette Filter in Thermal Insulation

Devesh Saxena, Shubham Srivastava, Nandan Kumar, and C. S. Malvi

1 Introduction

The massive rise of rail transportation, aircraft, and other industries has resulted in a contributed to increased demand for engineering. The dependability of composite materials is determined by a variety of parameters, including installation performance, thermal stability, mechanical qualities, ageing, and so on. Temperature is one of the most crucial elements determining the life of polymer-coated paper. The fundamental characteristics of insulating material sheets, including voltage breakdown, resistance of electric current, dielectric constant, and loss, may change as a result of high temperatures, and certain negative impacts will cause the materials to age and deteriorate. For example, with the increase in temperature, the breakdown voltage lowers, leading to an increase in leakage capacities, hence posing a safety issue. As a result, the future generation of fittings required to have great strength and effective heat dissipation. Due to its stable chemical composition and superior mechanical characteristics, aramid papers are well recognized for its electrical installation performance. They've been employed in transformers and electrics as electrical insulation materials, as well as high-temperature protection materials. However, aramid sheets, which are very effective polymers, show little thermal conductivity, especially in para-(poly-m-phenylene isophthalamide) and meta-(poly-m-phenylene isophthalamide) aramid products. As a result, it is an essential to create improved sheets that are more superior electrically insulating, having high thermal conductivities, remarkable thermal stabilities, and increased heat dissipation efficiency to widen their applications. Vacuum filtration of aramid pulp (AP), aramid nanofibers (ANFs) is used for processing and fabricating the composite sheets having high

D. Saxena (✉) · S. Srivastava · C. S. Malvi
Mechanical Engineering Department, MITS Gwalior, Gwalior, Madhya Pradesh, India
e-mail: deveshsaxena2308@gmail.com

N. Kumar
High Performance Textiles Pvt. Ltd, Panipat, Haryana, India

A. Kumar et al. (eds.), *Recent Advances in Manufacturing and Thermal Engineering*,
Lecture Notes in Mechanical Engineering, https://doi.org/10.1007/978-981-19-8517-1_24

thermal conductivities, excellent electrical insulation, and thermal thermodynamic stability. It also possesses excellent mechanical strength and flexibility, indicating that they have a lot of potential actually uses in flexible electronics and other fields [1]. It was also observed that uniform dispersion of aramid pulp is challenging task and effect was made to form the composite sheet by using varied amounts of acrylonitrile butadiene rubber (NBR). The result of this study indicated that improved dispersion might improve the homogeneity of the worn surface and reduce fracture development [2]. Further heterocyclic aramid nanofibers (HNFs) were generated by polymerization-induced self-assembly technique. These generated HNFs were then converted into nanopaper which has improved mechanical and electrical qualities due to more homogeneous structures [3]. ANFs was also used with Ag nanowires that have application in wearable's, artificial intelligence, and extremely efficient heating systems [4]. It was also addressed by researchers that industrial scale production of ANFs involves the use of harmful acids regardless of fantastic performance features and functionality. It was suggested to use other polymerization techniques for industrial production of ANFs [5]. It is evident that pulping and uniform dispersion are challenging in case of aramid fibres and drawing of ANFs utilized critical processes. In order to overcome this problem spinning waste of para-aramid (P-aramid) fibres were used for fabricating the insulation sheet and uniform dispersion was tackled by utilizing cigarette filter fibre waste. Cigarette filter fibres are made-up of 95% cellulose acetate and 81.3 billion cigarettes were sold in India that accounts for generation of same number of cigarette filter [6]. This study was also focused on recycling of both industrial p-aramid waste fibres and cigarette filter waste fibres.

2 Materials and Method

The thermal insulation sheets were fabricated by utilizing industrial spinning waste p-aramid fibres and cigarette filter fibres as shown in Fig. 1a, b. The industrial spinning waste p-aramid fibres were supplied by High Performance Textiles Pvt. Ltd. Panipat Haryana and cigarette filter fibres were collected from different shops. The supplied waste p-aramid was 1.7 dtex and fibre size of 51 mm (Fig. 2).

At first, supplied waste p-aramid fibres and collected cigarette filter fibres were manually cut into very fine pieces with the help of scissors as shown in Fig. 3a. The waste p-aramid fibres were kept in the range (5–8 mm) and cigarette filter fibres were in the range (2–3 mm). Fibre size of both materials was checked after reaming, and long fibres were cut again in to small pieces. Small fibres of were dipped in acetone for removing the impurities as shown in Fig. 3b. After that they were soaked into distilled water so that it could be turned into the pulp form. Further polyvinyl alcohol was used as a binder for binding of the pulps of both the fibres. Binder was prepared in ratio of 95% (distilled water) and 5% (PVA powder), in a beaker and magnetic stirrer machine was used for this purpose. Magnetic stirred and prepared PVA binder are shown in Fig. 3c, d, respectively. Fine fibres were dipped in PVA binder and

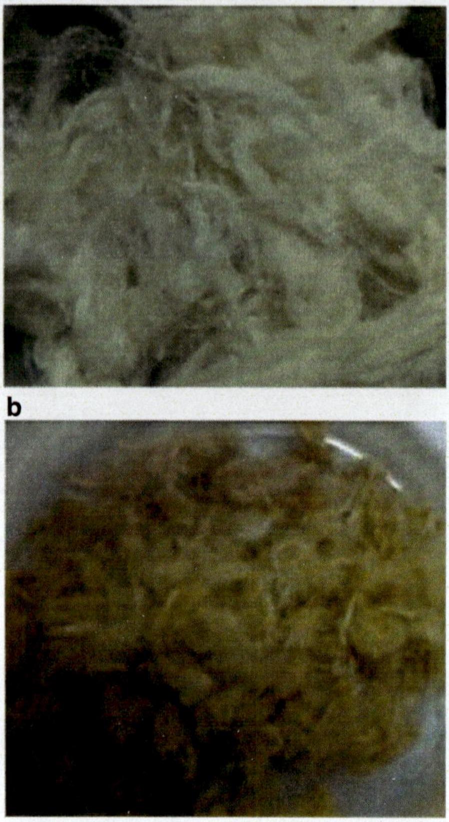

Fig. 1 **a** Industrial spinning waste p-aramid fibres, **b** cigarette filter fibres

spread over a glass sheet for uniform compression and for developing the insulation sheet as shown in Fig. 3e. The final prepared sample is shown in Fig. 3f.

The thermal insulation sheets were fabricated with waste p-aramid fibres and cigarette filter fibres along with the blended fibres of these two materials (in ratio of 50:50) as shown in Fig. 4a, b, c.

3 Result and Discussion

3.1 Physical Properties and Thermal Conductivity

The thermal conductivity of prepared samples was measured with the help of Quick thermal conductivity meter (QTM-710). Thickness and weight of the sample were

Fig. 2 Stepwise procedure of sample preparation

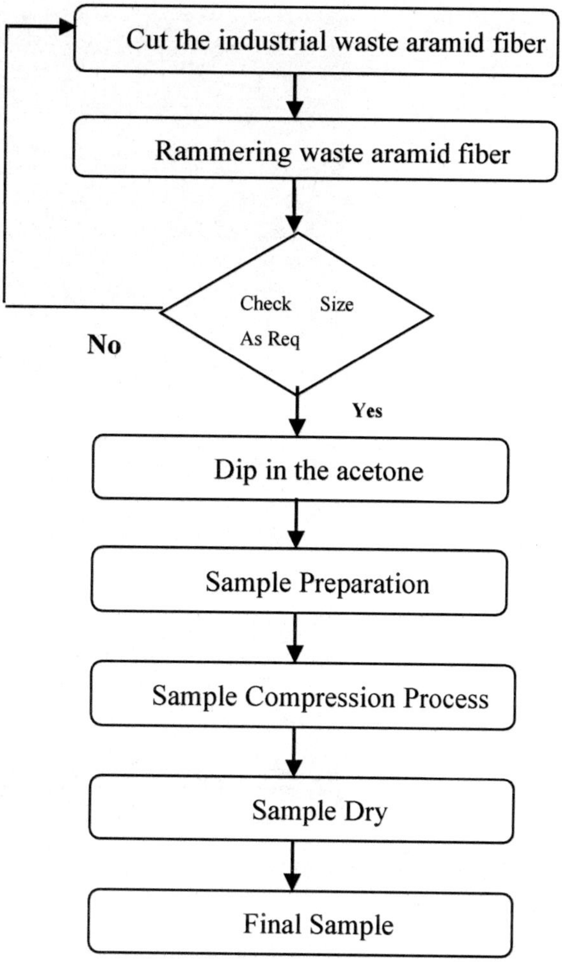

measured with the help of Mitutoyo thickness tester and weighing machine respectively. Corresponding values are shown in Table 1. It was observed that due to the long size of p-aramid fibres the insulation sheet made-up of 100% p-aramid fibres are thicker and heavier than the other two sheets. Data obtained from waste p-aramid fibre insulation sheet was used as a reference data for comparing other two insulation sheets.

4 Microstructure

Microstructure images of all three samples are shown in Fig. 5. Whitish colour of aramid waste is clearly visible in Fig. 5a whereas blackish colour of cigarette filter

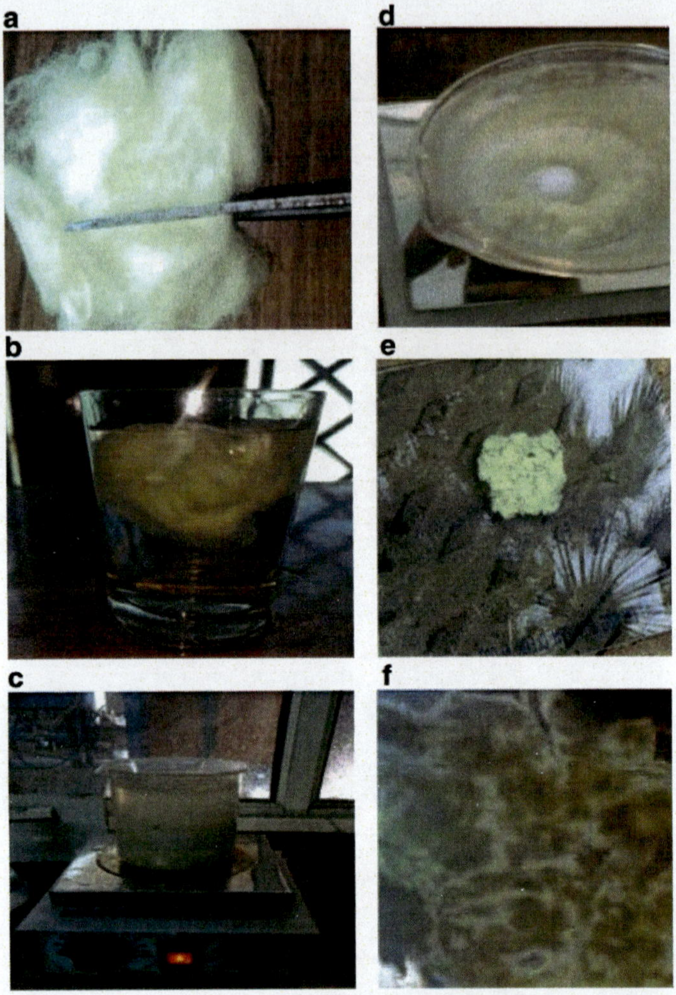

Fig. 3 **a** Cutting of fibres, **b** removing impurities, **c** binder preparation in magnetic stirrer, **d** PVA binder, **e** compressing of pulp, **f** final prepared sample

waste is visible in Fig. 5b. Mixture of aramid fibre waste and cigarette filter is also of white colour with black spot present everywhere in whole sheet.

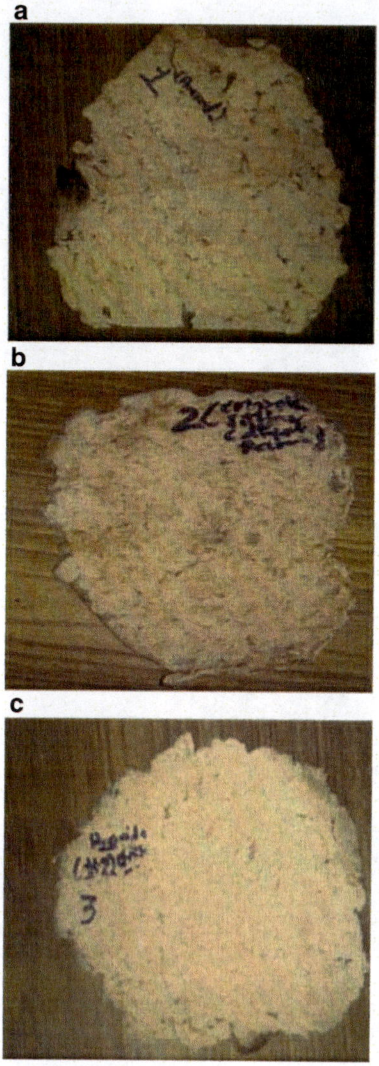

Fig. 4 **a** Waste p-aramid fibre insulation sheet, **b** cigarette filter fibre insulation sheet, **c** waste p-aramid fibre blended with cigarette filter fibre sheet

Table 1 Thickness weight and thermal conductivity of prepared insulation sheet

S. No.	Material	Thickness (mm)	Weight (g/m^2)	Thermal Conductivity (W/m-K)
1	Aramid + PVA	2.30	444.44	0.1721
2	Cigarette Filter + PVA	1.48	177.77	0.2262
3	Aramid + Cigarette Filter + PVA	2.30	278.18	0.1098

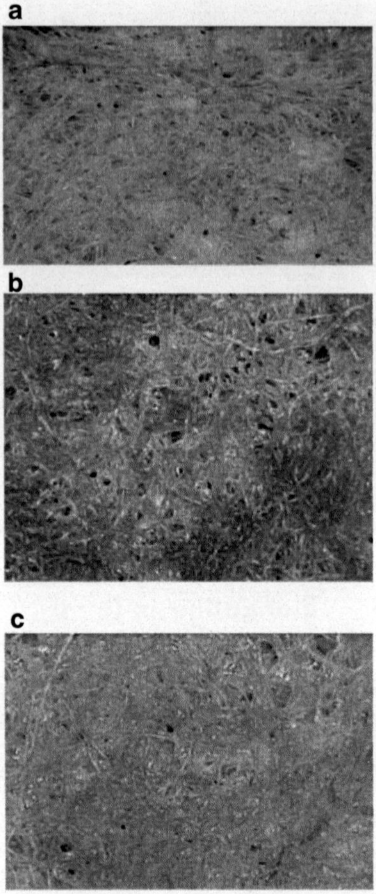

Fig. 5 **a** Microstructures of waste p-aramid fibres insulation sheet, **b** Microstructures of cigarette filter fibres insulation sheet, **c** Microstructures of p-aramid blended with cigarette filter fibres insulation sheet

5 Tensile Strength

Tensile test was carried out on Tensile Strength Test Machine *(TM2101-7)*, having capacity 0–500 N load. Figure 6a, b shows the tensile strength and maximum elongation of the samples. Tensile strength of the cigarette fibre was less as compared to other two samples. This is because of low strength of cellulose acetate (0.9–1.4 g/den) [7]. When waste p-aramid was blended with equal proportion of cigarette filter fibres, the tensile strength of blended insulation sheet was increased due to high strength of p-aramid (17 g/den). It could also be deduced that the uniform dispersion of small cigarette filter fibres increased the binding capacity of waste p-aramid fibres.

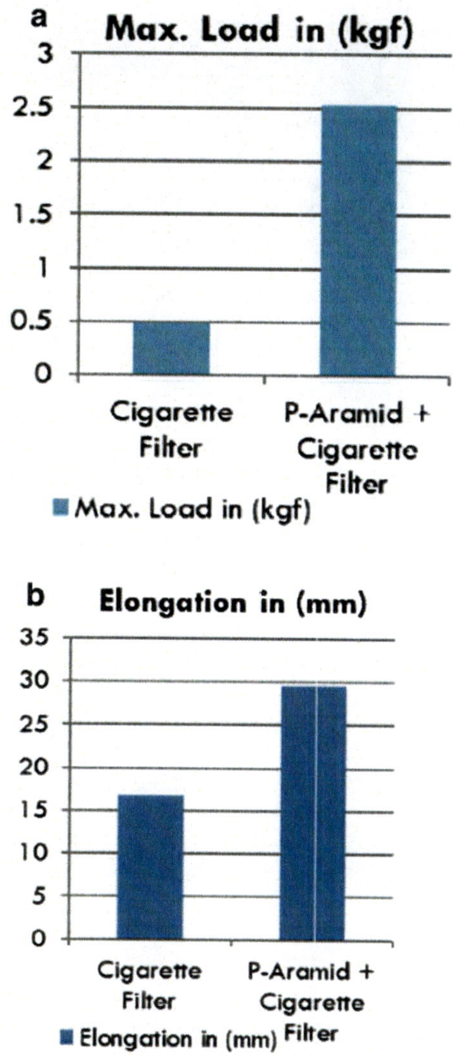

Fig. 6 **a** Maximum load, **b** maximum elongation

6 Contact Heat Transmission Testing

Contact heat transmission testing was performed on *T424* contact heat transmission tester as per ISO 12127-1 at 100 °C heat source temperature. In this test time taken to increase the temperature of the samples by 10 °C was noted. Waste p-aramid blended with cigarette filter fibre insulation sheet has performed well under contact heat due to low thermal conductivity as shown in Table 2, and it may also be possible that the pores present in waste p-aramid insulation sheet were filled with small fibres of

Table 2 Behaviour of different insulating sheet under 100 °C contact heat

S. No.	Sample	Heat source temperature (°C)	Atmospheric temperature (°C)	Time (Time taken to 10 °C rise) in sec
1	100% p-aramid waste	100	32.97	77.38
2	Cigarette filter waste	100	35.65	51.2
3	p-aramid blended with cigarette filter	100	34.92	105

cigarette filter fibres. This ultimately decreased the heat transmission. The time for 10 °C rise for insulating sheet of waste p-aramid, cigarette filter fibre, and blend of these two were found to be 77.38, 51.2, and 105 s, respectively. It can be used for insulating engines, boilers, etc. However, it may be tested at high temperature for further applications.

7 Vertical Flammability Test

It is evident that these samples have potential in insulating any system that is why these were also tested for direct flame exposure test on T419 vertical flammability testing machine. In this test flame was made to contact with the sample at 30° from the vertical and flame was made to strike at the edge of the insulation sheets. All the sheets were started burning in direct flame test which could be due to presence of PVA on surface. Shrinkage was also observed in case of cigarette filter fibre insulation sheet. The burned samples are shown in Fig. 7.

This study showed that blending of waste p-aramid and cigarette filter fibres in an equal proportion enhances the results and lowers thermal conductivity which is desirable property in insulation. Owing to the higher cost of p- aramid spinning waste, the mixing of cigarette filter fibres reduces the cost of insulation sheet. Further study is required by blending p-aramid waste to much lower levels, e.g. 10, 25, and 35% and also by using more sophisticated sample preparation methodology for better results.

8 Conclusions

The insulating behaviour of insulating sheets made of industrial spinning waste p-aramid, cigarette filter, and blend of these two fibres were studied. It was observed that insulating sheet prepared by blending foresaid fibres possess low thermal conductivity and have good performance against contact heat transmission and tensile load. The contact heat transmission results proved that it has potential of providing insulation. The prepared samples were burned in direct flame exposure so the binders

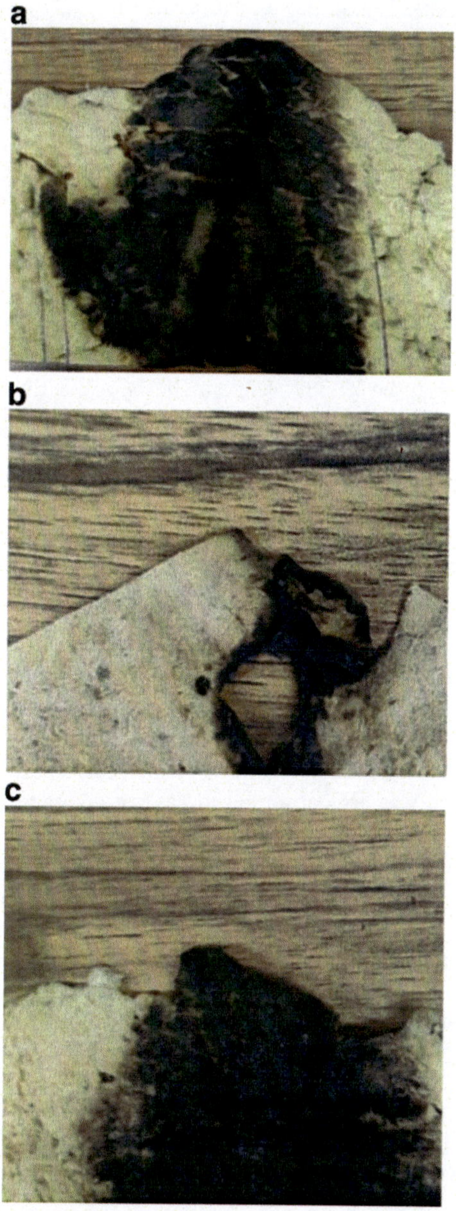

Fig. 7 **a** Insulating sheets after direct flame test waste p-aramid fiber, **b** insulating sheets after direct flame test cigarette filter waste, **c** insulating sheets after direct flame test waste aramid and cigarette filter waste insulating sheet

having fire-resistive properties could be utilized and behaviour of PVA binder in combination with phosphorous and bromine could be determined in future. Moreover, both p-aramid and cigarette filter fibre waste could be recycled by commercially developing these products which may be more economical; however, further study is required by changing blend ratio of p-aramid and cigarette filter fibres.

References

1. Li M, Zhu Y, Teng C (2020) Elsevier Compos Commun 21
2. Zhang R, Li Y, Du Z, Li Z, Wan S, Yuan X, Wang Y (2020) RSC Adv
3. Shi Y, Tuo X (2020) Mater Adv RSC
4. Ma Z, Gu J, Kang S, Yang B, Ma J, Dong D, Shao L, Wei L, Wei A, Ji Z, Liang C (2019) ACS Nano 13(7):7578–7590
5. Koo JK et al (2019) American Chemical Society
6. https://www.tobaccofreekids.org/problem/toll-global/asia/india.
7. https://textilefashionstudy.com/physical-and-chemical-properties-of-cellulose-acetate-rayon/

Experimental Study of Mechanical and Thermal Properties of Nano-carbon Areca Fiber Powder Reinforced Epoxy Composites

Alok Singh, Savita Singh, and Sudhir Kumar Sharma

1 Introduction

The early stages of human technological progress made use of bio-inspired natural materials including bone, wood, and shells. These materials gradually gave way to synthetic ones as antiquity developed in order to increase performance. Today, scientists and engineers explored distinctive qualities of natural materials as lightweight with desired mechanical properties can be engineered [1–3]. The modern civilization has a challenging issue of developing novel and propelled innovative techniques to utilize solid natural agro waste materials in polymer composites. Recently, a broader application of cellulose has been proposed at the nano-structure level for developing various biocompatible products and a variety of commercial cellulose derivatives [4, 5].

The most cutting-edge solution to many complex issues, including energy conversion, energy storage, and material science, is nanotechnology [6, 7]. To improve the mechanical, chemical, and thermo-physical properties of composite materials, nanoparticles are disseminated in matrix materials such metals, ceramics, or polymers [8, 9]. Polymer nano-composite materials are used in major industries as the automotive and aviation sectors [10, 11]. Polymer nano-composites can be produced using a variety of techniques. All methods intended to create nano-composite materials that dispersed evenly or randomly and didn't aggregate. The most popular methods for creating polymer nano-composite are melt-mixing, mixing, in-situ polymerization, electro spinning, and selective laser sintering [12, 13]. The similar function is also played by sonication at high frequencies for mixing procedures [14]. The

A. Singh (✉) · S. Singh · S. K. Sharma
Department of Physics, HBTU, Kanpur 208002, India
e-mail: alokphy@hbtu.ac.in

S. K. Sharma
e-mail: hodphy@hbtu.ac.in

A. Kumar et al. (eds.), *Recent Advances in Manufacturing and Thermal Engineering*,
Lecture Notes in Mechanical Engineering, https://doi.org/10.1007/978-981-19-8517-1_26

manufacture of thermodynamically stable nano-composites is made possible by in-situ polymerization [15]. An efficient technique that works well for creating porous materials is electro spinning. Additionally, the creation of nano-composites using selective laser sintering has apparent advantages to solve the aggregation issue. Each technique's operating principles include certain benefits and drawbacks [16, 17].

In this research work, an experimental developed different weight percentage NCAFP reinforced epoxy composites and its factual mechanical and thermal properties were discussed. In future, the analytical prediction models for density, mechanical strength, and thermal conductivity of NCAFP reinforced epoxy composites can be explored.

2 Experimental Setup

2.1 Materials and Methods

The Areca fiber was burned into closed furnace at 300 °C to develop nano-Carbon Areca Fiber Powder (NCAFP). The epoxy resin (LY 556) and the proportionate hardener (HY-917) were mixed to form neat epoxy matrix. The synthesized NCAFP was randomly dispersed into epoxy matrix in different ratios as 5, 10, 15, 20, and 25% by weight percentage to develop sample laminate sheets separately using casting method. The NCAFP composite sheets were formed in a mold with dimensions of 300 × 160 × 6 mm, with a base of glass and wood on either side. In order to quickly and easily remove the NCAFP composite sheet, the mold release sheet was kept on top of the glass plate. The NCAFP by weight percentage (i.e., 5, 10, 15, 20, and 25 wt%) was mixed with the matrix material consisting of epoxy resin and hardener in the ratio of 5:4. By curing for 24 h at room temperature in a vacuum chamber, air bubble generation was reduced. The cured laminate was then trimmed to the required size for a variety of mechanical tests.

These sample laminate sheets were machined as per ASTM D638-03 shown in Fig. 1. Five identical samples of each set of weight percentage of NCAFP reinforced epoxy composite were machined to validate reproducibility of measured data. The tensile strength of the developed samples was measured by INSTRON 1195 testing machine. The density and thermal properties of these developed NCAFP composite samples were estimated by LSA 1000 (Linseis Germany). The surface morphology of fractured NCAFP composite samples were analyzed by scanning electron microscopy (JEOL EDS System) of ACMS Laboratory at Indian Institute of Technology, Kanpur.

Fig. 1 Developed NCAFP composite sample as per ASTM D638-03

3 Results and Discussion

The density and thermal conductivity of these developed different weight percentage NCAFP composite samples were experimentally obtained by LSA 1000 (Linseis Germany) as shown in Figs. 2 and 3.

Figure 2 illustrates a merest increase in density of developed different weight percentage NCAFP composite samples as compared to neat epoxy composite. Figure 3 reports the gradual reduction in thermal conductivity and become steady in further loading of NCAFP in the epoxy matrix. The experimental percentage error in measuring density, thermal conductivity, and tensile strength is less than 5% (PE > 5%).

Figure 4 displays experimentally obtained tensile strength of developed different weight percentage NCAFP composite samples. An enormous increase in tensile strength is noticed in developed different weight percentage NCAFP composite samples which become optimum for 15% NCAFP composite sample. Further, higher loading of NCAFP in epoxy polymer matrix causes substantial decrease in NCAFP composite samples as shown in Fig. 4.

In the tensile test experiment, the samples were fractured by INSTRON 1195 testing machine to obtain tensile strength. Now, these fractured sample surfaces was analyzed by scanning electron microscope imaging technique to probe the reason behind the optimum tensile strength.

Fig. 2 Plots of density versus composite samples

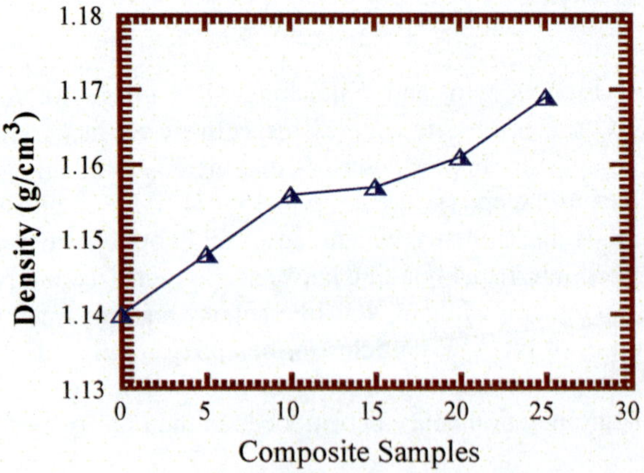

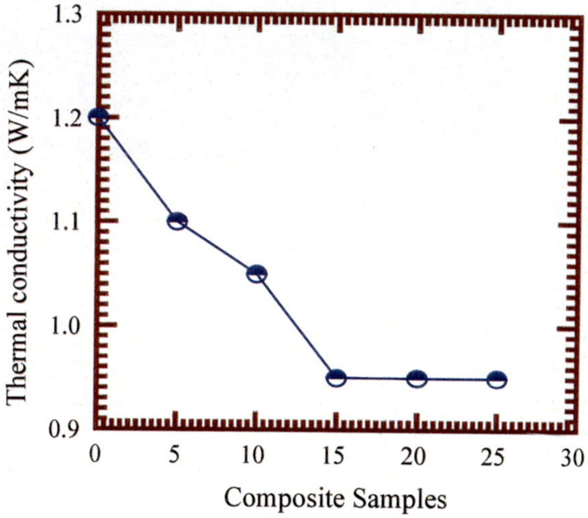

Fig. 3 Plots of thermal conductivity versus composite samples

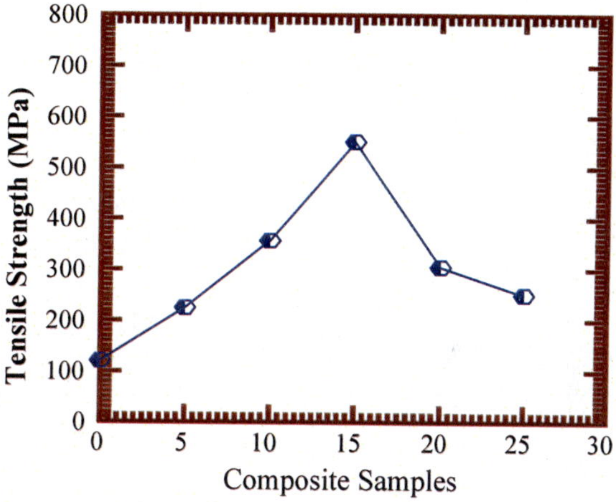

Fig. 4 Plots of tensile strength versus composite samples

Figures 5, 6, and 7 illustrate SEM images of fractured 15%, 20%, and 25% NCAFP composite samples, respectively. Surface analysis of these fractured NCAFP composite samples describes that there is minimal agglomeration of NCAFP particles in the epoxy matrix up to 15% NCAFP composite samples (Fig. 5) with least voids and porosity which causes an enormous increase in tensile strength having good interfacial bonding between epoxy matrix and NCAFP.

Higher loading of NCAFP in epoxy matrix greater than 15% produces agglomeration of NCAFP particles in the epoxy matrix which causes a decrease in tensile strength having low interfacial bonding between epoxy matrixes. The localized agglomeration area can be noticed in the Figs. 6 and 7 shown by black circles.

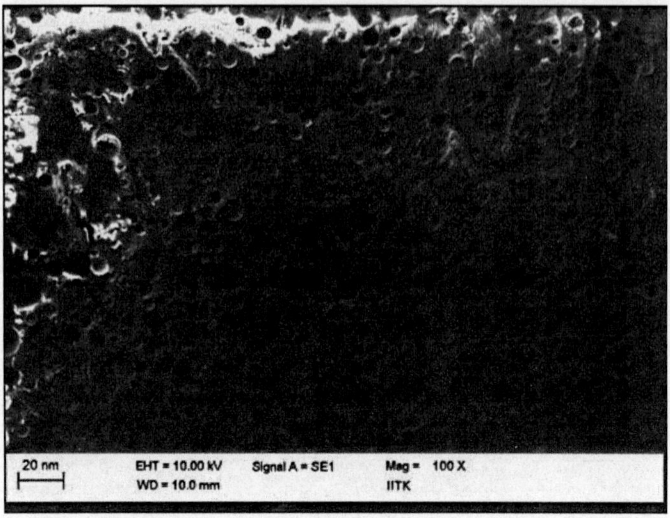

Fig. 5 Fractographic SEM images of 15% NCAFP composite sample

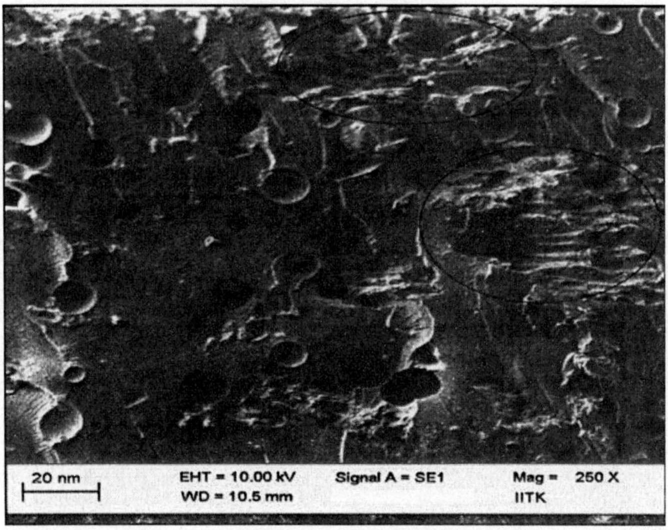

Fig. 6 Fractographic SEM images of 20% NCAFP composite sample

4 Conclusion

In this research work, an experimental developed different weight percentage NCAFP reinforced epoxy composites and its mechanical and thermal properties were discussed. The overall discussion leads to conclude that the optimum tensile

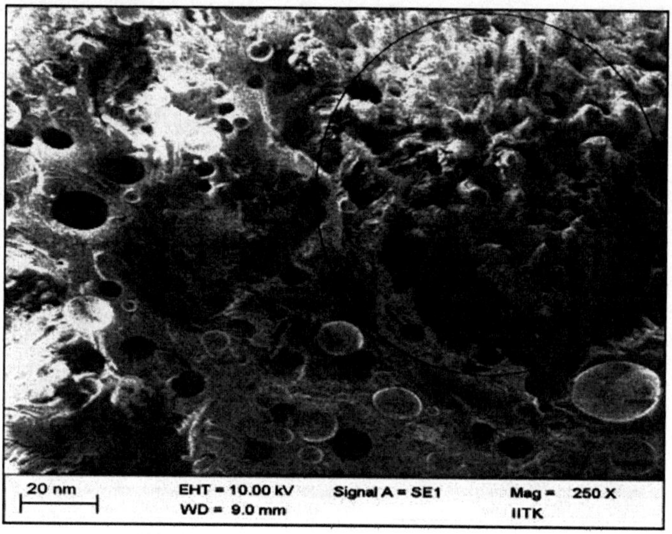

Fig. 7 Fractographic SEM images of 25% NCAFP composite sample

strength 550 MPa (see Fig. 4) with low density and thermal conductivity 0.95 W/mK (see Fig. 3) of developed 15% NCAFP reinforced epoxy composite were obtained experimentally. Absence of localized agglomeration, porosity, and voids in fractography SEM image (see Fig. 5) provide the justification for optimum tensile strength of 15% NCAFP composite. Consequently, the developed 15% NCAFP reinforced epoxy composites can be used in microelectronics packaging, aeronautics, and automobile engineering due to its lightweight (low density), optimum mechanical strength, and low thermal conductivity.

Future research can explore and compare the analytical prediction models for density, mechanical strength, and thermal conductivity of NCAFP reinforced epoxy composites with the current models for nano-composite materials.

Acknowledgements We are grateful to acknowledge AICTE sponsored quality improvement program for facilitating all experimental and measurements service.

References

1. Mirkhalaf M, Dastjerdi AK, Barthelat F (2014) Overcoming the brittleness of glass through bio-inspiration and micro-architecture. Nat Commun. https://doi.org/10.1038/ncomms4166
2. Singh S, Singh A, Sharma SK (2017) Analytical modeling for mechanical strength prediction with Raman spectroscopy and fractured surface morphology of novel coconut shell powder reinforced: epoxy composites. J Inst Eng (India) Ser C. https://doi.org/10.1007/s40032-016-0254-9

3. Singh S, Singh A, Sharma SK (2020) Analytical prediction models for density, thermal conductivity and mechanical strength of micro-scaled areca nut powder-reinforced epoxy composites. J Inst Eng (India) Ser C. https://doi.org/10.1007/s40032-019-00535-9
4. Shah SS, Shaikh MN, Khan MY, Alfasane MA, Rahman MM, Aziz MA (2021) Present status and future prospects of jute in nanotechnology: a review. Chem Rec. https://doi.org/10.1002/tcr.202100135
5. Ates B, Koytepe S, Ulu A, Gurses C, Thakur VK (2020) Chemistry, structures, and advanced applications of nanocomposites from biorenewable resources. Chem Rev. https://doi.org/10.1021/acs.chemrev.9b00553
6. Mehta KP, Sharma R, Haldar S, Kumar A (2021) Advancement in treatment of wastewater with nano technology. Mater Today: Proc. https://doi.org/10.1016/j.matpr.2021.07.253
7. Ageed ZS, Ahmed AM, Omar N, Kak SF, Ibrahim IM, Yasin HM, Salim NO (2021) A state of art survey of nano technology: implementation, challenges, and future trends. Asian J Res Comput Sci. https://doi.org/10.9734/AJRCOS/2021/v10i330245
8. Raj CR, Suresh S, Bhavsar RR, Singh VK (2020) Recent developments in thermo-physical property enhancement and applications of solid solid phase change materials. J Therm Anal Calorim. https://doi.org/10.1007/s10973-019-08703-w
9. Mendes JF, Martins JT, Manrich A, Luchesi BR, Dantas APS, Vanderlei RM,... Martins MA (2021) Thermo-physical and mechanical characteristics of composites based on high-density polyethylene (HDPE) e spent coffee grounds (SCG). J Polym Environ. https://doi.org/10.1007/s10924-021-02090-w
10. Kamal A, Ashmawy M, Algazzar AM, Elsheikh AH (2022) Fabrication techniques of polymeric nanocomposites: a comprehensive review. Proc Inst Mech Eng C J Mech Eng Sci. https://doi.org/10.1177/09544062211055662
11. Rajak DK, Pagar DD, Kumar R, Pruncu CI (2019) Recent progress of reinforcement materials: a comprehensive overview of composite materials. J Market Res. https://doi.org/10.1016/j.jmrt.2019.09.068
12. Lawal AT (2020) Recent progress in graphene based polymer nanocomposites. Cogent Chem. https://doi.org/10.1080/23312009.2020.1833476
13. Sun J, Shen J, Chen S, Cooper MA, Fu H, Wu D, Yang Z (2018) Nanofiller reinforced biodegradable PLA/PHA composites: current status and future trends. Polymers. https://doi.org/10.3390/polym10050505
14. Asgharzadehahmadi S, Raman AAA, Parthasarathy R, Sajjadi B (2016) Sonochemical reactors: review on features, advantages and limitations. Renew Sustain Energy Rev. https://doi.org/10.1016/j.rser.2016.05.030
15. Baniasadi H, Borandeh S, Seppälä J (2021) High-performance and biobased polyamide/functionalized graphene oxide nanocomposites through in situ polymerization for engineering applications. Macromol Mater Eng. https://doi.org/10.1002/mame.202100255
16. Yu WH, Sing SL, Chua CK, Kuo CN, Tian XL (2019) Particle-reinforced metal matrix nanocomposites fabricated by selective laser melting: a state of the art review. Prog Mater Sci. https://doi.org/10.1016/j.pmatsci.2019.04.006
17. Ni J, Ling H, Zhang S, Wang Z, Peng Z, Benyshek C, Khademhosseini A (2019) Three-dimensional printing of metals for biomedical applications. Mater Today Bio. https://doi.org/10.1016/j.mtbio.2019.100024

Numerical Study on Momentum and Heat Transfer Phenomenon from a Sphere Under Force Convection Environment

Numan Siddique Mazumder, Pradip Lingfa, and Asis Giri

Nomenclature

C_P	Specific heat
d	Diameter of the sphere
k	Thermal conductivity
Re	Reynolds number
Pe	Peclet number
ψ	Stream function
ω	Vorticity
r	Radial coordinate
Θ	Azimuthal coordinate
T	Temperature
P	Pressure
ϕ	Non-dimensional temperature
$T\infty$	Free stream temperature
T_s	Temperature at droplet surface
C_{DF}	Drag due to friction
C_{DP}	Drag due to pressure
C_D	Total drag coefficient
Nu_Θ	Local Nusselt number
Nu	Average Nusselt Number
$U\infty$	Free stream velocity
Z	Stretch coordinate

N. S. Mazumder (✉) · P. Lingfa · A. Giri
Department of Mechanical Engineering, North Eastern Regional Institute of Science and Technology, Nirjuli, Arunachal Pradesh 791109, India
e-mail: siddiquesil@gmail.com

A. Kumar et al. (eds.), *Recent Advances in Manufacturing and Thermal Engineering*, Lecture Notes in Mechanical Engineering, https://doi.org/10.1007/978-981-19-8517-1_27

Abbreviations

FEM	Finite Element Method
FVM	Finite Volume Method
FDM	Finite Difference Method
NS	Navier-Stokes
SOR	Successive Over Relaxation

1 Introduction

The solution of Navier–Stokes (NS) equation of spherical coordinate system is one of the most demanding engineering calculations. It has many practical applications, especially droplet combustion and evaporation, bubble dynamics, etc. Many studies [1–9] were performed on the fundamental aspect of flow and heat transfer phenomenon over a solid sphere. Abramzon and Elata [1] and Feng and Michaelides [2] performed numerical studies on transient heat transfer behavior from a solid sphere, but the Re was considered low. However, Feng and Michaelides [3] performed numerical study on solid sphere at higher Re ($0.1 \leq \text{Re} \leq 4000$) and Pe ($0.2 \leq \text{Pe} \leq 2000$). They solved the NS equation of the spherical polar coordinate system in terms of vorticity and stream function formulation via Finite Volume Method (FVM). But, the details of analysis of the flow pattern (streamline), vorticity contour, and the isotherm contour were not discussed at various Re and Pe, whereas Lee [4] also performed numerical study on solid sphere at the relatively higher Re, ($100 \leq \text{Re} \leq 500$). He solved the three-dimensional NS equation with the aid of Finite Element Method (FEM). He was mainly concerned with the vortex shedding which forms behind a solid sphere at higher Re. Subsequently, Bagchi et al. [5] performed a 3D study on solid sphere using Direct Numerical Simulation scheme. They observed that the flow over a solid sphere remains axisymmetric up to $\text{Re} \leq 210$. Moreover, Melissari and Argyropoulos [6] developed a co-relation between Nu, and Re and Pr for heat transfer phenomenon. Sekhar et al. [7] successfully implemented a 4th order higher compact scheme for the finite difference method (FDM) which can accurately solve the stream and vorticity function equation of spherical polar coordinate system for a widespread range of Re.

Moreover, it was perceived that the drag coefficients (C_D) of a solid sphere mainly depend on Re, and in practice, many correlations between C_D and Re exist for engineering applications. However, Duan et al. [8] successfully established a formulation to relate the appropriate drag coefficient with the heat transfer coefficients of a solid sphere based on experimental data. Will et al. [9] experimentally analyzed the force convection heat transfer over a solid sphere for a higher range of Re. They considered the radiation effects in their study. A co-relation between Nu and Re was also formulated based on the experiment data. The focus of their study was on the heat transfer

characteristics of sphere for higher range of Re. However, the flow and temperature profile over the solid sphere have not been discussed in their study.

By going through the previously reported studies on the solid sphere, the detailed analysis on flow and temperature pattern, vorticity contour over the sphere, and the influence of convective strength on the heat transfer coefficient at moderate Re and Pe is still required. Hence, in this study, we have solved the NS equation of the spherical polar coordinate system in vorticity and stream function formulation for moderate Re ($0.01 \leq \text{Re} \leq 100$) to capture the momentum transfer. The velocity field information generated after the solution of NS equation is utilized to solve the energy equation for moderate range of Pe ($0.1 \leq \text{Pe} \leq 200$). In this article, we have discussed about the streamline, vorticity contour, and isotherm contour over the solid sphere at various Re and Pe. Also, the C_D and Nu are also estimated at various Re and Pe.

2 Governing Equation

The non-dimensional governing equation of motion, the NS equation for viscous, and incompressible flow in terms of vorticity and stream function formulation in spherical polar coordinates may be written as [10].

$$\frac{\text{Re}}{2}\left[\frac{\partial\psi}{\partial r}\frac{\partial}{\partial\theta}\left(\frac{\omega}{r\sin\theta}\right)-\frac{\partial\psi}{\partial\theta}\frac{\partial}{\partial r}\left(\frac{\omega}{r\sin\theta}\right)\right]\sin\theta = E^2(\omega r\sin\theta) \tag{1}$$

$$E^2\psi = \omega r\sin\theta \tag{2}$$

Here, the operator E^2 is defined as

$$E^2 = \frac{\partial^2}{\partial r^2}+\frac{1}{r^2}\frac{\partial^2}{\partial\theta^2}-\frac{\cot\theta}{r^2}\frac{\partial}{\partial\theta} \tag{3}$$

The dimensionless Reynolds number (Re) is defined as

$$\text{Re} = \frac{\rho D U\infty}{\mu} \tag{4}$$

Here, ρ is the density of the ambient environment, D is the diameter of the sphere, $U\infty$ is the free stream velocity, and μ is the dynamic viscosity of the ambient fluid.

The non-dimensional heat transfer equation of spherical polar coordinate system may be expressed as [11].

$$\frac{\text{Pe}}{2}\left[\frac{\partial\psi}{\partial r}\frac{\partial\phi}{\partial\theta}-\frac{\partial\psi}{\partial\theta}\frac{\partial\phi}{\partial r}\right] = \sin\theta\left[r^2\frac{\partial^2\phi}{\partial r^2}+2r\frac{\partial\phi}{\partial r}+\frac{\partial^2\phi}{\partial\theta^2}+\cot\theta\frac{\partial\phi}{\partial\theta}\right] \tag{5}$$

Here, ϕ is the dimensionless temperature, and it has been converted as

$$\phi = \frac{T - T\infty}{T_S - T\infty} \tag{6}$$

Here, T_S is the temperature at the sphere surface, and $T\infty$ is the ambient temperature fluid at the far away from the droplet surface.

Also, the dimensionless Peclet number (Pe) is defined as

$$\mathrm{Pe} = \frac{\rho c_p D U\infty}{k} \tag{7}$$

Here, c_p is the specific heat, k is the thermal conductivity, and $U\infty$ is the free stream velocity.

2.1 Boundary Condition

At the free stream location ($r \to \infty$)

$$\psi = \frac{1}{2} r^2 \sin^2 \theta, \omega = 0, \phi = 0 \tag{8}$$

Axisymmetric location ($\theta = 0°, \theta = 180°$)

$$\psi = 0, \omega = 0, \frac{\partial \phi}{\partial \theta} = 0 \tag{9}$$

On the surface of the sphere ($r = 1$)

$$\psi = 0, \omega = \frac{1}{\sin \theta} \frac{\partial \psi}{\partial r}, \phi = 1 \tag{10}$$

2.2 Nusselt Number

The local Nusselt number (Nu_θ) is calculated as

$$\mathrm{Nu}_\theta = -2\left(\frac{\partial \phi}{\partial r}\right)_{|r=1} \tag{11}$$

The average Nusselt number (Nu) is calculated by integrating the local Nusselt number Nu_θ over the sphere surface

$$\mathrm{Nu} = \int_0^{\pi} \left(\frac{\partial \phi}{\partial r}\right)_{|r=1} \sin\theta \partial\theta \tag{12}$$

2.3 Drag Coefficient

Integration of the Θ-component of frontal stagnation point over the surface gives the surface pressure distribution

$$P_\theta = P_o + \frac{4}{\mathrm{Re}} \int_0^{\theta} \left(\frac{\partial \omega}{\partial r} + \omega\right)\Big|_{r=1} \partial\theta \tag{13}$$

Here, P_o is the frontal stagnation pressure, is obtained by integrating the r-component of the equation of motion along the line, and can be expressed as

$$P_o = 1 + \frac{8}{\mathrm{Re}} \int_1^{\infty} \left(\frac{\partial \omega}{\partial \theta}\right)\Big|_{\theta=0} \frac{\partial r}{r} \tag{14}$$

Now, Pressure Drag Coefficient (C_{DP}) can be obtained by integrating the surface vorticity over the sphere surface

$$C_{DP} = \int_0^{\pi} P_\theta|_{r=1} \sin 2\theta \partial\theta \tag{15}$$

Integration of surface vorticity distribution over the sphere surface gives the friction drag coefficient (C_{DF}) as

$$C_{DF} = \frac{8}{\mathrm{Re}} \int_0^{\pi} \omega|_{r=1} \sin^2\theta \partial\theta \tag{16}$$

And, the total drag coefficient is (C_D)

$$C_D = C_{DF} + C_{DP} \tag{17}$$

3 Method of Solution

Before solving of the governing equations motion (1)–(2) and energy Eq. (5), the radial coordinate 'r' is transformed to 'e^z' as it would generate denser mesh in the vicinity of the sphere surface. The governing equations are discretized using FDM. The discretization of the stream function and vorticity equations is carried out by following the research work of Hamielec et al. [10] and that of energy equation is carried out by the method of Woo and Hamielec [12] Now, these nonlinear partial differential equations are cast into set of algebraic equations after discretization. These are solved by the method of iteration with SOR method. As reported by Woo and Hamielec [12], constant relaxation parameters of 0.05 are used for all the three cases. The momentum transfer is first captured iteratively, and the velocity fields obtained from the solution of momentum equation are utilized to solve the energy equation. The maximum range of z-coordinate, $z_{\max} = 2.5$ is chosen for Re = 100, and for rest of the Re, $z_{\max} = 4$ is selected. A uniform grid size of 60 × 100 along Θ and z direction is used for all cases in the present simulation. The iteration is stopped when the relative difference of any quantity between two successive iterations is less than or equal to 10^{-06}, i.e.,

$$\frac{\left|\phi^{(n+1)} - \phi^{(n)}\right|}{\max\left(\left|\phi^{(n)}, 1\right|\right)} \leq 10^{-6} \tag{18}$$

4 Validation

An in-house code in Fortran-90 programming language is developed to visualize the momentum and heat transfer phenomenon over a rigid viscous sphere. Hence, a meticulous validation is performed with the standard existing works for both momentum and energy transfer equations. To validate the momentum equation, the shape and size of the vortex which forms in the downstream location of a sphere are compared first. Along with this, the total drag coefficient (C_D) is compared with literature at various Re. The simulation of the energy transfer is validated by comparing the numerically predicted Nu with existing literature at various Re and Pe.

In Fig. 1a, the separation angle (Θ_S) of the vortex and, in Fig. 1b, dimensionless width (L/D) of the vortex have been compared with the existing literature at various Re, whereas in Fig. 2, the total drag coefficient (C_D) which is estimated in the present numerical study is compared. Jenson [13] performed numerical study by considering the NS equation of spherical polar coordinate system in vorticity and stream function formulation via FDM to capture momentum transfer over a solid sphere for Re $\leq$ 40. The similar numerical work was also carried out by Hamielec et al. [10] by extending the Re range. Moreover, Rhodes [14] and Taneda [15] experimentally investigated the wake produced behind a sphere at moderate Re. Moreover, Feng and Michaelides

[3] performed numerical study for transient heat transfer analysis from a sphere by solving the stream function and vorticity equation via FVM. Along with the heat transfer analysis, they predicted C_D at various Re. The separation angle (Θ_S) of the vortex, which forms in the downstream location of the sphere at moderate Re, is compared with the earlier works [10, 13–15], whereas the vortex width (L/D) is also compared with literatures [10, 15]. In addition, predicted C_D from our study has been compared with [3, 10]. By observing the comparison graphs of Figs. 1 and 2, it can be said that numerically predicted C_D, separation angle, and vortex size are aligned with the existing experimental and theoretical works.

However, the average Nu, which was evaluated at various Re (=100, 50, 20, and 10) and Pe (=0.5, 33.6, 200, and 6.72), is arranged in Table 1. The estimated Nu is compared with the existing works of Feng and Michaelides [3] and Sayegh and Gauvin [11]. Both of them numerically studied the heat transfer phenomenon from a solid sphere by solving NS equation (in stream and vorticity formulation) and energy equation of spherical polar coordinate system. Moreover, it is observed from Table 1 that predicted Nu from our study is identical with the earlier works. Hence, heat formulation and method of its solution may be accurate.

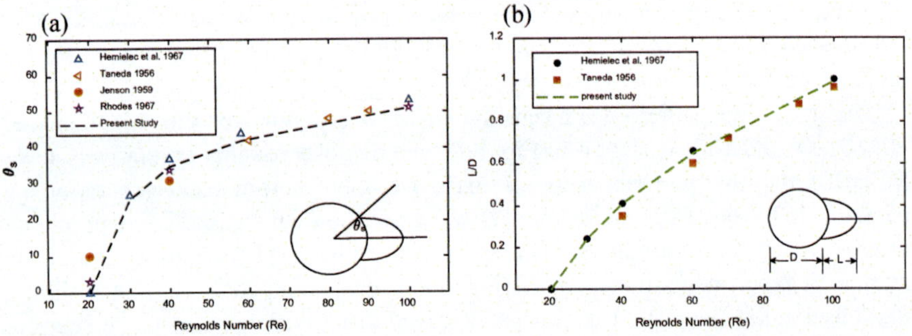

Fig. 1 Comparison of **a** separation angle (Θ) and **b** length (L/D) of the vortex at various Re with existing literature

Fig. 2 Comparison of drag coefficient (C_D) with existing literature at various Re

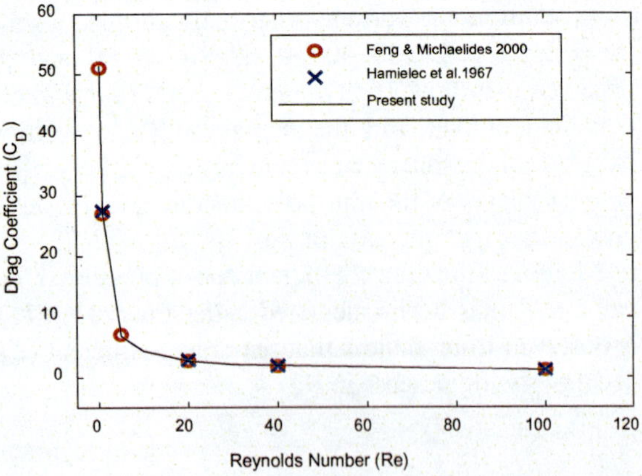

Table 1 Comparison of Nu with existing literature

Re	Pe	Source	Nu
100	0.5	Feng and Michaelides [3]	2.340
		Current work	2.330
50	33.6	Sayegh and Gauvin [11]	5.346
		Current work	5.362
20	200	Feng and Michaelides [3]	8.470
		Current work	8.347
10	6.72	Sayegh and Gauvin [11]	3.323
		Current work	3.308

5 Results and Discussions

5.1 Momentum Transfer Phenomenon

In this section, the flow over a solid sphere is captured and visualized in terms of streamline, vorticity contour, and the surface vorticity contour at various Re. Also, the drag coefficients (C_{DP}, C_{DF}, and C_D) have been estimated at various Re and presented in tabular format.

The streamline pattern over a rigid sphere at six different Re (=20, 30, 40, 50, 60, 70, 80, 90, and 100) is shown in Fig. 3. It can be observed that at relatively lower Re (=20), the fluid particles smoothly pass over the sphere without generating any backflow in the downstream location. However, as the Re increases, the recirculation motion of the fluid particle in back side of the sphere starts to grow. During the numerical experimentation, a vortex ring in the downstream location of the sphere is first observed at Re = 22. It is further observed that the vortex dimension keeps on increasing with Re. In Fig. 4, vorticity profiles along the surface of the sphere have been shown at various Re. The magnitude of the surface vorticity increases with Re; however, there is a sharp change of sign (from positive to negative) which is observed at an angle about 100° at a relatively higher Re.

In addition, the vorticity distribution around a solid sphere is shown in Fig. 5 at Re = 2, 5, 10, 20, 30, 40, 50, 60, 70, 80, 90, and 100. At low Re (=2), it can be observed that the vorticity contour exists in the whole region of the computational domain. However, with the increase in Re, the vorticity field slowly diffuses to the back side of the sphere. At a relatively higher Re > 50, the vorticity fields resemble the boundary layer of the transport phenomenon. However, the vorticity field thickness gets shorter with increase in Re.

Moreover, the drag coefficient due to pressure (C_{DP}), friction (C_{DF}), and the total value (C_D) has been calculated using Eqs. (13)–(17) and presented in Table 2. It is observed from Table 2 that the drag coefficient (C_D) is highest at low Re, and it decreases with increase in Re.

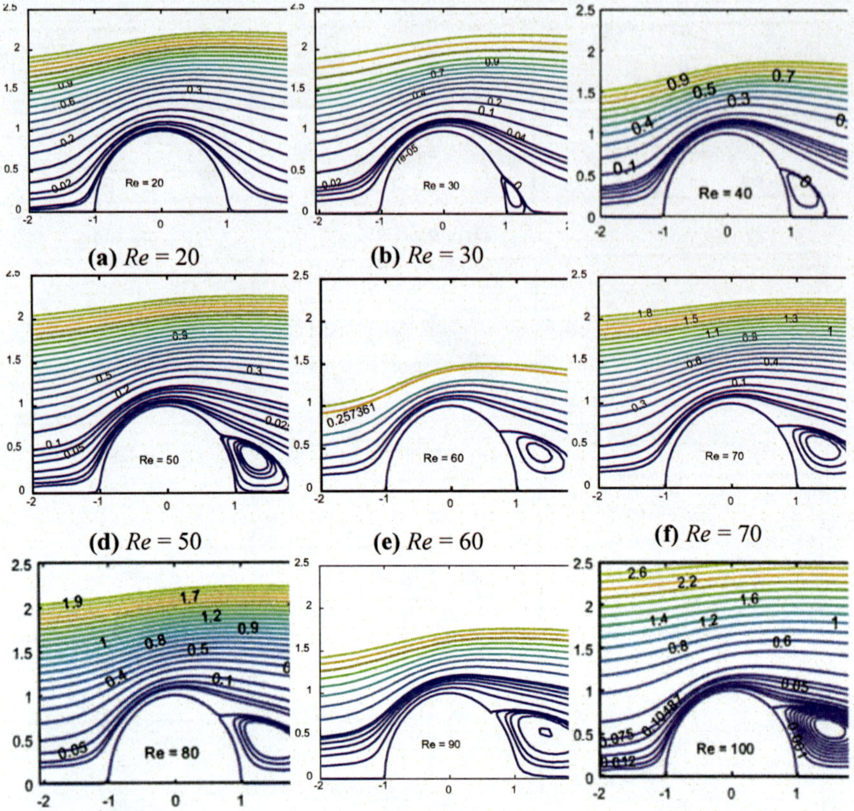

(a) $Re = 20$ (b) $Re = 30$ (c) $Re = 40$

(d) $Re = 50$ (e) $Re = 60$ (f) $Re = 70$

(g) $Re = 80$ (h) $Re = 90$ (i) $Re = 100$

Fig. 3 Streamline plots over a solid sphere at different Re

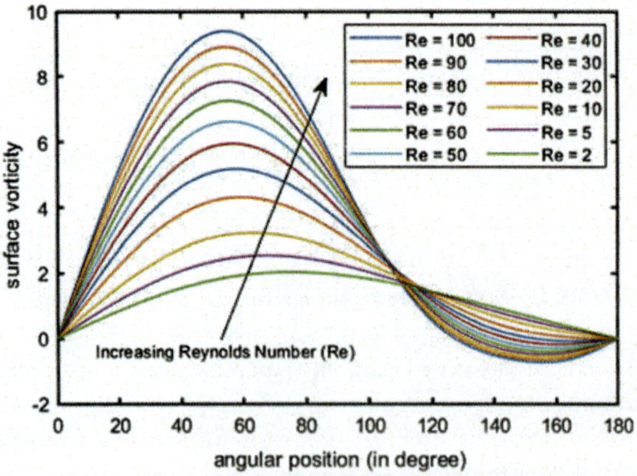

Fig. 4 Vorticity along the surface of the solid sphere at various Re

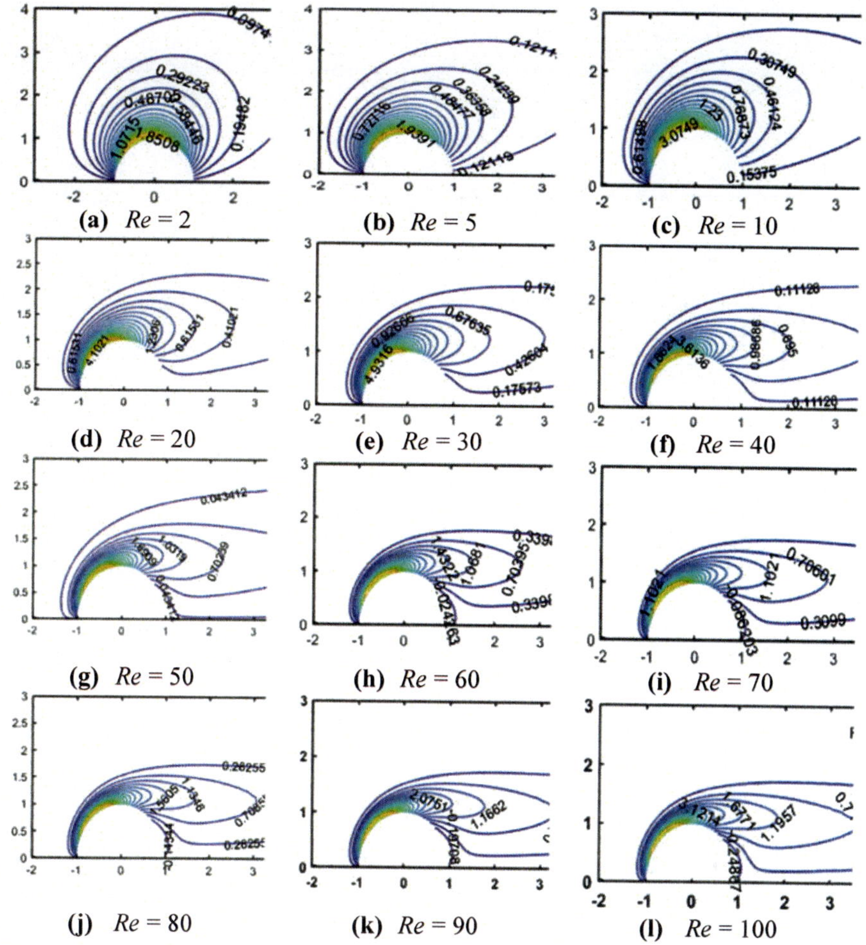

Fig. 5 Vorticity plots over a solid sphere at twelve different Re

5.2 *Energy Transfer Phenomenon*

In Fig. 6, the variation of local Nusselt number (Nu_Θ) along the surface of a solid sphere is shown at various Re at low and high Peclet numbers. In Fig. 6a, Nu plots are shown for low Pe (=0.2), whereas in Fig. 6b, relatively higher Pe (=200) is used. It is observed that, with increase in Re, noticeable differences in local Nusselt number have occurred at higher Pe (=200) compared to the lower counterpart. So, the convective strength (i.e., Reynolds number) has little effects on local Nusselt number at lower Pe. However, with increase in Pe the effects of Re on heat transfer coefficient (Nu) become dominant. Moreover, the effects of vortex ring on the heat transfer phenomenon are observed noticeable at moderate Re at relatively high Pe

Table 2 Calculated drag coefficients of solid sphere at various Re

Re	C_{DF}	C_{DP}	C_{DT}
0.1	162.92	78.12	241.04
0.2	83.750	40.127	123.877
1	17.923	7.926	25.849
5	4.7170	2.1463	6.8633
10	2.8565	1.5741	4.4306
20	1.7376	1.0462	2.7838
30	1.3149	0.8512	2.1661
40	1.0832	0.7457	1.8289
50	0.9335	0.6782	1.6117
60	0.8272	0.6306	1.4579
70	0.7472	0.5948	1.3420
80	0.6842	0.5667	1.2509
90	0.6331	0.5438	1.1770
100	0.5906	0.5248	1.1154

as the local Nusselt number curve starts to increase again once it crosses the vortex region (see Fig. 6b).

In Table 3, the estimated average Nusselt number (Nu) values from this study are presented at various Re (=2, 5, 10, 20, 30, 40, 50, 60, 70, 80, 90, and 100) and Pe (=0.2, 0.5, 1, 2, 10, 50, 100, 150, and 200). From the last column of Table 3, some noticeable changes of average Nu values with increase in Re are observed compared to the second column of Table 3. The second column of Table 3 refers to Nu values at low Pe (=0.2), whereas last column refers to Nu values at higher Pe (=200) at various Re. As previously discussed, Re has dominant effects on both local and average

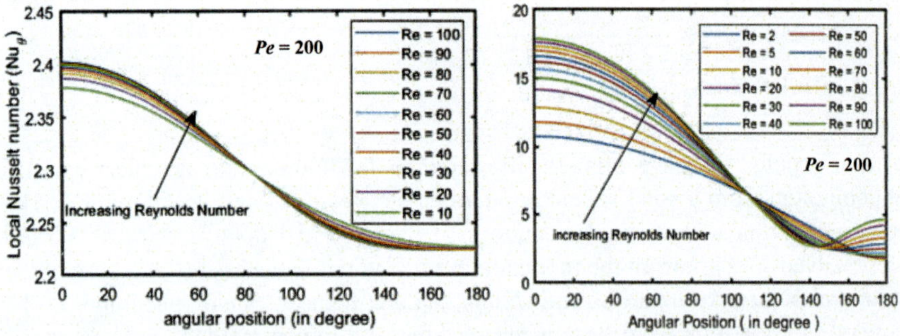

(a) *Pe = 0.2* **(b)** *Pe = 200*

Fig. 6 Local Nusselt number profile along the surface of a solid sphere at **a** low Pe (=0.2) and **b** high Pe (=200) at various Re

Table 3 Average Nusselt number at various Re and Pe

Re/Pe	0.2	0.5	1.0	2.0	10	50	100	150	200
2	2.291	2.315	2.399	2.585	3.424	5.011	6.029	6.748	7.323
5	2.291	2.319	2.411	2.614	3.516	5.190	6.260	7.014	7.617
10	2.292	2.322	2.422	2.641	3.612	5.388	6.514	7.308	7.943
20	2.292	2.325	2.432	2.667	3.721	5.631	6.830	7.674	8.347
30	2.292	2.326	2.438	2.681	3.786	5.789	7.039	7.916	8.615
40	2.292	2.327	2.441	2.690	3.832	5.905	7.195	8.098	8.818
50	2.292	2.328	2.443	2.696	3.865	5.998	7.321	8.247	8.984
60	2.292	2.328	2.445	2.701	3.892	6.074	7.427	8.374	9.128
70	2.292	2.329	2.446	2.705	3.913	6.139	7.520	8.486	9.257
80	2.292	2.329	2.447	2.708	3.931	6.195	7.602	8.588	9.376
90	2.293	2.329	2.448	2.710	3.946	6.245	7.677	8.683	9.488
100	2.293	2.330	2.449	2.712	3.959	6.289	7.746	8.771	9.593

Nusselt numbers for the case of heat transfer phenomenon over a solid sphere at relatively higher Pe only.

Now, in Fig. 7 the temperature profile in terms of isotherm contour line around a solid sphere is shown at different Re (=5, 20, 30, 50, 60, 70, 80, 90, and 100) at fixed Pe (=200). It is observed that the thermal boundary layer thickness at the frontal side of the droplet is more at low Re (=5) compared to the high Re (=100). Also, the pattern of the isotherm contour line at the extreme edge of the thermal boundary layer is little curve at low Re (=5); however, this contour line is straight at higher Re (=100). This may occur because of convection effect, at high Re, convective strength is more dominant, and rapid transfer of energy occurs from colder place to the hotter region.

6 Conclusions

In this article, we have numerically captured the momentum and heat transfer phenomenon from a solid sphere at intermediate Reynolds and Peclet number. The stream function and vorticity equations of the spherical polar coordinate system were solved for capturing the momentum transfer, while the energy equation of the spherical polar coordinate was utilized to capture the heat transfer phenomena. The governing equations were discretized by finite difference method, and successive over-relation method was used to solve the algebraic sets of discretized equations. An in-house code in Fortran-90 programming language was developed to perform the required computational iterations. The validation of the momentum transfer equation was carried out by comparing the vortex separation angle, dimension of the vortex, and the total drag coefficients with the literature at various Reynolds numbers. The

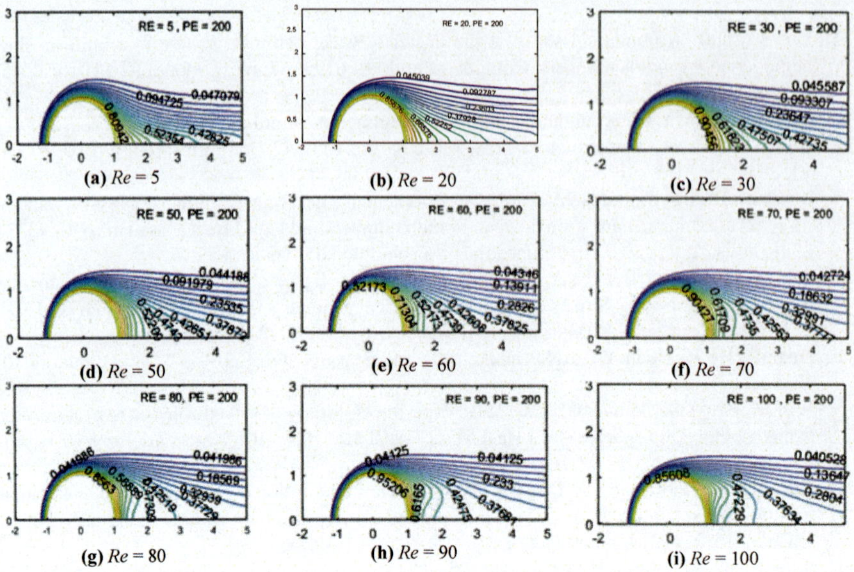

Fig. 7 Isotherm plot over a solid sphere at different Reynolds number at Pe = 200

heat transfer portion of the simulation was authenticated by matching the average Nusselt number with the existing works. The flow patterns over the solid sphere were visualized in terms of streamline, and also, total vorticity contour and surface vorticity line along the sphere were shown at various Reynolds numbers. During the present numerical experimentation, it was observed that a vortex ring in the downstream location of the sphere first appears at Re = 22. In this study, the maximum Reynolds number and Peclet number were curtained at 100 and 200, respectively. The energy transfer phenomenon from the solid sphere was analyzed based on local and average Nusselt numbers at various Reynolds and Peclet numbers. It was observed that the Reynolds number has significant effects on heat transfer coefficients at high Peclet number only. Moreover, the flow and temperature profile in the vicinity of the solid sphere in terms of streamline and isotherm contour plots are shown at different Reynolds numbers.

References

1. Abramzon B, Elata C (1984) Unsteady heat transfer from a single sphere in stokes flow. Int J Heat Mass Transf 27:687–695. https://doi.org/10.1016/0017-9310(84)90138-8
2. Feng Z-G, Michaelides EE (1996) Unsteady heat transfer from a sphere at small Peclet numbers. J Fluids Eng 118:96–102. https://doi.org/10.1115/1.2817522
3. Feng ZG, Michaelides EE (2000) A numerical study on the transient heat transfer from a sphere at high Reynolds and Peclet numbers. Int J Heat Mass Transf 43:219–229. https://doi.org/10.1016/s0017-9310(99)00133-7

4. Lee S (2000) A numerical study of the unsteady wake behind a sphere in a uniform flow at moderate Reynolds numbers. Comput—Fluids 29:639–667. https://doi.org/10.1016/s0045-7930(99)00023-7
5. Bagchi P, Ha MY, Balachandar S (2000) Direct numerical simulation of flow and heat transfer from a sphere in a uniform cross-flow. J Fluids Eng 123:347–358. https://doi.org/10.1115/1.1358844
6. Melissari B, Argyropoulos SA (2005) Development of a heat transfer dimensionless correlation for spheres immersed in a wide range of Prandtl number fluids. Int J Heat Mass Transf 48:4333–4341. https://doi.org/10.1016/j.ijheatmasstransfer.2005.05.025
7. Sekhar TVS, Raju BHS, Sanyasiraju YVSS (2012) Higher-order compact scheme for the incompressible Navier-Stokes equations in spherical geometry. Commun Comput Phys 11:99–113. https://doi.org/10.4208/cicp.171010.030311a
8. Duan Z, He B, Duan Y (2015) Sphere drag and heat transfer. Sci Rep 5. https://doi.org/10.1038/srep12304
9. Will JB, Kruyt NP, Venner CH (2017) An experimental study of forced convective heat transfer from smooth, solid spheres. Int J Heat Mass Transf 109:1059–1067. https://doi.org/10.1016/j.ijheatmasstransfer.2017.02.018
10. Hamielec AE, Hoffman TW, Ross LL (1967) Numerical solution of the Navier-Stokes equation for flow past spheres: part I. Viscous flow around spheres with and without radial mass efflux. AIChE J 13:212–219. https://doi.org/10.1002/aic.690130206
11. Sayegh NN, Gauvin WH (1979) Numerical analysis of variable property heat transfer to a single sphere in high temperature surroundings. AIChE J 25:522–534. https://doi.org/10.1002/aic.690250319
12. Woo SE, Hamielec AE (1971) A numerical method of determining the rate of evaporation of small water drops falling at terminal velocity in air. J Atmos Sci 28:1448–1454. https://doi.org/10.1175/1520-0469(1971)028%3c1448:anmodt%3e2.0.co;2
13. Jenson VG (1959) Viscous flow round a sphere at low Reynolds numbers (<40). Proc R Soc Lond. Ser A. Math Phys Sci 249:346–366
14. Rhodes JM (1967) Ph.D. thesis
15. Taneda S (1956) Experimental investigation of the wake behind a sphere at low Reynolds numbers. J Phys Soc Jpn 11:1104–1108. https://doi.org/10.1143/jpsj.11.1104

The Role of Quasi-static Strain Rate on the Mechanical Behaviour of WE43 Magnesium Alloy

Abdul Rahman, Md. Murtuja Hussain, and Naresh Prasad

1 Introduction

Magnesium alloys are now the next generation of ultra-lightweight metallic alloys for aerospace, transportation, and implants. The new magnesium alloys' light weight, excellent mechanical properties, and outstanding machinability make them ideal for these applications [1–6]. The influence of strain rates on mechanical characteristics and heat treatments has been widely researched in the AZ series magnesium alloys incorporating aluminium and zinc [5–8]. The Mg-Y-Nd-Zr system's WE43 magnesium alloy is a prominent one among magnesium alloys [9]. WE43 is a magnesium alloy having several advantageous properties, which include excellent mechanical strength, creep resistance, chemical inertness, and combustion and heat retardant properties, among others [10–14]. This alloy, in addition to its inherent low weight, has the potential to be used in medical implants and body armour applications [10, 11]. Recent animal investigations have demonstrated that WE43 magnesium alloys, in both casting and thermal forms, can be used for orthopaedic implants without generating cytotoxicity [9]. It has been discovered that the existence of certain rare-earth (RE) elements may alter the mechanical characteristics of magnesium (Mg) alloys, including the solid solution strengthening and precipitation strengthening processes [15]. It is stated that yttrium (Y) can be used to strengthen solid solutions. This is attributed to the conventional mismatch theory in terms of size and/or stiffness, as well as the effect of valence in the magnesium matrix [16]. Neodymium has the

A. Rahman (✉) · N. Prasad
National Institute of Technology, Jamshedpur, India
e-mail: 2019rsme012@nitjsr.ac.in

N. Prasad
e-mail: nprasad.me@nitjsr.ac.in

Md. Murtuja Hussain
National Metallurgical Laboratory, Jamshedpur 831014, India
e-mail: murtuja@nmlindia.org

A. Kumar et al. (eds.), *Recent Advances in Manufacturing and Thermal Engineering*, Lecture Notes in Mechanical Engineering, https://doi.org/10.1007/978-981-19-8517-1_28

potential to improve creep resistance because it can be strengthened by both precipitation and solid solution [17]. It has been revealed that crystalline strengthening and precipitation hardening may be used to magnesium–rare-earth alloys to improve their mechanical characteristics. Precipitation of a finely dispersed intermetallic solution results in the formation of high-temperature strengthening materials [18]. Yttrium-containing Mg alloys offer excellent creep and corrosion resistance compared to other Mg alloys up to 300 °C. Yttrium is an element that can endure ageing at low temperatures and dissolves readily in magnesium. Magnesium (Mg) and rare-earth alloys produce some secondary β phases based on the temperature of ageing [18, 19]. At quasi-static strain rates, Noradila et al. [20] showed that the yield stress and tensile strength of AZ31 were strain rate dependent, whereas those of AZ61 were not. Shen et al. [21] studied the influence of strain rate on AMX602's mechanical characteristics and found that it exhibits excellent mechanical behaviour but modest anisotropy throughout its length in strength and compressive ductility. Jahedi et al. [22] examined the temperature and strain rate dependence of direct-chill cast WE43 deformation characteristics and observed no early yield tension and compression asymmetry at ambient temperature and slip-dominated deformation. Feather et al. [23] used deformed twinning mechanism to study WE43's deformation as a function of strain rate and found that WE43 hardens more quickly when the strain rate is high than when the strain rate is almost the same. When it comes to alloys made of aluminium (Al) [24] and magnesium (Mg) [25, 26], plastic deformation before ageing treatment has been found to be an effective means of promoting the dispersion of precipitates. In addition, mechanical characteristics and creep investigations of cast [27, 28] and heat-treated [29–31] alloys indicate that hot-rolled WE43 alloys have greater strength and hardness compared to cast alloys. Furthermore, a T5-tempered WE43 alloy rolled at 250 °C exhibited superior mechanical performance. This research examined the mechanical and micro-structural properties of a hot-rolled, T5-tempered WE43 alloy. Different strain rates were used in tensile and compression tests to figure out the strength, ductility and strain hardening properties of the specimens.

2 Methodologies

In this study, a hot-rolled-T5 heat-treated WE43 magnesium alloy was used to determine the effect of quasi-static strain rates on the tensile, compressive, and work-hardening properties of the alloy. As received, the material was supplied by Suzhou Chuan Mao Material Co., Ltd., China. The initial thickness of the material was 80 mm, and it was T5-tempered and hot-rolled down to 35 mm plates on a reversing mill at falling temperatures beginning in the range of 450–500 °C with a pass reduction of 20–25% prior to reheating. Table 1 provides information on the nominal elements in the sample. For optical microscopy standard metallographic procedures were applied. Each sample was ground to 2500 grit using Silicon Carbide abrasive paper, and then polished with a diamond suspension of 3 μm particles, followed by

Table 1 Constituents of WE43

Constituents	Yttrium (Y)	Gadolinium (Gd)	Neodymium (Nd)	Zirconium (Zr)	Magnesium (Mg)
Weight (%)	4.04	0.70	2.17	0.46	Balance

1 μm particles. After cleaning the sample in an acetone bath, 2.1 g of picric acid, 5 ml of acetic acid, 35 ml of ethanol, and 5 ml of distilled water were used to etch it. Optical microscopy was done with an LEICA DM 2500 M microscope. Flat specimens of 30 mm gauge length were WEDM machined in the direction perpendicular to rolling for tension testing according to ASTM B557. Samples for compression testing were 8 mm dia. × 13 mm long cylinders. At room temperature, uniaxial tensile tests were conducted on the alloy using a JINAN TE WDW (20 KN) testing machine to determine its tensile characteristics, while compression tests were performed at H50KS, Hounsfield (United Kingdom). In the quasi-static tests, the cross-head speed determined the strain rate, and an extensometer was used to measure the strain so that any elastic deformation beyond the gauge length of the sample would not be recorded. For tensile and compressive tests, the nominal strain rates were 3×10^{-5}/s, 1×10^{-4}/s, and 1×10^{-3}/s, respectively. Using the findings of a quasi-static tensile test conducted at ambient temperature, the actual stress and strain were computed. Using Eqs. (1) and (2), [32] the actual stress and strain were determined.

$$\varepsilon_t = \ln(1 + \varepsilon_{eng}) \quad (1)$$

$$\sigma_t = \sigma_{eng}(1 + \varepsilon_{eng}) \quad (2)$$

In light of the formulae shown above, the strain hardening rate may be computed using Eq. (3) [32]. Where n is hardening exponent.

$$n = \frac{d(\log \sigma_t)}{d(\log \in_t)} = \frac{d(in\sigma_t)}{d(\ln \in_t)} = \frac{\in_t d\sigma_t}{\sigma_t d \in_t}$$

$$\text{Or, strain hardening} = \frac{d\sigma_t}{d \in_t} = n\frac{\sigma_t}{\in_t} \quad (3)$$

where ε_t is true strain, σ_t is true stress; σ_{eng} is engineering stress and ε_{eng} is engineering strain.

3 Outcomes and Discussion

3.1 Microstructural Properties

The microstructure of the treated WE43 alloy is shown in Fig. 1. The microstructure of WE43 alloy as it is cast is composed of large quantities of α-Mg phases with eutectic mixing at the boundaries between grains. The intermetallic phases observed in the eutectic mixture include $Mg_{41}Nd_5$, $Mg_{12}Nd$ and, $Mg_{22}Y_5$ [31]. Figure 1b shows the eutectics after heat treatment disintegrating into the matrix completely. The typical grain diameter of the rolled alloy as determined by optical microscopy was 22 μm. Kun Yu et al. [31] reported that the matrix contains both distorted dislocations and scattered precipitates. The dislocations can be pinned since the second phases are so tiny. When treated to these conditions, the alloy's high strength is influenced by two distinct processes, namely strain hardening and second-phase hardening. The second-phase precipitation, such as Mg-Nd phases, has been discovered in recent studies of the breakdown of WE alloys at 523 K, with the metastable β (fcc) phase serving as the primary hardening phase [33, 34]. Temperature has a role in determining the precipitates of various phases, which would in turn influence the alloy's yield strength. As a consequence of this, and in close agreement with the alloy's microstructure, an increase in tensile strength will be shown by a hot-rolled WE43 alloy that has undergone artificial ageing but has not been subjected to any kind of pre-solution treatment.

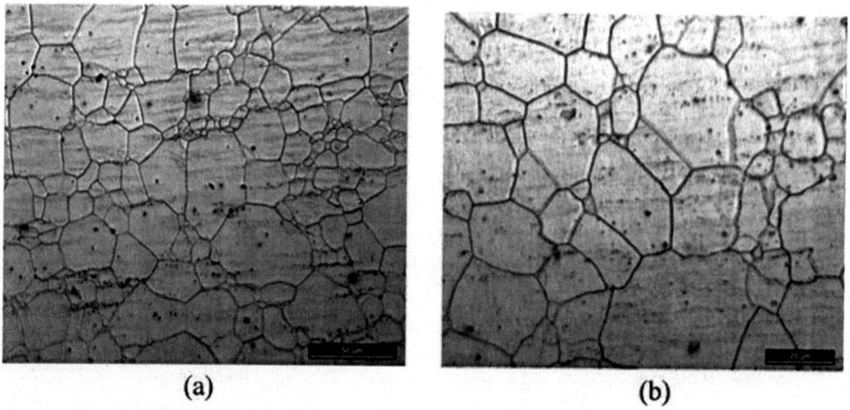

Fig.1 Microstructure of the WE43 treated condition **a** 50 μm and **b** 20 μm

3.2 Mechanical Properties

We first tested the material at room temperature under a variety of quasi-static compressive and tensile conditions until the fracture of the sample. The shattered sample is shown in Fig. 2. In tension, the fracture plane is rather flat and normal to the loading direction, however in compression, there is a shear fracture. This is immediately discernible from Fig. 2a, b. In his investigation, Jahedi et al. [35] came to comparable conclusions. Figure 3 displays the stress–strain graphs, while Table 2 provides a summary of the tensile and compressive characteristics and behaviour of WE43 magnesium (Mg) alloys. Figure 3 shows that both the ultimate tensile strength (UTS) and the yield strength (YS) go up as the strain rate goes up in both tensile and compression conditions. When the rate of strain goes up from 3×10^{-5} to 1×10^{-3}, the UTS and YS in tension increase by 28 and 65%, respectively, from 192 (MPa) to 246 (MPa) and 115 (MPa) to 190 (MPa), respectively, whereas the ultimate compressive strength (UCS) in compression increases by approximately 9%, from 370 (MPa) to 402 (MPa) (MPa).This trend was comparable to that seen by Jahedi et al. [22] for rolled-T5 heat-treated WE43 alloys made by direct-chill casting. It is hypothesised that this is connected to the strain rate dependence of critical resolved shear stress (CRSS) in magnesium (Mg) alloy non-basic slip systems [36].

The increased compressive strength of WE43-T5 may be ascribed to the smaller grain size and eutectic mixture exclusion in the microstructure [6]. At 1×10^{-3} strain rate, the 0.2% offset says that the TYS is 190 (MPa) and CYS is 245 (MPa). As a consequence, the yield stress (YS) of the alloy displays an imbalance between tension and compression. Saeede Ghorbanpoura et al. obtained comparable results [37]. Due to strain hardening, a tension–compression imbalance is seen in the plastic zone. A little amount of compression-versus-tension asymmetries are generated by slip-dominated distortion with some twining, as found by Jahedi et al. [22]. As seen in Fig. 3 and Table 2, strain increases as strain rate increases under both compressive and tensile loading conditions. Compressive strain grows to 50% and tensile to 29% as the strain rate goes from 3×10^{-5}/s to 1×10^{-3}/s. Asgari et al. [38] investigated the cause of this phenomenon and discovered that increasing the strain rate leads to the highest work-hardening rate, improves the material's strength and elongation, and reduces the contribution of twining to the material's distortion. Figure 4a, b depict the link between work-hardening rate and actual strain after yielding in tension (T) and compression (C) for the WE43-T5 alloy. It is plotted within a true strain range of 0.05 (mm/mm) to fracture in tension and 0.009 (mm/mm) to fracture in compression (C). It is evident that it is a strong, hardening material under stress and compression, and that hardening continues until the material reaches its full strength. As previously found by Savage et al. [39], as the strain rate increases, so does the hardening rate in tension (T) and compression (C). It is also found that the strain hardening rate reduces over the full spectrum of plastic deformation. This phenomenon is explained by the absence of twin-based stiffening at room temperature [37, 39]. The hardening rate curve is also shown to be steeper in tension than in compression as a result of nonlinear loading. This outcome is comparable to [37].

(a)

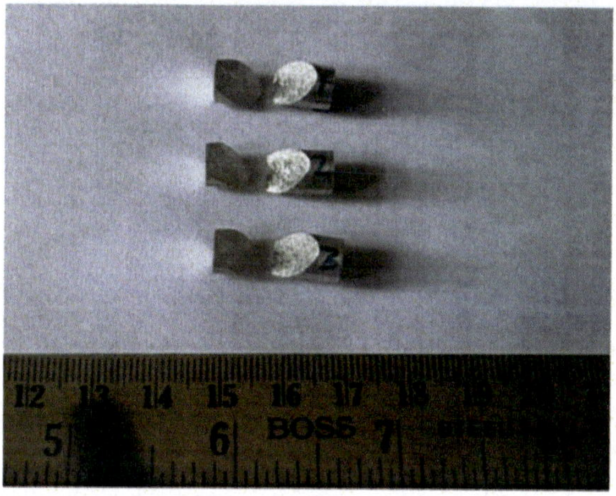

(b)

Fig. 2 Tension and compression tested specimens. **a** Tensile strain rate is 1×10^{-3} /s for 1–1, 1×10^{-4} /s for 2–1 and 3×10^{-5} /s for 3–1. **b** Compressive strain rate is 1×10^{-3} /s for 1, 1×10^{-4} /s for 2 and 3×10^{-5} /s for 3

4 Conclusions

WE43 is a magnesium alloy with a number of attractive features. This research was conducted to find out how quasi-static strain rate affects rolled, heat-treated WE43 Mg alloy mechanical properties at ambient temperature.

- According to reports, the microstructure of WE43 alloy as it is cast is composed of large quantities of α-Mg phases with eutectic mixing at the boundaries between grains. Following the application of heat, all eutectics disintegrate into the matrix.

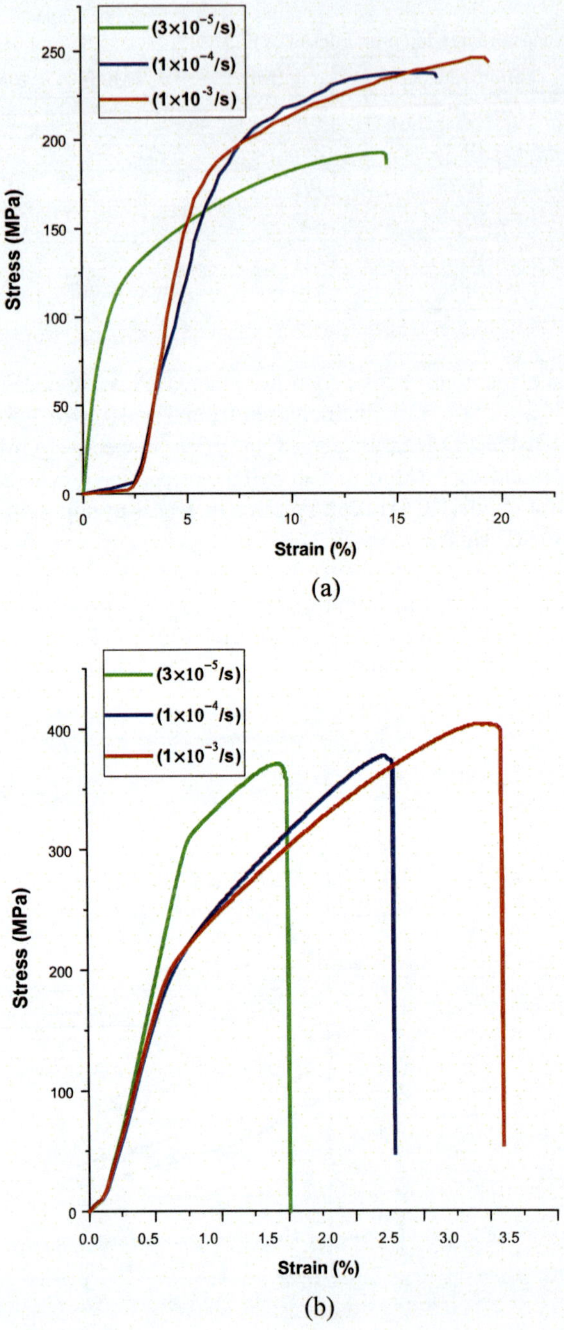

Fig. 3 WE43 rolled plate stress and deformation response **a** Graph of engineering stress versus engineering strain in tension, **b** engineering stress-engineering strain graph in compression

Table 2 Tensile and compression properties of WE43

	Strain rate (s^{-1})	σ_{uts} (MPa)	σ_y (MPa)	Elongation (%)
Tension	1×10^{-3}	246	190	18.6
	1×10^{-4}	237	165	16.2
	3×10^{-5}	192	115	14.4
Compression	1×10^{-3}	402	245	3.1
	1×10^{-4}	377	235	2.3
	3×10^{-5}	370	304	1.5

- In both tensile and compressive condition, increased strain rates increase tensile and YS. When the strain rate is increased from 3×10^{-5} to 1×10^{-3}, the UTS and YS in tension (T) increase by 28 and 65%, respectively, from 192 (MPa) to 246 (MPa) and 115 (MPa) to 190 (MPa), respectively, whereas the ultimate compressive strength (UCS) in compression increases by approximately 9%, from 370 (MPa) to 402 (MPa).
- The alloy's yield stress is asymmetric between tension and compression, and strain hardening causes a tension–compression imbalance in the plastic zone.
- Increased strain rate increases strength and ductility while decreasing the role of twinning to deformation. Strain rate increases under both compressive and tensile

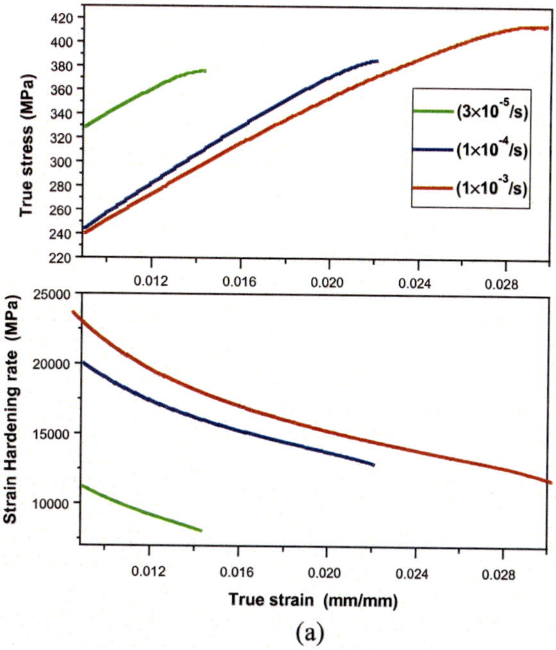

Fig. 4 Rate of work hardening versus true strain **a** tension **b** compression

Fig. 4 (continued)

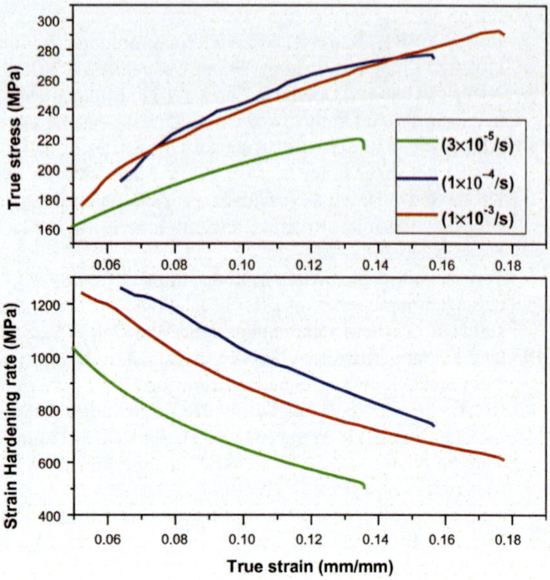

(b)

loading conditions. Compressive strain grows to 50% and tensile to 29% as the strain rate goes from 3×10^{-5}/s to 1×10^{-3}/s.

- As the strain rate increases, so does the hardening rate in tension (T) and compression (C). It is also found that the work-hardening rate is reduced over the full spectrum of plastic deformation. This phenomenon is explained by the absence of twin-based stiffening at room temperature.

Acknowledgements The first author would like to give his sincere thanks to Mr. Tahir Khan (Incharge, LAB, Mechanical engineering, Al-Kabir Polytechnic, Jamshedpur) and Mr. Mumtaz RIizwee (Lecturer, Al-Kabir Polytechnic, Jamshedpur) for his immense support and valuable suggestions.

References

1. Mordike BL (2001) Development of highly creep resistant magnesium alloys. J Mater Process Technol 117(3):391–394
2. Lambri OA, Riehemann W (2005) Damping due to incoherent precipitates in commercial QE22 magnesium alloy. Scripta Mater 52(2):93–97
3. Xiaoqin Z, Qudong W, Yizhen L, Yanping Z, Wenjiang D, Yunhu Z (2001) Influence of beryllium and rare earth additions on ignition-proof magnesium alloys. J Mater Process Technol 112(1):17–23
4. Friedrich H, Schumann S (2001) Research for a "new age of magnesium" in the automotive industry. J Mater Process Technol 117(3):276–281

5. Prasad YVRK, Rao KP (2009) Effect of homogenization on the hot deformation behavior of cast AZ31 magnesium alloy. Mater Des 30(9):3723–3730
6. Xiang C, Gupta N, Coelho P, Cho K (2018) Effect of microstructure on tensile and compressive behavior of WE43 alloy in as cast and heat treated conditions. Mater Sci Eng, A 710:74–85
7. Ahmad IR, Shu DW (2010) Tensile properties of die-cast magnesium alloy AZ91D at high strain rates in the range between 300 s-1 and 1500 s-1. Appl Mech Mater 24:325–330
8. Ulacia I, Dudamell NV, Gálvez F, Yi S, Pérez-Prado MT, Hurtado I (2010) Mechanical behavior and microstructural evolution of a Mg AZ31 sheet at dynamic strain rates. Acta Mater 58(8):2988–2998
9. Torroni A, Xiang C, Witek L, Rodriguez ED, Coelho PG, Gupta N (2017) Biocompatibility and degradation properties of WE43 Mg alloys with and without heat treatment: In vivo evaluation and comparison in a cranial bone sheep model. J Cranio-Maxillofac Surg 45(12):2075–2083
10. Cho K, Sano T, Doherty K, Yen C, Gazonas G, Montgomery J, DeLorme R (2009) Magnesium technology and manufacturing for ultra lightweight armored ground vehicles. ARMY RESEARCH LAB ABERDEEN PROVING GROUND MD
11. Ye CH, Zheng YF, Wang SQ, Xi TF, Li YD (2012) In vitro corrosion and biocompatibility study of phytic acid modified WE43 magnesium alloy. Appl Surf Sci 258(8):3420–3427
12. Mengucci P, Barucca G, Riontino G, Lussana D, Massazza M, Ferragut R, Aly EH (2008) Structure evolution of a WE43 Mg alloy submitted to different thermal treatments. Mater Sci Eng, A 479(1–2):37–44
13. Liu Y, Zheng S, Li N, Guo H, Zheng Y, Peng J (2016) Study on the in vitro degradation behavior of pure Mg and WE43 in human bile for 60 days for future usage in biliary. Mater Lett 179:100–103
14. Asgari H, Odeshi AG, Szpunar JA (2014) On dynamic deformation behavior of WE43 magnesium alloy sheet under shock loading conditions. Mater Des 63:552–564
15. Bhattacharyya JJ, Wang F, McQuade PJ, Agnew SR (2017) Deformation and fracture behavior of Mg alloy, WE43, after various aging heat treatments. Mater Sci Eng, A 705:79–88
16. Gao L, Chen RS, Han EH (2009) Effects of rare-earth elements Gd and Y on the solid solution strengthening of Mg alloys. J Alloy Compd 481(1–2):379–384
17. Yan J, Sun Y, Xue F, Xue S, Tao W (2008) Microstructure and mechanical properties in cast magnesium–neodymium binary alloys. Mater Sci Eng, A 476(1–2):366–371
18. Biswas S, Suwas S, Sikand R, Gupta AK (2011) Analysis of texture evolution in pure magnesium and the magnesium alloy AM30 during rod and tube extrusion. Mater Sci Eng, A 528(10–11):3722–3729
19. Suwas S, Gottstein G, Kumar R (2007) Evolution of crystallographic texture during equal channel angular extrusion (ECAE) and its effects on secondary processing of magnesium. Mater Sci Eng, A 471(1–2):1–14
20. Noradila AL, Sajuri Z, Syarif J, Miyashita Y, Mutoh Y (2013) Effect of strain rates on tensile and work hardening properties for Al-Zn magnesium alloys. IOP Conf Ser: Mater Sci Eng 46(10):012031
21. Shen J, Kondoh K, Jones TL, Mathaudhu SN, Kecskes LJ, Wei Q (2016) Effect of strain rate on the mechanical properties of magnesium alloy AMX602. Mater Sci Eng, A 649:338–348
22. Jahedi M, McWilliams BA, Kellogg FR, Beyerlein IJ, Knezevic M (2018) Rate and temperature dependent deformation behavior of as-cast WE43 magnesium-rare earth alloy manufactured by direct-chill casting. Mater Sci Eng, A 712:50–64
23. Feather WG, Ghorbanpour S, Savage DJ, Ardeljan M, Jahedi M, McWilliams BA, Knezevic M (2019) Mechanical response, twinning, and texture evolution of WE43 magnesium-rare earth alloy as a function of strain rate: experiments and multi-level crystal plasticity modeling. Int J Plast 120:180–204
24. Yassar RS, Field DP, Weiland H (2005) The effect of predeformation on the β'' and β' precipitates and the role of Q′ phase in an Al–Mg–Si alloy; AA6022. Scripta Mater 53(3):299–303
25. Kang YH, Huang ZH, Wang SC, Yan H, Chen RS, Huang JC (2020) Effect of pre-deformation on microstructure and mechanical properties of WE43 magnesium alloy II: aging at 250 and 300 °C. J Magnes Alloys 8(1):103–110

26. Jamili AM, Zarei-Hanzaki A, Abedi HR, Minárik P, Soltani R (2017) The microstructure, texture, and room temperature mechanical properties of friction stir processed Mg-Y-Nd alloy. Mater Sci Eng, A 690:244–253
27. Wang JG, Hsiung LM, Nieh TG, Mabuchi M (2001) Creep of a heat treated Mg–4Y–3RE alloy. Mater Sci Eng, A 315(1–2):81–88
28. Marrow TJ, Ahmad AB, Khan IN, Sim SMA, Torkamani S (2004) Environment-assisted cracking of cast WE43-T6 magnesium. Mater Sci Eng, A 387:419–423
29. Agnew SR, Mehrotra P, Lillo TM, Stoica GM, Liaw PK (2005) Texture evolution of five wrought magnesium alloys during route A equal channel angular extrusion: experiments and simulations. Acta Mater 53(11):3135–3146
30. Lambri OA, Riehemann W, Lucioni EJ, Bolmaro RE (2006) Mechanical spectroscopy of deformed WE43 magnesium alloys. Mater Sci Eng, A 442(1–2):476–479
31. Yu K, Li W, Wang R, Wang B, Li C (2008) Effect of T5 and T6 tempers on a hot-rolled WE43 magnesium alloy. Mater Trans 49(8):1818–1821
32. Zhang S, Pedersen PT, Villavicencio R (2019) 5—Collision damage assessment by nonlinear finite element simulations. Probability and mechanics of ship collision and grounding
33. Kamado S, Ashie T, Yamada H, Sanbun K, Kojima Y (2000) Improvement of tensile properties of wrought magnesium alloys by grain refining. In: Materials science forum, vol 350. Trans Tech Publications Ltd., pp 65–72
34. Nie JF, Muddle BC (2000) Characterisation of strengthening precipitate phases in a Mg–Y–Nd alloy. Acta Mater 48(8):1691–1703
35. Jahedi M, McWilliams BA, Knezevic M (2018) Deformation and fracture mechanisms in WE43 magnesium-rare earth alloy fabricated by direct-chill casting and rolling. Mater Sci Eng, A 726:194–207
36. Barnett MR, Keshavarz Z, Nave MD (2003) Metall Mater Trans A 34:1799–1806
37. Ghorbanpour S, McWilliams BA, Knezevic M (2019) Low-cycle fatigue behavior of rolled WE43-T5 magnesium alloy. Fatigue Fract Eng Mater Struct 42(6):1357–1372
38. Asgari H, Szpunar JA, Odeshi AG (2014) Texture evolution and dynamic mechanical behavior of cast AZ magnesium alloys under high strain rate compressive loading. Mater Des 61:26–34
39. Savage DJ, McWilliams BA, Vogel SC, Trujillo CP, Beyerlein IJ, Knezevic M (2020) Mechanical behavior and texture evolution of WE43 magnesium-rare earth alloy in Split-Hopkinson pressure bar and Taylor impact cylinder testing. Int J Impact Eng 143:103589

A Novel Scheme for Enhancing Quality of Pictures

Vineeta Singh and Vandana Dixit Kaushik

1 Introduction

Everyone wants to make sure they are using the best quality photo for their project, but there is a lot of technical terms and misinformation around image quality. Picture improvement is the most common way of further developing the image quality with next to no data misfortune so the outcomes are more appropriate for portrayal such as style, shading, and color as well as desired resolution or get ready pictures for additional examination in different machine vision applications, including object identification, scene understanding, picture classification, and significantly more. Enhancement of images usually consists of several transformations like image denoising, deblurring, up scaling, contrast enhancement, lighting up low-light pictures, removing optical distortion, etc. Image enhancement algorithms are not limited to digital photography [1]. They can be as well useful for processing satellite images [2], for processing medical images[3], e.g., in radiology, where AI not only helps enhance the quality of the picture itself but also helps detect certain medical conditions. Image enhancement of infrared images is crucial for smart city applications; forensic video and image processing software like Amped FIVE allows to restore a moving car, measure objects in crime scene photos, detect faces in blurred/dark photos, sharpen text-just as we have seen it in FBI and CSI movies; haze visibility enhancement and defogging of images can be used to interpolate and measure the distance between objects under different atmospheric conditions for vehicle visual systems development. Deep learning (DL) is a relatively new field of machine learning (ML), and this may be effectively imposed to image processing. A variety of neural networks can be utilized for solving different image enhancement tasks, for example, successful denoising, producing high-resolution images

V. Singh (✉) · V. D. Kaushik
Department of Computer Science and Engineering, Harcourt Butler Technical University, East Campus, Nawabganj, Kanpur, Uttar Pradesh 208002, India
e-mail: vineeta.singh.cs@gmail.com

A. Kumar et al. (eds.), *Recent Advances in Manufacturing and Thermal Engineering*,
Lecture Notes in Mechanical Engineering, https://doi.org/10.1007/978-981-19-8517-1_29

from low-resolution images by training super-resolution, and much more. Deep learning algorithms allow you to train a neural network to zoom in the images at 2× or even 4× ratio, making it possible to enhance the quality of your low-resolution pictures. Deep neural networks (GANs, deeply-recursive convolutional networks (DRCN)) can recuperate photograph realistic textures out of vigorously down sampled pictures: NN is 'hallucinating' details based on its training from example images. It is not a 100% accurate reconstruction of photos in HD; however, the results are quite impressive [4, 5].

2 Related Work

Wang J et al. (2022) [5] devised a platform for image enhancement. Here, the Residual Hybrid Attention Block (RHAB) was considered for processing the images. Here, multi-scale feature maps were combined amongst each other using nested sub-networks. In addition, a Hybrid Attention Network (HAN) was considered to attain good performance. However, the technique did not use loss functions [5]. Researchers during recent years have proposed various fusion frameworks based on advanced strategies such as deep learning and machine learning for improving picture quality involving satellite images, medical images, multi-exposure images, multi-focus images. In [6–11], researchers have analyzed and studied different fusion frameworks for enhancing picture quality based on certain parameters, and they have critically analyzed the existing current fusion schemes and given future directions, merits as well as demerits [6–14]. Singh and Kaushik proposed atom search optimization and sine cosine algorithm-based combination plan to intertwine multi-focus pictures, yet the created strategy utilizes just multi-focus images [15]. Singh and Kaushik [16] grew a lot of upgraded adaptation of advancement calculation in view of Taylor series and holoentropy alongside mixture of ASO and SCA yet proposed technique could utilize just multi-focus pictures for improving the quality [16]. Singh and Kaushik proposed a two level combination weighted model in view of DCNN for creating multi-focus pictures with a superior quality yet clinical pictures were not the piece of the research [17]. Singh and Kaushik fostered a streamlining-based combination calculation which was additionally used in preparing by means of DCNN for delivering great quality clinical pictures which could lead in better determination in clinical field, yet multi-focus pictures were not the piece of the exploration here [18]. Singh and Kaushik proposed an original combination conspire including Haar wavelet change for isolating the pictures into subgroups, and further created combination rule was used for delivering combination factors; here, clinical picture dataset was taken for exploratory and recreation work, however, strategy does not include multi-focus images [19].

3 Proposed Framework

Proposed HBA-based DCNN for picture quality improvement: The upgrade of picture quality is finished with created HBA-based DCNN. The HBA is used to prepare the DCNN to tune ideal loads which subsequently help to expand the nature of picture and creates upgraded picture. The DCNN training is performed with HBA. The HBA is formulated considering insightful scavenging elements of honey badger. The energetic search nature of honey badger along with honey and digging discovery methods are adapted. The HBA actually addresses the streamlining issues with search strategy. The algorithmic course of concocted HBA model is delineated as:

(a) Initially, measure of honey badger is introduced with population size alongside comparing positions.

(b) Assessment of fitness measure is finished finding ideal arrangement, and the fitness with lowest value is imprinted as best answer for picture upgrade. Plan of fitness function for considering error is shown as

$$\xi = \frac{1}{s}\sum_{\sigma=1}^{s}\left(M_s^* - O\right)^2 \tag{1}$$

where s shows aggregate sum of information, M_s^* determines target output, O is output from DCNN, and ξ depicts fitness function.

(c) Intensity is relating to uncover prey power and distance among honey badger as well as prey. There is quick movement for high smell, representation of which is done via inverse square law and expressed as

$$P_c = i_2 \times \frac{U}{4\pi v_c^2} \tag{2}$$

In above equation, i_2 stands for random value in the interval 0 and 1, U shows strength of concentration, as well as v_c shows distance from cth honey badger to prey. At the last, refer below representation to see interconnection in between U and v_c:

$$U = (O_c - O_{c+1})^2 \tag{3}$$

Here, O_c is the cth honey badger location, while O_{c+1} stands for $c+1$th honey badger location.

$$v_c = O_{\text{prey}} - O_c \tag{4}$$

In above equation, O_{prey} represents the position of prey.

(d) Here, density factor is responsible for managing the time fluctuating randomization for ensuring smoother headway from exploitation to exploration. The

density factor decreases close by cycles for diminishing randomization with time by below numerical portrayal.

$$\vartheta = Z \times \exp\left(\frac{-b}{b_{\max}}\right) \quad (5)$$

Here, Z stands for constant, as well as $b_{\max}$ stands for highest number of iterations.

(e) This procedure utilizes banner, which changes scan heading for remunerating raised opportunities for experts for separating search space harshly.

(f) The position updation association of HBA fundamentally consolidates two phases, specifically digging as well as honey phase, which are as follows,

(i) **Digging Phase**: The honey badger executes the action like Cardioid shape and the Cardioids headway which is convinced via,

$$O_{\text{new}} = O_{\text{prey}} + A \times \varepsilon \times E \times O_{\text{prey}} + A \times i_3 \times \vartheta \times v_c \times |\cos(2\pi i_4) \times [1 - \cos(2\pi i_5)]| \quad (6)$$

In Eq. (6), O_{prey} stands for location of prey, $\varepsilon \geq 1$, i_3, i_4 as well as i_5 refers to random numbers among 0 and 1. A acts as flag responsible for varying search direction denoted by Eq.(7),

$$A = \begin{cases} 1, & \text{if } i_6 \leq 0.5 \\ -1, & \text{else} \end{cases} \quad (7)$$

where i_6 implies random integer among 0 and 1. A honey badger mainly depends on small intensity E of prey O_{prey}, and badger may gain any trouble R in digging action that allows for detecting location of optimal prey.

(ii) **The Honey Phase**: In this phase, source stimulation is accompanied via Eq. (8) at the point when honey badger goes through honey guide bird for arriving at bee colony,

$$O_{\text{new}} = O_{\text{prey}} + R \times i_7 \times \vartheta \times v_c \quad (8)$$

In Eq. (8), O_{new} indicates new location of honey badger, and i_7 defines random integer among 0 and 1.

(g) The optimal solution is gained via means of fitness, denoted via Eq. (1), and here, optimum solution for fitness function having least possible value is considered.

(h) Step (a)–(h) is repeated until maximum solution is gained.

Table 1 Performance assessment of proposed technique

Images	Metrics	DRL [20]	GAN [21]	CNN [22]	Proposed technique
1	MI	1.14	1.32	1.20	**1.47**
2	MI	1.14	1.32	1.16	1.41
3	MI	1.21	1.25	1.13	1.35
1	UQI	0.62	0.66	0.55	**0.75**
2	UQI	0.57	0.55	0.52	0.71
3	UQI	0.54	0.54	0.50	0.67
1	SSIM	0.49	0.54	0.55	**0.76**
2	SSIM	0.49	0.46	0.54	0.67
3	SSIM	0.45	0.45	0.44	0.61

4 Simulation Results

Simulation work of the proposed technique has been done through MATLAB software, and proposed algorithm has been compared with three enlisted recent techniques DRL [20], GAN [21], and CNN [22] for three quality indicators MI, SSIM, UQI. Here, the effectiveness of created HBA-based DCNN with MI, SSIM, and UQI is shown. Here, absolute five pictures are thought of. The created model is demonstrated through MATLAB, and working framework taken was Windows 10, Intel I3 center processor as well as 2 GB RAM. The assessment of the proposed method has been tried on Lytro multi-center image [23]. The developed HBA-based DCNN model is contrasted with recently developed strategies like DRL [20], GAN [21], CNN [22], and HBA[24]. Experiment-related results are demonstrated through Table 1.

5 Conclusions

Proposed framework has shown better results for image quality improvement as compared to recent picture quality enhancement techniques with assessing key indicators mutual information, universal quality index, and structural similarity index measurement. For all the three assessment parameters, highest value denotes the better quality of the picture, as we can see from Table 1, our proposed algorithm is showing better result for all these three quality indicators.

Proposed framework is demonstrating better result in contrast with state-of-the-art schemes. Based on comparative assessment, it may be effectively inferred that our proposed method is providing better results in terms of max MI, max UQI, max SSIM, with a value of 1.47, 0.75, and 0.76 respectively. In this research, a novel honey badger optimization allied framework has been devised for enhancing the pictures in terms of quality. As we can sense the need of quality pictures in various research and

application domains such as photography, remote sensing, medical imaging, brain tumor detection, skin lesion detection, video surveillance, object tracking, undersea object detection, and so on. Our proposed framework is producing good quality pictures for multi-focus images. In future, we can also try this proposed framework on different image datasets for expanding the diversity of proposed framework.

References

1. Liu X, Pedersen M, Wang R (2022) Survey of natural image enhancement techniques: classification, evaluation, challenges, and perspectives. Dig Signal Process 103547
2. Wang R, Xiao W (2022) Adaptive enhancement algorithm of high-resolution satellite image based on feature fusion. J Math 2022
3. Hou H, Jin Q, Zhang G, Li Z (2022) CT image quality enhancement via a dual-channel neural network with jointing denoising and super-resolution. Neurocomputing
4. Arce GONZALOR, McLoughlin MICHAELP (1987) Theoretical analysis of the max/median filter. IEEE Trans Acoust Speech Signal Process 35(1):60–69
5. Wang J, Yang Y, Hua Y (2022) Image quality enhancement using hybrid attention networks. IET Image Proc 16(2):521–534
6. Amanlou A, Suratgar AA, Tavoosi J, Mohammadzadeh A, Mosavi A (2022) Single-image reflection removal using deep learning: a systematic review. IEEE Access 10:29937–29953. https://doi.org/10.1109/ACCESS.2022.3156273
7. Ramarao G, Bindu CH, Murthy TSN (2022) A critical review of image fusion methods
8. Xu F, Liu J, Song Y, Sun H, Wang X (2022) Multi-exposure image fusion techniques: a comprehensive review. Remote Sens 14(3):771
9. Singh V, Kaushik VD (2022) A study of multi-focus image fusion: state-of-the-art techniques. In: Advances in data and information sciences. Springer, Singapore, pp 563–572
10. Azam MA, Khan KB, Salahuddin S, Rehman E, Khan SA, Khan MA, Gandomi AH (2022) A review on multimodal medical image fusion: compendious analysis of medical modalities, multimodal databases, fusion techniques and quality metrics. Comput Biol Med 144:105253
11. Amanlou A, Suratgar AA, Tavoosi J, Mohammadzadeh A, Mosavi A (2022) Single-image reflection removal using deep learning: a systematic review. IEEE Access
12. Singh V, Kaushik VD (2022) A typical hybrid optimization-based image quality enhancement technique. In: Proceedings of international conference on computational intelligence, Data Sci Cloud Comput (pp. 225–233). Springer, Singapore
13. Singh V, Kaushik VD (2022) Study and performance analysis of image fusion techniques for multi focus images. In: Proceedings of second doctoral symposium on computational intelligence (pp. 247–259). Springer, Singapore
14. Singh V, Dixit Kaushik V (2021) A brief study and overview of image fusion methods. In: Proceedings of international conference on computational intelligence, Data Sci Cloud Comput (pp. 355–367). Springer, Singapore
15. Singh V, Kaushik VD (2021) Renyi entropy and atom search sine cosine algorithm for multi focus image fusion. Signal Image Video Process 15:903–912
16. Singh V, Kaushik VD (2021) HoEnTOA: Holoentropy and Taylor assisted optimization based a novel image quality enhancement algorithm for multi-focus image fusion. J Sci Ind Res 80(10):875–886. http://nopr.niscair.res.in/handle/123456789/58230
17. Singh V, Kaushik VD (2021) WeAbDeepCNN: weighted average model and ASSCA based two level fusion scheme for multi-focus images. J Sci Ind Res 80(10):905–914. http://nopr.niscair.res.in/handle/123456789/58227
18. Singh V, Kaushik VD (2022) DTCWTASODCNN: DTCWT based weighted fusion model for multimodal medical image quality improvement with ASO technique & DCNN. J Sci Ind Res (JSIR) 81(08):850–858. https://doi.org/10.56042/jsir.v81i08.56203

19. Singh V, Kaushik VD, Haar adaptive Taylor-ASSCA-DCNN: some novel fusion model for image quality enhancement. J Sci Ind Res (in press)
20. Zhang R, Guo L, Huang S, Wen B (2021) ReLLIE: deep reinforcement learning for customized low-light image enhancement. In: Proceedings of the 29th ACM international conference on multimedia, pp 2429–2437
21. Guo Y, Li H, Zhuang P (2019) Underwater image enhancement using a multiscale dense generative adversarial network. IEEE J Oceanic Eng 45(3):862–870
22. Tao L, Zhu C, Song J, Lu T, Jia H, Xie X (2017) Low-light image enhancement using CNN and bright channel prior. In: Proceedings of IEEE international conference on image processing (ICIP), pp 3215–3219
23. Lytro Multi-focus Image Dataset taken from. https://www.researchgate.net/publication/291522937_Lytro_Multi-focus_Image_Dataset. Accessed Jan 2022
24. Hashim FA, Houssein EH, Hussain K, Mabrouk MS, Al-Atabany W (2022) Honey badger algorithm: new metaheuristic algorithm for solving optimization problems. Math Comput Simul 192:84–110

Analysis of Lattice-Based Cranial Implant

Mohammad Zahid Khan, Jitendra Bhaskar, and Anand Kumar

1 Introduction

Brain is the main processing unit which handles the functionality and metabolic demands of the human body. It is covered inside the skull which protects it. In certain cases, such as accidents, birth defects, tumor removal, or in absence of cranial vault in children, these defects need to be restored and the process of repairing it is known as cranioplasty [1]. It is a surgical procedure to restore the defects of skull in both cosmetic and functional needs. Cranial implants are fitted on the place of defects by surgery. This surgery helps to increase the confidence of person and to reduce psychological disadvantages in social scenario. It is mostly performed after undergoing craniectomy [2].

Cranioplasty is also performed to control alterations in cerebral spinal fluid (CSF) and blood flow. It maintains metabolic demands, reduces the risk of sinking flap, maintains the intracranial pressure and reduces the risk of trephined syndrome after craniectomy. Trephined syndrome is basically caused due to direct atmospheric pressure on the open skull area which can vary the intracranial pressure, blood flow and CSF of brain leading to worsening of motor skills [3].

A set of 8 bones (Fig. 1a) meant for the protection of brain, when combined are called cranium. It contains and protects the brain, meninges and cerebral vasculature [1]. Meninges (Fig. 1b) is formed by three layers of membranes which are dura mater (tougher outer layer), arachnoid (Web like structure for cushioning of brain) and pia mater (cover layer on brain) [4]. Cerebral vasculature is very important specialized structures responsible for fulfilling the high demand of oxygen and glucose by neurons [5].

M. Z. Khan (✉) · J. Bhaskar · A. Kumar
Department of Mechanical Engineering, Harcourt Butler Technical University, Kanpur 208002, India
e-mail: 200204006@hbtu.ac.in

A. Kumar et al. (eds.), *Recent Advances in Manufacturing and Thermal Engineering*, Lecture Notes in Mechanical Engineering, https://doi.org/10.1007/978-981-19-8517-1_30

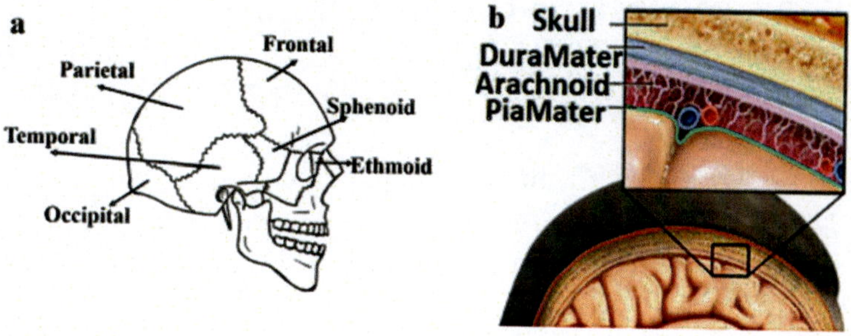

Fig. 1 **a** Types of cranial bones, **b** dissected view of different membranes [4]

Different materials used for the implants have their own merits and demerits. Cranial implant materials are classified into two categories: (1) biological, (2) synthetic [3]. Figure 2 shows the whole classification of cranial implants.

Biological implants are categorized into three categories known as autografts, allografts and xenografts. The bones are harvested from the same host body and carved into the desired shape to use as an implant known as autografts. If the bone is harvested from cadaver know as allografts and bones harvested from animals known as xenografts. Biological implants are very difficult to harvest and lead to high rate of infection and the esthetic outcomes are also not realistic [1, 3, 6].

Synthetic implants are classified according to the type of material, i.e., metal, polymers or ceramics. Synthetic materials are used as they have less risk of infection and provide better esthetics and cosmetic results [2, 6]. Due to advancement in the

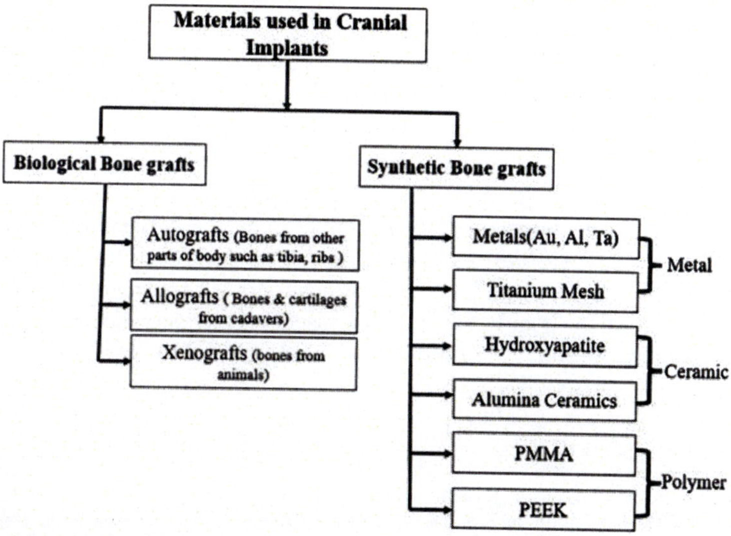

Fig. 2 Classification of cranial implants materials

technologies, these materials continued to develop and are suitable for the cranial implants for better and optimum results.

Currently, PEEK and titanium are the most favorable materials as they have shown better results and less complication rates. A retrospective study with single center experience compared data of patients (2001–2012) who underwent decompressive craniectomy has shown that there was little difference in the complication rates of PEEK and titanium implants after the removal of their previous implants due to complications [7]. PEEK and titanium implants are fairly costly. PMMA material is quite cost effective. A long-term clinical study on patients having PMMA-based cranial implants and found the complication rate to be 24% (70 patients, 2003–2010) which is less in comparison with the other materials [8].

Earlier the cranial implants were manufactured using traditional methods and during surgery, the alterations needs to be done in the implant which increases the surgery time. With recent advancement in technologies, 3D model of patient skull can be generated with the help of CT/MRI data, and patient specific implant can be modeled which gives more dimensional accuracy [9]. 3D printing technology can print the patient skull model, and designed implant fitment can be checked directly by placing on the defect and surgeon can pre-plan the surgery which leads to decrement in the time of surgery [10].

For improving the performance of implant in terms of osteointegration, the surface of the implant was roughly coated, but it was found that porous implants provide better results in term of interfacial adhesion with the bone [11, 12].

Lattice structures is basically a periodic or stochastic arrangements of unit cells in the particular volume which can help in the reduction of weight, alteration of mechanical properties. It is widely being developed and utilized in automobile, aerospace, infrastructure, energy and medical fields [13, 14]. These are classified into stochastic and non-stochastic cell structures.

Stochastic structures have closed cell and open cell foam type lattice structures. Non-stochastic are also two types: (1) 2D lattice such as honeycomb structure and (2) 3D lattice such as truss-based lattice structure. All these structures are developed over the years by researchers inspired from nature and environment [15].

In medical sector, lattice structures are widely being used to fabricate implants as these structures are nature and bio-inspired and can mimic bone topology. Earlier these structures were not integrated in the implant designs as these were impossible to manufacture with traditional methods, but 3D printing technology may print the most complex structures [14]. As per the previous conducted studies, these structures had better performance than foam and honeycomb structures. It gives more flexibility to alter the design variables as it consists of unit cell which means the design and parameters of each unit cell can be altered as per the requirement to get the desire results [16]. It also helps in the reduction of two major issues, i.e., osseointegration and stress shielding [17].

In this paper, three types of unit cell structures are considered for mimicking bone architecture which fall under polyhedral square geometry, which are [14]:

- Diamond (Fig. 3a)

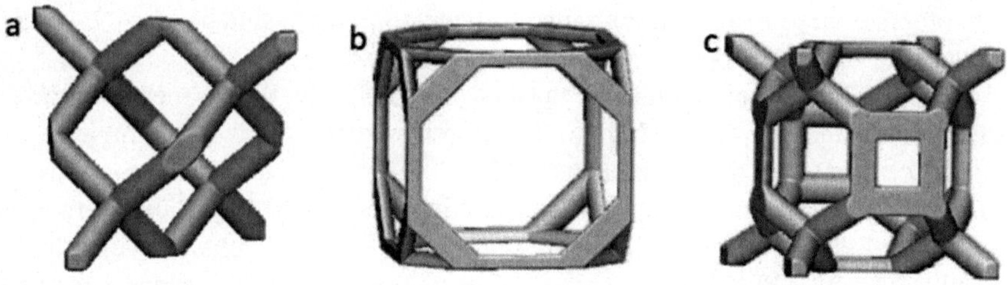

Fig. 3 Unit cell of lattice structures, **a** diamond, **b** truncated cube, **c** truncated octahedron

- Truncated cube (Fig. 3b)
- Truncated octahedron (Fig. 3c).

Fewer studies have been done using 3D lattice structure-based metallic and polymer cranial implants than the 2D lattice structure used in earlier studies. The main goal of this study was to attempt the integration of 3D lattice structures in cranial implant and to determine the displacement and von Mises stresses produced in an implant made of three different bio-compatible materials, namely Ti6Al4v, PEEK and PMMA. Finite element analysis study was done by applying force of 400 N as this force is similar to the effect of an adult man striking the implant site with a blow [18] whether the designed implant is safe or not. This study also demonstrated the significance of beam thickness in developing implants with lattice structures using different materials.

2 Methodology

2.1 Modeling

The geometry of implant is designed using SolidWorks 2021. Generation of lattice structure is done using nTopology software. Implant geometry is taken as elliptical dome to replicate the skull bones as they are curved in nature. Dimension of the implant is taken as 120 mm × 75 mm × 8 mm, and the cranial size defects are divided into 3 categories [19].

- Small size (>25cm^2)
- Medium size (25–200 cm^2)
- Large size (<200 cm^2).

As per the dimension taken above, the calculated area is 148.97 cm^2. Hence, a medium size implant is required. Figure 4 shows the geometry of implant with respective dimensions. Thickness of the implant is taken 1.5 mm [20]. After generation of geometry in SolidWorks, the model is imported into nTopology software, and workflow is generated to insert the selected lattice into the geometry. Cell map

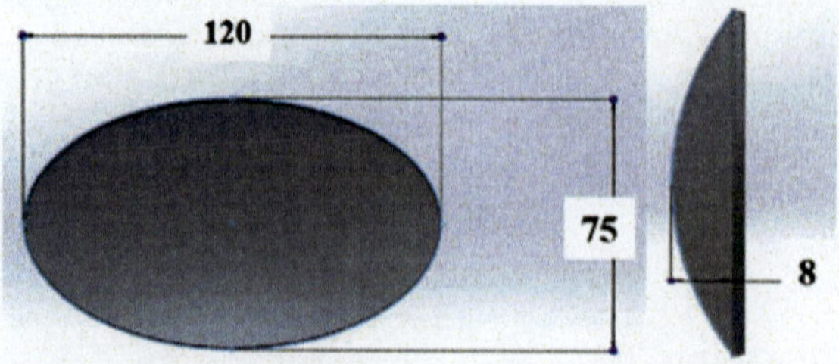

Fig. 4 Implant geometry and dimensions (all in mm)

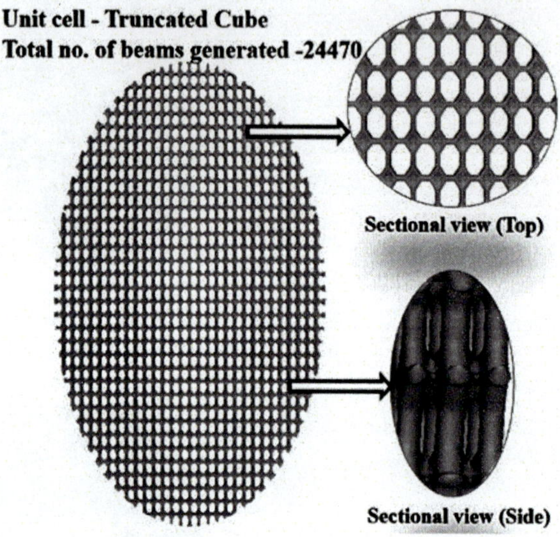

Fig. 5 Final implant geometry with truncated cube unit cell

is generated between upper and lower face of the geometry. Cell map is basically the division of total volume into multiple spaces where each unit cell will be generated and it gets connected to each other to generate the whole lattice covering the geometry. Final generated lattice structures with truncated cube, diamond and truncated octahedron unit cells are shown in Figs. 5, 6, and 7, respectively.

2.2 *Analysis*

Titanium alloy (TiAl64v), polyetherketone (PEEK) and polymethylmethacrylate (PMMA) have been considered for analysis. Titanium alloy is bio-compatible, corrosion resistance, high specific strength and chemically stable [20]. PEEK material is taken as it is also bio-compatible, light, strong and elastic in nature [21]. PMMA is biologically inert and maintains mechanical properties similar to bone [22].

Properties of materials considered are shown in Table 1. It is assumed that the material is isotropic linearly elastic in nature [23].

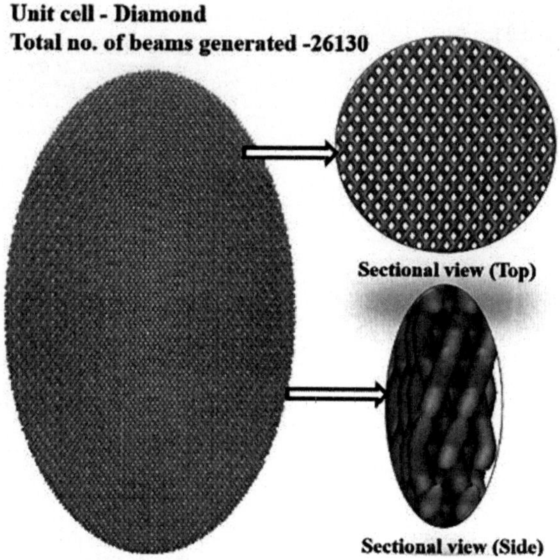

Fig. 6 Final implant geometry with diamond unit cell

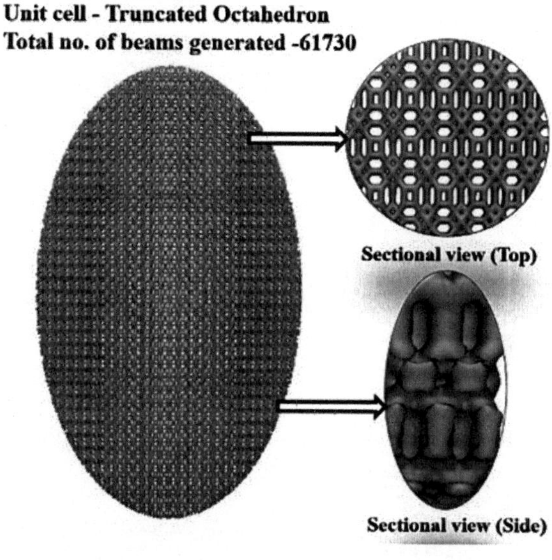

Fig. 7 Final geometry with truncated octahedron unit cell

Table 1 Material properties [24]

	TiAl64V	PEEK	PMMA
Young's Modulus (MPa)	110,000	4000	3000
Poisson's ratio	0.3	0.4	0.38
Density (kg/mm^3)	4430×10^{-9}	1360×10^{-9}	1180×10^{-9}
Yield strength (MPa)	880	95	83

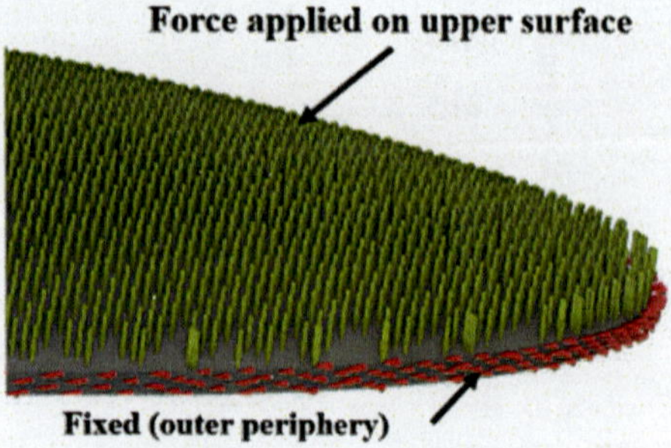

Fig. 8 Boundary conditions considered for simulation

Fig. 9 Lattice bean thickness

A force of 400 N magnitude is considered. Force is applied on the upper surface and outer periphery of the implant attached to the skull and may be considered fixed shown in Fig. 8. Two cases of lattice beam thickness of 0.5 mm and 0.75 mm, respectively, have been considered. Beam thickness is shown in Fig. 9.

Lattice integrated design files were very complex in nature and require high computational power for analyzing. We are using nTopology software finite element lattice mesh module which makes it easier to generate elements and nodes which can be further simulated with the boundary conditions to get the desired results. For the results to be accurate and error free, grid independence test conducted by varying the subdivision which basically divide one beam element into several 1D elements, so the values does not get affected by the number of element and nodes. It was seen that from value 16, there is very minimal difference in values of results so value 18 have been considered as an input in subdivision. Graphical representation of the test is shown in Figs. 10 and 11.

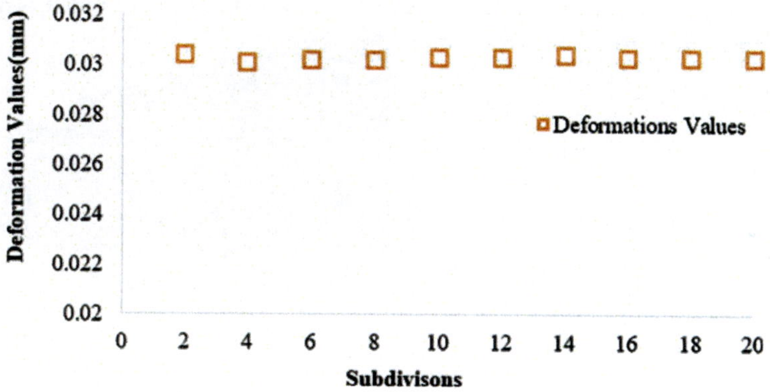

Fig. 10 Grid independence test results for deformation values

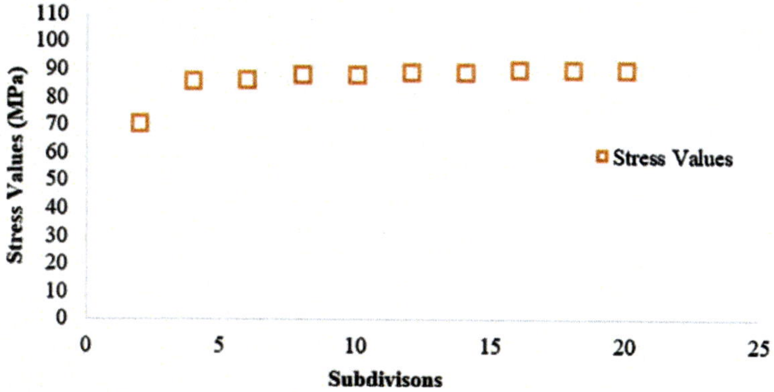

Fig. 11 Grid independence test results for stress values

3 Results and Discussion

A total of three lattice structures are used which are generated with two different beam thickness. Three different types of materials are used. Loading and boundary conditions are same. Total displacement and von Mises stress have been investigated. Higher stresses are observed in the mid region of the implant.

3.1 Case 1 (Beam Thickness = 0.5 mm)

- From Table 2, it can be observed that minimum displacement is observed in titanium alloy lattice structure ranging from 0.018–0.034 mm, PEEK from 0.536 to 0.996 mm and PMMA from 0.714 to 1.327 mm.

Table 2 Total displacement and stress analysis report (case 1)

Analysis report	Diamond			Truncated cube			Truncated octahedron		
	Ti6Al4v	PEEK	PMMA	Ti6Al4v	PEEK	PMMA	Ti6Al4v	PEEK	PMMA
Total displacement (mm)	0.03	0.866	1.153	0.034	0.996	1.327	0.018	0.536	0.714
Von Mises stress (MPa)	99.336	100.01	99.984	149	149.304	149.300	50.885	50.868	50.874

- Minimum von Mises stress observed in PEEK lattice structure ranging from 50.868–149.304 MPa, PMMA from 50.874 to 149.3 MPa and titanium alloy from 50.885 to 149 MPa.
- Truncated octahedron lattice structure implant with all three materials shows minimum equivalent von Mises stress ranging from 50.868–50.885 MPa following with diamond from 99.936 to 100.01 MPa and truncated cube ranging from 149 to 149.304 MPa.
- In this case, truncated octahedron unit cell lattice structure has shown the best results with minimum displacement (0.018–0.714 mm) and equivalent von Mises stress (50.868–50.885 MPa) with all three materials. Stress generated is less than the yield strength of materials used for implant. It confirms that the implant will elastically deform and regain its shape subsequently. Similarly, diamond and cube unit cell structure are safe with titanium alloy material but exceeding the yield strength and displacement values with PEEK and PMMA material. Stress comparison with respect to yield strength and displacement results comparison is shown in Fig. 12.

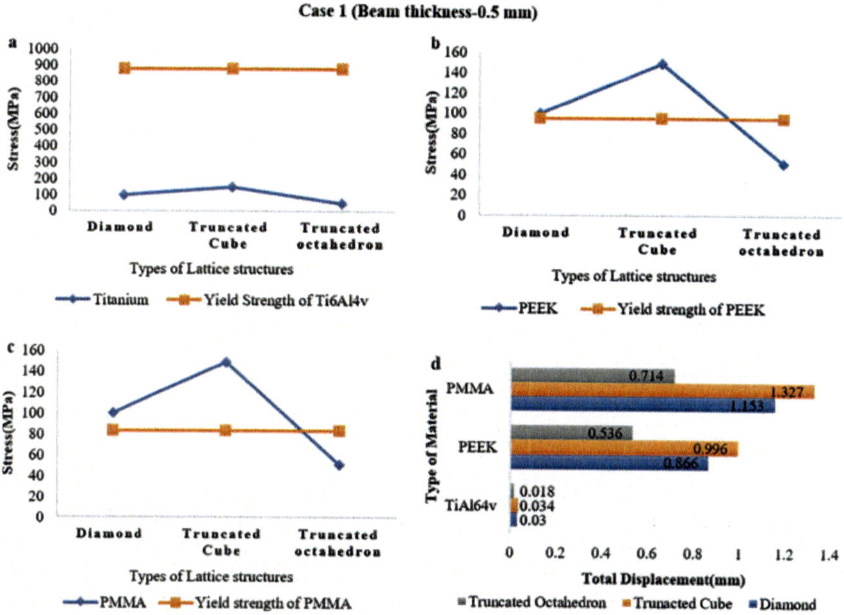

Fig. 12 **a** Ti6Al4v stress result; **b** PEEK stress result; **c** PMMA stress result comparison with yield strength of material; **d** total displacement results

Table 3 Total displacement and stress analysis report (case 2)

Analysis report	Diamond			Truncated cube			Truncated octahedron		
	Ti6Al4v	PEEK	PMMA	Ti6Al4v	PEEK	PMMA	Ti6Al4v	PEEK	PMMA
Total displacement (mm)	0.012	0.353	0.471	0.015	0.431	0.574	0.007	0.218	0.291
Von Mises stress (MPa)	37.429	37.476	37.460	65.405	65.393	65.397	19.972	19.973	19.973

3.2 *Case 2 (Beam Thickness = 0.75)*

- From Table 3, it can be observed that little displacement shown by all three lattice structures using titanium alloy. Minimum displacement value is ranging from 0.007–0.015 mm. PEEK displacement value ranges from 0.218–0.431 mm, and PMMA displacement value ranges from 0.291–0.574 mm.
- Minimum von Mises stress observed in titanium alloy lattice structures ranging from 19.972–65.405 MPa, PEEK and PMMA has shown similar stresses ranging from 19.973–65.397 MPa.
- In this case, all lattice structures shown minimal displacement and their equivalent von Mises stresses are less than the yield strength of materials. Best results are given by truncated octahedron unit cell structure followed with diamond and truncated cube structure, respectively. Stress comparison with respect to yield strength and displacement results comparison is shown in Fig. 13.

From overall study, the maximum displacement and stress values which are not exceeding the limits of yield strength and displacement of more than 1 mm are safe. As per the previous study conducted with solid, implants have shown the maximum displacement of 1 mm [24].

4 Conclusions

In this paper, a cranial implant based on three lattice structures is designed, and the geometries are analyzed. The study was carried out by varying the beam thickness of structures made of three different materials in each case. The current study's findings show that Ti6Al4v material provides minimal displacement in both cases. PEEK and Ti6Al4v demonstrated lower stress values in cases 1 and 2, respectively. In Case 1, two of the three lattice structures, diamond and truncated cube with PEEK and PMMA materials, failed the criteria. Similarly, in Case 2, all structures made from three different materials produce satisfactory results. Truncated octahedron lattice structure with all three materials produced the best results. Both case studies demonstrated the significance of the beam thickness parameter in the design of lattice structure-based cranial implants using specific materials.

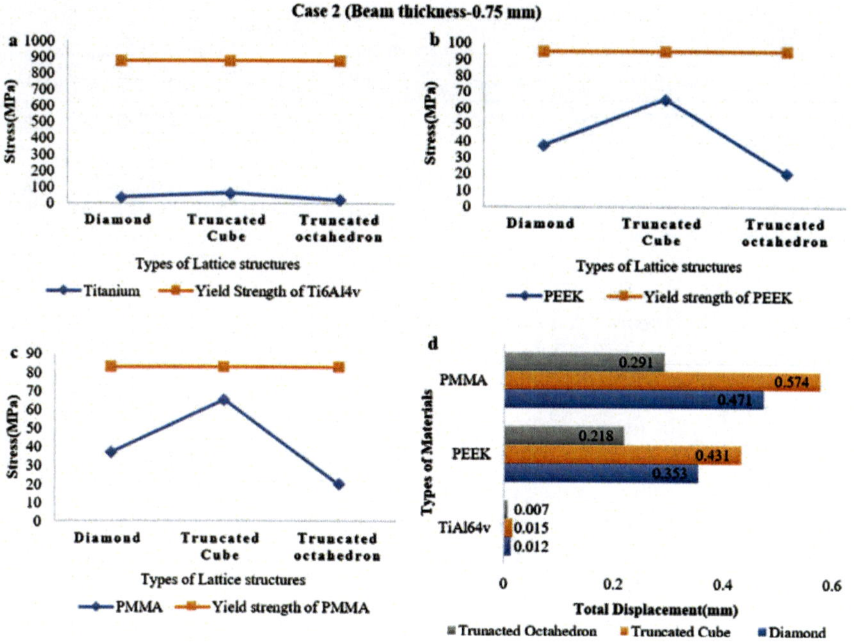

Fig. 13 **a** Ti6Al4v stress result; **b** PEEK stress result; **c** PMMA stress result comparison with yield strength of material; **d** total displacement results

Selection of reasonable lattice structures in cranial implants is of significant importance because they allow the implants to be lightweight, porous and able to withstand reasonable loads. Polymeric materials such as PMMA and PEEK are inexpensive, and this research could help to reduce the cost of implants. This research can help to advance the development of alternative designs for cranial implants that are more compatible with the human body by ensuring adequate strength to protect the brain.

Acknowledgements The authors would like to express their gratitude to nTopology team for providing the educational license for the research work and to Dr. Pawan Kumar Verma, Assistant Professor, SGPGI Lucknow for providing information related to cranial implants.

References

1. Khader BA, Towler MR (2016) Materials and techniques used in cranioplasty fixation: review. Mater Sci Eng, C 66:315–322
2. Shah AM, Jung H, Skirboll S (2014) Materials used in cranioplasty: a history and analysis. Neurosurg Focus 36(4)
3. Alkhaibary A, Alharbi A, Alnefaie N, Oqalaa Almubarak A, Aloraidi A, Khairy S (2020) Cranioplasty: a comprehensive review of the history, materials, surgical aspects, and complications. World Neurosurg

4. Meninges (2022) Mayo Foundation for Medical Education and Research. https://www.mayoclinic.org/diseases-conditions/meningioma/multimedia/meninges/img-20008665. Accessed 3 June 2022
5. Agarwal N, Carare RO (2021) Cerebral vessels: an overview of anatomy, physiology, and role in the drainage of fluids and solutes. Front Neurol 11
6. Aydin S, Kucukyuruk B, Abuzayed B, Aydin S, Sanus GZ (2011) Cranioplasty: review of materials and techniques. J Neurosci Rural Pract 2(2)
7. Thien A, King NKK, Ang BT, Wang E, Ng I (2015) Comparison of polyetheretherketone and titanium cranioplasty after decompressive craniectomy. World Neurosurg 83(2):176–180
8. Jaberi J, Gambrell K, Tiwana P, Madden C, Finn R (2013) Long-term clinical outcome analysis of poly-methyl-methacrylate cranioplasty for large skull defects. J Oral Maxillofac Surg 71(2):e81–e88
9. Bogu VP, Ravi Kumar Y, Asit Kumar K (2018) 3D printed, customized cranial implant for surgical planning. J Inst Eng India Ser C 99:255–259
10. Moiduddin K, Darwish S, Al-Ahmari A, Elwatidy S, Mohammad, Ameen W (2017) Structural and mechanical characterization of custom design cranial implant created using additive manufacturing. Electron J Biotechnol 29
11. Pilliar RM (1998) Overview of surface variability of metallic endosseous dental implants: textured and porous surface–structured designs. Implant Dentistry 7(4)
12. Nguyen HQ, Deporter DA, Pilliar RM, Valiquette N, Yakubovich R (2004) The effect of sol–gel-formed calcium phosphate coatings on bone ingrowth and osteoconductivity of porous-surfaced Ti alloy implants. Biomaterials 25(5):865–876
13. Raheem AA, Hameed P, Whenish R, Elsen RS, Aswin G, Jaiswal AK, Prashanth KG, Manivasagam G (2021) A review on development of bio-inspired implants using 3D printing. Biomimetics
14. Libonati F, Graziosi S, Ballo F, Mognato M, Sala G (2021) 3D-printed architected materials inspired by cubic Bravais lattices. ACS Biomater Sci Eng
15. Park KM, Min KS, Roh YS (2021) Design optimization of lattice structures under compression: study of unit cell types and cell arrangements. Materials (Basel)
16. Chen Pan, Yafeng Han, Jiping Lu. Design and Optimization of Lattice Structures: A Review. Applied Sciences, Volume 10(18):6374, 2020.
17. Maconachie T, Leary M, Lozanovski B, Zhang X, Qian M, Faruque O, Brandt M (2019) SLM lattice structures: properties, performance, applications and challenges. Mater Des 183
18. Jagoda Maciejewska and Hubert Jopek (2020) An analysis of auxetic cranioplasty implant. Vib Phys Syst 31:2020315-1–2020315-8
19. Uygur S, Eryilmaz T, Cukurluoglu O, Ozmen S, Yavuzer R (2013) Management of cranial bone defects: a reconstructive algorithm according to defect size. J Craniofac Surg 24(5)1606–1609
20. Moiduddin K, Mian SH, Umer U, Alkhalefah H (2019) Fabrication and analysis of a Ti6Al4V implant for cranial restoration. Appl Sci
21. Zhang J, Tian W, Chen J, Yu J, Zhang J, Chen J (2019) The application of polyetheretherketone (PEEK) implants in cranioplasty. Brain Res Bull
22. Huang GJ, Zhong S, Susarla SM, Swanson EW, Huang J, Gordon CR (2015) Craniofacial reconstruction with poly (methyl methacrylate) customized cranial implants. J Craniofac Surg 26(1)
23. Phanindra Bogu V, Ravi Kumar Y, Khanra A (2017) Modelling and structural analysis of skull/cranial implant: beyond mid-line deformities. Acta Bio Eng Biomech
24. Marciána P, Narrab N, Boráka L, Chamrada J, Wolffc J (2019) Biomechanical performance of cranial implants with different thicknesses and material properties: a finite element study. Comput Biol Med 109:43–52

Modeling and CFD Analysis of Hepatic Veins of Liver

Prabhat Agnihotri and Jitendra Bhaskar

1 Introduction

The liver is situated above the stomach region. The geometry of the liver is somewhat similar to the cone, and the weight of the liver is around 1.5 kg. The function of the liver in the human body is to produce bile juice and store vitamins and minerals. It breakdowns the blood coming from the stomach and intestine into nutrients and toxic parts are removed through the kidney [1]. The liver works as a dual-channel blood flow system in which it takes blood, nutrition through the portal vein, and oxygen-rich blood through hepatic arteries which comes from the heart and gives deoxygenated blood to the right portion of the heart through the inferior vena cava (IVC). Now, the blood from the heart is pumped to the lungs which absorb oxygen and removes carbon dioxide during the breathing process. The anatomy of the liver is shown in Fig. 1. It is a very challenging assignment [2].

The liver consists of hepatic veins, portal vein, hepatic arteries, and bile ducts. Hepatic veins are of three types, i.e., right hepatic vein (RHV), left hepatic vein (LHV), and middle hepatic vein (MHV), and these three veins combine and result in IVC. RHV is the largest vein out of the three. MHV divides the liver into two parts known as the left and right lobes of the liver [3]. The amount of blood carried by the liver is 1/8th part of the total blood present in the human body. Blockage of any arteries and veins in the liver can cause liver damage which may lead to several serious diseases such as Budd-Chiari syndrome and cirrhosis, and it can also create hindrance in the blood flow to the other parts of the body. Weeks or months may pass before symptoms appear, you may have severe pain or you may not experience any symptoms until the condition degrades. These diseases can be treated by surgeons through surgery in which the diameter of veins can be increased or they can change

P. Agnihotri (✉) · J. Bhaskar
Department of Mechanical Engineering, Harcourt Butler Technical University, Kanpur 208002, India
e-mail: prabhat2350@gmail.com

A. Kumar et al. (eds.), *Recent Advances in Manufacturing and Thermal Engineering*,
Lecture Notes in Mechanical Engineering, https://doi.org/10.1007/978-981-19-8517-1_31

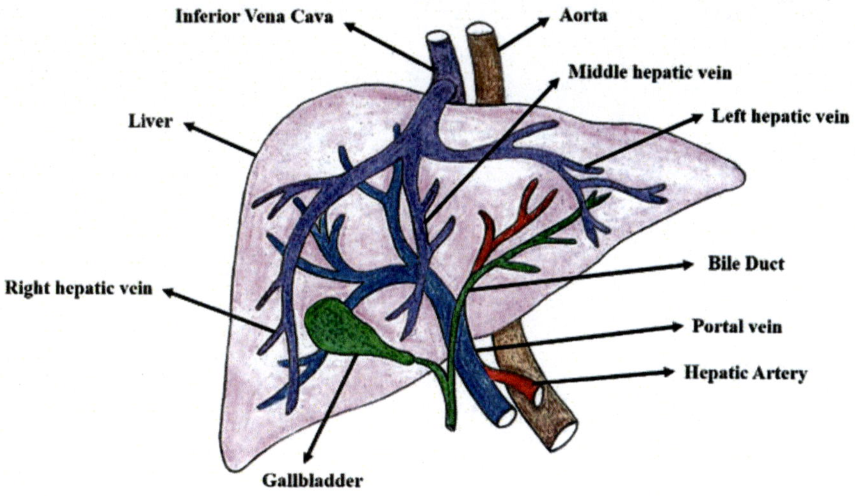

Fig. 1 Anatomy of the liver

the blood flow from one blood vessel to another. If the condition gets serious patient may require liver transplantation [4].

As there are certain challenges in directly using the STL file for analysis, the DICOM-generated STL model is converted into a 3D solid model using different software packages and the RE technique in this paper. The STL file typically contains the surface geometry of an object represented as an uneven distorted surface mesh, which does not meet the requirements for FEM/CFD applications. The conversion of a STL file into a solid model is very rigorous and complex because it depends on the geometry's complexity and requires a lot of computation power. We can obtain data such as outer dimensions and surface geometry from the STL file. For using this geometry in a CFD application, 3D solid mesh elements such as tetrahedral and hexahedral elements are required. Solid mesh elements contain an overall surface and discrete meshes inside the geometry. Because of these requirements, it was necessary to replicate the geometry using different modeling software packages as it will generate 3D solid model which can be further meshed into solid elements, and the required analysis can be performed.

The use of CFD in biomedical research and studies is increasing as it can analyze fluid flow systems with accuracy. With this technique, doctors can diagnose different types of parameters related to diseases. Virtual simulation and the patient-specific prototyped model may be beneficial in understanding the disease-related issue in a better way which will help in reducing the time and complications. In the past few years, researchers have shown that computational modeling and analysis have given better results than the old traditional methods used in the medical domain [5].

In this recent area of research, several challenges are to be addressed before integrating CFD models. This computer-based research strategy has been used to study other liver-related diseases. Most researchers have done their work on portal

vein-related diagnosis which is able to provide hemodynamic relation with the portal vein and provided assistance in developing software-based diagnoses for patients. This paper is focused on the study related to hepatic veins. The primary focus of this research paper is to develop a methodology to handle the DICOM data for the construction of 3D model which will be similar to real geometry and to study the simulation of blood flow in hepatic veins present in the liver with the use of computational fluid modeling techniques by applying real geometric boundaries.

2 Methodology

2.1 Modeling of Liver and Hepatic Veins

There are various steps involved to make a 3D geometrical model using DICOM data. MIMICS software was used for segmentation, 3-MATIC software for STL repairing as well as dimension measuring, SOLIDWORKS software for modeling the 3D CAD model, ANSYS for volumetric meshing as well as for computational fluid dynamics of hepatic veins [6]. CT images were used in form of DICOM and with the help of this DICOM data 3D CAD model using reverse engineering technique. All steps involved in creating a 3D geometrical model of hepatic veins are listed in Fig. 2.

Image acquisition and importing DICOM data: The CT image of the patient is uploaded in MIMICS software in DICOM format. There are three views present in MIMICS software, Axial, Sagittal, and Coronal. The greater number of slides gives a

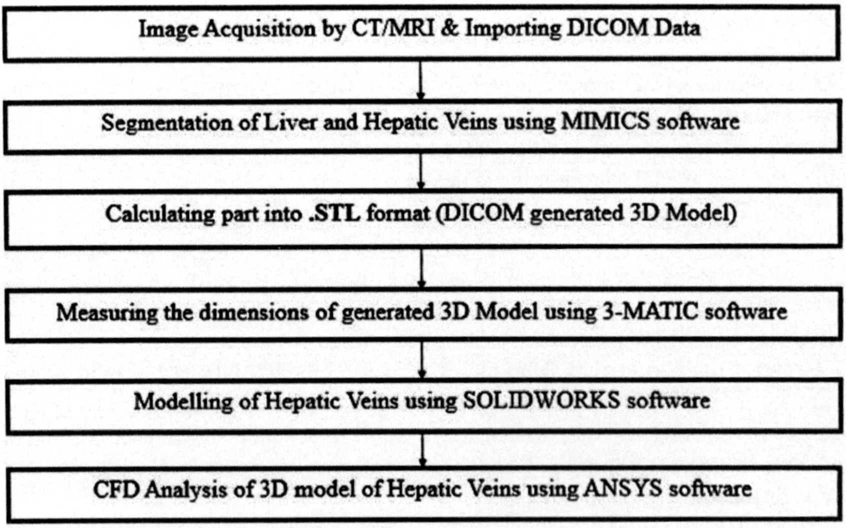

Fig. 2 Schematic study of making 3D model of liver and hepatic veins using dedicated software

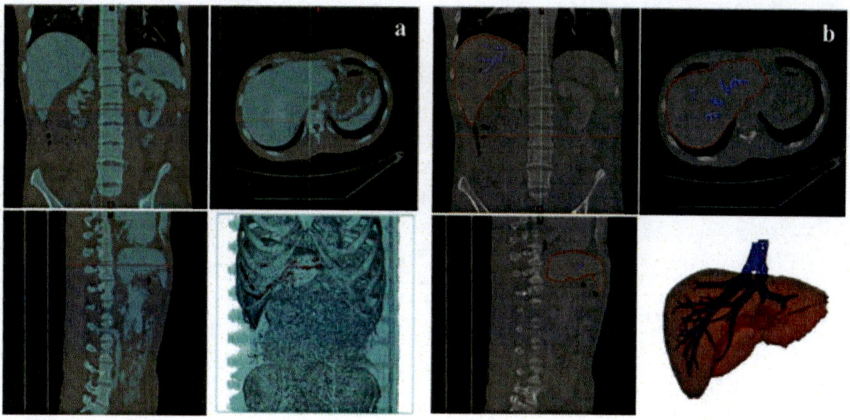

Fig. 3 **a** Coronal, axial, sagittal, and 3D generated views of venous phase of DICOM data; **b** coronal, axial, sagittal, and 3D generated views of venous phase of liver and hepatic veins after segmentation and thresholding

more accurate resultant model of HV. Image acquisition and DICOM data are shown in Fig. 3a.

Segmentation and Thresholding: The separation of soft tissues, bone, and hard tissues is known as segmentation. With the help of MIMICS software, liver and hepatic veins were separated from whole DICOM data using various inbuilt software tools. Thresholding is a method of image processing that creates a Bitonal image based on the thresholding value of the pixel intensity. Thresholding values for the liver vary from (100–1484) HU and for HV vary from (198–1626) HU. For finding the liver and HV, the venous phase was chosen. Segmented liver and HV are shown in Fig. 3b.

Calculation of 3D model: It is a method in which 3D model is obtained using optimal quality of DICOM data and creates the 3D model. Figure 3b represents the 3D model of the liver and HV (in STL format). It can also measure the surface area and volume of the liver from this 3D model.

Repairing and Dimensions Measurement: In this method, extra noisy parts were removed by tools, and then measurements (like length, diameter, and angle) were done using the 3-point method with the help of 3-MATIC software. In Fig. 4a–d, dimension measurements of each branch of hepatic veins are shown. After repair of DICOM, generated 3D model (in STL format) is shown in Fig. 5a.

Creation of geometrical 3D model: The measurements of the HV were measured using the 3-Matic software, and the 3D model was created using SOLIDWORKS, 3D modelling CAD software has variety of tools to produce a geometric 3D model and an output CAD file in the SLDPRT format. The 3D geometric solid model of HV is displayed in Fig. 5b.

Analysis of 3D model: For analysis, it required IGES format CAD file because ANSYS software supports only IGES format, so it was converted from SLDPRT file

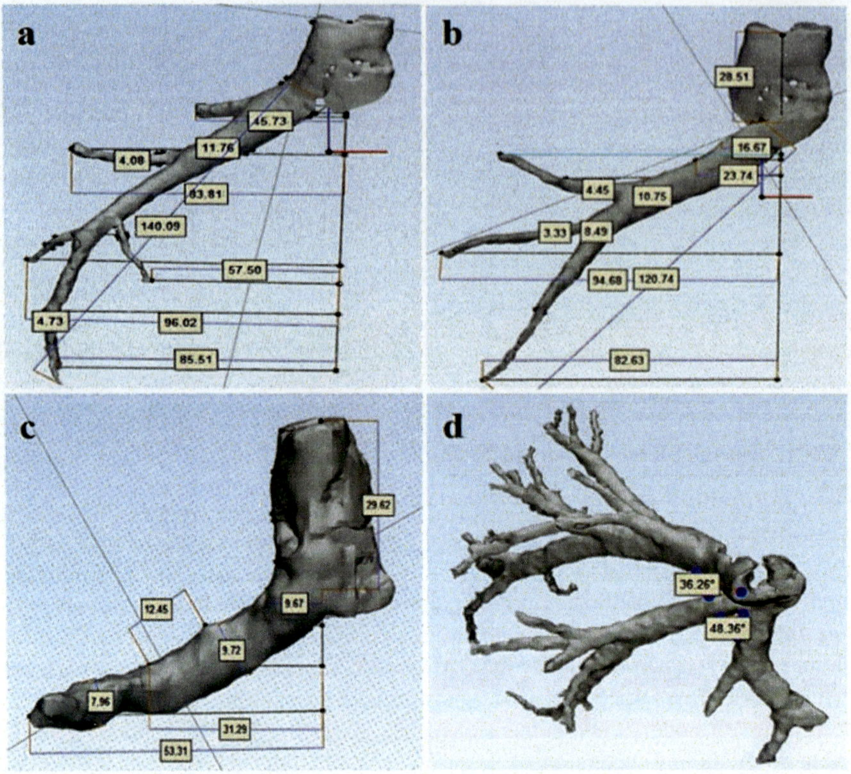

Fig. 4 **a** RHV dimensions; **b** MHV dimensions; **c** LHV dimensions; **d** angle measurement of all branches of HV, dimensions measured by using 3-MATIC software

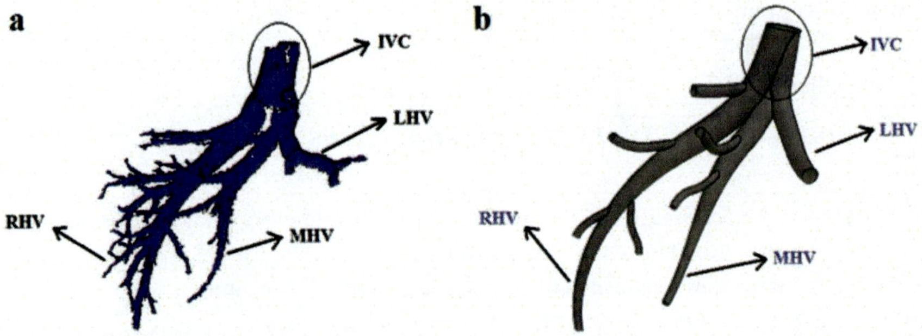

Fig. 5 **a** 3D model repaired and generated by 3-MATIC software; **b** 3D CAD solid model created by SOLIDWORKS software

Table 1 Geometrical 3D model dimensions

Description of 3D model	Dimensions
RHV diameter	3.5–11 mm
MHV diameter	3.5–10 mm
LHV diameter	7.5–10 mm
Angle between RHV and MHV	36.26°
Angle between MHV and LHV	48.36°
IVC diameter	17 mm

format to IGES format, that 3D CAD model uses various boundary conditions and governing equations to perform CFD.

2.2 Approach for 3D Geometrical Modeling

This study is focused on specific selections made by researchers when defining the model (governing equations, flow domain, and boundary conditions) because this selection affects the range, to which the model represents the aspects of real life to be analyzed [7]. There was many 3D CAD software available in which 3D modeling software SOLIDWORKS was used. In this step, the part design module was used to design the 3D model of HV. Dimensions taken from the 3-MATIC software given in Table 1 were in millimeters (mm), so during designing, the geometry all dimensions were taken in mm. Using all dimensions, the desired 3D geometrical model of HV was made. This 3D model was obtained in by default SLDPRT format and shown in Fig. 5b. Further, SLDPRT format was converted into IGES format, which is easily accepted by analysis software ANSYS.

2.3 Approach for Analysis of Hepatic Veins

For analysis, ANSYS FLUENT software was used. The geometry was converted into the fluid domain using the design modeler. The tetrahedron elements were used to create the volumetric mesh with an element size of 0.5 mm [8]. The skewness and orthogonal quality were mentioned in Table 2, values are under range as the skewness value should be near 0, and orthogonal quality should be about 1 for ideal condition. After volumetric meshing, the geometry was ready for applying boundary conditions, and CFD has been conducted.

Inlets of flow domains were taken as steady velocity or pressure. The outlet of the domain was considered as pressure-based boundary. The inlet and outlet of the flow domain are shown in Fig. 6a. Assumptions that wall of HV were taken as solid with no-slip condition for the reduction in simulation time [9, 10]. For simulation, the semi-implicit method for pressure linked equations (SIMPLE) algorithm was

Table 2 Parameters obtained using meshing of geometry

Parameters of geometry	Values
Skewness	0.00006–0.83643
Orthogonal quality	0.16357–0.99589
Nodes	2,921,671
Elements	2,106,363

used which had shown the correlation between pressure and velocity, and it is a pressure-based coupling [11].

The CFD of the blood flow has been done considering three-dimensional incompressible Navier–Stokes equations for mass and momentum. The flow of blood in HV is considered as laminar after calculating the Reynolds number. The calculated value ranges from 286–687 which is less than 2000. The conservation equations for laminar flow were used for the incompressible flow of non-Newtonian fluid. Blood density was taken as 1060 kg/m^3 [11]. The wall of the geometry was taken as rigid [12], and the elastic properties of the wall are neglected [13]. Other flow parameters are shown in Table 3. The HV simulation has been initialized by taking static pressure of 2731.1 pascals, and inlet blood flow velocity was assumed to be 35 cm/s for all inlets.

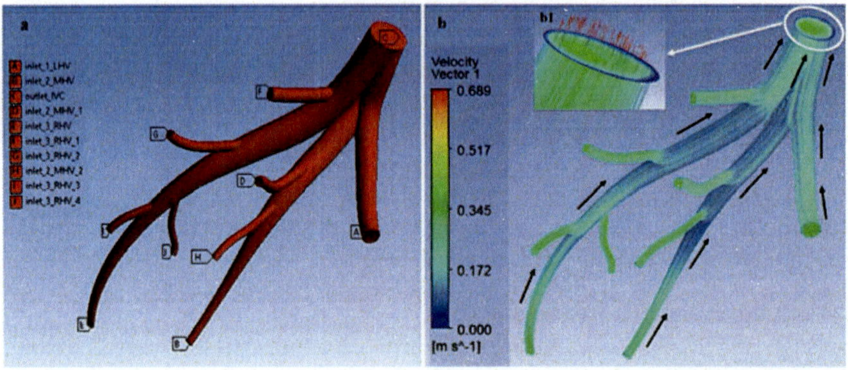

Fig. 6 **a** Inlet and outlet of hepatic veins are shown; **b** velocity vector generated by ANSYS FLUENT software; **b1** enlarged view of the direction of the velocity vector of blood flow of hepatic veins is shown

Table 3 Flow parameters [9]

Description of parameters	Values
Reference temperature (°K)	310
Maximum viscosity limit (η max) (kg/m-s)	0.003
Minimum viscosity limit (η min) (kg/m-s)	0.00125

3 Results and Discussion

After the analysis, it was found that the velocity of hepatic veins at the inferior vena cava (outlet) is 22.73 cm/s. This result is shown in Fig. 6b, b1. This study model is validated by comparing the blood flow velocity of hepatic veins with the Eco-Doppler data. As per Eco-Doppler data, the normal velocity of the hepatic vein in the healthy liver is less than or equal to 60 cm/s, and for a diseased liver patient, the velocity is greater than 60 cm/s [14]. Thus, according to the statement, output velocity at IVC obtained using the CFD technique shows an indication that there is no such significant problems with the liver caused by hepatic veins. Other veins that are present inside the liver could be the cause. A good connection between simulation results and clinical data has been obtained by comparing the simulation results with the Eco-Doppler data. The boundary conditions used in the numerical simulations as well as the time-variable velocity and pressure characteristics were taken from the literature and were not modified during this study.

4 Conclusions

In this paper, monitoring of the non-invasive hepatic veins approach is shown. The technical approach is shown for the conversion of the STL format file generated from DICOM data to the solid CAD model using dedicated software. As per the results obtained from the simulation, this study has shown the importance of a blood flow study as this approach may be useful for the analysis of hepatic vein disorders like diffuse venous obstruction and is also helpful for understanding the hemodynamic changes of hepatic veins such as pressure, velocity, and wall shear force. With this approach, diagnosis may become more effective and can help in the treatment of patients effectively. This approach can help researchers to utilize it in the diagnosis of diseases in different medical domains. The importance of 3D CAD modeling and computational fluid dynamics from a medical perception is shown in the study. Computational models can be used more correctly and effectively to diagnose patients with liver illnesses, identify potential treatment targets, and provide them with personalized care.

Acknowledgements The authors would like to thank Dr. Vivek Gupta and Dr. Sanjay Tripathi for sharing data and offering advice, as well as KGMU, Lucknow for giving access to Materialize software and other support.

References

1. Elobu AE, Kweyamba V, Rai R (2018) Surgical anatomy of the liver. Anat Sci 1
2. Christ B, Dahmen U, Karl-Herrmann, Koing M, Reichenbach JR, Ricken T, Schleicher J, Ole Schwen L, Sebastian, Waschinsky N (2017) Computational modelling in liver surgery. Front Physiol, Gastrointest Sci
3. Bashir O, Pianalto (2022) Hepatic veins. Radiopaedia.org. Available online: https://radiopaedia.org/articles/hepaticveins?lang=us. Accessed 06 June 2022.
4. Rath L (2022) Hepatic veins. WebMD. Available online: https://www.webmd.com/a-to-z-guides/hepatitis-hepatic-veins-anatomy; Accessed 06 June 2022
5. Tanja Cvitanovic T, Moskon M, Mraz M, Rozman D (2018) Computational modelling of liver metabolism and its applications in research and the clinics. Acta Chim Slov; 2018
6. Sahu YK, Anburajan M (2014) CAD model of patient specific portal vein for analysis of liver disorder. Int J Mech Ind Eng 4(1), article 12
7. Aramburu J, Anton R, Rodriguez-Fraile M, Sangro B, Bilbao JI (2021). Computational Fluid dynamics modelling of liver radioembolization: a review. In: CIRSE; 2021
8. Botar CC, Vasile T, Sfrangeu S, Clichici, Agachi PS, Badea R, Mircea P, Cristea MV, Moldovan R (2009) CFD simulation of the portal vein blood flow. In: IFMBE proceedings; 2009
9. Petkova S, Hossain A, Naser J, Palombo E (2003) CFD modelling of blood flow in portal vein hypertension with and without thrombosis. CSIRO
10. Taebi A, Pillai RM, Roudsari BS, Vu CT, Roncali E (2020) Computational modelling of the liver arterial blood flow for microsphere therapy: effect of boundary conditions. Bioengineering 2020
11. Khorseda Akhtar MST, Tajul Islam Md. (2017) A computational analysis of blood flow through portal vein under normal and extrahepatic obstructions. J Bangladesh Acad Sci 41(2):183–199
12. Botara CC, Vasileb T, Strangeub S, Clichicic S, Agachia PS, Badeab R, Mircead P, Cristeaa MV (2010) Validation of CFD simulation results in case of portal vein blood flow. In: 20th European symposium on computer aided process engineering ESCAPE 20
13. Vimmr J, Jonasova A (2010) Non-Newtonian effects of blood flow in complete coronary and femoral bypasses. Math Comput Simul 80(6):1324–1336
14. Afif AM, Chang JP-E, Wang YY, Lau SD, Deng F, Goh SY, Pwint MK, Ooi CC, Venkatanarasimha N, Lo RHG (2017) A Sonographic Doppler study of the hepatic vein, portal vein, and hepatic artery in liver cirrhosis. Ultrasound

Influence of Various Geometric Parameters on Sandwich Panel Under Ballistic Impact Using Finite Element Approach

Vinay Kumar and Mohammad Talha

1 Introduction

A sandwich panel is a type of composite structure that consists of layers of different materials such as metals, plastics and ceramics. Sandwich structures are known for their lightweight, high stiffness, high specific strength and moduli. These laminated composites are widely used as the structural components in military armour, aerospace industries, marine, sports, civil structures, etc. The ballistic impact load is one of the critical loads that is undergone by the composite structure during their service life but these are susceptible to damage due to their low toughness. Therefore, impact failure critical is a must to consider during the fabrication, operation and repairing processes. Along with it, minimum impact of bullet and less weight of the material are also important, and hence, there is a need to develop a lightweight and high ballistic performance material for armour composites [1]. Ceramic materials, because of the main energy absorption layer, played a serious role within the impact resistance tanks to their characteristics of high hardness and high compressive strength [2]. Meanwhile, the sandwich construction of the composite backplate is considered to be the ideal design strategy [3–5]. The laminated composite plate under impact load has received extensive attention from researchers. Significant reduction of the strength of the material has been observed under the low energy impact of FRP laminated plate; however, the damage on the impacted surface was not observable [6–8]. Due to this reason, many researchers are attracted to the low energy impact because in this case, prediction of the extent of damage is not easy to predict as the penetration of the impactor does not take place and after impact, it rebounds.

V. Kumar · M. Talha (✉)
School of Mechanical and Materials Engineering, IIT Mandi, Mandi, Himachal Pradesh 175075, India
e-mail: talha@iitmandi.ac.in

V. Kumar
e-mail: T20073@students.iitmandi.ac.in

A. Kumar et al. (eds.), *Recent Advances in Manufacturing and Thermal Engineering*, Lecture Notes in Mechanical Engineering, https://doi.org/10.1007/978-981-19-8517-1_32

The behaviour of laminated plate by three-dimensional hybrid stress finite element method along with the Newmark direct integration technique has been studied by Sun and Liou [9]. Under transverse impact, load damage on the composite laminated plate was analysed by Wang and yew [10]. Further, the analytical model proposed for woven fabric composite under impact with four edges of plate simply supported was studied by Mishra and Nayak [11]. Maximum force thresholds in three different types of composites, unidirectional E-Glass, woven E-Glass and woven Aramid composite, and their impact damage have been studied by Evci and Gülgeç [12].

The influence of skins and the core on the energy absorption capabilities of the sandwich panel during high-velocity impact was studied [13]. The high-velocity impact performance of glass reinforced polyester (GRP) resin with different types of reinforcements has been studied by Sabet et al. [14]. However, very few numbers of studies have been reported in the literature that highlights the effect of impact on deflection, contact force having different thickness, stacking sequence and different velocities along with boundary conditions. In the present work, Abaqus/Explicit (17.0) finite element software is used to evaluate the transient response of sandwich plate under impact with different thicknesses, different velocities and stacking sequence. Hashin's failure model has been used for the failure model to predict the damage initiation of composite sandwich plates.

2 Methodology

2.1 Model Configuration

A small ball of mass of 0.014175 kg and 5 mm radius is impacted on the laminated plate. A comparative study has been done using numerical simulation impact, and the results are compared with Ansari et al. [1]. The model consists of a plate that is composed of graphite–epoxy AS-3501-6. The dimensions of the plate are 140 mm × 140 mm with three layers of equal thickness and varying. The laminated plate is composed of three different ply orientations including cross-ply (0°/90°/0°) and angle plies as (45°/−45°/45°) and (30°/−30°/30°). And the laminated plate is clamped at all edges (Figs. 1 and 2).

2.2 Material Properties

See Table 1.

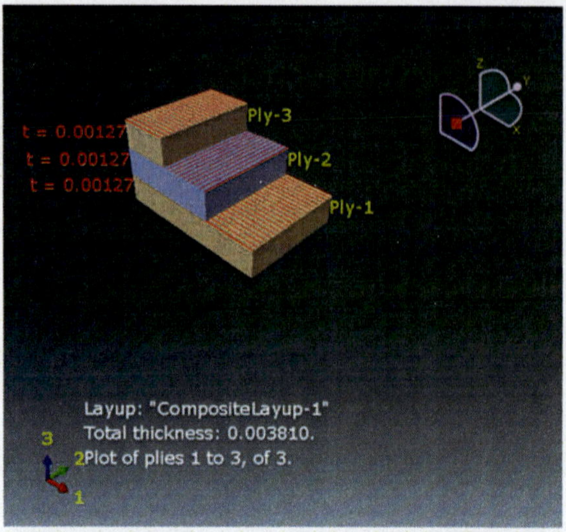

Fig. 1 Boundary condition applied to impact problems

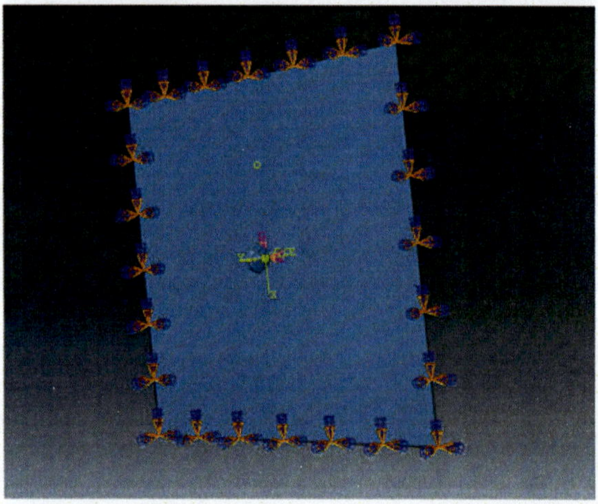

Fig. 2 Stack plot of plies of composite

Table 1 Mechanical properties of graphite–epoxy AS-3501-6 [1]

Young's modulus along the fibre direction, E_{11}	142.73 GPa
Young's modulus normal to the fibre direction, E_{22}	13.79 GPa
Poisson's ratio, ϑ_{12}	0.3
Shear modulus, G_{12}	4.64 GPa
Shear modulus, G_{13}	4.64 GPa
Shear modulus, G_{23}	3.03 GPa
Density, ρ	1.61×10^3 kg/m^3

(continued)

Table 1 (continued)

Strength of lamina	
Tensile strength in the fibre direction, X^T	1447 MPa
Compressive strength in the fibre direction, X^C	1447 MPa
Tensile strength in the direction perpendicular to fibres, Y^T	51.7 MPa
Compressive strength in the direction perpendicular to fibres, Y^C	206 MPa
Longitudinal shear strength, S^L	93 MPa
Transverse shear strength, S^T	103 MPa
Volume fraction	0.66

2.3 *Failure Criteria*

Failure of structural elements occurs if the required margin of safety for the same is up to mark what it is intended for. The failure of the composite material is very complex as it exhibits anisotropic and microscopic nature. However, in the case of laminated composites consisting of several layers of stacking and different fibre orientations and properties possesses more complexity. In different layers of laminate local stresses and strengths may be different so it may be possible that for most of the part overall loading is increased, and hence, stresses and strains may reach their critical values way before than the other layers. In this work, Hashin's failure criteria [15] have been chosen as it is the simplest failure criterion for unidirectional fibre composites. Hashin's criteria are an interacting failure criterion since more than one stress can be evaluated in different failure modes. Generally, Hashin's criteria are applied for two-dimensional classical lamination approaches where stress is calculated on the different plies; however, if required, it may also be extended to three-dimensional problems. In the present work, Hashin's criteria in two-dimensional form have been implemented. It is known that the criteria involve two failure mechanisms in which one is associated with fibre failure and other with the matrix failure. Four different failure modes included in Hashin's criteria [15] are as follows:

- Fibre tension initiation criteria (HSNFTCRT) [1]
- Fibre compression initiation criteria (HSNFCCRT) [1]
- Matrix tension initiation criteria (HSNMTCRT) [1]
- Matrix compression initiation criteria (HSNMCCRT) [1].

Fibre tension ($\sigma_{33} \geq 0$)

$$F_f^t = \left(\frac{\sigma_{11}}{X^T}\right)^2 + \left(\frac{\tau_{12}}{S^L}\right)^2 \tag{1}$$

Fibre compression ($\sigma_{11} < 0$)

$$F_f^C = \left(\frac{\sigma_{11}}{X^C}\right)^2 \tag{2}$$

Matrix tension ($\sigma_{22} \geq 0$)

$$F_m^t = \left(\frac{\sigma_{22}}{Y^T}\right)^2 + \left(\frac{\tau_{12}}{S^L}\right)^2 \tag{3}$$

Matrix compression ($\sigma_{22} < 0$)

$$F_m^C = \left(\frac{\sigma_{22}}{2S^T}\right)^2 + \left[\left(\frac{Y^C}{2S^T}\right)^2 - 1\right]\left(\frac{\sigma_{22}}{Y^C}\right) + \left(\frac{\tau_{12}}{S^L}\right)^2 \tag{4}$$

where F_f^t, F_f^C = failure indexes for fibre in tension and compression, respectively;
F_m^t, F_m^C = failure indexes for matrix in tension and compression, respectively;
X^T, X^C = tensile and compressive strength in fibre direction;
Y^C = compressive strength perpendicular to fibre direction;
S^L, S^T = longitudinal and transverse shear strength.

If the value of the indices is 1 or higher, then it will indicate that Hashin's initiation criteria are met and composite material elements will start to fail.

A value of 1.0 or higher for failure indices indicates that the initiation criterion has been met and elements in composite start to fail.

2.4 Finite Element Model

In this work, different stacking of plies along with the thickness of the plate is studied to optimise the layup of the composite plate to sustain the ballistic impact load. Graphite–epoxy is used as a composite material to validate the numerical results at low-velocity impact. To get accurate results at the contact areas, the mesh should be fine enough to represent the real failure modes during ballistic impact. To reduce the CPU time, coarse mesh has been used for the areas away from the collision region of the laminated plate. ABAQUS/Explicit version 17.0 is used to simulate the impact of plate with a striking object such as a ball or a bullet at different velocities. Continuum shell elements of type SC8R are used for composite plates (lamina). The laminated plate consists of 51,840 elements, and in the element library, Explicit continuum shell with geometric order linear has been chosen. Also, for the element controls second order accuracy with enhanced hour glass control along with the element deletion has been used. However, the impactor is modelled by using rigid elements (R3D4). For the boundary condition purpose, the laminated plate is considered as fixed at its edges and impactor is given the varying velocity. After the damage for the model

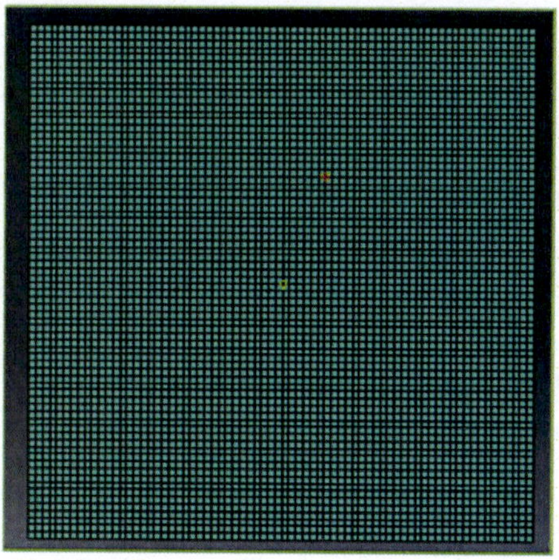

Fig. 3 Meshing details of laminated plate

Fig. 4 Central deflection variation (at thickness $t = 3.81$ mm) under $V = 22.6$ m/s

of low-velocity impact, the maximum stresses and contact force are also calculated (Figs. 3 and 4).

2.5 *Validation of the FE Models*

To validate the results, first the laminated plate made of graphite–epoxy is considered and the impactor is considered as a rigid body. The transverse central deflection of the laminated plate during impact has been considered as the parameters that are to be compared with the reference paper by Ansari et al. [1]. Apart from this, the various parameters such as maximum deflection, contact forces and Mises (equivalent stress) are obtained for the two velocities: one is for low-velocity impact (22 m/s) and the other is for medium-velocity impact (100 m/s). Also, these values are obtained at different stacking sequences of plies and thickness of the plate. For the different sizes of meshing elements, a convergence study has also been done to get the grid

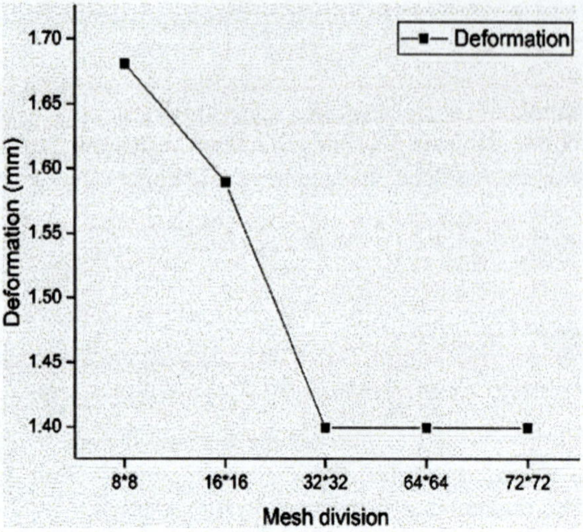

Fig. 5 Central deflection with mesh division

independency of the results such as maximum deformation, maximum contact force and stresses on the laminated plate as shown in Fig. 5.

The maximum deflection on the centre of the plate is validated with the paper [1]. It has been found that the values are matching very close enough to the paper results as shown in Fig. 6, and then, a parametric study has been performed to find out the effect of thickness on deflection, stresses and contact forces.

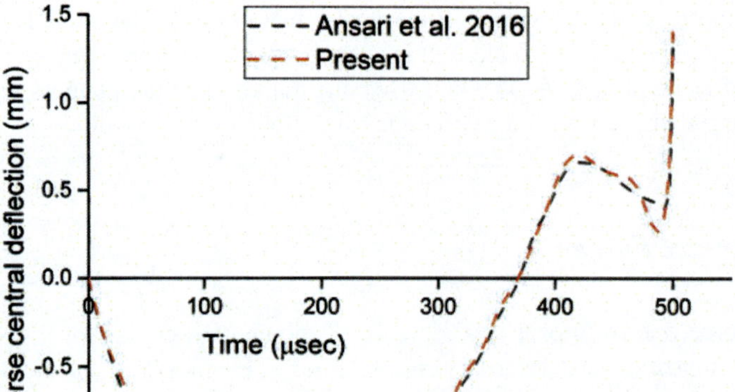

Fig. 6 Comparison of transverse central deflection of cross-ply laminated at (0°/90°/0°) at $V = 22.6$ m/s

3 Results and Discussion

In this section, laminated composite plate subject to impact with two velocities (i.e. low and medium impact velocities) and a stacking sequence is presented. Also, the thickness of the plates has been changed to get the effect of thickness on the other parameters.

3.1 Transverse Deflection

For the best suitability of angle plies to the laminated composite plate. Further, the study has been extended to observe the maximum transverse deflection for the laminated plate and it is found that the maximum deflection is less for angle ply (45°/−45°/45°) as compared to cross-ply (0°/90°/0°) and another angle ply (30°/−30°/30°) under both the low-velocity and medium-velocity impact. Variation of transverse central deflection of the plate is shown in Fig. 6. So, it can be shown from Table 2 that the angle ply laminate (45°/−45°/45°) is stiffer than the other laminates.

3.2 Stresses

As stated by Hashin [15], different types of stresses arise due to impact load on the composite plate which causes damage. Hence by taking different parametric changes in the composite plate are studied. Numerical values of von Mises stress for different ply stacks are presented in Table 2. It has been observed that the stresses are less in the case of angle ply (45°/−45°/45°) when the thickness of the plate, as well as the velocity of the impactor, is changed under the fully clamped condition.

3.3 Contact Force

Contact force due to impact is a measure of momentum change of impactor and clamped composite plate. To observe the variation of contact forces on the composite plate, two different velocities are applied to the impactor (low and medium velocity). It is seen that under low velocity (22.6 m/s), there is a clear difference between the contact forces for different plies and the minimum contact force is observed at the maximum thickness of the plate which is for the angle ply (45°/−45°/45°) as shown in Fig. 6. However, in the case of medium-velocity impact (150 m/s), there is not much change observed in the contact force in both cross-ply and angle ply laminates as shown in Fig. 7. But still, the contact force is slightly less in angle

Table 2 Maximum stresses and deflections of laminated composite plates at low velocity (V = 22.6 m/s) and medium velocity (V = 150 m/s)

Velocity (m/s)	Thickness (mm)	References	Equivalent stress (MPa)	Maximum deformation (mm)
0°/90°/0°				
22.6	3.81	Present	768.9	1.399
		Ansari et al. [1]	565.232	1.32
	5.715	Present	2107	1.399
	7.62	Present	2329	1.3
150	3.81	Present	1146	2.049
	5.715	Present	2188	1.95
	7.62	Present	2421	1.96
45°/−45°/45°				
22.6	3.81		766.8	1.389
	5.715		1989	1.28
	7.62		2047	1.29
150	3.81		1145	2.034
	5.715		1810	1.92
	7.62		2312	1.93
30°/−30°/30°				
22.6	3.81		710.4	1.409
	5.715		2091	1.3
	7.62		2228	1.28
150	3.81		1058	2.064
	5.715		2159	1.95
	7.62		2360	1.92

ply (45°/−45°/45°) as shown in Fig. 8. So, it may be observed that increasing the thickness and the velocities of the impactor leads to an increase in the contact force.

3.4 *Effect on Laminated Plate at Varying Bullet Velocities*

See Figs. 9, 10, 11, 12, 13 and 14 and Tables 3, 4.

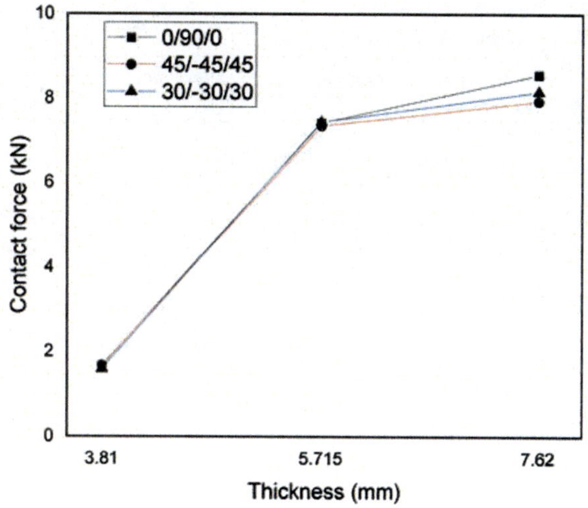

Fig. 7 Contact force variation with thickness at $V = 22.6$ m/s

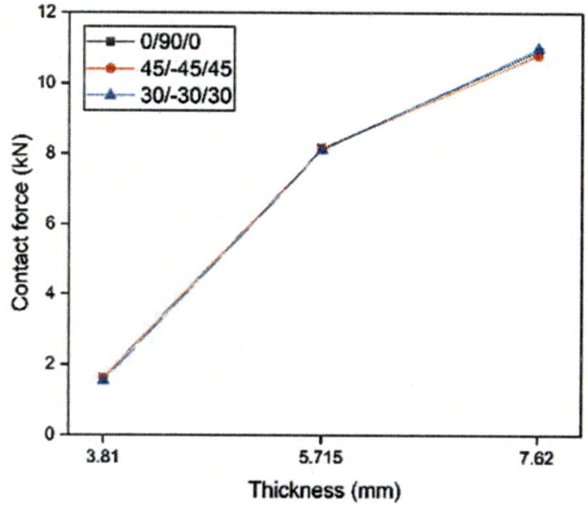

Fig. 8 Contact force variation with thickness at $V = 150$ m/s

3.5 Effect on Laminated Plate When Impacted by Multiple Bullets Simultaneously with Varying Velocities

See Figs. 15, 16, 17, 18 and 19.

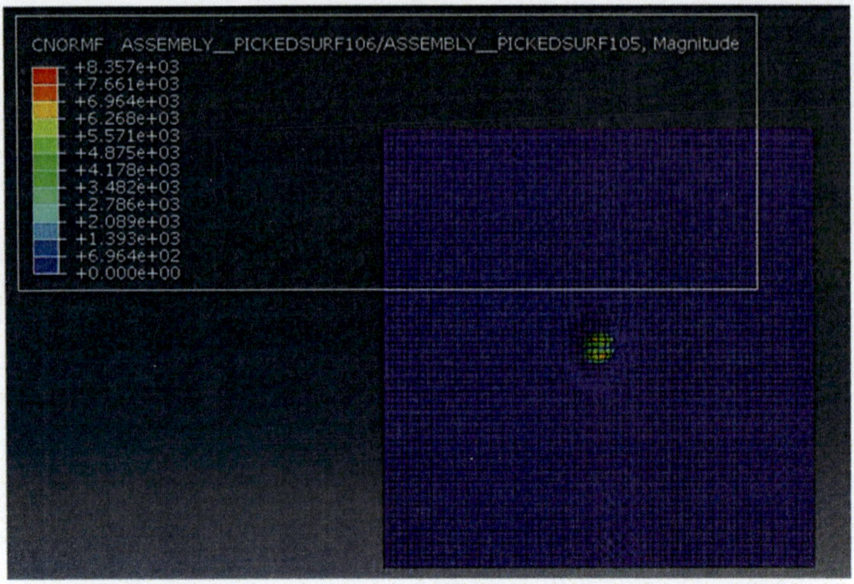

Fig. 9 Maximum contact force at bullet velocity of $V = 500$ m/s

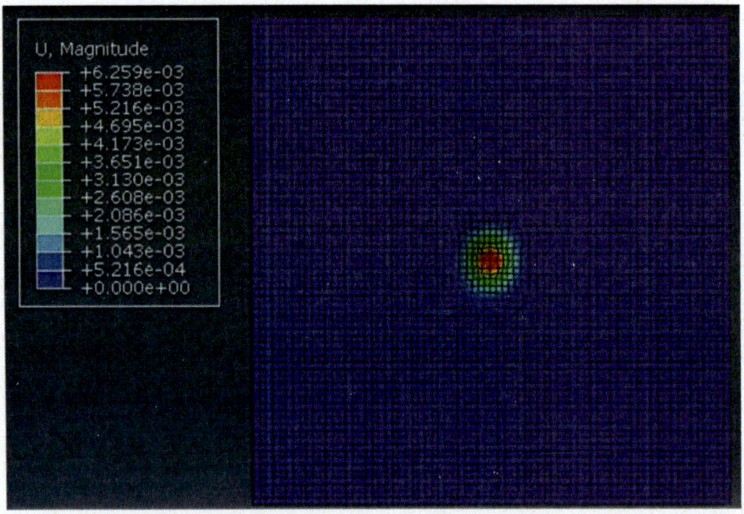

Fig. 10 Maximum central deflection at bullet velocity of $V = 500$ m/s

Fig. 11 Equivalent (von Mises) stress at bullet velocity of $V = 500$ m/s

Fig. 12 Equivalent (von Mises) stress variation for different ply arrangements in high-velocity impacts

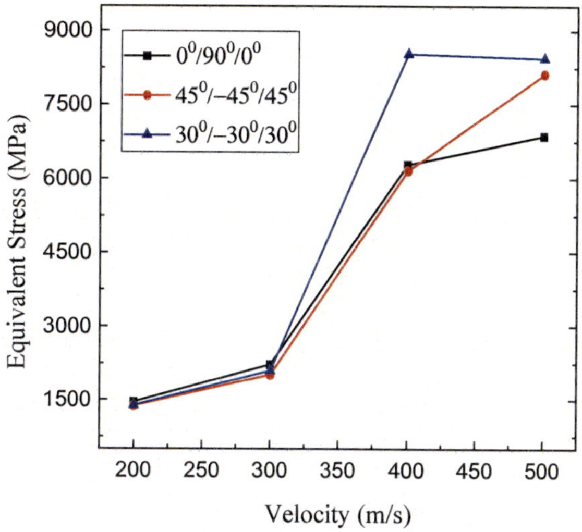

4 Conclusions

Sandwich composite plate subjected to impact load is analysed by changing different parameters of composite as well as the impactor. Different parameters such as maximum deflection, stresses and normal contact force in laminated composite with stacking sequence, different thickness of plate and different velocities of impactor are presented. The nature of the failure of the laminated composite plate has been discussed for different types of laminates and stacking sequences.

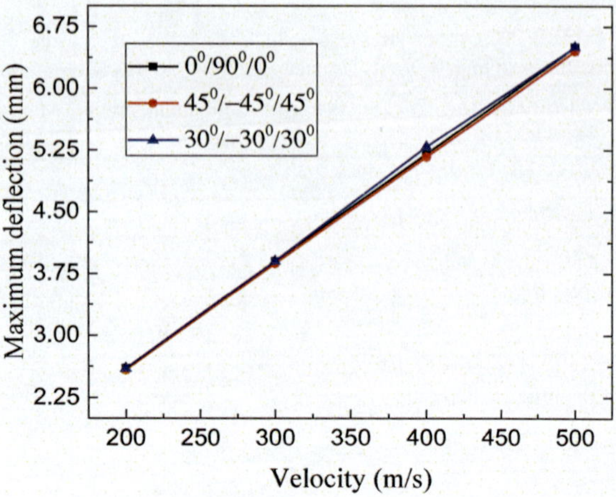

Fig. 13 Maximum central deflection variation for different ply arrangements in high-velocity impacts

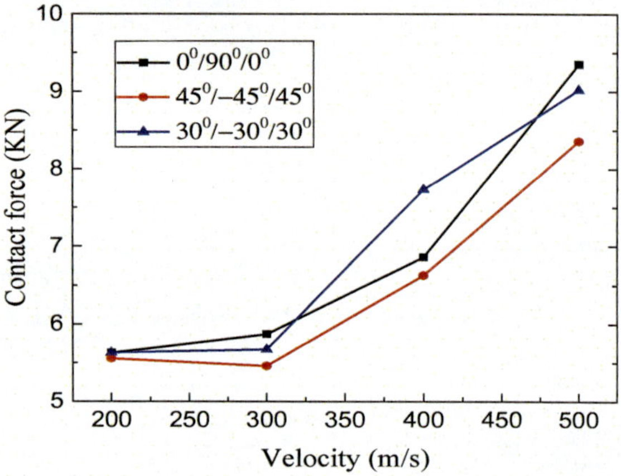

Fig. 14 Contact force variation for different ply arrangements in high-velocity impacts

Some important conclusion can be drawn from the present works which are as follows:

- In low- and medium-velocity deformation, the damage in the composite plate is majorly due to matrix failure in tension and compression.
- From the observation, it is evident that among the three laminates that are cross-ply (0°/90°/0°) and angle plies (45°/–45°/45°) and (30°/–30°/30°), the angle ply laminate (45°/–45°/45°) is stiffer against the deflection as compared to other.
- On account of the contact force, again the angle ply (45°/–45°/45°) is more suitable as it shows lower values under low- and medium-velocity impacts.
- In the range of high-velocity impact, we can see that still, the stiffer behaviour is shown by the plate having angle ply arrangement as (45°/–45°/45°).
- Impact of graphite/epoxy considering multiple spherical bullets hitting on the laminated plate has been studied, and it was found that the angle ply arrangement

Table 3 Maximum stresses, deflection and contact force of laminated composite plates from medium- to high-velocity impact, i.e. 200–500 m/s keeping thickness of plate fixed

Velocity (m/s)	Thickness (mm)	Equivalent stress (MPa)	Maximum deformation (mm)	Contact force (kN)
0°/90°/0°				
200	3.81	1454	2.6	5.628
300		2227	3.9	5.868
400		6286	5.2	6.86
500		6874	6.5	9.351
45°/−45°/45°				
200	3.81	1377	2.58	5.551
300		2018	3.87	5.454
400		6165	5.16	6.623
500		8134	6.45	8.357
30°/−30°/30°				
200	3.81	1381	2.6	5.627
300		2098	3.9	5.668
400		8534	5.287	7.732
500		8452	6.5	9.019

Table 4 Maximum stresses, deflection and contact force of laminated composite plates (graphite/epoxy) from medium- to high-velocity impact by considering multi-shot simultaneously

Velocity (m/s)	Thickness (mm)	Equivalent stress (MPa)	Maximum deformation (mm)	Contact force (kN)
0°/90°/0°				
200	3.81	1639	3.18	5.68
300		2410	4.77	6.576
400		2834	6.36	8.025
500		8272	7.95	10.22
45°/−45°/45°				
200	3.81	1469	2.98	4.699
300		2164	4.47	5.499
400		6790	6.033	7.064
500		7981	7.45	10.04
30°/−30°/30°				
200	3.81	1589	3.1	5.227
300		2220	4.68	5.625
400		8532	6.599	7.73
500		8777	7.75	10.29

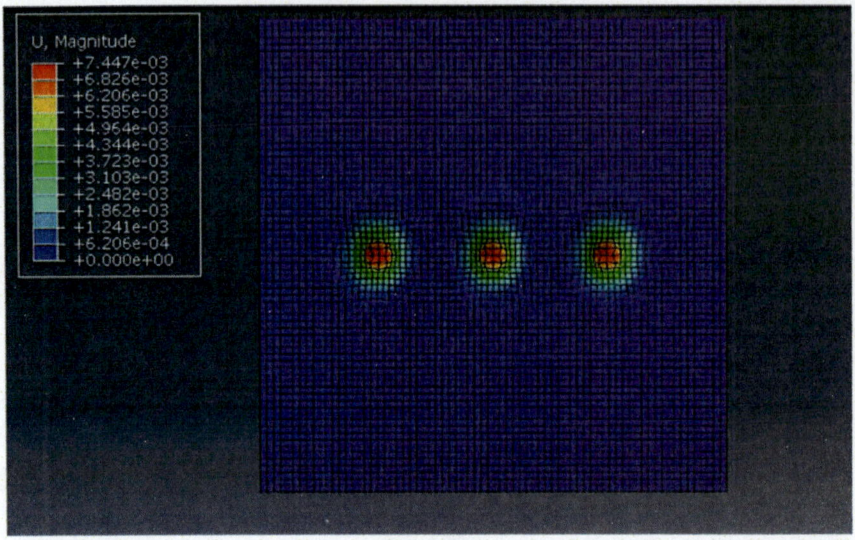

Fig. 15 Maximum central deflection at bullet velocity of $V = 500$ m/s at angle ply arrangement of (45°/−45°/45°)

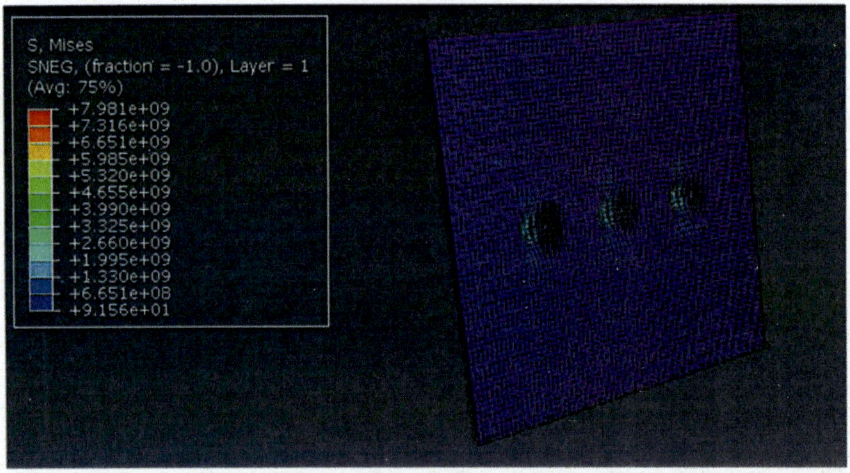

Fig. 16 Equivalent (von Mises) stress at bullet velocity of $V = 500$ m/s at angle ply arrangement of (45°/−45°/45°)

(45°/−45°/45°) for the laminate is more suitable as it possesses lesser values for the deflection, contact forces and maximum stresses. Hence, we can say that for the multiple bullets hitting on the target plate, angle ply arrangement shows more stiffer behaviour towards the deformation of the plate.

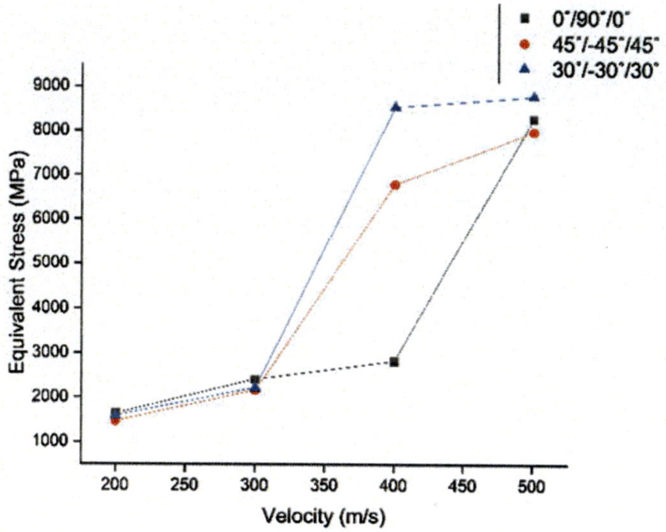

Fig. 17 Equivalent (von Mises) stress variation for different velocity of bullets

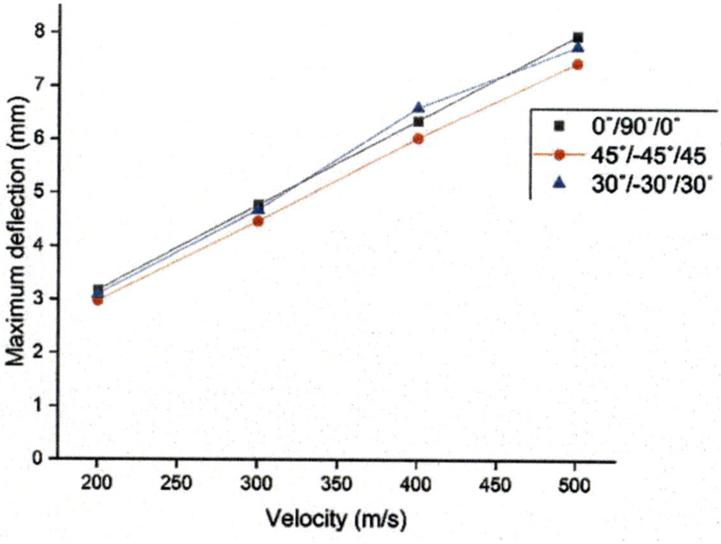

Fig. 18 Maximum central deflection variation for different velocity of bullets

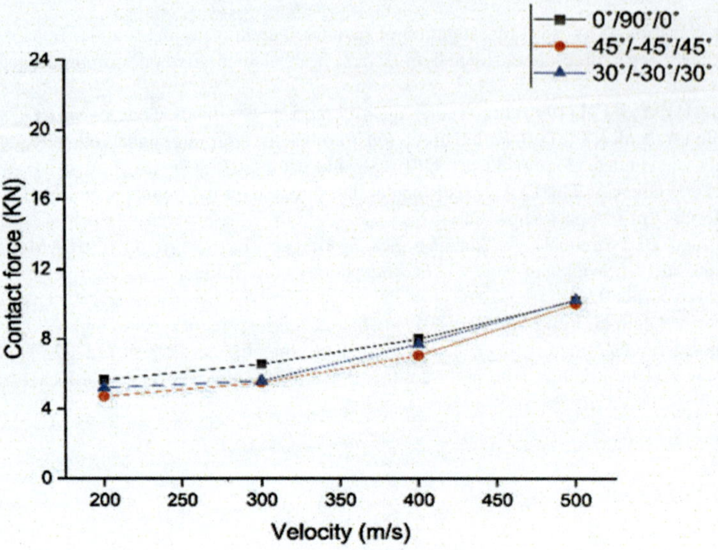

Fig. 19 Maximum contact force variation for different velocity of bullets

Acknowledgements The support provided by Dr. Mohammad Talha (Associate professor school of Engineering IIT Mandi), Computational and design Lab Staff of IIT Mandi: Mohammed Shakir (Ph.D. Scholar), Vikram Singh Chandel (Ph.D. Scholar) and Mohammad Fahed (M.S. Scholar) is gratefully acknowledged.

References

1. Ansari MM, Chakrabarti A, Iqbal MA (2016) Effects of impactor and other geometric parameters on impact behavior of FRP laminated composite plate. Model, Measur Control A 89(1):25–44
2. Krell A, Stranburger E (2007) Hierarchy of key influences on the ballistic strength of opaque and transparent armor. Ceram Eng Sci Proc 28:45e55
3. Tan P (2014) Numerical simulation of the ballistic protection performance of a laminated armor system with pre-existing debonding/delamination. Com Part B Eng 59:50e9
4. Gogotsi GA (2010) Fracture resistance of elastic ceramics in edge flaking: EF base line. J Eur Ceram Soc 30:1223e8
5. Lawn BR (1998) Indentation of ceramics with spheres: a century after Hertz. J Am Ceram Soc 81:1977e94
6. Tiberkak R, Bachene M, Rechak S, Necib B (2008) Damage prediction in composite plates subjected to low velocity impact. Compos Struct 83:73–82
7. Aslan Z, Karakuza R, Okutan B (2003) The response of laminated composite plates under low velocity impact loading. Compos Struct 59:119–127
8. Li CF, Hu N, Chen YG (2002) Fukunaga H, Sekine H. Low velocity impact induced damage of continuous fiber-reinforced composite laminates. Verification and numerical investigation. Compos A 33:1063–1072

9. Sun CT, Liou WJ (1989) Investigation of laminated composite plates under impact dynamic loading using a three-dimensional hybrid stress finite element method. Comput Struct 33:879–884
10. Wang CY, Yew CH (1990) Impact damage in composite laminates. Compos Struct 37:967–982
11. Mishra A, Naik NK (2010) Failure initiation in composite structures under low-velocity impacts analytical studies. Compos Struct 92(2):436–444
12. Evci C, Gülgeç M (2012) An experimental investigation on the impact response of composite materials. Int J Impact Eng 43:40–51
13. Buitrago BL, Santiuste C, Sánchez-Sáez S, Barbero E, Navarro C (2010) Modelling of composite sandwich structures with honeycomb core subjected to high-velocity impact. Compos Struct 92(9):2090–2096
14. Sabet A, Fagih N, Beheshty MH (2011) Effect of reinforcement type on high velocity impact response of GRP plates using a sharp tip impactor. Int J Impact Eng 38:715–722
15. Hashin Z (1980) Failure criteria for unidirectional fiber composites. J Appl Mech 47:329–334

Mathematical Modeling: Magnetic Field Effect on Oscillatory MHD Couette Two Dimensional Flow Regime

Alok Singh, Savita Singh, and Sudhir Kumar Sharma

1 Introduction

Over the past decade or so, the problem of time-dependent oscillatory flow in which fluctuating part of the motion is superimposed on mean steady parts was studied by several research groups [1–4]. The magnetic analog of light hill's investigation led to an increase of the oscillatory flow-based research on small harmonic oscillation in a channel of gap 'h'. The analysis of unsteady flow of the stream with its conjugate complex part concluded that the amplitude of the fluctuating velocity gradient varies with respect to transverse magnetic field [5, 6].

In recent years, oscillatory free convective flow and heat transfer problem in the presence of a magnetic field through a porous composite or medium have attracted the attention of many researchers due to their possible engineering application [7–11]. In real practice, the liquid or gas can be forced through solid capillaries to cool porous structures. Actually, they were utilized to keep a heated body's temperature constant by insulating it. Study of origin of flow through a porous medium is heavily based on Darcy's experimental law. In view of Darcy's law, Yamamoto and Yoshida, for the flow outside a vertex layer, Hooper et al. took into consideration injection and suction flow with convective acceleration via a planar porous wall [12–14]. By accounting for free convection currents, the oscillatory flow past a porous bed has demonstrated the study of two-dimensional flow of viscous fluid through a porous medium bordered by a porous surface subject to a constant suction velocity [15–17].

A smart fluid called a magneto-rheological fluid is typically an oil. The magnetic field of electromagnets can control magneto-rheological fluids filled in dampers for shock observer. A magneto-rheological fluid has three main modes of operation: flow mode, shear mode, and squeeze flow mode. This fluid is traveling between two

A. Singh (✉) · S. Singh · S. K. Sharma
Harcourt Butler Technical University, Kanpur, UP, India
e-mail: alokphy@hbtu.ac.in

A. Kumar et al. (eds.), *Recent Advances in Manufacturing and Thermal Engineering*,
Lecture Notes in Mechanical Engineering, https://doi.org/10.1007/978-981-19-8517-1_33

parallel plates, as well as between two plates that are moving perpendicular to each other's planes and have perpendicular magnetic fields [18].

The objective of this presented research work is the development of mathematical modeling to fluid flow in order to understand and predict the behavior of magneto-rheological fluid. In this presented research work, a mathematical modeling with quantitative analysis of oscillatory MHD Couette flow between two parallel plates (upper plate is moving relative to fixed lower plate) has been performed in terms of constitutive parameters using perturbation theory. The dependency of constitutive parameters and its variations is mathematical calculated and analyzed.

2 Mathematical Analysis

Consider a two-dimensional (2D) oscillating incompressible fluid with sturdy qualities where a magnetic field is given to the fluid's surface plates typically. The top surface plates remain stationary, while the other plates move at a slow speed, as shown in Fig. 1. The constraint in this scenario is that the velocity component normal to the wall is zero, and the flow is independent of the distance along the wall.

The momentum of fluid flow or inertial force of fluid can be neglected in comparison to the viscous force at low speed fluid in the Navier–Stokes expression of motion. The Couette flow motion is introduced here due to the movement of the upper plate which can be represented as [19],

$$\frac{\partial u}{\partial t} = -\frac{1}{\rho}\frac{\partial p}{\partial x} + \vartheta\frac{\partial^2 u}{\partial y^2} - \frac{\sigma B_0^2}{\rho}u \tag{1}$$

$$0 = -\frac{1}{\rho}\frac{\partial p}{\partial y} \tag{2}$$

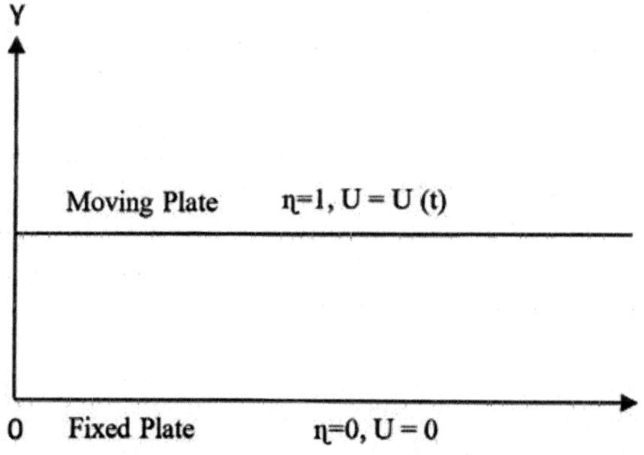

Fig. 1 Systematic of an oscillating incompressible fluid flow in 2D

The fixed plate is along x-axis and y-axis which is normal to it. Plates are at distance 'h' apart. The expression of unsteady velocity between the plates is represented by [19],

$$U(t) = U_h\left[1 + \frac{t}{2}(e^{iwt} + e^{-iwt})\right] \tag{3}$$

where ω is the frequency, and U_h and t are constants.

We considered the solution of Eq. (1) as [19],

$$u = U_h\left[u_0(\eta) + \frac{t}{2}\{u_1(\eta)e^{i\omega t} + \overline{u_1}(\eta)e^{-i\omega t}\}\right] \tag{4}$$

where $\eta = \frac{y}{h}$. On substituting Eqs. (3) and (4) in Eq. (1) and equating steady and unsteady parts, we have [19],

$$u_0'' = 0 \tag{5}$$

$$u_1'' - (m + ik)u_1 = -(m + ik) \tag{6}$$

The boundary conditions are [19],

$$\eta = 0, u_0 = u_1 = 0; \eta = 1, u_0 = u_1 = 1 \tag{7}$$

where $m = \frac{\sigma B_0^2}{\rho}\frac{h^2}{\upsilon}$ and $k = \frac{\omega h^2}{\upsilon}$ are magnetic and frequency parameters, respectively.

Equations (5) and (6) are simple linear differential equations, whose complementary and particular solutions are constituted as:

The solution of Eq. (5) is

$$u_0 = \eta \tag{8}$$

The solution of Eq. (6) is [19],

$$u_1 = 1 - \cosh(m + ik)^{\frac{1}{2}}\eta + \coth(m + ik)^{\frac{1}{2}} \sinh(m + ik)^{\frac{1}{2}}\eta \tag{9}$$

The magnitude of the fluctuating velocity gradient at lower wall is [19],

$$|u_1'| = \frac{2}{(\cosh 2\theta_1 - \cosh 2\theta_2)}(\theta_1^2 + \theta_2^2)^{\frac{1}{2}}(\sinh^2 2\theta_1 + \sinh^2 2\theta_2)^{\frac{1}{2}} \tag{10}$$

The magnitude of the fluctuating velocity gradient at mid plane is [19],

$$|u_1'| = \frac{2}{(\cosh 2\theta_1 - \cosh 2\theta_2)}(\theta_1^2 + \theta_2^2)^{\frac{1}{2}}(\sinh^2 2\theta_1 + \sinh^2 2\theta_2)^{\frac{1}{2}}$$

$$\times \left(\cosh^2 \frac{\theta_1}{2} \cos^2 \frac{\theta_2}{2} + \sinh^2 \frac{\theta_1}{2} \sin^2 \frac{\theta_2}{2} \right)^{\frac{1}{2}} \quad (11)$$

The magnitude of the fluctuating velocity gradient at upper wall is [19],

$$\left| u_1' \right| = \frac{2}{(\cosh 2\theta_1 - \cosh 2\theta_2)} (\theta_1^2 + \theta_2^2)^{\frac{1}{2}} (\sinh^2 \theta_1 \cos^2 \theta_2 + \cosh^2 \theta_1 \sin^2 \theta_2)^{\frac{1}{2}} \quad (12)$$

where $\theta_1 = (m^2 + k^2)^{\frac{1}{4}} \cos \alpha$, $\theta_2 = (m^2 + k^2)^{\frac{1}{4}} \sin \alpha$ and $\alpha = \frac{1}{2} \tan^{-1} \frac{k}{m}$.

The shear stress at $y = 0$ is given by

$$T_w = \mu \left(\frac{\partial u}{\partial y} \right)_{y=0} \quad (13)$$

$$T_w = \frac{\mu U_h}{h} \left\{ 1 + \frac{t}{2} \left(A e^{i\omega t} + \overline{A} e^{-i\omega t} \right) \right\} \quad (14)$$

where $A = (m + ik)^{\frac{1}{2}} \coth(m + ik)^{\frac{1}{2}}$.

An expression of the amplitude of shear stress of lower wall is given by

$$|A| = \frac{2}{(\cosh 2\theta_1 - \cosh 2\theta_2)^{\frac{1}{2}}} (\theta_1^2 + \theta_2^2)^{\frac{1}{2}} (\sinh^2 2\theta_1 + \sinh^2 2\theta_2)^{\frac{1}{2}} \quad (15)$$

An expression of the phase angle of shear stress of lower wall is given by

$$\psi = \tan^{-1} \left[\frac{\theta_2 \sin 2\theta_1 - \theta_1 \sinh 2\theta_2}{\theta_1 \sin 2\theta_1 - \theta_2 \sinh 2\theta_2} \right] \quad (16)$$

2.1 *Temperature Field*

Under the flow conditions, the equations for the temperature distribution with negligible heat generation due to Joule's effect are [19],

$$\frac{\partial T}{\partial t} + u \frac{\partial T}{\partial x} = K \left(\frac{\partial^2 T}{\partial x^2} + \frac{\partial^2 T}{\partial y^2} \right) + \frac{\upsilon}{c} \left(\frac{\partial u}{\partial y} \right)^2 \quad (17)$$

In this case, x is the distance beside the wall, T is the temperature, K is the thermal diffusivity, and C is the specific heat.

The following boundary conditions were used to solve equation,

$$\mathrm{T} = \mathrm{T_w} \text{ or } \frac{\partial T}{\partial y} = 0 \text{ at } y = 0$$
$$T = T_h \text{ at } y = h \tag{18}$$

The above obtained solutions are independent of x with applied boundary conditions. In the present case, the heat transferred due to convection is negligible. Temperature distribution is caused only due to conduction and viscous dissipation.

Thus, we have

$$\frac{\partial T}{\partial t} = K\frac{\partial^2 T}{\partial y^2} + \frac{\nu}{c}\left(\frac{\partial u}{\partial y}\right)^2 \tag{19}$$

Looking for the solution of Eq. (19) in the form as,

$$T(\eta) = T_0(\eta) + \frac{1}{2}\varepsilon\left[T_1(\eta)e^{i\omega t} + \overline{T_1}(\eta)e^{-i\omega t}\right] + \frac{1}{2}\varepsilon^2\left[T_2(\eta)e^{2i\omega t} + \overline{T_2}(\eta)e^{-2i\omega t}\right] \tag{20}$$

Applying harmonic coefficients equating to zero, we have

$$T_0'' = -\frac{P_r U_h^2}{c} \tag{21}$$

$$T_1'' - iP_r k T_1 = -\left(\frac{P_r U_h^2}{c}\right)u_0' u_1' \tag{22}$$

where $P_r = \frac{\mu c}{k}$ is the Prandtle number, $k = \frac{\omega h^2}{\nu}$ is frequency parameter.

The boundary conditions reduce to as

$$T_0 = T_\omega,\ T_1 = T_2 = 0 \text{ at } \eta = 0$$
$$T_0 = T_h,\ T_1 = T_2 = 0 \text{ at } \eta = 1 \tag{23}$$

The solution of Eq. (22) and (23) is

$$\frac{T_0 - T_h}{T_\omega - T_h} = -P_r E\frac{\eta}{2}(\eta - 1) + \eta + 1 \tag{24}$$

$$\frac{T_1}{T_\omega - T_h} = P_r E\frac{l}{(l^2 - I^2)}[\sinh l\eta(\mathrm{cosechIcoshl} + 1)$$
$$+ \coth l(\mathrm{cosechIcoshI}(1 - \eta) + \cosh l\eta)] \tag{25}$$

where $l^2 = m + ik$, $I^2 = iP_r k$ and $E = \frac{U_h^2}{c(T_\omega - T_h)}$ is Eckert number.

The heat transfer rate is given as

$$\frac{T_1'(\eta)}{T_\omega - T_h} = P_r E \frac{l}{I(l^2 - I^2)}\left[\frac{\cosh l\eta}{l}(\text{cosechIcoshl} + 1) - \coth l\left(\text{cosechIcoshI}\frac{(1-\eta)}{\text{I}} + \frac{\sinh l\eta}{l}\right)\right] \quad (26)$$

The heat transfer rate at lower plate is given by

$$\frac{T_1'(0)}{T_\omega - T_h} = \frac{P_r E}{I(l^2 - I^2)}[I + \text{cosechIcoshl}(I - l\text{cosechloshI})] \quad (27)$$

The heat transfer rate at upper plate is given by

$$\frac{T_1'(1)}{T_\omega - T_h} = \frac{P_r E}{I(l^2 - I^2)} \frac{\coth l}{2\sinh l}[I \sinh 2l - 2l] \quad (28)$$

3 Results and Discussion

The significance of magnetic field is that it controls the flow of fluid between two parallel plates, while frequency parameter of pulsating flow perturbs the other constitutive parameters. The magnitude dispersal of velocity gradient variation $|u_1'|$ is shown in Table 1 and Fig. 2 for several values of m and at fixed value of frequency parameter. Variation of the phase lag in the shear stress and amplitude at lower wall with respect to magnetic field parameter for fixed value of frequency parameter has been shown in Table 2 and variation of the amplitude and phase lag of the shear stress shown in Fig. 3. The amplitude of shear stress decreases rapidly between $m = 1$ and $m = 2$, then after it becomes constant with magnetic field parameter.

Several values of frequency parameter for lower wall with fixed value of magnetic field parameter have been shown in Table 3. We observe that amplitude of the shear stress at lower plate increases with increasing frequency parameter for fixed magnetic field parameter, while phase lag increases with magnetic field parameter increases as shown in Figs. 5 and 4, respectively. Moreover, we observe that the phase lag decreases with increasing frequency parameter as shown in Fig. 6.

The amplitude and phase lag of heat transfer rate at lower plate increase with increasing magnetic field parameter. However, the amplitude and phase lag of rate of heat transfer at upper plate are same at lower plate. These variations are shown in Tables 4 and 5.

Figure 2 illustrates that the higher rate of increase in amplitude of fluctuating velocity gradient with increasing magnetic field parameter at lower plate than upper plate at fixed frequency parameter. A significant increase in velocity gradient occurs above magnetic field parameter $m = 4$.

Table 1 Variation of fluctuating velocity gradient with η

S. No.		$\lvert u_1' \rvert$		
	m	$\eta = 0.0$	$\eta = 0.5$	$\eta = 1.0$
1	0	0.09	0.25	0.20
2	1	2.70	0.61	0.52
3	2	3.33	0.94	0.73
4	3	3.79	1.15	0.83
5	4	4.20	1.98	0.95
6	5	5.01	2.32	1.02
7	6	5.98	3.01	2.31
8	7	6.02	3.99	3.12
9	8	6.99	4.20	4.05
10	9	7.73	5.39	5.13
11	10	8.08	6.17	6.09

For fixed value $K = 1$

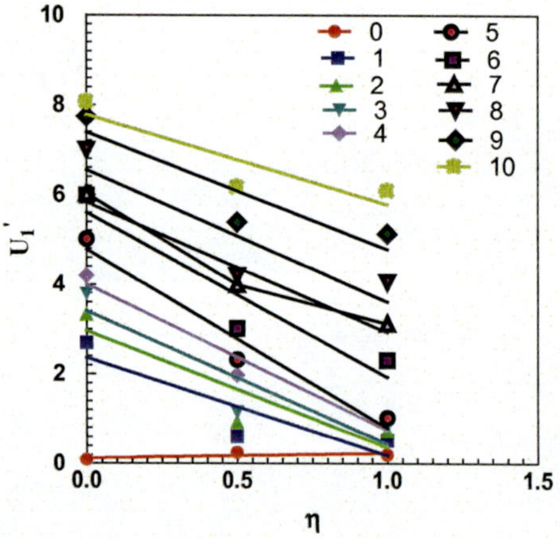

Fig. 2 Distribution of fluctuating velocity gradient verses η

Table 2 Variation of phase lag in the shear stress and amplitude at lower wall for fixed value of m = 2

S. No.	K	$\lvert A \rvert$	Ψ
1	1	0.3880	72.8051
2	2	1.5694	60.6989
3	3	2.3435	52.8321
4	4	4.3413	49.3235
5	5	5.8898	40.3135

For fixed value $m = 2$

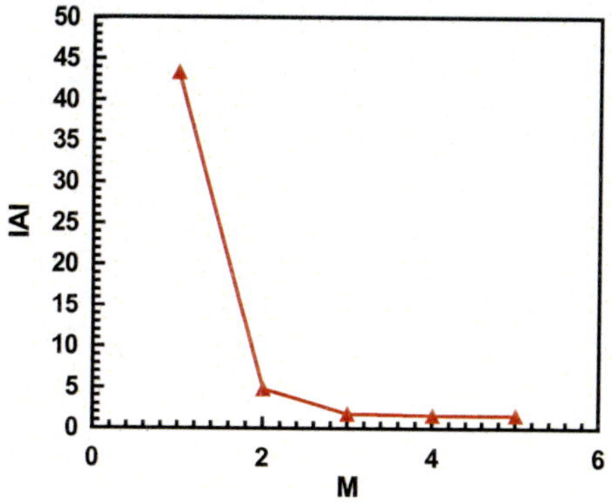

Fig. 3 Distribution of amplitudes of shear stress with magnetic field parameter

Table 3 Variation of phase lag in the shear stress and amplitude at lower wall for fixed value of K = 2

S. No.	m	$\frac{T_1'(0)}{T_w - T_h}$	$\lvert \beta_1 \rvert$	$\emptyset_1$
1	1	9.2631 – i 1.5032	9.3842	10.2416
2	5	5.2140 + i 10.5660	11.7824	70.8167
3	10	1.2315 + i 15.3216	15.3710	94.8940

For fixed value, $K = 2$

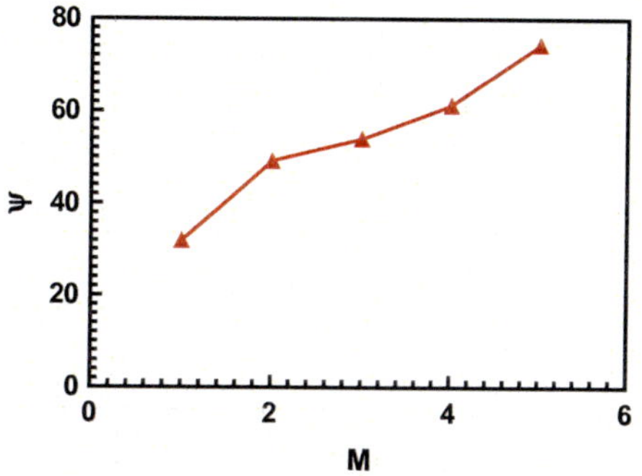

Fig. 4 Variations of phase lag of shear stress with magnetic field parameter

Figure 3 shows that the amplitude of shear stress decreases rapidly with increasing magnetic field parameter between 0 and 2 value, but beyond the value, $m = 3$ becomes constant.

Figure 4 indicates almost linear increment with increasing magnetic field parameter.

Figure 5 illustrates that amplitude of shear stress with increasing frequency parameter beyond the value of unity.

Figure 6 illustrates a linear decrement of the phase lag with increasing frequency parameter.

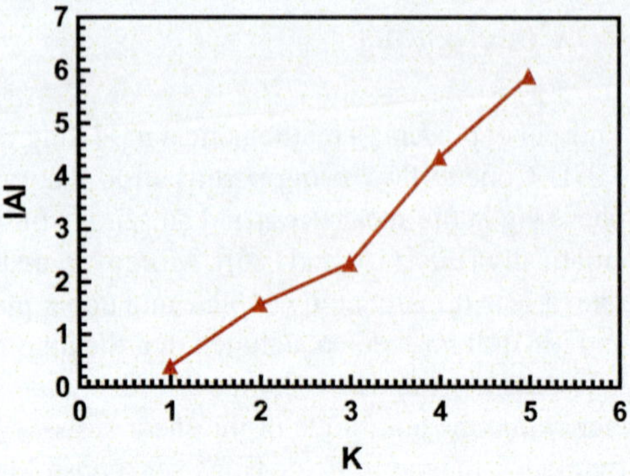

Fig. 5 Variations of amplitude of shear stress with frequency parameter

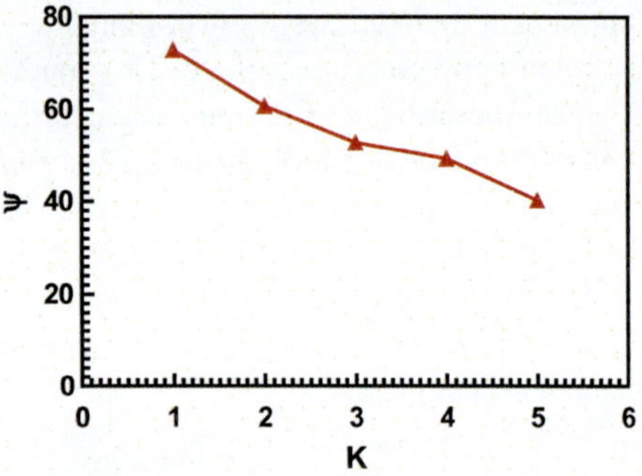

Fig. 6 Variations of phase lag of shear stress with frequency parameter

Table 4 Heat transfer variation at lower wall with respect to magnetic field parameter *m*

S. No.	*m*	|A|	Ψ
1	1	43.2995	31.9641
2	2	4.8000	49.2559
3	3	1.8000	53.9830
4	4	1.6000	61.3257
5	5	1.6000	74.5323

$\mathrm{Pr} = 2, E = 10, K = 5$

Table 5 Heat transfer variation at upper wall with respect to magnetic field parameter *m*

S. No.	m	$\frac{T_1'(1)}{T_w - T_h}$	$\|\beta_2\|$	$\emptyset_2$
1	1	7.6567 – i 0.3548	7.6649	2.9478
2	5	10.5204 + i 8.6241	13.6034	43.7146
3	10	14.5109 + i 13.0990	19.5486	46.7473

$\mathrm{Pr} = 2, E = 10, K = 5$

4 Conclusions

This paper presents a mathematical modeling with quantitative analysis of oscillatory MHD Couette flow using perturbation theory. A set of modeled expression of the phase lag in the shear stress and amplitude of lower wall have been developed mathematically [Eqs. (15) and (16)]. Moreover, mathematical modeled expression for the rate of heat transfer at lower plate and upper plate has been also developed [Eqs. (27) and (28)]. It has been concluded that the amplitude of shear stress decreases rapidly between $m = 1$ and $m = 2$, then after it becomes constant with magnetic field parameter, while the amplitude of the shear stress at lower plate increases with increasing frequency magnetic field. However, phase lag increases with increasing magnetic field, but it decreases with increasing frequency parameter. The amplitude and phase lag of heat transfer rate at lower plate increase by increasing magnetic field which is equal at upper and lower plate. This mathematical modeling with quantitative analysis can be utilized in various engineering fields for instance magneto-rheological dampers, magneto-rheological finishing in optics, enhanced body armor in military and defense, automotive, aerospace, human prosthesis, petroleum industry, polymer technologies, and magneto-hydrodynamic power pumps.

In future, these developed mathematical models can be analyzed for validation in nano-fluid dynamics, polymer fluid dynamics, paraffin oil fluid dynamics, oil and paint technology, etc.

References

1. Lighthill MJ (1954) The response of laminar skin friction and heat transfer to fluctuations in the stream velocity. Proc R Soc Lond A. https://doi.org/10.1098/rspa.1954.0137
2. Scotti A, Piomelli U (2001) Numerical simulation of pulsating turbulent channel flow. Phys Fluids 10(1063/1):1359766
3. Soundalgekar VM, Wavre PD, Takhar HS (2001) Effects of free convection currents on the oscillatory flow of an electrically conducting fluid past a stationary or steadily moving vertical plate with variable suction and constant heat flux. Int J Appl Mech Eng 6(3):639–658
4. Riley N (2001) Steady streaming. Annu Rev Fluid Mech 33:43
5. Sidlof P, Zorner S, Huppe A (2015) A hybrid approach to the computational aeroacoustics of human voice production. Biomech Model Mechanobiol. https://doi.org/10.1007/s10237-014-0617-1
6. Hanratty TJ, Campbell JA (2017) Measurement of wall shear stress. In: Fluid mechanics measurements. Routledge, pp 575–648
7. Varsakelis C, Papalexandris MV (2014) A numerical method for two-phase flows of dense granular mixtures. J Comput Phys. https://doi.org/10.1016/j.jcp.2013.10.023
8. Noranuar WNIN, Mohamad AQ, Shafie S, Khan I, Jiann LY, Ilias MR (2021) Non-coaxial rotation flow of MHD Casson nanofluid carbon nanotubes past a moving disk with porosity effect. Ain Shams Eng J. https://doi.org/10.1016/j.asej.2021.03.011
9. Mhamed B, Sidik NAC, Yazid MNAWM, Mamat R, Najafi G, Kefaati GHR (2016) A review on why researchers apply external magnetic field on nanofluids. Int Commun Heat Mass Transfer. https://doi.org/10.1016/j.icheatmasstransfer.2016.08.023
10. Singh S, Singh A, Sharma SK (2017) Analytical modeling for mechanical strength prediction with raman spectroscopy and fractured surface morphology of novel coconut shell powder

reinforced: epoxy composites. J Inst Eng (India): Ser C. https://doi.org/10.1007/s40032-016-0254-9
11. Singh S, Singh A, Sharma S K (2020) Analytical Prediction Models for Density, Thermal Conductivity and Mechanical Strength of Micro-scaled Areca Nut Powder-Reinforced Epoxy Composites. Journal of The Institution of Engineers (India): Series C, doi: https://doi.org/10.1007/s40032-019-00535-9
12. Yamamoto K, Yoshida ZI (1974) Flow through a porous wall with convective acceleration. J Phys Soc Jpn, pp 774–779. https://doi.org/10.1143/JPSJ.37.774
13. Yamamoto K, Iwamura N (1976) Flow with convective acceleration through a porous medium. J Eng Math. https://doi.org/10.1007/BF01535425
14. Zeghari K, Louahlia H, Masson Le S (2019) Experimental investigation of flat porous heat pipe for cooling TV box electronic chips. Appl Therm Eng https://doi.org/10.1016/j.applthermaleng.2019.114267
15. Ahmed S, Kalita K (2013) Analytical and numerical study for MHD radiating flow over an infinite vertical surface bounded by a porous medium in presence of chemical reaction. J Appl Fluid Mech https://doi.org/10.36884/jafm.6.04.21201
16. Sharma BK, Yadav K, Mishra NK, Chaudhary RC (2012) Soret and Dufour effects on unsteady MHD mixed convection flow past a radiative vertical porous plate embedded in a porous medium with chemical reaction. Sci Res Publishing https://doi.org/10.4236/am.2012.37105
17. Sharma PK (2005) Simultaneous thermal and mass diffusion on three-dimensional mixed convection flow through a porous medium. J Porous Media. https://doi.org/10.1615/JPorMedia.v8.i4.70
18. Hua D, Liu X, Li Z, Fracz P, Hnydiuk A, Stefan Li Z (2021) A review on structural configurations of magnetorheological fluid based devices reported in 2018–2020. Front Mater https://doi.org/10.3389/fmats.2021.640102
19. Batchelor CK, Batchelor GK (2000) An introduction to fluid dynamics. Cambridge University Press

Investigation of Novel Epoxy/Mixed Plastic Waste Char Nano-composites: Mechanical, Thermal, and Morphological Properties

Parul Dwivedi, Ashwani Kumar Rathore, Deepak Srivastava, Jitendra Bhaskar, Kavita Srivastava, Deepa Agrahari, and Shilpi Tiwari

1 Introduction

Plastic garbage has increased dramatically all over the world as a result of economic development and changing patterns of production and consumption. Plastic trash disposal into open areas or dumping yards is harmful to the environment and poses a health risk to human beings. This issue has attracted many researchers to work in this field and recommend possible solutions. Completely eliminating plastics from our day-to-day lives is not possible because of their various properties and the vast range of products already available on the market. When elimination is not possible, search for the recycling techniques that have been started and suggested by the scientists. Education is critical for reducing plastic waste because it can improve people's understanding, attitudes, and actions about plastic waste management [1]. However, arguably the most pressing problem, which runs through this whole world, is that our existing practices are unsustainable. Around 4% of global oil production is utilised as a feedstock for the production of plastics, with a similar amount of energy consumed in the process. Despite this, more than a third of current output is used to create packaging, which is subsequently quickly discarded. This linear use of hydrocarbons via packaging and other short-lived applications of plastic is simply not sustainable, given our diminishing fossil fuel supplies and finite capacity for waste disposal to landfill. There are options, such as material reduction, design for end-of-life recyclability, enhanced recycling capacity, bio-based feedstock development, and litter prevention techniques [2].

If we talk about the numbers, then in 2019, India generated roughly 9.4 million tonnes of plastic garbage, compared to a global creation rate of more than 380 million tonnes per year. As a result, India's contribution to global plastic garbage output

P. Dwivedi · A. K. Rathore (✉) · D. Srivastava · J. Bhaskar · K. Srivastava · D. Agrahari · S. Tiwari
Chemical Engineering Department, Harcourt Butler Technical University, Kanpur, India
e-mail: akrathore75@gmail.com

A. Kumar et al. (eds.), *Recent Advances in Manufacturing and Thermal Engineering*, Lecture Notes in Mechanical Engineering, https://doi.org/10.1007/978-981-19-8517-1_34

was roughly 3.1%. The packaging industry consumes 42% of all plastics produced globally, while the construction industry consumes 17%. In India, the packaging industry consumes 35% while the construction industry consumes 23%. Bioplastic production accounts for 1% of the annual 300 million tonnes of conventional plastics produced [3].

Following waste avoidance, the highest focus is given to the recovery of materials for recycling and composting. Recycling differs from reuse and remanufacturing in that it refers to the use of collected waste materials as a raw material in the production of a new compound. It allows valuable resources to be reclaimed and landfill waste to be reduced. Plastic recycling may or may not be possible depending on the qualities of the plastics, but because of the possibility of contamination, additional attention is required [4].

2 Literature Review

One of the most important categories of thermosetting polymers is epoxy resins (ERs). They can react with a variety of curing agents to produce solid, cross-linked materials with high strength, excellent adhesion to a variety of substrates, and improved heat and chemical resistance. As a result of these facts, a wide range of epoxy resin applications have emerged. They've been used as sealants, adhesives, protective coatings, and dielectric materials with great success [5]. However, their brittleness in the cure stage has shown severe disadvantages, limiting their use in many industries that demand much higher ductility from applied materials. As a result, ER hardening has been a difficult problem for decades [6]. Epoxy-based composites are used in aerospace, automotive, and marine applications. Epoxy is used as a filler or a polymer matrix in composite applications. Filler refers to the epoxy modification of the fibre. Epoxy as a polymer matrix is enhanced by natural and synthetic fibres. The process of fabrication as well as the performance of created epoxy-based composites (mechanical and thermal properties, for example) were investigated [7].

It was created a one-of-a-kind nanocomposite with phase separation characteristics, enhanced strength, and reduced stiffness. When hexahydrophthalic acid (anhydride-cured bisphenol A) diglycidyl ether (BADGE) was used as a matrix material for the manufacture of the epoxy nanocomposite, these features emerged [8]. Plastic [polyethylene terephthalate (PET)] waste was used as a raw material to make diglycidyl ether of bisphenol A-type epoxy composite materials [9]. Wood shaving char and pine cone char (PCC), both obtained from natural resources, were also used to prepare the composites. Plastic waste char (PWC) and pine cone char (PCC) can significantly improve the thermal stability of neat epoxy resin at temperatures above 300 °C, according to thermogravimetric analysis [9]. The pyrolysis of the liquid, gas, and solid fractions (char) of high-density polyethylene (HDPE) and low-density polyethylene (LDPE) plastic wastes in the temperature range of 300–700 °C was investigated, and the pyrolysis char was employed as an additive in the

production of epoxy composite. The LDPE and HDPE char additions increased the hardness of neat epoxy from 79.8 Shore D to 86.3 and 87 Shore D, respectively. The paper proposes a novel method for reusing HDPE and LDPE-type plastic trash [10].

Polypropylene (PP) plastic waste pyrolysis was studied [11]. For the preparation of epoxy composite, solid pyrolysis product (char) was used as a filler material. The ultimate pyrolysis temperatures were chosen to be 300, 400, 500, 600, and 700 °C. The most optimal behaviour was seen in an epoxy composite made from PP char heated to 300 °C. The investigations by Sogancioglu et al. [11] were carried out on char obtained from the pyrolysis of metallised food packaging plastic waste (MFPWs), which are the most complex type of plastic waste and have high aluminium content in their char. Vacuum-assisted resin transfer and curing methods were used to create the composite panels. When compared to a pure sample, 0.75 wt% of food-based char (FBC) improved panel modulus of elasticity by 22%. Furthermore, with the same concentration of FBCs, thermal stability and energy impact absorption were increased by 21% (in the major decomposition zone) and 38%, respectively. The bio-based epoxy resin was synthesised using the esterification mechanism of sebacic acid and epichlorohydrin and modified by the coconut shell and converted to an epoxy composite at the end. They have claimed that the composite showed more elastic moduli compared to prepared epoxy resin [12].

In this study, a novel epoxy composite reinforced with mixed plastic pyrolyzed char has been prepared. Also, the mechanical properties of the synthesised composite have been studied.

3 Research Gap

This research area has not been explored much by the researchers as there is very limited work present. The epoxy-based nano-composite was prepared first to improve the strength and other mechanical properties of the composites, which will help the aerospace, automotive, and many more industries for their growth. Scientists used biomass as a filler for the epoxy nano-composite preparation by scientists and found excellent results [13, 14]. Similar to biomass filling into the epoxy, plastic waste was also used for the nano-composite production and found tremendous results.

Researchers in previous studies have utilised one type of plastic only as a filler for the epoxy nano-composite synthesis [9–11]. Municipal solid waste contains mixed plastic waste materials, and selection of one type of plastic char will take an extra step and increase the cost of production due to the addition of the segregation cost of the particular plastic from the mixed plastic waste.

The present work focuses on the production of mixed plastic char-based nano-composite, which excludes the segregation step in the synthesis process and reduces the cost of production. Study of the mechanical properties and thermal stability of the epoxy polymer when reinforced with mixed plastic char was not performed elsewhere and it is the need of the hour.

4 Objectives

The objective of this research work is described as follows:

To reuse the plastic waste and reduce the environment's pollution.

Synthesis of the epoxy nano-composite with mixed plastic waste reinforcement.

Reduction in the cost of segregation from municipal solid waste for the production of epoxy-nano-composites.

Prepare a nano-composite with improved mechanical and thermal properties.

5 Material and Method

The mixed plastic waste-based char was purchased from a local industry. Epoxy and hardener (B11 LAPOX) were purchased from ATUL India limited and maintained epoxy to hardness ratio 4:1. Curing agent Triethylenetetramine $[CH_2NHCH_2CH_2NH_2]_2$ and Chemicals Diglycidyl ether, Acetone and bisphenol-A (2 ethylene oxide) were used in the study.

6 Specimen Preparation

A mixed plastic char-based nano-composite was prepared using a standard method. Mixed plastic char was measured using weighing balance and mixed with the appropriate weight fraction of epoxy and hardener. The epoxy-to-hardener ratio was maintained at 4:1 with varying char weight fractions. The mixture was then ultrasonicated by which it was properly mixed, and a uniform mixture was formed. A small air bubble was detected on the surface of the uniformly mixed liquid mixture, which was removed by putting the beaker under a vacuum oven for half an hour. This process will result in an air-free solution. A steel mould pained with silicon wax was taken, and the final solution was poured into it. It was left for 72 h to set. Final samples with varying char concentrations (Table 1) were then unmolded and taken for the characterisation and testing as given in Fig. 1.

Table 1 Compositions for preparation of mixed plastic char-based epoxy composite

Sample code	Mixed plastic char weight %	Epoxy weight %
EC0	0	100
EC1	2.5	97.5
EC2	5	95
EC3	7.5	92.5
EC4	10	90

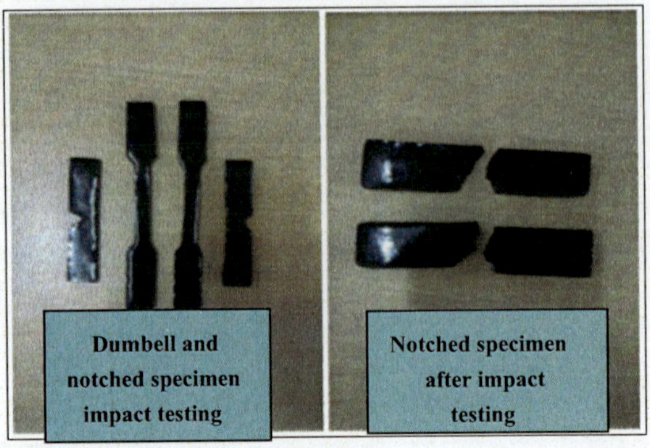

Fig. 1 Image shows moulded samples EC1, EC2, EC3, EC4 (from left to right)

7 Characterisation of Epoxy Composites

7.1 Mechanical Testing

Mechanical behaviour of specimen was examined by calculating its ultimate tensile strength and impact strength. For ultimate tensile strength analysis sample was flattened first and then tested via screw-driven machine known as INSTRON 1195. This testing was performed at 5 mm/min. strain rate with 50.14 mm gauge length and 62 mm^2 cross-sectional area by stroke mode.

For the impact strength analysis samples were converted to V notch geometry with dimensions such as 63.99 mm length, 12.68 mm width, and 3.15 mm thickness. Impact absorption of samples was measured using swinging pendulum with reference to standard method, i.e. ASTM D256. Mechanism of this method is that the machine is initially set with each of the samples and when pendulum moves and hits the V-notch side of the sample. Kinetic energy generated at the time of hit was measured in joule and measured in regard to sample thickness. Impact strength is known as proportional to kinetic energy and inversely proportional to sample thickness. This calculation will result in impact strength of the different samples.

7.2 Field-Emission Scanning Electron Microscope (FESEM)

The morphology of the material was observed using a field-emission scanning electron microscope (FE-SEM, MIRA 3 LMH (TESCAM)). The instrument was operated at 10 kV voltage and 50.13 mm width. The prepared sample showed very low conductivity properties. Therefore, to increase its conductivity properties, we have to clean the flat surface with acetone and apply a very thin coating of carbon. Scanning of the

fracture specimen is done by AU or gold metal sprayed on the micro-thin surface of the sample. A high vacuum evaporator was used for the coating of au metal on the surface.

7.3 *Fourier Transform Infrared Spectroscopy*

A Fourier transform infrared spectrum (FTIR) (Thermo Scientific Nicolet iD5, ATR) was obtained for the functional group analysis. Sample analysis was done for wavenumber ranges from 500 to 4000 cm^{-1}.

7.4 *Water Absorption Test*

Moisture absorbed on the surface of samples was examined by the standard method, i.e. ASTM D590. For moisture content absorption analysis, all samples with a known weight were dipped in a separate flask filled with distilled water for 24 h. After 24 h, take out the submerged sample and dry it on tissue paper immediately. Then samples were weighed after drying. This process continued for 6 days. In this way, we prepare 6 wet samples of all specimens. The percent of wet content of the sample is measured by the following equation:

$$M_t = \frac{(W_f - W_i)}{W_i} \times 100 \quad (1)$$

where

W_i is initial weight of sample before immersion
W_f is final weight of wet sample after 24 h
M_t is the rate of wet absorption of samples.

7.5 *Thermal Gravimetric Analysis (TGA)*

TGA MODAL SDT Q600 type of TA was utilised for the thermal stability test of the samples. The flow rate of 99.99 mL/min and ramp heating rate 10 °C/min over the temperature 800° in the presence of 100 kPa N_2 atmospheric gas were used during analysis. TGA is a function of change in weight or decomposition of sample with respect to change in temperature or time at sufficient condition.

Table 2 Stress versus strain data for tensile strength study of varying char concentration-based nano-composite samples

Sample code	Tensile strength (stress, MPa)	Elongation at break (strain %)
EC0	10.24	0.642
EC1	14.20	0.416
EC2	18.25	0.682
EC3	31.16	1.096
EC4	18.89	0.774

8 Results and Discussion

8.1 Mechanical Testing

The tensile strength of the samples EC0-EC4 (Table 1) was examined. The tensile strength of samples with plastic char EC1-EC4 was compared with a neat composite with zero plastic waste char (EC0). It was found that the neat composite had a tensile strength of 10.24 MPa at 1.963 KN load (Table 2). Tensile strength increases to a maximum of 31.16 MPa at 1.196 KN load as plastic char composition increases from 2.5 to 7.5 weight percentage. However, at a 10 weight percentage concentration of mixed plastic char, tensile strength suddenly reduced due to necking. The maximum tensile strength reached at 31.16 MPa is also known as the ultimate strength of a prepared mixed plastic char-filled nano-composite (Fig. 2).

This study illustrates that mixed plastic waste char has been completely dispersed in epoxy composite and prepared into a high-tensile material.

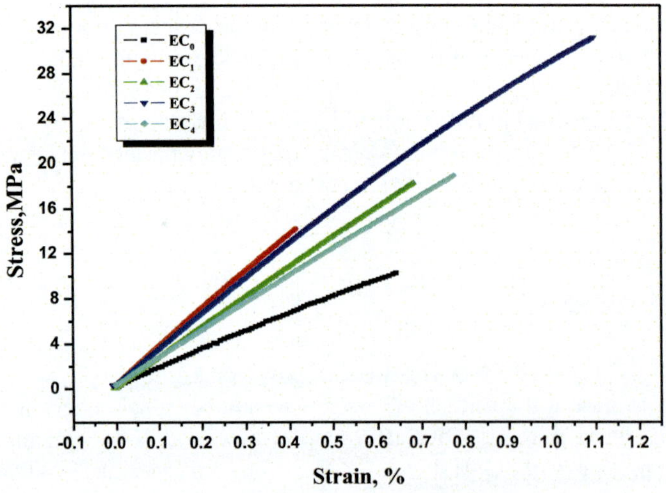

Fig. 2 Stress versus strain curve for varying char concentration in nano-composites

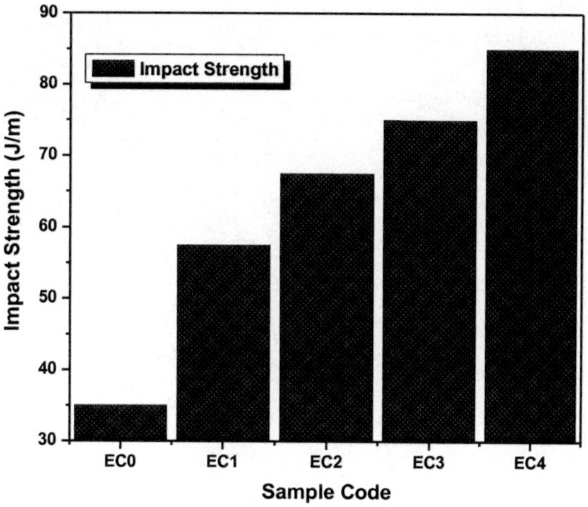

Fig. 3 Impact strength shown in Y-axis for varying char concentration ranges from 0–10 weight %

Table 3 Impact strength data for epoxy nano-composite with varying plastic char concentration (weight %)

The content of Char wt%	0	2.5	5	7.5	10
Impact strength (J/m)	35	57.5	67.5	75	85

The effect of waste plastic char on the impact behaviour of epoxy is shown in Fig. 3. It was discovered (Fig. 3) that as the amount of char in epoxy increased, the impact strength increased from 35 to 85% for char content ranges of 0 to 10%. Maximum energy was absorbed by specimens during the impact test, which indicates maximum impact strength. For epoxy/char nano-composite reinforced with (0, 2.5, 5, 7.5, and 10) wt% plastic waste nano-char, the impact strength was 35 and 57.5 J/m, 67.5, 75, and 85 J/m (shown in Table 3). This aforementioned enhancement in impact strength can be credited to the improved interfacial bonding of char with the epoxy matrix, which led to transferring the stresses appreciably through the intermolecular structures and promoting the higher impact strength.

9 FTIR Analysis

FTIR analysis of a cured 5 wt% char loaded epoxy composite sample was performed and found to have a transmittance versus wavenumber graph as given in Fig. 4. The strong sharp peak at 3750.41087 cm^{-1} shows the N–H stretching and at 3649.438 cm^{-1} shows the O–H stretching. The strong broad peak at 2924.08408 cm^{-1} is because of the aliphatic C-H and the medium broad peak at 2359.46178 is due to C–C stretching. The presence of C=O is responsible for the little strong intensity sharp peak at 1683.56356 cm^{-1}. Aromatic nitro has a high-intensity sharp peak at

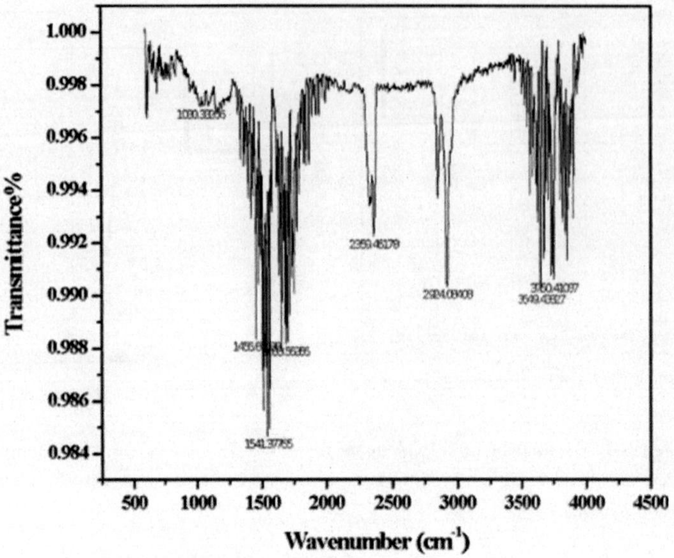

Fig. 4 FTIR image of cured EC3 sample

1541.37765 cm^{-1}, and CH3 asymmetrical and aromatic binding has a medium sharp peak at 1456.89038 cm^{-1} [15].

10 Thermal Analysis

The TGA analysis of the samples ranging from 0 to 10 wt% char concentration has been performed, and the results are summarised in Table 4. This analytical technique is basically used to calculate a substance's thermal stability and the corresponding fraction of volatile constituents by measuring the weight change that takes place as a sample is substantially heated at a constant rate.

Table 4 Sample temperatures for various weight loss during TGA analysis

Sample code	Initial temperature at 5% weight loss	Temperature at 50% weight loss	Solid residue in wt. % at 800 °C	Tangent at middle point (temperature, °C)
EC0	312.86	399.28	3.906	349.01
EC1	299.92	398.54	8.468	342.58
EC2	309.98	399.68	0.1161	342.58
EC3	310.70	397.30	11.56	338.75
EC4	307.11	400.25	2.170	343.00

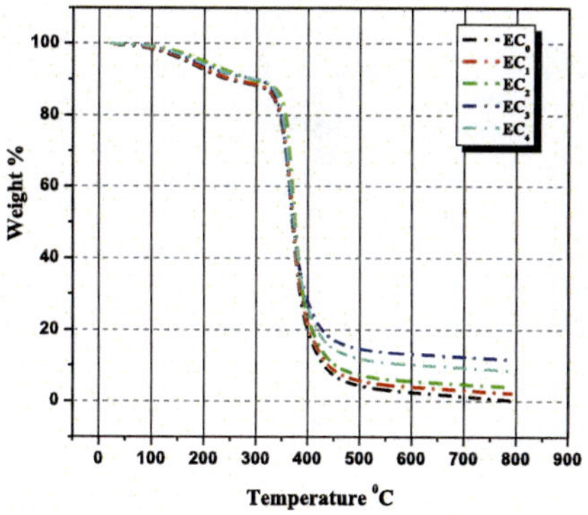

(a)

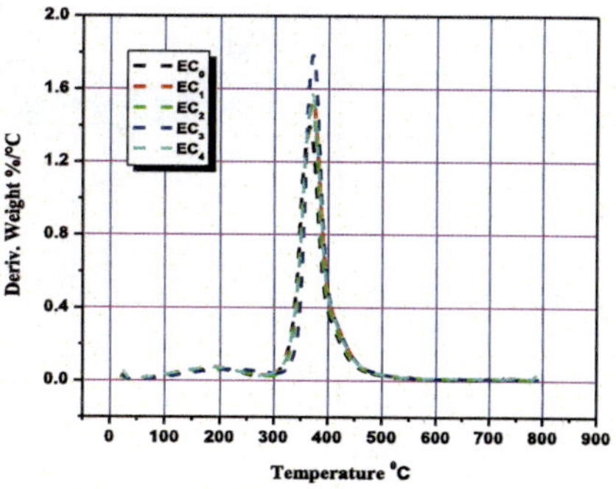

(b)

Fig. 5 TGA curve of samples EC0-EC4 denoting **a** Weight loss percent with temperature, **b** deviation in weight percent with temperature

The TGA curve between weight loss percent and temperature was plotted by the TGA analyser as shown in Fig. 5. Thermal degradation temperature can be calculated by drawing a tangent line in Fig. 5a between the initial inflection point and the middle point. Thermal degradation temperature is summarised in Table 4. From the data, this can be observed that 10 wt% of waste mixed plastic char reinforced with epoxy composite compared with others shows an increase in thermal degradation temperature, which tells that there is sufficient binding occurring between waste mixed plastic char molecules and epoxy molecules. Because the mechanical strength of this composite is optimal, the intermolecular force between waste mixed plastic char and resin component is excellent. We can also see that as the waste plastic char fraction in the sample increases, so does the thermal stability. At 50% weight loss, thermal degradation of samples EC0, EC1, EC2, EC3, and EC4 was found at temperatures given in Table 4 by the 2nd column. At 800 °C, the remaining solid residue was calculated and its data has been shown in the 3rd column of Table 4. Figure 5b denotes the deviation wt.% curve with a change in temperature. The DSC curve of the samples, which shows heat flow variation with temperature, is shown in Fig. 6.

11 FESEM

The morphology behaviour of the dispersion of waste mixed plastic char on an epoxy surface was investigated by field emission scan microscopy (FESEM). Figure 7 depicts a high-resolution image of neat epoxy and 2.5, 5, 7.5, and 10% waste mixed plastic char loaded epoxy composite. The image of a neat epoxy composite surface in Fig. 7a represents a bubbly texture due to the absence of a waste mixed plastic char molecule. Figure 7b depicts a 2.5 wt% char loaded epoxy nano-composite. By looking

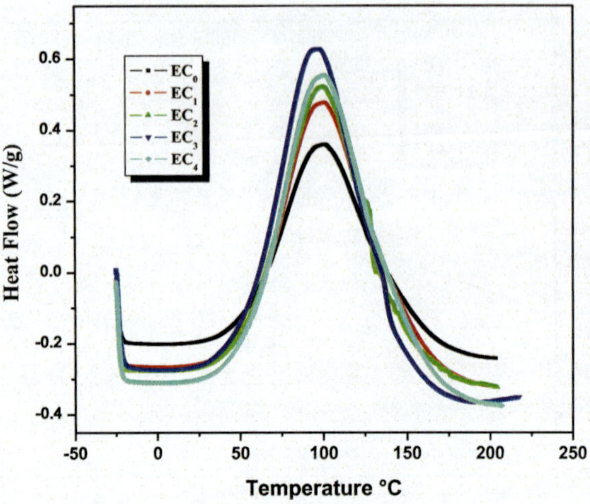

Fig. 6 DSC curve of samples ranging from EC0-EC4

at the surface image, it is clearly visible that particles of waste mixed plastic char are gathered in the form of small pebbles and spread over the entire surface, which makes the surface appear to be rough. This means there is a weak intermolecular force present between char and epoxy particles, which results in lower mechanical strength. Moving on, Fig. 7c, which depicts a 5 wt% char-based composite, has the smoothest surface of all. This shows that in this case, uniform dispersion occurred and mechanical properties increased from the previous one. Other images (Fig. 7d, e) show a rough surface of agglomerated char particles that contribute to the mechanical strength of the nano-composite [11].

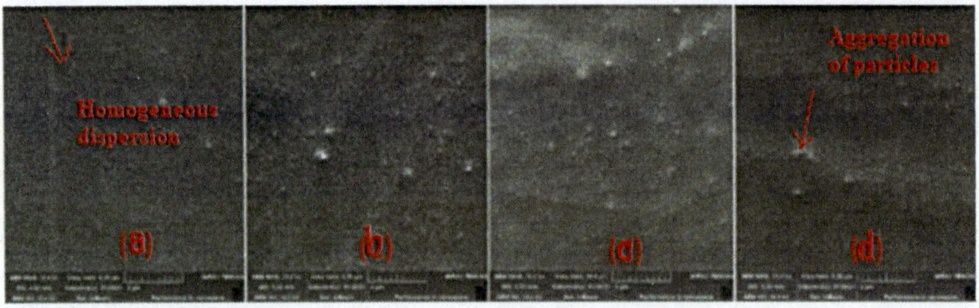

Fig. 7 FESEM images of plastic char reinforced epoxy nano-composite **a** EC1-2.5 wt%, **b** EC2-5 wt%, **c** EC3-7.5 wt%, **d** EC4-10 wt%

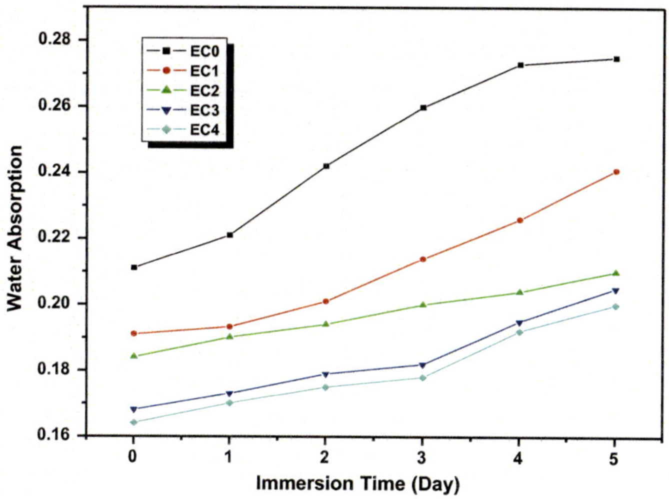

Fig. 8 Water absorption capacity of char leaded epoxy composite versus immersion time of the samples in water

12 Water Absorption Capacity

Composite samples' water absorption capacity was tested by immersing the different samples in distilled water for varying time periods and calculating water absorption capacity using Eq. (1). Figure 8 shows that with an increase in the immersion time, water absorption capacity increases. However, with an increase in char concentration starting from neat composite, nano-composite moves from a highly hydrophilic to a hydrophobic character. Therefore, due to this property, this type of nano-composite can be utilised for oil/water or organic solvent/water separation applications.

13 Conclusion

A novel mixed plastic char-filled nano-composite was prepared by the standard composite making method. The synthesised nano-composites were physically analysed and characterised for their physical and mechanical study by various means. FESEM results successfully interpreted that mixed plastic char has been uniformly distributed at the surface of epoxy, which has imparted very good characteristics in mixed char loaded epoxy composite compared to neat epoxy composite.

After observing its properties, it can be claimed that the prepared materials have shown high thermal stability, high tensile strength, and high impact strength by increasing char concentration in the epoxy composite. However, stress–strain data

show that 10 wt% plastic char-based composites were unable to withstand additional strain and broke after reaching a certain stress limit.

A 7.5 wt. percent char loaded epoxy nano-composite with a maximum tensile strength of 31.16 MPa was developed, which is roughly three times stronger than neat epoxy resin (10.24 MPa). Similarly, increasing the char concentration in epoxy resin boosted the tensile strength, impact strength, and thermal stability several times. These nano-composites' increased qualities suggest that they could be used in a variety of industries, including automotive, aerospace, and many others. However, more research is needed to determine the precise applications of these new nano-composites.

References

1. Chow CF, So W, Cheung YTY, Yeung SK (2017) Plastic waste problem and education for plastic waste management. Emerg Pract Scholarsh Learn Teach a Digit Era pp 125–140. https://doi.org/10.1007/978-981-10-3344-5_8
2. Thompson RC, Moore CJ, vom Saal FS, Swan SH (2009) Plastics, the environment and human health: current consensus and future trends. Philos Trans R Soc Lond B Biol Sci 364:2153–2166. https://doi.org/10.1098/rstb.2009.0053
3. Rafey A, Siddiqui FZ (2021) A review of plastic waste management in India—Challenges and opportunities. Int J Environ Anal Chem 1–17. https://doi.org/10.1080/03067319.2021.1917560
4. Banerjee T, Srivastava DR, Hung YT (2014) Plastics waste management in India: an integrated solid waste management approach
5. Pizzi A, Mittal KL (1994) Handbook of adhesive technology. CRC Press
6. Januszewski R, Dutkiewicz M, Nowicki M, Szołyga M, Kownacki I (2021) Synthesis and properties of epoxy resin modified with novel reactive liquid rubber-based systems. Ind Eng Chem Res 60:2178–2186. https://doi.org/10.1021/acs.iecr.0c05781
7. Zhao X (2022) Epoxy as filler or matrix for polymer composites. In: Lu S (ed) IntechOpen, Rijeka, p Ch 1. https://doi.org/10.5772/intechopen.102448
8. Fröhlich J, Thomann R, Mülhaupt R (2003) Toughened epoxy hybrid nanocomposites containing both an organophilic layered silicate filler and a compatibilized liquid rubber. Macromolecules 36:7205–7211. https://doi.org/10.1021/ma035004d
9. Ahmetli G, Kocaman S, Ozaytekin I, Bozkurt P (2013) Epoxy composites based on inexpensive char filler obtained from plastic waste and natural resources. Polym Compos 34. https://doi.org/10.1002/pc.22452
10. Sogancioglu M, Yel E, Ahmetli G (2017) Pyrolysis of waste high density polyethylene (HDPE) and low density polyethylene (LDPE) plastics and production of epoxy composites with their pyrolysis chars. J Clean Prod 165:369–381
11. Sogancioglu M, Yel E, Ahmetli G (2020) Behaviour of waste polypropylene pyrolysis char-based epoxy composite materials. Environ Sci Pollut Res Int 27:3871–3884. https://doi.org/10.1007/s11356-019-07028-3
12. Oral I (2021) Preparation and ultrasonic characterization of modified epoxy resin/coconut shell powder biocomposites (2021) 1–16. https://doi.org/10.1002/app.51772
13. Matykiewicz D (2020) Biochar as an effective filler of carbon fiber reinforced bio-epoxy composites. Processes 8. https://doi.org/10.3390/pr8060724

14. Giorcelli M, Khan A, Pugno N, Rosso C, Tagliaferro A (2018) Biochar as a cheap and environmental friendly filler able to improve polymer mechanical properties. Biomass Bioenergy 120. https://doi.org/10.1016/j.biombioe.2018.11.036
15. Doblies A, Boll B, Fiedler B (2019) Prediction of thermal exposure and mechanical behavior of epoxy resin using artificial neural networks and Fourier transform infrared spectroscopy. Polymers 11. https://doi.org/10.3390/polym11020363

TOPSIS Approach for Selecting the Most Appropriate Brand of Spark Plug

Kiran Pal, Naveen Kumar Garg, Rajan Kumar Jha, and Mumtaz Ahmad Khan

1 Introduction

The most important part affecting the efficient operation of the spark ignition engine is the spark plug [1]. It must be able to withstand tremendous temperatures, pressures, and the high voltage that must pass through it [2]. Spark plug manufacturers recommend replacing spark plugs every 20,000 to 40,000 miles, depending on how frequently you use your car and the type of vehicle you drive, in order to achieve good mileage per litre. The negative effects of spark plug failure are reduction in vehicle comfort and increase in travel costs; it also causes more damage to environment [3]. Three major and popular spark plug brands (DENSO, NGK, and BOSCH) are considered in the proposed study. TOPSIS decision-making method has been utilized in this study to determine which brand of spark plug is best for use of consumers. The consumer has several options when it comes to deciding which product is the best [4]. In this work, the selection of a spark plug is based on seven parameters (cost, voltage, and strength of metal shell, heat range, life of spark plug, service period, and deterioration of ground electrode).

K. Pal (✉) · M. A. Khan
Delhi Skill & Entrepreneurship University, DITE DSEU Okhla-II Campus, New Delhi 110020, India
e-mail: kiranpaldite@gmail.com

N. K. Garg
Delhi Skill & Entrepreneurship University, G B Pant DSEU Okhla-III Campus, New Delhi 110020, India

R. K. Jha
Sujan Group of Industries, IMT Manesar, Gurugram, Haryana, India

A. Kumar et al. (eds.), *Recent Advances in Manufacturing and Thermal Engineering*, Lecture Notes in Mechanical Engineering, https://doi.org/10.1007/978-981-19-8517-1_35

1.1 High Heat Range

The right heat range prevents the spark and inducing the thermal auto-ignition that could lead to engine damage.

1.2 Life of Spark Plug and Voltage

The first spark plug was expected to ignite 15–25 times per second. Today's spark plugs must ignite five times as often. The upper temperature limits as risen from 600 °C to approximately 900 °C and the ignition voltage from 10,000 V to up to 30,000 V [3]. Whereas today's spark plugs must function for at least 30,000 km.

Deterioration of Ground Electrode
The electrode wears from areas that quickly discharge with spark discharge. Because the core electrode is exposed to high temperatures, it oxidizes and wears. The degree of electrode wear varies depending on the melting point, strength, hardness, and other properties of the electrode material. Nickel alloys, platinum, iridium, and other similar materials are utilized for the electrodes to minimize the amount of wear, and service lifetimes are also prolonged with fine electrodes [5, 6]. In present case, only one attribute, i.e. cost is fit to non-beneficial criteria, and rest of all attribute is fit to benefit criteria.

2 Research Background

The consumer may see an advertisement on social media in which the corporation claims that their product is the greatest [7]. The consumer may be told by distributors of the product which one is best, but they do not know why. The consumer may depend on a mechanic which is best, but the chances are the mechanic does not know why this product is best. These are few ways the consumer has to use for determine which brand spark plug to buy. In this section, the authors have provided a brief review of spark plug with different material which are used and make a decision-making for various material used in ground electrodes. Iridium spark plugs were chosen for this investigation because they have a lifespan of roughly 125,000 miles and are made of a stronger, more durable metal than platinum. Hristov and Bogdanov [8] applied TOPSIS approaches to determine the best material of spark plug if more effective.

3 Research Methodology

3.1 TOPSIS Method

Hwang and Yoon [9] devised the TOPSIS approach. The approach has been focussed on the assertion that the selected option ought to be the farthest away from the negative ideal solution and the least Euclidean distance from the ideal answer. The positive appropriate outcome has been the hypothetical work, able alternative for which all model parameters correspond to the maximum attribute values in the database containing the satisfying solutions. The negative ideal solution is the hypothetical solution for which all attribute values correspond to the minimum attribute values in the database. Thus, TOPSIS provides a response that is not only the furthest from the supposedly worst but also the answer that is nearest to the presumably best. The primary approach of the TOPSIS method for choosing the best option out of those that are accessible is described. TOPSIS is a numerical method for solving multi-criteria decision-making problems [10]. The collected data/information contains a significant degree of ambiguity and imprecision. The TOPSIS technique has the major benefits: it is straightforward, reasonable, understandable, highly computer-efficient, and able to express the relative performance of each option in a concise mathematical form. The MCDM TOPSIS approaches refer to selecting the optimal alternative from a finite number of decision alternatives based on various, often conflicting criteria [11].

3.2 Assumptions Considered for Modelling

- All brands of spark plug consist a set of 4 numbers.
- The average life of all brands of spark plug assumed to be 125,000 km.
- All brands of spark plug facilitate reliable cold starts.
- All brands of spark plug ensure consistent operation with no ignition miss throughout its service life.
- All brands of spark plug do not overheat under extended operation at or near top speed.
- Engine conditions and driving conditions assumed to be ideal.

In Fig. 1, comparison of general specification the three top spark plug manufacturing companies can be seen.

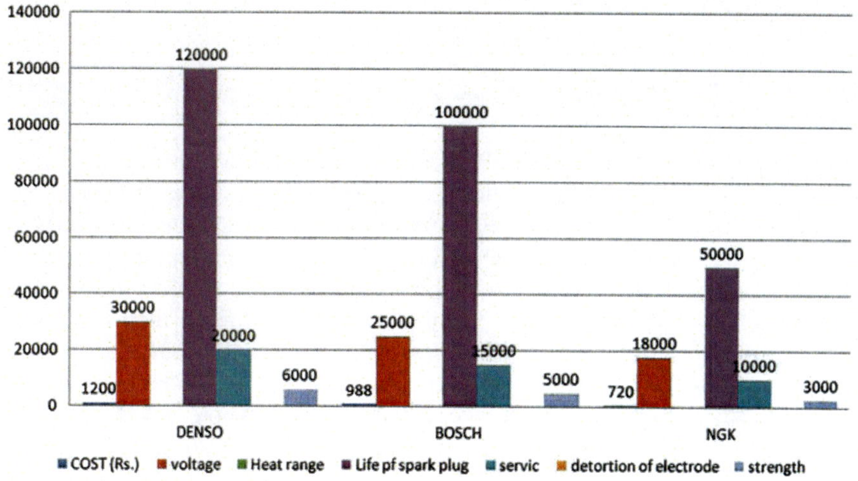

Fig. 1 Comparison of general specification the three spark plug manufacturing companies

4 Calculations Using TOPSIS Method

- Step 1. Constructing the decision matrix

  Create a decision matrix that includes alternative and attributes for choosing the best decision

$$\text{Alternatives: } i = 1, 2, 3, 4 \ldots m$$
$$\text{Attributes: } j = 1, 2, 3 \ldots n$$
$$\text{Decision matrix: } X = \left[x_{ij}\right]_{m*n} \quad (1)$$

  The decision matrix of general specification with three spark plug manufacturing companies, alternatives is shown on rows, and attributes are shown on columns matrix rank: $m * n$ can be seen in Table 1.
- Step 2. Calculate the normalized decision matrix

Table 1 Decision matrix of general specification the three spark plug manufacturing companies

Brand/manufacturer	Cost (Rs.)	Voltage (V)	Strength of metal shell (PSI)	Heat range	Life of spark plug (km)	Deterioration of ground electrode	Service (km)
Denso	1200	30,000	6000	24	120,000	50	20,000
BOSCH	988	25,000	5000	4	100,000	90	15,000
NGK	720	18,000	3000	8	50,000	80	10,000

Elements of the matrix are divided by the square root of the sum of each squared element in a column (attributes). Lai and Hwang [10] introduced linear normalization formula

$$n_{ij} = \frac{x_{ij}}{\sqrt{\sum_{i=1}^{m} x_{ij}^2}}, \quad (2)$$

After doing the necessary calculations, normalized decision matrix is obtained and can be seen in Table 2.

On the basis of different attributes, a weightage matrix has been developed and can be seen in Table 3.

- Step 3. Calculating the weighted normalized decision matrix

  The weighted normalized value V_{ij} is calculated in the following way

$$V_{ij} = w_{ij} * n_{ij}, \quad i = 1, 2, 3, 4 \dots m, \quad j = 1, 2, 3, 4 \dots n \quad (3)$$

where $V_{ij} = W_{ij} \times n_{ij}$

Obtained normalized weighted matrix has been given in Table 4.

- Step 4. Determine the positive ideal and negative ideal solutions

  In the case of beneficial attributes (i.e. those of which higher values are desirable for the given application), V_j^+ indicates the higher value of the attribute

$$V_j^+ = \left\{ \left(\sum_i^{\max} v_{ij} / j, \sum_i^{\min} v_{ij} / j \varepsilon J' \right) \quad i = 1, 2, \dots N \right\}$$
$$= \left\{ V_1^+, V_2^+, V_3^+ \dots V_M^+ \right\} \quad (4)$$

In the case of non-beneficial attributes (i.e. those of which lower values are desired for the given application), V_j^- indicates the lowert value of the attribute

$$V_j^- = \left\{ \left(\sum_i^{\min} v_{ij} / j, \sum_i^{\max} v_{ij} / j \varepsilon J' \right) \quad i = 1, 2, \dots N \right\}$$
$$= \left\{ V_1^-, V_2^-, V_3^- \dots V_M^- \right\} \quad (5)$$

where $J = (j = 1, 2, \dots, M) / j$ is associated with beneficial attributes, and $J' = (j = 1, 2, \dots, M) / j$ is associated with non-beneficial attributes.

The positive and negative ideal solution are given in Tables 5 and 6, respectively.

- Step 5. Calculate the difference measure (S_i^+ and S_i^-) for each alternative A_j from PIS and NIS Difference measure is given by Euclidean for each alternative which has been calculated using the formulae given below, and obtained results are shown in Table 7.

$$S_i^+ = (v_{ij}, v_j^+) = \sqrt{(v_{ij}, v_j^+)^2} \quad (6)$$

Table 2 Normalized decision matrix

Brand/manufacturer	Cost (Rs.)	Voltage (V)	Strength of metal shell (PSI)	Heat range	Life of spark plug (km)	Deterioration of ground electrode	Service (km)
Denso	0.700505	0.697674	0.717137166	0.937043	0.731653	0.383482	0.742781
BOSCH	0.576749	0.581395	0.597614305	0.156174	0.609711	0.690268	0.557086
NGK	0.420303	0.418605	0.3585685583	0.312348	0.304855	0.613572	0.371396

Table 3 Weightage matrix

W	Cost (Rs.)	Voltage (V)	Strength of metal shell (PSI)	Heat range	Life of spark plug (kms)	Deterioration of ground electrode	Service (km)
Weightage	0.2	0.2	0.1	0.1	0.1	0.1	0.2

and

$$S_i^- = d(v_{ij}, v_j^-) = \sqrt{(v_{ij}, v_j^-)^2} \text{ for } i = 1, 2, 3 \ldots m \tag{7}$$

- Step 6. Calculate the corresponding closeness coefficient (CC_i) of the alternatives

  Corresponding closeness coefficient (CC_i) of the alternative has been obtained using the formulae as under:

$$CC_i = S_i/(S_i^+ + S_i^-) \text{ for } i = 1, 2, 3 \ldots m \tag{8}$$

  Obtained result is given in Table 8.
- Step 7. Rank the preference order or select the alternatives closest to 1

  Preference order using the TOPSIS method has been obtained and given as under in Table 9 and shown in Fig. 2.

5 Results and Discussion

The determined finding implies that Denso will be the highest priority (Rank 1), Bosch will be the second highest (Rank 2), and NGK will be the lowest (Rank 3). Denso has the lowest maintenance cost per 100,000 km, whilst NGK has the most. On 100,000 km, NGK will be more expensive than Denso, and Bosch will be an intermediate selection. As a result, the best alternative for the consumer is Denso. The authors carried out research on a method (TOPSIS) which suggested spark plug utility/analysis for procuring best suitable spark plug out of the available brands for using in a spark ignited internal combustion engine. By this method, the spark plug voltage becomes measured by voltage sensors, installed (non-contact) on the high voltage zone, in the vicinity of the spark plugs.

6 Conclusions

This article was undertaken to choose the best product from a vendor. The spark plugs in this analysis were examined for a number of factors, including price, voltage, metal shell strength, heat tolerance, spark plug life (km), ground electrode deterioration,

Table 4 Normalized weighted matrix (V_{ij})

Brand/manufacturer	Cost (Rs.)	Voltage (V)	Strength of metal shell (PSI)	Heat range	Life of spark plug (kms)	Deterioration of ground electrode	Service (km)
Denso	0.140101	0.139535	0.071714	0.093704	0.073165	0.038348	0.148556
BOSCH	0.11535	0.116279	0.059761	0.015617	0.060971	0.069027	0.111417
NGK	0.084061	0.083721	0.035857	0.031235	0.030486	0.061357	0.074278

Table 5 Positive ideal solution (PIS)

V_j^+	0.084061	0.139535	0.071414	0.093704	0.073165	0.069027	0.148556

Table 6 Negative ideal solution (NIS)

V_j^-	0.140101	0.083721	0.035857	0.015617	0.030486	0.038348	0.074278

Table 7 Difference measure (S_i^+ and S_i^-) for each alternative A_j from PIS and NIS

S_i^+	S_i^-
0.06389	0.13356
0.12372	0.1247
0.09503	0.06404

Table 8 Corresponding closeness coefficient (CC_i) of the alternatives

Alternatives	*CCi*
Denso	0.676425
BOSCH	0.50196
NGK	0.402605

and service. The research component suggests important data that correlate with the theoretical part based on data and observation findings between the theoretical and experimental parts. Modern automobiles demand appreciably more durable products at the lowest possible cost as their effectiveness increases. Automobile manufacturers

Table 9 Rank the preference order

Rank	Brand/manufacturer
1	Denso
2	BOSCH
3	NGK

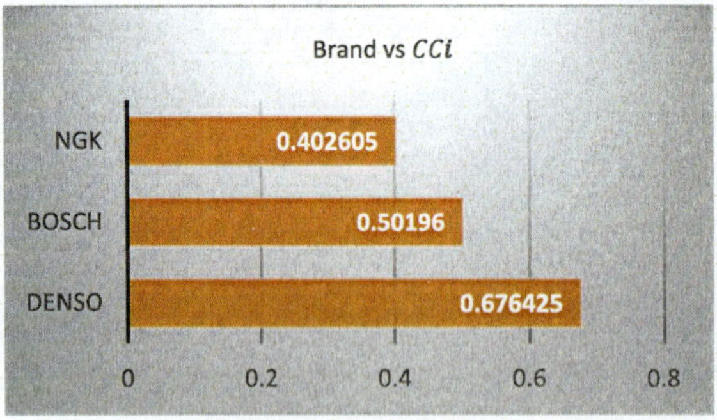

Fig. 2 Preference order

nowadays entirely rely on technological approaches to choose the best product from a vendor. This aids in locating the best solution at the lowest possible per kilometre maintenance cost.

The choosing of a spark plug for a certain engine is not an easy affair. In this research article, the TOPSIS approach was used to identify an appropriate spark plug for a given engine and rank all options or choices. The many procedures that must be taken in order to employ this strategy were discussed. Triptych software is used to do the analysis. To explain how to utilize the technique, a case study demonstrating its implementation was considered.

TOPSIS and Triptych can be utilized as effective tools for selecting and rating spark plugs.

Conflict of Interests All contributors' attests that they are not connected to or a part of any group or organization that has a financial or non-financial stake in the topics or materials covered in this work.

References

1. Azrin AA, Yusri IM, Sudhakar K, Mohd Nor CW, Zainal A, Majeed APPA (2021) An overview of the spark plug engine profile in a spark ignition engine. IOP Conf Ser: Mater Sci Eng 1092:012030. https://doi.org/10.1088/1757-899X/1092/1/012030
2. Özçelik Z, Gültekin N (2019) Effect of iridium spark plug gap on emission, noise, vibration of an internal combustion engine. Int J Energy Appl Technol. https://doi.org/10.31593/ijeat.561307
3. Shimasaki Y, Kanehiro M, Baba S, Maruyama S, Miyata THS (1993) Spark plug voltage analysis for monitoring combustion in an internal combustion engine. In: International congress and exposition, Detroit, MI, March 1–5, SAE Paper 930461
4. Ahmed A (2013) Impact of spark plug number of ground electrodes on engine stability. Shams Eng J 4(2):307–316. https://doi.org/10.1016/j.asej.2012.09.006
5. Osamura H, Abe N (1999) Development of new iridium alloy for spark plug electrodes. SAE paper 1999-01-0796
6. Yorita H, Okabe H, Hiroya I, Shibata M (2007) Ignition simulation and visualization for spark plug electrode design. SAE paper 2007-01-0940. https://doi.org/10.4271/2007-01-0940
7. Hristov R, Bogdanov K, Dimitrov R (2018) Research the influence of spark plugs types on the performance of the engine operating on gaseous fuels. Mob Vehicle Mech. https://doi.org/10.24874/mvm.2018.44.01.05
8. Papathanasiou J, Ploskas N (2018) TOPSIS. Springer optimization and its applications, pp 1–30. https://doi.org/10.1007/978-3-319-91648-4_1
9. Hwang CL, Yoon K (1981) Multiple attribute decision making: methods and applications. Springer-Verlag, New York. http://dx.doi.org/10.1007/978-3-642-48318-9
10. Lai Y-J, Liu T-Y, Hwang C-L (1994) TOPSIS for MODM. Eur J Oper Res 76:486–500. https://doi.org/10.1016/0377-2217(94)90282-8
11. Basakayi J, Ilinca A (2019) TOPSIS method for supporting the selection and ranking of phase change materials for a latent heat storage. Int J Eng Res Technol (IJERT) 08(04)

Investigating Membrane Degradation in Low-Temperature Proton Exchange Membrane Fuel Cell (PEMFC)

Jay Pandey

1 Introduction

Proton exchange membrane fuel cell (PEMFC) is an electro-chemical device with compact configuration, zero noise, fast start-up, high efficiency and zero emission. PEMFC has several applications in many fields such as automobiles, power generation, consumer electronics, defense and space applications [1–3]. In last few decades, the PEM fuel cell technology has made tremendous progresses in enhancing performance, lowering costs and improving stability. However, the electro-chemical performance, cost and durability of PEM fuel cells are still underlying for global commercialization [4–6]. There are two the most prevalent performance degradation processes in the PEM fuel cell operation that includes two categories, one is recoverable performance degradation in the short term, such as the cell in the cold start, dead-ended anode (DEA) and anode recirculation operation [7, 8], and another one is unrecoverable performance degradation in the long term such as materials' degradation (membrane, electro-catalysts and bipolar plates) [8, 9]. The key parameters causing PEMFC degradation are many, and the degradation mechanism is very complex in nature. Therefore, at this juncture, the detailed analysis and prediction of the degradation behaviors of real PEMFC system are very crucial and must be addressed carefully [10, 11].

Proton exchange membrane (PEM), one of the components in PEMFC, is primarily used to allow the proton transport from anode to cathode as well as for separating the both the half cells to reduce the fuel/oxidant crossover [11]. An ideal PEM, in low-temperature PEM fuel cell, should exhibit high proton conductivity, high mechanical strength, low electronic conductivity, less feed/oxidant crossover [3, 5]. However, design and development of PEMs having all intrinsic properties are very tedious task; therefore, the optimization of the desired properties as per the required

J. Pandey (✉)
Department of Chemical Engineering, BITS Pilani, Pilani, Rajasthan 333031, India
e-mail: jay.pandey@pilani.bits-pilani.ac.in

A. Kumar et al. (eds.), *Recent Advances in Manufacturing and Thermal Engineering*, Lecture Notes in Mechanical Engineering, https://doi.org/10.1007/978-981-19-8517-1_36

application is very challenging. Till today, Nafion, discovered by DuPont USA, has proven as the best polymeric membrane for low temperature PEM fuel cell. The Nafion series membranes (N-112/115/117) have demonstrated excellent oxidative stability and proton conductivity below 80 °C and under fully hydrated (98%) conditions. However, Nafion series membranes are very expensive due to poly(tetrafluoro ethylene) (PTFE) backbone, high fuel crossover and declined performance at elevated temperature (~100 °C) and lower humidity (below 30%) [12]. In order to minimize, these issues many efforts were made to design and develop Nafion-based hybrid membranes by doping Nafion suspension with other polymers such as PVDF, PVA, SPEEK and PBI to reduce the fuel crossover but declined fuel cell performance was observed due to lower proton conductivity compared to Nafion-based membranes. The one of the key reasons of poor fuel cell performance is the membrane degradation under fuel cell environment and thereof lower efficiencies.

In this direction, several efforts have been also made to focus on non-Nafion-based inorganic–organic ion-exchange membranes by embedding inorganic ion-exchangers such as heteropoly acids (HPAs), zirconium phosphate (ZrP), solid metal oxides, graphene oxides (GO) and metal–organic frameworks (MOFs) into stable polymer substrate to enhance thermo-mechanical and electro-chemical properties of membrane [13–15]. Unfortunately, in most of cases, the long-term stability of these proton conducting membrane is not up to DOE expectations mainly due to degradation of PEMFC components (membrane and electro-catalyst) and thus not meeting the norms for commercialization.

Therefore, the present research paper aims at mainly emphasizing on different types of membrane degradation such as mechanical/physical, chemical, thermal and electro-chemical degradation and their impact on fuel cell performance, i.e., polarization curve (I-V curve) and power density curves. The root-cause analysis and its mitigation are one of challenges with handling proton conducting membrane for low-temperature fuel cell. So far very limited efforts have been made on degradation analysis of PEMs and their real-time investigations. Therefore, the aim of the present study was to understand the influence of operational parameters (physical, operational and electro-chemical) on the degradation of PEM membranes (both polymeric and nanocomposite) in order to evaluate the rate of the membrane degradation under fuel cell environment.

2 Investigation and Mitigation of Membrane Degradation

Polymer electrolyte membrane fuel cell (PEMFC) also called proton exchange membrane fuel cell possesses high power/energy densities with small size, low weight and volume. PEMFC, commonly operating at room temperature 50–80 °C, mainly composed of a polymer electrolyte membrane (PEM) as a proton conducting medium and electro-catalysts for redox reactions such as hydrogen oxidation reaction (HOR) at anode and oxygen reduction reactions (ORR) at cathode. PEMFC is typically fueled with pure hydrogen as feed at anode side and atmospheric air or pure oxygen

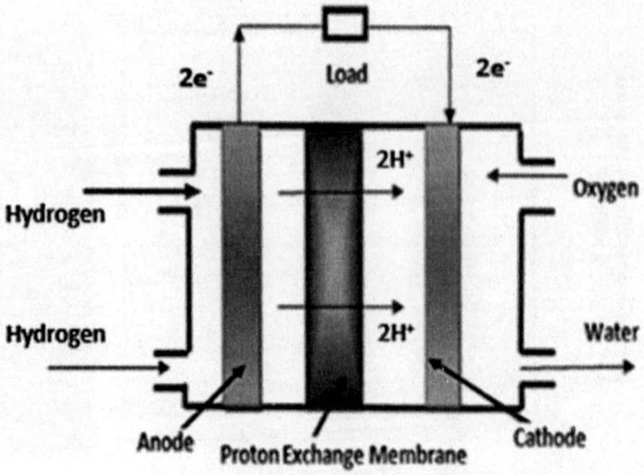

Fig. 1 Working of proton exchange membrane fuel cell (PEMFC)

as oxidant at cathode side as shown in Fig. 1. PEMFCs are well suited for transportation applications and some stationary applications. However, the problem with PEMFC is high cost of precious catalysts, CO poisoning and high fuel crossover though solid electrolyte.

The electro-chemical reactions occurring in a PEMFC are as following [16]:

Reaction at anode $H_2 \rightarrow 2H^+ + 2e^-$
Reaction at cathode $1/2O_2 + 2H^+ + 2e^- \rightarrow H_2O$
Overall reaction $H_2 + 1/2O_2 \rightarrow H_2$

Role of PEM membrane is very critical in PEM fuel cell such as proton conduction, inhibiting the electronic conductions and separation of feed and oxidants. The membranes used in PEMFC are also quite prone to fuel (hydrogen/water) crossover from anode to cathode causing loss of cell potential and thus poor power density performance. These membranes also cannot also sustain higher temperatures and also very prone to degradability. The proton conductivity of PEMs used in PEMFC reduces significant at lower humidity. Management of operational humidity requires additional supporting components in PEMFC that is not only complex but also adds extra cost to fuel cell. This is the reason that the membrane's life has a great effect on both fuel cell performance and long-term stability. In reality for instances during portable, vehicular and stationary applications the proton conducting membrane experiences chemical, mechanical, electro-chemical and thermal stresses that eventually deteriorate its structural configurations of PEMs. There are few studies on detailed investigation and curtailment of membrane degradation in real PEM fuel cells [15–17]. A very interesting work on proton conducting membrane degradation was done by Lim et al. [18], where a cyclic accelerated stress test (AST) was designed to understand the combined effect of chemical, mechanical and electrochemical membrane degradation in PEM fuel cells. Figure 2 shows the SEM images of the membrane degradation after operating in PEM fuel cell as reported in literature [18].

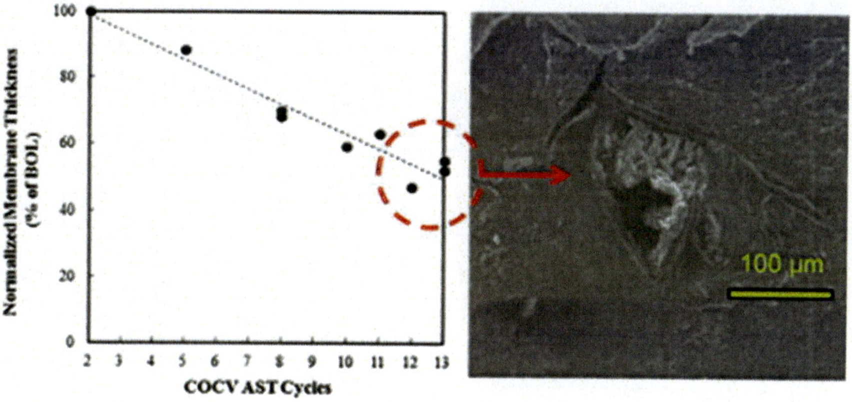

Fig. 2 Cyclic accelerated stress test and SEM image showing the membrane degradation in PEM fuel cell [18]

PEM degradation is typically initiated by the attack of peroxide radical species that are formed during half-cell reactions at PEM fuel cell and attack the functional membrane, thus causing polymeric backbone loss and deterioration of electrochemical properties. The thermal degradation in membrane happens during MEA formation at high temperature as well as the high temperature (>120 °C) operation of PEM fuel cell/ causing dimensional disintegration and also pin-hole formation. Whereas the mechanical membrane degradation is induced by applied stresses and failures/defects generated in the membrane during synthesis and fuel cell operational environments. Other than the chemical, thermal, mechanical, there is also significant role of electro-chemical degradation in inducing the degradation in PEM membranes such as electro-osmosis drag (EOD) [19]. The EOD mainly allows the water molecules to pass through the membrane along with protons during PEM fuel cell operation and causes water drag through membrane. The high EOD imparts flooding of membrane and causes membrane's dimensional instabilities. It is worth mentioning that all the degradation types in the proton conducting membranes are very sensitive to the fuel cell working environment such as temperature, pressure and feed/oxidant compositions and exhibit combined effects in membrane degradation over operating time span. Table 1 gives a summary of different types of membrane degradation and possible impact on membrane.

Table 1 Key factors affecting membrane degradation

Degradation type	Identification
Thermal	Expansion, pin holes,
Mechanical	Pin holes, shrinking, film thinning
Chemical	Polymer decomposition, film detachments
Electro-chemical	Electro-osmotic drag causing membrane swelling

Understanding of the fuel cell membrane degradation is much complex and has been extensively studied and reported in literature by now [11, 12]. Thermal durability can be improved by optimized process conditions. Chemical degradation can be curtailed using high strength polymeric backbone with good membrane functionality. On the other hand, the mechanical durability of membranes can be increased by increasing the thickness of the membrane and also by choosing high strength polymeric materials; however, at the same time, it has an adverse effect on fuel cell performance such as high ionic resistance and cost. To mitigate the electro-chemical degradation, it is advisable to optimize the operating current densities in PEM fuel cell to avoid water flooding in PEM fuel cells.

There are many electro-chemical diagnostic tools for analysis of fuel cell performance in batch and/or real-time operating fuel cell such as electro-chemical impedance analyzer (EIS), cyclic voltametery (CV), linear sweep voltametery (LSV) and polarization curve [20]. The polarization curve analysis is one of such very useful tools to investigate the thermodynamic irreversibilities in PEM fuel cell under real operation. The polarization curve has several zones such as activation, ohmic and concentration overpotential. There are several methods for investigation and curtailment of the membrane degradation in PEM fuel cells.

1. Accelerated stress testing (AST)
2. Fluoride release rate (FRR) test
3. OCV loss test
4. Image analysis.

During real-time operation, the PEM fuel cell undergoes to different types of irreversibilities (losses) due to reactions at electro-catalyst, transport of ions through membrane and feed/oxidant flows in bipolar flow channels. Figure 3 shows the different zones in fuel cell polarization curve such as activation overpotential, ohmic overpotential and concentration overpotential. These over potentials arise due to several electro-chemical and physical phenomena at different zones of fuel cells, i.e., electro-catalysts, membranes and flow channels in bipolar plates. The ohmic overpotential region in polarization curve is mainly showing membrane degradation in PEM fuel cell.

3 Conclusion

Cost, electro-chemical activity and stability are the key parameters to evaluate the techno-economic feasibility of a PEM fuel cell device for EV applications. Longer durability of PEM fuel cell's components such as membrane, electro-catalyst and bipolar plates is one of major challenges in PEM fuel cell. In our present work, the rigorous investigations on major operating parameters (physical and electro-chemical parameters) causing membrane degradation were carried out. The membrane thinning, pin-hole formation, polymer backbone detachment and membrane buckling

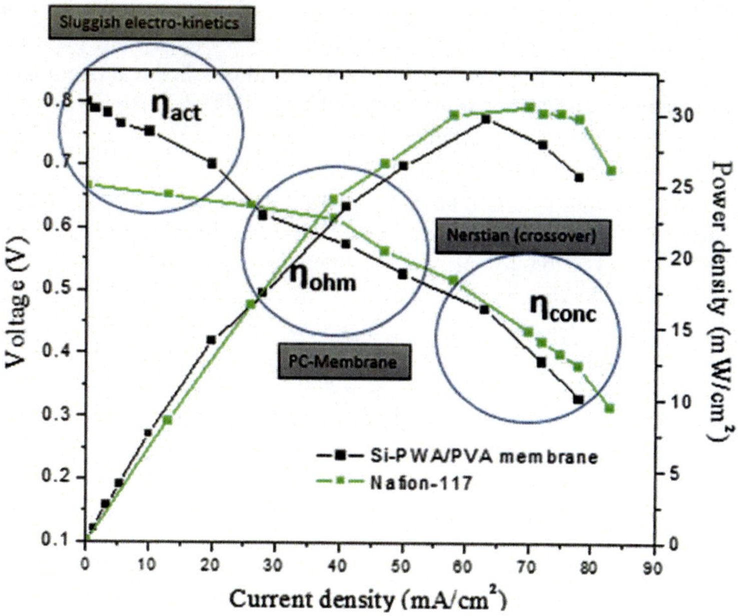

Fig. 3 Different overpotential zones in polarization curve of a PEM fuel cell

due to mechanical stress and peroxide radical attacks are some of factors causing faster membrane degradation and thus affecting PEM fuel cell life.

References

1. Pandey J (2015) Synthesis and characterization for membrane. Mater Lett 2:99–110
2. Knights S, Colbow K, St-Pierre J, Wilkins D (2004) Aging mechanisms and lifetime of PEFC and DMFC. J Power Sources 127:127–134
3. Panha K, Fowler M, Yuan X, Wang H (2012) Accelerated durability testing via reactants relative humidity cycling on PEM fuel cells. Appl Energy 93:90–97
4. Okada T (2002) Ion and water transport characteristics of perflurosulfonated ionomer membranes with H^+ and alkali metal cations. J Phys Chem B 106:1267–1272
5. Pandey J (2015) Synthesis and characterization for membrane. Int J Hydrogen Energy 2:99–110
6. Carter D, Wing J (2013) The fuel cell industry reviews 2013. Fuel Cell Today 1:1–46
7. Pan R, Yang D, Wang Y, Chen Z (2020) Performance degradation prediction of proton exchange membrane fuel cell using a hybrid prognostic approach. Int J Hydrogen Energy 45:30994–301008
8. Chen B, Cai Y, Shen J, Tu Z, Chan SH (2018) Performance degradation of a proton exchange membrane fuel cell with dead-ended cathode and anode. Appl Thermal Eng 132:80–86
9. Kamarudin SK, Ahmad F, Daud WRW (2009) Overview on application of direct methanol fuel cell (DMFC) for portable electronic devices. Int J Hydrogen Energy 34:6902–6916
10. Zelenay P, Piela P (2004) Researchers redefine the DMFC roadmap. The Fuel Cell Review 1:17–23

11. U.S. Department of Energy (2000) Fuel cell handbook, 5th edn. Science Applications International Corporation E&G Services Parson Inc.
12. Wee JH (2007) Feasibility study on direct methanol fuel cells for laptop, computers based on a cost comparison with lithium ion batteries. J Power Sources 173:424–436
13. Strathmann H (2004) Ion-exchange membrane separation processes. Elsevier Sci Technol 9:10–30
14. Nagarale RK, Gohil GS, Shahi VK, Rangarajan R (2004) Organic-inorganic hybrid membrane: thermally stable cation-exchange membrane prepared by sol-gel method. Macromolecules 37:10023–10030
15. Ahmad H, Kamarudin SK, Harsan UA, Daud WRW (2010) Overview of hybrid membranes for direct methanol fuel-cell applications. Int J Hydrogen Energy 35:2160–2175
16. Slade S, Campbell SA, Ralph TR, Walsh FC (2002) Ionic conductivity of an extruded nafion 1100 EW series of membranes. J Electrochem Soc 149:A1556–A1564
17. Nagarale RK, Shin W, Singh PK (2010) Progress in ionic organic-inorganic composite membranes for fuel cell applications. Polym Chem 1:388–408
18. Lim C, Ghassemzadeh L, Van Hove F, Lauritzen M, Kolodzie J, Kjeang E (2014) Membrane degradation during combined chemical and mechanical accelerated stress testing of polymer electrolyte fuel cells. J Power Sources 257:102–110
19. Peighambardoust SJ, Rowshanzamir S, Amjadi M (2010) Review of the proton exchange membranes for fuel cell applications. Int J Hydrogen Energy 35:9349–9384
20. Smitha B, Sridhar S, Khan AA (2005) Solid polymer electrolyte membrane for fuel cell applications. J Membr Sci 259:10–26

Rational Efficiency Analysis of Laminar Nanofluid Flow in the Heat Exchanger

K. Manjunath

Abbreviations

d_p	Nanoparticle diameter (m)
$\dot{E}$	Rate of exergy (W)
$\dot{I}$	Irreversibilities (W)
N_s	Dimensionless entropy generation
$\dot{S}_{gen}$	Entropy generation rate (W/K)

Subscripts

c	Cold stream
h	Hot stream
in	Fluid inlet
out	Fluid outlet
bf	Base fluid
nf	Nanofluid
p	Nanoparticle

K. Manjunath (✉)
Department of Mechanical Engineering, Delhi Technological University, Bawana Road, New Delhi 110042, India
e-mail: manjukmys@gmail.com

A. Kumar et al. (eds.), *Recent Advances in Manufacturing and Thermal Engineering*, Lecture Notes in Mechanical Engineering, https://doi.org/10.1007/978-981-19-8517-1_37

Greek Symbols

ψ Rational or exergetic efficiency
φ Volume concentrations

1 Introduction

From the point of view of energy conservation, reduction of losses happening in heat exchanger is of importance in thermal engineering. The miniature achievement of energy systems is resulting in various heat transfer appreciation techniques used in the case of heat exchanger. In these days, nanofluid usage is carried out to appreciate the heat transfer in equipment. The nanoparticles from the sizes ranging from 80 to 120 nm made of different materials are uniformly mixed with the base fluids in volume fractions from 1 to 6% and used as nanofluids. The properties of nanofluids will be different from that of base fluids and to be calculated from different empirical expressions and experimental techniques.

Various losses happening in processes in terms of irreversibilities can be obtained by second-law thermodynamic analysis. Energy lost in different ways will lead to decrease in world's energy resources. Second-law thermodynamic analysis considers finding of irreversibilities in the heat transfer equipment. From these analyses, we are able to find out losses happening and find the ways to reduce it [1]. Second-law thermodynamic method of analyses is reviewed in [2] along with constructal theory.

Some of the literature reviews are carried out based on the analysis of this manuscript to understand the nanofluid research in heat exchangers. Theoretical study of the entropy generation to analyze nanofluid heat transfer and fluid flow is investigated by [3]. Second-law analysis of the nanofluid flow in a pipe is theoretically analyzed in [4]. Entropy generation of Al_2O_3–water nanofluid in terms of constant wall heat flux in turbulent forced convection flow is performed in [5]. Exergetic performance of electronic devices by using exergy-entropy analysis applied to nanofluids by environment effect was investigated by [6]. Numerical study of the second law for the flow of the water–Al_2O_3 nanofluid in a mini-channel of triangular shape in terms of constant wall heat flux is performed in [7]. Second-law performance of counterflow plate heat exchanger having corrugation with the use of hybrid nanofluid in a milk chilling plant is applied in [8].

Numerical simulations have been performed by [9] on analysis of turbulent nanofluid flow in heat exchanger having twisted tapes. Energetic and exergetic analysis in a corrugated counterflow plate type heat exchanger by hybrid nanofluid with different particle volume ratios was carried out experimentally by [10]. A novel configuration of turbulator was analyzed by [11] to model the nanomaterial turbulent flow to improve the performance by irreversibility analysis. Recently, the hybrid nanofluid laminar flow analysis is carried out for double pipe heat exchanger by [12] incorporating entropy generation and exergy analysis. Numerical investigation of

irreversibility characteristics of shell and tube heat exchanger having helical baffles is analyzed by [13] having different shapes of nanoparticles in nanofluid. Improvement in performance by using hybrid nanofluid flow in microplate heat exchanger is reported by [14] using first- and second-law analysis. A ternary hybrid nanofluid was used to analyze heat transfer improvements in radiator using exergetic method [15]. Experimental comparison with numerical performance for nanofluid flow in mini-channels was investigated by [16] for heat transfer and pressure drop characteristics. The influence of nanoparticle shape considering spherical and non-spherical types on thermal performance of printed circuit heat exchanger is investigated by [17].

Considering the nanoparticle shape and fin effects of a mini-shell and tube heat exchanger, the irreversibility analysis is performed by [18] to decide whether the fins provide improvements along with nanofluids. First- and second-law analysis was employed by [19] to investigate heat transfer and pressure irreversibilities for the nanofluid flow in hair pin heat exchanger having annulus helical baffle. Novel nanofluid flow in heat exchanger was investigated experimentally and numerically by [20], and also, analysis was carried out using hybrid nanofluids. In the double pipe heat exchanger, the nanofluid parametric optimization is done by [21] incorporating exergetic and entropy generation performance techniques.

From the literature review, it can be found out that there exist very less manuscripts which use the performance methods, namely rational or exergetic efficiency analysis which includes entropy generation. By using nanofluids to know whether there are advantages and disadvantages is not easily able to investigate by using first-law analysis (heat transfer and pressure drop). But by using second-law analysis, the irreversibilities found in the heat exchangers can be investigated. The second-law analysis of heat exchanger using nanofluid and considering the imbalance irreversibilities is not yet carried out in literatures. Optimum values of operating parameters of nanofluid particles present in the base fluids are not obtained for laminar flow in heat exchanger by the earlier investigators.

Also, there is less work of research carried out for heat exchanger based on rational or exergetic analysis with the use of laminar flow nanofluid. This paper concentrates its research based on exergetic efficiency which includes entropy generation as performance parameters to analyze the outcome of laminar nanofluid used in a counterflow configuration. The parameters which are varied for the purpose of study are the volume concentration and diameter of nanoparticles in the based fluid. Water–alumina (Al_2O_3–water) is considered for the analysis of nanofluid by varying the Reynolds number in laminar flow conditions. Also, the variations of ratio of heat capacity rate heat exchanger are carried out to study the imbalance irreversibility along with pressure drop and heat transfer irreversibilities. Heat transfer and pressure drop study is also carried out for the variations of volume concentration and diameter of nanoparticles.

2 Formulations

In this analysis, same type of nanofluid is considered in both the streams of a counterflow heat exchanger. The assumptions used in this analysis are as follows: Fully developed laminar fluid flows in both the streams; perfect heat transfer exists between hot and cold streams; in the constant cross section of tubes, the pressure drop is due to friction loss; and material resistance between the streams is neglected [22].

2.1 *Exergetic Efficiency Expression*

The second-law efficiency is found for heat exchanger as [23],

$$\psi = \frac{\dot{E}_c^T}{\dot{E}_c^T + \dot{I}} \tag{1}$$

The exergy of the cold side is given by the following expression [23],

$$\dot{E}_c^T = C_c\left[\varepsilon\left(T_{\mathrm{h,in}} - T_{\mathrm{c,in}}\right) - T_o \ln\left\{1 + \varepsilon\left(\frac{T_{\mathrm{h,in}}}{T_{\mathrm{c,in}}} - 1\right)\right\}\right] \tag{2}$$

Gouy-Stodola expression is calculated with respect to reference temperature and entropy generation rate as,

$$\dot{I} = T_{\mathrm{o}}\dot{S}_{\mathrm{gen}} \tag{3}$$

Equation (3) is re-written in terms of entropy generation number and minimum of heat capacity.

$$\dot{I} = T_{\mathrm{o}}(N_s C_{\mathrm{min}}) \tag{4}$$

where the N_s is calculated as follows [1].

$$N_S = \frac{\dot{S}_{\mathrm{gen}}}{(\dot{m}c_P)_{\mathrm{min}}} \tag{5}$$

The S_{gen} is defined for heat exchanger as [2],

$$\dot{S}_{\mathrm{gen}} = \left(\dot{m}c_{\mathrm{p}}\right)_c \ln\left(\frac{T_{\mathrm{c,out}}}{T_{\mathrm{c,in}}}\right) + \left(\dot{m}c_{\mathrm{p}}\right)_{\mathrm{h}} \ln\left(\frac{T_{\mathrm{h,out}}}{T_{\mathrm{h,in}}}\right) + \left(\frac{\dot{m}}{\rho T_{\mathrm{ave}}}\right)_{\mathrm{c}}(\Delta P)_{\mathrm{c}} + \left(\frac{\dot{m}}{\rho T_{\mathrm{ave}}}\right)_{\mathrm{h}}(\Delta P)_{\mathrm{h}} \tag{6}$$

The S_{gen} is expressed as follows [1],

$$\dot{S}_{gen,T} = (\dot{m}c_p)_c \ln\left(\frac{T_{c,out}}{T_{c,in}}\right) + (\dot{m}c_p)_h \ln\left(\frac{T_{h,out}}{T_{h,in}}\right) \quad (7)$$

$$\dot{S}_{gen,P} = \frac{2\dot{m}^3 f L}{\rho T_{ave} D A_c^2} \quad (8)$$

Heat capacity for hot and cold streams is given as

$$(\dot{m}c_p)_c = C_c \text{ and } (\dot{m}c_p)_h = C_h$$

Ratio of heat capacity rate is found as,

$$C = \frac{C_{max}}{C_{min}} \quad (9)$$

Considering C_{max} as C_h and C_{min} as C_c,

$$C = \frac{C_h}{C_c} \quad (10)$$

Using Eqs. (7) and (8) in Eq. (5), N_s for unbalanced heat exchanger is found as,

$$N_S = \ln\left[1 + \varepsilon\left(\frac{T_{h,in}}{T_{c,in}} - 1\right)\right] + C\ln\left[1 - \frac{1}{C}\varepsilon\left(1 - \frac{T_{c,in}}{T_{h,in}}\right)\right] + \left[\frac{2\dot{m}^3 f L}{\rho T_{ave} D A_c^2}\right]_c + C\left[\frac{2\dot{m}^3 f L}{\rho T_{ave} D A_c^2}\right]_h \quad (11)$$

The effectiveness and NTU = heat transfer unit expression for the counterflow configuration heat exchanger are as follows,

$$\varepsilon = \frac{1 - e^{-NTU(1 - C_{min}/C_{max})}}{1 - (C_{min}/C_{max})e^{-NTU(1 - C_{min}/C_{max})}} \quad (12)$$

NTU is obtained as,

$$NTU = \frac{UA}{C_{min}} \quad (13)$$

Heat exchanger thermal resistance is obtained by the addition of heat exchanger resistances of both sides [24],

$$\frac{1}{UA} = \frac{1}{(hA)_c} + \frac{1}{(hA)_h} \quad (14)$$

Here, U = heat exchanger overall heat transfer coefficient and h = heat transfer coefficients for cold and hot streams. Here, in the analysis, the material resistance of conduction separating the two streams is assumed very small and can be neglected. Using NTU expression from Eq. (13), Eq. (14) becomes

$$\frac{1}{\mathrm{NTU}} = \frac{1}{(\mathrm{NTU})_{\mathrm{c}}} + \frac{1}{(\mathrm{NTU})_{\mathrm{h}}} \tag{15}$$

Stanton number is defined as,

$$\mathrm{St} = \frac{h}{c_{pG}} \tag{16}$$

Here, G = mass velocity. Considering the Stanton number and NTU expressions, the combination of NTU Eq. (14) used for cold and hot sides is obtained as,

$$\frac{1}{\mathrm{NTU}} = \left(\frac{A_{\mathrm{c}}}{A}\right)_{\mathrm{c}} \frac{1}{(\mathrm{St})_{\mathrm{c}}} + \frac{C_{\mathrm{c}}}{C_{\mathrm{h}}} \left(\frac{A_{c}}{A}\right)_{\mathrm{h}} \frac{1}{(\mathrm{St})_{\mathrm{h}}} \tag{17}$$

Here, the tube cross sectional area is represented as A_c and surface area is given as A. Considering the same geometrical dimensions of both stream tubes and assuming same fluid is flowing in both tubes' sides, Eq. (17) reduces to,

$$\mathrm{NTU} = \frac{4L}{D}\mathrm{St}\left[\frac{C}{C+1}\right] \tag{18}$$

For the assumptions done as above, the total N_s defined for unbalanced heat exchanger in respect of Eq. (11) is given as,

$$\begin{aligned} N_S = {} & \ln\left[1 + \varepsilon\left(\frac{T_{\mathrm{h,in}}}{T_{\mathrm{c,in}}} - 1\right)\right] \\ & + C\ln\left[1 - \frac{1}{C}\varepsilon\left(1 - \frac{T_{\mathrm{c,in}}}{T_{\mathrm{h,in}}}\right)\right] + \frac{2fL}{\rho D A_{\mathrm{c}}^2}\left[\left(\frac{\dot{m}^3}{T_{\mathrm{ave}}}\right)_{\mathrm{c}} + C\left(\frac{\dot{m}^3}{T_{\mathrm{ave}}}\right)_{\mathrm{h}}\right]_{\mathrm{c}} \end{aligned} \tag{19}$$

As the considered nanofluid flow is fully developed laminar, the ratio of heat exchanger length to diameter will be more than 100. Therefore, we will obtain constant heat transfer coefficient. For uniform circular duct which is a conventional channel uniform heat flux, the Nusselt number will be 4.364 [3, 22, 24]. The Stanton number is given as,

$$\mathrm{St} = \frac{\mathrm{Nu}}{\mathrm{Re}\cdot\mathrm{Pr}} \tag{20}$$

With the use of Reynolds number, the friction factor is defined as,

$$f = \frac{16}{\mathrm{Re}} \tag{21}$$

Heat exchanger heat transfer rate is defined with effectiveness as,

$$\dot{Q} = C_{\min}\varepsilon\left(T_{\mathrm{h,in}} - T_{\mathrm{c,in}}\right) \tag{22}$$

Heat exchanger pressure drop equation is obtained as,

$$\Delta P = \left(\frac{fL\rho V^2}{2Dg}\right)_{\mathrm{c}} + \left(\frac{fL\rho V^2}{2Dg}\right)_{\mathrm{h}} \tag{23}$$

2.2 Thermophysical Properties of Nanofluid

Referring to [5], thermophysical properties considered in the investigation are as follows. Nanofluid density is calculated as,

$$\rho_{\mathrm{nf}} = (1-\varphi)\rho_{\mathrm{bf}} + \varphi\rho_{\mathrm{p}} \tag{23}$$

Specific heat of nanofluid is calculated as,

$$c_{p,\mathrm{nf}} = \frac{(1-\varphi)\rho_{\mathrm{bf}}c_{\mathrm{p,bf}} + \varphi\rho_{\mathrm{p}}c_{\mathrm{p,p}}}{\rho_{\mathrm{nf}}} \tag{24}$$

Nanofluid thermal conductivity is calculated as,

$$\frac{k_{\mathrm{nf}}}{k_{\mathrm{bf}}} = 1 + 4.4\mathrm{Re}_d^{0.4}\,\underset{\mathrm{bf}}{\mathrm{Pr}}^{0.66}\left(\frac{T_{\mathrm{nf}}}{T_{\mathrm{fr}}}\right)^{10}\left(\frac{k_p}{k_{\mathrm{bf}}}\right)^{0.03}\varphi^{0.66} \tag{25}$$

Here, base fluid freezing point is T_{fr}. Reynolds number of nanoparticles is calculated as,

$$\mathrm{Re}_d = \frac{\rho_{\mathrm{bf}}u_{\mathrm{b}}d_{\mathrm{p}}}{\mu_{\mathrm{bf}}} \tag{26}$$

where Brownian velocity is calculated as,

$$u_b = \frac{2k_bT_{\mathrm{nf}}}{\pi\mu_{\mathrm{bf}}d_{\mathrm{p}}^2} \tag{27}$$

where k_b is Boltzmann's constant given as 1.38066×10^{-23} J/K. Boltzmann's constant is $k_b = 1.38066 \times 10^{-23}$ J/K.

Nanofluid viscosity is obtained by the following relation,

$$\frac{\mu_{\mathrm{nf}}}{\mu_{\mathrm{bf}}} = \frac{1}{1 - 0.10255 d_{\mathrm{p}}^{-0.264} \varphi^{1.028}} \tag{28}$$

3 Results and Discussion

In this analysis, the formulae considered are empirical which are obtained from open literature. These empirical relations are formulated and validated by experimental results and numerical procedure by earlier researchers.

For the investigation, the input values considered as followed in [5, 22, 24] are as follows: Hot and cold stream temperature ratio at inlet ($T_{h,in}/T_{c,in}$) is 1.5; cold stream temperature at inlet ($T_{\mathrm{c,in}}$) is 301 K; and length of heat exchanger tube (L) is 1 m. The reference temperature considered as (T_{o}) is also equal to $T_{\mathrm{c,in}}$. Water and alumina (Al_2O_3–water) nanofluid properties are obtained from [3]. Water (base fluid): density, ρ_{bf} is 1000 kg/m^3; thermal conductivity, k_{bf} is 0.6 W/m K; viscosity, μbf is 0.001 N s/m^2; and specific heat, $c_{\mathrm{p,bf}}$ is 4180 J/kg K.

Alumina (nanoparticle): density, ρ_{p} is 3900 kg/m^3; thermal conductivity, k_p is 40 W/m K; and specific heat, $c_{\mathrm{p,p}}$ is 880 J/kg K. The analysis program is solved in EES software [25]. For laminar flow which is fully developed, the two values of Reynolds number considered are 1500 and 2000. The capacity ratio values are 1 and 2 considered to study the variations of imbalanced reversibility. The volume concentration value as ($\varphi = 0.03$) and particle diameter value as ($d_{\mathrm{p}} = 20$ nm) are considered for analysis. The fluid properties are referred at average temperatures.

To validate the present work, comparison with previous work [3] is carried out by considering same input values. The results of validation and comparison are shown in Fig. 1 where the performance parameter entropy generation ratio is plotted by varying volume fractions (φ). Here, entropy generation ratio is defined as ratio of entropy generation rate with nanofluid versus entropy generation rate with base fluid (without nanofluid) in heat exchanger. The deviation of results observed is about 5–7%, and it is having acceptable range. This is due to differences in thermophysical property variations of nanofluid.

Figure 2 shows the variations of ψ with change in the values of φ for different values of Reynolds number (Re) and heat capacity rate ratio (C). As φ increases, ψ will initially increase, establishes maximum value and further decreases. Corresponding to the attained higher value of ψ, it provides optimum value of φ for lower losses and increased performance for the considered input values. This is because of increase in heat transfer due to increase in φ which leads to ψ increase initially. After certain φ value, there is decreasing trend because of pressure drop increase due to increase in nanofluid viscosity which decrease ψ after attending peak value. Also, we can observe from Fig. 2 that as the value of heat capacity ratio changes

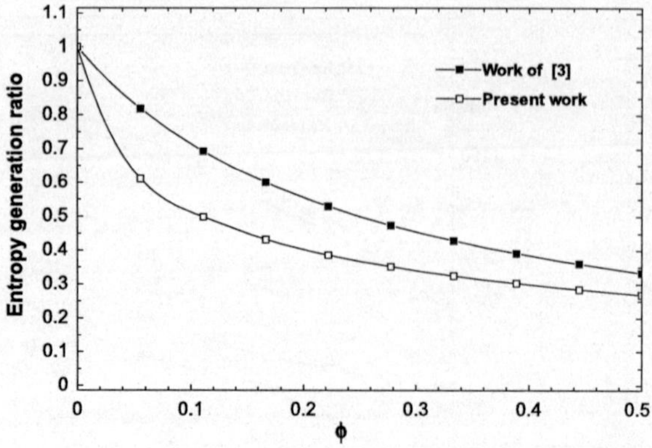

Fig. 1 Validation of present work with previous work [3]. Entropy generation ratio versus volume fraction (φ)

from 1 to 2, the exergetic efficiency increases. This is due to increase in randomness of nanoparticles when there is more of imbalance between the fluid streams. As Re decreases in the range of laminar flow, ψ increases, indicating lowering of randomness and losses.

Figure 3 gives the behavior of Q and ΔP for different values of nanoparticle volume concentration (φ). It is observed that as φ increases, Q increases indicating the advantages of nanofluid usage in heat exchangers leading in increase of heat transfer. When $\varphi = 0$, the value of Q will be lowest. This corresponds to the base fluid only, that is the conventional working fluid. For the given value of φ, the value of Q will increase as the value of Re increased in the given laminar region. This is because of increase in randomness of nanoparticles. As C reduced from 2 to 1, Q reduces because of decrease in imbalance irreversibility. Also seen in Fig. 3 is the increase in ΔP for increase in φ because of increase in viscosity. As Re reduces, the ΔP reduces accordingly because of reduction in randomness of fluid. There is no effect for changes in the value of C for Re = 1500 in the case of laminar flow.

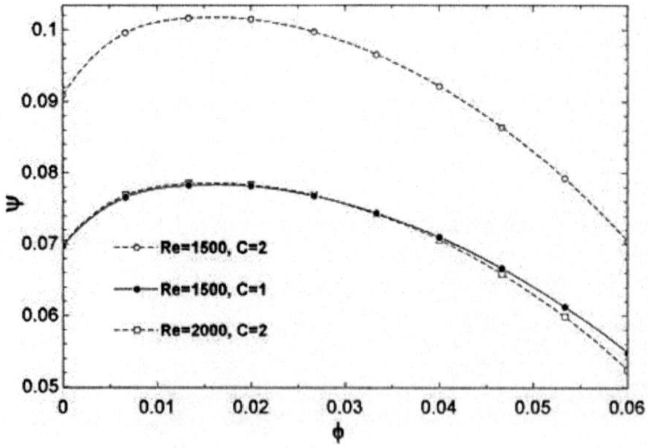

Fig. 2 Exergetic efficiency (ψ) versus volume fractions (φ)

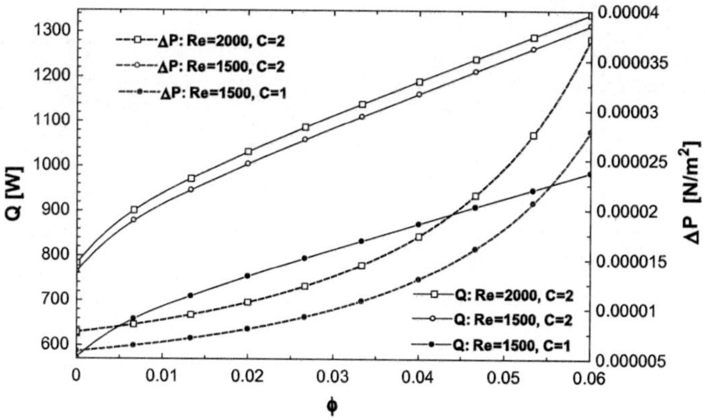

Fig. 3 Heat transfer rate (Q) and pressure drop (ΔP) versus volume fractions (φ)

In Fig. 4, the relationship of ψ with the nanoparticle diameter (d_p) variation is presented. As d_p increases in nanofluid, the ψ increases with more slope initially and with less slope then afterward. This is due to increase in effectiveness of heat exchanger. This indicates that after certain value of d_p, there will be no appreciable increase in ψ. As Re decreased, the value of ψ increases for a particular value of d_p. This is due to reduction of heat transfer irreversibility in laminar flow. As the value of C decreased, the ψ decreases for a particular value of d_p in the case of laminar flow. This is due to reduction of randomness.

Figure 5 provides the behavior of Q and ΔP for changes in the value of nanoparticle diameter (d_p). There is decreasing trend for Q for increase in the value of d_p. This is due to that as nanoparticle diameter increases, the thermal conductivity decreases.

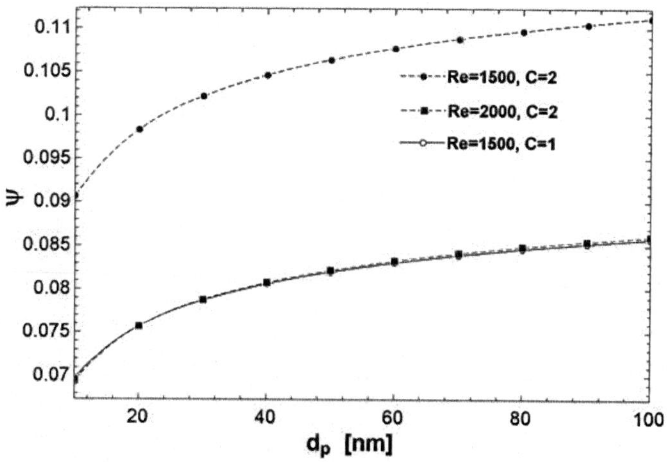

Fig. 4 Exergetic efficiency (ψ) versus nanoparticle diameter (d_p)

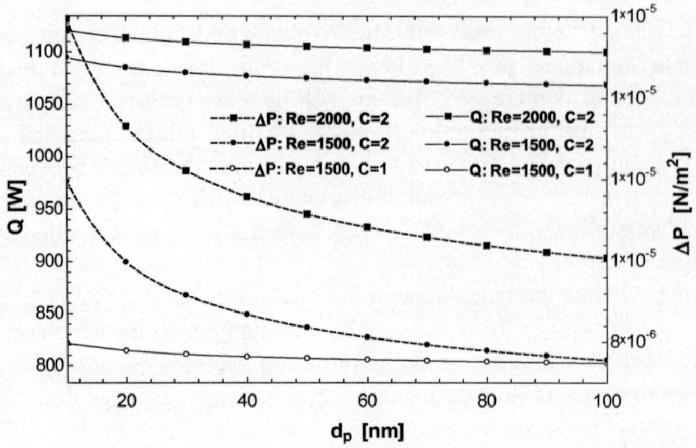

Fig. 5 Heat transfer rate (Q) and pressure drop (ΔP) versus nanoparticle diameter (d_p)

This is because of decrease in Brownian velocity as the nanoparticle size increases which leads to decrease in heat transfer [26]. In the same trend, the value of ΔP decreases as the nanoparticle size increases due to increase in viscosity. As the Re increased, the Q and ΔP increase. This is due to increase in randomness in the fluid. The value of Q decreases when the value of C reduces due to operating conditions. This is due to reduction of imbalance between the fluid streams. But as value of C decreased for Re 1500, there are no effects for ΔP in the case of laminar flow.

4 Conclusions

For a counterflow heat exchanger, exergetic analysis is performed with the use of nanofluid considering laminar flow. In this study, performance methods used are rational efficiency or exergetic efficiency which includes entropy generation. To analyze heat exchangers operated by nanofluids, the irreversibilities considered are heat transfer, pressure drop and imbalance between the fluid streams in a single closed form equation. With the method of exergetic analysis, the optimum value of nanoparticle volume concentration of nanofluid is obtained for higher thermal efficiency. This provides realistic operating parameters of nanofluid heat exchangers in heat transfer equipment with reduced losses leading to energy conservation. The present work is validated with the earlier work, and the variations in values are within 7%.

Water and alumina nanofluid combination is used for parametric study by varying nanofluid volume fraction and particle diameter. There is appreciation of heat transfer rate in heat exchangers considering nanofluid compared to conventional fluids for variations in volume fraction. But also, it is observed that the pressure drop irreversibility increases in the nanofluids. Therefore, the second-law analysis provides

trade-off between different irreversibilities to obtain optimum operating parameters. That is in heat exchangers, to achieve higher heat transfer rate, temperature difference between the streams is increased. But this will increase heat transfer irreversibility. To increase heat transfer rate, mass flow rate of fluid will be increased. This will lead to increase in pressure drop irreversibility. Also, if imbalance between the fluid streams increased, then also, irreversibility increases. So, exergy methods include all these irreversibilities, by which we can have trade-off between these losses to achieve higher output.

For increase in nanoparticle diameter, rational efficiency continuously increases. This is due to increase in effectiveness of heat exchanger. As the nanoparticle diameter increases, the heat transfer and pressure drop decrease because of decrease in Brownian motion in nanofluid. In the considered laminar nanofluid flow type, it will be advantageous to operate the counterflow heat exchanger with Reynolds number equal to 1500 and imbalance heat capacity rate ratio.

References

1. Bejan A (1996) Entropy generation minimization. CRS Press, Boca Raton
2. Manjunath K, Kaushik SC (2014) Second law thermodynamic study of heat exchangers: a review. Renew Sustain Energy Rev 1(40):348–374
3. Singh PK, Anoop KB, Sundararajan T, Das SK (2010) Entropy generation due to flow and heat transfer in nanofluids. Int J Heat Mass Transfer 53(21–22):4757–4767
4. Moghaddami M, Mohammadzade A, Esfehani SA (2011) Second law analysis of nanofluid flow. Energy Convers Manage 52(2):1397–1405
5. Bianco V, Manca O, Nardini S (2014) Entropy generation analysis of turbulent convection flow of Al_2O_3–water nanofluid in a circular tube subjected to constant wall heat flux. Energy Convers Manage 1(77):306–314
6. Khaleduzzaman SS, Sohel MR, Mahbubul IM, Saidur R, Selvaraj J (2016) Exergy and entropy generation analysis of TiO_2–water nanofluid flow through the water block as an electronics device. Int J Heat Mass Transf 1(101):104–111
7. Bahiraei M, Majd SM (2016) Prediction of entropy generation for nanofluid flow through a triangular minichannel using neural network. Adv Powder Technol 27(2):673–683
8. Bhattad A, Sarkar J, Ghosh P (2019) Energetic and exergetic performances of plate heat exchanger using brine-based hybrid nanofluid for milk chilling application. Heat Transfer Eng 19
9. Nakhchi ME, Rahmati MT (2021) Entropy generation of turbulent Cu–water nanofluid flows inside thermal systems equipped with transverse-cut twisted turbulators. J Therm Anal Calorim 143(3):2475–2484
10. Bhattad A, Sarkar J, Ghosh P (2020) Heat-transfer characteristics of plate heat exchanger using hybrid nanofluids: effect of nanoparticle mixture-ratio. Heat and Mass Transfer, 1–16
11. Shafee A, Sheikholeslami M, Jafaryar M, Babazadeh H (2021) Irreversibility of hybrid nanoparticles within a pipe fitted with turbulator. J Therm Anal Calorim 143(1):715–723
12. Manjunath K (2021) Hybrid nanofluid laminar flow analysis in double pipe heat exchanger (February 28, 2021). In: Proceedings of the international conference on systems, energy & environment (ICSEE). https://doi.org/10.2139/ssrn.3794702
13. Bahiraei M, Naseri M, Monavari A (2021) A second law analysis on flow of a nanofluid in a shell-and-tube heat exchanger equipped with new unilateral ladder type helical baffles. Powder Technol 1(394):234–249

14. Garud KS, Hwang SG, Lim TK, Kim N, Lee MY (2021) First and second law thermodynamic analyses of hybrid nanofluid with different particle shapes in a microplate heat exchanger. Symmetry 3(8):1466
15. Sahoo RR (2021) Heat transfer and second law characteristics of radiator with dissimilar shape nanoparticle-based ternary hybrid nanofluid. J Therm Anal Calorim 146(2):827–839
16. Wang H, Pang M, Diao Y, Zhao Y (2022) Heat transfer characteristics and flow features of nanofluids in parallel flat minichannels. Powder Technol 1(402):117321
17. Nogueira É (2022) Influence of nanoparticle shapes of boehmite alumina on the thermal performance of a straight microchannel printed circuit heat exchanger. J Metal Mater Res 5(01)
18. Bahiraei M, Monavari A (2022) Irreversibility characteristics of a mini shell and tube heat exchanger operating with a nanofluid considering effects of fins and nanoparticle shape. Powder Technol 7:117117
19. Yang L, Baghaei S, Suksatan W, Barnoon P, Davidyants A, El-Shafay AS (2022) Numerical assessment of the influence of helical baffle on the hydrothermal aspects of nanofluid turbulent forced convection inside a heat exchanger. Sci Rep 12(1):1–5
20. Irshad K, Islam N, Zahir MH, Pasha AA, Abdel Gawad AF (2022) Thermal performance investigation of Therminol55/MWCNT+ CuO nanofluid flow in a heat exchanger from an exergy and entropy approach. Case Stud Thermal Eng 1(34):102010
21. Manjunath K (2022) Optimization of nanofluid parameters for double pipe heat exchanger. In: Recent advances in manufacturing, automation, design and energy technologies. Springer, Singapore, pp 803–814
22. Manjunath K, Kaushik SC (2015) Second law efficiency analysis of heat exchangers. Heat Transfer—Asian Res 44(2):89–108
23. Kotas T-J (2013) The exergy-method of thermal plant analysis. Elsevier
24. Manjunath K, Kaushik SC (2014) The second law analysis of an unbalanced constructal heat exchanger. Int J Green Energy 11(2):173–192
25. Klein SA (2008) Engineering-equation-solver, version 8.158., F-Chart Software, Middleton, WI
26. Apmann K, Fulmer R, Scherer B, Good S, Wohld J, Vafaei S (2022) Nanofluid heat transfer: enhancement of the heat transfer coefficient inside microchannels. Nanomaterials 12(4):615

Investigation on Lead Free Ferroelectric $[(Ba_{0.825+x}Ca_{0.175-x})(Ti_{1-x}Sn_x)O_3]$ Ceramics for Energy Storage Density and Thermal Energy Harvesting Capacity

Shatrughan Singh, Ashok Kumar Yadav, Niraj Kumar, Umakanta Choudhury, and Mukesh Kumar

1 Introduction

Lead-based ceramics are very harmful, and it is the origin for various undesirable problems like damage of brains and cancer, etc. Ferroelectric materials which are environment friendly and without lead are having more affection in regards to researchers as the evolution of new materials. Nowadays, mostly this kind of materials have been using mainly as long as vitality repository as well as heated vitality accumulation [5, 10]. Therefore, the investigation of pyroelectric materials is in trend for mostly vitality accumulation applications. This kind of materials change heated vitality into electrical signals [4]. Various researchers have been proposed a lot of unique patterns for power storage and power harvesting utilization. Olsen was explored the basics of energy harvesting concepts by adopting ferroelectric hysteresis loop. Moreover, pyroelectric coefficient is not straight touching power accumulation effectiveness regards to materials along with electric field-convinced polarization is utilized for output power against ferroelectric materials [1, 6]. Further, ferroelectric materials explored for huge power density capacitor functions [8].

S. Singh
Mechanical Engineering Department, IILM College of Engineering & Technology, Greater Noida, India

A. K. Yadav (✉) · U. Choudhury
Department of Mechanical Engineering, Raj Kumar Goel Institute of Technology, Ghaziabad 201003, India
e-mail: ashokme015@gmail.com

N. Kumar
Department of Mechanical Engineering, Ghani Khan Choudhury Institute of Engineering & Technology, Malda, India

M. Kumar
Department of Mechanical Engineering, Gautam Buddha University, Greater Noida, India

A. Kumar et al. (eds.), *Recent Advances in Manufacturing and Thermal Engineering*, Lecture Notes in Mechanical Engineering, https://doi.org/10.1007/978-981-19-8517-1_38

Environmental friendly and lead free ferroelectric materials have been studied in details. Therefore, those kinds of material compositions are manufactured nearby the region of morphotropic phase boundary; so, the arrangement may display the huge pyroelectric, piezoelectric as well as ferroelectric properties [3]. After all, $Ba_{0.85}Ca_{0.15}Ti_{0.9}Zr_{0.1}O_3$ (BCT-BZT) ceramics get described as long as big pyroelectric or piezoelectric properties just as correlated into alternative lead free or along with lead components on minimum electric field. Therefore, these materials show great values of all parameters [8]. So, BCT-BZT displays the big dielectric properties in this area [2] and having less dielectric damage about 0.001 that is teeny with respect of more ferroelectric materials [9]. This material having huge pyroelectric coefficient of 5.84×10^{-8} C/(cm^{2}K) nearby room temperature and that is noted bigger value with respect to $BaTiO_3$-dependent materials. There are the possibilities of enhancement in piezoelectric, pyroelectric and ferroelectric properties with the presence about MPB nearby room temperature as regards to ferroelectric rhombohedral or tetragonal phases. BCT-BZT has focused in power shift or electrocaloric by cause of, it displays less electric field congestion. So, it has been proved that a big EC desire a huge entropy (S) and thermal (T) revolution along electric field. Thus, demands as long as catching of huge voltage are heavy, high-priced and contribute to be risky. Therefore, most of the researchers focused on freezing machines could be regulated through less voltage. For experimental designs, improvement of dielectric materials and able to develop giant EC through comparatively less implemented electric field shift E needed [7]. This kind of power changeover also needed less electric field. Accordingly, Sn doped $[(Ba_{0.825+x}Ca_{0.175-x})(Ti_{1-x}Sn_x)O_3$ [abridge such as BCT-Sn after this] looks feasible substitute for the replacement of Pb dependent oxides for dielectric applications. Furthermore, BCT-Sn can also be utilized as well as solid state refrigeration applications. Therefore, we can say BCT-Sn may be a specialized required material as the counterfeit of lead-dependent materials for vitality repository as well as heated vitality accumulation potentiality application.

2 Materials and Method

Polycrystalline $[(Ba_{0.825+x}Ca_{0.175-x})(Ti_{1-x}Sn_x)O_3]$ ceramics were made in reach a solid state reaction path. Three distinct patterns were prepared through varying the portion as regards Sn ($x = 0.0625$, 0.0875 and 0.100). Pure crumb of $BaCO_3$, CaO $SnCl_2$ and TiO_2 were utilized. According to stoichiometric ratio, the powders were prepared after weighing. By using the mortar and pestle for the better physical homogeneity, the mixture was ground in acetone. The crumb calcined at 1200 °C as long as 4 h with atmosphere. Thereafter, the PVA binder (2% through weight) was deposited in calcined crumb to provide the density to the sample. The pellets were prepared by using hydrostatic press and apply an uniaxial pressure about 6 ton/cm^2. Sintering is done at 1450 °C among 4 h bide time. Silver paste was utilized for providing the connections. The phase condition was firmly established via X-ray diffraction facility. Microstructure has been checked adopting scanning electron microscope

(SEM) as concern the processed patterns. Evaluate the density through Archimedes principle of samples which is covered during examination. The polarization (P) versus electric field (E) hysteresis shifts on separate electric field along with temperatures determined by using Sawyer Tower circuit.

3 Results and Discussion

Figure 1 displays the X-ray diffraction movement same as lead free $[(Ba_{0.825+x}Ca_{0.175-x})(Ti_{1-x}Sn_x)O_3]$ ceramics along $x = 0.0625$, 0.0875 and 0.100 patterns. It verifies the single state formation along perovskite architecture. During the XRD, no inferior crest was recorded. So, Sn ions have mixed with the B site to convert a uniform solid mixture.

Figure 2 expositions the images which were arranged by the sintered surface for all the compositions through scanning electron microscopy (SEM). In the course of investigation, we have to found that no any big porosity are visibly on the surface area of all samples. Moreover, through checking the density by Archimedes principle and found the relative density ~95% with respect to theoretical density.

Figure 3 displays the hysteresis curve of P–E loops as long as compositions with $x = 0.0625$, 0.0875 and 0.100 patterns at 293 K temperature and 25 kV/cm and also shows P–E loops as long as all patterns under examinations.

Figure 4a, b show the characteristic as long as vitality repository as well as heated vitality accumulation potentiality. So, it permits the concept about the investigation way by which we can vitality repository as well as heated vitality accumulation potentiality.

In Fig. 4b, the data recorded that the reversible polarization may be build in the form of a clockwise twist in the middle of two distinct temperatures. The cycle

Fig. 1 X-ray diffraction pattern of calcined powder

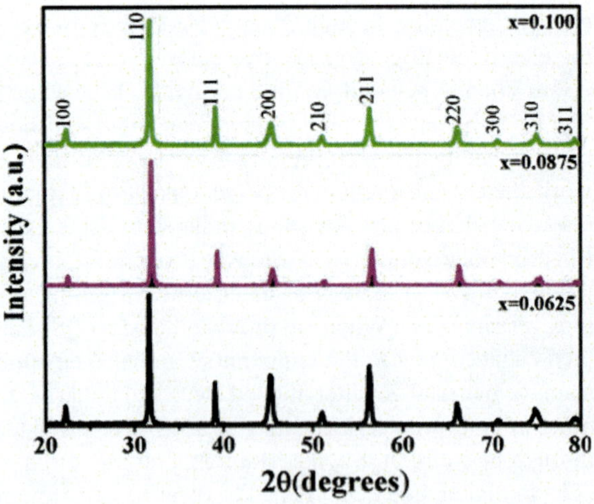

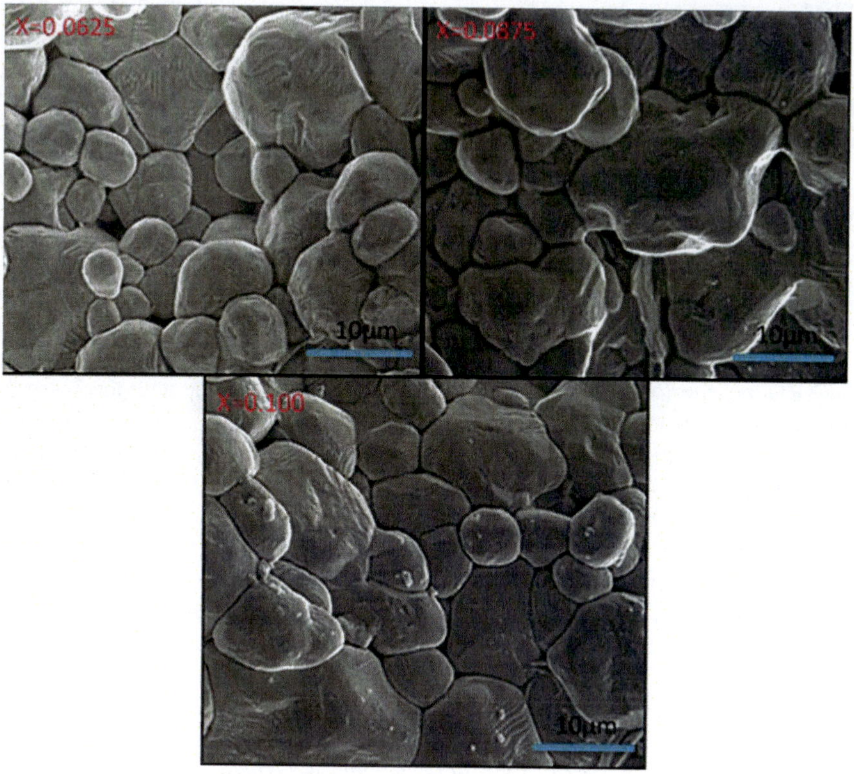

Fig. 2 SEM image of sintered pellets

form is called Olsen cycle and through it can be change the form of energy from thermal energy to electricity. Hence, Fig. 4b displays the complete processes for energy conversion through Olsen cycle. In the Olsen cycle, the steps 1–2–3–4–1 show the electric analog, so respective embedded area also shows the maximum electric power change per unit volume per cycle. Whenever the Olsen cycle is running in bipolar P–E twist then the inside area (1–2–1–1) shows the hysteresis loss. So, we can say that hysteresis forgetting is very less in single polar P–E twist. In the present work, the enhancement in the power changes is explained through the relocated twist (1–2–3–4–1) because for single polar state hysteresis forgetting may be neglected. Material polarization rises, when the process 1–2 is conduct on a fixed downside heat temperature (TL) as well as electric field rises from the downside restrict (EL) to long restrict (EH). When we provide the heat (QS) during steps 2–3 to the material, then enhancement in the temperature and also developed lattice fluctuation. Therefore, the material get alter at fixed electrical field (EH). So, there is a development of much alteration current that may be utilized or saved to drive the electronic machines through best circuit. So, the heat temperature of material rises from lower side heat (TL) to higher side heat (TH). The electric field minimizes from EH to EL on fixed

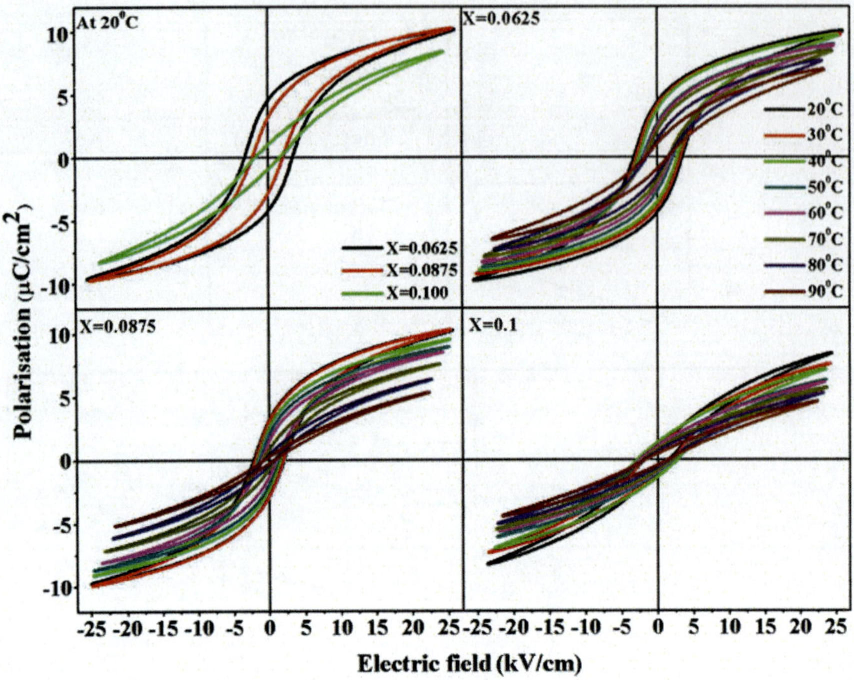

Fig. 3 P–E hysteresis loops at 293 K as long as all patterns and P–E hysteresis loops along with in heat change from 200 to 900 °C

heating temperature TH, during the process 3–4. Without electric field, the polarization minimizes and which developed a weak depolarization current. After that, during the process 4–1, extracting the temperature from the body on fixed electric field (EL) is completed, then the pattern material approaches to its starting position or cycle is completed. By utilizing these steps, we can estimate the power density per unit quantity of the pattern material [3].

Figure 5 shows the vitality repository and vitality repository efficiency as long as all compositions under study. For the vitality repository, it has been seen that the power repository weight rises along the increase in a quantity of Sn up to $x = 0.0875$ and then decreases afterward. The results of maximum vitality repository potentiality were 62.04 kJ/m^3 for $x = 0.0875$ pattern.

Figure 6 shows that vitality accumulation vs electric field as regards $x = 0.0625$, 0.0875 and 0.100 samples at 25 kV/cm and at also with heat changes. During investigation, EL and TL are counted as 25 kV/cm electric field and 293 K temperature. The maximum vitality accumulation 90 kJ/m^3 is observed. At the minimum value of electric field and change in temperature, BCT-Sn having a very good candidature for thermal vitality accumulation.

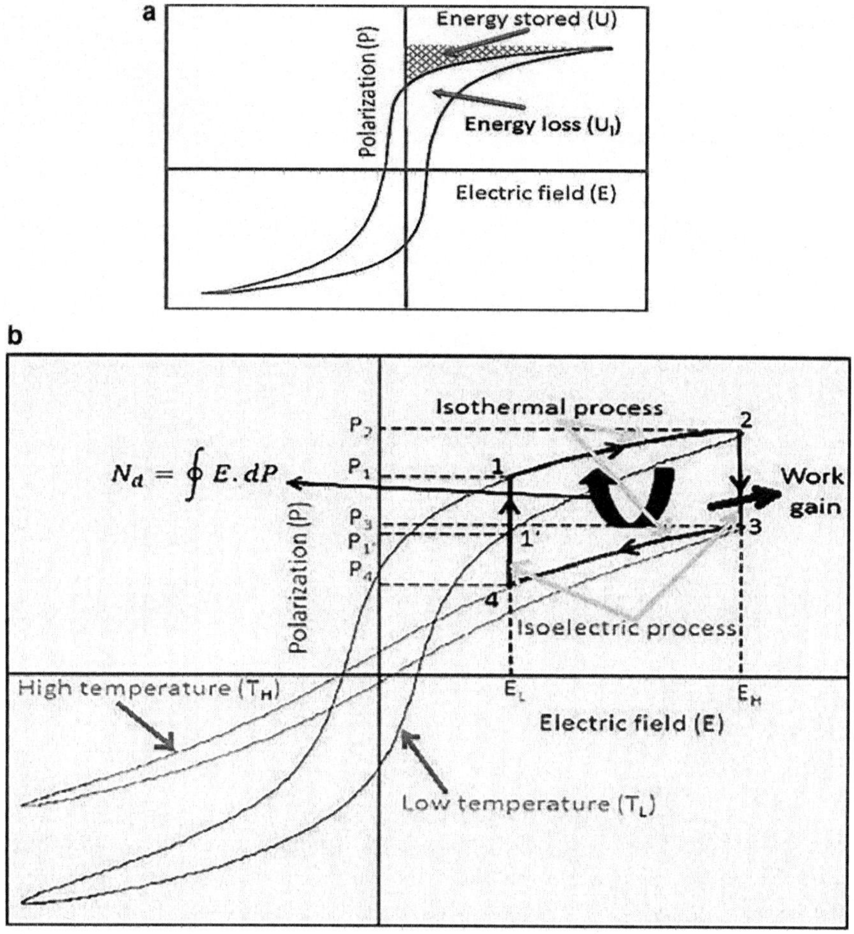

Fig. 4 **a** Vitality repository characteristic. **b** Heated vitality accumulation cycle

4 Conclusions

In the present work, lead free [$(Ba_{0.825+x}Ca_{0.175-x})(Ti_{1-x}Sn_x)O_3$] ceramics along $x = 0.0625$, 0.0875 and 0.100 samples were explored in detail with the help of Olsen cycle. The specimen was sintered on 1450 °C as long as time interval 4 h to study the vitality repository as well as heated vitality accumulation applications. The maximum vitality repository as well as heated vitality accumulation are found 62.04 kJ/m^3 and 90 kJ/m^3 at $x = 0.0875$.

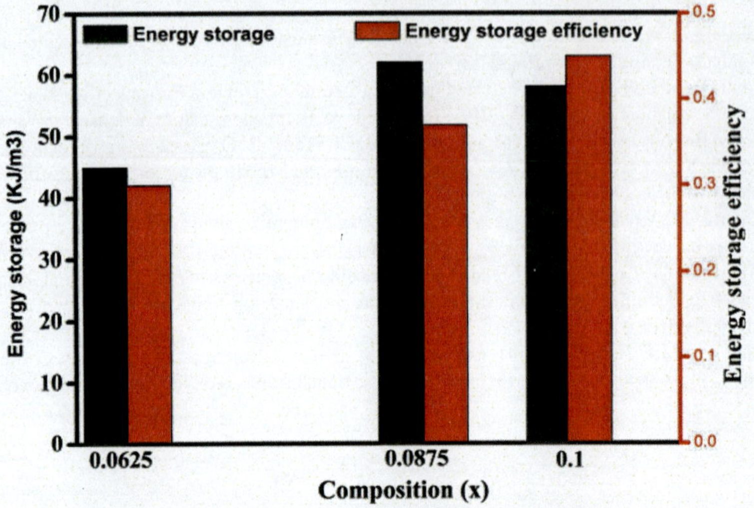

Fig. 5 Plot for vitality repository and vitality repository efficiency

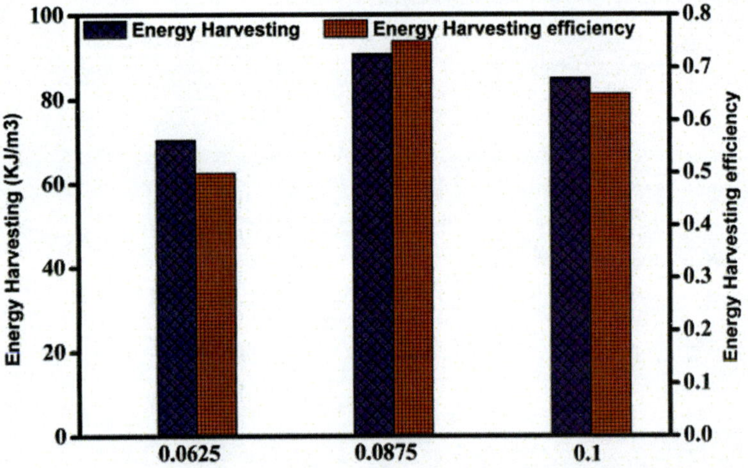

Fig. 6 Plot for comparison of heated vitality accumulation potentiality in compositions under study

References

1. Chauhan A, Patel S, Vats G, Vaish R (2014) Enhanced thermal energy harvesting using Li, K-doped $Bi_{0.5}Na_{0.5}TiO_3$ lead-free ferroelectric ceramics. Energy Technol 2:205–209
2. Coondoo I, Panwar N, Amorin H, Alguero M, Kholkin A (2013) Synthesis and characterization of lead-free $0.5Ba(Zr_{0.2}Ti_{0.8})O_3$–$0.5(Ba_{0.7}Ca_{0.3})TiO_3$ ceramic. J Appl Phys 113:214107
3. Kutnjak Z, Petzelt J, Blinc R (2006) The giant electromechanical response in ferroelectric relaxors as a critical phenomenon. Nature 441(22):956–959

4. Lee FY, Navid A, Pilon L (2012) Pyroelectric waste heat energy harvesting using heat conduction. Appl Therm Eng 37:30–37
5. Panda P, Sahoo B (2015) PZT to lead free piezo ceramics: a review. Ferroelectrics 474:128–143
6. Patel S, Chauhan A, Vaish R (2015) Improved electrical energy storage density in vanadium-doped $BaTiO_3$ bulk ceramics by addition of $3BaO–3TiO_2–B_2O_3$ glass. Energy Technol 3:70–76
7. Saadon S, Sidek O (2011) A review of vibration-based MEMS piezoelectric energy harvesters. Energy Convers Manage 52:500–504
8. Sharma M, Vaish R, Chauhan VS (2016) Development of figures of merit for pyroelectric energy harvesting devices. Energy Technol 4:1–9
9. Wang P, Li Y, Lu Y (2011) Enhanced piezoelectric properties of $(Ba_{0.85}Ca_{0.15})(Ti_{0.9}Zr_{0.1})O_3$ lead-free ceramics by optimizing calcination and sintering temperature. J Eur Ceram Soc 31:2005–2012
10. Zhang S, Li F, Jiang X, Kim J, Luo J, Geng X (2015) Advantages and challenges of relaxor-$PbTiO_3$ ferroelectric crystals for electroacoustic transducers—a review. Prog Mater Sci 68:1–66

Performance Enhancement and Emissions Reduction in a DME Fueled Compression Ignition Engine Using Hydrogen Under Dual-Fuel Mode

Anilkumar Shere and K. A. Subramanian

Nomenclature

CI	Compression ignition
DICI	Direct injection compression ignition
DME	Dimethyl ether
H_2	Hydrogen
DME ES	Dimethyl ether energy share
HES	Hydrogen energy share
BTE	Brake thermal efficiency
BSEC	Brake specific energy consumption
BSFC	Brake specific fuel consumption
BMEP	Brake mean effective pressure
ECU	Electronic control unit
CN	Cetane number
LHV	Lower heating value
NDIR	Non-dispersive infrared sensor
HRR	Heat release rate
CHR	Cumulative heat release
LTR	Low-temperature region
HTR	High-temperature region
HCCI	Homogeneous charge compression ignition
RCCI	Reactivity controlled compression ignition
PM	Particulate matter
HC	Hydrocarbon
CO	Carbon monoxide

A. Shere · K. A. Subramanian (✉)
Engines and Unconventional Fuels Laboratory, Department of Energy Science and Engineering, Indian Institute of Technology Delhi, New Delhi, India
e-mail: subra@dese.iitd.ac.in

A. Kumar et al. (eds.), *Recent Advances in Manufacturing and Thermal Engineering*,
Lecture Notes in Mechanical Engineering, https://doi.org/10.1007/978-981-19-8517-1_40

CO_2	Carbon dioxide
NO_x	Oxides of nitrogen

1 Introduction

The developing countries are looking for the indigenous renewable energy source due to the fluctuation in crude oil prices and the increase in the demand for fuels. The concern over the environmental protection and depletion of crude oil move the efforts of researchers toward energy security by finding alternative fuels. The researchers are focused on the achievement of high performance as well as ultra-low emissions to meet the emission standard norms.

The compression ignition (CI) engines have significant advantages in high power, high fuel economy, durability, and low carbon dioxide emissions. CI engines are used globally for trucking, shipping, rail, electricity generation, and other tasks requiring reliable power. The diesel engine has the highest thermal efficiency compared to the spark-ignition engine. The dependency on internal combustion engines does not decrease in the next few decades. However, the higher smoke and NO_x emissions from diesel engines are still a serious concern. So, to meet the standard emission norms, exhaust emissions must be reduced without compromising the engine's efficiency. Therefore, a lot of research is focused on high efficiency and ultra-low emissions to develop advanced engine systems. The alternative fuels such as dimethyl ether (DME), biodiesel, and Fischer–Tropsch diesel, having higher cetane numbers, are more suitable for CI engines. Among the alternative fuels, dimethyl ether (DME) is one of the best suitable fuels for CI engines. Under atmospheric conditions, it can be stored in a liquid state and converted to a gaseous state.

Dimethyl ether (DME) fuel has been widely studied in recent years because of its higher cetane number (~58) and is considered a promising alternative fuel for CI engines. DME belongs to the ether family (CH_3–O–CH_3) and can be produced from coal or biomass through gasification and the Fischer–Tropsch synthesis process. It can also be produced via catalytic dehydration of methanol. According to the Niti Ayog, Government of India, DME has the potential to play a significant role in improving India's energy security (petroleum oil import is about 87%). Niti Ayog also stated that DME is a more efficient fuel that emits less NO_x and produces no smoke than diesel fuel. It is a non-toxic fuel that can be handled safely. India has the world's fifth largest coal reserves, which, if used wisely, can contribute significantly to DME production [1]. DME fuel is typically stored in liquid form in a pressurized cylinder, where it can quickly evaporate into gaseous form (less than 5 bar) under atmospheric conditions. Under dual-fuel mode, DME fuel can be used in the CI engines without requiring significant changes to the engine's design or hardware system. The intake system of the engine needs to be modified to integrate a DME injector to inject DME in a gaseous form [2]. However, DME-diesel dual-fueled CI engines suffered from problems like marginal increase in CO and HC emissions

compared to base diesel operations [3]. This problem can be addressed by utilizing hydrogen gas as a secondary fuel (additive) in the CI engine under DME-diesel mode, as hydrogen is a clean fuel that possesses a wide flammability range with higher flame speed and no C–C bond, which produces water vapor as a by-product.

Researchers investigated and examined the utilization of different alternative fuels as secondary fuel under various dual-fuel modes and analyzed its impact on performance of the engine. Bo et al. studied the effect of DME and ethanol on fuel consumption in a single-cylinder CI engine. The emulsified fuel (diesel emulsified with water) was injected into the combustion chamber, and DME or ethanol fuel was first heated to about 1000 K, and then the product gas was inducted into the intake manifold of the engine. They reported that BSFC and diesel fuel consumption could be decreased by 10% and 18%, respectively [4]. Saleh studied the effect of LPG fuel with five different compositions in a two-cylinder CI engine under dual-fuel mode and compared the performance and emissions characteristics with conventional diesel mode. Their finding indicates that the LPG fuel with higher butane content reduces NO_x emission and higher propane content leads to lower CO emission. 40% of LPG energy share was obtained under dual-fuel mode. The LPG fuel with 30% butane content was considered a better fuel as the SO_x and NO_x emissions were reduced by 69% and 27% and CO emission rises to 15.7% at full load with better engine performance than neat diesel fuel [5]. Kakoee et al. examined the influence of hydrogen as an additive in a single-cylinder RCCI engine under diesel-natural gas and DME-natural gas dual-fuel mode. Three different percentages of hydrogen (3, 6, and 9%) were inducted in diesel-natural gas and DME-natural gas dual-fuel mode and compared the combustion, performance, and emission characteristics. The results showed that 9% of hydrogen addition advanced the start of combustion by 2 CA degree in DME-natural gas mode, whereas it was observed to be 0.3 CA degree in diesel-natural gas mode. The emissions of CO and HC increased, and NO_x emission decreased in the case of DME-natural gas as compared to diesel-natural gas by adding 9% of hydrogen as an additive [6].

Chen et al. investigated and reviewed the emissions of automobiles fueled with dimethyl ether (DME), methanol, natural gas, ethanol, biodiesel, and polyoxymethylene dimethyl ether (PODME). They reported that natural gas-diesel dual-fuel produces lower BTE, soot, and NO_x emissions, whereas higher HC and CO emissions than base diesel. An engine fueled with blending of methanol or ethanol in diesel improved the BTE and lower soot emissions, whereas emissions of NO_x, HC, and CO are uncertain compared to diesel fuel. As DME does not contain a C–C bond and pertains to higher oxygen content lower the soot emissions and improves the BTE in diesel engine [7]. Liu et al. investigated the influence of hydrogen addition on a single-cylinder four-stroke CI engine with DME-CNG dual-fuel mode. The author observed the advancement in ignition time and increase in peak cylinder pressure by hydrogen addition. The emissions of CH_4 and CO were reduced, whereas NO_x emission was raised with the addition of hydrogen [8]. The effect of H_2 addition on the engine performance and emissions characteristics have been analyzed by Lounici et al. by performing the experimental work on the CI engine under CNG-diesel mode at low loads. The results reported that 10% H_2 addition under CNG-diesel mode showed a

considerable reduction in BSFC and significant improvement in BTE as compared to 20% and 30% H_2 blends under CNG-diesel mode. The author also observed that HC, CO, and CO_2 emissions were reduced, whereas NO_x emission increased marginally by H_2 addition under CNG-diesel dual-fuel mode [9].

Wang et al. carried out an experimental investigation on the effect of DME-diesel dual-fuel on emission characteristics of two-cylinder, four-stroke, direct injection (DI) compression ignition (CI) engine using LPG as ignition inhibitor to reduce the detonation at higher DME ES. The experimental tests were conducted at 1700 rpm at two brake mean effective pressures (BMEPs) of 0.24 MPa and 0.48 MPa. The study revealed that the BTE and NO_x emission reduced slightly with an increase in the mixture of LPG in DME under DME-diesel mode [10]. The influence of hydrogen as a secondary fuel in a four-cylinder CI engine under LPG-diesel dual-fuel mode has been examined by Lata et al. and compared the combustion characteristics of the CI engine with diesel, hydrogen-diesel, LPG-diesel, and hydrogen-LPG-diesel modes at a constant speed of 1500 rpm. It was observed that the maximum rate of pressure rise by 0.88 bar/°CA and combustion duration reduced by 4°CA, whereas maximum combustion pressure and HRR increased by 5.25 bar and 35.24 J/°CA, respectively, with the addition of hydrogen as secondary fuel under LPG-diesel mode as compared with the remaining three cases [11]. Khatri et al. studied the performance and emission characteristics of a single-cylinder CI engine under hydrogen-biogas-diesel fuel mode and compared it with the base diesel mode at 1500 rpm. The hydrogen and biogas were used as secondary fuel and inducted into the suction stroke, whereas diesel was used as a pilot fuel inducted directly into the engine's combustion chamber. It was observed that BSFC was reduced by 12% and BTE improved by 3.1% with hydrogen-biogas-diesel mode compared to diesel mode. NO_x, CO, and HC emissions were reduced by 83%, 88%, and 5.6%, respectively, with H_2-biogas-diesel mode compared to diesel mode [12]. Verma et al. performed and analyzed the experimental tests on a single-cylinder CI engine under biogas-diesel dual-fuel mode and compared the performance and emission characteristics with hydrogen addition (5–20% with the increment of 5%) in biogas-diesel mode at 1500 rpm. It was observed that up to 5% H_2 addition, the performance, and emission characteristics were insignificant as compared to biogas-diesel mode whereas, further increase in the percentage of H_2 addition showed improvement in performance and reduction in emissions characteristics of the engine under hydrogen-biogas-diesel mode as compared to biogas-diesel dual-fuel mode [13]. The combined effect of hydrogen fuel (inducted into the intake system) and EGR application was investigated by Shere et al. using the CI engine in diesel-hydrogen mode. The authors studied the enhancement in hydrogen energy share (HES) by suppressing the knocking tendency to some extent using a cooled EGR strategy. The NO_x emission was reduced to a near-zero level, with a slight increase in the smoke emission at higher EGR ratios [14].

The aforementioned literature reflects that relatively few research was reported on the use of alternative fuels as an additive (supplementary) for performance improvement and emissions reduction in CI engines, but no information is available on the influence of utilization of hydrogen supplementation as a secondary fuel for performance enhancement and emissions reduction with DME-diesel dual-fuel mode. The

purpose of this research is to investigate the influence of hydrogen gas supplementation on the performance, combustion, and emission characteristics of a DME-diesel fueled CI engine in dual-fuel mode at 2200 rpm and various load conditions (5, 10, 15, and 18 Nm).

2 Experimental Setup Details and Methodology

2.1 Engine Setup

The single cylinder, four-stroke direct injection compression ignition (DICI) engine was used to conduct the experimental tests. The dynamometer was coupled with the output shaft of the engine to load the engine, whereas a dyno-controller was used to regulate the engine's speed and torque. Figure 1 shows the schematic layout of the setup, and Table 1 shows the specification of the CI engine used for the experimental investigation. The conventional diesel engine was converted into DME-diesel and hydrogen-DME-diesel mode with minor modifications in the inlet manifold of the engine. DME gas (secondary fuel) and hydrogen gas (secondary fuel as additive) were inducted during the suction stroke into the engine's inlet manifold, and diesel fuel (pilot fuel) was injected at the end of the compression stroke into the engine's cylinder. Two electronic injectors were mounted on the intake manifold to inject the DME and hydrogen fuels. The start and end of DME fuel injection timing were controlled by using an electronic control unit (ECU) driver. The hydrogen fuel injection timings were controlled using an Arduino-based programmable injector driver developed in the laboratory. The diesel, DME, and hydrogen flow rate were measured using Coriolis-based mass flow meters, whereas air flow rate was recorded using a thermal-based flow meter. Table 2 listed the properties of fuels (diesel, DME, and hydrogen) used in the experimental study. The flame-trapped system was integrated in the hydrogen fuel supply line to prevent any risk due to propagation of backfire. The smoke discharge from the exhaust of a diesel engine was measured by using a smoke opacity meter, whereas NDIR-based AVL di-gas analyzer was used to record the exhaust discharge such as oxides of nitrogen, carbon monoxide, and unburned hydrocarbon. The piezoelectric pressure sensor recorded the pressure in the combustion chamber of the engine. An optical encoder was used to record the crank angle in degree.

2.2 Measured Data Uncertainty Analysis

The error analysis of the measured parameters can be estimated based on the root mean square method by using Eq. (1).

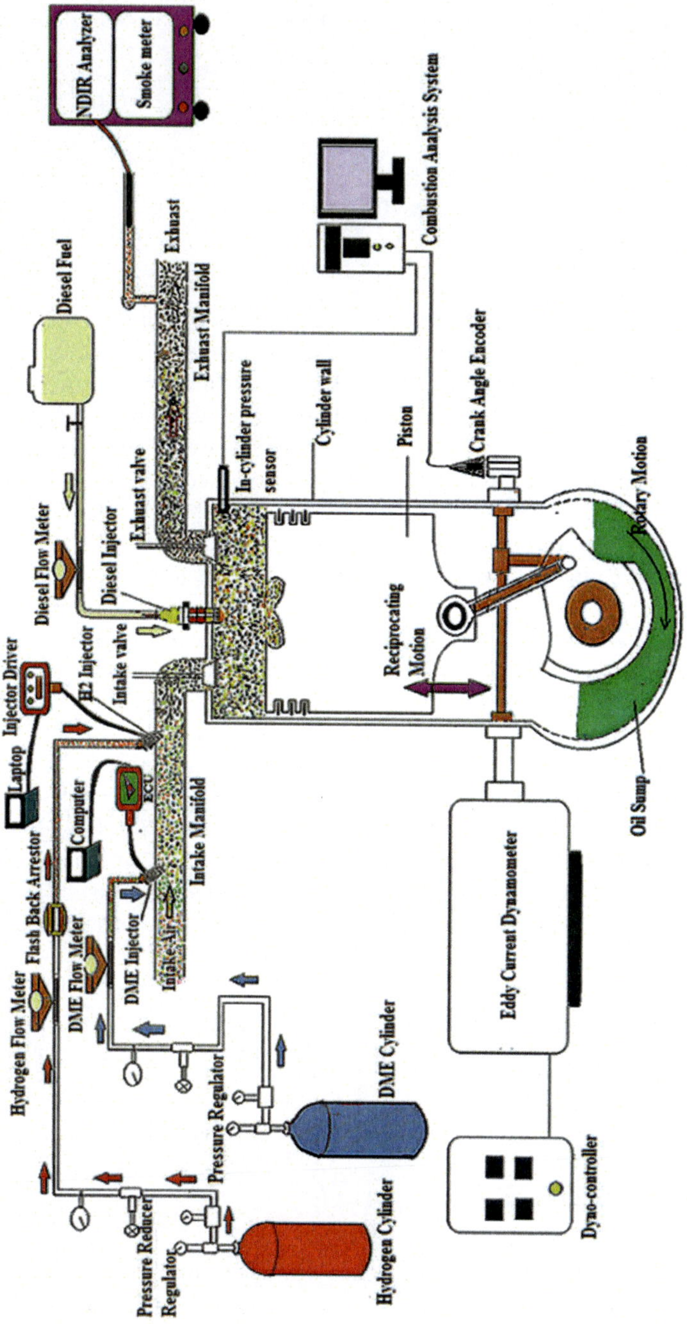

Fig. 1 Schematic layout of the engine setup

Table 1 Technical specifications of the DICI engine

Parameters	Specifications
Aspiration mode	Natural
Engine details	Four-stroke DICI automotive engine
Number of cylinders	One
Bore and stroke	0.086 m and 0.075 m
Displacement volume	0.000435 m^3
Maximum torque	18 Nm at 2200 rpm
Engine's compression ratio (CR)	19:1
Power output (rated)	7.4 HP (5.5 kW) at 3000 rpm

Table 2 Properties of fuels (diesel, DME, and hydrogen) used in the experimental investigation

Characteristics	Diesel	DME	Hydrogen
Molecular structure	$C_{16}H_{34}$	CH_3OCH_3	H_2
Cetane number (CN)	50–55	55–60	–
Calorific value (MJ/kg)	43.6	28.4	120
Self-ignition temperature (K)	532	508.15	858
Density (kg/m^3)	854	1.98	0.083
Stoichiometric A/F ratio (kg of air/kg of fuel)	14.6	9.0	34.3
Boiling point (°C)	150–400	−24.85	−252.9

$$Y_R = \left[\left(\frac{\partial U}{\partial x_1} y_1\right)^2 + \left(\frac{\partial U}{\partial x_2} y_2\right)^2 + \cdots + \left(\frac{\partial U}{\partial x_n} y_n\right)^2\right]^{1/2} \quad (1)$$

where Y_R denotes the total value of uncertainty; U implies the given function; x_1, x_2, ... x_n indicates the independent measured variables in function; y_1, y_2, ... y_n shows the uncertainty values of independent measured variables.

2.3 *Experimental Test Procedure*

The experimental tests were conducted on a single-cylinder automotive DICI engine fueled with diesel, DME-diesel, and DME-diesel with hydrogen addition as secondary fuel (H_2-DME-diesel) under dual-fuel mode at 2200 rpm with varied load conditions (5, 10, 15, and 18 Nm), respectively. The experimental tests were performed in three phases. During the first phase, the engine was set to run under diesel mode at 2200 rpm and varied load conditions. The second phase implies the tests under DME-diesel dual-fuel mode (induction of DME during suction stroke in the intake system and diesel injection during the end of the compression stroke in

the engine's cylinder) under the same conditions substituting the diesel percentage by the amount of DME percentage. The DME energy share (DME ES) was kept constant at 30%. The third phase utilized hydrogen gas as a secondary fuel (additive) in the CI engine under DME-diesel mode. The hydrogen gas was injected along with DME fuel during the suction stroke. The energy share of the DME and hydrogen (DME + H_2) mixture was kept constant at 30% (15% DME and 15% hydrogen gas) during the experimental tests under the required conditions. Further, the results of H_2-DME-diesel mode were compared with DME-diesel and base diesel mode, respectively.

2.4 Data Reduction

The DME energy share (ES_{DME}) and hydrogen energy share (ES_{H2}) can be estimated using Eqs. (2) and (3).

$$ES_{DME} = \left[\frac{\left(m_{g_{DME}} * C.V._{DME}\right)}{\left(m_{l_D} * C.V._D\right) + \left(m_{g_{DME}} * C.V._{DME}\right) + \left(m_{g_{H_2}} * C.V._{H_2}\right)}\right] \times 100 \quad (2)$$

$$ES_{H_2} = \left[\frac{\left(m_{g_{H_2}} * C.V._{H_2}\right)}{\left(m_{l_D} * C.V._D\right) + \left(m_{g_{DME}} * C.V._{DME}\right) + \left(m_{g_{H_2}} * C.V._{H_2}\right)}\right] \times 100 \quad (3)$$

where m_{l_D} denotes mass flow rate of diesel (pilot fuel) in a liquid state (g/s), $m_{g_{DME}}$ indicates the flow rate of DME (main fuel) in a gaseous state (g/s), $m_{g_{H_2}}$ implies the flow rate of hydrogen main fuel) in a gaseous state (g/s), $C.V._D$ denotes the lower calorific value of diesel (43.6 MJ/kg), $C.V._{DME}$ indicates the lower calorific value of DME (28.4 MJ/kg), and $C.V._{H_2}$ implies the lower calorific value of hydrogen (120 MJ/kg), respectively.

Thermal efficiency (η) and brake specific energy consumption (BSEC) of the CI enginender DME-diesel dual mode with hydrogen supplementation as secondary fuel can be analyzed using Eqs. (4) and (5).

$$\eta = \frac{BP\ (kW)}{\left(m_{l_D} * C.V._D\right) + \left(m_{g_{DME}} * C.V._{DME}\right) + \left(m_{g_{H_2}} * C.V._{H_2}\right)} \times 100 \quad (4)$$

$$BSEC = \frac{\left(m_{l_D} * C.V._D\right) + \left(m_{g_{DME}} * C.V._{DME}\right) + \left(m_{g_{H_2}} * C.V._{H_2}\right)}{BP\ (kW)}\ (MJ/kWh) \quad (5)$$

3 Results and Discussion

Figure 2 implies the influence of H_2 addition as a secondary fuel on the brake thermal efficiency (BTE) under H_2-DME-diesel fueled CI engine and compared with DME-diesel and diesel operations at 2200 rpm with different loads. The BTE of the CI engine was estimated using Eq. (4). The figure displays that at 10 Nm (55% load), the BTE of the CI engine with DME-diesel mode found as higher by 2.5% as compared to diesel mode. Similarly, at 15 Nm (83% load) and 18 Nm (100% load), the efficiency with DME-diesel mode was 5.22% and 6.07%, respectively, higher than in diesel mode. The increased homogeneity in the DME-air mixture and decreased brake specific energy consumption (BSEC) lead to rapid combustion (reduced ignition delay due to higher DME cetane number), which raises the maximum combustion pressure. Further, enhancement in the performance of the engine can be obtained by using H_2 gas as a supplementary fuel along with DME during suction stroke under DME-diesel mode at required engine conditions.

It can be observed from Fig. 2 that when the engine was operated with H_2-DME-diesel mode at 18 Nm (100% load), BTE was found as 9.0% and 15.6% higher than DME-diesel and neat diesel operation, respectively. Similarly, at other loads, BTE was improved with H_2-DME-diesel as compared with DME-diesel and diesel modes. This improvement in the performance of the engine may be due to the high self-ignition temperature (585 °C) and high diffusivity (0.63 cm^2/s) of hydrogen gas which was advantageous to form a homogeneous mixture with air leads to enhancing the combustion rate.

Figure 3 depicts the impact of H_2 addition on brake specific energy consumption (BSEC) under H_2-DME-diesel fueled CI engine and compared with DME-diesel and diesel operations at 2200 rpm with various loads. The BSEC of the engine was determined using Eq. (5). The BSEC for H_2-DME-diesel, DME-diesel, and diesel

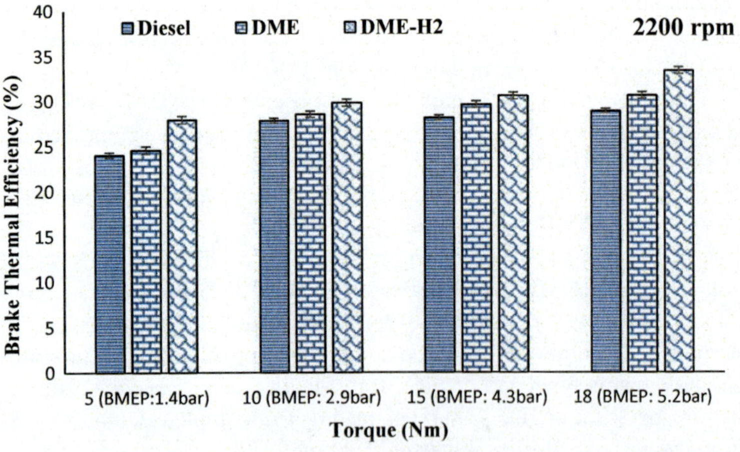

Fig. 2 Influence of H_2 addition on the BTE under DME-diesel mode

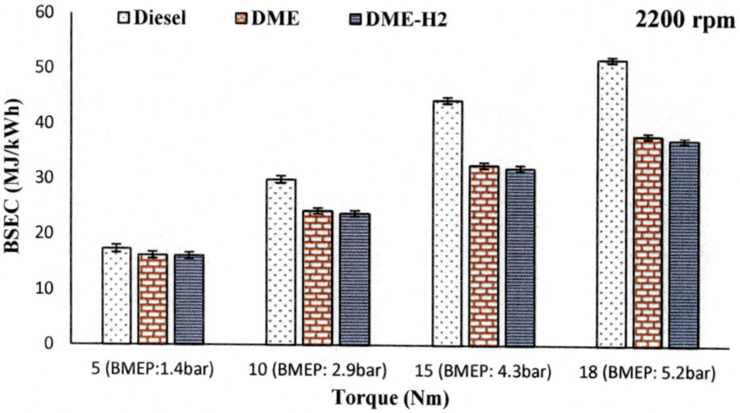

Fig. 3 Influence of H_2 addition on BSEC of the engine under DME-diesel mode

operations was observed to be 37.2 MJ/kWh, 38 MJ/kWh, and 51.8 MJ/kWh at the load of 18 Nm (100% load). The BSEC of the engine was reduced by 2.10% and 28.18% for H_2-DME-diesel mode at 100% load compared to DME-diesel and diesel modes, respectively. The uniform mixing of hydrogen and DME fuels leads to a reduction in BSEC. The high flame speed (3.25 m/s) and wide range of flammability (4–75% volume in air) of hydrogen fuel lead to better mixing of charge resulting in quick and complete combustion.

Figure 4 displays the variation in CO emission under diesel, DME-diesel, and H_2-DME-diesel modes in CI engine at 2200 rpm with various loads. At 18 Nm (100% load), CO emission was observed to be 22.09 g/kWh and 7.95 g/kWh for DME-diesel and diesel modes. CO emission was raised by 64.01% for DME-diesel mode compared to diesel operation. CO emission rises under DME-diesel mode due to the incomplete combustion of the trapped mixture of DME and air present around the piston bowl and crevice of the cylinder. This trapped mixture slows down the oxidation reaction and results in the formation of CO. When the engine was operated with H_2-DME-diesel mode, CO emission was observed to be 0.86 g/kWh at 18 Nm. CO emission dropped by 96.01% for H2-DME-diesel operation compared to DME-diesel mode at 100% load. The rapid oxidation of the mixture due to the higher flame velocity (3.25 m/s) of hydrogen could be the possible reason for the reduction in CO emission under H_2-DME-diesel mode.

Figure 5 presents the variation in HC emission at different loads with diesel, DME-diesel, and H_2-DME-diesel modes in CI engine at 2200 rpm. At 18 Nm (100% load), unburned HC was 61.5% higher for DME-diesel mode than diesel operation. The increase in HC emission could be due to the trapped DME fuel mixture inside the cylinder's piston bowl and crevice. It results in a heterogeneous distribution of fuel mixture inside the engine's chamber, leading to incomplete combustion. The HC emission under H_2-DME-diesel was reduced by 92.30% compared to DME-diesel

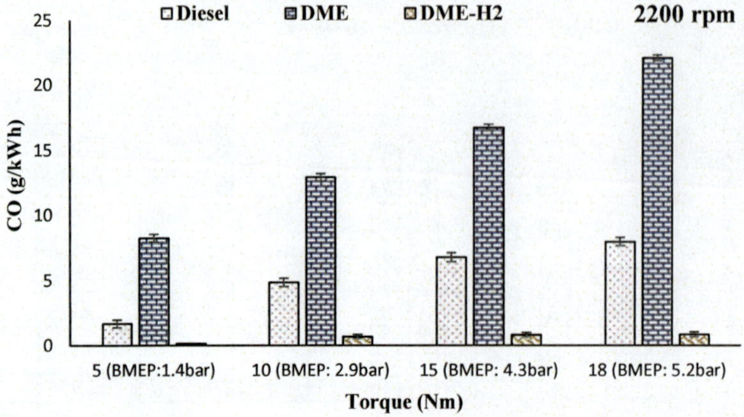

Fig. 4 Influence of H_2 addition on CO emission under DME-diesel operation

mode at 100% load. This curtailment in the formation of HC emission under H_2-DME-diesel mode could be due to the absence of carbon availability, higher flame velocity (3.25 m/s), and shorter quenching distance (0.64 mm) of hydrogen fuel.

Figure 6 indicates the impact of H_2 addition on the NO_x emission under DME-diesel mode at 2200 rpm with various loads. At 100% load, NO_x emission for H_2-DME-diesel, DME-diesel, and diesel modes were observed to be 4.76 g/kWh, 4.53 g/kWh, and 3.55 g/kWh, respectively. NO_x emission for H_2-DME-diesel was 5.07% and 34.08% higher than DME-diesel and neat diesel operation. The formation of NO_x emission mainly depends upon the residence time, combustion temperature, and, oxygen and nitrogen concentration. The increase in combustion temperature and higher flame speed of H_2 could be the possible reason for the increase in NO_x emission with H_2-DME-diesel mode.

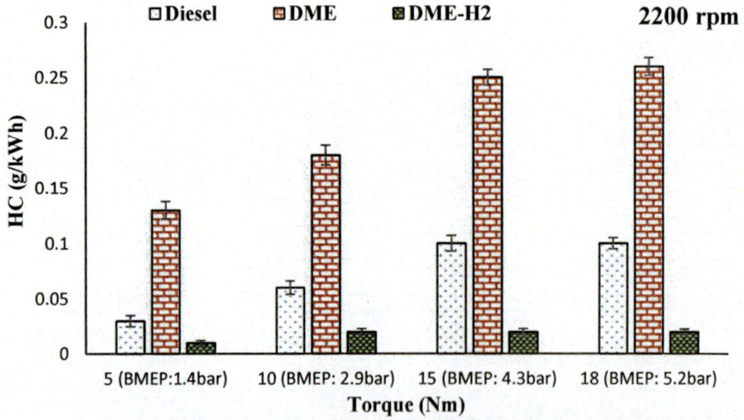

Fig. 5 Influence of H_2 addition on HC emission under DME-diesel operation

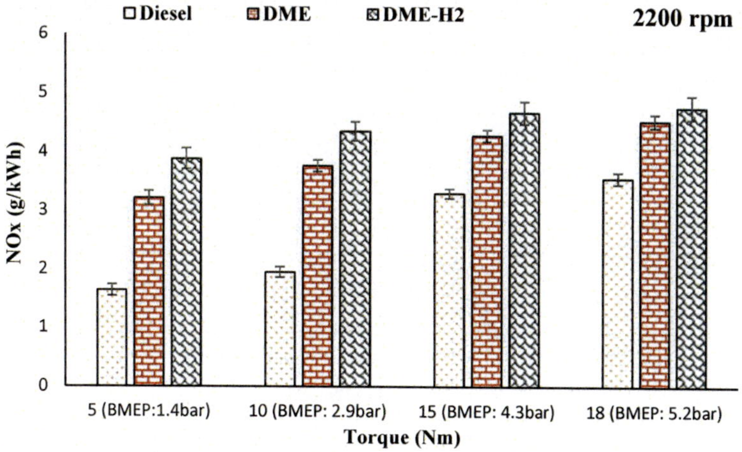

Fig. 6 Impact of H_2 addition on the NO_x emission under DME-diesel operation

Figure 7 presents the effect of H_2 addition on smoke emission under DME-diesel mode at 2200 rpm with various loads. At 18 Nm, smoke emission for DME-diesel mode was reduced by 83.09% which is lower than diesel operation. The better homogeneity of the DME-air mixture due to higher diffusivity and absence of the C–C bond in the DME molecular structure results in the reduction of smoke emission. Further, the smoke emission with H_2-DME-diesel mode was reduced drastically by 93.33% which was lower than DME-diesel mode at 100% load due to better homogeneity of the H_2-DME-air mixture, which results in quick and complete combustion.

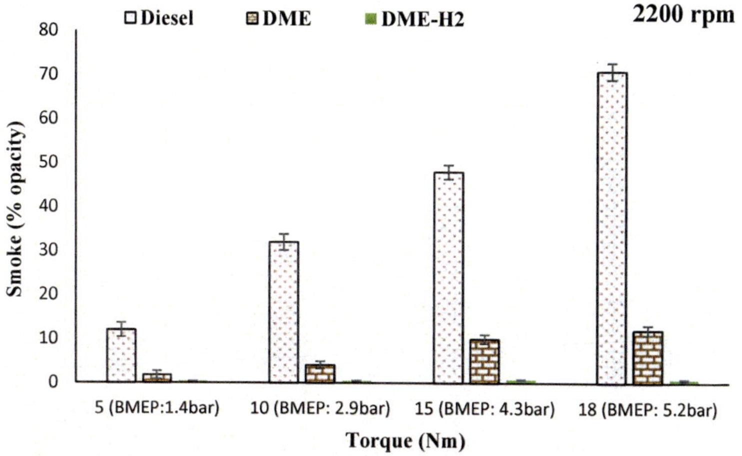

Fig. 7 Influence of H_2 addition on smoke emission under DME-diesel operation

Figure 8 demonstrates the maximum combustion pressure, and Fig. 9 displays the variation in the plots of combustion pressure (bar)—theta (θ) with diesel, DME-diesel, and H_2-DME-diesel modes at 2200 rpm with various loads. The experimental result showed that as the load increases, more amount of fuel gets consumed which subsequently leads to an increase in the maximum combustion pressure and temperature in the combustion chamber of the engine. It was observed that maximum combustion pressure with DME-diesel and H_2-DME-diesel system was 67.3 bar and 64.5 bar, respectively, higher than with base diesel operation of 59.6 bar at 100% load condition. The DME fuel due to its higher CN (~58) and lower auto-ignition temperature (508 K), gets ignition prior to diesel and initiates the combustion process for the DME-diesel system in the form of homogeneous charge compression ignition (HCCI) mode resulting in homogeneous mixture formation between air and DME fuel during the period from suction stroke to compression stroke (prior to diesel injection) and, leads to complete oxidation and better combustion of the mixture resulting in the advancement of position and, increased in maximum combustion pressure and temperature significantly. Whereas, when the engine operates with H_2-DME-diesel case, the higher-octane number (Hydrogen) fuel mixed with DME fuel and air, and postpones the start of combustion time, which shifts the rise of combustion pressure toward a later crank angle and hence, the maximum combustion pressure got reduced marginally as compared to DME-diesel mode.

Figures 10 and 11 depict the influence of diesel, DME-diesel, and H_2-DME-diesel modes on heat release rate and cumulative heat release at 2200 rpm with various loads. It can be observed from the figure that DME-diesel and H_2-DME-diesel combustion shows a complex nature as compared to diesel combustion. The heat release curve clearly shows three-stage combustion, i.e., low-temperature reaction

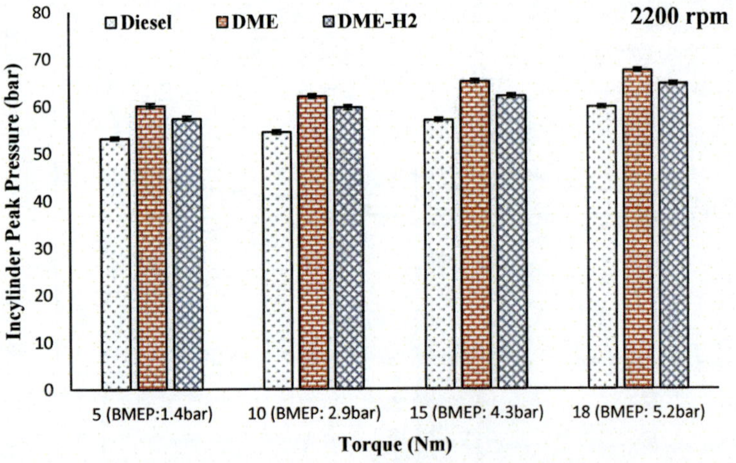

Fig. 8 Variation in maximum combustion pressure under diesel, DME-diesel, and H_2-DME-diesel modes

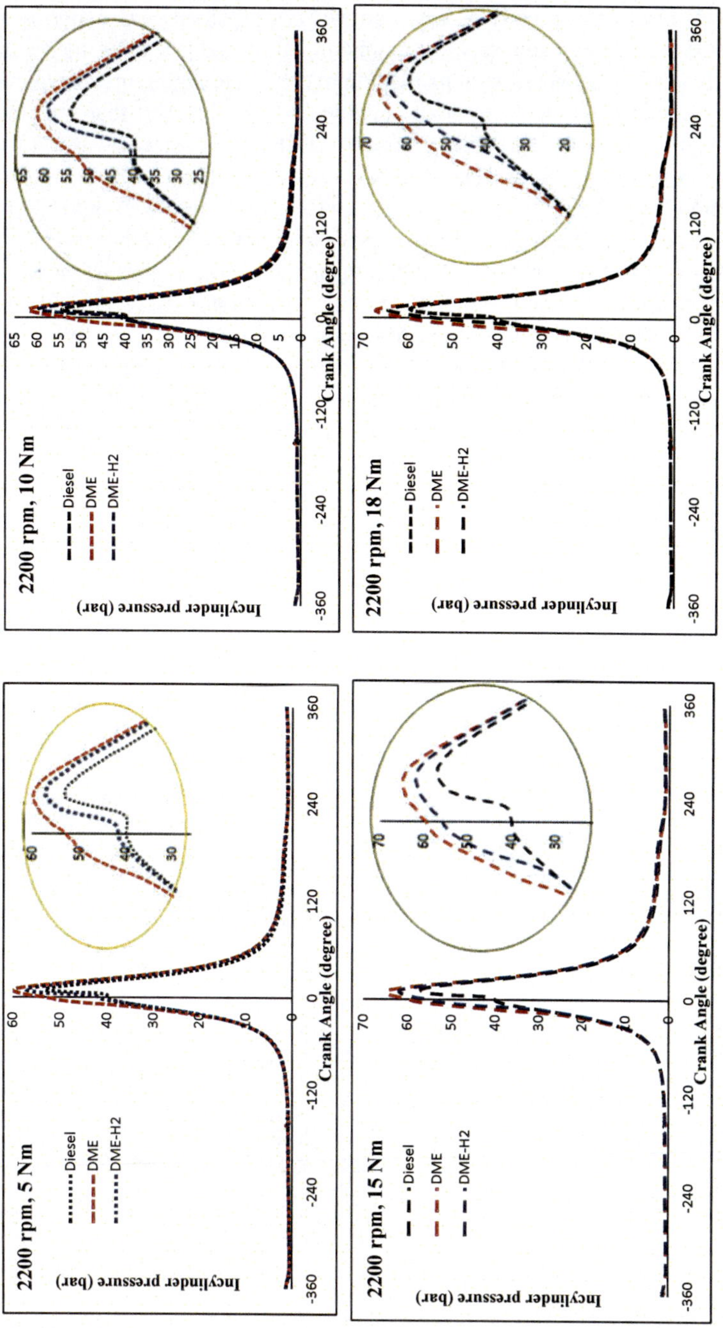

Fig. 9 Influence of hydrogen addition on combustion pressure under DME-diesel dual mode

(LTR—cool flame) region, high-temperature reaction (HTR) region, and diffusion-controlled CI combustion. The DME-diesel mode shows combustion similar to DME HCCI combustion and exhibits the LTR and HTR regions [15]. When the temperature is in the range of 650–850 K, then the LTR region promotes the ignition by breaking the C–H bond of DME, formation of $CH_3O–CH_2O_2$ with subsequent intermolecular isomerization. Further, when the temperature reached above 1200 K, the mixture gets ignited in the HTR region. In this region, a large amount of energy gets released with the formation of H_2O, CO_2, and CO.

Due to the higher-octane number, high flame speed (3.25 m/s), wide flammability range (4–75% volume in air), and high self-ignition temperature (858 K) of hydrogen fuel, the H_2-DME-diesel combustion eliminates the cool flame combustion region and the start of combustion started in HTR combustion region with the mixture of DME and H_2 (DME + H_2 HCCI) where the temperature reached above 1200 K.

4 Conclusions

The impact of hydrogen addition as a secondary fuel on the performance, emission, and combustion characteristics in a DME-diesel fueled engine is scrutinized. The conclusions that emerged from the experimental investigation are as follows:

- The BTE of the CI engine enhanced from 28.8% (diesel) to 30.5% (DME-diesel) and further improved to 33.3% (H_2-DME-diesel) at maximum torque conditions.
- The smoke emission decreased drastically from 71% with diesel to 12% with DME and further reduced to 1.2% of opacity with hydrogen induction along with DME.
- CO and HC emissions increased from 7.9 and 0.1 g/kWh with diesel to 22.1 and 0.26 g/kWh with DME-diesel, whereas they further decreased drastically to 1.8 and 0.02 g/kWh with H_2-DME-diesel mode at maximum torque condition.
- NO_x emission raised to 4.76 g/kWh with DME + H_2-diesel from 4.53 and 3.55 g/kWh with DME-diesel and diesel mode.
- DME-diesel mode shows multi-staged combustion, i.e., low-temperature region (LTR-cool flame), high-temperature region (HTR), and diffusion-controlled region, whereas H_2-DME-diesel mode eliminates the cool flame combustion region and the start of combustion occurred in HTR combustion region.

It is demonstrated through this study that hydrogen induction along with DME leads to higher BTE with a drastic reduction in CO, smoke, and HC emissions.

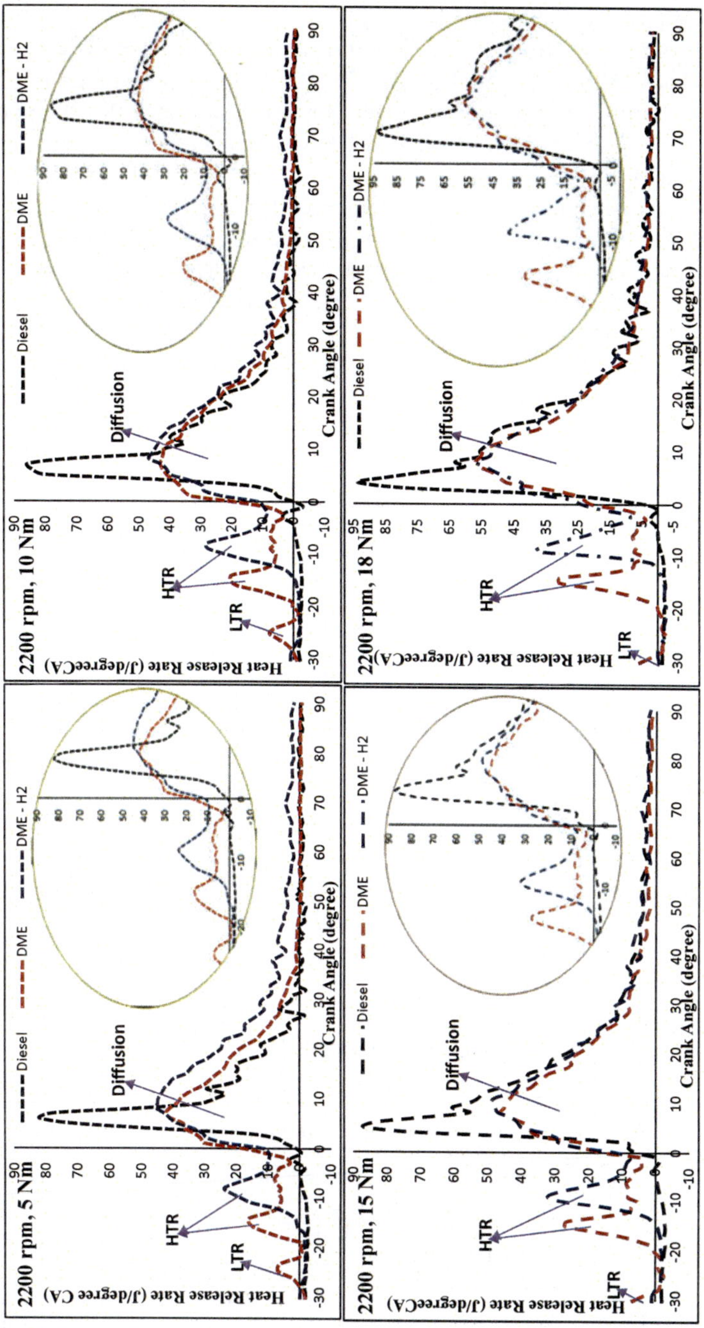

Fig. 10 Influence of hydrogen addition on heat release under DME-diesel dual mode

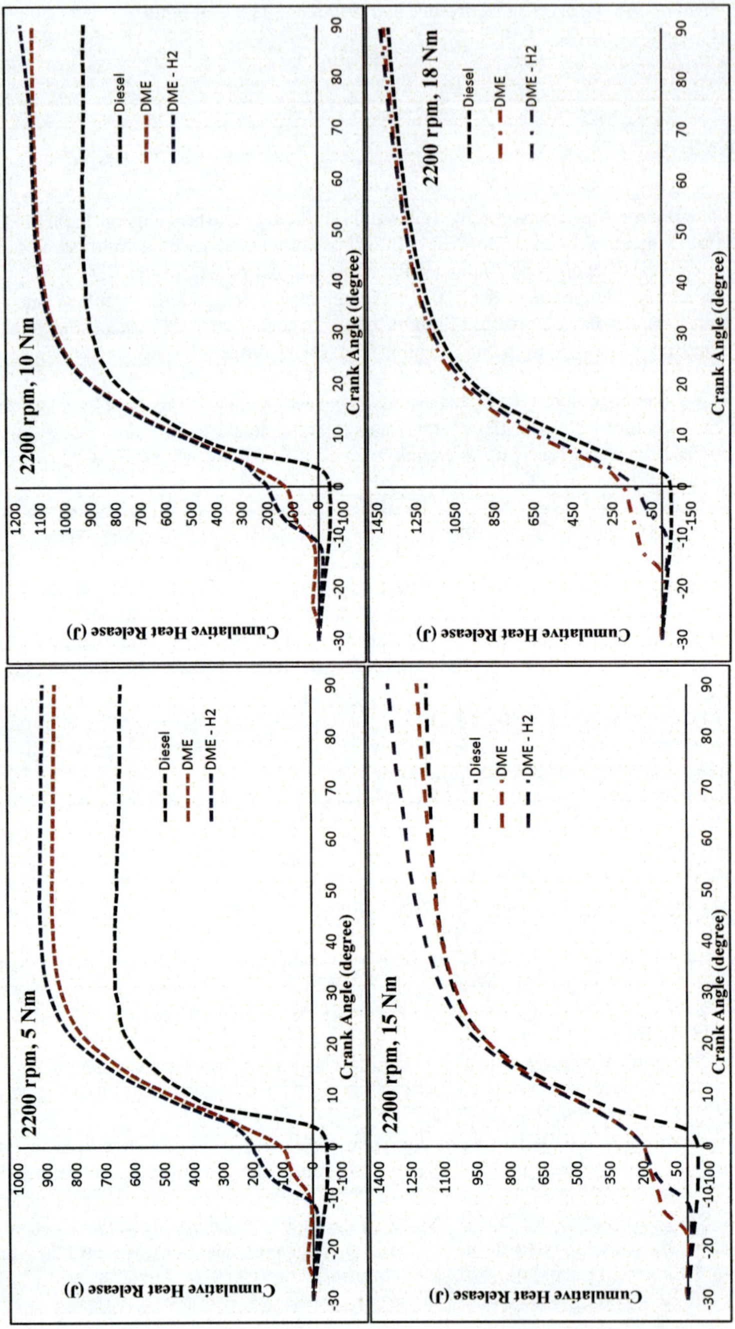

Fig. 11 Influence of hydrogen addition on cumulative heat release under DME-diesel dual mode

Acknowledgements The authors wish to express thanks to IOCL R&D, India, for providing the DME fuel funded by the Department of Science and Technology (DST), India.

References

1. National Institution for Transforming India (NITI) Aayog, Government of India. India's leapfrog to methanol economy. https://niti.gov.in/methanol-economy https://niti.gov.in/writereaddata/files/Indias-Leapfrog-to-Methanol-Economy.pdf. Accessed 27 Sept 2021
2. Sinha S, Shere A, Subramanian KA (2022) Experimental investigations on enhancement of DME energy shares in compression-ignition engine under dual fuel mode using reduced compression ratio. Int J Autom Sci Technol 6(1):83–97. https://doi.org/10.30939/ijastech.972029
3. Shere A, Subramanian KA (2022) Experimental investigation on effects of fuel injection timings on Dimethyl Ether (DME) energy share improvement and emission reduction in a dual-fuel CRDI compression ignition engine. Int J Autom Sci Technol 6(2):98–112. https://doi.org/10.30939/ijastech.999261
4. Bo Z, Weibiao F, Jingsong G (2006) Study of fuel consumption when introducing DME or ethanol into diesel engine. Fuel 85:778–782. https://doi.org/10.1016/j.fuel.2005.08.042
5. Saleh HE (2008) Effect of variation in LPG composition on emissions and performance in a dual fuel diesel engine. Fuel 87:3031–3039. https://doi.org/10.1016/j.fuel.2008.04.007
6. Kakoee A, Gharehghani A (2019) Comparative study of hydrogen addition effects on the natural-gas/diesel and natural-gas/dimethyl-ether reactivity-controlled compression ignition mode of operation. Energy Convers Manage 196:92–104. https://doi.org/10.1016/j.enconman.2019.05.113
7. Chen Y, Ma J, Han B, Zhang P, Hua H, Chen H, Su X (2018) Emissions of automobiles fueled with alternative fuels based on engine technology: a review. J Traffic Transp Eng 5(4):318–334. https://doi.org/10.1016/j.jtte.2018.05.001
8. Liu J, Yang F, Wang H, Ouyang M (2012) Numerical study of hydrogen addition to DME/CH_4 dual fuel RCCI engine. Int J Hydrogen Energy 37:8688–8697. https://doi.org/10.1016/j.ijhydene.2012.02.055
9. Lounici MS, Boussadi A, Loubar K, Tazerout M (2014) Experimental investigation on NG dual fuel engine improvement by hydrogen enrichment. Int J Hydrogen Energy 39:21297–21306. https://doi.org/10.1016/j.ijhydene.2014.10.068
10. Wang Y, Liu H, Huang Z, Liu Z (2016) Study on combustion and emission of a dimethyl ether-diesel dual-fuel premixed charge compression ignition combustion engine with LPG (liquefied petroleum gas) as ignition inhibitor. Energy 96:278–285. https://doi.org/10.1016/j.energy.2015.12.056
11. Lata DB, Misra A, Medhekar S (2011) Investigations on the combustion parameters of a dual fuel diesel engine with hydrogen and LPG as secondary fuels. Int J Hydrogen Energy 36:13808–13819. https://doi.org/10.1016/j.ijhydene.2011.07.142
12. Khatri N, Khatri KK (2020) Hydrogen enrichment on diesel engine with biogas in dual fuel mode. Int J Hydrogen Energy 45(11):7128–7140. https://doi.org/10.1016/j.ijhydene.2019.12.167
13. Verma S, Das LM, Kaushik SC, Tyagi SK (2018) An experimental investigation of exergetic performance and emission characteristics of hydrogen supplemented biogas-diesel dual fuel engine. Int J Hydrogen Energy 43:2452–2468. https://doi.org/10.1016/j.ijhydene.2017.12.032

14. Shere A, Subramanian KA (2022) Emissions reduction in an automotive compression ignition engine using hydrogen and exhaust gas recirculation. Int J Ambient Energy, 1–20. https://doi.org/10.1080/01430750.2022.2091032
15. Shere A, Subramanian KA (2022) Experimental investigation on effects of equivalence ratio on combustion with knock, performance, and emission characteristics of dimethyl ether fueled CRDI compression ignition engine under homogeneous charge compression ignition mode. Fuel 322:124048. https://doi.org/10.1016/j.fuel.2022.124048

Energy and Exergy Analysis of the Automotive Spark-Ignition Engine Fueled with Ethanol, Methanol, and Gasoline

Sachin Kumar Gupta and K. A. Subramanian

1 Introduction

The reserve of conventional fuels is limited, and the demand for internal combustion engines is increasing faster [1]. The engine emits net nonzero carbon oxide emission from the tail pipe while utilizing conventional fuels. Irrespective of the advancement in vehicle technologies such as the slight penetration of electric vehicles but still generation of electricity in the world is mostly from coal. Thus, renewable fuels such as bio-ethanol and bio-methanol are attractive over conventional fuels when utilized in engines due to their carbon neutral characteristics [2]. Since the CO_2 emitted from its utilization in the engine can be captured back while growing trees and easily replace gasoline usage in spark-ignition (SI) engines with minor modifications. Organic matters such as wood residue, waste food, agriculture waste, and crop residue can be used to produce bio-ethanol and bio-methanol through thermochemical methods. Therefore, it has been initiated to add a volume of ethanol in gasoline and increase its volume.

The performance of SI engines fueled by different fuels can be evaluated based on energy and exergy analysis. These techniques allow to quantify various performance indicators and energy fluxes of the engine [3]. Exergy is the maximum work produced by a system or stream while bringing it into equilibrium with the reference environment. Exergy analysis based on the second law of thermodynamics can determine the magnitude of combustion source and irreversibility in heat transfer during the various engine processes. Various research on the energy and exergy analysis of the spark-ignition engine fueled with ethanol and methanol are present in the literature, as discussed ahead.

S. K. Gupta · K. A. Subramanian (✉)
Department of Energy Science and Engineering, Indian Institute of Technology Delhi, New Delhi 110016, India
e-mail: subra@dese.iitd.ac.in

A. Kumar et al. (eds.), *Recent Advances in Manufacturing and Thermal Engineering*,
Lecture Notes in Mechanical Engineering, https://doi.org/10.1007/978-981-19-8517-1_41

Sezer et al. [4] applied the exergy analysis on a 2D thermodynamic cycle model of SI engine fueled with oxygenated fuels. They found that the irreversibility for methanol and ethanol was lowered by 7.44% and 4.29%, respectively, and exergy transfer with heat and work for methanol was 9.47% and 7.35% which decreased for ethanol to 6.45% and 3.24%. Caton [5–7] developed a thermodynamic model of the engine for alcohol and hydrocarbon fuels and performed energy and exergy analysis. During the combustion process of iso-octane in an engine, the fuel exergy destruction was higher than in alcohol fuels. Haq and Morshed [8] analyzed the exergy and exergy of the engine for various fuels including methanol and ethanol. It was found that the useful work is about 34–39% of the fuels' energy, which increased with the engine speed. Dogan et al. [9] operated the engine using the blends of ethanol and gasoline at full load conditions and utilized the experimental data to evaluate the exergy of the system. They found that the exergy efficiency of the engine decreased with an increase in the percentage of ethanol in the blend. Further, Dogan et al. [10] prepared the blend of 10% ethanol and methanol with gasoline (i.e., E10 and M10) for exergy analysis. Their study showed that the exergy is highest for pure gasoline (i.e., 21%) as compared to M10 (16.8%) and E10 (17.9%). Kul and Ciniviz [11] prepared various blends of ethanol and gasoline (E5, E10, E15, E85, and E100) to determine the engine characteristics and perform exergy analysis. The results showed that the exergetic efficiency of the engine was the highest for E100 (i.e., 14.7%) as compared to other blends.

The exhaustive literature review has not evaluated the effect of different throttle openings of the engine on exergy and energy. Thus, in the present study, the energy and exergy analysis of the port-fueled spark-ignition engine was performed for different throttle openings for ethanol, methanol, and gasoline fuels.

2 CFD and Exergy Models

2.1 CFD Model

A CAD model of the engine's combustion chamber was prepared by making the silicon rubber mold and scanning it using the laser scanner. Then, the computational fluid dynamics (CFD) model of the spark-ignition engine was developed using converge CFD, as shown in Fig. 1. All the geometrical parameters of the engine (bore & stroke : 74 mm & 58 mm), compression ratio (9.8:1), engine speed (4000 rpm), and measured valves profile were provided to the model. Grid scaling, adaptive mesh refinement, and fixed embedding were used to generate the structured mesh [12]. In the present study, a base grid size of 4 mm was selected for a good balance between computational time and accuracy. The conservation equations for mass, momentum, energy, and species were solved including heat transfer and RANS-based RNG $k-\varepsilon$ turbulence models. A detailed chemical kinetics for ethanol, methanol, and iso-octane (i.e., gasoline surrogate) was solved using SAGE.

The CFD model of the spark-ignition engine was simulated for four complete cycles, and the results of fourth cycle were used for further calculation. The model was validated with the experimental data obtained while running the engine at 30% WOT and 4000 rpm fueled with ethanol, methanol, and gasoline, as shown in Fig. 2. The predicted peak cylinder pressure and IMEP through the CFD simulation showed a good agreement with the experimental results for the various fuels.

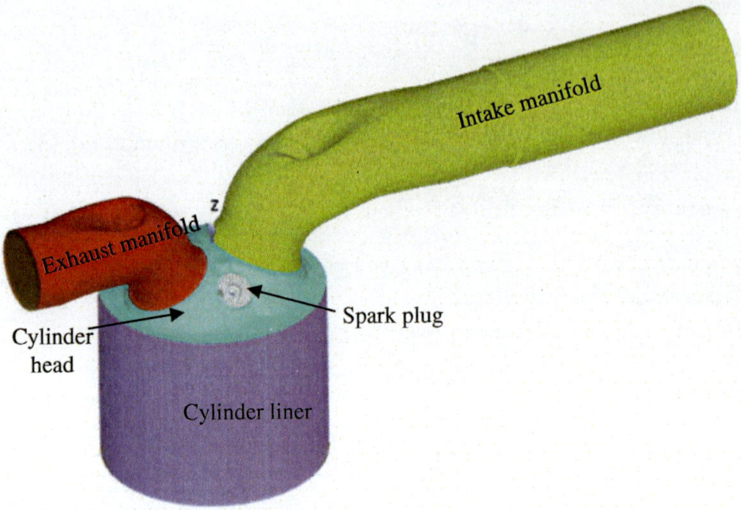

Fig. 1 Computational fluid dynamics model of the spark-ignition engine

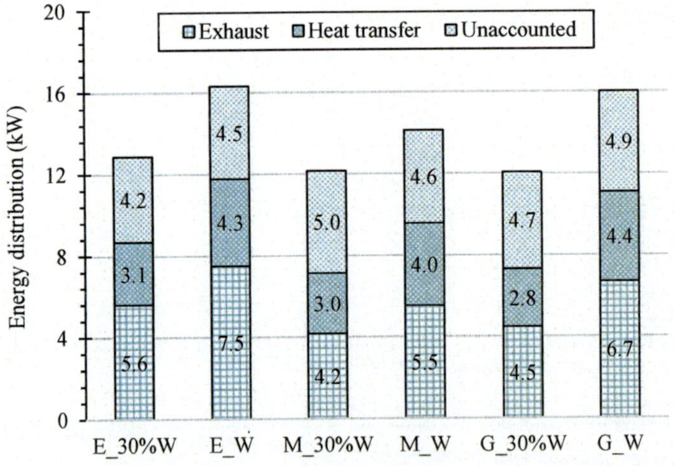

Fig. 2 Energy distribution for ethanol, methanol, and gasoline at 30% WOT and WOT

2.2 Energy Analysis

The mass and energy through the IC engine should be conserved, can be written as Eqs. (1) and (2), respectively.

$$\dot{m}_{\text{air}} + \dot{m}_{\text{fuel}} - \dot{m}_{\text{exh}} = 0 \quad (1)$$

$$\dot{Q}_{\text{loss}} - \dot{W} + \dot{E}_{\text{air}} + \dot{E}_{\text{fuel}} - \dot{E}_{\text{exh}} = 0 \quad (2)$$

In the engines, fuel provides energy to the system, which is mass flow rate of fuel ($\dot{m}_{\text{fuel}}$) times lower heating value of the fuel (LHV_{fuel}), as written in Eq. (3).

$$\dot{E}_{\text{fuel}} = \dot{m}_{\text{fuel}} \times \text{LHV}_{\text{fuel}} \quad (3)$$

The indicated work and useful work of the four-stroke engine can be calculated using Eqs. (4) and (5), respectively.

$$\dot{W}_i = \frac{P_{i\text{mep}} V_{\text{disp}} N}{60 * 2} \quad (4)$$

$$\dot{W} = \frac{2\pi NT}{60} \quad (5)$$

The thermo-physical properties of the air–fuel mixture in the engine during various processes were precisely determined from the CFD model result. The indicated and brake thermal efficiency of the engine can be calculated using Eqs. (6) and (7), respectively.

$$\eta_{\text{th},i} = \frac{\dot{W}_i}{\dot{E}_{\text{fuel}}} \quad (6)$$

$$\eta_{\text{th}} = \frac{\dot{W}}{\dot{E}_{\text{fuel}}} \quad (7)$$

2.3 Exergy Analysis

The destruction rate of energy from the control volume of the engine can be calculated using Eq. (8).

$$\dot{E}x_{\text{dest}} = \sum \dot{m}_{\text{in}}\text{ex}_{\text{in}} - \dot{E}x_w - \sum \dot{m}_{\text{out}}\text{ex}_{\text{out}} - \dot{E}x_{\text{cw}} \quad (8)$$

where $\dot{E}x_{\text{dest}}$ is the irreversibility or exergy destruction rate, $\dot{E}x_{\text{w}}$ is the shaft exergy or work exergy output. $\dot{E}x_{\text{cw}}$ is the heat transfer exergy, $\dot{m}$ is the rate of mass flow, and ex is the specific flow exergy. The fuel's exergy ($\dot{E}x_{\text{in}}$) is converted to various exergy forms such as shaft exergy, exhaust gas exergy, and exergy destroyed. This exergy can be destroyed by the engine due to mixing, combustion, and mechanical friction. The input exergy of various fuels can be evaluated using Eq. (9), whereas exergy power output is given by Eq. (10).

$$Ex_f = \dot{m}_f \times \text{LHV}_f \tag{9}$$

$$\dot{E}x_W = \dot{W} \tag{10}$$

The heat transfer through the engine results in exergy loss, as represented by Eq. (11).

$$\dot{E}x_{\text{loss}} = \dot{Q}_{\text{loss}}\left(1 - \frac{T_0}{T_s}\right) \tag{11}$$

The exhaust gases leaving out the engine result in physical exergy losses and can be calculated using Eq. (12).

$$\begin{aligned}\dot{E}x^{\text{th}}_{\text{exh}} &= \dot{Q}_{\text{exh}} + \left(\dot{m}_f + \dot{m}_a\right) \times T_{\text{amb}} \\ &\times \left\{C_{p,\text{exh}} \ln\left(\frac{T_{\text{amb}}}{T_5}\right) - R_{\text{exh}} \ln\left(\frac{p_{\text{amb}}}{p_{\text{exh}}}\right)\right\}\end{aligned} \tag{12}$$

where R_{exh} is the specific gas constant of the exhaust gases. The exergy destruction can be determined by Eq. (13)

$$\dot{E}x_{\text{dest}} = \dot{E}x_w - \left(\dot{E}x_w + \dot{E}x_{cw} + \dot{E}x_{\text{exh}}\right) \tag{13}$$

The exergetic efficiency can be calculated using Eq. (14).

$$\eta_{\text{ex}} = \frac{\dot{E}x_w}{\dot{E}x_f} \tag{14}$$

3 Results and Discussion

The throttle opening and spark timing of the engine model was simulated and the CFD results were utilized for energy and exergy analysis. The fuel energy in the engine was converted to various forms of energy, i.e., power output, heat carried by exhaust gases, and heat lost to the environment. Figure 2 shows the distribution of

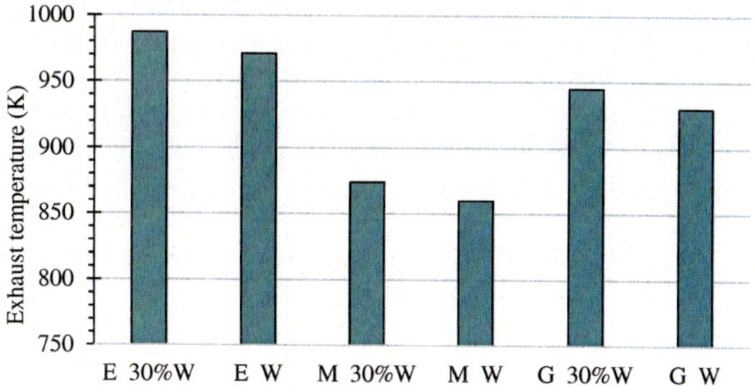

Fig. 3 Exhaust temperature for ethanol, methanol, and gasoline at 30% WOT and WOT

fuel energy converted to exhaust, heat transfer, and unaccounted energy. It can be observed that the heat transfer and exhaust energy losses were more for the engine operating at wide-open throttle (WOT) as compared to the part throttle opening (i.e., 30% WOT) for the different fuels. The energy lost by engine exhaust for ethanol, methanol, and gasoline increased from 5.6 kW, 4.2 kW, and 4.5 kW to 7.5 kW, 5.5 kW, and 6.7 kW, respectively, when the throttle opening increased from 30 to 100%. It may be noted that the energy lost through exhaust gases of the engine increased by 24% for ethanol and decreased by 7% for methanol fuel as compared to gasoline, which is mainly due to the trend of exhaust gas temperature, as shown in Fig. 3.

Figure 4 shows the variation of net and gross indicated thermal efficiency of the engine while operating it with ethanol, methanol, and gasoline at different throttle openings. The indicated thermal efficiency of the engine was the highest for methanol as compared to ethanol and gasoline while operating at both the throttle positioning. Figure 5 shows the variation of the fuel exergy input to the engine while operating the engine with ethanol, methanol, and gasoline at different throttle openings.

Further, the exergy analysis was performed on the CFD results for the different fuels and throttle positions. Figure 6 shows the exergy lost in the exhaust, work output, heat transfer, and destructed during various engine processes. A better understanding of exergy destruction during the combustion phenomenon due to chemical and thermal interactions of reactants and products. The work output at the shaft increased for methanol by 22.7% as compared to gasoline at both the throttle openings. In the case of ethanol, the work output increased by 13.6% as compared to gasoline, which decreased by 15% under WOT conditions. This is mainly due to an improvement in the combustion characteristics of the engine when methanol and ethanol were used as compared to gasoline at 30% WOT, which degraded for ethanol at WOT.

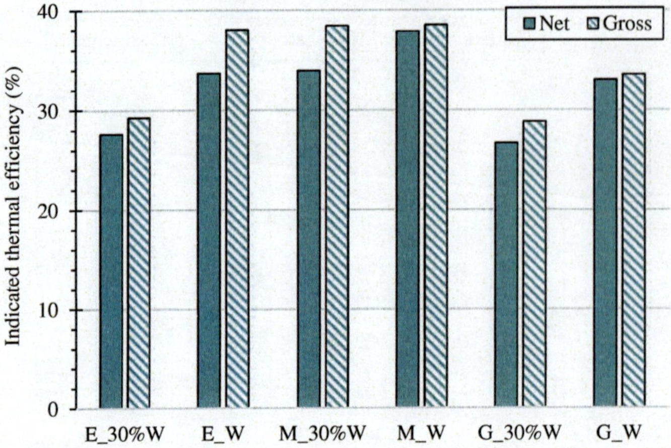

Fig. 4 Net and gross indicated thermal efficiency for ethanol, methanol, and gasoline at 30% WOT and WOT

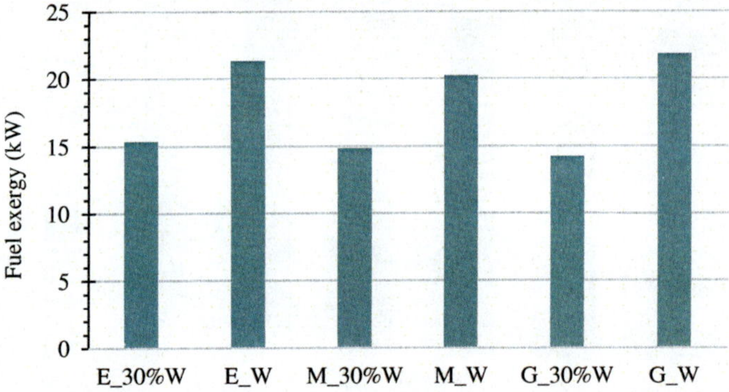

Fig. 5 Fuel exergy input to the engine for ethanol, methanol, and gasoline at 30% WOT and WOT

4 Conclusions

The thermodynamics-based analysis was performed in the present study based on the first and second law of thermodynamics for ethanol, methanol, and gasoline fuels at wide-open throttle (WOT) and part throttle (30% WOT). The study was performed on the air-cooled port-fuel injection, single-cylinder, and spark-ignition engine. The major conclusions that can be drawn from the energy and exergy investigation are as follows:

1. The maximum power output of 6.1 kW was obtained at WOT while operating the engine on methanol fuel, which was about 22% and 4% higher than ethanol and gasoline.

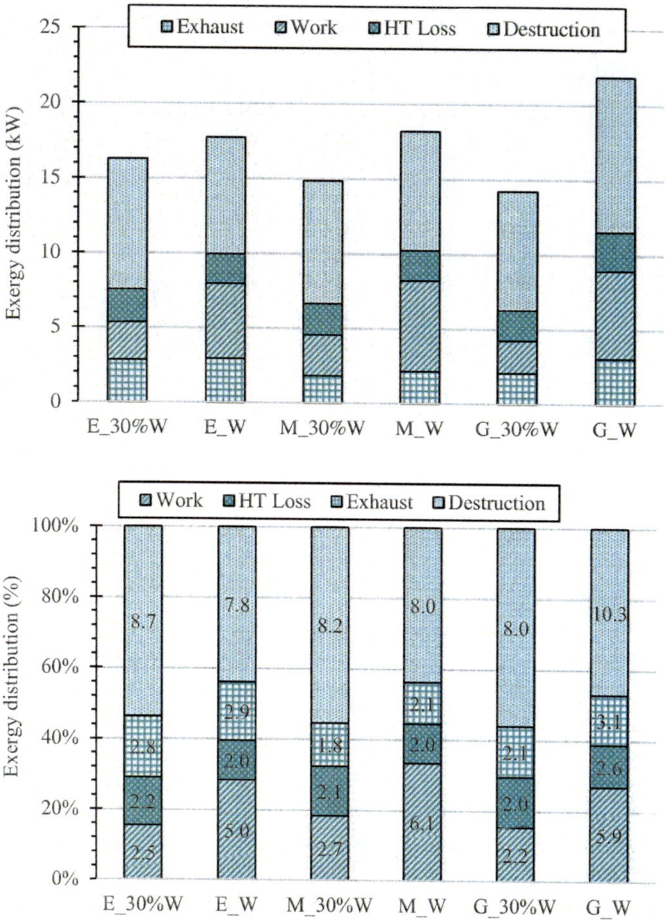

Fig. 6 Exergy balance for ethanol, methanol, and gasoline at 30% WOT and WOT

2. Fuel's exergy was the highest for gasoline at wide-open throttle and ethanol at part throttle condition.
3. The energy lost through a heat transfer and exhaust gases were the lowest for methanol fuel as compared to ethanol and gasoline. At WOT, combined heat transfer and exhaust losses contributed about 4.1 kW for methanol as compared to gasoline (5.7 kW) and ethanol (4.9 kW).
4. The energy destructed was maximum for gasoline while operating at WOT condition, whereas maximum for ethanol at part throttle.
5. The exergy efficiency of the engine was the highest for methanol (i.e., 33.4%) as compared to gasoline and ethanol.

References

1. Bhatti SS, Verma S, Tyagi SK (2019) Energy and exergy based performance evaluation of variable compression ratio spark ignition engine based on experimental work. Thermal Sci Eng Progr 9:332–339. https://doi.org/10.1016/j.tsep.2018.12.006
2. Caton JA (2010) A thermodynamic evaluation of the use of alcohol fuels in a spark-ignition engine. SAE Int J Fuels Lubr 2(2):1–19
3. Caton JA (2011) Exergy destruction during the combustion process as functions of operating and design parameters for a spark-ignition engine. Int J Energy Res 36(3):368–384. https://doi.org/10.1002/er.1807
4. Caton JA (2011) The destruction of exergy during the combustion process for a spark-ignition engine. In: Internal combustion engine division fall technical conference, pp 717–731.https://doi.org/10.1115/ICEF2010-35036
5. Dogan B, Erol D, Yaman H, Kodanli E (2017) The effect of ethanol-gasoline blends on performance and exhaust emissions of a spark ignition engine through exergy analysis. Appl Thermal Eng 120(25):433–443
6. Dogan B, Yesilyurt MK, Dervis E, Cakmak A (2020) A Study toward analyzing the energy, exergy and sustainability index based on performance and exhaust emission characteristics of a spark-ignition engine fuelled with the binary blends of gasoline and methanol or ethanol. Int J Eng Res Dev 12(2):529–548
7. Gupta SK, Mittal M (2019) Analysis of cycle-to-cycle combustion variations in a spark-ignition engine operating under various biogas compositions. Energy Fuels 33(12):12421–12430. https://doi.org/10.1021/acs.energyfuels.9b02344
8. Gupta SK, Subramanian KA (2022) Analysis of combustion and cycle to cycle variations of an ethanol (E100) fueled spark-ignition engine. Int J Autom Sci Technol 6(1):68–74. https://doi.org/10.30939/ijastech..999559
9. Haq MZ, Morshed A (2013) Energy and exergy based analyses of a multi-fuelled SI engine. In: ASME power conference. https://doi.org/10.1115/POWER2013-98279
10. Kul BS, Ciniviz M (2021) An evaluation based on energy and exergy analyses in SI engine fueled with waste bread bioethanol-gasoline blends. Fuel 286:119375. https://doi.org/10.1016/j.fuel.2020.119375
11. Sezer I, Altin I, Bilgin A (2009) Exergetic analysis of using oxygenated fuels in spark-ignition (SI) engines. Energy Fuels 23(4):1801–1807. https://doi.org/10.1021/ef8002608
12. Statistical Review of World Energy. BP. Retrieved Jan 2022, from https://www.bp.com/content/dam/bp/business-sites/en/global/corporate/pdfs/energy-economics/statistical-review/bp-stats-review-2021-oil.pdf